2001. A Clay Odyssey

2001. A Clay Odyssey

Proceedings of the 12th International Clay Conference

Bahía Blanca, Argentina, July 22-28, 2001

Editors

Eduardo A. Domínguez   Graciela R. Mas   Fernanda Cravero

Universidad Nacional del Sur

Organized by

The Argentine Local Committee
In Collaboration with the Clay Mineral Society
under the auspices of
Association Internationale pour l'Etude des Argiles (AIPEA)

2003
Elsevier

Amsterdam – Boston – Heidelberg – London – New York – Oxford
Paris – San Diego – San Francisco – Singapore – Sydney – Tokyo

ELSEVIER SCIENCE B.V.
Sara Burgerhartstraat 25
P.O. Box 211, 1000 AE Amsterdam, The Netherlands

© 2003 Elsevier Science B.V. All rights reserved.

This work is protected under copyright by Elsevier Science, and the following terms and conditions apply to its use:

Photocopying
Single photocopies of single chapters may be made for personal use as allowed by national copyright laws. Permission of the Publisher and payment of a fee is required for all other photocopying, including multiple or systematic copying, copying for advertising or promotional purposes, resale, and all forms of document delivery. Special rates are available for educational institutions that wish to make photocopies for non-profit educational classroom use.

Permissions may be sought directly from Elsevier's Science & Technology Rights Department in Oxford, UK: phone: (+44) 1865 843830, fax: (+44) 1865 853333, e-mail: permissions@elsevier.com. You may also complete your request on-line via the Elsevier Science homepage (http://www.elsevier.com), by selecting 'Customer Support' and then 'Obtaining Permissions'.

In the USA, users may clear permissions and make payments through the Copyright Clearance Center, Inc., 222 Rosewood Drive, Danvers, MA 01923, USA; phone: (+1) (978) 7508400, fax: (+1) (978) 7504744, and in the UK through the Copyright Licensing Agency Rapid Clearance Service (CLARCS), 90 Tottenham Court Road, London W1P 0LP, UK; phone: (+44) 207 631 5555; fax: (+44) 207 631 5500. Other countries may have a local reprographic rights agency for payments.

Derivative Works
Tables of contents may be reproduced for internal circulation, but permission of Elsevier Science is required for external resale or distribution of such material.
Permission of the Publisher is required for all other derivative works, including compilations and translations.

Electronic Storage or Usage
Permission of the Publisher is required to store or use electronically any material contained in this work, including any chapter or part of a chapter.

Except as outlined above, no part of this work may be reproduced, stored in a retrieval system or transmitted in any form or by any means, electronic, mechanical, photocopying, recording or otherwise, without prior written permission of the Publisher. Address permissions requests to: Elsevier's Science & Technology Rights Department, at the phone, fax and e-mail addresses noted above.

Notice
No responsibility is assumed by the Publisher for any injury and/or damage to persons or property as a matter of products liability, negligence or otherwise, or from any use or operation of any methods, products, instructions or ideas contained in the material herein. Because of rapid advances in the medical sciences, in particular, independent verification of diagnoses and drug dosages should be made.

First edition 2003

Library of Congress Cataloging in Publication Data
A catalog record from the Library of Congress has been applied for.

British Library Cataloguing in Publication Data
A catalogue record from the British Library has been applied for.

ISBN: 0 444 50945 3

∞ The paper used in this publication meets the requirements of ANSI/NISO Z39.48-1992 (Permanence of Paper).
Printed in The Netherlands.

Organizing Committees

CONFERENCE ORGANIZING COMMITTEE

Eduardo Dominguez	Chair	Rodolfo Salomón	Internet Coordinator
Pedro J. Maiza	Vice Chair and Program Chair	Silvina Marfil	
		Wanda Alló	Posters and Exhibits
Fernanda Cravero	Secretary-General	Cristina Frisicale	Logo Design
Leandro Bengochea	Treasurer	Hebe Peral	
Eduardo Dominguez	Fund Raising	Mirta Garrido	Social Program
Graciela Mas	Program and Publication Chair	Cristina Gómez	
		Liliana Luna	
		Hebe Peral	

FIELD EXCURSION LEADERS

Precambrian saprolitized basement rocks an Precambrian and Lower Paleozoic sedimentary sequences: kaolinitic, pyrophillitic and illitic clays.	Patricia Zalba and Renato Andreis
Kaolin deposits of Patagonia	Claudio Iglesias and Eduardo Dominguez
Semid-arid and Humid soils of Buenos Aires Province	Nilda Amiotti and Ma. del Carmen Blanco
Precambrian-Paleozoic clay deposits. Sierras Septentrionales	Jorge A. Dristas and María C. Frisicale
Bentonitic and Kaolin deposits: Extra-Andean Patagonia	Jorge Vallés
Lateritic-Tropical soils. Misiones Province	Gabriel Píccolo

SYMPOSIA AND GENERAL THEMES CONVENORS

Symposia		General Sessions	
Teaching Clay Mineralogy	Darell Schulze	Clays in Geology	Patricia Zalba
Clay Barriers and Waste Managements	Jorge Vallés	Clay minerals and environment	Silvana Bertolino
Clays in Hydrothermal Deposits	Jorge Dristas	Soil Mineralogy	Nilda Amiotti Ma. del Carmen Blanco
Clays in Ceramics	Michele Dondi	Crystal Chemistry	Silvia Acebal Elsa Rueda
Clays in Petroleum Exploration and production	Daniel Poiré	Clays in Industry	Haydn H. Murray
Clay Resources in the Mercosur	J.C. Factorovich	Methods	Pedro J. Maiza

3rd INTERNATIONAL SYMPOSIUM ON ACTIVATED CLAYS (3 ISAC): Cristina Volzone

PREFACE

The 12th International Clay Conference of the Association Internationale pour l'Etude des Argilles (AIPEA) was held at Universidad Nacional del Sur, Bahía Blanca, Argentina in July 22-28, 2001. Bahía Blanca is located 686 km South of the most widely known Argentine city: Buenos Aires. The Conference also included the participation of the International Society of the Soil Science (Commission VII) and a special symposium: the Third International Symposium on Activated Clays (ISAC).

The theme of the conference was "2001 a Clay Odyssey". When we decided upon this title for the Conference, we believed that it would reflect our hope for the development of thought in the new century. Several years ago, we believed that by 2001 we would be using spacecrafts instead of automobiles. High tech computers and the landing on the moon and Mars are some of our dreams that came true. But, on Earth, a lot of problems still remain to be solved and we should emphasize the importance of clay science in improving the standard of living, both in developed and developing countries.

By 1997, at the time of the 11th Clay Conference, in Ottawa, I believed that my country was finally in the right way towards development. Unfortunately, that was not the case when the Conference was held. Then, by the end of 2001, we found that after the terrorist attack of September 11 the world had radically changed.

The meeting was organized by a local University Committee and 205 delegates from 35 countries took part. European participation was regretfully reduced by the economic crisis experienced by National Air Lines. During the Conference, the AIPEA medals to Gerhard Lagaly and Tom Pinnavaia were awarded.

This Volume of the Conference Proceedings contains 85 out of a total of 235 oral presentations and posters presented at the following symposia: Teaching Clay Mineralogy, Clays in Hydrothermal Deposits, Clays in Ceramics, Clays in Petroleum Exploration and Production, Clay Barriers, and Waste Management, as well as in the following general sessions of the Conference: Clays in Geology, Clay Minerals and Environment, Soil Mineralogy, Methods, Crystal Chemistry Structure and Synthesis, and Clays in Industry.

This was the second time an AIPEA Conference was held in the Southern hemisphere and the first time in Latin America. The field trips organized gave visitors the opportunity to see various Argentine landscapes, to be in touch with the local people, and to know how we live: improvising every day in order solve difficulties. However, the Conference was successful in spite of the hard economic and political circumstances in our country.

Another thing we should not fail to mention is that the Editors gratefully acknowledge to all reviewers for their useful corrections. And finally, on behalf of the Conference Organizing Committee, the editors would like to express their appreciation for the generous support granted by Universidad Nacional del Sur, Agencia Nacional de Promoción Científica y Tecnológica, and the following industrial organizations:

Piedra Grande SA,
Piedra Grande La Toma,
Canteras Cerro Negro,
Cerámica Ilva,
Rothenberger Minerales
Minera Ameghino

E. Domínguez, G. Mas, and M.F. Cravero
Departamento de Geología – Universidad Nacional del Sur
CONICET

Table of Contents

Conference Organizing Committees V
Preface.. VII
Table of Contents... IX
List of Technical Referees.. XV

I. Special Lecture and Keynotes
Clays in Industry
    H.H. Murray... 3

Keynotes
The Tandilia System, Province of Buenos Aires, Argentina: its sedimentary successions
    R. Andreis.. 15
Technological and compositional requirements of clay materials for ceramic tiles
    M. Dondi.. 23
Activation of Clays for environmental uses-Can we improve on nature, and can it pay?
    G.J. Churchman, C. Volzone... 31

II. Clays in Geology
Na-bentonites and K-bentonites from Argentina: composition, origin and age
    R.R. Andreis, P.E. Zalba... 41
Clay mineralogy and construction problems in volcanic soils near Mount St. Helens: Sediment Retention Structure, southwestern Washington, USA
    M.L. Cummings... 49
Neoformed mineral parageneses in acid weathering systems: sedimentary vs. volcanic environments
    Th. De Putter, A. Bernard, A. Perruchot, J. Yans, Fr. Verbrugghe, Ch. Dupuis.............. 57
Geochemical study of clay materials in Fornos de Algodres Region (Central Portugal) in an archaeometric view
    M.I. Dias, M.I. Prudêncio, M.A. Gouveia... 65
Fibrous clay minerals as lithostratigraphic markers in a Tertiary continental deposit (Malpica do Tejo, Portugal)
    M.I. Dias, F. Rocha... 71
Cr-bearing chlorites in low-grade metapelites of the Puncoviscana Formation (Neoproterozoic), Northwestern Argentina
    M. Do Campo, F. Nieto... 79
Two types of hydrothermal clay deposits in the south-east area of Tandilia, Buenos Aires Province, Argentina
    J.A. Dristas, M.C. Frisicale.. 85
Occurrence of bentonites in Southern South America
    M.L.L. Formoso, L.M. Calarge, A.A.M. Misusaki, A. Meunier, A.D. Alves, P. Zalba, R. Andreis, J. Bossi... 93
Clay mineralogy, illite crystallinity and polytypes in the Campana Mahuida Porphyry Cu Deposit, Neuquén, Argentina
    A. Impiccini, M. Franchini, G.H., Grathoff, A. Schalamuk, L. Meinert............. 101
A neoformed kaolinitic mineral in the Upper Pleistocene of northeastern Argentina
    M. Iriondo, D. Kröhling... 109
Surface microtopography of pyrophillite from different modes of occurrence
    M. Jige, R. Kitagawa, V. Zaykov, I. Sinyakovskaya, V. Udachin, J.Y. Hwang............ 117
Geochemistry of the hydrothermal kaolins in the SE area of Los Menucos. Prov. of Río Negro, Argentina
    P.J. Maiza, D. Pieroni, S.A. Marfil.. 123

Chlorite-smectite geothermometry of two wells from the Copahue Geothermal Field, Argentina
    G.R. Mas, L. Bengochea, L.C.Mas.. 131

Clay Mineral diagenesis of a Pleistocene volcanogenic sequence, Mexican Basin
    L. de Pablo-Galan, J.J. De Pablo, M.L. Chavez-Garcia.. 139

Analysis of colour rhythmites in sensitive marine clays (Leda Clay) from Eastern Canada
    J.B. Percival, J.M. Aylsworth, A. Fritz.. 147

Berthierine formation in spotted slates
    M.D. Ruiz Cruz, E. Galán.. 155

The application of clay mineralogical analysis to the reconstruction of a Greek Bronze Age coastal environment
    C.M. Shriner and H.H. Murray.. 163

Facies distribution and pedogenetic evolution of clayey deposits in Caxambu Hill, Quadrilátero Ferrífero, Minas Gerais, Brazil
    A.F.D.C. Varajão, M.C. Santos.. 171

The effect of deformation on illite crytallite sizes
    K. Wagner (née Hetfeld)... 179

II.   Soil Mineralogy

X-Ray analysis of clay and silt fractions of soils developed in allochthonous materials of Aeolian (loess) and alluvial origin in the southwestern Pampa, Argentina
M. del C. Blanco, I.M. Natale, N. Amiotti, E.A. Ferreiro and M.E. Mandolesi................... 189

Coastal dune soils in Oregon, USA, forming allophane, imogolite and gibbsite
    G.H. Grathoff, C.D. Peterson, D.L.Beckstrand.. 197

Deep and actively forming illuvial clay in the regolith and on bedrock
    D.L. Johnson, D.N. Johnson, D.M. Moore... 205

Pedogenic mineral formation as environmental indicator in a paleosol sequence from the Miocene to the Holocene at Mergelstetten, Southwest-Germany
    K. Stahr, P. Kallis, M. Zarei, K. Bleich.. 211

Experiences with selective extraction procedures for iron oxides
    J.K Torrance, J.B. Percival... 219

Neoformed halloysite in podzols developed on the Bärhalde granite, Southern Black Forest, Germany
    M. Zarei, M. Sommer, K. Stahr... 227

III. Applied Clay Science
Ceramics, Petroleum and Waste Managements

Clay mineralogy of raw materials and ancient pottery from archaeological sites in the Ambato valley, Catamarca, Argentina
    S.R. Bertolino, M. Fabra... 237

Transformation of chlorinated aliphatic compounds by ferruginous smectite
    J. Cervini-Silva, R.A. Larson, J. Wu, J.W. Stucki... 241

Porous alumina-nano-clay functionally gradient membrane structures
    K. Darcovich, F.N.Toll, L. Kotlyar.. 247

Effect of chemical modification of *stilbite* zeolite on removing lead from wastewaters
    A.C.P. Duarte, M.B.M. Monte, A.A. Neto, A.B. Luz.. 255

Pore lining chlorites in hydrocarbon reservoirs: structure and composition related to their origin
    C. Durand, B. Rebours, E. Rosenberg, E. Brosse... 261

Mechanochemical activation of kaolinite surfaces
    R.L. Frost, É. Makó, J. Kristóf, Z. Ding, J.T. Kloprogge... 269

The role of hydrophobic solids in the separation and upgrading of bitumen from Athabasca oil sands
    L.S. Kotlyar, B.D. Sparks, J. Woods, J. Kung, K.H. Chung... 277

The physico-chemical and rheological characterizations of two industrial bentonites
    C. Malfoy, A. Besq, A. Pantet, P. Monnet 285

Application of bentonite and organoclay in stabilization/solidification of tannery waste
    C.A. Pinto, S.M. Toffoli, L.T. Hamassaki, F.R. Valenzuela-Diaz, P.M. Büchler 293

Plasticity of bricks clays: comparison of several empirical tests and correlation with mineralogical composition and particle size distribution
    M. Raimondo, M. Cocchi, G. Dircetti, M. Dondi, A. Ulrici, P. Zannini 301

Adsorption of phenol by organo-clays
    M.M.G Ramos Vianna, C.L.Vieira José, F.R Valenzuela Diaz, P.M. Büchler 309

Portuguese clays used in Geomedicine: study of their relevant properties
    J. Silva, C. Gomes, F. Rocha 315

Industrial clays of Brazil: a review
    P. de Souza Santos 323

White bentonite from Patagonia, Río Negro Province, Argentina
    J.M. Vallés, A. Impiccini 331

Corrensite, a stratigraphic marker in the Quintuco Formation, Neuquén Basin, West-Central Argentina, and Mississipian carbonates of the Illinois Basin, Illinois, USA
    J.M. Vallés, G.R. Pettinari, D.M. Moore 339

Study of the structural order degree of Brazilian kaolinites by X-ray diffraction
    A.C. Vieira Coelho, P. Souza Santos 347

Geology and evaluation of the Yarmouth Kaolin Deposit, Nova Scotia, Canada
    I.R. Wilson, H.H. Murray, G. MacGillivray, J. Keating 355

Demonstration of the existence of subcritical growth cracks in sintered kaolin
    C. Xavier, P.K. Kiyohara, P. Souza Santos 361

Geology and physical properties of palygorskite from Central China and Southeastern United States
    H. Zhou, H.H. Murray 369

V. Mineral Structure and Investigational Methods
a) Crystal Chemistry

Distribution and characterization of forms of Fe and Al in particle-size fractions of an Entic Haplustoll by selective dissolution techniques, X-ray powder diffraction and Mössbauer spectroscopy
    S.G. Acebal, M.E. Aguirre, R.M. Santamaría, A. Mijovilovich, S. Petrick, C. Saragovi 379

Thermal transformation of synthetic bayerite and nordstrandite as studied by electron-optical methods
    M.L.P. Antunes, H. Souza Santos 387

Proton binding at clay surfaces in aqueous media
    M.J. Avena, M.M. Mariscal, C.P. De Pauli 395

Hydrothermal transformation of kaolinite into 2:1 expandable minerals
    M. Bentabol, M.D. Ruiz Cruz, F.J. Huertas, J. Linares 403

Synthesis and characterization of novel inorganic/ organic hybrid materials prepared from layered double hydroxides
    G.A. Caravaggio, A. Moser, C. Detellier 411

EPR characterisation of iron in various clay minerals: montmorillonites and layered double hydroxides
    F. Dijoux, B. Deroide, D. Tichit 419

Synchrotron x-ray study of hydration dynamics in the synthetic swelling clay Na-fluorohectorite
    E. DiMasi, J.O. Fossum, G. J. da Silva 427

Experimental weathering of biotite, muscovite and vermiculite: a Mössbauer spectroscopy study
    E.A. Ferrow 435

Syntheses of smectite-analogue/coumarin composites
    K. Fujii, S. Hayashi 443

Thermogravimetric analysis-mass spectrometry (TGA-MS) of hydrotalcites containing CO_{32}^-, NO_3^-, Cl^-, SO_{42}^- or ClO_4^-
    J.T. Kloprogge, J. Kristóf, R.L. Frost.. 451
Spectroscopic analysis and X-ray diffraction of zinnwaldite
    J.T. Kloprogge, S. van der Gaast, P.M. Fredericks, R.L. Frost........................... 459
Clay mineralogy and tensile strength of different soils from the southwestern Buenos Aires Province, Argentina
    M. Del P. Moralejo, O.E.Soulages, S.G. Acebal, M.E. Aguirre, R.M. Santamaría................ 465
Molecular and particulate organisation in dye-clay films prepared by the Langmuir-Blodgett method
    R.H.A. Ras, B. Van Duffel, M. Van der Auweraer, F.C. De Schryver, R.A. Schoonheydt... 473
Si-Al-Mg and Si-Al-Zn montmorillonite: synthesis and characterization by EXAFS and quantitative ^{27}Al MAS-MNR
    M. Reinholdt, J. Miehé-Brendlé, L. Delmotte, M.H. Tuilier, R. Le Dred....................... 481
Electron optical study of star shaped gibbsite microcrystals
    H. Souza Santos, P. Souza Santos, P.K. Kiyohara... 489
A new method for the preparation of microcrystals of boehmite
    P. Souza Santos, H. Souza Santos, P.K. Kiyohara... 497
Studies of Synthetic Kaolinites Containing Copper and Zinc.
    A.R. Tong, B.J. Kennedy, B. Singh... 505
Adsorption of quinoline on Na-sepiolite and Na-palygorskite
    L.I. Vico, S.G. Acebal... 513

b)   Methods
Controlled rate thermal analysis of formamide intercalated kaolinites
    R.L. Frost, J. Kristof, Z. Ding, E. Horvath... 523
Cathodoluminescence of kaolin group minerals
    M. Plötze, J. Götze... 531
FT-IR photoacoustic spectroscopy of kaolinite and gibbsite surfaces
    H.D. Ruan, R.L. Frost, J.T. Kloprogge, L. Duong, D.G. Schulze 537
FT-Raman spectroscopy and SEM of gibbsite, bayerite, boehmite and diaspore in relation to the characterization of bauxite
    H.D. Ruan, R.L. Frost, J.T. Kloprogge, D.G. Schulze, L. Duong......................... 545

VI.   Teaching Clay Mineralogy
Evaluation of colloidal properties using clay minerals: experiments for undergraduate and introductory chemistry students
    M.B. Lombardi, M. Baschini... 555
Three simple experiments to visually demonstrate the impact of clay minerals on the behavior of organic pollutants in soils
    G. Rytwo.. 561
Microbes associated with clay minerals – Formation of bio-halloysite –
    K. Tazaki, R. Asada... 569

3^{ed} International Symposium on Activated Clays (ISAC)
Acid activation and bleaching capacity of bentonites from the Troodos ophiolite, Cyprus
    G. Christidis, S. Kosiari, E. Petavratzi... 579
Synthesis and characterization of copper-loaded Titania pillared clays.
    Z. Ding, R.L. Frost, J.T. Kloprogge, G.Q. Lu, H.Y. Zhu...................................... 587
Synthesis and characterization of $SiO_2 - Cr_2O_3$ pillared montmorillonites
    E.M. Farfán-Torres, A.G. Mercado, E.L. Sham, M.A. Blesa.............................. 595
Castor, cottonseed, and soybean oil bleaching by activated bentonites
    E.L. Foletto, C. Volzone, A.F. Morgado, L.M. Porto... 603

OH-Al complexes on kaolinite surface and their effect on suspension properties
    L.B. Garrido, C. Volzone 611

Remotion of chromium from water by treated altered-tuffaceous materials
    A. Hülsken, D. Hall Gómez, J. Venaruzzo, J. Ortiga, C. Volzone 619

Clay iron oxide magnetic composites for the adsorptions of contaminants in water
    L.C.A. Oliveira, R.M. Lago, R.V.R.A. Ríos, J.D. Fabris, C. Solar, K. Sapag 625

Thermal decomposition of layered Co-Al hydrotalcite: an *in situ* study
    J. Pérez-Ramírez, G. Mul, J.B. Taboada, F. Kapteijn, J.A. Moulijn 631

Some structural properties of an Al-PILC influenced by grinding the starting montmorillonitic clay
    M. Sergio, S. Cardozo, S. Froche, M. Bentancor, M. Musso, J. Medina, W. Diano 639

Synthesis and characterization of pillared clays from Argentinean bentonites
    C. Solar, E. Fernández, E. Perino, E. Strasser, K. Sapag 647

Influence of the smectite type on the basal spacing after polyhydroxy-aluminum pillaring in some Brazilian and Argentinian clays
    A.C. Vieira Coelho; P. Souza Santos; C. Volzone, L.D.V. Abreu 655

List of Technical Referees

The following individuals gave their time and talent as technical referees of manuscripts submitted for publication in this volume. Their devotion is greatly appreciated.

Alves Barbosa, D.	Fujii, K.	Ramos Vianna, M.
Bergaya, F.	Galán, E.	Rocha, F.
Berry, R.	Gomes, C.	Rytwo, G.
Campbell, A.	Grandjean, J.	Sandler, A.
Christidis, G.	Grathoff, G.	Schoonheydt, R.
Churchman, J.	Guggenheim, S.	Schulze, D.
Cummings, M.	Harvey, C.	Shriner, C.
Czurda, K.	Huff, W.	Souza Santos, H.
DePutter, T.J.M.	Kasbohm, J.	Souza Santos, P.
Dias, M.I.	Kelm, U.	Stahr, K.
DiMasi, E.	Lalonde, A.	Stucki, J.
Ding, Z.	Lessovaia, S.	Theng, B.
Dondi, M.	Merriman, R.	Torrance, J. K.
Duarte, M.	Michot, L.	Tsutsumi, S.
Dupuis, C.	Mihe-Brendle, J.	Van der Gaast, S.
Durand, C.	Moore, D.	Vieira Coelho, A.C.
Emmerich, K.	Murray, H.H.	Volzone, C.
Ferrell, R.	Ottner, F.	Wilson, I.
Ferrow, E.	Pantet, A.	Xavier, C
Forano, C.	Percival, J.	Yuan, Y.
Formoso, M.	Perez Ramirez, J.	Yuretich, R.
Fridriksson, T.	Pinto, C.A.	Yvon, J.
Frost, R.	Plötze, M.	

I
Special Lecture and Keynotes

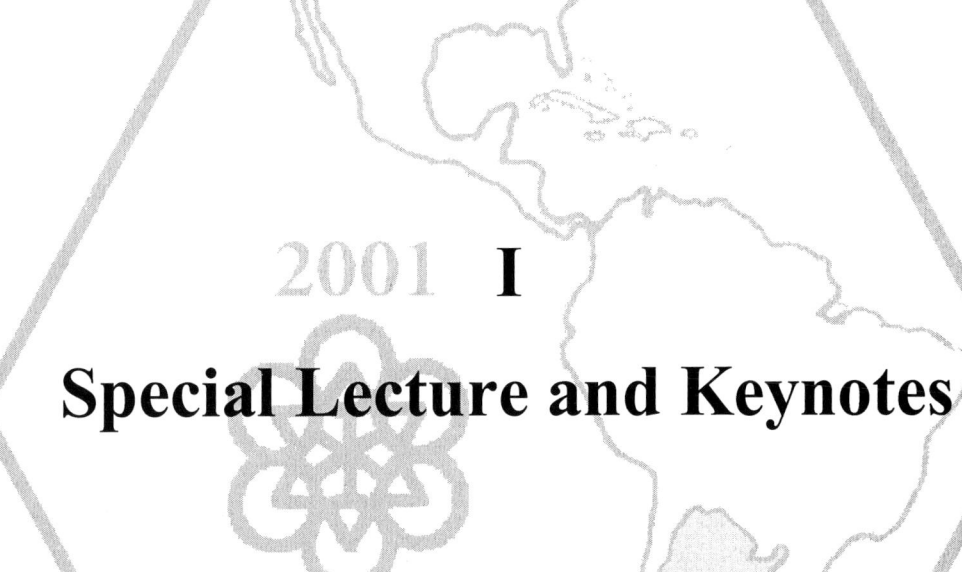

Clays in Industry

Haydn H. Murray

Department of Geological Sciences. Indiana University, Bloomington, Indiana 47405. USA

Clays are one of the more important industrial minerals. Clays are comprised largely of clay minerals and have applications in many facets of today's society including agriculture, geology, construction, engineering, process industries, and protecting the environment. Research and development by clay scientists in academia, government, and industry contribute to our fundamental understanding of these extremely small particle size minerals. This research and development results in new, innovative, and improved clay products.

It is important to understand that a fundamental knowledge of the structure and composition of clay minerals - kaolins, smectites, palygorskite-sepiolite, illite, and chlorite - is necessary to develop industrial applications. The various clay minerals have major and minor differences in their physical and chemical properties which are related to their structure and composition. These structural and compositional attributes of clay minerals are correlated with their physical and chemical properties in this paper. The correlation between the properties and the structure and composition explains why kaolinite is an excellent paper coating material and a necessary component in many ceramic products; why sodium montmorillonite is a major ingredient in drilling muds; why calcium montmorillonite is an excellent sorbent clay; and why palygorskite is a superior gelling clay in sys__tems with high contents of electrolytes and salts. Other applications related to the structure, composition, and physical and chemical properties of clay minerals are described. Processing techniques and some new and improved products that have recently been developed are also described. As our fundamental knowledge concerning clay minerals increases, they will become even more important industrially and scientifically.

1. INTRODUCTION

The clay minerals kaolinite, smectite, and palygorskite-sepiolite can occur in substantially pure deposits and/or in common clays and shales associated with illite and chlorite along with common non-clay minerals such as quartz, feldspar, micas, ilmenite, rutile, anatase, goethite, and trace amounts of zircon, tourmaline, and other heavy minerals. These clay materials are very important and useful industrial minerals. Millions of tons of these clays are used annually in a large variety of industrial applications. The relatively pure deposits of kaolin, smectite, and palygorskite-sepiolite are relatively rare in comparison to common clay and shales which are abundant. Kaolin rich deposits are formed by weathering and hydrothermal alteration of feldspathic rocks which are called primary kaolins or by deposition in lagos, lakes or oxbows, most often in a deltaic environment which are called secondary or sedimentary kaolin (Murray and Keller 1993). The most common kaolin mineral is kaolinite. Other kaolin minerals are halloysite, dickite, and nacrite all of which are rare relative to kaolinite. Smectite deposits are most commonly altered volcanic tuffaceous rocks or volcanic ash deposits called bentonites (Grim and Guven, 1978). Bentonites are comprised most commonly of sodium and/or calcium montmorillonite. A relatively rare bentonite used commercially is hectorite, a lithium montmorillonite. Saponite, another relatively rare bentonite, is a magnesium montmorillonite. Palygorskite and sepiolite are magnesium rich clay minerals which were deposited in marine lagoons (Patterson, 1974) or in magnesium rich lacustrine environments (Zhou 1996). The mineral terms palygorskite and attapulgite are synonymous but palygorskite is the preferred term. Ball clays and fireclays are kaolinitic clays used by the ceramic industry. Common clays and shales are generally mixtures of illite, and chlorite along with some kaolinite and smectite.

Important physical properties of clays that are related to their application are particle size and shape, surface chemistry, area, and charge, color and brightness, viscosity, absorption, plasticity, green, dry, and fired strength, casting rate, permeability, bond strength, thermal characteristics, and other special properties. A summary of the mineralogical properties of clay minerals that relate to their applications is summarized in Table 1.

Table 1 - Some important properties that relate to their applications.

Kaolin	Smectite	Palygorskite
1:1 layer	2.1 layer	2.1 layer inverted
White or near white	Tan, olive green, bluish gray or white	Light tan or gray
Little substitution	Octahedral and tetrahedral substitutions	Octahedral substitution
Minimal layer charge	High layer charge	Moderate layer charge
Low base exchange capacity	High base exchange capacity	Moderate base exchange capacity
Pseudo-Hexagonal flakes	Thin flakes and laths	Elongate
Low surface area	Very high surface area	High surface area
Very low absorption capacity	High absorption capacity	High absorption capacity
Low viscosity	Very high viscosity	High viscosity

The structure and composition of the clay minerals largely determine their physical and chemical properties (Murray 2000). This explains the significant differences in the physical and chemical properties and thus in their industrial applications. Improved processing techniques are continually evolving and also have a significant effect on traditional and new applications. Some of the more important applications of the clays are discussed in this paper, along with new and improved processing.

2. KAOLINS

As mentioned above, the most common mineral in the kaolin group of minerals is kaolinite which is by far the most used industrially. As noted by Murray and Keller (1993) there are kaolins, kaolins and kaolins because most every kaolin deposit may have different physical and chemical properties which must be fully tested and evaluated to determine its utilization. The occurrence of kaolinite is common but commercially useable and relatively pure deposits are few in number. Some deposits such as the primary kaolin in the Cornwall district of Southwestern England have a very low kaolinite content but can be beneficiated to produce a relatively pure product. The two most well known sedimentary kaolin areas are in Georgia and South Carolina in the United States and in the lower Amazon in Brazil (Pickering and Murray, 1994).

As shown in Table 1, kaolin is white or near white in color. There is very little substitution in the structural lattice so therefore it has a minimal layer charge and a very low base exchange capacity. The kaolinite crystals are pseudo-hexagonal in shape and occur as thin plates and large books and vermicular stacks.

Particle size and size distribution are very important properties in determining the industrial application. A coarse particle size kaolin has very different physical and optical properties compared to a fine particle kaolin. A coarse particle kaolin is one that has a particle size of 60±10 percent finer than two microns and a fine particle kaolin is one that has 90±10 percent finer than two microns. The Cretaceous kaolins mined in middle Georgia and the Rio Capim kaolin in Brazil are examples of coarse kaolin. The Tertiary kaolins from East Georgia and the Rio Jari kaolin in Brazil are examples of fine particle kaolins. The fine particle kaolins do not contain large book, stacks, or vermicular crystals whereas the coarse particle kaolins have a multitude of stacks and vermicular crystals.

Another important physical property of kaolin is that it has a low viscosity at high solids concentration. This is particularly important in paper coating and paint applications. The viscosity of kaolin products is dependent on several factors including mineral content, particle size, shape, and size distribution, presence of soluble salts, and dispersability. For example, the presence of a small amount of smectite, illite, halloysite, and soluble salts can adversely affect the low and high shear viscosity.

Some important industrial applications of kaolin are shown in Table 2.

Table 2 - Some important applications of kaolins

Paper Coating	Plastic filler	TiO_2 extender (calcined)
Paper filling	Ink extender	Adhesives
Paint extender	Cracking Catalysts	Enamels
Ceramic raw material	Fiberglass	Pharmaceuticals
Rubber filler	Cement	Molecular sieves

The largest use of kaolin is for coating paper. The physical properties that are important to the paper coater are dispersion, viscosity (both low and high shear), brightness, whiteness, gloss, smoothness, adhesive demand, film strength, ink receptivity, and print quality. The properties shown in Table 1 are very important factors that relate to why kaolinite is an excellent paper coating material. It is white or near white in color, has a low surface area and charge, a low base exchange capacity, a low absorption capacity, and a near two dimensional platy particle shape.

Its applications in paint, plastics, rubber, and ink are again related to its good physical properties and the fact that it is nearly chemically inert between pH3 and 9. A very large industrial application for kaolins and kaolinitic materials such as ball clays and fire clays is as a ceramic raw material. They are used as important ingredients in the manufacture of whitewares, sanitary ware, insulators, pottery, and refractors. Many kaolins that do not have good viscosity for paper coating do have good ceramic properties. These include plasticity, shrinkage, modulur of rupture, absorption, fired color, casting rate, and pyrometric cone equivalent (PCE). The PCE test determines the fusion or melting point of the kaolin.

A few industries use kaolins because of their chemical composition. These include fiberglass, cement, molecular sieve, and cracking catalysts where the kaolin is converted to a zeolite type structure. This is accomplished by using a strong alkali base such as potassium, sodium, calcium, or magnesium hydroxide and heating the slurry to a temperature of about 100EC (Murray, 1994).

Calcined kaolins are value-added products used primarily by the paper and paint industries to extend titanium dioxide. There are three general types of calcined kaolins; meta-kaolins calcined at about 650EC, mullite/spinel calcined at about 1050EC, and mullite refractory grog calcined at about 1300EC. The partially calcined or meta-kaolin type of product is used as a filler in plastics or rubber used for wire coatings. The meta-kaolin has a high dielectric capacity and good thermal insulating properties. Partially calcined kaolins are used as an additive in cement (5 to 10 percent) which doubles the compressive and tensile strength of concrete. The amorphous meta-kaolin reacts with excess calcium ions to form a calcium alumino silicate mineral which is elongate in shape and therefore improves strength characteristics. The mullite/spinel calcined product has by far the largest use. The cost of titanium dioxide is very high (around $1.00 per lb.) Whereas the cost of regular mullite/spinel calcined products sell for 10 to 20 cents per lb. Up to 60 percent of T_iO_2 can be replaced with calcined clay with little or no loss in opacity and brightness. The high temperature (1300EC) granular calcined product is used as a refractory grog to reduce shrinkage and to increase the refractoriness. A pulverized high temperature calcined product is also used in foundrys as a mold release and in some cases, sand sized granules are used in high temperature mold applications.

In the past five to ten years many tailored or engineered products have been developed for paper coating to enhance specific properties such as opacity, gloss, ink holdout, brightness, whiteness, and print quality. These tailored products are produced by special processing such as high speed centrifuges, flotation, selective flocculation, and blending of particle sizes and with calcined kaolins. These are called engineered products and are increasingly important because of the needs of the coated paper manufacturers to improve specific properties of the coated paper. Another area that is increasingly important is the production of surface modified products. Kaolins are naturally hydrophilic and are easily dispersed in water by
using small amounts of a chemical dispersing agent such as sodium hexametaphosphate, sodium silicate, or sodium polyacrylate. Ionic and/or polar non-ionic surfactants can be applied to the surface of the kaolin to make it hydropholic or organophilic (Iannicelli, 1991). These surface-modified kaolins are used in the plastics, rubber, and ink industries to improve dispersion and thus produce a more functional filler.

Many properties of kaolin can be altered and improved by processing which is covered in the last section of this paper.

3. SMECTITES

The smectite group of clay minerals is comprised of several clay minerals but the four most important industrially are sodium montmorillonite, calcium montmorillonite, hectorite (lithium montmorillonite), and saponite (magnesium montmorillonite) (Elzea and Murray, 1994). The most common occurrence of the smectites is in bentonite, a clay altered from glassy igneous rocks such as volcanic ash and tuff (Grim and Guven, 1978). As shown in Table 1, the smectites have very different physical and chemical properties than kaolins. Furthermore, there are significant differences in the properties of sodium, calcium, lithium, and magnesium montmorillonites.

The most important and largest deposits of sodium bentonite occur in Montana, South Dakota, and Wyoming in the United States. The deposits are Cretaceous in age and the bentonite beds are part of the Mowry formation (Elzea and Murray, 1994) Calcium bentonite deposits are much more common than sodium bentonites and occur in Nevada, Arizona, Texas, Mississippi, Alabama, and Georgia in the United States. In Europe calcium bentonites are produced in England, Germany, Italy, Greece, Hungary, Republic of Georgia, and Yugoslavia (Murray, 2000). In Asia calcium bentonites are produced in India, Japan, Malaysia, and China. Hectorite, a lithium bentonite, is produced in California and Nevada and saponite a magnesium bentonite is produced in Arizona and Nevada.

Both the octahedral and tetrahedral sheets can have ionic substitutions which creates a charge imbalance in the 2:1 layer, which Grim (1962) reported was about 0.66 per unit cell. This net positive charge deficiency is balanced by exchangeable cations which are absorbed between the 2:1 layers. If the dominant cation is sodium, then the smectite mineral is sodium montmorillonite, and if the exchangeable cation is predominantly calcium, it is calcium montmorillonite. The high charge on the layer, the very fine particle size, the very thin flakes, the high base exchange capacity, and the large surface area result in physical and chemical properties that make these smectite clays very useful in many industrial applications. The more important industrial applications are shown in Table 3.

Table 3 - Some important industrial applications of smectites

Drilling mud	Bleaching clays	Emulsion stabilizers
Foundry bond clay	Agricultural carrers	Dessicants
Pelletizing iron ores	Cat box absorbents	Oil and grease absorbents
Sealants	Adhesives	Slurry trenches
Animal feed bonds	Pharmaceuticals	Organoclays

The fine particle size, high layer charge, large surface area, and high swelling capacity of sodium montmorillonite makes this clay the most important and necessary ingredient in fresh water drilling muds (Elzea and Murray, 1990). Sodium montmorillonite has a high viscosity and is thixotropic. It becomes a gel at 5 percent solids in contrast to low viscosity kaolins which flow readily at 70 percent solids.

Both sodium and calcium montmorillonites are used to bond foundry sands. Calcium montmorillonite has a higher green strength, lower dry strength, lower hot strength, and better flowability than sodium montmorillonite (Grim and Guven, 1978). In order to gain the optimum properties for a molding sand needed for a particular molten metal, blends of sodium and calcium montmorillonites are used.

Sodium montmorillonite because of its high dry strength is used to pelletize pulverized iron ore concentrates. This is done for ease of handling for shipment and to produce a better furnace feed. Because many iron and other metal ores need to be beneficiated to remove impurities, they must be pulverized to free the impurities which can then be removed by flotation, magnetic separation, selective flocculation, etc. The pulverized and beneficiated metal concentrate is then pelletized with about 1 weight percent or less of sodium montmorillonite.

Sodium montmorillonite is used in applications requiring barriers to water movement because of its high swelling capacity. Examples are liners for irrigation ditches, to prevent leakage in earthen dams, to prevent seepages and leaks in ponds, and to contain chemicals and toxic wastes in landfills (Keith and Murray, 1994). Blending sodium montmorillonite with local clays and soils in landfills to prevent contamination of water supply and aquifers is a growing application. Another growing application of high swelling sodium montmorillonite is in the slurry-trench or diaphragm-wall method of excavation in construction when unconsolidated materials are encountered (Lang, 1971). In this application, the trench being excavated is filled with bentonite slurry and the material being excavated is lifted through it. A thin filter cake on the walls of the trench prevents loss of fluid and the hydrostatic head of the slurry prevents caving which makes expensive shoring unnecessary.

Both sodium and calcium montmorillonites are used in binding or pelletizing animal feed. In addition to their binding ability, these minerals absorb bacteria and certain enzymes which when removed promote the health and growth of the animal. A large and growing use for both calcium and sodium montmorillonite is as a pet waste absorbent. This market uses granules of calcium montmorillonite which may be treated with deodorants, fungicides, and bacteriacides. The sodium montmorillonite is blended with the calcium montmorillonite granules to produce clumping cat litter. The sodium montmorillonite swells when the waste moisture contacts the litter and soon dries and hardens to produce a clump which can easily be removed from the litter box.

Another sizeable market for calcium montmorillonite is to produce acid activated clay for decolorizing vegetable, animal, and mineral oils. Acid treatment with sulfuric or hydrochloric acid removes the calcium ions on the clay particle which increases the negative charge. This makes the acid activated clay more effective in removing anionic color bodies from the oil (Odom, 1984). Natural sodium or calcium montmorillonite is used to remove colloidal impurities in wine. These colloidal impurities are positively charged and are attracted and coagulated by the negatively charged clay particles. Sodium montmorillonites having a light or white color with high dispersability are preferred for use in the clarification of wine, beer, vinegar, and fruit juices. A new market being developed for acid activated clays is in animal feed where it absorbs mycotoxins.

Calcium montmorillonite is used as a carrier in the agricultural business for insecticides, herbicides, and fertilizers. It is also used as an oil and grease absorbent in factories, garages, etc. where oil and grease are spilled on the floor.

Three important value added applications for sodium montmorillonites are organoclays, pillared clays, and nanoclays. In the organoclay processing, the exchangeable sodium ions are replaced with organic cationic molecules such as alkylamines (Rausell-Colom and Serratosa, 1982) which produces a hydrophobic or organophyllic surface. These organic clad montomorillonites are used as thickeners in oil based paints, oil-based drilling fluids, to gel some organic liquids, to stabilize the gel properties of lubricating greases, and to thicken and gel certain cosmetics. For the paint and cosmetic markets a white organoclay would be advantageous. Pillared clays tailored for specific catalyst and absorbent uses is a developing market. Interlamellar reactions using sodium montmorillonite are restricted to low temperatures (less than 200EC) because at high temperatures, the layers collapse. By varying the size of the pillar and/or the spacing between pillars, the pore size and space can be adjusted to suit a particular application. Hydroxyaluminum and zirconium cations yield thermally stable clays with surface areas of 200 to 500 m^2/g and interlayer free spacing near 9A (Rupert et.al, 1987). This relatively recent development will result in many new industrial applications because of the need to absorb and catalyze toxic ions, anions, and chemicals to prevent their release into groundwater, soils, and the atmosphere. Another recent development using sodium montmorillonite is the separation of the unit layers into unit cell thicknesses for use in plastic compositions called nanocomposites. These very thin platy particles are exchanged using organic molecules which interact with the plastic polymers to produce very strong and heat resistant products. Currently these nanoparticles are being utilized in certain automotive plastic components and in plastic food wrappings and packaging (Beall, personal communication).

4. PALYGORSKITE AND SEPIOLITE

Palygorskite and attapulgite as mentioned in the introduction are synonymous terms for the same hydrated magnesium aluminum silicate mineral. The preferred name as specified by the International Nomenclature Committee is palygorskite. However, the name attapulgite is so well entrenched in trade circles that it continues to be used by some producers and users. Sepiolite is structurally and chemically almost identical to palygorskite except it has a slightly larger unit cell and may have a higher magnesium content.

The term fuller_s earth is used to describe clays which have sorptive and bleaching properties so palygorskite, sepiolite, and some smectite clays (particularly calcium montmorillonite) are called fuller_s earth. This term has no compositional or mineralogical meaning. The word fuller in early times in England was one whose trade was that of scouring and felting woolen cloth. A slurry of fuller_s earth was used to remove grease and dirt in the wool thus the name fuller_s earth (Robertson, 1986).

Palygorskite and sepiolite are both elongate in shape, which is a controlling physical property (Table 1). Some of the more important applications of these minerals are listed in Table 4.

Table 4 - Some important industrial applications of palygorskite and sepiolite

Drilling fluids	Cat box absorbents	Paper
Paint	Suspension Fertilizers	Pharmaceuticals
Agricultural carriers	Animal feed bondents	Anti-caking agent
Industrial floor absorbents	Catalyst supports	Reinforcing fillers
Tape joint compounds	Adhesives	Environmental Absorbents

There is considerable substitution of aluminum by magnesium and iron in the octahedral layer which gives the elongate particles a moderately high layer charge. The layer charge and high surface area gives palygorskite an intermediate cation exchange capacity of about 30 to 40 meg/100 gms. The high surface area, the charge on the lattice, and the inverted structure, which leaves parallel channels through the lattice, give palygorskite and sepiolite a high absorption capacity. These properties along with the elongate crystal habit of these minerals, make them very useful in many industrial applications. The high viscosity that these minerals impart to liquids is a physical rather than a chemical viscosity which makes a chemically stable suspension viscosity.

Galan (1996) reviewed the properties and applications of palygorskite-sepiolite clays and showed the relationship between the structure, composition, and physical properties and their industrial applications. The most important areas where these minerals are produced are sepiolite in Spain and palygorskite in South Georgia - North Florida in the United States, Senegal in West Africa, Ukraine in Europe, and in Anhui and Jianesi Provinces in China.

The largest use for palygorskite and sepiolite is for viscosity builders in salt water drilling muds because of their insensitivity to flocculation in salt water and other electrolytes. Sepiolite is stable at high temperatures and is used in drilling muds for geothermal wells. The colloidal or gel grades are used in paints, adhesives, sealants, fertilizer suspensions, and cosmetics. A large growth market for the future is the suspension fertilizer application. Suspension fertilizer solutions, which are gel-like, are stable over extended time periods of one to two months. These gels can be readily redispersed with mild agitation to be fluid enough to be pumped and applied uniformly to the soil.

Palygorskite and sepiolite are excellent absorbents which are used in many applications that utilize this physical property. These include industrial floor absorbents, agricultural carriers, cat litter, anti-caking agents, and environmental absorbents (Murray, 1995). The good absorbent properties and elongate shape make palygorskite and sepiolite very useful as barrier clays in landfills and toxic waste impoundments (Keith, 2000). Mixtures of palygorskite and sodium montmorillonite work well to prevent movements of liquids through a barrier and to absorb heavy metals and toxic pollutants. Sodium montomorillonite flocculates easily and shrinks in the multiple drying and wetting cycles that the barrier undergoes. Palygorskite does not flocculate so it maintains a constant volume that prevents permeability changes.

Another larger and growing use for the gel grades of palygorskite is in tape joint compounds used to fill cracks and joints in wallboard. The tape joint compound must form a level and smooth surface that does not shrink on drying. The adhesive filled with palygorskite is very good for this application because it forms a smooth surface and does not shrink. Other uses for palygorskite and sepiolite are reinforcing fillers in rubber and plastics, suspending agents in pharmaceutical products, animal feed bondants and supplements, catalyst supports, and as a coating clay on carbonless copy paper. In these applications, the elongate shape and the high absorbency are the prime physical properties of palygorskite and sepiolite that make them particularly functional.

5. COMMON CLAYS AND SHALES

Clays and shales occur in many types of rocks ranging in age from Precambrian to present. They include glacial clays, soils, alluvium, loess, shale, claystone, and slates. The most common clay mineral present in these common clays is illite, a hydrated potassium aluminum silicate. Other clay minerals that are often present are chlorite, kaolinite, and smectite along with mixed layer clays. Quartz is usually the most common non-clay mineral present. It is estimated that worldwide over 200,000,000 tons of these common clays are used as raw materials for use in producing structural clay products, lightweight aggregate, and cement. These materials are economically very important in the construction industry. Generally, they are locally available and are an inexpensive raw material.

Clays and shales are used in so many different structural clay products that they necessarily have a wide range of physical properties (Murray, 1994). The properties that are important and are tested include plasticity, green strength, dry strength, dry and fired shrinkage, vitrification range, and fired color. The properties desired vary with the type of structural clay product being manufactured.

Most clays are plastic when naturally wet or mixed with water. The type of clay minerals, particle size, particle shape, organic matter content, soluble salts, absorbed ions, and the amount and type of non-clay minerals influence the plastic properties of the clay or shale. Green and dry strength properties are important because most structural clay products must be strong enough to be handled by hand or by machine and still maintain their shape. The presence of a small amount of smectite generally increases the plasticity and dry strength. In fact, if the plasticity and dry strength are too low then a small addition of sodium bentonite can improve both of these properties. Both drying and firing shrinkage are critical properties of clay and shale used for structural clay products. The presence of rather large amounts of smectite (10 to 20%) commonly causes excessive shrinkage, cracking, and slow drying. The temperature range of vitrification on glass formation during firing is a very important property because if this temperature range is too short then it is difficult to regulate the kiln temperature so that the product does not completely melt to a glass. Uniform color is also an essential property of many structural clay products and must be tested.

Clays and shales used as a raw material to make lightweight aggregate must contain a mineral or minerals which will produce a gas at the same temperature that the material reaches the vitrification temperature. The most common gas is CO_2 which is emitted when calcite or dolomite breaks down at temperatures in the range of 800EC. If the clay or shale vitrifies at this temperature then the gas forms small vesicules and looks like a pumice. The finished product must be strong and lightweight in the range of 1281 to 1602 kg/m^3 or less. Crushed stone and gravel weighs 2002 kg/cm^3 or more. Other properties that are important are the thermal and acoustical insulation, high fire resistance, and toughness (Mason, 1994). For use in cement it is the alumina and silica content which are important. Lightweight aggregate is produced and used when the weight of a structure is critical and needs to be relatively low. Long bridge spans and tall buildings commonly use lightweight aggregate materials.

6. IMPROVED BENEFICATION RESULTING FROM IMPROVED PROCESSES

Over the past two or three decades, many new and/or improved processes have been developed which has resulted in higher quality and more uniform clay products. This has resulted in increased industrial utilization and has also resulted in new products. Another important factor is that these improved processing technologies has permitted the use of marginal and sub-marginal clays which substantially increased the reserves of these industrial clays.

The process equipment manufacturers and the process engineers in the clay companies have made some outstanding innovations in process equipment and technology over the past 30 years. Some of these more important special and improved processes are listed in Table 5.

Many of these processes have been described in the literature (Pickering and Murray, 1994; Prusad et.al., 1991; Pruett, 2000). Therefore, only air classification, calcination, flocculation, flotation, and magnetic separation will be described here. Air classification is used in the dry processing of kaolin, bentonite, and palygorskite-sepiolite to remove quartz, mica, and other non-clay minerals and to produce various size fractions. Size classifications of minus 44, 20, 10, and 5 microns can now be produced using air classification.

Table 5 - Special and improved processes applicable to clay beneficiation

Acid activation	Dewatering	High intensity blunging
Air classification	Dispersion	Hydrocyclones
Blending	Drying	Magnetic separation
Calcination	Electrostatic separations	Organocladding
Centrifuging	Extrusion	Oxidation

Chemical leaching	Flocculation	Pulverization
Computer controls	Flotation	Surface treatments
Delamination	Granular sizing	West and dry screening

The calcination process is used in the kaolin industry to produce value added special grades for use in filling plastic and rubber coatings on electrical wire, in paint, in both paper coating and filling to extend T_iO_2 and to improve brightness and opacity, and for producing very fine abrasives for polishing compounds. Calciners with very good temperature controls have been developed. High heat drying is used by the absorbent clay producers to harden granules so that they will not slake in water or other liquids. The temperature is controlled so that the mineral structure is maintained. For example, a calcium montmorillonite is still a calcium montmorillonite, but is completely dehydrated but the internal structure has not changed.

Flocculation is an important process used in wet processing of both kaolin and bentonites in order to improve filtration rates. Common flocculants are sulfuric acid, alum, and long chain polyacrylates. Special selective flocculation processes have been developed in the kaolin industry to separate contaminants like iron and titanium minerals (Attia, 1982).

Flotation is now a common process used by kaolin producers to make high brightness products by removing iron and titanium mineral impurities (Prasad et.al. 1991). Improved flotation cells and column flotation columns have been developed which are more effective in improving brightness and reducing losses.

High intensity wet magnetic separation is now a standard process used in the kaolin industry and has been pilot tested to produce a very high brightness white bentonite. Significant design improvements have resulted in a more uniform field strength and more efficient flow distribution through the canisters. Large cryogenic magnetic separators with field strengths as high as 10 Tesla are now available. All these design improvements have resulted in reduced power requirements, a higher brightness product, and better ability to remove ultrafine paramagnetic minerals.

New and improved process equipment for both wet and dry processing is continually being developed. This enables clay producers to make new and improved products (Murray, 1999; Harvey and Murray, 1997).

7. CONCLUSIONS

Clays in industry are exceedingly important industrial minerals. The diversity of use, value, and location make clays an essential raw material in the manufacture of many products as enumerated in this paper. Applied research and development and innovative processes are continually needed in order to produce new and improved products and to enable the clay raw materials to be beneficiated to meet uniform and high quality standards. The important clyas in industry are kaolins, smectites, palygorskite-sepiolite, and common clays.

REFERENCES

1. H.H. Murray and W.D. Keller, Kaolins, Kaolins and Kaolins. In. H.H. Murray, W.M. Bundy, and C.C. Harvey (eds) Kaolin Genesis and Utilization. Special Pub. 1, (1993) 1-24
2. R.E. Grim and H. Guven, Bentonites. Developments in Sedimentology 24 Elsevier, N.Y.

(1978) 256

3. S.H. Patterson, _Fuller_s Earth and Other Industrial Mineral Resources of the Meigs - Attapulgus-Quincy District, George and Florida , Geol. Surv. Prof. Pap. (U.S.) 828 (1974) 45

4. H. Zhou, Mineralogical and Industrial Evaluation of a Palygorskite Deposit from Guanshon, Anhui Province, P.R. China. Ph.D. Thesis, Indiana Univ. (1996) 197

5. H.H. Murray, _Traditional and New Applications for Kaolin, Smectite, andPalygorskite: a General Overview Applied Clay Science 17 (2000) 207-221

6. S.M. Pickering and H.H. Murray, _Kaolin in D.D. Carr (ed): Industrial Minerals and Rocks, 6th ed. SME (1994) 233-246

7. H.H. Murray, _Catalysts in D.D. Carr (ed): Industrial Minerals and Rocks, 6th ed. SME (1994) 191-194

8. J. Iannicelli, _Polymer Reinforcement with Amino and Mercaptosilane Grafted Kaolin Miner. Metall Process 8 (1991) 135-138

9. J.M. Elzea and H.H. Murray, _Bentonite In D.D. Carr (ed) Industrial Minerals and Rocks, 6th ed. SME (1994) 233-246

10. R.G. Grim, Applied Clay Mineralogy, McGraw-Hill, N.Y. (1962) 422

11. J.M. Elzea and H.H. Murray, _Variation in the Mineralogical, Chemical and Physical Properties of the Cretaceous Clay Spur Bentonite in Wyoming and Montana, USA, App. Clay Sci. 5 (1990) 229-248

12. K.S. Keith and H.H. Murray, _Clay Liners and Barriers. in D.D. Carr (ed). Industrial Minerals and Rocks, 6th ed. SME (1994) 435-452

13. W.J. Lang, _Bentonite: The Demand and Markets for the Future SME Preprint 71-H-29 Littleton, CO 1971 3

14. I.G. Odom, _Smectite Clay Minerals: Properties and Uses Philos. Trans. R. Soc. London, Series A 311 (1984) 391-409

15. J.A. Rausell-Colom and J.M. Serratosn, _Reactions of Clays with Organic Substances in A.C.D. Newman (ed) Chemistry of Clays and Clay Minerals Miner. Soc, London (1987) 371-422

16. J.P. Rupert, W.T. Granguist, and T.J. Pinnavaia. _Catalytic Properties of Clay Minerals in A.C.D. Newman (ed) Chemistry of Clays and Clay Minerals Miner. Soc, London (1987) 275-318

17. G. Beall, Personal Communication

18. R.H.S. Robertson, Fuller_s Earth-A History, Volturna Press, Kent., U.K. (1986) 421

19. E. Galan, _Properties and Applications of Palygorskite-Sepiolite Clays, Clay Miner. 31, (1996) 443-445

20. K.S. Keith, Charactarization and Permeability of Sepiolite, Palygorskite, and Other Commercial Clays and Their Applicability for Use as Impermeable Barriers inWaste Disposal. Ph.D. Thesis, Indiana Univ. (2000) 274

21. M.S. Prasad, K.J. Reid, and H.H. Murray, _Kaolin: Processing, Properties, and Applications. App. Clay Sci. 6 (1991) 87-119

22. A.J. Pruett, Georgia Kaolin: Development of a Leading Industrial Mineral.Mining Egr. 52 (2000) 21-27

23. V.A. Attia, _Fine Particle Separation by Selective Flocculation Sep. Sci. Technol., 17 (1982) 485-493

24. H.H. Murray, _Applied Clay Mineralogy Today and Tomorrow Clay Miner. 34 (1999) 39-49

25. C.C. Harvey and H.H. Murray, _Industrial Clays in the 21st Century: A perspective of Exploration, Technology, and Utilization. Appl. Clay Sci. , (1997) 285-310

26. B.H. Mason, _Lightweight Aggregates in D.D. Carr (ed); <u>Industrial Minerals and Rocks</u> 6th ed. SME (1994) 343-350
27. H.H. Murray, _Common Clay in D.D. Carr (ed): <u>Industrial Minerals and Rocks</u> 6th ed. SME (1994) 247-

The Tandilia System, Province of Buenos Aires, Argentina: its sedimentary successions

Andreis, R. R. – Museo Paleontológico "Egidio Feruglio" – Av. Fontana 140 (9100) Trelew, Chubut, Argentina – E-mail: rrandreis@sinectis.com.ar

Over Paleoproterozoic to Mesoproterozoic hydrothermalised and weathered basement rocks (2200-1000 My), named "Complejo Buenos Aires", several sedimentary cycles of Neoproterozoic to early Ordovician ages were deposited. A slow sea level rise allow the first Neoproterozoic sedimentation related to the Villa Monica/ La Juanita Formations quartzites, that covers an probable embayment at Sierra Bayas (Olavarría) and extends to Cuchilla de las Aguilar (Barker) and San Manuel areas. The same areas also show a regressive event produce the deposition of the Cerro Largo Formation quartzites under nearshore-tidal conditions, followed by a second rise deposited a single parasequence, the Olavarrría Formation (Upper Rhyphean). A slow transgression allows the sedimentation of black, grayish to brownish calcareous sediments (Loma Negra Formation), followed by a falling sea level that formed an karstic paleorelief on this carbonatic rocks (Loma Negra and El Infierno quarries). A fourth transgression deposited the Cerro Negro Formation including shaly, heterolithic and sandy facies (Rhyphean?, 680 My), with two depositional systems separated by an ondulate surface produced by a slow sea rise. The successions represent an epeiric sea under subtidal conditions followed by an open sea sedimentation. The correlated Las Aguilas Formation (Cuchilla de las Aguilas) includes a cherty limestone breccia, reddish claystones and a heterolithic cycle affected by longshore and tidal currents.

During the early Ordovician, a fifth transgression occurred with the deposition of frequent cross-bedded quartzites, some clast-supported breccias, and minor massive siltstones. Bioturbations are common. Paleocurrents are mainly oriented to SW to WNW, but also NE-SW to N-S tidal currents can be found.

1. INTRODUCTION

The Tandilia System is a non continuous NW-SE range which extends 300 km in the Province of Buenos Aires, from Mar del Plata to Blanca Grande localities, with heights varying from 50 to 490 m over sea level and comprising Paleo to Mesoproterozoic crystalline basement rocks (2200-1000 Ma). This old basement was later covered by different sedimentary cycles of Neoproterozoic to early Ordovician age (8, 11, 26, 37, among others).

The Sierras de Tandilia is the first mining area of the Province of Buenos Aires, with important reserves on limestones and shales as the principal mining commodities. Located near great consumer centers (Tandil, Olavarría, Buenos Aires, Mar del Plata), the deposits have underwent hydrothermal, weathering and diagenetic processes superimposed, which led to different clay assemblages depending on the old hydrothermalised and weathered crystalline basement paleorelief and the geological time elapsed.

The effects of the horst and graben structure developed during the Tertiary led, for more than a hundred years, to the wrong idea of the existence of a unique basin and sedimentary cycle. Recently, most of the evolution history of the Sierras has been partially enlighten. Yet, some stratigraphical problems are still to be solved. The old basins, only preserved here as "patches" of the original ones are, however, well developed and preserved in South Africa, Namibia (Nama Group), after the continental break during Mesozoic times (13). The geological correlation between these deposits is a very important clue to the understanding of the variation, lithological differences and fossiliferous content of the depositional cycles involved.

2. TRANSGRESSSIVE - REGRESSIVE PROCESSES AND PALEO ENVIRONMENTAL CONDITIONS

The crystalline basement rocks, named "Complejo Buenos Aires" (16) are mainly composed of granitoids, migmatites, ectinites, mylonites, amphibolites, tonalites and dike rocks (38), with ages from 2000 to 1400 My (37). They also include a 90 m thick metapelite unit (Punta Mogotes Formation), recognized through subsurface studies in Mar del Plata (9, 28). Metamorphic processes occurred during Neoproterozoic-Cambrian times (11, 585 to 515 ± 14 My).

A <u>slow and continuous sea level rise</u> would have been linked to a regional tectonic subsidence, which led to the development of the <u>first Neoproterozoic depositional basin</u>. At the beginning of this transgression the " Complejo Buenos Aires" remained as a positive relief in the San Manuel, Barker and Estación López (Constante 10 and El Cañón quarries) areas, undergoing hydrothermal and superimposed weathering processes. Mainly at San Manuel and Estación López quarries it is possible to recognize in the basement bedrock, saprock and saprolite zones, with superimposed paleosols (22, 43), favored by restricted geomorphic conditions as well as tectonic stability. In San Manuel area the basement is characterized by reddish weathered migmatitic rocks, while Constante 10, the rocks are of tonalitic composition. Here, the tonalites reached the maximum stage of weathering, suggested by abundant detrital kaolinite and rare illite in the paleosols, also confirmed by the existence of remainder diagenetic anatase. This old depositional sequence is represented by the <u>Villa Mónica Formation</u> and the correlated <u>La Juanita Formation</u>, cropping out in Sierras Bayas, Olavarría (29, 30), in Barker (27, 39, 46), and San Manuel areas (4, 43). The sequences probably covered a possible embayment at Sierras Bayas and extended from Cuchilla de Las Aguilas (near Barker) to San Manuel. Their extension to the SE (Mar del Plata region) is still doubtful.

In the Olavarría (29, 30) and Barker areas (2, 3, 12, 22, 27, 35), both correlated units show two lithofacies of about 16 m in thickness and comprise: a) basal quartzites, somewhat arcosian at the base, and including intraclasts derived from the erosion of the underlying basement (saprolite/saprock), with through cross-bedding and thinning upward tendency, and b) tabular laminated or domed stromatolitic dolomites and greenish to reddish illitic shales of 36 m thick (Olavarría) or also fine sandstones and ferruginous shales of 10 m thick (Barker). The first association suggests littoral to shoreface sedimentation, sometimes covering an eroded undulate basement paleorelief (San Manuel, 43). The second one may represent a single parasequence with progressive rise of sea level and deposition of shales followed by shallow sandstones (shoaling) on top. In Barker, the succession is characterized by algal barriers or even narrow platforms, with deposition under subtidal to intertidal conditions, while the increase in sandstones and reddish shales in near outcrops may also suggest supratidal conditions, probably near the ancient coast (21). Paleocurrents are mainly oriented to E-NE and SE and symmetric straight ripple-marks show NE-SW orientation, parallel to the near paleocoast (Barker), while in Olavarría, paleocurrents are oriented to SW or to NW (30), probably perpendicular to the supposed embayed coast. The origin of the clayey sediments is related to the erosion of the near basement and are, in general, detrital (42), including illite, kaolinite, and scarce smectite (see also 27).

A continuos regressive event is now related to the deposition of about 25 m thick quartz sands (Cerro Largo Formation, 27, 29, 43), followed by a weak second transgression and subsequent aggradation of sandy sediments (44). The lowering of the sea produced an irregular regional paleosurface, which separates the underlying muddy-sandy succession from the quartzites. This paleosurface has also been identified in San Manuel area (43), in Cuchilla de Las Aguilas and other hills located to the west of Barker (3). The unit measures 22 m in Olavarría and 33 m in San Manuel (29, 43). The quartzites are lenticular, with frequent trough cross-bedding, scarce ripple-marks and glauconite. Bioturbation was also observed (*Paleophycus, Didymaulichnus, Phycodes and Skolithos*, 31, 34, 43, 46). This unit was deposited in nearshore conditions, on tidal channels with NW and SE orientation, perpendicular to the coast line, under meso to macrotidal influence (43). Also, frequent lateral migration, probably related to storms, was observed. Paleocurrents oriented to NE and NNE are linked with littoral currents flowing along the coast (44). The sandy sediments could have been deposited in 5 to 10 m shallow sea water, but a possible fluvial influence in the depositional history is not discarded.

Later on, a renewed second transgression occurred, covering the quartzites with a single parasequence of 38 m thick, starting with massive illitic muddy sediments, showing an upward progressive increase in glauconitic heterolithic (wavy) and sandy facies. The Olavarría Formation represents the deposition in an open sea platform (50 m in depth), changing upward to a subtidal platform related to a lowering in the transgression velocity (2). In some dark-gray shales, an abundant pyritised microflora (cyanophyceous algae and *Acritharca*) of Upper Rhiphean age (700-800 My) were found (32). Similar Rb/Sr ages (756-797 My) were also obtained (8). Paleocurrents flew mainly to NW, SW, E and SE.

At the Aust, LOSA and Cerro Negro quarries (Sierras Bayas, Olavarría), illite, minor chlorite and illite-smectite and chlorite-smectite interstratified clay minerals, with variable degree of crystallinity were identified (24). Illitic material (I+ISII), with less than 15 % expansive layers and 1M/1Md polytypes were identified. Detrital micas (1M) + illite derived from the erosion of the weathered granitic basement rocks. The 1Md polytype reflects diagenetic processes which led to the formation of illitic material (44).

A third slow Neoproterozoic transgressive event (29), is related to the deposition of up to 45 m of black, dark-gray or brownish massive micritic limestone, included in the Loma Negra Formation, that covered the Olavarría Formation fine sediments (Olavarría and Barker). This unit shows basal reddish calcipelites with ripple-marks and some terrigenous sandy material over a carbonatic platform (Olavarría locality, 29). At Loma Negra (near Barker) and El Infierno quarries an extended undulate karstic plain developed under warm, high humidity and frequent rain conditions, producing the dissolution and formation of towers and dolines, and which represents a long subaerial exposition related to the lowering of the sea level (6, 7). The regression process favored the formation of phosphates and also the precipitation of silica, dolomite and chamosite. As the sea retreated rapidly, the limestones suffered partial reworking, producing lenticular mixed phosphate and limestone breccias over a wide an irregular paleosurface (25). The lowering of the sea has been related to a glacimarine event (25), corroborated by the presence of glacimarine debris between the basement and the Balcarce Formation, to the East of Balcarce locality (36). Also was suggested the existence of a shallow epeiric or perhaps an epicontinental sea, with slopes of 3cm/km, with restricted water circulation which provoked hypersaline conditions (25).

A <u>fourth transgressive event</u> is represented by the deposition of the <u>Cerro Negro</u> <u>Formation</u> sediments. At Loma Negra (Barker), El Infierno and Cerro Negro quarries it is mainly composed of gray-olive shales, illitic in composition (1Md and 1M polytypes), of good crystallinity, with random chlorite-smectite and scarce chlorite, smectite and kaolinite (7, 41). These fine sediments partially cover lenticular intraformational calcareous breccia deposits of radial distribution along a wide bay (5, 14). Also, some whitish, yellowish or reddish shales of this unit (7), together with quartzite and limestone fragments have fallen into the dolines by roof collapsing. To the NW of the Loma Negra quarry, thin basal quartzites with disperse limestone blocks and pebbles, may appear. To the top, the shales are substituted by fine rippled sandstones (Cerro Chato, 7), forming a unique parasequence.

In some interbedded black shales *Acritharca*, (*Spheromorphitae*) fossils have been identified and K-Ar datation showed an average age of 680 My (12). The existence of basal shale deposits suggests a higher paleorelief for the limestone sediments.

At Sierras Bayas, this formation was also studied (24, 49). Considered of Neoproterozoic age (723 ± 21 My)(8), this unit is up to 150 m thick. Two depositional systems (LDS and UDS) were recognised, separated by an undulate paleosurface (5). The lower one (LDS), of 50 m thick, covers a wide irregular gentle karstic paleorelief with excavated channels and open to the SE. It is composed of basal greenish to reddish siltstones and claystones with chert nodules, phosphate breccias, oolites, oncolites and some limestone blocks followed by massive, laminated or with linsen to wavy structures, ripple marls, calcareous siltstones and scarce breccias. These deposits cover the Loma Negra Formation limestones, filling channels with upward coarsing or thinning metric thick cycles. Paleocurrents are oriented to WNW and less to SE. The absence of biogenic structures suggests precipitation of aragonitic mud during low sediment input from the basement rocks.

The UDS (50 to 89 m thick), includes muddy and heterolithic facies (90%) and subordinate sandy facies (10%), of greenish, brown-olive to reddish hues. Centimetric to decimetric tabular beds are frequent. Sandstones bear sole marks, current or wave ripple-marks. Paleocurrents flew to SE and W and wave ripples are oriented NW-SE or NS, parallel to a distant coast. The LDS was deposited during a transgressive phase event in an epeiric basin, under subtidal conditions and rather muddy and warm waters. The UDS sediments were deposited in an open epicontinental sea, probably in nearshore areas related to <u>slow sea level rise</u> that occupied the valley and peripheral channels. The sequences reflect deposition under subtidal to intertidal conditions, with wave action, periodically intensified by storms. Thin localized low-density turbidites are attributed to this storm action or may be associated to westward oriented mass gravity slumps, probably related to earthquakes along active faults located toward the northern basin border (21).

The LDS contains I+ISII, with <15% expansive layers and 1Md/1M polytypes, but the UDS includes random chlorite-smectite and scarce smectite and most of the clay minerals recognize a detrital origin related to the erosion of the basement rocks (49).

At Cuchilla de Las Aguilas, 3 km NW from Barker, a new unit was recognized (45), the Las Aguilas Formation, of about 20 m thick, and later studied in more detail at Sierra La Juanita east from Barker (4, 28, 45). Three lithofacies were identified: the lower one represented by silicified calcareous breccia, with silicified oolites and pelloids; the middle one, of reddish to whitish claystones, and the upper one, a coarsening upward sandstone-heterolithic-shale sequence with ripple-marks, lamination, hummocky structures and syneresis and dessication cracks (3, 33, 45). The basal lithofacies may represent a transgressive-regressive event, and fills wave action formed channels. The reddish claystones were deposited under intertidal conditions favored by abundant debris derived from the weathered basement, while the upper lithofacies was deposited continuously in upper intertidal conditions, in extended tidal plains with currents and wave action influenced by periodic storms. Paleocurrents are mainly oriented to SW and NE (tidal).

Based on K-Ar datation on shales (11, 600 My), as well as on the stratigraphical position of this unit and the presence of oolites in the silicified basal breccia, it is possible to suggest a geological correlation between the Las Aguilas Formation and the Cerro Negro Formation (21, 45). The clay content includes: A. Pyrophyllite, illitic material (I+ISII) and traces of kaolinite (basal subunit); B. Dominant kaolinite and variable proportion of illitic material and pyrophyllite, lenses of alunite, diaspore and halloysite (10Å), the last three ones of diagenetic origin (middle subunit) and C. Illitic material, with less proportions of kaolinite, halloysite (10Å) and pyrophyllite (upper subunit), the latter diminishing toward the top of the deposit (44, 45). Kaolinite, pyrophyllite and illite are of detrital origin (42), but other authors postulate a hydrothermal alteration of pyroclastic sediments for the Las Aguilas Formation (17, 20). The incorporation of pyroclastic material to the shales is somewhat doubtful and even mysterious because the "basic and vesiculate pumice fragments" described by these authors are cogenetic with the microgabroid diabase that intruded the Balcarce Formation during Ordovician times (17, 20) and no one else has found any evidence of concomitant explosive volcanic processes occurred there. On the contrary, this intrusion occurred after the deposition of the Las Aguilas Formation (Neoproterozoic)(33). Moreover, the hypothesis that the pumices were deposited in littoral to marine conditions, with depths up to 50 m and covering great extensions is absolutely improbable. Another two questions arise from this statement: Where was the emission center located? and Which were the mechanisms that permitted the pumice fragment to reach the local Proterozoic coast? The other problem is related to the hydrothermal modifications of the clay deposits (18, 19). No hydrothermal veins have been recognized in the sequence and many of the clay minerals are detrital (39), coming from the erosion of the basement rocks and transported by currents during the different transgressive-regressive events occurred.

After a long-termed erosive process, a fifth transgressive event occurred, with the deposition of the Balcarce Formation sediments. The unit measures up to 90 m between Mar del Plata and Balcarce covering the weathered basement, the Punta Mogotes metapelites (28) and a glacimarine diamictite (36). Thickness is reduced at Cuchilla de Las Aguilas and Sierra La Juanita and may cover the Las Aguilas Formation (2, 27) and the weathered basement rocks to the SE, NW and central areas (3, 21, 24, 47). The Balcarce Formation includes basal clast-supported breccias (northern Cuchilla de Las Aguilas), fine to medium quartzites, well sorted sabulites with frequent through cross-bedding, lamination and ripple-marks, and scarce lenticular, massive siltstones, as well as rare heterolithic facies. At Mar del Plata, Constante 10 and La Verónica quarries, the quartzites show frequent bioturbation (*Cruziana* and *Skolithos* biofacies, 1, 2, 10, 15. 45, 48), considered of Ordovician age. This age was confirmed by the datation on a basic intrusive body (33, 450-498 My, Early to Middle Ordovician). The major transport occurred towards NNE-SSW or NNW-SSE by tidal currents (21, 23).

Clay minerals include abundant kaolinite, scarce illite and illitic material at Constante 10, La Veronica and in Balcarce-Mar del Plata region (1M, 2M)(44). All these material are detrital (including very fin|e micas), originated from the erosion of the weathered basement, and with diagenetic processes superimposed (39, 42).

The interbedded sandstones (with trough cross-bedding) and some heterolithic facies (with mud-drapes, reactivation surfaces, sigmoid structures, channels and bipolar palcocurrents) suggest slow oscillation of the sea level, changing from open sea to lagoonal paleoenvironments. The cherty breccias at Cuchilla de Las Aguilas suggest the existence of a wide stony beach whose materials were derived from the erosion of the underlying Las Aguilas Formation (3).

REFERENCES

1. **Alfaro, M., 1981.** Estudio geológico de la zona comprendida por las Hojas La Numancia, Licenciado Matienzo y Estancia San Antonio, en las Sierras Septentrionales de la Provincia de Buenos Aires. Resumen V Reunión Científica Informativa CIC, La Plata.
2. **Andreis, R.R. & P.E. Zalba, 1986.** La transgresión del Ordovícico inferior y la evolución paleogeográfica del basamento cristalino en el sector Barker-Chillar, provincia de Buenos Aires, Argentina. I Reunión Argentina de Sedimentología, La Plata, 1986. Resúmenes Expandidos, 189-192.
3. **Andreis, R.R. & P.E. Zalba, 1989.** Estratigrafía y paleogeografía de las secuencias cuarciticas al oeste de Barker (Buenos Aires, Argentina). I Jornadas Geologicas Bonaerenses, Tandil, 1985. Actas, 909-930.
4. **Andreis, R.R. & P.E. Zalba, 1998.** El basamento cristalino y eventos transgresivos y regresivos en las sucesiones silicoclásticas proterozoicas y eopaleozoicas aflorantes entre Chillar y San Manuel, Sierras Septentrionales, Buenos Aires, Argentina. VII Reunión Argentina de Sedimentología, Salta, 1998, Actas de Resúmenes, 101-103.
5. **Andreis, R.R.; P.E. Zalba & A.M. Iñiguez Rodriguez, 1992.** Paleosuperficies y sistemas depositacionales en el Proterozoico superior de Sierras Bayas, Sistema de Tandilia, Provincia de Buenos Aires, Argentina. IV Reunión Argentina de Sedimentología, La Plata, 1992. Actas, 1: 283-290, La Plata.
6. **Barrio, C.A.; Poiré, D. & Iñiguez Rodriguez, A.M., 1991.** El contacto entre la Formación Loma Negra (Grupo Sierras Bayas) y la Formación Cerro Negro: un ejemplo de Paleokarst, O)lvarría, provincia de Buenos Aires. Revista Asociación Geológica Argentina, 46(1-2): 69-76, Buenos Aires.
7. **Bertolino, S.R.A., 1988.** Estratigrafía, mineralogía y geoquímica de la Formación Cerro Negro en la zona de Villa Cacique, partidos de Juárez y Necochea, provincia de Buenos Aires. Revista Asociación Geológica Argentina, 43(3): 275-286, Buenos Aires.
8. **Bonhomme, M.G. & C.A. Cingolani, 1980.** Mineralogía y geocronología Rb-Sr y K-Ar de fracciones finas de la "Formación La Tinta", Provincia de Buenos Aires. Asociación Geológica Argentina, Revista, 35(4): 519-538, Buenos Aires.
9. **Borrello, A.V., 1962.** Formación Punta Mogotes (Eopaleozoico, Provincia de Buenos Aires). Notas CIC, provincia de Buenos Aires, 1(1), La Plata.
10. **Borrello, A.V., 1966.** Trazas, restos tubiformes y cuerpos fósiles problemáticos de la Formacion La Tinta, Sierras Septentrionales de la Provincia de Buenos Aires. Paleontografía Bonaerense, Fasc. 5, CIC, PBA, La Plata.
11. **Cingolani, C.A. & Bonhomme, M.G., 1982.** Geocronology of La Tinta upper Proterozoic sedimentary rocks, Argentina. Precambrian Research, 18(1-2), Elsevier Science Publ. Co, Amsterdam.
12. **Cingolani, C.A. & Rauscher, R., 1985.** Datos geocronológicos en las sedimentitas del Grupo La Tinta de Villa Cacique, partido de Juarez, provincia de Buenos Aires. I Jornadas Geológicas Bonaerenses, Tandil, 1985, Resúmenes, 128.

13. **Dalla Salda, L. H., 1979.** Nama and La Tinta groups: a common Southern Africa-Argentine Basin?. Chamber of Mines, 16th An. Rep., 1978, Univ. Cape Town, South Africa.
14. **Dalla Salda, L.H.; M. Guichon & C. Rapela, 1972.** Hallazgo de una brecha de talud en el techo de las calizas de Barker, provincia de Buenos Aires. Asoc.Min, Petr. y Sedimentología, III(4): 133.
15. **Del Valle, A., 1987.** Nuevas trazas fósiles en la Formación Balcarce, Paleozoico Inferior de las Sierras Septentrionales. Su significado cronológico y ambiental. Rev. Museo de La Plata (Nueva Serie), Sec. Paleont., 9: 19-41, La Plata.
16. **Di Paola, E.C. & Marchese, H.G., 1974.** Relación entre la tecto-sedimentación, litología y mineralogía de arcillas del Complejo Buenos Aires y la Formación La Tinta (Prov de Buenos Aires). Revista Asociación de Mineralogía, Petrología y Sedimentología, 5(3-4): 45-58, Buenos Aires.
17. **Dristas, J.A. & Frisicale, M.C., 1996.** Geochemistry of an altered pyroclastic suite interbedded in the sedimentary cover of the Tandilia Area, Buenos Aires, Argentina. Zentralblatt f. Geologie u. Paleontologie, 1(7-8): 659-675.
18. **Frisicale, M.C., 1991.** Estudio de algunos yacimientos de arcilla originados por actividad hidrotermal, en las Sas. Septentrionales de la Prov. De Buenos Aires. Tesis Doctoral, Univ. Nacional del Sur, Bahia Blanca, 217 pp. (inédito).
19. **Frisicale, M.C. & J.A. Dristas, 1993.** Alteración hidrotermal en el contacto entre el basamento y la secuencia sedimentaria en el Cerrito de la Cruz, Tandilia. XII Congreso Geológico Argentino y II Congreso de Exploración de Hidrocarburos, Actas, 5: 222-228, Mendoza.
20. **Frisicale, M.C. & J.A. Dristas, 2000.** Génesis de los niveles arcillosos de Sierra de la Tinta, Tandilia. Asociación Geológica Argentina, Revista, 55(1-2): 3-14, Buenos Aires.
21. **Iñiguez Rodriguez, A.M.; A. del Valle; D. Poire; L.A. Spalletti & P.E. Zalba, 1989.** Cuenca precámbrica-paleozoica inferior de Tandilia, Provincia de Buenos Aires. In: Chebli, G. & Spalletti, L.A. (Eds.), Cuencas Sedimentarias Argentinas, Universidad Nacional de Tucumán, Instituto Superior de Correlación Gcológica, Publ. 1430, Serie Correlación Geológica 6: 245-263, Tucumán.
22. **Iñiguez Rodriguez, A.M.; Zalba, P.E & Andreis, R.R., 1990.** Mineralogy and Chemistry of Cambrian (¿) paleosols, Tandilia System, Buenos Aires Province, Argentina. Proceed. 9 th Intern. Clay Conference, Strasbourg, 1989, V.C. Farmer & Y. Tardy (Eds.), Sci. Géol. Mémoir, 85: 175-184, Strasbourg, France.
23. **Iñiguez Rodriguez, A.M.; M.J.Manassero; D. Poiré & J. Maggi, 1996.** Génesis y procedencia de sedimentitas cuarzosas del área de Olavarria, provincia de Buenos Aires, Argentina. VI Reunión Argentina de Sedimentologia, Bahía Blanca, 1996. Actas, 61-66, Bahia Blanca.
24. **Iñiguez Rodriguez, A.M. & P.E. Zalba, 1974.** Nuevo nivel de arcilitas en la zona de Cerro Negro, Partido de Olavarria, Provincia de Buenos Aires. LEMIT, Anales, Serie II, 264, p. 95-100, La Plata.
25. **Leanza, C.A. & C.A. Hugo, 1987.** Descubrimiento de fosforitas sedimentarias en el Proterozoico Superior de Tandilia, Buenos Aires, Argentina. Revista Asociación Geológica Argentina, 42(3-4): 417-428, Buenos Aires.
26. **Linares, E., 1977.** Catálogo de edades radimétricas determinadas para la República Argentina. Publ. Especial, Asociación Geológica Argentina, Serie Didactica y Complementaria, 4, pp……, Buenos Aires.
27. **Manassero, J.M., 1986.** Estratigrafía y estructura en el sector oriental de la localidad de Barker, provincia de Buenos Aires. Asociación Geológica Argentina, Revista, 41(3-4): 375-384, Buenos Aires.
28. **Marchese, H.G. & Di Paola, E.C., 1975.** Reinterpretación estratigráfica de la Perforación Punta Mogotes N° 1, Provincia de Buenos Aires. Revista Asociación Geológica Argentina, 30(1): 17-44, Buenos Aires.
29. **Poiré, D.G., 1987.** Mineralogía y sedimentología de la Formación Sierras Bayas en el Núcleo Septentrional de las sierras homónimas, partido de Olavarria, provincia de Buenos Aires. Tesis Doctoral 494, FCN y Museo, Universidad Nacional de La Plata (inédito).
30. **Poiré, D.G., 1996.** La Formación Villa Mónica en el Precámbrico Sedimentario de Olavarría: Implicancias sedimentológicas y paleogeográficas. Actas 13 Congreso Geológico Argentino y III Congreso de Exploración de Hidrocarburos, Buenos Aires, 1996. 2: 107. Buenos Aires.
31. **Poiré, D.G. & A.M. Iñiguez Rodriguez, 1984.** Miembro Psamopelitas de la Formación Sierras Bayas, partido de Olavarría, Provincia de Buenos Aires. Revista Asociación Geológica Argentina, 39(3-4): 276-283, Buenos Aires.
32. **Pothe de Baldis, E.D.; Baldis, B. & Cuomo, J., 1983.** Los fósiles precámbricos de la Formación Sierras Bayas (Olavarría) y su importancia intercontinental. Revista Asociación Geológica Argentina, 38(1): 73-83, Buenos Aires.

33. **Rapela, C.; L. Dalla Salda & C. Cingolani, 1974.** Un intrusivo básico ordovícico en la Formación La Tinta (Sierra de los Barrientos, Provincia de Buenos Aires). Revista Asociación Geológica Argentina, 29(3): 319-331, Buenos Aires.
34. **Regalia, G.M. & H.M. Herrera, 1981.** *Phycoides aff pedum* (traza fósil) en estratos cuarcíticos de San Manuel, Sierras Septentrionales de la provincia de Buenos Aires. Revista Asociación Geológica Argentina, 36(3): 257-261, Buenos Aires.
35. **Schauer, C. & J. Venier, 1967.** Observaciones geológicas en la zona de Barker, Sierra de la Tinta, Provincia de Buenos Aires. Notas CIC, PBA, 5(6), La Plata.????
36. **Spalletti, L. A. & del Valle, A.,1984.** Las diamictitas del sector oriental de Tandilia: caracteres sedimentológicos y origen. Revista Asociación Geológica Argentina, 39(3-4): 188-206, Buenos Aires.
37. **Stipanicic, P. & E. Linares, 1969.** Edades radimétricas determinadas para la República Argentina y su significado geológico. Bol. Acad. Nac. Cienc., Córdoba, Argentina, 43: 51-96.
38. **Teruggi. M.E. & Kilmurray, J.O., 1980.** Sierras Septentrionales de la Provincia de Buenos Aires. En: J. Turner (Ed.), Geología Regional Argentina. Academia Nacional de Ciencias, Córdoba, Argentina. 2: 919-956.
39. **Zalba, P.E., 1978.** Estudio geológico-mineralógico de los yacimientos de arcillas de la zona de Barker, partido de Juarez, provincia de Buenos Aires y su importancia económica. Museo de La Plata, tesis Doctoral 362, 75 pp, La Plata (inédita).
40. **Zalba, P.E., 1979.** Clay deposits of Las Aguilas Formation, Barker, Buenos Aires province, Argentina. Clays and Clay Minerals, 27(6): 433-439.
41. **Zalba, P.E., 1981.** Nuevo nivel de arcilitas sobre la caliza en la Cantera Loma Negra, Barker. Asociación Geológica Argentina, Revista, 36(1): 99-102, Buenos Aires.
42. **Zalba, P.E., 1988.** Clasificación de arcillas de las Sierras Septentrionales de la Provincia de Buenos Aires. CETMIC, Publ. Especial 1, 62 pp., Buenos Aires.
43. **Zalba, P.E. & R.R. Andreis, 1998.** Basamento saprolitizado y secuencia sedimentaria suprayacente en San Manuel, Sierras Septentrionales de Buenos Aires, Argentina. VII Reunión Argentina de Sedimentologia, Salta, 1998, Actas, 143-153, Salta.
44. **Zalba, P.E. & Andreis, R.R., 2001.** Stratigraphy, Sedimentology and Mineralogy of Neoproterozoic clay deposits, Sierras de Tandilia, Province of Buenos Aires, Argentina. Economical Importance. 12th International Clay Conference (AIPEA), Bahía Blanca, Argentina, 2001. Pre-Simposium Field Trip, 79 pp.
45. **Zalba, P.E.; R.R. Andreis & A.M. Iñiguez Rodriguez, 1988.** Formación Las Aguilas, Barker, Sierras Septentrionales de la Prov de Buenos Aires, nueva propuesta estratigráfica. Asociación Geológica Argentina, Revista, 43(2): 198-209. B.Aires.
46. **Zalba, P.E.; R.R. Andreis & F. Lorenzo, 1982.** Consideraciones estratigráficas y paleoambientales de la secuencia basal eopaleozoica en Cuchilla de las Aguilas, Barker, Argentina. V Congreso Latinoamericano de Geología, Buenos Aires, 1982. Actas, 2: 389-409, Buenos Aires.
47. **Zalba, P.E. & Garrido, L.B., 1984.** Estudio de yacimientos de arcilla de El Ferrugo y Constante 10, provincia de Buenos Aires. I. Geología, mineralogía y clasificación textural, genética y tecnológica de las arcillas. Actas IX Congreso Geológico Argentino, Bariloche, 1984, 5: 575-588.
48. **Zalba, P.E.; D. Poiré; R.R. Andreis & A.M. Iñiguez Rodriguez, 1992.** Precambrian and Lower Paleozoic records and paleosurfaces of the Tandilia System, Buenos Aires Province, Argentina. In: Schmitt, J.M. & Gall, Q. (Eds.), Mineralogical and Geochemical Records of Paleoweathering. Memoire des Sciences de la Terre, 18: 93-113, Paris, France.
49. **Zalba, P.E.; C. Volzone; L.B. Garrido; M. Morosi & E. Pereira, 1994.** Mineralogical composition and diagenetic processes in the two depositional systems of the Cerro Negro Formation, Buenos Aires, Argentina: industrial application. Revista Geológica de Chile, 21(2): 303-311, Santiago, Chile.

Technological and compositional requirements of clay materials for ceramic tiles

M. Dondi

CNR-IRTEC, Istituto di Ricerche Tecnologiche per la Ceramica,
Via Granarolo 64, 48018 Faenza, Italy

The ceramic tile industry has been experiencing for the last three decades a dramatic process and product innovation, leading to a drastic upgrading of the technological cycle and the development of new typologies of wall and floor tiles. The evolution of body formulations and the changing requisites of composition, particle size distribution, and technological behaviour of clay materials are overviewed. The effects of this technological innovation are discussed, with particular reference to the applications and the technical requirements of clay raw materials.

1. INTRODUCTION

In the last two decades the tilemaking industry has registered the highest innovation rate of the entire ceramic sector. This circumstance has brought about a drastic renewal of the manufacturing cycle, which turned to be faster, more flexible, with considerable energy savings and a reduced environmental impact.

This process of technological upgrading of tilemaking plants has required, in the same time, an adjustment of composition and characteristics of the bodies, in order to meet the requisites of completely new manufacturing stages, such as wet grinding, spray drying and fast firing [1-2]. These changes in body formulation accomplished a real product innovation, with development of brand new typologies, such as *porcelain stoneware* or *monoporosa* tiles [3-4]. As a matter of fact, the traditional majolica or stoneware tiles produced in the Seventies (at maximum 15x15 cm2 in size) have evolved to present-day porcelain stoneware slabs (up to 120x120 cm2) with greatly enhanced technical performances and aesthetic appearance [4].

The variations in body composition have especially concerned both the amount and the technological properties of clay materials. The percentage of clays has been progressively lowered, both in wall and floor tile formulations [1-2]. But also the role of the clay component has been gradually varied from the single raw material of traditional majolica or stoneware tiles, to merely inorganic binder of the advanced bodies for porcelain stoneware tiles.

This paper is basically aimed at outlining the technological evolution in tile manufacturing, in order to highlight the most important repercussions that process innovation had on body design and raw materials selection, with particular emphasis on clays.

2. PROCESS INNOVATION IN TILEMAKING

The innovative trend in tilemaking plants has not been gradual, but more properly a stepwise process, with a rapid success of a specific innovation, followed by a period of consolidation and upgrading of the technology, prior to a further step.

There have been several driving forces of innovation in the last twenty years, mostly concerning economic, technological, environmental and marketing aspects. The starting-point was the mid Seventies energetic crisis, which forced the ceramic plant and tile manufacturers to revisit thoroughly the thermal process, introducing the roller kiln and the fast single firing [5]. As a consequence, during the Eighties the entire manufacturing cycle was progressively adapted to the new requirements (wet grinding, spray drying, fast drying, etc.). In the same period, ceramic districts with a high concentration of tilemaking plants, such as Sassuolo-Scandiano in Italy, had to face a restrictive environmental policy which: i) put very low limits for gaseous emissions (especially F and SOx), and ii) obliged every producer to recycle in the ceramic body all liquid and solid manufacturing residues [6]. In the Nineties, the driving force has mainly been the growing international competition, which urged the Italian industry to develop a high performance tile (porcelain stoneware) through an innovated technology [4, 7].

This picture can be simplified defining three broad technological stages: i) *traditional single- and double-firing*, ii) *fast single firing*, and iii) *porcelain stoneware*, more or less corresponding to the leading tilemaking plants in the '70s, the '80s and the '90s respectively. A further stage can be identified with the more recent technological advancements, and it could be referred to as iv) *advanced porcelain stoneware* (Table 1).

Table 1
Typical technological cycles used in tilemaking

Slow double-firing	Fast single-firing	Porcelain stoneware	Advanced stoneware
Dry grinding with hammer mills	Wet grinding with drum mills (silica balls)	Wet grinding with drum mills (alumina balls)	Wet grinding with drum mills (alumina balls)
Powder pelletization	Spray drying	Spray drying	Spray drying
Mechanical pressing (10-15 MPa)	Hydraulic pressing (20-35 MPa)	Hydraulic pressing (40-50 MPa)	Pressing (multiple loading), slip casting, compaction
Drying in tunnel driers (~24 h)	Drying in vertical driers (~1 h)	Drying in vertical driers (~1 h)	Drying in microwave, infrared driers (~30')
Biscuit firing (~1000°C, 30-36 h)	Glazing and decoration		Decoration (soluble salts, glass ceramics, glazing, etc.)
Glazing and decoration	Single firing (1050-1150°C, 30-45')	Single firing (1200-1250°C, 50-70')	Single firing (1200-1250°C, 50-70')
Gloss firing (~950°C, 12-16 h)		Polishing	Polishing

3. PRODUCT INNOVATION IN TILEMAKING

The enhancement of the technical performances of ceramic tiles followed different routes, starting from the classical body formulations of the double-firing technology, separate for the wall and the floor tiles, and for white and red coloured bodies. The frequent coexistence of different tilemaking technologies, even in the same district or company, brought about a complex mosaic of product typologies (Tables 2 and 3).

The traditional single- and double-firing technology used simple bodies made up of one clay, plus non-plastic materials (white bodies) or a small amount of scraps (red bodies) [2, 8]:

a) marly clay, typically illite-chlorite with some smectite and abundant carbonates, for red wall tiles (majolica) and glazed floor tiles (cottoforte);
b) red shale, generally illite+chlorite±kaolinite±smectite and carbonate-free, for unglazed red stoneware floor tiles;
c) kaolinite-illite ball clay, together with quartz sand and carbonate additions for white wall tiles (earthenware) or quartz and feldspars for white stoneware floor tiles.

These bodies, especially the red ones, cannot be utilized with the single firing technology, for several drawbacks, in particular the occurrence of drying cracks, black core, glaze pinholes, deformation during firing [9-10].

In order to overcome these drawbacks, a wide range of non-plastic materials, such as basalt, granite, aplite, feldspar or arkosic sand [11], have been introduced into the body in relevant amounts. The consequent reduction of plasticity have been compensated by the stronger forming pressure, and the finer particle size of powders, achieved through wet grinding.

Table 2

% weight	Red bodies			White bodies		
	Classic Majolica	*Birapida*	*Mono porosa*	Classic Earthenware	*Mono porosa*	Multi-purpose
Raw material	Slow double firing	Fast double firing	Fast single firing	Slow single firing	Fast single firing	
Red shale		20-30	20-30			
Marly clay	100	70-80	40-50			
Ball clay				45-50	20-40	20-30
Plastic clay					10-20	10-20
Feldspars						35-40
Granite, aplite					10-30	
Quartz sand			30-40	30-40	10-30	10-15
Carbonates				15-20	10-15	5-10

Body formulations for floor tiles

Table 3

% weight	Red bodies			White bodies			
	Red stoneware	*Cotto-forte*	*Mono-cottura*	Classic Stoneware	*Mono-cottura*	Porcelain stoneware	Advanced stoneware
Raw material	Slow single firing	Slow double firing	Fast single firing	Slow single firing	Fast single firing		
Red shale	100		70-80				
Marly clay		100					
Kaolin loam					20-40		
Ball clay				40-50	10-20	20-30	0-10
Plastic clay					15-20		20-30
Feldspars				30-40	0-10	40-45	45-50
Granite, aplite			20-30		30-40		
Quartz sand				10-20	10-20	10-15	5-10
Sintering promoters						0-5	
Pigments						0-5	5-10

Body formulation for wall tiles

In such a way, new products were developed and gradually replaced the classic ones: red *monoporosa* for majolica and white *monoporosa* for earthenware, while the traditional red and white stoneware tiles became glazed products (red and white *monocottura*). A technical improvement of classic majolica, achieved adopting a fast double firing, required a change in body composition, with red shale partially substituting marly clay *(birapida* tiles) [12].

The search for always better mechanical and tribological properties was addressed to a strong reduction of porosity, as low as 0.1% water absorption, with conspicuous vitrification of tiles, that acquire a *porcelainized* appearance (porcelain stoneware). This result was pursued improving the sintering kinetics: higher firing temperature (up to 1240 °C), finer particle size (prolonged grinding), denser unfired tiles (increased forming pressure). But It was necessary to modify the body formulation, starting from the white *monocottura* recipes, reducing the total amount of clay (rejecting kaolinite loams) and introducing large amounts of sodic and sodic-potassic feldspars instead of quartz-feldspathic materials, such as granite or aplite, and even adding sintering promoters (talc, dolomite, chlorite, wollastonite, etc.) [13].

Porcelain stoneware tiles were initially unglazed and their body had to be coloured in bulk, using pure raw materials (low iron and titanium contents) and 'whitening' agents (zircon or corundum) to enhance the effect of ceramic pigments [13]. A continuously growing fraction of porcelain stoneware production is now decorated on the surface (glazing, soluble salts, glass-ceramic coatings, etc.); this circumstance makes the use of expensive pigments and raw materials to a large extent unnecessary.

Multipurpose formulations have been designed in order to simplify the manufacturing cycle [14]: different typologies (e.g. white *monoporosa, monocottura* and porcelain stoneware tiles) can be actually produced with the same body, just modifying some processing conditions and adding calcium carbonate in the case of wall tiles (Table 3).

The most recent technological innovation has been focused on shaping, introducing new processes, such as slip casting, multiple loading pressing, compaction or stiff extrusion, which require a significant change of rheological properties of slips and powders [15]. The trend of this *advanced stoneware* is toward a further reduction of the clay percentage in the body, compensated by the use of inorganic (i.e. bentonite) and organic binders.

4. REQUIREMENTS OF CLAY MATERIALS

In the traditional single- and double-firing technology, the clay must meet all the technological requisites of processing and the product specifications. As a matter of fact, being the single raw material constituting the body, the clay has to be at the same time:
- fine-grained and easily disagglomerated by dry grinding (jaw crusher and hammer mill);
- very plastic, in order to get dense unfired tiles even with the low pressure of mechanical equipments; in particular, the mechanical strength must be high enough to withstand the weight of tiles, which were laid one upon the other during drying and firing;
- capable to consolidate during firing without significant shrinkage, coupling a certain mechanical resistance with a remarkable porosity (20-25% water absorption) necessary for the glazing process (majolica);
- capable to vitrify (water absorption below 1%) at relatively low temperature (~1000 °C) with good mechanical strength, quite uniform shrinkage and tolerable deformations on small sizes, typically 5x10 cm2 (stoneware).

During the transition to the fast firing technology, the above-mentioned requirements were put on one side by the new process [1-2]: i) particle size is actually determined by a more effective wet grinding, ii) plasticity is balanced with stronger pressing of spray dried powders, iii) tiles are no longer overlaid but arranged in monolayers, and iv) sintering behaviour is controlled by the clay-to-feldspar ratio. On the other hand, attention was paid to the occurrence of minor components, generally well tolerated in the traditional processing, but undesired in the new manufacturing cycles:
- smectite, interstratified I/S, soluble salts: interfere with the rheological behaviour of slips during wet milling and spray drying, increasing viscosity (less grinding efficiency) and thixotropy (difficult and slow emptying of ball mills) [16];
- iron oxi-hydroxides, Fe-rich chlorite, organic matter: enhance the fusibility, promoting both 'black core' and 'overfiring' phenomena (deformation, softening, bloating) in vitrified tiles [9];
- sulfides, rock fragments: cause defects to the integrity or colouring of glazes [17];

- large amounts of calcite and dolomite: lead to excessive porosity and formation of pinholes at the glaze surface [10].

These new criteria of clay selection brought about the rejection of several raw materials, and a drastic reduction in the use of many others, particularly red shales and marly clays. Typical ball clays for whitewares met satisfactorily the new requisites and entered in use for white *monocottura*, together with a different class of unconventional clay materials: white-firing, coarse-grained, illite-kaolinite loams with a considerable quartz content [18].

The progressive shift of the fast firing technology toward the porcelain stoneware brought about some changes in the requirements of clay selection. Clay purity was furtherly stressed: current acceptance limits have been lowered to 1.5% for both Fe_2O_3 and TiO_2, in order to avoid

Table 4
Composition and technological properties of the main clay types used in tilemaking

Property	Unit	Marly clays	Red shales	Kaolinitic loams	Ball clays (glazed)	Ball clays (unglazed)	Plastic clays
Kaolinite	% wt	<5	<5	10-30	40-55	40-55	35-55
Illite + I/S		20-25	30-35	10-20	20-30	15-25	5-10
Chlorite		5-15	5-10	<5	absent	absent	traces
Smectite		5-10	5-10	<5	<5	<5	10-20
Quartz		20-25	25-30	30-50	15-25	20-35	25-40
Feldspars		5-10	5-15	0-20	<5	<5	traces
Carbonates		15-25	<5	absent	absent	absent	absent
Fe-oxihydroxides		<5	5-7	<5	<2	<1	<2
SiO_2	% wt	46-52	56-60	64-75	57-62	60-67	61-68
TiO_2		<1	<1	<1	1-2	<1.5	<1.5
Al_2O_3		12-15	17-20	14-20	24-28	22-27	18-24
Fe_2O_3		4-6	6-8	1-6	1-2	<1.5	2-3
MgO		3-4	2-4	<1	<1	<1	<1
CaO		8-13	<3	<1	<0.5	<0.5	<1
Na_2O		<1.5	<1.5	<1.5	<0.5	<0.5	<0.5
K_2O		2-3	3-4	2-4	1-3	1-2	<1
L.o.I.		11-15	6-8	3-6	7-10	6-9	9-10
Fraction >63μm	% wt	<5	<5	15-65	<5	<10	<5
Fraction <2μm		40-50	55-65	10-40	40-75	45-75	60-70
Atterberg IP	% wt	20-30	40-50	5-15	15-30	15-30	35-45
Methylene Blue	meq	5-10	20-25	<5	5-10	5-10	15-20
Dry MOR	MPa	3-4	3-5	1-2	2-3	2-3	3-5
Firing shrinkage	cm/m	<1*	5-10*	0-5+	3-8+	1-9+	2-4+
Water absorption	% wt	20-25*	3-13*	5-20+	3-8+	3-12+	8-16+
Fired MOR	MPa	15-20*	25-40*	10-20+	20-40+	20-40+	10-20+

* Firing 1000 °C - 36 h. + Firing 1200 °C - 1 h.

undesired colouring after firing. On the other hand, plasticity returned to be an important criterion, as the clay amount in porcelain stoneware body has been lowered below 40%. As a matter of fact, fine-grained, plastic clays with high dry modulus of rupture are now preferred (Table 4). The occurrence of smectite and interstrafied I/S in these ball clays is tolerated and to some extent required; moreover, special Ca-Mg exchanged bentonites have found a certain application, particularly for the 'advanced' porcelain stoneware tiles.

The evolution of the tilemaking technology has dramatically modified the role of the clay component of ceramic bodies. Moving from the traditional double-firing to the current porcelain stoneware production, clays gradually lost their importance in the firing process, that has been transferred to the balance of the complex mixture of clays, feldspars, quartz and various additives which made up current bodies.

Even the role of mullite precursor, played by kaolinite-bearing clays, has taken a backseat, being the amount of mullite formed in porcelain stoneware usually around 10% [13].

Ceramic clays are now basically seen as 'inorganic binders', which must ensure to the unfired tiles a sufficient mechanical strength to withstand handling, drying shrinkage and glazing stresses before firing. The restricted requisites of purity and whiteness applied to ball clays (essentially a kaolinite-quartz-illite system) grew these materials ever refractory, with water absorption values as high as 5-10% after fast firing at 1200 °C, without significant repercussions, since their sinterabiliS is no longer considered as a discriminant requirement.

5. CONCLUSIONS

The evolution of the manufacturing technology has a considerable influence on both body composition of ceramic tiles and technical requirements that clay raw materials have to meet. The technological development during the last two decades has dramatically changed the role of the clay into the ceramic tile bodies:
- single and fundamental component of traditional majolica and stoneware tiles;
- main component (70-80%) with addition of non-plastic materials, of fast single fired red bodies *(monoporosa* and *monocottura* tiles);
- basic component, with a prevailing cohesive function, of fast single fired white stoneware tiles (about 50:50 with quartz-feldspathic materials);
- minor component (30-40%) of porcelain stoneware bodies, where it plays essentially the role of inorganic binder.

Different economic, technological and market factors have determined with time a superposition of various trends:
a) red-firing bodies have been gradually replaced by white-firing ones, leading to rejection of clays with significant content of iron oxi-hydroxides, Fe-chlorite, etc.;
b) dry grinding has been progressively substituted by wet milling, with rejection of raw materials containing significant amounts of smectite or expandable clay minerals, which worsen the rheological behaviour of slips, or mica, that is difficult to be ground;

c) firing has been almost completely transferred to roller kilns, with very fast cycles (30-50') that hinder the use of clays rich in carbonates, organic matter or sulfides, which cause defects to the glaze and promote 'black core' phenomena;
d) porcelain stoneware bodies are mostly decorated in bulk by addition of ceramic pigments, that require a white base for a reliable and economic colouring (clays must have very low iron and titanium contents);
e) always larger tile sizes (over 1 m2) and new shaping techniques demand for more plastic clays, somehow acting as 'inorganic binders' (fostering raw materials with finer particle-size and wider specific surface).

In this framework, it is possible to sketch some short-term tendencies of clay application in ceramic tile production:
- the demand for <u>very plastic and low-Fe, low-Ti ball clays</u> will increase with porcelain stoneware growing output;
- <u>conventional ball clays</u> will be confined to bodies for fast fired glazed tiles (*monocottura*);
- <u>marly clays and red shales</u> will be used just in red coloured bodies for glazed tiles;
- <u>kaolinitic loams</u> and other unconventional white-firing clays will be rejected or restricted to fast fired glazed tiles.

These trends point out some unresolved problems and subjects for future research:
- new mineralurgic treatments are needed to lower iron and titanium percent in ball clays;
- better comprehension of slip rheology is required to allow a wider use of smectite and I/S interstratified;
- special additives have to be developed to reduce the colouring effect of iron and titanium in porcelain stoneware tiles;
- exploitation of the huge amounts of kaolinitic loams, red shales and other clay materials no longer used in tilemaking.

REFERENCES

1. P.G. Burzacchini, Ceram. Inf., 20 (1985) 283
2. M. Bertolani, Fabbri B., Fiori C., Loschi Ghittoni A.G., Ceram. Inf., 21 (1986) 333
3. P.G. Burzacchini, Ceram.World Rev., 11 (2001) 114.
4. P.G. Burzacchini, Ceram. World Rev., 10 (2000) 96.
5. M. Poppi, La Ceramica, 32 (1979) 1
6. C. Palmonari, F. Cremonini, A. Tenaglia, G. Timellini, La Ceramica, 34 (1981) 1
7. A. Brusa, Int. Ceram. J., October (1999) 73.
8. M. Dondi, Appl. Clay Sci., 15 (1999) 337.
9. A. Escardino, A. Barba, A. Blasco, F. Negre, Br. Ceram. Trans., 94 (1995) 103
10. J.E. Enrique, A. Escardino, J. Garcia-Ten, V. Cantavella, 4th QUALICER, 2 (1998) 69
11. M. Dondi, Tile & Brick Int., 10 (1994) 77
12. M. Dondi, G. Guarini, S. Guicciardi, C. Melandri, M. Raimondo, Ceram. Inf., 35 (2000) 985
13. M. Dondi, G. Ercolani, M. Marsigli, C. Melandri, C. Mingazzini C., Interceram, 48 (1999) 75
14. A. Brusa, A. Bresciani, Ceram. Eng. Sci. Proc., 17 (1996) 50
15. M. Brezina, Ceram. Eng. Sci. Proc., 20 (1999) 113
16. A. Albertazzi, E. Rastelli, Ceram. Acta., 9, No.4 (1997) 5
17. V. Capucci, Int. Ceram. J., 59 (1993) 14
18. Fabbri, C. Fiori, I. Venturi, Ceramurgia, 11(1981) 131

Activation of clays for environmental uses - can we improve on Nature, and can it pay?

G.J. Churchman[a] and C. Volzone[b]

[a]CSIRO Land and Water, PMB 2, Glen Osmond, South Australia 5064, AUSTRALIA

[b]CETMIC, C.C. 49, Cno Centenario y 506, (1897) M.B. Gonnet, Prov. Buenos Aires, ARGENTINA

Examples are given of two different methods for the activation of clays to enhance or enable their uptake of a variety of both solutes and gases. In one, complexes of a smectite with organic polycations varied in their ability to remove toluene from water, with an aromatic polycation showing more uptake than others. Only small additions of highly cationic polycations resulted in positively-charged complexes for anion uptake. In the other, acid-activated smectites showed enhanced gas adsorption and also some selective adsorption, including that of CO_2. A kaolinite after acid-activation, followed by mechanical and thermal treatments, showed similar characteristics, while amorphous derivatives of kaolinite had a high capacity for gas sorption. Results of related studies are also discussed, along with other methods for the activation of clays for environmental uses.

1. INTRODUCTION

While possessing the capability to remove, immobilise or degrade contaminants, soils and other natural reactive materials each have only limited capacities for these purposes. It has been claimed that Nature is able to deal with virtually all wastes given enough time [1]. Nonetheless, other materials and technologies are required to provide effective adsorbents or barriers for contaminants, or as catalysts for their degradation in the shorter term.

As the minerals within the <2 μm material, clays form the most reactive inorganic component of soils. They provide soils with virtually all of their surface area [2] and, apart from a contribution from "humus" i.e. processed organic matter in surface soils [3] generally all of their electrical charge. Charge-bearing minerals such as zeolites occur in coarser particles within some rare soils as exceptions. The use of clays in both their natural and activated forms to remove, contain or alter contaminants represents the augmentation of these useful constituents of the natural system in order to alleviate its overloading or poisoning. Among clay minerals, aluminosilicates are most commonly mined or marketed for use as environmental materials.

Their extensive surfaces and electrical charge encourage the adsorption of other materials by aluminosilicate clay minerals. Generally, they are predominantly negatively charged, so most have a much lower affinity for anions than for cations. They also have strong affinity for water. On the other hand, aluminosilicate clays have little attraction for non-ionic or non-polar compounds, including many organic contaminants of concern such as hydrocarbons. Some, e.g. smectites, sepiolite and palygorskite, have extensive internal surfaces and pores that are useful for the adsorption of gases, including their selective adsorption, which has a potential for environmental applications [4]. Aluminosilicate minerals or their activated products often act environmentally as adsorbents. However, they are also used commonly as barriers, which may be for the containment and/or adsorption or immobilisation of contaminants [5, 6] and as catalysts, either for the breakdown [7] of pollutants or in chemical processes to minimise effluent pollutants [8].

Activation can extend the capabilities of clays for environmental uses both quantitatively and qualitatively. Clays may become activated by a variety of means, including sodium saturation, organic modification, various forms of pillaring, acid treatment, heating and alteration with different chemicals. Some will be discussed here, but no doubt many others await development.

This paper presents particular examples of methods of the activation of clays for providing solutions to some environmental problems that are illustrated by summaries of results from our laboratories.

2. ACTIVATION BY POLYCATIONS TO ENABLE UPTAKE OF NON-IONIC ORGANIC COMPOUNDS AND ANIONS FROM WATER

Quaternary alkylammonium cations (QACs) have been used most commonly for the organophilic activation of clays, to enable their uptake of sparingly soluble non-ionic organic compounds such as hydrocarbons and phenols from aqueous solutions [9-12]. However, it is worthwhile for economic, among other reasons, to broaden the range of compounds suitable for this purpose. Like QACs, cationic polyelectrolytes comprise organic cations containing non-polar groups and recent work has shown that they are able to attract non-ionic organic compounds [13]. Some, in particular poly-diallyldimethylammonium chlorides (poly-DADMACs), have been recognised as safe materials for use in potable waters [14].

Complexes of smectites with organic polymeric cations can develop a net positive charge [13, 15]. In our work [16], a comparison has been made of the effect of different types of polycation on the development of positive charge on smectites in relation to their loading by the polymers. The focus has been on the possibility of the use of the resulting complexes for anion uptake. The study involved the activation of a Wyoming montmorillonite clay with different polycations from polyelectrolytes, viz. poly-DADMAC (PD), a poly-vinylbenzyltrimethyl ammonium chloride (polystyrene) (PS), a epichlorohydrin based polymer (polyamine) (PAm) and a polyacrylamide (PAc). PD, PS and PAm respectively have molecular weights of 2×10^4, 2.6×10^4, and $\sim 10^4$, and also cation densities of 100, 74 and $\geq 80\%$. PAc had a much higher molecular weight (3×10^6) and much lower cation density (8%) than the other polyelectrolytes.

The complexes formed with the poly-DADMAC (PD) and the cationic polystyrene (PS) showed substantially more uptake than those with the other polymers [16]. The amount of uptake varied with the polymer content of the complex. maximum uptake occurred when the polymer content was 0.5g g^{-1} clay for both the polystyrene and poly-DADMAC complexes [16]. Comparisons of isotherms for toluene uptake (Fig. 1) show that the complexes of the clay with PS had a greater affinity for toluene than the PD-clay complexes.

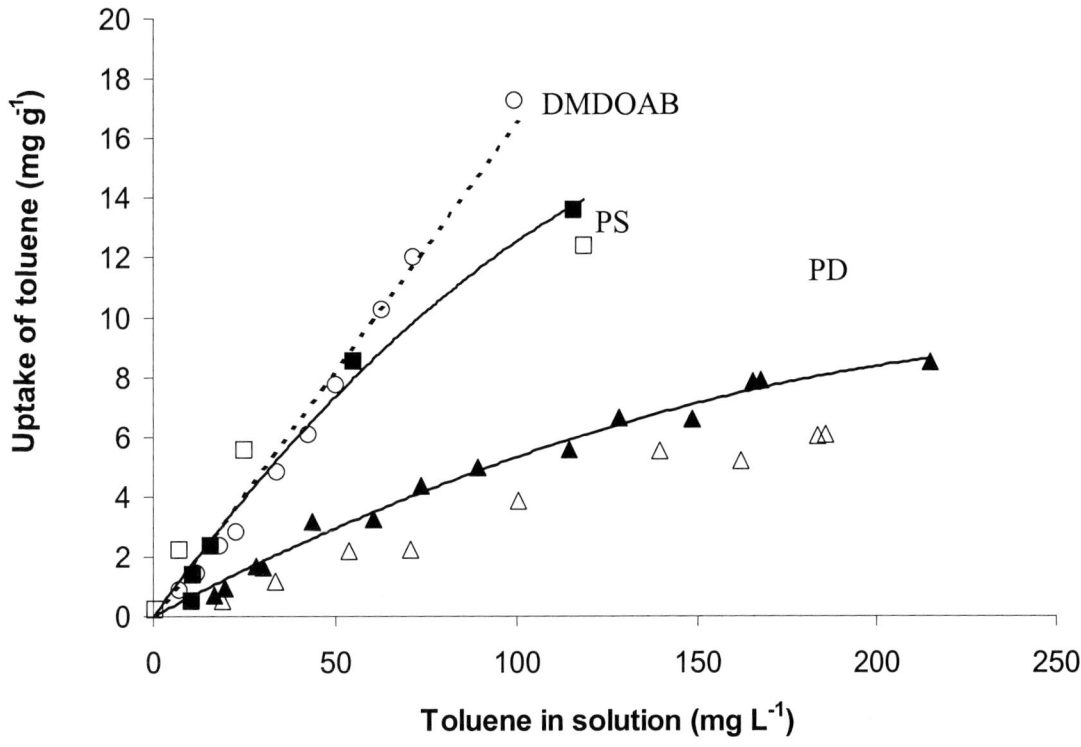

Figure 1. Isotherms for the uptake of toluene by complexes formed by Wyoming montmorillonite clay with a cationic polystyrene (PS) and a poly-DADMAC (PD) in comparison with its uptake on a complex between the clay and dimethyldioctadecyl ammonium bromide DMDOAB. Filled symbols are for 0.5 g polymer g^{-1} clay loading; open symbols for a 0.2 g polymer g^{-1} clay loading. The data is from [16].

The carbon contents of the two complexes at a loading of 0.5 g polymer g^{-1} clay were both 13.3 ± 0.2% [16]. Toluene uptake by the PS-complexes (with loadings of both 0.5 and also 0.2 g polymer g^{-1} clay) was similar to that by the complex with the double-chain quaternary ammonium cation DMDOAB over much of the concentration range. The relatively strong attraction of the polystyrene complex for toluene may indicate an enhanced attraction for toluene by the aromatic benzyl group in the polystyrene cation.

Fig. 1 also indicates that the isotherms for uptake by the complexes of the clay with both PD and PS are curved, whereas that for uptake by the DMDOAB complex is linear. A linear isotherm, indicating that all sorbate-sorbent interactions are equal in energy, is characteristic for the uptake of non-ionic organic molecules by smectites modified by large QACs [12]. It is consistent with a partition of the organic molecules from solution into the organic phase in the clay interlayers [17]. On the other hand, a curved isotherm indicates there are different sites, offering differing attractive energies for uptake.

Only small additions (<~0.2g polymer g^{-1} clay) of highly cationic polyelectrolytes produced a complex with a net positive charge (Fig. 2). However, a low-charge cationic polymer (a polyacrylamide) was unable to yield a positively charged complex regardless of amount added.

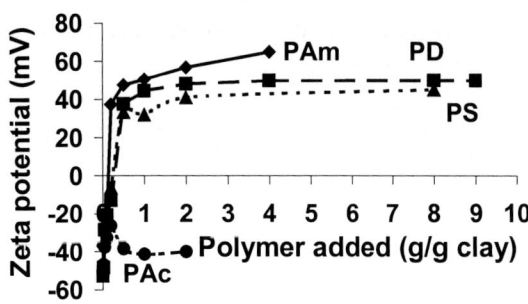

Figure 2. Changes in zeta potential of products with increasing addition of polymers to Wyoming montmorillonite clay. Cationic polyelectrolytes are: PAm: a polyamine; PD: a poly-DADMAC; PS: a polystyrene; PAc: a polyacrylamide. The data is from [16].

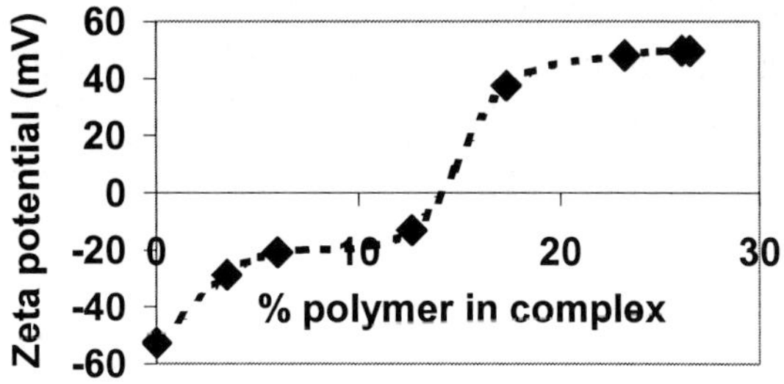

Figure 3. Zeta potential in relation to the actual percentage content of the poly-DADMA$^+$ polycation (PD) as calculated from the measured carbon contents of the complexes as a proportion of the carbon content in the chemical formula of the polycation. The data is from [16].

When related to the actual polycation in the complexes, (e.g. Fig. 3), zeta potential showed an initial gradual rise with polymer additions, prior to the major rise at the point of zero charge (PZC). The addition of a large QAC to a Na-saturated Wyoming montmorillonite had shown only one sharp rise (at the PZC) in electrophoretic mobility, i.e. zeta potential [18], indicating that adsorption of the QAC on external surfaces occurred only near the PZC. By contrast, polycations apparently adsorb on external surfaces, as well as within interlayers at the initial stage. The occurrence of two types of sites is consistent with the curved shape of the isotherms (Fig. 1).

Other work in one of our laboratories has shown that complexes of both smectites and palygorskites with polycations removed anionic dyes from water (P. Self, unpub. data). They showed a high affinity for phosphate in concentrations of 20-200 µg P kg^{-1} (T. Kleinig, unpub. data). After activation with appropriate cationic polymers, clays could be used to remove anionic pollutants from water.

3. ACTIVATION TO ENABLE THE SELECTIVE ADSORPTION OF GASES

Both the surface areas and also the porosity of clays can be increased by pillaring, which results in pores being created by addition within interlayers of expandable clays. Both surface area and porosity can also be increased by the creation of pores within the framework of the clays by subtraction as a result of the removal of constituents of the aluminosilicate layers at different rates from one another.

When a dioctahedral smectite (a montmorillonite Ch) and a trioctahedral smectite (a saponite S) were each treated with 5M sulphuric acid., both showed large increases in surface area, and particularly that of their micropores, especially after 60 min. treatment (Table 1). Table 1 shows that the various acid activated clays resulting from these treatments showed increases in the adsorption of each of O_2, CH_4 and CO_2.

Table 1. Total, and micropore surface, total volume, and adsorption of oxygen, methane and carbon dioxide by a dioctahedral smectite (Ch) and a trioctahedral smectite (S) before, and also after, treatments with 5M sulphuric acid for different lengths of time (from [19])

Sample	Time (min.)	Total surface (m^2g^{-1})	Micropore surface (m^2g^{-1})	Total volume $cm^3 g^{-1}$	O_2 adsorption (mmol g^{-1})	CH_4 adsorption (mmol g^{-1})	CO_2 adsorption (mmol g^{-1})
Ch	0	78	71	0.175	0.004	0.009	0.065
	15	276	222	0.288	0.014	0.049	0.206
	60	467	419	0.441	0.020	0.045	0.217
S	0	8	6	0.031	0.009	0.030	0.214
	15	322	288	0.257	0.028	0.063	0.329
	60	437	390	0.347	0.030	0.062	0.357

The acid activated trioctahedral smectite showed a greater adsorption of gas than the acid activated dioctahedral smectite. For both samples, however, retention of CO_2 was higher than that of O_2 or CH_4. Differences in gas retention values could be attributed to physicochemical characteristics of their molecules including their molecular size and quadruple moments.

The adsorption of CO_2 gas by acid smectites decreased by ~10 % and ~20% for acid trioctahedral and dioctahedral smectites, respectively, when they were heated up to 600 °C.

Furthermore, retention of CO_2 by the acid-activated trioctahedral smectite samples was higher than that of Al-, Zr-, Cr- and Ti-pillared clays and modified Al-pillared clays [20], although lower than that of smectites treated with a tetra-methyl ammonium cation [21] (Fig. 4). The effectiveness of a molecular sieve with a high retention of CO_2 depends upon the degree of separation of mixtures of CO_2 and CH_4. Since a ratio of ≥3 for the retention of two different gases is considered commercially viable for gas separation [22], the results obtained here suggest that acid-activated smectites could be useful for the separation of CO_2 and CH_4.

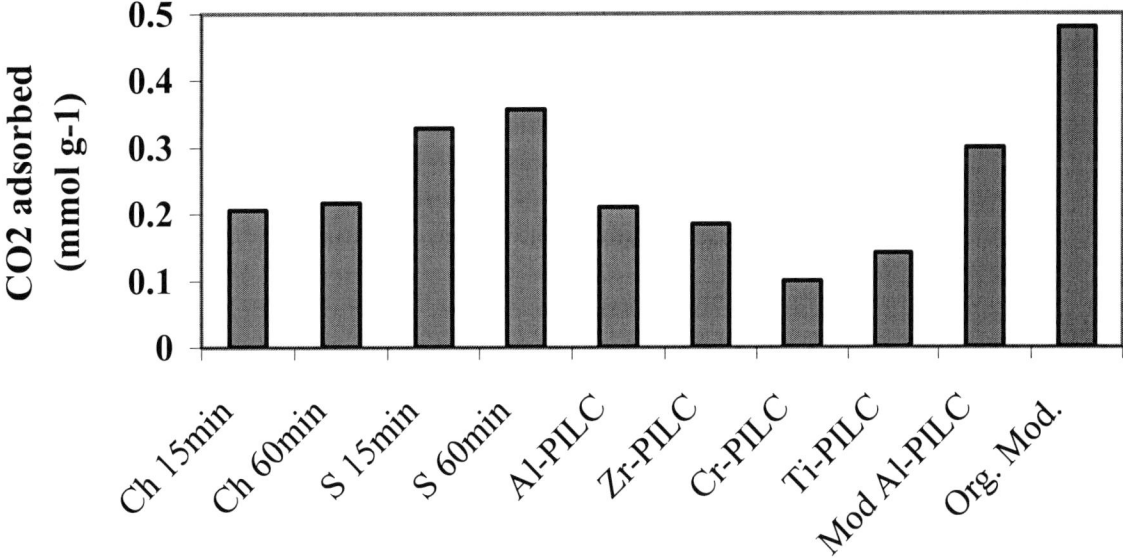

Figure 4. The adsorption of CO_2 by smectites (dioctahedral, Ch, and trioctahedral, S) after treatment with 5M H_2SO_4 for 15 and 60 min in comparison with that of various pillared clays and also a clay that had been modified organically with a tetra-methyl ammonium cation. The data is from [19-21].

The adsorption of a number of gases (N_2, O_2, CO, CO_2, CH_4, C_2H_2) by amorphous clay derivatives [23] are compared with those by Al-, Zr-, Cr- and Ti-pillared clays [20]. The metal exchanged amorphous clay derivatives showed that the highest gas retention levels per unit area were achieved for alkali metal exchanged-derivatives or alternately for clays exchanged with cations with the largest ionic radii [24]. A high CO_2 retention value (0.606 mmol/g) was obtained after an acid treatment of kaolinite that had been mechanically and thermally pre-treated [C. Volzone and J. Ortiga, unpub. data].

4. DISCUSSION

Only two among the many methods of activation of clays have been discussed in any detail here. However, these two methods (the use of polycations and that of acid activation) represent two quite different general approaches to activation. They each provide effective alternatives to more established methods of activation i.e. the use of QACs and pillaring respectively; polycations with aromatic groups for adsorbing aromatic hydrocarbons and acid activated clays for gas adsorption. They also extend the capabilities of clays; highly charged polycations for anion adsorption and acid activation for selectivity of gas sorption.

In the use of polycations, activation is brought about because of the capacity of clays for cation exchange. One of the simplest means, yet often a particularly effective one, is that of exchange of interlayer inorganic cations. In another study in one of our laboratories, the uptake of proteins from the wastewaters of abattoirs apparently increased when smectites became saturated with Na^+ (G.J. Churchman, unpub. data). Maximum amounts adsorbed were ~10% of the maximum for protein uptake by clays [25]. The different smectitic clays showed a contrast of >5 times in protein uptake between the most adsorbent (a saponite) and least adsorbent of the 8 Australian smectitic samples, used as mined or as marketed. However, their exchangeable sodium contents, among other properties, covered a wide range of values. Nonetheless, when Na-saturated with NaCl to give ESPs of 92±4%, 7 of the 8 samples removed amounts of protein that were within 40% of each other in spite of differences of up to 220% between their CECs. Even so, one exceptional sample required ultrasonic treatment, following sodium saturation, to break down microaggregates apparent in transmission electron micrographs before adsorption of protein could occur to a comparable extent to the other samples. Activation of clays was necessary to ease access to interlayers by bulky protein molecules. Generally, this could be achieved by sodium saturation.

The use of quaternary alkylammonium cations (QACs) for the activation of clays [9-12] clearly depends upon cation exchange. Long chain QACs form a hydrophobic and organophilic phase for the uptake of nonionic organic compounds by partitioning [12] while smaller QACs form pillars for adsorption of gases [11,] and solutes [9-11]. Pillaring of clays generally, to increase micropore volumes and surface area (Section 3), and also for catalysis [7, 8,], also occurs as a result of cation exchange. Pillaring can be carried out with hydrolysed salts of many different cations (e.g. Section 3), with the hydroxides in the interlayers being calcined to give poly-oxo ions. Other approaches to pillaring have included the addition of cationic surfactants either after or during the synthesis of metal-pillared smectites [26, 27]. The products in these cases are effective sorbents of uncharged organic compounds. A further approach using surfactants with a source of silica to create a silica matrix within the clay interlayers produces highly acidic catalysts [28].

In another type of approach, activation results from the chemical alteration of the aluminosilicate framework of the clay minerals. This includes acid activation to increase micropore volume and surface areas (Section 3), the formation of amorphous derivatives with high surface areas and/or substantial cation exchange capacities [29], as well as the capacity to intercalate alkaline earth metal salts [30] and treatment with alkalis and mild heating to convert clays to zeolites, with their variety of environmental uses [31]. It also includes a vast number of treatments that affect the oxidation state of Fe, by oxidation, including heating, and by reduction.

The work described here has been largely limited to the activation of smectites. However, the results have shown variations between different smectites (Section 3). Other work in our laboratories [32-34] has shown differing physical and chemical behaviours of different smectites, especially in relation to the distribution of layer charge between octahedral and tetrahedral layers e.g. for adsorption of Cr [34]. Studies with a structural basis should enable better understanding of the phenomena and prediction of the behaviour of different clays following different types of activation and also better choices of materials for environmental uses. The aim of making the use of activated clays economic poses a challenge that may be achieved partly by the re-use of products from the use of clays and activated clays for treatments of wastes. It may also be met partly by the re-use of some of the millions of tonnes of clay wastes produced each year from a number of industries e.g. paper manufacture, wine making and edible oil processing [35, 36].

REFERENCES

1. P. Hawken, A. Lovins and L.H. Lovins, Natural Capitalism: Creating the Next Industrial Revolution, Little, Brown and Company, Boston, 1999
2. M. Robert and C. Chenu, in G. Stotzky and J-M Bollag (eds.), Soil Biochemistry, Marcel Dekker, New York, 7 (1992) 307.
3. R.G. Burns, in P.M. Huang and M. Schnitzer (eds.) Interactions of Soil Minerals with Natural Organics and Microbes, Special Publication No. 17 Soil Science Society of America, Madison, Wisconsin, (1986) 429.
4. C. Volzone, J.G. Thompson, A. Melnitchenko, J. Ortiga and S.R. Palethorpe, Clays Clay Min., 47 (1999) 647.
5. G.R. Alther, Engineering Geology, 23 (1987) 177.
6. J. Arch, in A. Parker and J.E. Rae (eds.) Environmental Interactions of Clays, Springer-Verlag, Berlin (1998) 207.
7. D.E.W. Vaughan, in R. Burch and B. Delmon (eds.) Catalysis Today 2, Elsevier, Amsterdam (1988) 187.
8. T.J. Pinnavaia, in by G.J. Churchman, R.W. Fitzpatrick and R.A. Eggleton (eds.) Clays Controlling the Environment, CSIRO Publishing, Melbourne (1995).
9. R.M. Barrer and D.M. MacLeod, Trans. Faraday Soc., 51 (1955) 1290.
10. J.F. Lee, M.M. Mortland, C.T. Chiou and S.A. Boyd, J. Chem. Soc. Faraday Trans. I, 85 (1989) 2953.
11. J.F. Lee, M.M. Mortland, C.T. Chiou, D.E. Kile and S.A. Boyd, Clays Clay Min., 38 (1990) 113.
12. W.F. Jaynes and S.A. Boyd, Soil Sci. Soc. Am. J., 55 (1991) 43.
13. C. Breen and R. Watson, J. Coll. Interf. Sci., 208 (1998) 422.
14. R.D. Letterman and R.W. Pero, J. Amer. Water Works Assoc., 82 (1990) 87.
15. T. Ueda and S. Harada, J. Appl. Polym. Sci., 12 (1968) 2395.

16. G.J. Churchman, Appl. Clay Sci. (in press).
17. S.A. Boyd, M.M. Mortland and C.T. Chiou, Soil Sci. Soc. Am. J., 52 (1988) 652.
18. S. Xu and S.A. Boyd, Langmuir 11 (1995) 2508.
19. C. Volzone and J. Ortiga, J. Mater. Sci., 35 (2000) 5291.
20. C. Volzone, L.B. Garrido, J. Ortiga and E. Pereira, in F. Rodríguez-Reinoso and P. Andreu (eds.) *Programa Iberamericano de Ciencia y Technología (CYTED) Subprogama V. Catálisis y Absorbentes. Proyecto V.3,* CYTED-España (1998) 97.
21. C. Volzone, J.O. Rinaldi and J. Ortiga, in Anais do 43rd Congresso Brasileiro de Cêramicado Mercosul. Florianópolis, SC Brasil, 2-5 June 1999, 32401.
22. M.S.A. Baksh and R. T. Yang, American Institute of Chemical Engineers J., 38 (1992) 1357.
23. C. Volzone, J. G. Thompson, A. Melnitchenko, J. Ortiga, S. R. Palethorpe. Clays Clay Min., 47, (1999) 647.
24. A. Melnitchenko, J. G. Thompson, C. Volzone, J. Ortiga. Appl. Clay Sci., 17 (2000) 35-53.
25. B.K.G. Theng, Formation and Properties of Clay-Polymer Complexes, Elsevier, Amsterdam, 1979.
26. K. R. Srinivasan and S.H. Fogler, Clays Clay Min., 38 (1990) 277.
27. L.J. Michot and T.J. Pinnavaia, Clays Clay Min., 39 (1991) 634.
28. A. Galerneau, A. Barodawalla and T.J. Pinnavaia, in H. Kodama, A.R. Mermut and J.K. Torrance (eds.) Clays for our Future, Published by the ICC97 Organizing Committee, Ottawa, 1999, 21.
29. J.G. Thompson, I.D.R. MacKinnon, A. Koun and N. Gabbitas, Kaolin derivatives, PCT patent application, WO 95/00441 AU94/00323, 1994.
30 B. Singh and I.D.R. MacKinnon,. in H. Kodama, A.R. Mermut and J.K. Torrance (eds.) Clays for our Future, Published by the ICC97 Organizing Committee, Ottawa, 1999, 489.
31. D.W. Breck, Zeolite Molecular Sieves. Structure, Chemistry and Uses, John Wiley & Sons, New York, 1974.
33. P.G. Slade, J.P. Quirk and K. Norrish, Clays Clay Min., 39 (1991) 234.
34. C. Volzone, Aust. J. Soil. Res., 36 (1998) 423.
35. C. Volzone and L. B. Garrido. Clay Min., 47 (2001) 115.
36. T. Kendall, in T. Kendall (ed.) Industrial Clays, 2nd. Edition, Industrial Minerals. London (1966) 53.
37. P. Crossley, Industrial Minerals, March 2001, (2001) 69.

II
Clays in Geology

Na and K-bentonites from Argentina: composition, origin and age.

Andreis, R. R.[a] and P.E. Zalba [b]

[a]Museo Paleontológico "Egidio Feruglio", Av. Fontana 140-(9100) Trelew, Chubut, Argentina - E-mail: rrandreis@sinectis.com.ar

[b]Comisión de Investigaciones Científicas de la Provincia de Buenos Aires – Centro de Tecnología de Recursos Minerales y Cerámica – Facultad de Ciencias Naturales y Museo – C. C. 49, (B1897ZCA) M. B. Gonnet, Argentina – E-mail: pezalba@netverk.com.ar

Almost all bentonites in Argentina are of the Na-rich type (Triassic, Cretaceous, Tertiary), but scarce K- (Ordovician), Mg- (Tertiary) or Ca-Mg-rich bentonites (Triassic) have been recognized. The genesis of the Na-bentonites is related to proximal to distal ash showers of rhyolitic, andesitic to dacitic nature associated to plinian and phreato-plinian explosive volcanism, produced at the Andean chain from the Mesozoic to the Tertiary. The parental magma of the Ordovician K-bentonites (San Juan Province) was of rhyolitic to trachyandesite composition and represents distal fall-out ashes related to a subduction zone with probable collisional implications.

The main area covered with Na-bentonites is situated in western Argentina at the Precordillera (Cuyana Basin) where the Triassic succession of the Potrerillos and Cacheuta formations (Province of Mendoza) and the Barreal Group (Province of San Juan) can be found. These deposits are rich in montmorillonite-beidellite or in montmorillonite, with minor beidellite and traces of illite and kaolinite, respectively. The other important area is related to the Upper Cretaceous bentonites (Allen Formation), covering the NW Patagonia extra-Andean region, Neuquén, Río Negro and La Pampa provinces. Associated to extended tidal flats related to lagoons, they are montmorillonitic with poor to excellent crystallinity. In Patagonia, Tertiary bentonites were found in the Collón-Curá (Miocene) and the Chichinales formations (Late Oligocene-Late Miocene). The former are rich in montmorillonite with less than 14% of illite-smectite, while the latter contains illite-smectite, but changing the frequency from 80-95% smectite to 70% illite.

Other areas with Na-Mg bentonite were identified at Sierra de Mogna (Miocene, San Juan Province). The deposits are montmorillonitic with minor beidellite. The Mg-bentonites of the Miocene Calchaquí Formation (Mendoza Province) and the Río Chico and Sarmiento Formations (Chubut Province) with Na-bentonites include montmorillonite of variable crystallinity.

1. INTRODUCTION

Almost all Argentinean bentonites are of the Na-rich type (Triassic to Tertiary), but scarce K- (Ordovician), Mg- (Tertiary) or Ca-Mg-rich bentonites (Triassic) have been recognized. The genesis of the Na-bentonites is related to proximal to distal ash-fall deposits of rhyolite, andesitic to dacitic nature, associated to explosive volcanism of plinian and phreato-plinian characteristics, produced at the Andean chain from the Mesozoic to the Tertiary (21).

The parental magma of the K-bentonites of Ordovician age was of rhyolitic to trachyandesite composition and represents distal fall-out ashes related to a subduction zone with probable collisional implications. The provenance of the ash-falls is doubtful, but further studies will probably associate the ash-falls with a magmatic arc located at Famatina (NW Argentina), northern Chile and SE Bolivia (8). The distribution of bentonites is shown in Figure 1.

2. BENTONITES OF ARGENTINA

2.1. Ordovician bentonites

The K-bentonites were recognized in the Eastern Precordillera (San Juan Province) and in the Los Azules Formation (Llanvirnian-Caradocian), in both cases covering conformably the Caliza San Juan (8). The bentonite beds show variable thickness, from a few centimeters up to 1m, and exceptionally reach 2 m at the profiles of Cerro Viejo and Cerro La Chilca. They are yellowish-brown to pale-orange or gray and the glass shards were modified by diagenetic processes to about 70-90% illite-smectite (ordered, R=1). Also minor kaolinite and authigenic anatase were found. Geochemical analysis show that the parental magma was of rhyolitic to trachiandesitic nature and the explosive eruptions of great scale were related to the formation of calderas associated with a subduction zone (collisional margin).

2.2. Triassic bentonites

In western Argentina, in the Precordilleran area, the Cuyana Basin (16) was originated by extensive reactivation of Eopaleozoic faults and thermal subsidence during the Lower to the Middle Triassic (9). It includes successions that crop out at Lujan de Cuyo and Las Heras Departments (Carrizalito District, north of Uspallata), Province of Mendoza. Also in the Province of San Juan, bentonites were studied in a large area comprising a 1-5 km-wide and 22 km-long belt east of Barreal-Sorocayense-Hilario region related to an anticlinal structure (12, 27). In Mendoza Province, the Triassic sequence comprises the Potrerillos, Cacheuta and Río Blanco formations, but only the first unit includes bentonites, while the Cacheuta Formation only shows centimeter thick gray to dark-gray layers, associated with black shales (9, 18, 26).

The basal part of the Potrerillos Formation shows an alluvial fan-braided river system, which changes upwards to a meandering one with extended flood plains, which is covered by a progressive inundation related to a lacustrine environment. Mainly during this lacustrine sedimentation, the gray to black shales include ash falls transformed to bentonites by

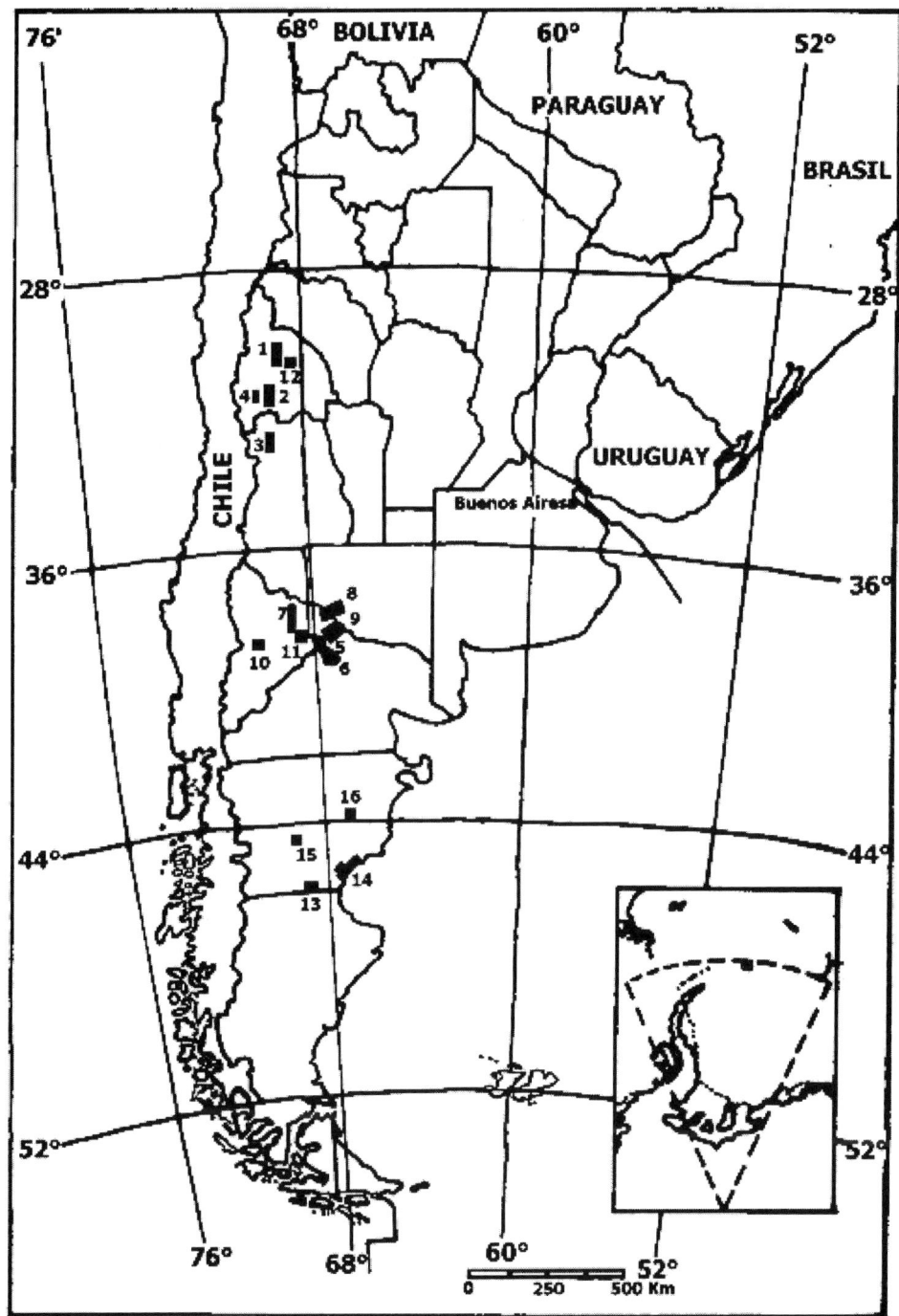

FIGURE 1 – Distribution and age of bentonites in Argentina. **Ordovician bentonites**: 1. Los Azules Formation (Llanvirnian-Caradocian) – **Triassic bentonites**: 2. Cortadera and Barrial Formations 3. Potrerillos Formation (Middle-Upper Triassic). 4. Ischichuca Formation (Triassic) – **Cretaceous bentonites**: Allen Formation (Campanian to Lower Maastrichtian). 5. Pellegrini Lake. 6. Allen and Gral. Roca localities. 7. Añelo. 8. Puelén. 9. Río Colorado – **Tertiary bentonites**: Collón-Curá Formation (Miocene). 10. Western Zapala, Chichinales Formation (Upper Oligocene-Upper Miocene). 11. Añelo. 12. Sierra de Mogna (Tertiary). 13. Sarmiento Formation (Early Eocene-Upper Oligocene). Río Chico Formation (Late Paleocene). 14. From Pico Salamanca to Bajada de Hansen. 15. Laguna Payahilé. 16. Gaiman (Chubut River valley).

diagenetic processes (1, 10, 16, 27). These bentonites, 0,20 to 2,00 m in thickness, are yellow, greenish to grayish-blue, and their composition is about 40-70% montmorillonite-beidellite but locally they can be rich in montmorillonite (76-95%) and contain minor kaolinite (18).

The San Juan bentonites are related to the Barreal Group sequence of Middle to Upper Triassic age (31). They appear only in the Barreal and Cortaderita formations and are composed of a sandy-conglomeratic basal alluvial fan and braided river sequences followed to the top by tuffaceous siltstones, bentonites and tuffs, and final braided-distal fluvial system and detrital flows (29, 31). Bentonites are lenticular, yellow, yellowish gray, gray, greenish-gray to reddish, of 1,50 m up to 7 m thick, and montmorillonitic in composition (27). They are of the Ca-Mg type bentonites. The Barreal-Sorocayense-Hilario bentonites, are 1 to 7, massive, yellow, greenish, rose, gray or grayish-white beds, of 0,60 m to 3,00 m thick. Their composition includes 60-75% montmorillonite, minor beidellite and traces of illite and kaolinite.

Also in the Quebrada de Ischichuca, located 90 km west of San Juan, the Triassic Ischichuca Formation contains incipient developed bentonites, associated with lacustrine black shales (6, 32).

2.3. Cretaceous bentonites

At the Neuquén Basin there are many bentonite deposits, which extend to the extra-Andean region at the provinces of Río Negro, La Pampa and Neuquén (14, 36). The most important one belongs to the clastic succession of the Allen Formation (Malargüe Group) of Late Campanian to Late Paleocene age (17) that covers long distances only disturbed by small faults. Historically, the production of bentonites started near the localities of Cinco Saltos and Pellegrini Lake (Río Negro Province). The last is located at the Vidal Basin where this lake is situated (12). These bentonites are included in the Middle Member (3, 15, 36), which is characterized by a lithological uniformity, including gray to greenish-gray claystones and siltstones with massive, laminated or wavy and linsen structures, thin lenses of fine sandstones and pale-green to whitish bentonites. These sediments were deposited in extended tidal flats related to lagoons. The same environment is found at the base of the Upper Member but the entrance of the sea also produced the deposition of stromatolithic limestones and, to the top, the deposition of gypsum is related to partial close of the lagoonal environment. The Middle Member is 20 to 32 m thick (3). Bentonites show tabular or lenticular bedding, 0,05 to 0,70 cm thick and are montmorillonitic with good to excellent crystallinity (19, 24).

Other related bentonites, also included in the Middle Member of the Allen Formation, crop out at Añelo (Neuquén Province, Bajada del Palo and El Caracol), the right margin of the Colorado River (Río Negro Province, Aguará, Vaca Mahuida, Catriel), north of Allen and General Roca localities, and also at Puelén (La Pampa Province). Bentonites are 95-99% rich in montmorillonite, but with moderate to low crystallinity (33, 34, 35, 36). The bentonite deposits of Puelén, Allen and General Roca localities are quite similar to those outcropping around the Pellegrini Lake, but they show good to excellent crystallinity (19).

2.4. Tertiary bentonites

Other bentonite deposits, of minor quality, are relatively common in Tertiary deposits (37).

These deposits were mined at the Collón-Curá Formation (Miocene), 25 km west from Zapala locality (Neuquén Province). The bentonites appear at the basal clastic unit, associated with greenish-gray to yellowish-gray, medium to very coarse sandstones or conglomeratic sandstones, representing a braided distal river system. They also appear in the middle unit (pyroclastic) (25). Bentonites are lenticular, 2 to 7 m (Cerro Bandera) and 3 to 10 m (Barda Negra) thick, olive-green, pale-green to greenish-gray and mixed with green andesitic to dacitic tuffs. They are montmorillonitic, with less than 14% illite-smectite (with 70-80% of illite, R=1), although some bentonites are only montmorillonitic in nature (37). The abundance of gypsum veins lowers the quality of these bentonites (12). Also, in the Miocene Pinturas Formation (Santa Cruz Province), some bentonites may be found (7).

At the base of the continental Chichinales Formation (Late Oligocene-Late Miocene), at Añelo (Róo Negro Province), covering an irregular surface excavated in the Jaguel Formation, some bentonites appear mixed with acid to intermediate tuffs and siltstones and are lenticular with variable degree of substitution of the glass shards. At Don Alfredo mine the thickness of the bentonites varies from 2,50 to 7,00 m, showing 76-83% of clay components, and are composed of illite-smectite but changing the frequency of illite and smectite from 80-95% smectite to 70% illite (12, 34, 35, 37).

At Cañón del Colorado (Sierra de Mogna, Jachal), 90-100 km northward from San Juan, Upper Miocene-Upper Pliocene (11) bentonites are related to an anticlinal structure, and they are interstratified with conglomerates, sandstones and claystones. The thickness of the two levels varies from 1,00 to 2,00 m. They are whitish to grayish-white, with sodium-magnesium montmorillonite and scarce beidellite (12, 13, 23, 27). In the area of Ramblón (Sarmiento Dept. Mendoza), near the border of San Juan Province, the Miocene Calchaquí Formation sandstones cover the San Juan limestones, and contain two levels of low quality bentonites, Mg-rich in composition (5, 12).

Many bentonite levels were described in the Sarmiento Formation (Early Eocene-Upper Oligocene) (4, 20, 21, 29, 30), cropping out at the southern cliffs of the lake Colhué-Huapi (Chubut Province). They are massive, greenish to brownish and bear montmorillonite with variable degree of crystallinity, which is higher at the basal member (Gran Barranca) and lower in the other members (Puesto Almendra and Colhué-Huapi). Their thickness varies from 0,70 m to 6,50 m and their SiO_2 content is around 60% (rhyolitic to dacitic composition). The Sarmiento succession represents ash falls deposited over wide undulate plains under temperate to warm climatic conditions and vague humidity. Ashes were partially removed by ephemeral rivers, and over the plains the development of paleosols is rather common. The tephras show a clear dominant dacitic composition with calc-alkaline affinity (21).

Other bentonite deposits were also described at the Río Chico Formation-Visser Member (2, 4) along a littoral area from Pico Salamanca, Puerto Visser and Bajada de Hansen, as well as in the Cañadón Hondo region (Chubut Province). They are massive, tabular, and yellowish-gray, greenish-gray, dark-gray to light red. Also, in the Chubut lower valley river, near Gaiman, other dark-gray to greenish montmorillonitic bentonites were mentioned in the same unit (22).

3. ENVIRONMENT OF DEPOSITION AND PALEOCLIMATIC CONDITIONS

Almost all Na- and K-bentonites were originated by explosive eruptions that produced the distal fallout of tephras (mainly glass shards and pumices) in shallow standing aqueous bodies with poor occurrence of currents (shallow interior and littoral lakes, small lakes in delta plains, swamps, or lagoons), and latter transformed to clay minerals by diagenetic processes (3, 8, 36). In some cases (Allen bentonites, 36), the high values of Na and Mg are related to the entrance of seawater, while in other cases, the transformation occurred in fresh water swamps or related to ascendant phreatic level (Añelo bentonites, 37). In other cases, the generation of bentonites is related to alteration of re-elaborated glass fragments deposited in fluvial channels with low gradients and associated with shallow lakes.

The Triassic bentonites from Mendoza and San Juan provinces (31), as well as the Sarmiento Formation bentonites (30), were deposited under temperate to warm and high humidity conditions.

4. USES AND APPLICATIONS

The Cretaceous Colorado River, Pellegrini Lake, Añelo, Allen, Roca and La Pampa sodium-rich bentonites are used for foundries, oil drilling, civil engineering and metallurgical purposes, but also they are qualified for applications on iron ore pelletizing, red ceramics, balanced foodstuff, asphaltic emulsifiers, paints, water proofing and landfills (19, 36). Also the Triassic bentonites of Mendoza and San Juan provinces are qualified for foundries, enology, iron ore pelletizing, balanced foodstuff, ceramics, cosmetics, paints, and 30 to 40% of them are available for oil drilling (38). The Tertiary Zapala and Añelo bentonites (Neuquén Province) are used in the metallurgical industry and iron ore pelletizing, while the Mogna bentonites (San Juan Province) are qualified for purifying greases and oils and for the paper industry and ceramics (37).

REFERENCES

1. Alfonso, R.; S. Alurralde; O. Mancilla; R. Manoni & R. Pombo, 1984. Análisis litoestratigráfico de las unidades triásicas del subsuelo en el sector septentrional de la Cuenca Cuyana en la provincia de Mendoza. *IX Congreso Geológico Argentino*, Bariloche, 1984. Actas, 1: 7-24, Buenos Aires.
2. Andreis, R.R., 1977. Geología del área de Cañadón Hondo, Depto. Escalante, Provincia de Chubut, República Argentina. *Obra del Centenario del Museo de La Plata*, 4: 77-102.La Plata.
3. Andreis, R. R.; M. A. Iñíguez Rodríguez; J. J. Lluch & D. A. Sabio, 1974. Estudio sedimentológico de las formaciones del Cretácico Superior del área del lago Pellegrini (Provincia de Río Negro, República Argentina). *Revista Asociación Geológica Argentina*, 29(1): 85-104, Buenos Aires.
4. Andreis, R.R.; M.M. Mazzoni & L.A. Spalletti, 1975. Estudio estratigráfico y paleoambiental de las sedimentitas terciarias entre Pico Salamanca y Bahía Bustamante, provincia de Chubut, República Argentina. *Revista Asociación Geológica Argentina*, 30 (1): 85103, Buenos Aires.
5. Angelelli, V. & R. Fernández, 1980. Los yacimientos no metalíferos y rocas de aplicación de la región Centro-Cuyo. Secretaría de Estado de Minería, Anales, 19, Buenos Aires.
6. Bossi, G. E, 1970. Asociaciones mineralógicas de las arcillas en la cuenca de Ischigualasto-Ischichuca. Parte II. Perfiles de la Hoyada de Ischigualasto. *Acta Geológica Lilloana*, 11(4): 73-100, Tucumán.

7. Bown, T.M. & C. L. Larriestra, 1990. Sedimentary paleoenvironments of fossil platyrrhine localities, Miocene Pinturas Formation, Santa Cruz Formation, Argentina. *Journal of Human Evolution*, 19: 87-119, Academic Press.
8. Cingolani, C. A.; W. Huff; S. Bergstrom & D. Kolata, 1997. Bentonitas potásicas Ordovícicas en la Precordillera de San Juan y su significación tectomagmática. *Revista Asociación Geológica Argentina*, 52(1): 47-55, Buenos Aires.
9. Criado Roque, P., 1979. Subcuenca de Alvear (Provincia de Mendoza). *II Simposio Geología Regional Argentina*, Academia Nacional de Ciencias, 1: 811-836, Córdoba.
10. Días, H. A. & A. C. Massabie, 1974. Estratigrafía y tectónica de las sedimentitas triásicas, Potrerillos, provincia de Mendoza. *Revista Asociación Geológica Argentina*, 29(2): 185-204, Buenos Aires.
11. Gallardo, C. A.; L. M. Martos & R. D. Weidmann, 1987. Presencia de la Formación Mogna en el área de La Rinconada, San Juan, República Argentina. *X Congreso Geológico Argentino*, Tucumán, 1987, Actas, 2: 105-108
12. Herrmann, C. & E. Menoyo, 2000. Bentonitas. Informe Integral. *Revista Panorama Minero*, 250, Suplemento Especial Investigación y Desarrollo, 3: 39-54.
13. Hevia, R. P., 1991. Bentonitas de Mogna. *Revista Panorama Minero*, 160
14. Impiccini, A., 1999. Correlación entre capas de bentonita de edad Cretácico Superior y su vinculación con las áreas volcánicas de origen, en el norte de la Patagonia, Argentina. *Boletim V Simpósio sobre o Cretáceo do Brasil y I Simpósio sobre el Cretácico de América del Sur*, Serra Negra, São Pablo, Brasil, 1999, 225-230, Rio Claro, Brasil.
15. Iñíguez Rodríguez, A. M., R. R. Andreis & J. J. Lluch, 1975. Estudio geológico y tecnológico de las bentonitas del lago Pellegrini, Depto. Gral. Roca, provincia de Río Negro. *Minería*, 143: 18-25, Buenos Aires.
16. Kokogian, D. A.; L. Spalletti; E. Morel; A. Artabe; R. N. Martínez; O. A. Alcober; J. P. Milana; A. M. Zavattieri & O. H. Papú, 1999. Los depósitos continentales triásicos. Instituto de Geología y Recursos Minerales, Geología Argentina, *Anales* 29(15): 377-398, Buenos Aires.
17. Legarreta, L. & C. A. Gulisano, 1989. Análisis estratigráfico secuencial de la cuenca neuquina (Triásico Superior – Terciario Inferior). En: Chebli, G. A. & Spalletti, L. A. (Eds.), *Cuencas Sedimentarias Argentinas*, Instituto Superior de Correlación Geológica, Universidad Nacional de Tucumán, Serie Correlación Geológica 6: 221-243.
18. Lluch, J. J., 1971. Sedimentología del Triásico en el área Papagayos-Divisadero Largo, provincia de Mendoza. *Revista Asociación Argentina de Mineralogía, Petrología y Sedimentología*, 2(3-4): 93-116, Buenos Aires.
19. Marconi, C. R., 1998. Bentonites in Argentina. *Proceedings International Workshop of Activated Clays*, La Plata, Argentina, 1998, 97-107.
20. Mazzoni, M. M., 1979. Contribución al conocimiento petrográfico de la Formación Sarmiento. *Revista Asociación Argentina de Mineralogía, Petrología y Sedimentología*, 10: 33-54, Buenos Aires.
21. Mazzoni, M. M., 1985. La Formación Sarmiento y el vulcanismo paleógeno. *Revista Asociación Geológica Argentina*, 40(1-2): 60-68, Buenos Aires.
22. Mendia, J. E. & A. Bayarsky, 1981. Estratigrafía del Terciario en el valle inferior del río Chubut. *VIII Congreso Geológico Argentino*, San Luis, 1981. Actas, 3: 593-606.
23. Miolano, A. D., 1977. *Estudio integral de las cuencas bentoníticas de San Juan*. Departamento de Minería, San Juan.

24. Natale, I. M. & M. E. Mandolesi, 1985. Caracterización de un mineral montmorillonítico de la provincia de Río Negro. *Revista Asociación Geológica Argentina*, 40: 290-292, Buenos Aires.
25. Nullo, F. E., 1979. Descripción Geológica de la Hoja 39c, Paso Flores, provincia de Río Negro. *Servicio Geológico Nacional,* Boletín 167, Buenos Aires.
26. Rolleri, E. O. & C. A. Fernández Garrasino, 1979. Comarca septentrional de Mendoza. II Simposio de Geología Regional Argentina, Academia Nacional de Ciencias, 1: 771-809, Córdoba.
27. Schalamuck, I. B. & M. C. Cábana, 1999. Bentonitas de la región precordillerana de San Juan y Mendoza. En: E. O. Zappettini (Ed.), *Recursos Minerales de la República Argentina*, Instituto de Geología y Recursos Minerales (SEGEMAR), Anales, 35: 921-927, Buenos Aires.
28. Spalletti, L., 1995. Los sistemas de acumulación fluviales y lacustres del Triásico en la región occidental de la Precordillera Sanjuanina, República Argentina (Resumen). *II Reunión Triásico del Cono Sur*, Bahía Blanca, 1995, 27-28.
29. Spalletti, L.A. & Mazzoni, M.M., 1977. Sedimentología del Grupo Sarmiento en un perfil ubicado al Sudeste del lago Colhué Huapí, provincia de Chubut. *Obra del Centenario del Museo de La Plata*, 4: 261-283, La Plata.
30. Spalletti, L. A. & M. M. Mazzoni, M.M., 1979. Estratigrafía de la Formación Sarmiento en la barranca sur del lago Colhué Huapí, provincia del Chubut. *Revista Asociación Geológica Argentina*, 34(4): 271-281, Buenos Aires.
31. Stipanicic, P. N., 1979. El Triásico del valle del río de Los Patos (provincia de San Juan). En: Turner, J. C. M. (Ed.), *Geología Regional Argentina*, Academia Nacional de Ciencias, 1: 695-744, Córdoba.
32. Stipanicic, P.N. & J. F. Bonaparte, 1979. Cuenca Triásica de Ischigualasto-Villa Unión (Provincias de La Rioja y San Juan). II Simposio de Geología Regional Argentina, Academia Nacional de Ciencias, Córdoba, Argentina, 1:523-575
33. Vallés, J. M., 1987. Posición estratigráfica y distribución de los horizontes de bentonita en Río Negro, Neuquén y La Pampa, Argentina. *Actas X Congreso Geológico Argentino*, II: 33-37, Tucumán.
34. Vallés, M. & A. Impiccini, 1994. Hallazgos de bentonita en los niveles basales de la F. Chichinales, Depto. Añelo, Pcia. del Neuquén. *II Reunión de Mineralogía y Metalogenia*, La Plata. Publicación 3: 423-428.
35. Vallés, J. M. & A. Impiccini, 1996. Geología, mineralogía y propiedades tecnológicas de las bentonitas de Pozo Cavado, Depto. Añelo, Neuquén. *Actas VI Reunión Argentina de Sedimentología*, Bahía Blanca, 1996, 317-324.
36. Vallés, J. M. & A. Impiccini, 1999a. Bentonitas de la Cuenca Neuquina, Río Negro, Neuquén y La Pampa. En: E.O. Zapettini (Ed.), *Recursos Minerales de la República Argentina*, Instituto de Geología y Recursos Minerales (SEGEMAR), Anales, 35: 1113-1125, Buenos Aires.
37. Vallés, J. M. & A. Impiccini, 1999b. Depósitos de bentonitas terciarias de Zapala y Añelo, Neuquén. En: E.O. Zapettini (Ed.), *Recursos Minerales de la República Argentina*, Instituto de Geología y Recursos Minerales (SEGEMAR), Anales 35: 1385-1390, Buenos Aires.
38. Zuleta, M. & C. G. Rudolph, 1992. Tipificaciones de las bentonitas de Cuyo. *II Jornadas Argentinas de Tratamiento de Minerales*, La Plata, 1992, 14-16.

Clay mineralogy and construction problems in volcanic soils near Mount St. Helens: Sediment Retention Structure, southwestern Washington, USA

Michael L. Cummings

Applied Mineralogy Laboratory, Geology Department, Portland State University, P. O. Box 751, Portland, Oregon, USA, 97207   CummingsM@pdx.edu

During the late 1980's, a 60 m earth-cored dam was constructed across the North Fork of the Toutle River in southwestern Washington to impound sediment released during the 1980 eruption of Mount St. Helens volcano. Unanticipated soil conditions during excavation, handling, stockpiling, and compaction were encountered in saprolite developed from flow-top breccia and volcanic debris flows and in less intensely weathered bedded sediment. Saprolite developed from glassy flow-top breccia and 35 m below the surface contains 100 percent low-charge smectite. In saprolite developed from dacite and andesite clasts and matrix in volcanic debris flow, the percentage of halloysite increases upward while the percentage of smectite and expandable mixed layer clays decreases. Increase in the percentage of fines, decrease in bearing strength and stability, and decrease in optimum moisture content from in place to remolded soils characterize saprolite used in the impervious core of the structure. The bedded sediment was disposed as waste due to high natural water content relative to optimum moisture content and loss of stability and bearing capacity upon handling.

1. INTRODUCTION

On May 18, 1980, Mount St. Helens erupted in the Cascade Range of southwestern Washington, USA (Fig. 1). At the start of the eruption, a debris avalanche with volume estimated at 2.8 km^3 covered 60 km^2 of the valley of the North Fork of the Toutle River to an average depth of 45 m. About 200 million metric tons of sediment were removed in a large volcanic mudflow that originated in the debris avalanche a few hours after emplacement (Costa, 1994). The debris entered the Columbia River via the Cowlitz River and blocked the dredge-maintained shipping channel trapping ships in upstream ports and disrupting international shipping. Sediment yield estimates made in 1985 predicted long-term dredging would be required to maintain shipping on the Columbia River.

To reduce the volume of sediment moving down the Toutle River from the debris avalanche deposits, a Sediment Retention Structure (SRS) was constructed. During construction the mechanical properties of site soils resulted in construction delays and an investigation into the characteristics of these soils was undertaken. In this paper, the geology of the construction site, the bulk and clay mineralogy of soils from the right abutment, and the relation between clay mineralogy and construction problems are discussed.

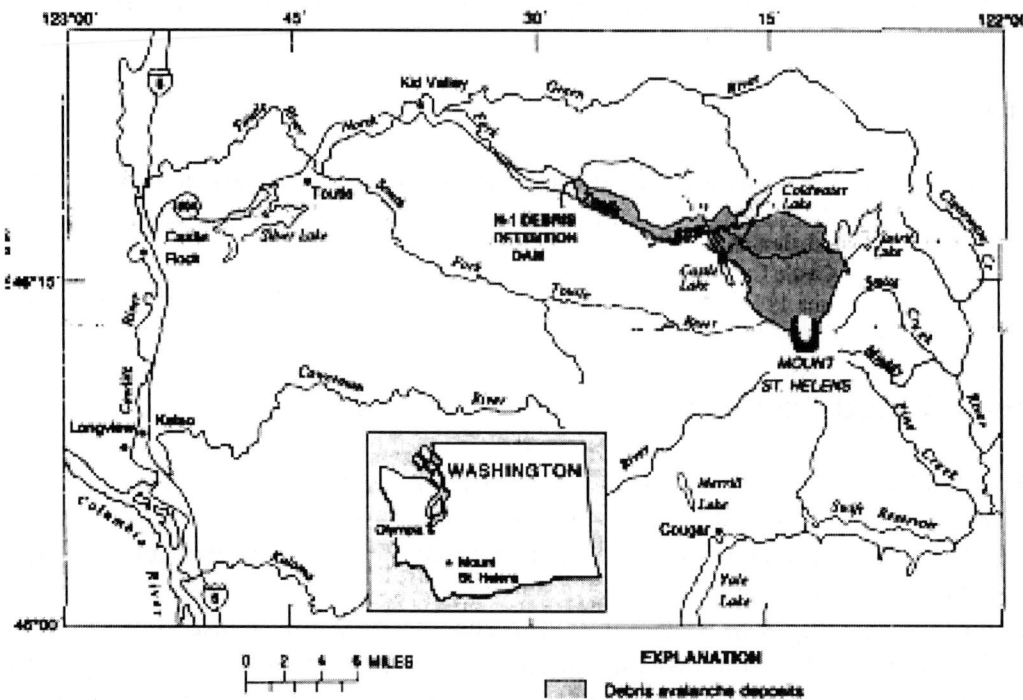

Figure 1. Location map of the Sediment Retention Structure (SRS) in relation to Mount St. Helens and debris avalanche deposits from the May 18, 1980 eruption.

2. GEOLOGIC SETTING AND SITE GEOLOGY

The Cascade volcanic arc, extending from southern British Columbia, Canada to northern California, USA, reflects subduction of oceanic plates of the Pacific basin beneath the North American plate. The Juan de Fuca plate, a remnant of the Farallon Plate, is subducting with a convergence rate of 4.5 cm yr^{-1} (Rogers, 1988).

The Quaternary edifice of the Mount St. Helens stratovolcano overlies volcanic and volcaniclastic rocks and shallow-level intrusions of the middle Tertiary core of the Cascade volcanic arc. The middle Tertiary section is several kilometers thick, strikes north-south and dips approximately 20 degrees east (Evarts and others, 1987).

The site geology of the SRS based on outcrops, excavations, and shallow exploration drill holes was determined by the U.S. Army Corps of Engineers (CoE) (1986). The Hatchet Mountain volcanics, the oldest unit, underlies the construction site. The formation consists of interlayered massive, jointed flow interior and rubbly flow-top breccia zones of basaltic andesite lava flows. Deeply weathered flow-top breccia on the right abutment was used in the impervious core of the SRS and moderately to weakly weathered platy jointed rock was used for rock fill.

Unconsolidated units of differing age and degree of weathering overlie the Hatchet Mountain volcanics. The deposits record repeated stages of deposition and erosion along the Toutle River valley. Units examined in this study include a saprolite developed from a volcanic debris flow and containing subrounded boulders of andesite and dacite in a cobble to silt matrix that overlies the Hatchet Mountain volcanics and a less weathered unit consisting of interbedded silt and sand with irregularly shaped volcanic rock fragments overlies the saprolite across an erosional unconformity, both on the right abutment. The debris flow saprolite was used for the impervious core of the SRS while the bedded sediments were excavated as waste.

3. RESULTS

The clay and bulk mineralogy of units on the right abutment are described below and stratigraphic relations are shown in Figure 2. Samples were collected during the site preparation, excavation, and construction phases of the SRS from excavations either freshly opened for the purpose of sampling or excavated within days of sampling. Samples were placed in plastic bags and the bags were sealed to maintain the natural moisture content.

3.1 Hatchet Mountain volcanics

The middle Tertiary Hatchet Mountain volcanics were exposed in a small CoE quarry on the right abutment at the start of construction and, as excavation continued, the flow-top breccia and upper contact of the unit were exposed. Below the level of the quarry, exploratory drill holes penetrated massive flows with lenses of breccia. The spillway of the SRS is cut into these massive flows at an elevation of 286.5 m.

The massive to platy jointed flow interior and the transition to flow-top breccia were sampled from the quarry. The flow interior ranged from hard to easily scratched with a rock hammer. In thin sections, plagioclase, orthopyroxene, and clinopyroxene phenocrysts and microphenocrysts are surrounded by plagioclase, pyroxene, Fe-Ti oxide, and glass-bearing groundmass. Smectite preferentially replaces orthopyroxene relative to clinopyroxene and plagioclase phenocrysts. Clay alteration in the groundmass is spotty and fractures are lined by clay. In the transition between the flow interior and the flow-top breccia, small cavities (< 3 mm) and the percentage of glass in the groundmass increase. Clay replaces minerals and glass and partially to fully infills cavities (foliated smectite). The extent of alteration of phenocrysts and groundmass increases as the abundance of cavities and microfractures increases. The clay mineral in the flow interior zone is low-charge smectite.

In the flow-top breccia, clasts are vesicular and the size of cavities between clasts increases toward the outer surface of the flow. Bulk and clay mineralogy (Fig. 2) indicate an increase in the percentage of clay in the bulk sample from the flow-interior (TR-1) to the flow-top breccia zone (OW2, OW3).

In the lower levels of the flow-top breccia (OW7, OW8), low-charge smectite or low-charge smectite and 7Å halloysite are present with plagioclase. Where the samples are completely clay, low-charge smectite is present with traces of vermiculite and expandable mixed layer clay (OW3) or low-charge smectite with 7Å halloysite and traces of expandable mixed layer clay (OW2). Two species of low-charge smectite are present in some samples.

The decomposed flow-top breccia was used in the impervious core of the SRS. It was very hard in-place, but broke down to a stiff, clayey silt during handling (Cornforth Consultants, Inc, 1991).

3.2 Debris-flow saprolite.

On the right abutment the Hatchet Mountain volcanics are overlain by up to 13.8 m of saprolite developed from a volcanic debris flow. The debris flow consists of rounded to subrounded cobbles to boulders (up to 1 m diameter) of andesite and dacite in a fine-grained matrix. The cobbles, boulders, and matrix are thoroughly weathered and can easily be cut by a pocket knife. Core stones are sparse. At the time of excavation, the otherwise uniform saprolite contained three color zones; lower yellow, beige, and upper yellow. Sample collection for mineralogical analysis was distributed relative to these color zones (Fig. 2).

The bulk mineralogy of the debris-flow saprolite (Fig. 2) is dominated by clay minerals and quartz. The percentage of feldspar is greatest in the beige and lower part of the upper yellow zones. At approximately 3.4 to 4.6 m beneath the upper contact, the bulk mineralogy changes significantly. In this depth interval, quartz and cristobalite increase, feldspar and magnetite are not detected, and the clay mineral content decreases.

The clay mineral suite is dominated by halloysite (7Å and 10Å), high- and low-charge smectite, and vermiculite (Fig. 2). Expandable mixed layer clay and kaolinite are present in some samples. In the lower yellow zone, 7Å and 10Å halloysite, smectite, and either vermiculite or expandable mixed layer clay comprise 52 and 64 percent of the bulk sample. In the beige zone and lower yellow zone, interlayered 7Å and 10Å halloysite and vermiculite are present, but smectite is not detected. At higher levels in the upper yellow zone where feldspar is not detected in the bulk sample, smectite is present with 7Å and 10Å halloysite and vermiculite. Kaolinite is present in a sample from near the upper contact.

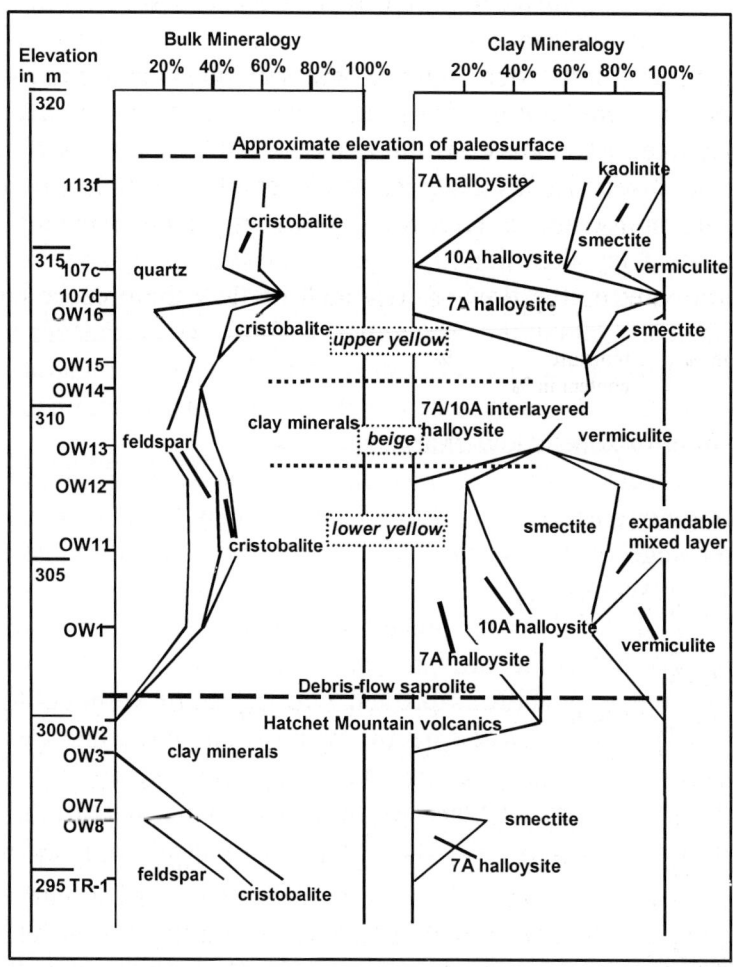

Figure 2. Bulk and clay mineralogy for samples from the flow-top breccia and debris flow saprolite from the right abutment of the SRS. Stratigraphic boundaries and color zones within the debris flow saprolite are indicated.

The engineering properties of the debris-flow saprolite were investigated thoroughly during excavation and emplacement in the impervious core of the SRS. The engineering properties changed from in place (undisturbed) to the impervious core (remolded) with respect to grain size, strength and optimum moisture content.

The primary texture of the debris flow saprolite is unsorted rounded cobbles and boulders of dacite and andesite in a fine-grained matrix that probably contained abundant glassy ash. Although boulders are up to 1 m in diameter, they are so deeply weathered that the mechanical size grading of in-place saprolite is a silty to sandy gravel. However, the saprolite broke down to clayey silt during placement in the impervious core. The majority of samples from the impervious core are silts of medium plasticity (MH) (Strid, 1988).

Undisturbed saprolite was stable in vertical cuts for months, but was no longer stable in moderate slopes after disturbance and showed loss of bearing capacity after working. Both the flow-top breccia and debris flow saprolites lost much of their in-place strength after excavation (Cornforth Consultants, Inc., 1991).

The natural water content of the in-place saprolite averaged 41.9 percent and the optimal moisture content (OMC) averaged 34.2 percent (Cornforth Consultants, Inc., 1991). This required a 4.7 percent reduction in water content to meet specifications of 3 percent above OMC. However, samples of remolded soil from the impervious core indicated OMC of 30.9 percent (Strid, 1988), requiring an additional 3.3 percent reduction in water content to meet specifications. Thus, the optimum moisture content (OMC) decreased measurably from undisturbed to remolded soil. As a result, the predicted effort to decrease the natural water content to OMC based on in-place samples was significantly underestimated.

Table 1. Average properties of in place and remolded volcanic debris flow saprolite.

	Natural water content in %	Optimum moisture content in %	Liquid limit in %	Plastic limit in %	Plasticity index	Dry density
In place (undisturbed)	42.1 (n=22) (32.6 to 49.5)	34.3 (n=21) (30.4 to 38.8)	48.9	37.1	11.8	1340.0 kg/m^3
Impervious core (remolded)	34.2 (n=60) (27.2 to 40.8)	31.0 (n=13) (29.2 to 32.7)				1384.9 kg/m^3 compacted to 1821.9 kg/m^3

(Strid, 1988; Cornforth Consultants, Inc., 1991).

3.3 Upper bedded unit

The upper bedded sediments consist of weathered interbedded clay/silt and silt/sand beds that were deposited upon an eroded surface cut into the debris-flow saprolite. Included in the unit are sediments deposited within a channel cut at least 13 m into the debris-flow saprolite. These bedded sediments were excavated for disposal.

Although generally composed of clay-rich silt and sand, partially weathered lithic fragments were present near the base of the 14 m thick unit. Vertical cuts remained wet for weeks without weeping and did not form shrinkage cracks during dry-weather conditions. The natural moisture content was 16.4 percent above OMC (Cornforth Consultants, Inc., 1991). The material released progressively more water with each pass of heavy equipment that eventually became mired.

Samples collected from relatively thin layers (3-5 cm) have different bulk and clay mineral suites with no systematic change in assemblage with stratigraphic position. Feldspar (trace to 42%), quartz (21-53%), and clay (18-46%) are the prominent minerals and are associated with cristobalite (trace to 22%) and amphibole (trace to 9%). The clay mineral suite displays considerable variability among samples. Halloysite (7Å and 10Å), high- and low-charge smectite and vermiculite are present in the most common assemblages and may be associated with chlorite, or kaolinite, or expandable mixed layer clay. Chlorite, smectite, and vermiculite have interlayers of other clay minerals. Chlorite and vermiculite are poorly crystalline in some layers.

4. DISCUSSION

Soil mineralogy, texture, and structure develop as rock weathers to saprolite. However, the relative contribution of the mineral suite, texture and structure to sensitivity in soils is poorly known. Below, the weathering history and engineering properties of soils at the SRS are discussed in relation to the genesis of the clay-mineral suites in the saprolite and overlying bedded sediment.

4.1 Genesis of clay-mineral suites:

Patches of deeply weathered volcanic soils are present throughout western Washington and Oregon. These soils are developed on a variety of volcanic parent materials including lava flows, debris flows, and sedimentary rocks. Locally, ferruginous bauxite is developed from flows of the Columbia River Basalt Group (Cummings and Fassio, 1990). The age of intense weathering is constrained by upper Miocene to lower Pliocene sedimentary rocks that overlie the lateritic soil (Wilson, 1997). However, all deeply weathered soils are not of the same age.

At the SRS, the oldest and most deeply weathered deposits are the flow-top breccia and volcanic debris flow saprolite of the right abutment. The contact and mineralogical patterns observed in the Hatchet Mountain volcanics and volcanic debris flow saprolite indicate a systematic variation in bulk and clay mineralogy with depth in the weathering profile. Patterns are consistent with chemical weathering of these units after juxtaposition. The changes in bulk and clay mineralogy in the lower part of the upper yellow zone (Fig. 2) are believed to reflect the water table at the time of weathering. The saprolite was partially eroded before burial by the bedded sediments unit.

The irregularly shaped hard volcanic fragments and variability in clay and bulk mineralogy among different beds suggest that the bedded sediments may have accumulated in a short-lived lake similar to those formed during the 1980 eruption of Mount St. Helens, when several lakes formed in tributary valleys to the North Fork of the Toutle River where the debris avalanche blocked their outlets. The original glass content in different beds may have varied substantially and is reflected in the bulk mineralogy (beds with low or trace feldspar) and clay mineralogy (kaolinite, smectite, 7Å halloysite assemblage). The relatively high percentage of cristobalite suggests that it was a component of the original volcanic material.

4.2 Engineering relations

The problems encountered at the SRS are similar to problems at other sites in the western slopes of the Cascades. The Mud Mountain Dam on the White River, near Seattle, Washington, was redesigned in 1941 because soils selected for the earthen core had residual moisture content that exceeded optimum and generated excess fines during mechanical grading (Anonymous, 1941a,b). Construction of the Hills Creek Dam on the Middle Fork Willamette River, Oregon started in 1957. The impervious fill soil could not be compacted as specified due to rutting by wheels of the rubber-tired roller. The dominant clay mineral was halloysite with as much as 15 percent smectite (USACE, 1966).

The construction problems encountered in soils of the right abutment of the SRS reflect the weathering history of these volcanic deposits. The well-preserved primary volcanic textures in the flow-top breccia and debris flow saprolite suggest chemical weathering occurred with little change in volume. As primary minerals and volcanic glass weathered, bulk density decreased while porosity and a succession of secondary minerals formed as replacement and/or precipitated phases. Velbel (1993) and Delvigne (1998) document the complex relations among chemical leaching, generation of porosity, and crystallization of a succession of minerals that are stable as the composition of the soil solution evolves.

The change in grain size, loss of strength, and change in optimum moisture content of saprolite between in place and remolded state reflect the weathering processes whereby the primary mineralogy and volcanic glass were converted to the bulk and clay suites of the residual soil. Vaughan et al. (1988) suggest that engineering characteristics of residual soils that are related to their geologic origins include: a component of strength and stiffness due to bonding, varying mineralogy and grain strength, and widely varying porosity.

Bonding in residual soils develops progressively as the soil evolves (Vaughan et al., 1988). In volcanic deposits, bonding may form very early, near the time of deposition, and evolve as weathering progresses. Approximately seven years after the 1980 eruption of Mount St. Helens, the gray, highly porous volcanic mudflow deposits were distinctly cemented, supported vertical cuts, and broke into large blocks. SEM photomicrographs show clay minerals and amorphous silica precipitated on pore walls and the fine-grained glass-rich matrix cemented to rock fragments. The bonding in the saprolite evolved during the weathering process to produce mechanically stable soils with considerable in-place strength that yielded water and changed physical properties during mechanical working.

Vaughan et al. (1988) argue that standard engineering tests that describe the mechanical properties of soils are inadequate to describe the essential characteristics of residual soils. The structural breakdown of the soil when manipulated is associated with the breaking of the bonds formed during evolution of the soil. Water trapped in pores within the bonded residual soil is released as mechanical working takes place.

The clay mineral suite develops in a residual soil concurrently with bonding between grains and evolution of porosity. The porosity of clasts of different texture and composition and of matrix in the debris flow influences the local water:rock ratio resulting in considerable variation in the bulk and clay mineral assemblage. Dissolution of primary minerals provide chemical compounds that are either flushed from the soil, incorporated into replacement minerals or precipitated as new minerals in other parts of the soil. The crystallization of halloysite, smectite, and other clay minerals and development of bonding between particles at the same time primary minerals are dissolving and porosity is evolving provides an opportunity for water to be trapped in the saprolite and bedded sediments. Box-work textures formed by replacement and precipitated minerals in sensitive soils developed from basalt lava flows are documented in SEM photomicrographs by Bednarz (2002). Mechanical working progressively disrupts the box-work and releases pore water.

5. CONCLUSIONS

The release of water during handling is related to the break down of the residual structure of the soil. Chemical weathering to form saprolite is an isovolumentric process. Leaching of chemical constituents and the formation of clay minerals go hand in hand. At constant volume, the density of the soil decreases as pore spaces develop. The contemporary development of pore space and the crystallization of soil clays traps water in a complex maze of variably connected pores and domains of variable inter-particle bonding. As these soils are disturbed, at least some of these pores are ruptured and water is released, however with continued working more and more pores are ruptured and the bonds that form the soil structure are increasingly disturbed. The processes that produce the structure in residual soils from volcanic materials also produces halloysite and smectite as prominent clay minerals. These minerals are present in the problem soils at the SRS and at the Hills Creek dam (USACE, 1966) where they line pores and help trap water within the residual soil.

6. ACKNOWLEDGEMENTS

Clay and bulk mineralogy for the SRS site were determined by Reka K. Gabor, Research Associate, Department of Geology, Portland State University. Richard A. Lewis of Granite Construction Company supported this research and encouraged its publication. Review by Colin Harvey and a second reviewer. Their comments are appreciated.

REFERENCES

Anonymous, 1941a, Engineering News-Record, p. 408-410, March 13, 1941.
Anonymous, 1941b, Engineering News-Record, p. 282-284, August 28, 1941.
Bednarz, S., 2002, Influence of halloysite on the engineering behavior of basaltic saprolites in northwestern Oregon and southwestern Washington [M.S. thesis] Portland State University, Portland, Oregon
Cornforth Consultants, Inc., 1991, Clay mineralogy claims analysis Sediment Retention Structure. Report to U.S. Army Corps of Engineers, Contract DACW57-88-C-0018.
Costa, J. E., 1994, Evolution of sediment yield from Mount St. Helens, Washington, 1980-1993: U.S. Geological Survey Open-File Report 94-313.
Cummings, M. L. and Fassio, J. M., 1990, Chemical Geology, v. 84, p. 40-41.
Delvingne, J. E., 1998, The Canadian Mineralogist Special Publication 3, 495 p.
Evarts, R. C., Ashley, R. P., and Smith, J. G., 1987, Journal of Geophysical Research, v. 91, no. 10, p. 10,155-10,169.
Rogers, G. C., 1988, Canadian Journal of Earth Sciences, v. 25, p. 844-852.
Strid, J. M., 1988, Unusual soil characteristics of the right abutment impervious material (Ancient Glacial Drift/Saprolitized Diamicton). Internal report prepared by Granite Construction Company employee.
U.S. Army Corps of Engineers, 1966, Hills Creek Reservoir Middle Fork Willamette River, Oregon: Design Memorandum No. 4, Embankment Design Completion Report.
U.S. Army Corps of Engineers, 1986, Mount St. Helens Sediment Control Sediment Retention Structure engineering plans and geologic site characterization.
Vaughan, P. R., Maccarini, M., and Mokhtar, S. M., 1988, Quarterly Journal of Engineering Geology, London, v. 21, p. 69-84.
Velbel, M. A., 1993, American Mineralogist, v. 78, p. 405-414.
Wilson, D. C., 1997, Post-middle Miocene geologic history of the Tualatin basin, Oregon with hydrogelologic implications [Ph.D. dissertation] Portland State University, Portland, Oregon.

Neoformed mineral parageneses in acid weathering systems: sedimentary *vs* volcanic environments

Th. De Putter[a], A. Bernard[b], A. Perruchot[c], J. Yans[a], Fr. Verbrugghe[b] and Ch. Dupuis[a]

[a]Faculté Polytechnique de Mons, Géologie Fondamentale et Appliquée, 9 rue de Houdain, B-7000 Mons, Belgium

[b]Université Libre de Bruxelles, CP 160/02, Géochimie et Minéralogie, 50 av. Fr. Roosevelt, B-1050 Brussels, Belgium

[c]Université de Paris XI – Orsay, UMR 8616, Géochimie des Roches Sédimentaires, bâtiment 504, F-91405 Orsay, France

Various weathering systems are compared in this paper, on both sedimentary and volcanic substrates. In all the systems, acid fluids generated by sulfur oxidation dissolved primary silicates and precipitated weathering minerals according to their decreasing acidity (due to progressive neutralization by the host sediment). The resulting mineral parageneses are fairly similar in both settings, confirming that changes in weathering mineralogy are governed by "normal" parameters of mineral stability (temperature and composition). The influence of the bulk composition of the protolith, and the role played by the availability/activity of alkalis and earth alkaline elements (Ca, Mg) in the profile, are especially emphasized.

1. INTRODUCTION

Recently, B. Velde suggested that clays were forming under conditions of extreme chemical disequilibrium in most weathering systems. Here, contrary to deep burial conditions, chemistry is that of an open system, with many solids entering into the aqueous phase, and eventually leaving the zone where the clays are forming (1). As a consequence, minerals phases formed mostly respond to chemical forces. This means that neoformed minerals "mirror" the chemical compositions of (i) the weathered protolith, (ii) the weathering fluid phase, and (iii) the variations in these chemistries, with time.

In this paper, we deal with weathering systems in rather different surface environments, i.e. on one hand karsts and weathering profiles developed on sedimentary rocks, and on the other hand the vicinity of volcanic fumaroles.

In karsts, acid fluids are generated mostly by the oxidation of sedimentary pyrite contained in the sedimentary filling of the depression, according to two possible oxidation reactions:

$FeS_2 + (15/4) O_2 + (7/2) H_2O \rightarrow Fe(OH)_3 + 2 H_2SO_4$ (2) (reaction 1)
$FeS_2 + 14 Fe^{3+} + 8 H_2O \rightarrow 15 Fe^{2+} + 2 SO_4^{2-} + 16 H^+$ (3) (reaction 2)

The fluids are then "funneled" through the sediments of the karst filling, that are mostly mature sediments, rich in phyllosilicates. Eventually, they reach the karst's wall, where alkaline earth elements (Ca, Mg) dominate. We also consider in this section a thick and "old" (probably Mesozoic, see below, 4) kaolinite regolith, though in this case, the observed minerals do not conclusively point to a massive sulfuric acid intervention.

In fumarolic environment, acid fluids are generated through the cooling of volcanic SO_2-rich gases, according to the sulfur disproportionation reaction, and through a direct oxidation of gaseous H_2S (g: gas; l: liquid; s: solid):

$4\ SO_2\ (g/l) + 4\ H_2O\ (g/l) \rightarrow 3\ H_2SO_4\ (l) + H_2S\ (g/l)$ (4) (reaction 3)

At low temperature (< 200°C), prevailing in the environments considered below, this reaction is written:

$3\ SO_2\ (g/l) + 3\ H_2O\ (g/l) \rightarrow 2\ H_2SO_4\ (l) + S\ (l/s) + H_2O\ (g/l)$ (reaction 4)
$H_2S\ (g/l) + 2\ O_2\ (g/l) \rightarrow H_2SO_4\ (l)$ (reaction 5)

These fluids are then percolating through primary magmatic materials (pyroclastic deposits and/or basalt), with high alkalis and alkaline earth elements contents.

The mineralogy of the studied samples was determined by X-ray diffraction (XRD) and SEM examination. Major element analyses were routinely performed at the "Centre de Recherches Pétrographiques et Géochimiques" (CNRS, Nancy, France), using emission-ICP.

2. BRIEF OUTLINE OF THE SEDIMENTARY SYSTEMS

2.1. Newhaven karstic aluminous minerals deposits (S. England)

The Newhaven karsts are metric pre-Thanetian depressions in the Upper Cretaceous chalk from Sussex. Chalk and the depressions, are unconformably overlain by the Thanetian/Ypresian "Woolwich Beds", and the uppermost dark silty/clayey sediments of the "Peaty Beds". The "Woolwich Beds" comprise a lower, marine, glauconite-bearing sand series, and an upper, freshwater (fluviatile), cross-bedded feldspathic sand series. The "Peaty Beds" contain lignite layers in which are abundant vegetal remains, and root imprints (5). Organic matter from the "Peaty Beds" is now highly oxidized (degradofusinite), while root imprints and small vertical fissures are filled with jarosite and gypsum, two common products of pyrite oxidation. The "Peaty Beds" are thus assumed to be the origin of acidity, through total oxidation of the pyrite they contained (6, 7).

Acid fluids percolated downward, through the Tertiary sediments overlying chalk. Karstic depressions acted as "funnels" where the downward percolating fluids concentrated, and extensive mineral neoformation occurred in the depressions, and in the voids created by karstification. The progressive neutralization of the fluids resulted in "sequentially" layered neoformation of minerals, i.e. from top to bottom: jarosite, iron oxides (yet in the Tertiary sands), aluminous minerals (namely: alunite, aluminite, basaluminite, hydrobasaluminite, gibbsite, and bayerite), gypsum, and Fe-Mn oxides (in the karsts) (7). The Newhaven neoformed sediments were tentatively dated by U/Th α-spectrometry, at 40-120 Kyr BP (8).

2.2. Counterpart to the Newhaven fossil system: Dhainaut quarry, Flines-Les-Raches (N. France)

A modern counterpart of the Newhaven profiles is presently studied in Dhainaut quarry, at Flines-les-Raches (9). The Thanetian Flines sand is stratigraphically, mineralogically, and chemically similar to the unweathered glauconite-bearing marine sand of the lower

"Woolwich Beds", though it still contains pyrite (contrary to the Newhaven sand). In Flines, the geological context is non karstic. Meteoric fluids percolate downward in the exposed Flines sand quarry walls, and flow out in the lower part of the walls, due to the presence of an underlying impervious substrate (Louvil Clay Fm). Fluids are characterized by their low pH (2 - 3), their high [SO_4^{2-}] contents (3-5000 ppm), and the predominance of ferrous (*vs* ferric) iron (Fe^{2+}/Fe^{3+} ratio ~10). The presence of dominant divalent iron in fluids, and of ferrous neoformed minerals (see below), could suggest that pyrite oxidation was controlled by reaction 2 (see above, 1).

Neoformed minerals include fairly common jarosite, iron amorphous compounds (ferrihydrite?), some gypsum, and rare aluminum sulfates. Besides these minerals, there are also in Flines "anecdotic" occurrences of alunogen efflorescence (indicating very low pH, <1), and a rich and diverse series of Fe^{2+} and Fe^{3+} sulfates (Fe^{2+}: melanterite, halotrichite, rozenite, and szomolnokite; Fe^{3+}: jarosite, copiapite, magnesiocopiapite, and coquimbite). Though sometimes spectacular (e.g. bright yellow copiapite mm-wide nodules), these minerals are quantitatively minor. Interestingly, Al-bearing neoformed minerals in Flines appear to be nor as abundant, nor as diverse as in Newhaven. This suggests that the fluids are only weakly neutralized, due to the fact that the release of acid fluids through pyrite oxidation is a fast process, when compared to their neutralization by the host sediment.

2.3. Transinne kaolinite regolith (E. Belgium)

The Transinne regolith is a thick (~65 m) kaolinite profile, superimposed on folded rocks from the Devonian Oignies Fm, comprised of shale and sandstone. The unweathered shale/sandstone series of the Oignies Fm is locally carbonaceous, and contains a few percent of pyrite, nowadays epigenized in iron oxides within the regolith.

Neoformed minerals are predominantly kaolinite (probably formed through the dissolution of pre-existing phyllosilicates, and especially of chlorite), and associated iron oxides. The acid(s) involved in the weathering process, whose age could be as old as Upper Jurassic-Lower Cretaceous (10), is/are a matter of debate. On one hand, the lack of sulfate minerals, indicative of low pH (e.g. jarosite), does not support a leaching by sulfuric acid originating from pyrite oxidation. On the other hand, epigenized pyrite cubes are observed in the profile, and, above all, kaolinite deposit is rather thick, suggesting a massive aluminum mobilization process (at pH < 4).

Schmitt and Thiry once suggested that thick kaolinite profiles could be efficiently formed through leaching of the protolith by meteoric fluids, especially in periods (such as the Cretaceous) where these fluids had high CO_2 fugacities (11). Preliminary geochemical modelling of the Transinne regolith indeed suggests that a "mixed" situation, in which pyrite was oxidized by (supposedly Upper Jurassic-Cretaceous) fluids with high CO_2 fugacities, could account for the observed mineralogy (gibbsite → kaolinite + hematite).

3. BRIEF DESCRIPTION OF THE VOLCANIC FUMAROLIC SYSTEMS

3.1. Mount Papandayan (Java, Indonesia)

Preliminary investigation of the weathering facies within the Papandayan volcano crater (W. Java, Indonesia) reveals a complex situation. The crater is made up from andesitic lava, though ash deposits, and (rare) obsidian blocks are also observed. Active fumaroles, "sulfur pools", and hot springs are in contact with a predominantly andesitic material. Native sulfur, and various polymorphs of silica, are neoformed in the immediate vicinity of the active fumaroles. "Sulfur pools" and hot springs are more interesting, with regard to clay neoformation. Spring fluids are hot (50-90°C), with moderate SO_4 content (typically 400-1200 ppm, though values as high as ~5000 ppm were occasionnally measured). Their typical pH range is 1-2. Fluids from the "sulfur pools" are generally more acid (pH as low as 0.35), and contains rather high halogen concentrations (up to 27,000 ppm Cl, and >200 ppm F), suggesting a more "direct" connection with volcanic gas vents.

In these environments, neoformed minerals are silica (*ab initio* sursaturated), alunogen, jarosite, iron oxide, Al-sulfate, and kaolinite. Interestingly, the presence of kaolinite in the immediate vicinity (a few meters) of the active "sulfur pool" indicates an advanced stage of neutralization of these strongly acid fluids. It is also noteworthy that some light coloured, earthy (though still coherent) samples collected in the lower part of the external flank of the volcano are fully transformed into kaolinite. The presence of an amorphous phase (preserved glass, allophanes ?) has not yet been investigated, but is suspected in several samples. More fieldwork is needed to get a better understanding of that complex weathering system, that remains poorly studied at the time.

3.2. *Faraglione di Vulcano* (Vulcano Island, Sicily)

The Faraglione hills are located on Vulcano Island, immediately to the north of the Fossa crater. Though these small hills are locally known as *Faraglione Ranni* (56 m), and *Faraglione Nicu* (27 m), they are in fact the residual parts of one and only small volcano, comprised of a porphyritic "neck" (still visible in the central part of *Faraglione Nicu*) surrounded by roughly concentric pyroclastic series, that constitutes most of the Ranni hill (12). The protoliths (porphyry, and pyroclastics, of subalkaline basaltic affinity) are always weathered to some extent. The Nicu hill has been almost completely strongly weathered, while the upper part of the Ranni hill seems to be less intensely weathered. The weathering process itself is presently undated. It could have been still active in the 18th century A.D., as copperplates by the Italian naturalist L. Spallanzani show active fumaroles in the presently collapsed Nicu cave.

Neoformed minerals are mostly found in the pyroclastics deposits, i.e. on the *Baia di Levante* beach, at the foot of the Nicu hill, and in the Ranni's *Grotta dell'Alume*. These minerals include a wide variety of sulfates: alunogen/meta-alunogen, jarosite/natrojarosite, jurbanite, alunite/natroalunite, minamiite, tamarugite, pickeringite, pentahydrite, chalcanthite, and gypsum/bassanite/anhydrite. Though weathering fluids can no longer be sampled, these sulfate minerals clearly point to a weathering process by fluids enriched in sulfuric acid (originating from reactions 4-5 above). Minor oxides (hematite) and sulfides (pyrite, greigite) are also observed. It is noteworthy that no clays have ever been described in this material, and that glass shards from the pyroclastic beds have often been unexpectedly spared by weathering (12).

3.3. "*Cave di caolino*" (Lipari Island, Sicily)

The "*Cave di caolino*" are disused kaolinite quarries located in the upper part of the seacliffs, on the northwestern coast of Lipari Island. These cliffs are comprised of thick series of interstratified pyroclastic beds, volcanic ashes, and obsidian flows. All these volcanic rocks pertain to the third volcanic phase recognized in the island, dated 20 – 13 Ka B.P. (13). Pyroclastic deposits are of overall andesitic affinity, though containing porphyric xenoliths of trachyandesitic affinity. These rocks were most probably weathered through secondary fumarolic activity. This hypothesis is supported by the overall "chimney-like" geometry of the deposit, and by the fact that low discharge fumarolic vents (fluids with pH~3) are still visible within the cave site. These fumaroles are unable to account for the extensive weathering of the protolith, but they could have been more active in past periods.

Neoformed minerals are: alunogen, jarosite/natrojarosite, iron oxides, alunite/natroalunite, gypsum/anhydrite, and kaolinite. These minerals sometimes emphasize the thin bedding of the pyroclastic deposits while in other places, they occur in very pure, bright masses, with no apparent structure. This is especially the case for kaolinite. This clay mineral, that is still extremely abundant in the disused parts of the quarry, was exploited within large (n.10m), snow-like white masses (coherent, earthy, or pulverulent, possibly after ashes?). The presence of kaolinite in Lipari recalls the observation made in Papandayan volcano (see 3.1. above), where this mineral was however neoformed from basaltic lava, and in much smaller quantities than in Lipari. Further work is planned on the "*Cave di caolino*" site.

4. COMPARISON OF NEOFORMED MINERALS AND DISCUSSION

The geochemical analyses (major elements) of the protoliths from the different studied systems are listed in table 1. But first of all, it is worth emphasizing an important difference between sedimentary and volcanic systems.

The source of acidity is indeed quite different between volcanoes, and the sediments. In volcanoes, there is an infinite source of acidity (volcanic gases), that allows quick, congruent (or nearly congruent), dissolution of primary magmatic minerals and/or rocks, and hence an almost immediate, and synchronous availability of all cations. The presence of halogens in the fluids (Cl, F) frequently results in the neoformation of Al-fluorides that form only under reducing conditions (14).

Weathering systems in sediments release "quickly" (as pyrite oxidation is a fast process) comparatively small quantities of acid fluids. These fluids can experience different scenarios:
- Quick neutralization by efficient buffers (carbonates→gypsum; chlorite→smectite-kaolinite); in this case, neutralization can successfully compete with acid release. Such reactions prevent any strong pH decrease, and thus limit the availability of some cations. Aluminum is one of these critical cations, because its solubility strongly decrease above pH~4.
- Partial neutralization on poorly reactive substrates, with precipitation of mineral species that in turn release H^+ ions (jarosite, ferrihydrite); in this case, the pH appears to be buffered to low values.
- Progressive and "slow" neutralization by reactive silicates. This is the "normal" scenario – i.e. the end-member of long-term processes (involving large quantities of fluids) that result in most weathering profiles within the geological timescale.

Notwithstanding this important difference in acidity source, field observations, and geochemical modeling show that neoformed minerals tend to precipitate "sequentially", according to the progressive neutralization of the acid fluids (7, 15): alunogen (pH<1) → (K, Fe) sulfates (1<pH<2) → iron oxides (2<pH<2.5) → K/Al-sulfates/hydroxides (pH>2.5) → silicates (pH>4-5), often via gels (as in Dieppe profile, see 16).

Let us now briefly consider how the observed weathering minerals, in the studied systems, fit within this chemical framework.

The simplest case is the Newhaven one: pH was low, efficient buffers were absent, and fluids were funneled, and concentrated, within karst pockets: we identified a sequential mineral assemblage, which matches fairly well the modeled one (7). The absence of any clay mineral

Table 1
Petrology and geochemical data (major elements, in wt%; < d.l. = below detection limit) of the protoliths weathered within the studied weathering systems

Location	Petrology	SiO_2	Al_2O_3	Fe_2O_3 + FeO	MgO	CaO	Na_2O	K_2O	Σ alkalis
Volcanic systems									
Mount Papandayan	andesite	56.3	19	7.9	3.5	8	3	1.2	4,2
Faraglione Hills	subalkaline basalt	57	15.7	6.3	3.2	5.2	4.2	5.6	9.8
Cave di caolino	(trachy)andesite	59.7	18.6	4.4	1.2	5.7	2.8	3.3	6.1
Sedimentary systems									
Newhaven Karsts	Glauconite-bearing sand	87	2.9	4.3	0.5	0.1	0.1	1.3	1.4
Flines-les-Raches	Glauconite-bearing sand	85.5	4.9	2	0.6	0.1	0.3	1.9	2.2
Transinne Regolith	interlayered shale and sandstone	65.4	17.4	6.6	2.2	<d.l.	0.4	3.2	3.6
Boom Clay Fm	pyrite-bearing clay	56.2	18.9	5.5	2.3	0.8	0.7	3.4	4.1

clay mineral in Newhaven is probably due to the fact that, when the fluid pH increased (at the karst wall), Al was short, and the precipitation of gypsum competed successfully with the (slow) precipitation of silicates.

In Flines, both the acidity source and the protolith were similar to the Newhaven ones. However, Flines differs from Newhaven by at least two important facts: Newhaven is an "old" (40 → 120 Kyr) system, where the fluids were concentrated within small volume pockets, while the Flines system is nowadays active, and fluids can freely diffuse within the quarry. Consequently, we identify in Flines only minor neutralization by the sandy substrate. Iron amorphous compounds (ferrihydrite?) and jarosite (with K^+ originating from feldspar dissolution) can form, but these solids both release H^+ ions while precipitating, leaving fluid acidity roughly unaffected. Fluids thus percolate within the quarry with no measurable pH decrease, which confirms that neutralization is a slow process with regard to the acid release.

Turning now to the Faraglione hills, we can observe a mineral paragenesis that shows some similarities with the Newhaven one, though there are some important differences in the protolith chemistry. The Al, alkalis and alkaline earth elements contents were significantly higher in the Faraglione than in Newhaven. These characteristics led to the precipitation of Al sulfates/hydroxides, and of many alkali and alkaline earth sulfates in which aluminum was included as a secondary cation. The absence of clay in this system is most probably due to the fact that the neutralization process did obviously not proceed to pH values higher than

~2.5. Maybe this results from the insufficient neutralization capacity of the volcanic rock (lack of Al?) and/or to a strong disequilibrium between (infinite) acid fluids generation, and the neutralization capacity provided by a restricted quantity of primary magmatic material.

In this respect, it is useful to compare the Faraglione hills with Mt Papandayan where kaolinite can form in the immediate vicinity of acid fluids venues. This suggests that, in the Papandayan active crater, the neutralization of the acid fluids by andesitic lava (rich in Al) has – at least locally – proceeded to pH conditions allowing silicates to form (pH>~4).

An interesting comparison confirms that the key factor determining the mineral output of acid weathering is the bulk chemistry of the protolith. Let us compare the andesitic lava of Mt Papandayan, and the Oligocene Boom Clay Fm from NE Belgium (see relevant data in table 1, and 17). Both rocks are at first quite different in origin, and in "maturity", though their major elements contents are quite similar (SiO_2 = 56%; Al_2O_3 = 19%; Σalkalis = 4,2%). Field data from the Papandayan, and the model output for the acid weathering of the pyrite-bearing Boom Clay Fm (local pyrite oxidation scenario, 17) consequently show remarkable similarities in the sequential neoformation of minerals, if one excepts the initial silica, and alunogen in the Papandayan, that both result from the stronger acidity of the initial volcanic fluids ("infinite" acidity source).

The situation in the "*Cave di Caolino*" is not yet fully investigated, though the abundance of kaolinite is a noticeable feature, which appears to be surprising in the case of a true "fumarolic" weathering system. Other scenarios, as a long lasting *per descensum* leaching of the protolith by meteoric fluids, should be considered also.

The weathering systems mentioned up to now shared one common, and important characteristic: the protolith never contained efficient mineral "buffers", chlorite or calcium carbonate, that continuously react with the acid fluids, and prevent any strong decrease in their pH value. On the contrary, chlorite was present in Transinne, where both field data, and the results of preliminary chemical modeling suggest that (at least) two scenarios could account for the observed paragenesis (kaolinite + iron oxide):

1. Pyrite oxidation by "Cretaceous" meteoric fluids (log $f(CO_2)$ = -2.45) + protolith → (transient gibbsite, later converted to) kaolinite + iron oxide
2. Chlorite + $2x$H$^+$ (also originating from pyrite oxidation) → smectite/kaolinite + x(Mg^{2+}, Fe^{2+}, later oxidized in Fe^{3+} to form iron oxides) + nH$_2$O (18)

Both pathways would account for (i) the absence of sulfate minerals, as acidity was never low enough (acid fluids are "quickly" neutralized by surrounding reactive mineral phases), and (ii) the massive neoformation of kaolinite. In both cases, it is suggested that the "product" of low acidity fluids by a long lasting weathering period (i.e. involving sufficient quantity of fluids) resulted in this thick kaolinite deposit.

The Transinne system could, in our opinion, be representative of what could be called the "slow way" to kaolinite, as opposed to the "fast way", involving infinite quantities of volcanic strongly acid fluids neutralized by primary magmatic protoliths.

5. CONCLUSIONS

The study of various weathering systems allowed us to address different aspects of mineral (including clay) neoformation. Obviously, mineral neoformation in acid weathering systems overall responds to chemical forces, as already stated by B. Velde (1). Neoformed minerals mirror fairly well the "simple" chemical interaction between a given protolith, and the acid fluids percolating through it, whatever the origin or nature of the protolith. However, in our opinion, clays are not forming "under conditions of extreme chemical disequilibrium" (1), but rather result from favourable – even if transient – physico-chemical conditions in the fluid/rock interaction. Clays, as solids, are thus always in equilibrium with the fluids whence they precipitate, even if these, as fluids, are chemically highly unstable.

Finally, an interesting consequence of the overall chemical "predictability" of the acid weathering systems, is that it is generally possible to model fairly easily the interactions between a given protolith and acid fluids, in most environments, using commercial chemical modeling codes (7, 11, 15).

ACKNOWLEDGEMENTS

This research was funded by the Belgian Agency for the Management of Radioactive Waste (ONDRAF/NIRAS), by the "Cimenteries Belges Réunies" (CBR), and by the "Commissariat Général aux Relations Internationales" (CGRI) of the "Communauté française de Belgique". The Dhainaut quarry (Flines-les-Raches, N. France) offered an easy access to his quarry.

REFERENCES

1. B. Velde, in G.D. Price and N.L. Ross (eds), The stability of minerals, Chapman and Hall, London (1992) 329-351.
2. D.K. Nordstrom, in J.A. Kittrick, D.S. Fanning and L.R. Hossner (eds), Acid Sulfate Weathering, Soil Science Society of America Special Publication 10, Madison (1982) 37-55.
3. K.J. Edwards, P.L. Bond, T.M. Gihring and J.F. Banfield, Science, 287 (2000) 1796-1799.
4. M. Kusakabe and Y. Komoda, Rept Geol. Surv. Japan, 279 (1992) 93-96.
5. Ch. Dupuis and C. Gruas-Cavagnetto, The London Naturalist, 75 (1996) 27-39.
6. R.D. Wilmot and B. Young, Proc. of the Geol. Ass., 96 (1985) 47-52.
7. Th. De Putter, A. Bernard, A. Perruchot, D. Nicaise and Ch. Dupuis, Clays and Clay Minerals, 48 (2000) 238-246.
8. Y. Quinif, Th. De Putter and Ch. Dupuis, Spec. Public. Ass. Sédim. Français, 30 (1998) 83-84.
9. F. Verbrugghe, Unpublished M. Sc. thesis, Brussels Univ., 2001.
10. Ch. Dupuis, J.-M. Charlet, L. Dejonghe and J. Thorez, Ann. Soc. Géol. Belgique, 119 (1996) 91-109.
11. J.-M. Schmitt and M. Thiry, 14th Intern. Sediment. Congress, abstract S8 (1994) 21-23.
12. A. Bernard and Th. De Putter, in prep.
13. H. Pichler, Rend. Soc. Ital. Mineral. Petrol. 36 (1980) 415-440.
14. F. Africano and A. Bernard, J. Volc. Geoth. Res. 97 (2000) 475-495.
15. P. Delmelle and A. Bernard, Geochim. et Cosmochim. Acta, 58 (1994) 2445-2460.
16. A. Perruchot, Ch. Dupuis, Th. De Putter, D. Nicaise and Fr. Arbey, C.-R. Acad. Sci. Paris, 332 (2001), 315-322.
17. A. Bernard, J. Jedwab, A. Van Moer and N. Yourassowsky, Unpubl. Rep. Geoch. and Mineralogy Dept, Brussels Univ. (1997).
18. D. Craw, Chem. Geol. 171 (2000) 17-32.

Geochemical study of clay materials in Fornos de Algodres region (Central Portugal) in an archaeometric view.

M. I. Dias[a], M. I. Prudêncio[a] and M. A. Gouveia[a]

[a]Instituto Tecnológico e Nuclear. Estrada Nacional 10, 2686-953 Sacavém, Portugal

Within an ongoing archaeometric project clay materials from Fornos de Algodres region (Central Portugal) have been studied. In this paper the results obtained will be presented with a geochemical characterisation. The main purpose of this research is to identify potential raw materials for pottery production in an archaeological context (3rd millennium BC). The geological context for this research is mainly granites, with quartz, doleritic and aplite-pegmatite veins, and also a schist-greywacke complex.

Chemical data of the whole sample and the < 2 μm fraction was obtained by instrumental neutron activation analysis.

Dolerites clearly are differentiated from the other lithologies by higher contents of Fe, Sc and Cr, also found in the finer ceramics. The Eu anomaly is higher in granites and aplite-pegmatite veins in the clay fraction and especially in the whole sample. Rb and K are related to the presence of feldspars, which have higher levels in granites and ceramics with coarser pastes. The aplite-pegmatite veins have lower iron content and appear not to be related with the pottery production.

Chemical differences between clays allowed the definition of geochemical fingerprints for each material. A comparative study with ceramics point generally to a ceramic production with local raw materials, using two main sources – weathered dolerite veins as the plastic component and weathered granites as the non-plastic addition.

1. INTRODUCTION

For a better understanding of provenance, production technology, nature and function of ancient ceramics a geochemical approach to the used raw materials assume importance, specially in the processes applied in clay mixing and/or tempering. Clay minerals are the plastic part of ceramics so changes in their chemistry will result obviously in different physical changes too. The chemistry of clays, at a certain point, can control their interaction with water, their plastic properties, and their behaviour under firing during the ceramic process and also the need of adding more or less tempering material. So, it's of crucial importance to know about clay mineral composition when we need to understand the choices of potters for a given clay type to make his pots.

In this case geochemical characterisation of clay materials are included in a project that aims to understand pottery production technologies and resource strategies in a settlement network during the 3rd millennium BC in Central Portugal, included in a broader archaeological research (Valera, 1997 a, b, 1999).

2. GEOLOGICAL CONTEXT: CLAY RESOURCES

The main petrological outcrops in the region are granites, in general porphiroid of medium to coarse grains of two micas ("Granito da Muxagata"), with quartz, doleritic and aplite-pegmatite veins, and in a very narrow area a schist-greywacke complex ("Complexo Xisto-Grauváquico do Grupo do Douro").

Sample collection was essentially done in granites and their veins, specially the doleritic ones, which contain the most clayey materials of the studied area. Twelve samples were collected in dolerites, four samples in a granite profile and two in the schist-greywacke complex. Sampling was performed at the different stages of material alteration. These samples were collected near the archaeological sites for all the geological materials.

3. METHODS

Chemical data of the whole sample and the < 2 µm fraction were obtained by instrumental neutron activation analysis and two reference materials were used as calibration standards, sediment GSD 9 and soil GSS 1. Samples and standards were irradiated together in the core grid of the Portuguese Research Reactor (Sacavém) at a flux of 4.4×10^{12} n cm^{-2} s^{-1} for 2 minutes (short irradiation) and seven hours (longer irradiation). Details concerning the measurement and processing of the gamma spectra can be found in Prudêncio *et al*, 1986 and Dai Kin *et al*, 1999. These analyses obtained the concentration of 30 elements: Na, K, Fe, Sc, Cr, Mn, Co, Zn, Ga, As, Br, Rb, Zr, Sb, Cs, Ba, La, Ce, Nd, Sm, Eu, Tb, Dy, Yb, Lu, Hf, Ta, W, Th, U.

The methodology used in this work aims to obtain chemical features with special attention to the role of trace elements as fingerprints, for later comparison studies with ceramics.

The chemical results obtained were submitted to cluster analysis (tree clustering method with the unweighted pair-group average – UPGMA – as linkage rule, the Pearson coefficient as distance measure, and also the Euclidean distances, using as variables the chemical elements) and other statistical analysis, such as correlations, bivariate graphics, factor analysis, etc, using the *Statistica* program (Statistica, 1997).

4. RESULTS AND DISCUSSION

The main petrological outcrops in the local region are granites, with quartz, doleritic and aplite-pegmatite veins, and in a very narrow area the schist-greywacke complex. The most clayey materials correspond to higher stages of alteration, above all dolerites.

The chemical characterisation of these potential raw materials reveals expected differences. The basic veins have higher levels of iron, chromium, manganese, cobalt and zinc, lower levels of sodium, potassium and thorium and lower europium anomaly, than the granitic samples.

Previous work considering the mineralogical diversity of the regional clay materials (Dias *et al*, 2000) emphasises a very well established correlation between the mineralogical and the chemical results obtained. The Na and K differences observed in all the samples agree with the plagioclase and alkali feldspar proportions, which is much lower in dolerites. Dolerites have higher contents of iron in both the whole sample and clay fraction, with a higher concentration in the < 2 μm fraction, stressing higher contents of iron oxides (mainly goethite) in the clay fraction. Among the other geological materials the aplite-pegmatite veins show the lower iron contents, except in one case where goethite was found. We also found iron enrichment from the whole to the clay fraction.

Sc, Cr, Co and Zn show the same general trend as iron in the whole sample. In granites Cr, Co and especially Zn are concentrated in the clay fraction when compared with the whole sample. In dolerite veins Cr and Co prevailed in the whole sample when compared with their clay fraction (Figure 1), which can be due to their presence in the ferromagnesian minerals widespread in coarser fractions. Rb has a similar behaviour as K, which is correlated with the proportion of alkali feldspars (Figure 2).

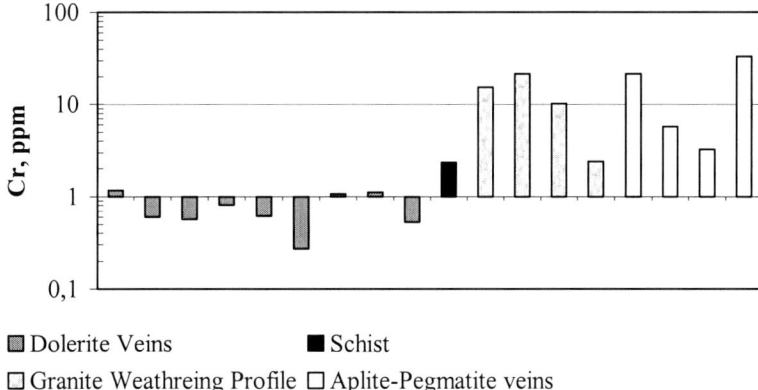

Figure 1. Cr contents in the < 2 μm fraction relative to the respective whole sample

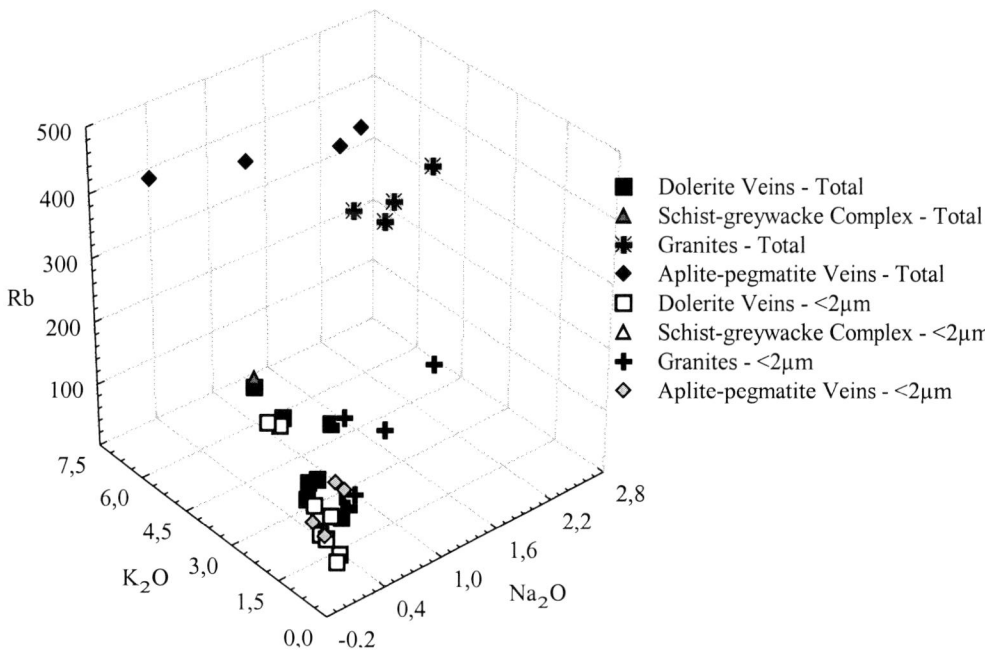

Figure 2 – Rb, Na e K distribution in the whole sample and in the clay fraction.

A few elements doesn't show any significant variations with the lithology and the grain size, like Cs and U, which only show a tendency of enrichment in the clay fraction only in the granite weathering profile.

In granites and aplite-pegmatite veins there is a higher distribution uniformity and lower concentrations in several elements, like Sc, Cr, Zr, Hf, Ta, Fe, Co, Cs and Ba. Nevertheless, when studying their behaviour according to grains size, Hf, Zr and Ta are depleted in the clay fraction of granites, aplite-pegmatite veins and schists (particularly Hf), whereas in dolerites they are enriched in the < 2 μm fraction (Figure 4).

Th is in general concentrated in the clay fraction, especially in granites and schist (Figure 3).

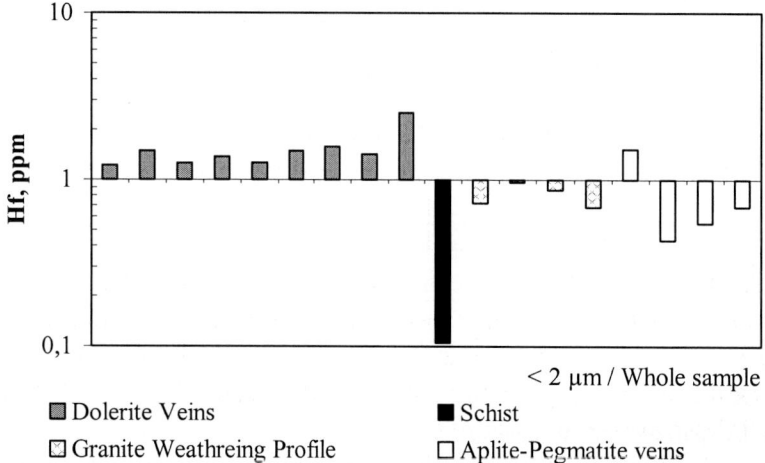

Figure 3. Hf contents in the clay fraction relative to the whole sample.

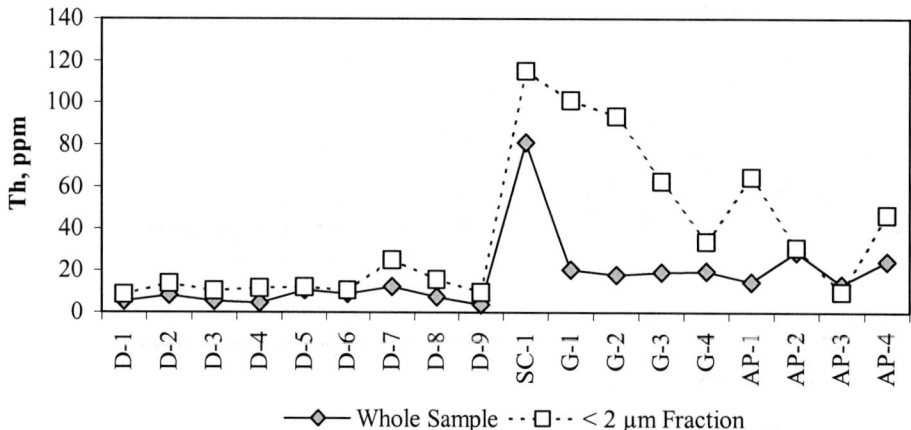

Figure 4. Th distribution in the whole sample and in the < 2 μm fraction.

Rare earth elements (REE) show in general a higher content in the clay fraction of all geological materials in the study, particularly in granites (Figure 5). The weathered schist is different from the other materials by a much higher La/Yb ratio in both whole and clay fraction. The Eu anomaly is higher in granites and aplite-pegmatite veins in the clay fraction and especially in the whole sample. Another important feature is the higher variety of REE behaviours present in dolerites.

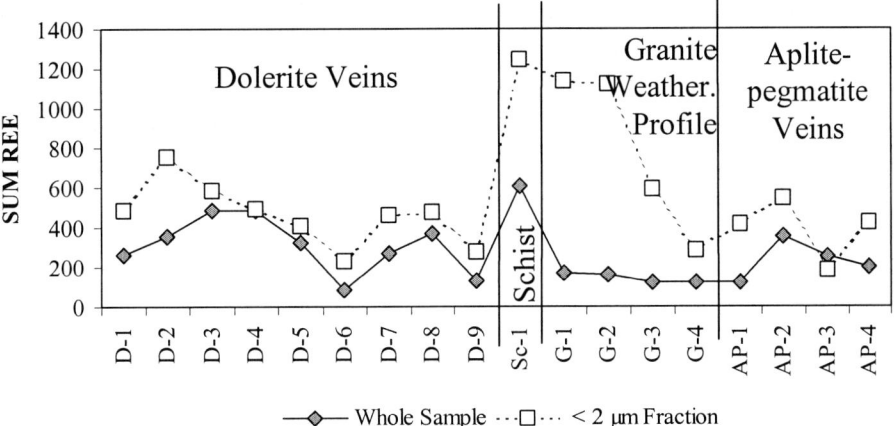

Figure 5 – REE Sum in the whole sample and respective < 2 µm fraction.

A more detailed geochemical study of the granite weathering profile from the bottom to the top (G4 to G1) shows differences in some elements, especially in the REE. An increase of Na and K and a slight decrease of Fe occurs from the bottom to the top. However, where the changes are more significant it's in the clay fraction with an increase of the REE (light and heavy, indicating mobility of these elements (Figure 6).

Figure 6. REE distribution in the < 2 µm fraction relative to the respective whole sample in the granite weathering profile.

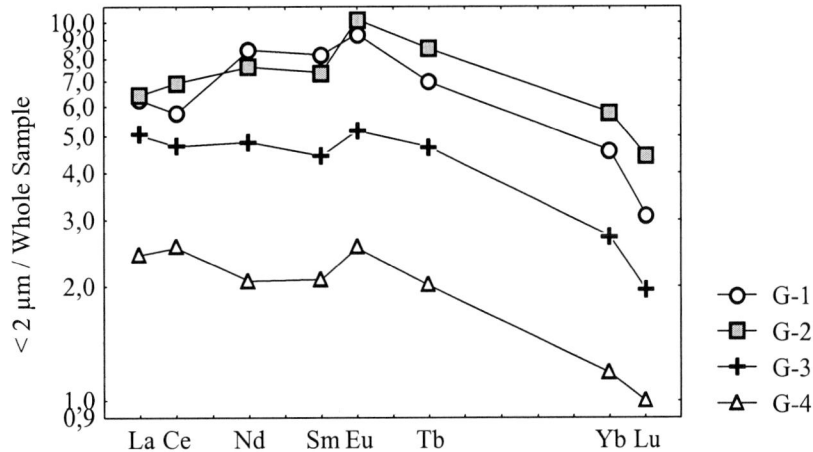

A Ce positive anomaly appears in the top of the profile pointing to the fact that Ce is less mobile than the other REE, which is explained by the partial oxidation of Ce^{3+} to Ce^{4+}. Similar behaviours have been found in weathering profiles of granites (Gouveia *et al*, 1993) and in basaltic rocks (Prudêncio *et al*, 1995).

5. CONCLUDING REMARKS

The chemical characterisation applied to the clay materials accessible in Fornos de Algodres region allowed the distinction between the different types of potential raw materials available for ceramic manufacture in the region.

Geochemical fingerprints were established aiming for better distinction between clays in the area, as well as to set up possible raw materials for archaeological ceramics in the surrounding archaeological sites.

The chemical characterisation of these potential raw materials reveals the expected differences. The basic veins have the higher levels of iron, chromium, manganese, cobalt and zinc, the lower levels of sodium, potassium and thorium, and lower europium anomaly, than the granitic samples, and correspond to the most clay materials in the region.

REFERENCES

Dai Kin, F., Prudêncio, M.I., Gouveia, M.A., and Magnusson, E., (1999), Determination of rare earth elements in geological reference mateirals: a comparative study by INAA and ICP-MS. *Geostandards Newsletter*, 23, 47-58

Dias, M.I.; Prudêncio, M.I.; Prates, S.; Gouveia, M.A. & Valera, A.C. (2000) – "Tecnologias de Produção e Proveniência de Matéra-Prima das Cerâmicas Campaniformes da Fraga da Pena (Fornos de Algodres – Portugal)". *Actas do 3º Congresso de Arqueologia Peninsular*, V.Real, Portugal, vol IV (2000) 253-268.

Gouveia, M.A., Prudêncio, M.I., Figueiredo, M.O., Pereira L.C.J., Waerenborgh, J.C., Morgado, I., Pena, T. and Lopes, a., (1993), Behaviour of REE and other trace and major elements during weathering of granitic rocks, Évora, Portugal. *Chemical Geology*, 107, p. 293-298.

Prudêncio, M.I., Gouveia, M.A. and Cabral, J.M.P., (1986), Instrumental neutron activation analysis of two french geochemical reference samples-basalt BR and biotite mica-Fe. *Geostandards Newsletter*, 10, 29-31.

Prudêncio, M.I., Gouveia, M.A. and Sequeira Braga, M.A., (1995), REE distribution in actual and ancient surface environments of basaltic rocks (central Portugal). *Clay Minerals*, 30, p. 239-248.

Statistica ® (1997), User Manual, Release 5, 3th Ed., Stat Soft Inc., Tulsa.

Valera, A.C. (1997a). O Castro de Santiago (Fornos de Algodres, Guarda): aspectos da calcolitização da bacia do Alto Mondego, Textos Monográficos, **1**, Fornos de Algodres.

Valera, A. C. (1997b). "Fraga da Pena (Sobral Pichorro, Fornos de Algodres): uma primeira caracterização no contexto da rede local de povoamento", *Estudos Pré-Históricos*, Vol. V, Viseu p. 55-84.

Valera, A. C. (1999). "The re-creation of territorialities and identities in the III millennium BC: research problems in Central Portugal", *Journal of Iberian Archaeology*, Vol **1**, Porto ADECAP, p.109-115.

Fibrous clay minerals as lithostratigraphic markers in a Tertiary continental deposit (Malpica do Tejo, Portugal)

M. I. Dias[a] and F. Rocha[b]

[a]Instituto Tecnológico e Nuclear. Estrada Nacional 10, 2686-953 Sacavém, Portugal
[b]Dep. Geociências, Univ. de Aveiro, Campo Santiago, 3810-193 Aveiro, Portugal

Fibrous clays are very common in the Iberian Peninsula, especially in the Tagus Tertiary Basin in several stratigraphic levels of deposits of Palaeogene age. We report here on a sedimentary profile located in a small tectonic basin (Castelo Branco, Eastern Portugal) in which the sediments, mainly feldspar and quartz, are detrital. A lithostratigraphic scheme is proposed for this region based mainly on statistically distinct clay mineral suites. The statistical approach was useful in differentiating among several lithostratigraphic levels, and their classification based on mineralogical parameters. This leads to a better clarification of the variables that discriminate the units, emphasising the importance of the clay mineralogical associations, especially palygorskite in the Palaeogene levels.

1. INTRODUCTION

The first sedimentological studies done on the Portuguese Tagus Tertiary basin were those of Carvalho (1967, 1968, 1969), in which palygorskite was shown to be associated with smectite at several stratigraphic levels of deposits attributed to the Palaeogene. In Portugal, palygorskite occurs almost always in a Tertiary context, being mainly associated with montmorillonite and more rarely with illite and kaolinite (Carvalho, 1964; Carvalho and Alves, 1970; Grade and Moura, 1981). Dias (1998) presented the "state of the art" concerning fibrous clay minerals in the Portuguese deposits with respect to the geological contexts as well as the main mineralogical associations.

In this paper, a sedimentary record of a Portuguese Tertiary deposit located in Beira Baixa is presented. In this deposit, the presence of palygorskite is important as a lithostratigraphic marker (Dias, 1998; Dias and Prates, 1997), and as a means for testing statistical analysis of the mineralogical parameters. These deposits fill two small tectonic basins, Sarzedas and Castelo Branco, and are mainly detrital, conglomerates rich in feldspar and quartz, sandstones or silty-sandstones. The studied profile is located in the Castelo Branco basin. The basement is a "schist-greywacke complex" dated from Precambrian-Cambrian, intruded in some places by the granite plutons of Castelo Branco and Penamacor, by the Pre-Hercynian of Salvaterra do Extremo, Zebreira and Segura, and also by veins of quartz, pegmatite, aplite, microgranite and basic rocks. Some higher upper units are covered with a "raña" deposit (Plio-Quaternary), which consists of a red fanglomerate of quartzite pebbles.

2. MATERIALS AND METHODS

Several lithostratigraphic sections have been analysed and thirty-six representative samples have been collected of the different beds of the Tertiary sediments of the studied profile, in Malpica do Tejo, south of Castelo Branco. Powdered samples were analysed by X-ray diffraction (bulk rock prepared as unoriented aggregates; <2μm fractions prepared as oriented aggregates under ambient conditions, after solvation with ethylene-glycol and after heating to 550°C). Semi-quantitative mineralogical contents were obtained according to Schultz (1964), Barahona (1974) and Galan (unpublished, for fibrous clay minerals). Multivariate data analysis (cluster analysis, R-mode factor analysis, and discriminant function analysis) of the mineralogical data has been carried out.

3. RESULTS AND DISCUSSION

According to Dias (1998) two mineralogical patterns (Table 1) can be defined for the bulk sample: 1 - quartz, phyllosilicates and goethite associated with the upper stratigraphic units and 2 - quartz, phyllosilicates, feldspars (K feldspars and Ca-Na feldspars), dolomite and sometimes calcite associated with the middle to lower stratigraphic units. In the clay-size fraction, four units were defined (Table 1): A - an upper unit in which kaolinite is associated with illite; B - downwards, one unit in which smectite is associated with illite and kaolinite; C - other unit with palygorskite associated with illite; and D - in the basement (closer to the schists) smectite associated with palygorskite and illite.

Table1: Lithostratigraphic units defined in the Malpica do Tejo region according to mineralogical parameters (Dias, 1998).

Unit	Clay Mineralogical Association	Bulk Sample Mineralogical Association
A	KT (+ ILL)	1) Qz (+Phyllosilicates + Goethite)
B	SM (+ ILL + KT)	2) Qz (+Phyllos. + Felds ± Calc ± Dol)
C	PAL + ILL	2) Qz (+Phyllos. + Felds ± Calc ± Dol)
D	SM (+ PAL + ILL)	2) Qz (+Phyllos. + Felds ± Calc ± Dol)

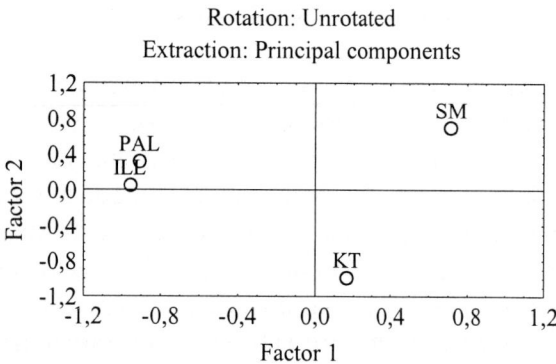

Figure 1. R-mode factor analysis of the mineralogical composition of the clay fractions.

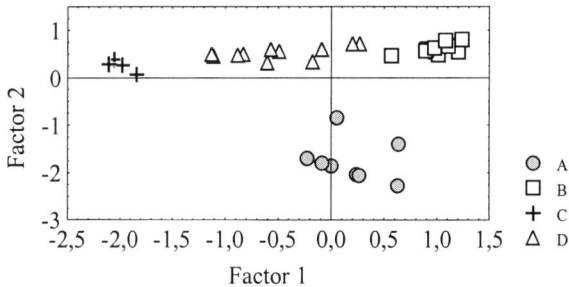

Figure 2. Factor scores of the mineralogical composition of the clay fractions.

The application of R-mode factor analysis (Figure 1) to the mineralogical composition of the clay fractions allow some interesting observations to be made: 1) smectite is positively weighted on factor 1 relative to palygorskite and illite; 2) kaolinite is negatively weighted on factor 2; and 3) there is a correlation between palygorskite and illite.

Considering the factor scores resulting from the application of R-mode factor analysis to the mineralogical composition of the clay fractions, it is important to highlight the clear differentiation obtained between the four groups defined according to the prevailing mineralogical association shown in Table 1 (Figure 2).

The application of R-mode factor analysis (Figure 3) to the mineralogical composition of the bulk samples, stresses the contribution of goethite with quartz to define one group and dolomite and feldspars another group (both groups explained by factor 1). Calcite and phyllosilicates plot in opposition, explained by another factor (factor 2).

Considering the factor scores resulting from the application of R-mode factor analysis to the mineralogical composition of the bulk sample (Figure 4), two groups are clearly defined, corresponding respectively to groups 1 and 2 of Table 1, although four samples are detached from the others, due to their higher amounts of calcite, or for sample 10B, higher amount of phyllosilicates. Two samples of group 1 pot close to group 2 as a result of their amounts of feldspars, in general absent in that group.

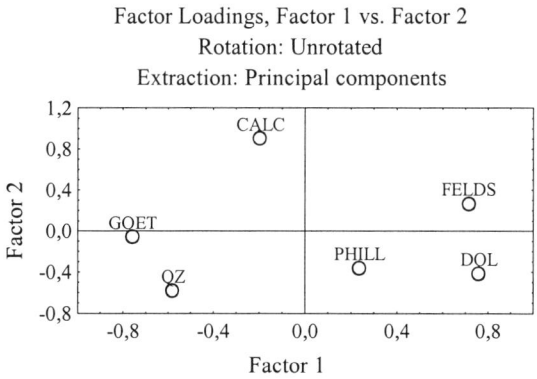

Figure 3. R-mode factor analysis of the mineralogical composition of the bulk samples.

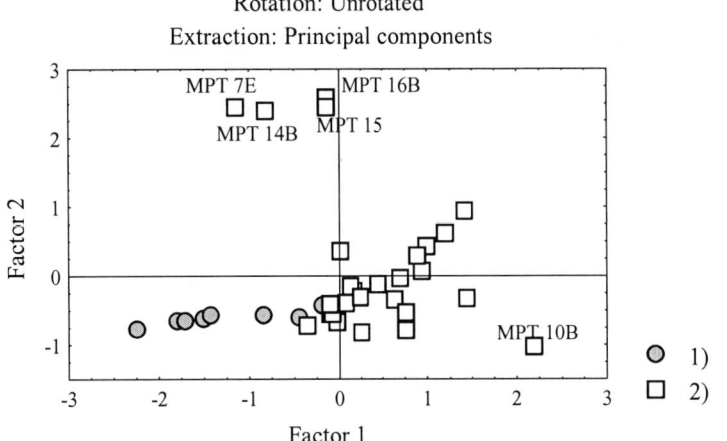

Figure 4. Factor scores of the mineralogical composition of the bulk sample.

Regarding the discriminant analysis results, it is important to notice that the groups previously defined through the mineralogical patterns of the clay fractions (and also distinguished by factor analysis) are now clearly confirmed, by Wilks' Lambda parameter (Table 2), which is very close to zero, as well as by Correlations Variables - Canonical Roots (Pooled-within-groups correlations) and classifications of cases (Figure 5 and Tables 3 and 4).

Table 2: Discriminant Function Analysis for the mineralogical composition of the clay fraction. Wilks' Lambda: .00412 approx. F (12.74)=43.223 p< .0000

	Wilks' Lambda	Partial Lambda	F-remove (3.28)	p-level	Toler.	1-Toler. (R-Sqr.)
SM	.004415	.931989	.681093	.571026	.015426	.984574
PAL	.004571	.900327	1.033270	.392985	.046992	.953008
ILL	.004519	.910568	.916683	.445516	.031957	.968043
KT	.004734	.869333	1.402869	.262655	.026776	.973224

Figure 5. Discriminant function analysis of the clay fraction mineralogy.

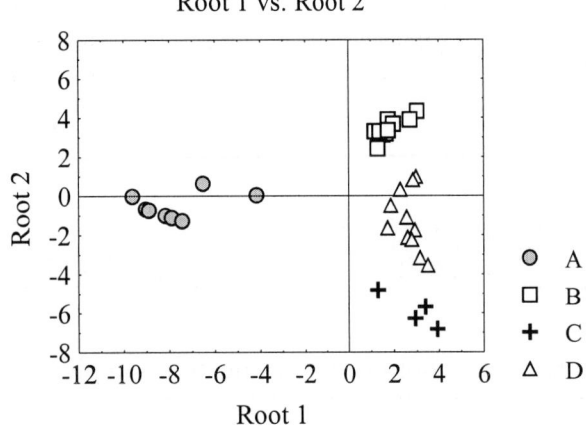

Table 3: Factor Structure Matrix – Clay fraction
Correlations Variables - Canonical Roots. (Pooled-within-groups correlations)

	Root 1	Root 2	Root 3
SM	.410296	.888525	.175690
PAL	.338766	-.809295	.225071
ILL	.102340	-.588814	-.381699
KT	-.913830	-.053155	.091114

Table 4: Classification Matrix – Clay fraction
Rows: Observed classifications. Columns: Predicted classifications

	Percent Correct	A p=.22857	B p=.34286	C p=.11429	D p=.31429
A	100.0000	8	0	0	0
B	100.0000	0	12	0	0
C	100.0000	0	0	4	0
D	100.0000	0	0	0	11
Total	100.0000	8	12	4	11

It is important to emphasize that no incorrect cases of classification occur as shown in the classification matrix, which shows the number of cases that were correctly classified according to their mineralogical pattern (clay fraction). Therefore 100 % of all cases are correctly classified (Table 4). Root 1 seems to discriminate unit A from all the other units (B, C and D). Root 2 discriminates unit C from unit B (Figure 5). Discriminant function analysis was useful to determine which variables discriminate the previous defined units. Kaolinite distinguishes unit A, smectite distinguishes unit B, and palygorskite unit C (Table 3).

Table 5: Discriminant Function Analysis – Bulk sample
Wilks' Lambda: .34540 approx. F (6.28)=8.8443 p< .0000

	Wilks' Lambda	Partial Lambda	F-remove (1.28)	p-level	Toler.	1-Toler. (R-Sqr.)
GOET	.393261	.878289	3.880159	.058824	.717435	.282565
QZ	.361125	.956447	1.275026	.268407	.000687	.999313
FELDS	.364175	.948436	1.522271	.227533	.001255	.998745
CALC	.364094	.948649	1.515649	.228520	.000845	.999155
DOL	.383239	.901258	3.067680	.090812	.224640	.775360
PHYLL	.362261	.953448	1.367087	.252171	.000756	.999244

Table 6: Classification Matrix – Bulk sample
Rows: Observed classifications. Columns: Predicted classifications

	Percent Correct	1) p=.22857	2) p=.77143
1)	75.0000	6	2
2)	100.0000	0	27
Total	94.2857	6	29

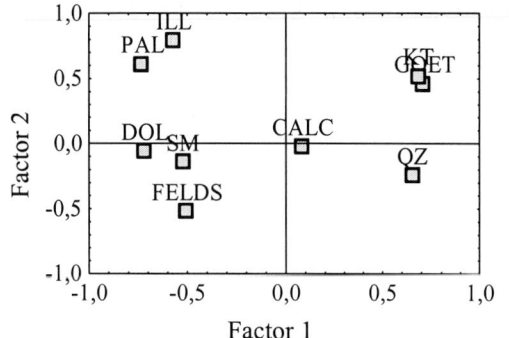
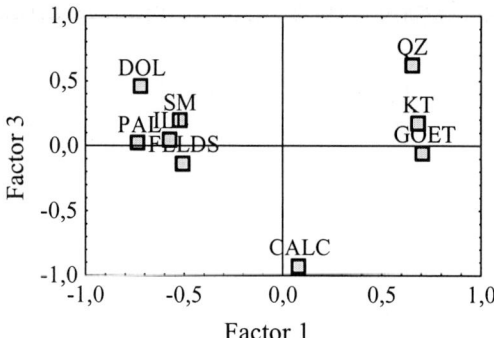

Figure 6 - R-mode factor analysis of the mineralogical composition of the global samples (with redistribution of phyllosilicates contents by kaolinite, illite, palygorskite and smectite).

Applying the discriminant analysis to the bulk sample mineralogy, the classification into two groups is mostly correct, although not with the same accuracy as obtained for the clay fraction. This is indicated by the higher Wilks' Lambda parameter (Table 5) and by the case classifications (Table 6), being 94 % of all cases correctly classified, 75 % of the samples of group 1 correctly classified, as well as 100 % of group 2. The samples from group 1, which appear distinct, correspond to those samples in Figure 4 that plotted close to group 2 (due to the presence of feldspars). Although some samples plotted distinctly away from the others in figure 4 due to differences in calcite or phyllosilicate content, these do not support this statistical approach. Nevertheless, the higher values of Wilks' Lambda parameter for bulk samples relative to the clay fraction, indicate less discrimination between groups. It was not possible to present an x-y plot, because only one canonical root was extracted.

The application of R-mode factor analysis (Figure 6) to the mineralogical composition of the global samples (bulk sample with redistribution of phyllosilicate content by kaolinite, illite, palygorskite and smectite) allows some interesting observations to be made: 1) Factor 1 shows kaolinite, goethite and quartz positively weighted and dolomite, smectite, feldspars, palygorskite and illite negatively weighted; 2) Factor 2 shows illite and palygorskite correlated and positively weighted but feldspars negatively weighted; 3) calcite is negatively weighted in factor 3.

Considering this global sample, it is clear that these statistical analysis allow, as a first step, to differentiate two main groups, directly connected to the previously defined 1 (units A and B) and 2 (units C and D) for the bulk sample.

Combining the results obtained in all these statistical analysis of the mineralogical data (clay fraction and bulk sample), we propose the subdivision of Palaeogene sediments of Malpica do Tejo (Castelo Branco Basin) not in four but only in three main units: A, B and C, subdividing unit C, in C2 (previous C) and C1 (previous D), based on their clay mineralogical features. The differentiation observed in the clay fraction, related to the association of palygorskite with illite in what is now subunit C2, as well as the behaviour of all the other minerals, lead us to the assumption that subunit C2 can be interpreted as an evolutionary products of subunit C1.

The discriminant analysis performed with the mineralogical composition of the global sample (after redistribution of phyllosilicates contents by kaolinite, illite, palygorskite and smectite) became useful for a proposal of dividing in mineralogical units the Palaeogene deposits of Castelo Branco, confirming the proposed division into three main groups (Figure 7).

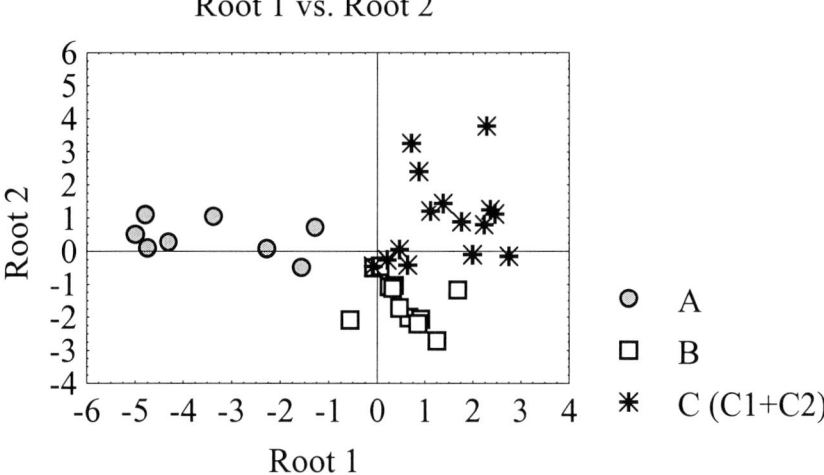

Figure 7 – Discriminant function analysis of the global sample (after redistribution of phyllosilicates contents by kaolinite, illite, palygorskite and smectite).

Table 7: Factor Structure Matrix – Global sample
Correlations Variables - Canonical Roots. (Pooled-within-groups correlations)

	Root 1	Root 2
GOET	-.443290	.045739
QZ	-.364925	-.156205
FELDS	.302531	-.261611
CALC	.117714	.102591
DOL	.237710	-.016252
SM	.265991	-.371426
PAL	.286839	.583956
ILL	.137508	.513872
KT	-.620503	.167000

Root 1 discriminates group A from groups B and C, and root 2 discriminates group B from group C. Looking at which variables discriminates each one of the defined groups, goethite associated with quartz and kaolinite discriminates group A, whereas feldspars, dolomite and smectite, in opposition to palygorskite and illite, discriminate group B (Table 7).

Table 8

Lithostratigraphic units defined in the Malpica do Tejo region according to mineralogical *vs* statistical parameters, considering fingerprints minerals of each unit

Unit	Subunit	Mineralogical Association (clay fraction crossed within bulk sample)
A		kaolinite + goethite + quartz
B		feldspars + smectite + dolomite + (calcite)
C	C2	palygorskite + illite + (+ calcite + feldspars + dolomite)
	C1	smectite + feldspars + palygorskite + illite + dolomite + calcite

Table 8 synthesise the results of the application of multivariate data analysis to mineralogical data of clay fractions and bulk samples, emphasising minerals fingerprints of each defined unit. Therefore, mineralogical patterns became quite useful to lithostratigraphic differentiation, namely the clay minerals suites combined with the bulk rock ones. Some mineral fingerprints were defined for each unit/subunit, emphasising the role of palygorskite as the main lithostratigraphic marker in the Palaeogene deposit of Malpica do Tejo. This means that only unit C (C1 and C2) are Paleogene in age and units A and B are younger.

4. CONCLUDING REMARKS

A lithostratigraphic scheme for the Tertiary deposits of Castelo Branco region (Beira-Baixa, Portugal) is proposed based mainly on the clay mineral suites. Statistical analysis was useful in differentiating the several lithostratigraphic levels. The sample's classification considering mineralogical parameters allowed a better clarification of which variables discriminate the previously defined units. The clay fraction is the best feature for differentiating between lithostratigraphic levels, emphasising the role of kaolinite for Pliocenic levels and of palygorskite for Palaeogene ones. Three main units were proposed, unit A in a Quaternary context and the other two (B and C) in a Tertiary context.

REFERENCES

1. Barahona E., Arcillas de ladrilleria de la provincia de Granada. Evaluacion de algunos ensayos de materias primas, PhD thesis, Univ. Granada, Spain (1974).
2. Carvalho A.M.G., Étude géologique et sédimentologique de la région de Ponte de Sor (bordure Est du bassin tertiaire du bas-Tage), Thèse de doctorat de 3e cycle, Paris (1964).
3. Carvalho A. M. G., Finisterra II (1967), Lisboa, 174-200.
4. Carvalho A. M. G., Memórias Serviços Geológicos de Portugal, nº 15 (1968)
5. Carvalho A. M. G., Estudos, Notas e Trabalhos, XVII (3/4) (1969), 297-307.
6. Carvalho A.M.G. and Alves C., Finisterra, vol. V, nº 10 (1970), Lisboa, pp. 282-291.
7. Dias M. I., Caracterização Mineralógica e Tecnológica de Argilas Especiais de Bacias Terciárias Portuguesas. Unpubl. PhD thesis, Univ. Lisboa, 333 p. (1998)
8. Dias M. I. and Prates S., Proc. of The 11th International Clay Conference (1997), Canada, pp.631-638.
9. Grade J. and Moura A.C., Memórias e Notícias, Coimbra, nºs 91-92 (1981), pp.173-182.
10. Schultz L.G., Geol. Survey Prof. Paper 391 – C (1964), 31 pp.

Cr-bearing chlorites in low-grade metapelites of the Puncoviscana Formation (Neoproterozoic), Nortwestern Argentina

M. D. Do Campo[a] and F. Nieto[b]

[a]Instituto de Geocronología y Geología Isotópica and Facultad de Ciencias Exactas y Naturales,U.B.A. Pabellón INGEIS, Ciudad Universitaria (1428) Buenos Aires. Argentina.

[b]Departamento de Mineralogía y Petrología and I.A.C.T., Universidad de Granada-CSIC, Avda. Fuentenueva s/n, 18002-Granada, Spain.

Chlorites containing appreciable amounts of Cr were recognized by AEM in an anchizonal metapelite of the Puncoviscana Formation during a TEM study. Cr-bearing chlorites are restricted inside a rounded crystal of quartz several microns diameter, they form a group of tabular crystals all with the same crystallographic orientation. AEM analysis reveals Cr contents from 0.07 to 0.25 a.f.u., and a low sum of octahedral cations. Average analysis is $Si_{3.07}Al_{1.92}Mg_{262}Fe_{2.07}Cr_{0.19}O_{10}(OH)_8$. Rock-forming chlorites do not contain Cr, a mean formulae based on AEM analyses is $Si_{2.76}Al_{2.46}Mg_{2.43}Fe_{2.29}Mn_{0.03}O_{10}(OH)_8$. This Cr-bearing chlorites are unusual in sediments; their presence in the rock together with chlorites having the usual composition in metapelites is significant regarding the absence of chemical equilibrium in low-grade metamorphic systems.

1. INTRODUCTION

Chlorites containing significant amounts of Cr were identified by AEM in a representative metapelite of the Puncoviscana Formation (1) during a TEM study; sample location is shown in Figure 1. Lattice-fringe images, selected area electron diffraction, and AEM analyses coupled with previous data of white mica crystallinity index (IC), indicate a *state of reaction progress* (2) for Puncoviscana slates consistent with medium anchizonal to epizonal grade metamorphism (3-6). The mineral assemblage of these rocks is mainly quartz, albite, phengite and chlorite.

2. RESULTS

Cr-bearing chlorites are restricted inside a rounded crystal of quartz three microns diameter; they form a group of tabular crystals 260 to 1300 Å thick and 500 to 4200 Å long (Fig. 2), all with the same crystallographic orientation. Micro-diffraction indicates semi-random or random stacking, while lattice-fringe images display scarce interleaved layers at 24 Å and 20 Å in several crystals (Fig. 3). AEM analysis (Table 1) reveals Cr contents from 0.07 to 0.25 a.f.u., and a low sum of octahedral cations. The mean formulae, based on AEM analysis, is $Si_{3.07}Al_{1.92}Mg_{262}Fe_{2.07}Cr_{0.19}O_{10}(OH)_8$.

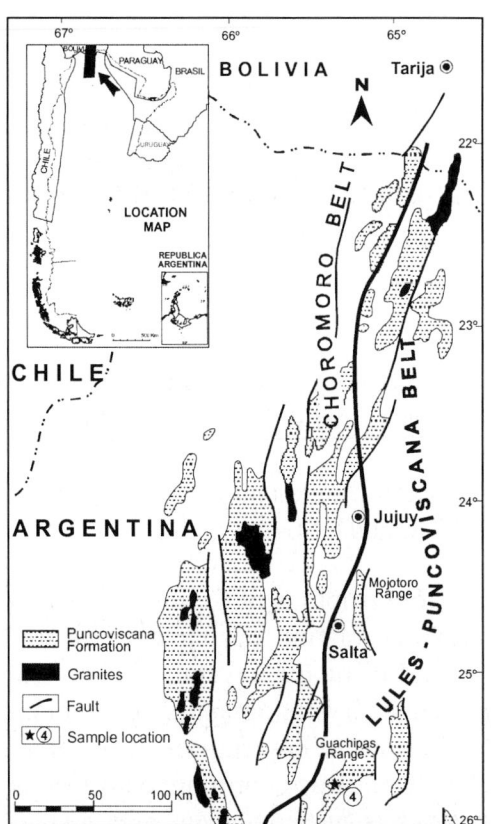

Figure 1. Outcrops of Puncoviscana Formation with sample location (see star).

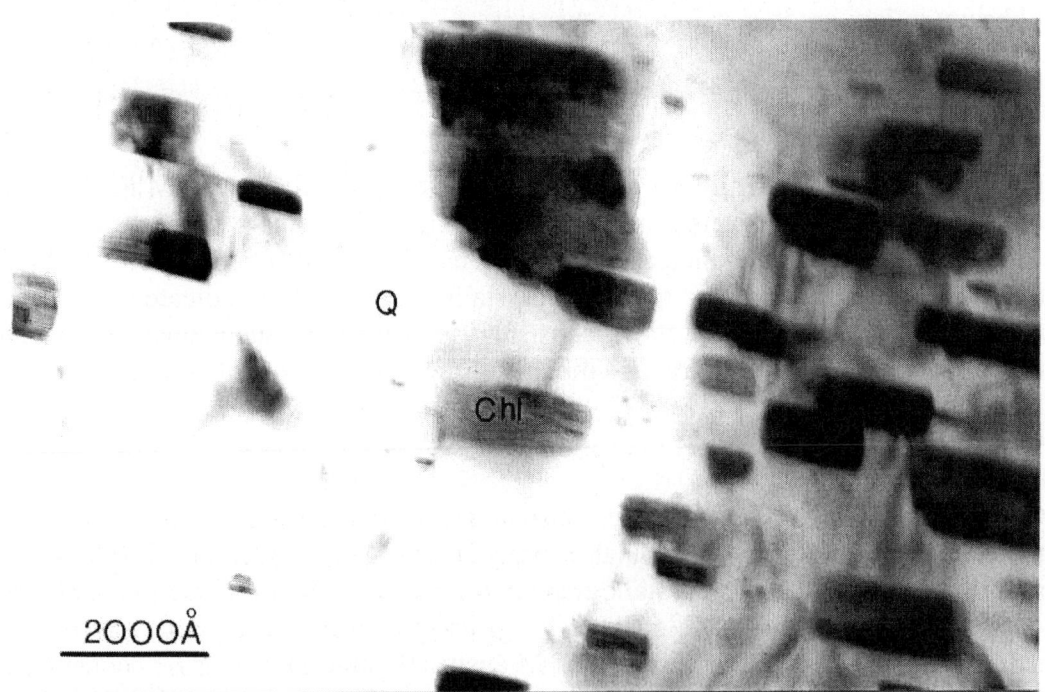

Figure 2. Textural image of Cr-bearing chlorites inside a quartz grain.

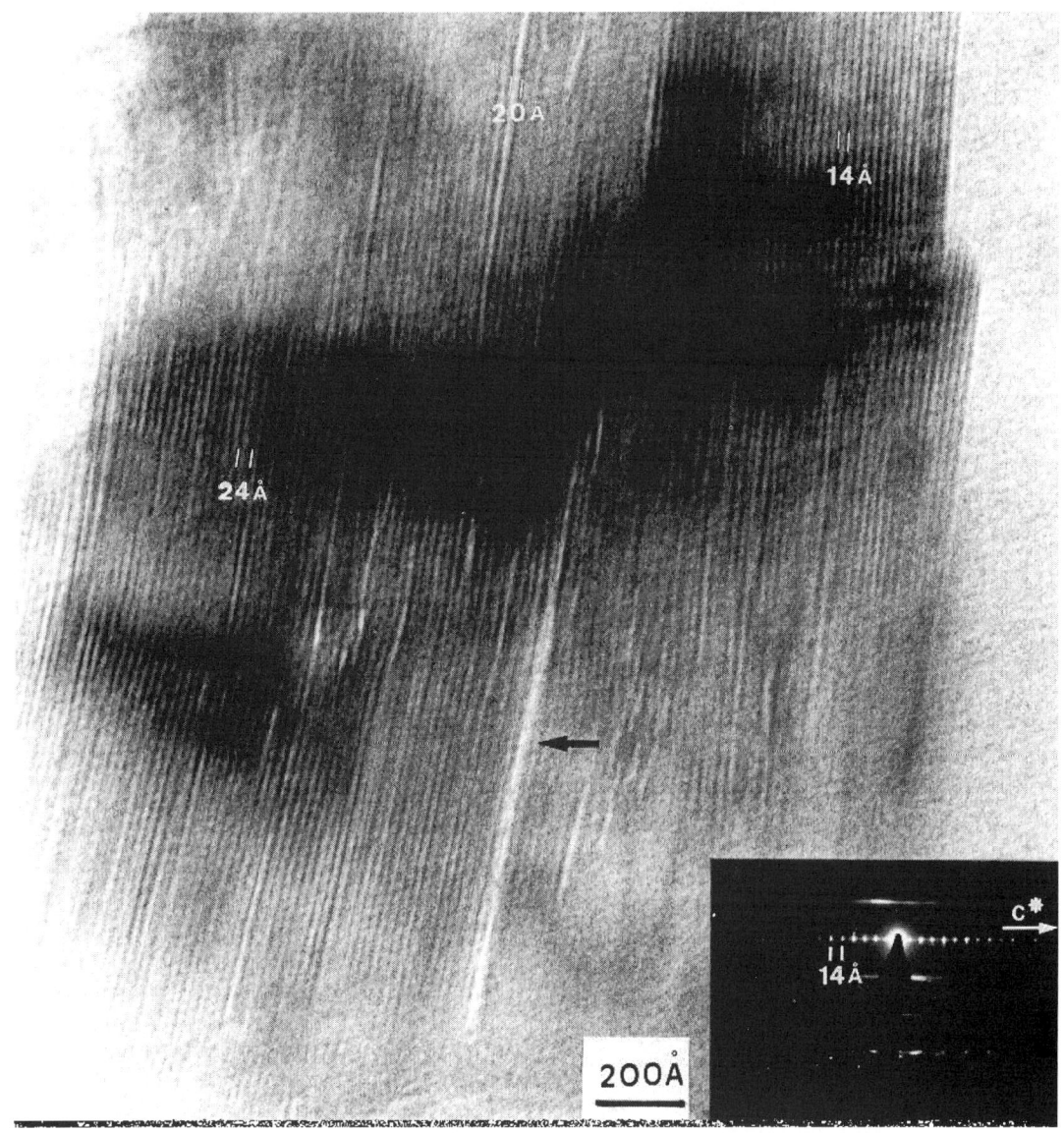

Figure 3. SAED pattern of semi-random Cr-bearing chlorites and corresponding lattice fringe image.

IC of 0.30 Δ°2θ and a **b** parameter of 9.035 Å were measured for metapelite PU4. Phyllosilicates of this sample appear in TEM images in packets with straight and continuous lattice fringes. More common crystalline defects are low-angle boundaries between packets of the same mineral (Fig. 4) or between mica and chlorite packets. 2M polytype was identified in all the dioctahedral micas analyzed. Muscovite occurs as defect-free packets some 40-350 layer thick. Bulk sample chlorite exhibit semi-random and ordered stacking in SAED patterns. Among ordered chlorites one (Fig.4, inset) and two layer sequences were recognized. Ordered as well as semi-random chlorite show straight and continuous 14 Å and occasionally 28 Å lattice fringes (Fig. 4). Chlorite of the bulk sample do not contain Cr (Table 1) and the mean formulae, based on AEM analyses, is $Si_{2.76}Al_{2.46}Mg_{2.43}Fe_{2.29}Mn_{0.03}O_{10}(OH)_8$. Smectite, identified at TEM scale, was interpreted as a retrogradation product of previous phyllosilicates (6).

3. DISCUSSION

It is well known that chlorites easily reach chemical equilibrium with the system; furthermore it is frequent that chemically equilibrated chlorites coexist in low-grade metamorphic rocks with metastable micas that are far from equilibrium (e.g. 7). Nevertheless Cr-bearing chlorites in PU4 metapelite are not equilibrated with the rock-forming chlorites. In this case the impossibility that fluids reached these detrital chlorites due to the preservation by the quartz grain has been determinant to allow them to maintain their original composition and texture. The concept of state of reaction progress is now fundamental in the understanding of the diagenetic and low-grade metamorphic processes (2) as it assumes the general lack of equilibrium and the metastable character of most of the mineral phases generated by these geological processes. Therefore the usual grade-indicators reflect more the kinetic than the thermodynamic state of evolution. From this point of view very local factors, as permeability or the existence of physical barriers for fluids, are highly determinant of the mineral chemistry in low-grade rocks.

The first indication of the coordination of Cr^{+3} in chlorites comes from crystal field theory that suggests that this cation has a large preference for octahedral coordination (8). Afterwards structural refinements of Cr-bearing chlorites carried out by Phillips and co-workers (9) determined that Cr^{+3} was concentrated in the interlayer octahedral site M(4). The studied Cr-bearing chlorites present a low sum of octahedral cations that are compatible with Cr^{+3} replacing dioctahedral cations in octahedral sites. This replacement is also compatible with the amount of positive charge required to balance the negative charge due to substitution of ^{IV}Al for Si in tetrahedral sheets.

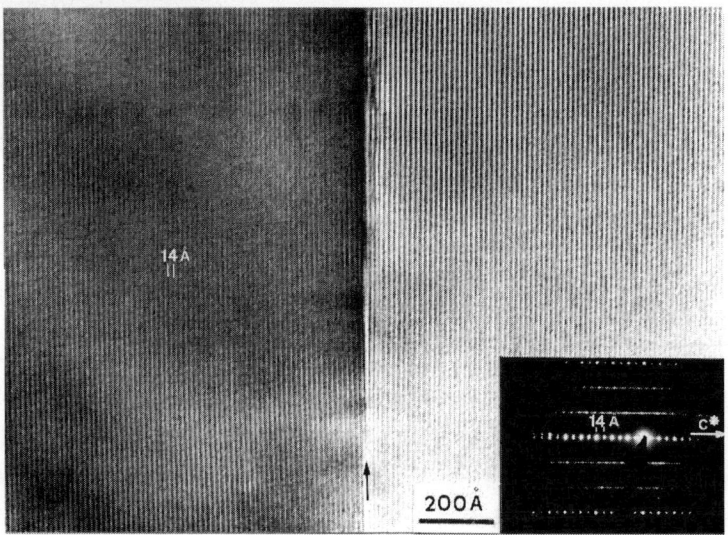

Figure 4. SAED pattern (inset) of 1-layer polytype chlorite and corresponding lattice fringe image showing a thick defect-free packet. Arrow indicates low angle boundaries.

Table 1

AEM analyses of Cr-bearing and bulk sample chlorites

	Cr-bearing			Bulk sample		
Element	PU4-3	PU4-4	PU4-5	PU4-6	PU4-7	PU4-10
Si	2.84	3.17	3.19	2.70	2.57	3.01
^{IV}Al	1.16	0.83	0.81	1.30	1.43	0.99
^{VI}Al	1.15	0.93	0.89	1.19	1.17	1.31
Mg	2.40	2.75	2.72	2.42	2.29	2.57
Fe	2.35	1.89	1.97	2.37	2.53	1.96
Ti	-	-	-	-	-	-
Mn	-	-	-	0.04	0.07	-
Cr	0.07	0.26	0.25	-	-	-
Σ oct.	5.97	5.82	5.83	6.02	6.06	5.84
K	-	-	0.01	-	-	-
Ca	-	-	-	-	-	-

REFERENCES
1. J.C.M. Turner, Estratigrafía de la Sierra de Santa Victoria y adyacencias. Boletín Acad. Nac. Ciencias, Cordoba, 41 (1960), 163.
2. R.J. Merriman and D.R. Peacor, Very low-grade metapelites: mineralogy microfabrics and measuring reaction progress. In: Low-grade metamorphism, Frey, M. & Robinson, D. (eds), Blackwell Science, Oxford, 1999.
3. M. Do Campo, F. Nieto and R. Omarini, Mineralogía de arcillas y metamorfismo de la Formación Puncoviscana en localidades de la Cordillera Oriental y Puna, Argentina. *X* Congr. Lat. Geología, Buenos Aires. Actas 2, 217 (1998)..
4. M. Do Campo, Metamorfismo del basamento en la Cordillera Oriental y borde oriental de la Puna. In: Relatorio de la Geología del Noroeste de Argentina, González Bonorino, G., Omarini, R. & Viramonte, J. (eds), Universidad de Salta, Salta, 1999.
5. M. Do Campo, Mineralogía, geoquímica y geocronología de la Formación Puncoviscana (Neoproterozoico) entre los 23°30' y 25°50' de Latitud Sur, Noroeste de Argentina. Ph Thesis (unpublished). Buenos Aires University, 1999.
6. M. Do Campo and F. Nieto, Transmission Electron Microscopy study of very low-grade metamorphic evolution in Neoproterozoic Pelites of the Puncoviscana Formation (Cordillera Oriental, NW Argentina), Clay Miner. (in press).
7. G. Giorgetti, I. Memmi, and F. Nieto,. Microstructures of intergrown phyllosilicate grains from Verrucano metasediments (northern Apennines, Italy). Contr. Mineral. Petrol., 128 (1997) 127.
8. R. G. Burns, Mineralogical Applications of Crystal Field theory. Cambridge University Press, Cambridge, 1970.
9. T.L. Phillips, J.K. Loveless and S.W. Bailey, Cr^{+3} coordination in chlorites: a structural study of ten chromian chlorites. Am. Miner., 65 (1980) 112.

Two types of hydrothermal clay deposit in the south-east area of Tandilia, Buenos Aires Province, Argentina.

J.A. Dristas[a] and M.C. Frisicale[b]

[a] Departamento de Geología, Universidad Nacional del Sur, San Juan 670, 8000 Bahía Blanca and CIC de la provincia de Buenos Aires, Argentina.
[b] Departamento de Geología, Universidad Nacional del Sur, San Juan 670, 8000 Bahía Blanca and CONICET, Argentina.

In the SE region of Tandilia two types of clay deposits originated by hydrothermal activity have been studied:
1a) Derived from alteration of original fall out pyroclastic material (mostly lapillistone) without material rework, deposited in restricted areas of a platform environment dominated by quartz sandstones (Cerro del Corral area) showing a kaolinite(3T)-rutile-anatase assemblage. The genesis of these deposits intercalated in the sedimentary cover of Tandilia, are clearly related to diabase intrusions.
1b) Deposits involving mainly reworked pyroclastic material are located in the west Barker, Taglioretti and east of Sierra de La Tinta areas displaying an advanced argillic alteration (AAA), which included pyrophyllite, kaolinite, dickite, rutile, alunite, diaspore, hematite and secondary quartz as mineral association.
2) Derived from alteration of deformed migmatites from the igneous-metamorphic basement rocks, showing an AAA represented by pyrophyllite, kaolinite, sericite, diaspore, hematite, alunite, turmaline and rutile. Though in a lesser extent, the sedimentary cover (orthoquartzites, wackes) of these deposits have been also affected (San Manuel and Cerrito de la Cruz areas).

Deposits1b) and 2) type show clear evidences of replacement of preexistent rocks and developed mineral zoning due to hydrothermal activity. Trails of secondary fluid inclusions have been found in the orthoquartzite wall rocks. In the secondary quartz associated to 1b) type deposits primary fluid inclusions and solid inclusions (pyrophyllite and/or sericite) can be recognized. Small amounts of aluminum phosphate sulfate (APS) minerals enriched in REE have been detected in these deposits.

1.INTRODUCTION

This paper presents a study of hydrothermal genesis of the two types of clay deposits and the usefulness of ashfall pyroclastics as a stratigraphic marker in the area SE of Tandilia. New localities of altered pyroclastic rocks have been recognized in the area SE of Tandilia, from Sierra de la Tinta to San Manuel areas, (figure 1), providing new evidences which support an hydrothermal origin for all these clay deposits.
According to Leveratto and Marchese (1983) the sedimentary cover of Tandilia comprises a unique stratigraphic sequence called La Tinta formation deposited on the igneous-

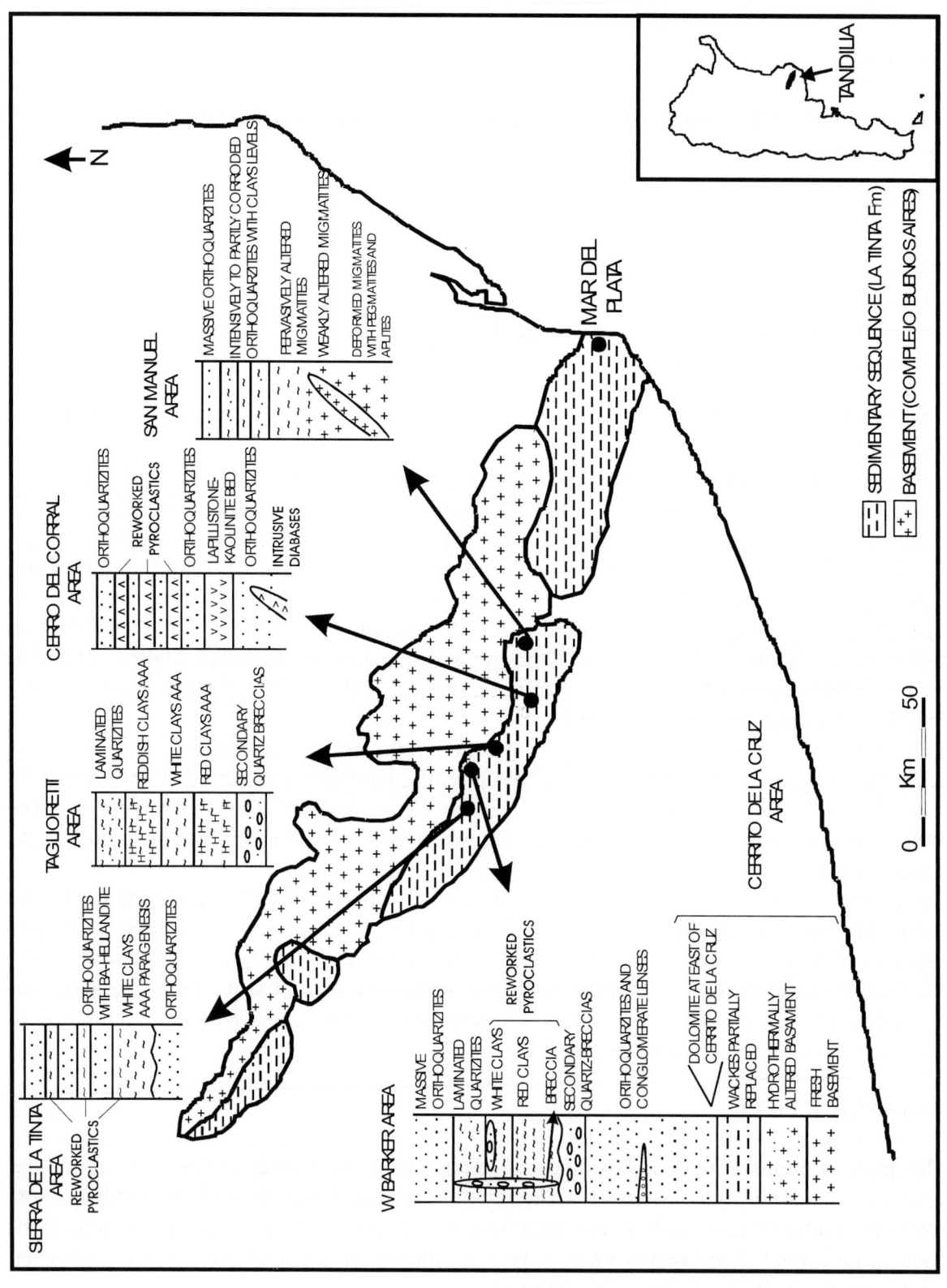

Figure 1. Location map

metamorphic basement named Complejo Buenos Aires. Dristas & Frisicale (1987) recognized for the first time, that the clay beds interbedded in the orthoquartzite sequence of La Tinta Fm. in Cerro del Corral area are altered pyroclastic rocks. The two types of clay deposits identified in the SE region of Tandilia are: 1) located at the contact between basement rocks and the sedimentary cover and 2) interbedded in the orthoquartzite sequence of the sedimentary cover.

2. METHODOLOGY AND ANALYTICAL THECHNIQUES

Sampling of several horizontal beds and various vertical profiles in each deposit was made. Bulk clays samples and wall rocks were analysed by XRD. The method of known additions (Brindley & Udagawa, 1960) checked by petrography and chemical bulk analyses were used in order to establish the zoning patterns. After the preliminary evaluation, some selected samples were studied by SEM and EMP in order to identify textures of clay minerals or the composition of accessory minerals, respectively. Double polished sections of secondary quartz associated to the clay deposits, were used for the analyses of primary fluid inclusions in secondary quartz and secondary fluid inclusions in orthoquartzite wall rocks.

A JEOL JXAS 900R electron microprobe analyser (EMP) from the Geochemisches Institut, Göttingen University, with the following operating conditions: 20 kV, 80 nA were employed for the mineralogical analyses. Bulk rock analyses for mayor elements in research quality were made in Activation Laboratories Ltd. (Canada) by ICP. XRD analyses were obtained using a Rigaku Geigerflex D max III C equipment, with 35kV and 15 mA as operating conditions. The sulfur isotope data from alunites were obtained in the Geochemisches Institut, Göttingen University, Germany. Fluid inclusions were studied by means of a Chaix-Meca heating-freezing stage (-180 +600°C) and crushing stage.

3. LOCALITIES OF CLAY DEPOSITS FORMED FROM PYROCLASTIC MATERIALS

3.1. Cerro del Corral area (San Ramón, Cerro Nuevo, Sierra de la Tigra and Sierra de Barrientos deposits)

Dristas & Frisicale (1987) gave details of the geology, mineralogy and petrography of these kaolinite deposits, which were mined to use in the refractory industry. In this area spectacular delicate replacement textures of the original lapillistone by aggregates of kaolinite (3T) and rutile, which delineate the original forms, have been preserved.

In addition, the original titanomagnetite crystals were transformed to an anatase-rutile aggregate preserving the unmixing textures (pseudomorphs) of the original mineral. The strongly vesicular nature of the lapillstone vitroclasts suggests a subaerial ejection process derived from alow viscosity magma. The presence of relatively abundant titanomagnetite, as the unique crystallized mineral from the eruption process together with the vitroclasts, indicate that the parent material was a basic magma. Well crystallized kaolinite(3T) is almost the unique clay mineral in the beds. No halloysites (10 Å and/or 7 Å) were detected by XRD and SEM studies. Small amounts of sericite and smectite have also been recognized. The genesis of all these deposits is related to the intrusion of diabase bodies, which generated the hydrothermal activity.

3.2. West Barker area (Carabelli, El Tinterito, Vittor and La Elisa deposits)

Dristas & Frisicale (1992) have extended the areas of ashfall pumiceous material to the west Barker region, (figure 1). This is an area of intensive mining for clays, (mainly pyrophyllite with subordinate kaolinite and sericite) where the clay beds are normally represented by a lower level of reddish clays, and an upper level of whitish clays without clear limits between them. Both types of clays are used in sanitary wear and in structural ceramic industry, respectively. The clay paragenesis is a typical AAA, as follows: pyrophyllite, kaolinite (dickite), sericite ($2M_1$), hematite, diaspore, alunite, rutile and secondary quartz, as the more representative minerals. The hydrothermal activity has been more aggressive than in the Cerro del Corral area, strongly affecting the original pyroclastic textures of the protolith through pervasive alteration. Also the quartz sandstones wall rocks, which underlie and overly the pyroclastics, show clear replacement textures. In the least altered zones pumiceous textures replaced by clay minerals have been identified. A clear vertical zoning from lower to upper levels has been determined in the clays deposits of these areas, as follows: a pyrophyllite-secondary quartz zone, kaolinite-diaspore zone, pyrophyllite-sericite zone and a sericite zone. The lower contact of the exploited clay beds shows typical concentrations of secondary quartz, which exhibits mainly a pseudo-spherulitic texture. As was previously suggested this type of texture represents secondary quartz, derived from the recrystallization of silica, dissolved from initial orthoquartzite wall rock, (Frisicale, 1991; Dristas & Frisicale, 1992). The presence of relict quartz grains with quartz cement in optical continuity, in a matrix of fine-grained secondary quartz supports this interpretation.

Breccia branches from the main sub-horizontal lenses cross-cut the upper orthoquartzite levels. Likewise, at the Carabelli deposit (west Barker area), occasionally vertical breccia-dykes up to 4 m thick including blocks up to 2 m^3 cross-cut the sedimentary sequence. At the contact between the laminated orthoquartzites and the underlying clay bed, breccia levels of lenticular form, are frequently present.

Sulfur isotopes

Sodian alunite veinlets representing a later event of the hydrothermal alteration process have been recognized, Dristas & Frisicale (1983). Samples for isotopic studies from different deposits of this area were chosen based on their abundance and purity. Consistent, homogeneous data were obtained ($\delta^{34}S‰$ = 21.7, 21.4 and 22.2) and plot in the area of primary hypogene alunite, figure 2, compared with the data from Julcani, Summitville and Goldfield deposits (Hayba et al., 1985). These results support previous studies of Dristas & Frisicale (1983) and invalidate the proposal of Zalba et al. (1988), assigning a diagenetic origin to the alunite.

Aluminum-phosphate-sulfate (APS) minerals

Dristas & Frisicale (1996) informed that the presence of considerable amounts of P_2O_5 (1.15 wt.%), CaO (0.25 wt. %), Sr (4340 ppm) and lanthanoids (Ln's Σ 1457 ppm) in alunites from this area suggests a probable solid solution between alunite and APS minerals: woodhouseite-svanbergite-crandallite-goyazite as previuosly have been documented in localities related to AAA, (Stoffregen & Alpers, 1987).

Fluid inclusions

The above-described levels of secondary quartz at the base of clay beds and secondary quartz breccias clasts contain liquid-rich and vapour fluid inclusions of primary origin. Also small solid inclusions of pyrophyllite (and/or sericite) and hematite have been found in the secondary quartz. The homogenization temperature (Th) is shown in figure 3.

Based on the scheme of Bodnar (1985) and the low contents of dissolved CO_2 detected in the liquid phase, by means of crushing stage, Frisicale (1991) and Dristas & Frisicale (1992) considered that the temperature of crystallization of the secondary quartz varies between 290-340°C, from a fluid phase containing about 2 NaCl wt. % equivalent.

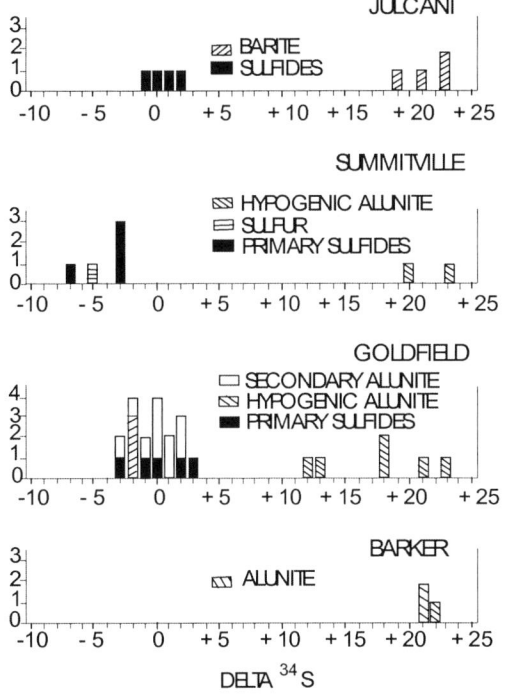

Figure 2. $\delta^{34}S$ histograms for Julcani, Summitville, Goldfield and Barker area. Data sources: Hayba et al.(1985) & Frisicale (1991).

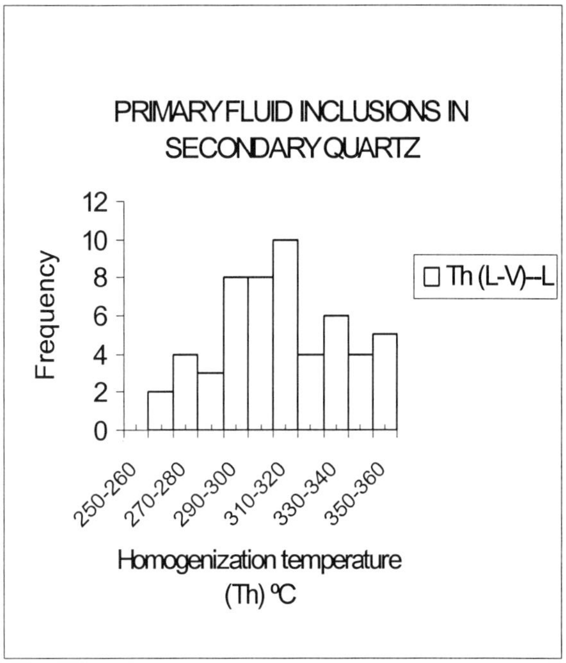

Figure 3. Th histogram of primary fluid inclusions in secondary quartz from Barker area. Dristas & Frisicale (1992).

3.3. Taglioretti area (El Diamante, El Infierno and Taglioretti deposits)

These deposits are located 4 km southeastwards from Barker, (Fig. 1). Here, iron rich clay beds with hematite and goethite minerals up 70-weight % of Fe2O3, were recognized. Some evidence of clay pseudomorphs of pyroclastic textures has been found. They are represented partly by lapillistone material including accretionary lapilli and agglomerates. An AAA mineral clay paragenesis similar to that described for the west Barker area is present here. The clay beds exhibit a typical mineral zoning and the corrosion of the orthoquartzitic wall rocks are similar to that described for the west Barker area. Bedded quarzites again overly the clay bed. Secondary quartz and breccias can be also found.

3.4. Sierra de La Tinta area (La Tinta, Cantera 1 and Cantera 2 deposits)

The most southwesterly clay deposits in the sedimentary sequence, derived form pyroclastic protoliths affected by hydrothermal activity, has been studied in the Sierra de la Tinta area, (Frisicale & Dristas, 2000), figure 1. This area represent marginal zones of alteration giving rise to low temperature and neutral paragenesis, which included minerals like zeolite, smectite, adularia and opal.

APS minerals have been found in La Tinta deposit as veins or aggregates in the massive kaolinite bed. Electron microprobe (EMP) analyses (Table I) indicate a composition close to goyazite for this mineral according to the scheme of Stoffregen & Alpers (1987). The EMP analyses indicate also very high content of lanthanoids (Ln's Σ: 16.23 wt. %) in these minerals, (Table I).

4. LOCALITIES WITH CLAY DEPOSITS DERIVED FROM HYDROTHERMALLY ALTERED BASEMENT ROCKS

4.1. San Manuel area (Cantera Grande, Misato and La Liebre deposits)

The main locality for these clay deposits is the San Manuel area, (figure 1), where intense AAA have affected basement rocks, mostly migmatites of granodioritic composition, (Dristas & Frisicale, 1984). These clay deposits are actually mined for structural ceramic uses.

Generally the pseudomorphism of the original banded structures of the migmatites can be macroscopically recognized. Orthoquartzites covered the original basement rocks in the Cerro Reconquista hill, San Manuel area.

The clay deposits of this area show the following mineral association: pyrophyllite, kaolinite, sericite, hematite, diaspore, turmaline, rutile and alunite. In wall rocks relicts, the intense replacement of quartz, K-feldspar, plagioclase and biotite by pyrophyllite in the migmatite wall rock, can be clearly observed microscopically. At the contact with the overlying sediments, a tourmalinization zone has been found showing curious whitish nodules filled with kaolinite-pyrophyllite including aggregates of radial arranged diaspore, giving rise to a green mottled rock.

Trails of secondary fluid inclusions cross-cutting quartz clasts have been recognized in the overlying orthoquartzites in the Cantera Grande deposit. This is a clear evidence that the

Table I
Electron microprobe analyses of APS minerals from hydrothermal clay deposits of the SE region of Tandilia, Argentina.

Oxides	C. Cruz deposit	C. Cruz deposit	C. Cruz deposit	C. Cruz deposit	La Tinta deposit	La Tinta deposit
	Anal. 1 wt. %	Anal. 2 wt. %	Anal. 6 wt. %	Anal. 7 wt. %	Anal. 9 wt. %	Anal.10 wt. %
K_2O	0.84	1.06	0.53	0.27	0.92	0.70
CaO	1.09	0.74	0.82	0.77	0.80	0.83
SrO	7.50	6.80	7.99	7.32	5.73	5.20
BaO	1.75	1.89	1.98	1.84	1.29	1.19
La_2O_3	2.26	2.39	2.27	2.67	5.65	5.54
Ce_2O_3	4.08	4.97	5.24	6.49	2.21	2.37
Pr_2O_3	0.67	0.80	0.91	1.02	1.74	1.58
Nd_2O_3	2.33	3.26	2.99	3.69	5.04	5.57
Sm_2O_3	0.35	0.52	0.52	0.58	0.78	0.77
Gd_2O_3	0.10	0.12	0.32	0.25	0.33	0.33
Al_2O_3	33.64	33.94	32.66	33.02	31.39	31.92
Fe_2O_3	1.38	0.59	0.63	0.50	0.29	0.24
SO_3	6.67	6.74	8.76	8.32	10.83	9.01
P_2O_5	20.68	20.72	22.21	23.11	22.43	22.42
Total	83.28	85.54	87.83	89.85	89.43	87.67
Σ Ln's	9.79	12.06	12.25	14.70	15.75	16.16

hydrothermal solutions affected both, the basement and the sedimentary cover. The Th range of these secondary fluid inclusions is similar, to that of the primary fluid inclusions in secondary quartz, from west Barker area.

Given the resistate nature of the sedimentary cover the intensity of the replacement texture developed in the orthoquarzite suggests this was throughly pervasive hydrothermal event.

4.2. Cerrito de la Cruz area (Cerrito de la Cruz Norte and Cerrito de la Cruz Sur deposits)

Hydrothermal fluids circulated throughout the contact between the basement rocks (granodioritic migmatites) and the overlying sediments (wackes-orthoquartzites). Both rocks were affected by the hydrothermal activity. An asymmetric hydrothermal alteration was evident, reflected in the mineral paragenesis and zones extension due to different original wall rocks reactivity. Both rocks types exhibit an AAA with clear replacement textures, (Frisicale & Dristas, 1993), (figure 1).

APS minerals

These minerals, which are close to goyazite composition, show anomalous high content of lanthanides (Ln´s Σ=14.70 wt. %, Table I).

5. DISCUSSION

The described clay deposits lying in the contact between the migmatitic basement and the sedimentary cover (San Manuel and Cerrito de la Cruz areas) or within the upper orthoquartzite sequences (west Barker, Taglioretti and La Tinta areas), occur along faults, fractures or breccias through which the hydrothermal fluids could be mobilized to horizons with higher porosity: 1) the paleosurface of the covered basement and 2) the reactive pumiceous material interbedded in the orthoquartzite sequence, respectively. This led to a selective hydrothermal replacement in both levels of the geologic scheme of the SE region of Tandilia. Such a phenomenon is clearly evident in the studied clay deposits. In addition to pervasive AAA of the clay deposits of these areas, there are several concurrent characteristics that include: 1) similar vertical mineral zoning of the clay beds, 2) quartz wall rock dissolution and 3) secondary quartz deposition with fluid and solid inclusions (sericite, pyrophyllite, hematite).

This would indicate the presence of a main horizontal fluid circulation throughout the more porous and reactive zones: 1) the contact basement-sedimentary cover and 2) the pyroclastic levels, that induced a lateral (vertical) diffusion giving rise to the vertical mineral zoning in both occurrences. The prograde reactions involved silica dissolution (quartz corrosion). The quartz–kaolinite patches, recognized in massive pyrophyllite, are interpreted as a retrograde reaction associated to silica fixation. Prograde and retrograde reactions are based on the stability diagrams of Hemley et al. (1980) modified by Bryndzia (1988). The associated breccias intrusions recognized in many deposits are probably related to the retrograde event, involving temperature and pressure decrease.

The consequent mixing of hydrothermal solutions with meteoric water gave rise to more oxidizing conditions and to the formation of alunite veinlets representing a later event of the hydrothermal process in these areas. These hydrothermal residual fluids probably contained high concentrations of lanthanides, which could not be retained by the crystallization of the phyllosilicates and oxides of the AAA mineral paragenesis. Consequently, the availability of the sulfate structure, but mainly the sulfate-phosphate mineral structure at this latter stage, could be suitable for the lanthanides retention in alunite and APS minerals.

Probably an unique hydrothermal event generated both types deposits described, because they have in common the following features: 1) a similar mineral paragenesis including the APS minerals, 2) equivalent alteration and zoning patterns, 3) wall rock replacement textures and 4) the similarity of fluid inclusion Th in both types of clay deposits.

6. CONCLUSIONS

Two types of clay deposits have been formed by a single hydrothermal event in the area SE of Tandilia:

I- In the Cerro Reconquista and Cerrito de la Cruz areas clay deposits formed by replacement of basement rocks (migmatites) displaying an AAA at the contact with the overlying sediments, which also are affected.

II- Selective replacement of pyroclastic rocks interbedded with orthoquarzites, lead to suggest two categories of clay deposits in this group:

a) Low temperature hydrothermal clays formed mainly by kaolinite replacement of ashfall pyroclastic rocks with minor or without reworking in the Cerro del Corral area including the: San Ramón, Cerro Nuevo, Sierra de la Tigra, and Sierra de Barrientos deposits.

b) Advanced argillic alteration assemblage (pyrophyllite, dickite, kaolinite, sericite, diaspore, alunite, tourmaline) of reworked and epiclastic contaminated ashfall pyroclastics in the west Barker, Taglioretti and the east of La Tinta areas.

REFERENCES

Bodnar, R.J., T.J. Reynolds & C.A. Kuenhn, Rev. Econ. Geol., 2 (1985) 73.

Brindley, G.W. & S. Udagawa, J.Am.Ceram.Soc., 43, (1960) 59.

Bryndzia, T.L., Econ.Geol. 83 (1988) 450.

Dristas, J.A. & M.C. Frisicale, Rev.Asoc.Arg.Min.Petr. 14, 1-2, (1983) 34.

Dristas, J.A & M.C. Frisicale, IX Congr. Geol.Arg., V, (1984) 507.

Dristas, J.A & M.C. Frisicale, Rev. Asoc.Arg.Min.Petr. 18,1-4, (1987) 33.

Dristas, J.A. & M.C. Frisicale, Zbl. Geol. Paläont., I,6, (1992) 1901.

Dristas, J.A. & M.C. Frisicale, 15 Geowissenschaftliches Lateinamerika Kolloquium, Terra Nostra, Heft 8/96, (1996) 37, Hamburg.

Frisicale, M.C., Ph.D. Thesis, Departamento de Geología, Universidad Nacional del Sur, Bahía Blanca (1991).

Frisicale, M.C. & J.A. Dristas, XII Congr.Geol.Arg., V,(1993) 222.

Frisicale, M.C. & J.A. Dristas, Rev.Asoc.Geol.Arg, 55, 1-2, (2000) 3.

Hayba, D.O., Ph. M. Bethke, P. Heald, & N.K. Foley, Rev.Econ. Geol., 2 (1985) 129.

Hemley, J.J., J.W. Montoya, J.W. Marinenko & R.W. Luce, Econ.Geol. 75, (1980) 210.

Leveratto, M.A. & H.G. Marchese, Rev.Asoc.Geol.Arg., 38, 2, (1983)23.

Stoffregen, R.E. & Ch.N. Alpers, Can.Mineral. 25 (1987) 201.

Zalba, P.E., R. Andreis & F.C. Lorenzo, V Congr. Lat.Geol. Act. 2, (1982) 389.

Zalba, P.E., R. Andreis & A.M. Iñiguez, Rev.Asoc.Geol.Arg. 43,2, (1988) 198.

Occurrence of bentonites in Southern South America

M. L. L. Formoso[a]; L. M. Calarge[a]; A. M. Misusaki[a]; A. Meunier[b]; D. B Alves[c]; P. Zalba[d]; R.Andreis[e]; J. Bossi[f]

[a] Inst. de Geociências – UFRGS Av. Bento Gonçalves, 9500 –91500 Porto Alegre, RS, Brazil
[b] HydrASA - UMR 6532 CNRS – Univ. de Poitiers - 40 av. du Recteur Pineau – 86022 Poitiers Cedex - France
[c] CENPES – PETROBRAS – Cidade Universitária, Rio de Janeiro, RJ, Brazil
[d] Comision de Investigaciones Cientificas de La Provincia de Buenos Aires – CETEMIC). (UNLP) - (LAC) - GONNET (1897), Procincia de Buenos Aires, Argentina
[e] Museo Paleontologico «Egidio Feruglio», Av da Fontana 140 – Chubut, Argentina
[f] Universidad de la Republica, Av da Garzon 780 – Montevideo, Uruguay

Bentonite and tonstein beds, derived from the alteration of volcanic ash, may be used as important stratigraphic markers in either a local or a regional scale, as they occur in several areas in southern South America.

This work highlights some findings in Permian rocks from the Paraná Basin, including the bentonite beds of Aceguá (Brazil) and Mello (Uruguay) and those from the Sierra de La Ventana (Argentina). Cretaceous bentonites from the Campos Basin ("3-Dedos" marker) and Permian tonsteins from the Candiota Coal Mine (Brazil) are also discussed in this paper.

The Mello bentonite comprises a single bed (1.5 to 3 m thick) with gradational upper and lower contacts; the Aceguá finding is composed of at least two successive beds (around 0.3 m thick) displaying sharp and/or gradational contacts; the Sierra de La Ventana occurrence presents five beds (less than 1 m thick) displaying sharp upper and lower boundaries; and the "3-dedos" marker includes three beds (about 1 m thick) presenting sharp lower contact.

This study aims to evaluating the potential use of these occurrences as local or regional stratigraphic markers.

1. INTRODUCTION

Knight (1898) defined bentonite as a swelling rock of the Cretaceous Fort Benton Formation from Wyoming, USA, and later Hewit (1917) described bentonite as an alteration product of volcanic ash. Today, the term bentonite is also used for any clay with dominantly smectitic composition, with special physical properties related to its smectitic composition (Grim & Güven, 1978).

Fallout volcanic ashes are mainly the precursors of bentonites. Volcanic ash deposited in marine enviroments forms smectitic clay minerals by alteration (smetitic bentonites). In continental enviroments, especially in palustrine ones, fallout volcanic ash goes to kaolinitic clay minerals (Bohor & Triplehorn, 1993).

However, the deposition of those volcanic ashes in a semi-arid continental environment (alkaline lakes) leads to the formation of smetitic clay minerals. The alteration of distal volcanic ash may generate very pure smectites.

The so-called K – bentonites or metabentonites (Grim & Güven, 1978) are described principally in many Lower Paleozoic formations. Commonly, they are formed by illite – smectite mixed layers and interpreted as altered volcanic ash to which potassium has been added later on. Bergstrom et al (1997), Huff et al (1997) and Cingolani et al (1997) describe K – bentonites in the Silurian (Arisaig) of North American and in the Ordovician (Argentina Precordillera) in South America.

Five occurrences of Southern South America bentonites (figure 1) were selected for this study. They represent the diversity of geologic types:

1 – Tonsteins in the Candiota Coal Mine (RS, Brazil), a kaolinitic bentonite;

2 – The Aceguá and Melo bentonites (RS, Brazil and Uruguay, very pure smectitic bentonites from a semi-arid, deltaic, lacustrine or lagoonal environments (in the Rio do Rastro and Yaguari Formations);

3 – Smectitic bentonitic material from the Permian Tunas Formation, located at Abra del Despeñadero (Sierras Australes), Provincia de Buenos Aires – Argentina (alteration of pyroclastic beds interbedded with detrital sedimentary rocks);

4 – The Campos Basin bentonites (RJ, Brazil), Cretaceous bentonite beds in turbidites, consisting of pure dioctahedral illite – smectite mixed layers.

2. PERMIAN TONSTEINS FROM THE PARANA BASIN, RIO GRANDE DO SUL, BRAZIL

Tonsteins beds are found within Permian strata of Sakmarian age (Alves, 1994) in the Candiota Coal Mine, near the town of Bagé (RS).

Two tonsteins beds (8 – 10 cm and 1 – 3 cm thick) were identified in the Upper Candiota Coal Seam, and one (1 – 2 cm thick) in the Lower Candiota Coal Seam. At the top of the Upper Candiota Coal Seam, an argillaceous bed 1.5 to 2 m thick occurs (the so-called "upper claystone"). Another claystone bed ("intermediate claystone", 0.75 – 1.20 m thick) separates the Upper and the Lower Candiota Coal Seams. The upper and intermediate claystones have been studied for comparison with the tonsteins (Calarge, 1997 and Formoso et al. 1997).

The tonsteins consist of 89 – 92% clay minerals and 8 – 11% quartz. Kaolinite and illite – smectite mixed layers (I – S) form about 90% of the tonsteins. The upper and intermediate claystones contain 65 – 84% clay minerals, 15 – 32% quartz and 1 – 3% feldspars. I/S constitutes 70 – 96% of the clay mineral composition of the claystones.

SEM study of tonsteins has shown that vermicular kaolinite crystals are dispersed within a fine argillaceous matrix. The crystals are platy, with a pseudo – hexagonal to hexagonal shape, and most range 2 to 10μm in size. Some "booklets" of kaolinite are, however, smaller than 5μm. The upper claystone displays platy kaolinite crystals with pseudo – hexagonal faces, rounded quartz grains and I – S grains that range from 0.5 to 5μm. Kaolinite in the tonsteins is well-ordered. SEM studies (figure 2A) show crystals with ß-quartz form, ranging between 5 – 20μm in size, and with a bimodal size distribution. They are generally euhedral (Spears and Lyons, 1995). Chemical data revealed that the SiO_2/Al_2O_3 ratios from tonsteins fall within a narrow range of 2.3 – 2.4, whereas those from the claystones vary from 5.5 to 7.7, thus indicating that tonsteins are almost pure kaolinite. Sulfur varies from 0.14 to 0.30, and it is associated with diagenetic pyrite.

The tonsteins have a lower minor element content than the claystones. However, U (8 – 19 ppm) and Th (67 – 87ppm) are higher in the tonsteins. On the other hand, Nb, Y, Cs, Rb, Ba and Zr are much higher in the upper and intermediate claystones.

REE values are higher in the claystones than in the tonsteins (figure 2B). A negative anomaly of Eu and the high value of La/Yb are conspicuous (Spears et Kanaris - Sotiriou, 1979).

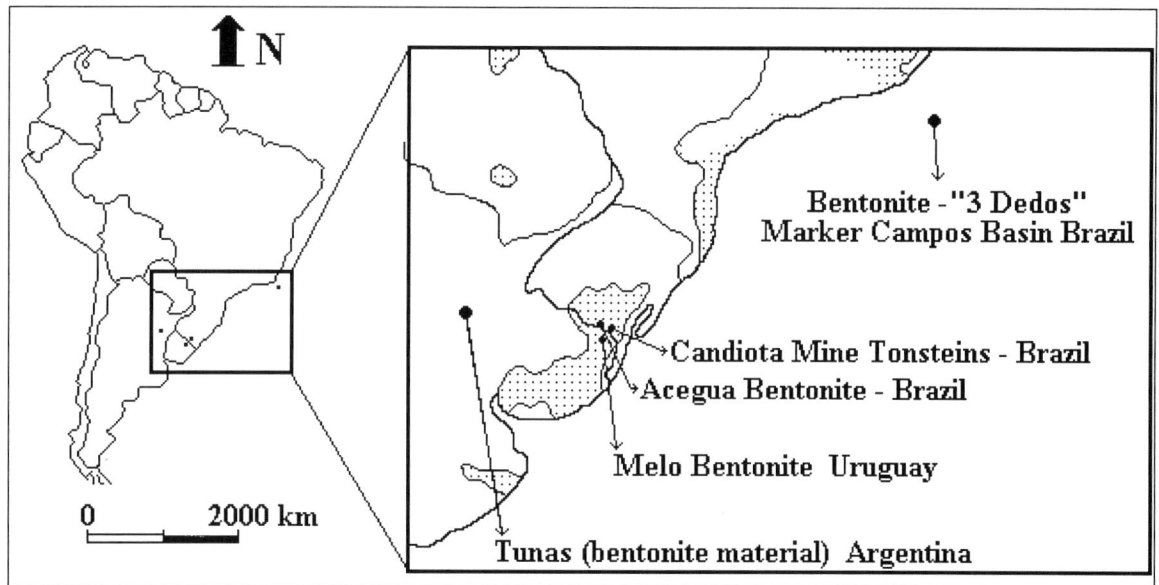

Figure 1 – Volcanic markers in the South American continent

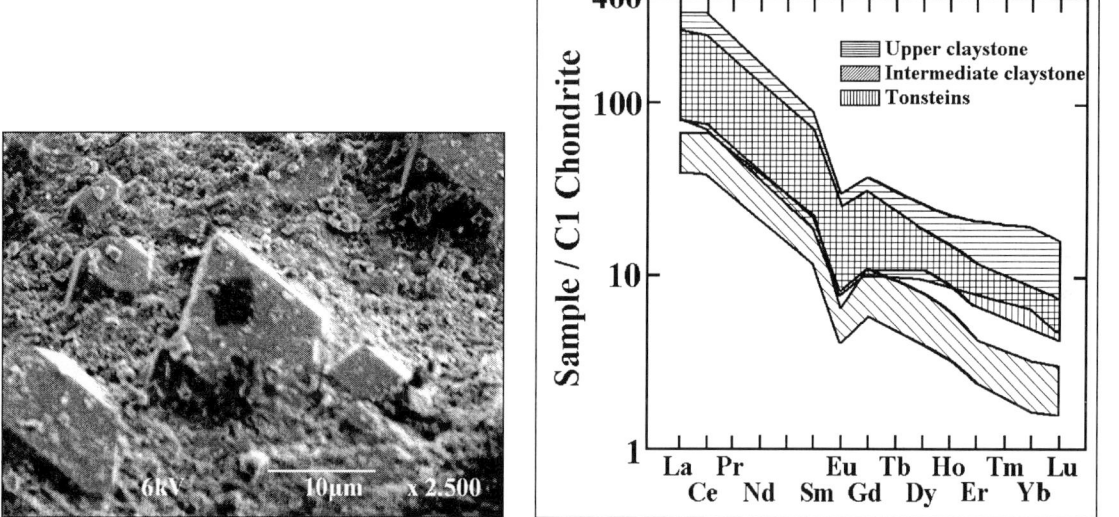

Figure 2 – A: SEM bipiramidal, ßform quartz; B – REE patterns of the upper and intermediate claystones, and of tonsteins from the Candiota Mine (RS), Brazil

The zircon from upper tonsteins within the Upper Candiota Coal Seam was dated by Matos (1999), by use of the U/Pb method developed by Basei et al. (1995). It has yield a radiometric age of 267.1 ±3.3 Ma.

Formoso et al (op cit.) concluded that the Candiota tonsteins display thin, widespread and continuous beds, which may indicate an air-fallout volcanic origin; tonsteins are nearly monomineralic (90% pure kaolinite), and the presence of euhedral, □-form quartz suggests a volcanic origin. The higher contents of U and Th in comparison with the upper and intermediate claystones, the negative Eu anomalies, and the high La/Yb ratio (figure 2B) (Winchester and Floyd, 1977) also indicate a volcanic origin.

3. PERMIAN SMECTITE BENTONITES

3.1. Bentonites of Aceguá and Melo

The bentonite beds of Aceguá occur on the Bage - Aceguá road, at about 3 km from the Brazil – Uruguay border. The Melo bentonite occurs on the Melo – Montevideo road – Uruguay, 20 km from Melo. The Aceguá (in the Rio do Rastro Formation) and Melo (in the Yaguari Formation) bentonites are stratigraphically equivalent. A fossil reptile (Pareiasaurus Americanus) has been identified in Aceguá, being correlated with the Daptocephalus Biozone of the Upper Permian (Tatarian) of the Karoo Basin (Barberena et al., 1985). The Rio do Rastro and Yaguari Formations are red beds formed during the Upper Permian regression. Andreis (1996) described the development of a volcanic arc in the southern part of Argentinian Patagonia, beginning during the Carboniferous and reaching its highest intensity in Triassic – Lower Jurassic..

Aceguá: bentonite beds are interbedded with continental sedimentary rocks (fluviat sandstones – eolian dune and lacustrine deposits) – thickness: 15 – 30 cm.

Melo: a lenticular bed – about 2 m thick – interbedded with fine-grained sandstones and siltstones in the upper part of the Yaguari Formation. The massive clay in the center of the bed is pink-colored material. X – ray diffraction patterns (figure 3) of the Aceguá and Melo bentonites show a very pure smectite with minor quartz and zeolites (Calarge et al., 2001, in preparation).

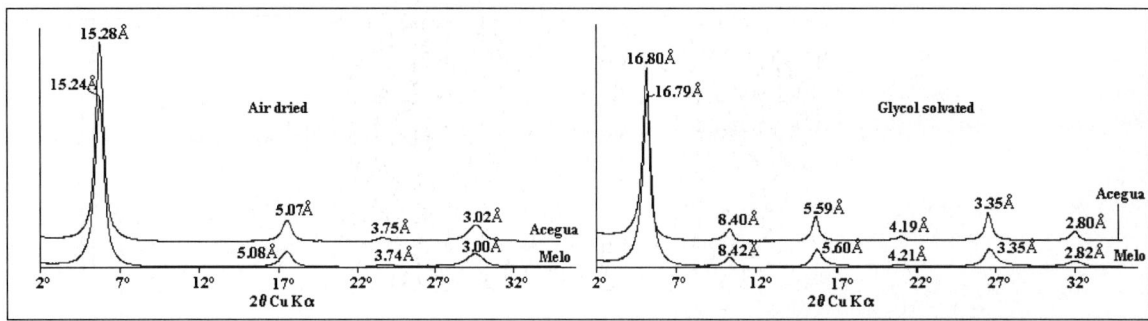

Figure 3: X – ray diffraction patterns of the Aceguá and Melo bentonites (< 2μm)

It was observed that smectite presents an elevated coherent scattering domain size (CSDS). The high crystallinity is found even in the fine-size fraction. The x-ray diffraction of treated samples using the Greene – Kelly method (Greene & Kelly, 1953, Li+ - heated to 300°C) shows that the smectite is dominantly a montmorillonite (layer charge originating from the octahedral sheet). Calarge et al. (op cit.), after saturation with different cations (Ca2+, K+ and K+–Ca2+) observed that the smectitic minerals are not homogeneous but heterogeneous. The apparently homogeneous, well-crystallized montmorillonite from Melo (Uruguay) is composed of layers where the composition varies from low-charge smectite to high-charge vermiculite. According to the expandability degree in the Ca2+, K+-Ca2+ and K+-saturated states, "layer types" have been identified. These layers, which are nearly fully expandable (2EG (Ethylene Glycol)) in the Ca-saturated state, form three-component mixed layer minerals (MLM) (2EG, 1EG and 0EG) after K+ adsorption. Based on increasing amounts of

K+ ions, the following sequence of crystal structure is observed: (a) 2EG layer-rich random MLM (Ca-saturated state); (b) 1EG-0EG layer-rich ordered MLM (K-Ca saturated state) presenting segregation of 0EG layers; (c) two separated phases 1EG layer-rich random MLM + 0EG layer-rich random MLM (K-saturated state). Hence high-charge layers are segregated or belong to distinct clay particles.

3.1.1. Chemical composition

The bentonite beds of Aceguá and Melo are nearly monomineralic. In Aceguá, the bulk and < 0.1 µm fractions are similar. The calculated structural formula: (bulk rock)
$[Si_{3.91} Al_{0.09}]O_{10} (Al_{1.46} Fe_{0.08} Mg_{0.51} Ti_{0.01}) (OH)_2 Na_{0.02} K_{0.02} Ca_{0.16}$; Σ oct = 2.06
Tetrahedral charge = - 0.09; Octahedral charge = - 0.32; Interlayer = 0.36
78% of the layer charge is originated from ionic replacement in the octahedral sheet.
The fraction < 0.01 µm has a chemical composition very close to the bulk rock:
$[Si_{3.87} Al_{0.13}] O_{10} (Al_{1.43} Fe_{0.08} Mg_{0.53} Ti_{0.01}) (OH)_2 K_{0.01} Ca_{0.24}$; Σ oct = 2.05
Tetrahedral charge = - 0.13; Octahedral charge = - 0.23; Interlayer charge = 0.49
74% of the layer charge is originated from ionic replacement in the octahedral sheet.

3.1.2. Geochemical data

The REE content in bulk rock is twice as much the content in the <1 µm fraction, what means that REE are higher in minerals coarser than 1µm (for example, zircon). A weak negative Eu anomaly and an enrichment in LREE characterize the REE patterns. Zr / TiO2 vs. Nb/Y diagram of the samples (Winchester & Floyd, op cit.) indicates that the original volcanic ash is rhyodacitic to dacitic in composition. The Zr/Y is about 12.3 (bulk sample), which indicates a calc-alkaline magmatic affiliation. The positive Ce anomaly is probably related to oxidation of Ce3+ to Ce4+ and precipitation of cerianite.

3.2. Bentonites of Argentina – Permian Bentonitic Clays of Abra del Despeñadero

Argentina bentonites occur from the Ordovician to the Tertiary (Zalba, personal communication).

Ordovician bentonites were studied by Huff et al. (1997); Cingolani et al. (1997) studied K-bentonites in the Argentina Precordillera, emphasizing their tectonomagmatic significance. They compared the radiometric and biostratigraphic ages. Iñiguez Rodrigues & Zalba (1994) described smectitic horizons in the Permian rocks of the Sierras Australes, where pyroclastic material is principally associated with mudstones. Marconi (1988) presented a summary of occurrence, properties and applications of the principal deposits in Argentina. Marconi (op cit.) described Upper Triassic, Upper Cretaceous and Tertiary bentonites.

3.2.1. Occurrence of bentonitic beds as alteration of pyroclastic material in Abra del Despeñadero (Sierras Australes in the Permian Tunas Formation).

Sedimentary sequences display an alternating succession of sandstones and mudstones. Py oclastic horizons are associated with clayey sediments, interbedded with sandy, fining-upward cycles and channelized bodies.

Mudstones occur as siliciclastic and pyroclastic clays. The latter displays glass shards dispersed in a fine smectitic matrix. X – ray diffraction patterns of the clay fraction are characteristic of illite and smectite. Li saturated heated to 300°C (Greene – Kelly, op cit.) samples show 001 reflection – 10Å and the 060 reflection at 1.504 to 1.502 Å (dioctahedral smectite – montmorillonite). K – saturated samples exhibited 10Å reflection (superimposed with illite peak) and broad 12Å reflection. Mg – saturated samples show basal 15Å and 17Å reflections (high charge and low charge smectites).

According to Iñiguez Rodrigues and Zalba, (op cit.), the pyroclastic events responsible for the generation of the Permian tephra falls would be related to a volcanic arc located to the SW of the studied area. Their origin would be associated with the collision between the Patagonia plate and the southern part of the Gondwana continent.

3.3. Cretaceous Campos Basin bentonite

Bentonite beds occur in the Santonian (Upper Cretaceous) of the Campos Basin, Brazil. Alves et al. (1995a) described the occurrence of three main claystone beds in the mud-dominated Ubatuba Formation at southern Campos Basin, offshore Brazil. The beds are usually 1 – 2 meters thick and can be differentiated from the surrounding shales and siltstones by their low resistivity and high porosity. The beds display sharp upper and lower contacts. Other characteristics include the lack of fissility, the local occurrence of spheroidal exfoliations, the tabular geometry (thin and laterally extensive beds), and the repetitive and episodic character.

The associated shales and siltstones are composed of quartz, plagioclase, potassium feldspar, calcite, micas and clay minerals, and eventually chlorite mixed layers. Clay minerals are I – S and C – S mixed layers, illite and chlorite. I – S consists of around 50 – 55% illite layer.The bentonites, on the other hand, consist almost exclusively of a homogeneous dioctahedral illite – smectite mixed layer (figure 4B) with disordered inter-stratification and around 85 – 90% of smectite layers.

The high porosity in electric logs is related to the high microporosity due to the random arrangement of the microscopic particles in the clay minerals . The lower resistivity is related to the high water content, and also to the absence of carbonates. Alves et al. (op cit.) also described diagenetic features thatindicate that the claystone has undergone intense recrystallization. The absence of fissility and the occurrence of local spheroidal exfoliation indicate that the clay minerals are not detrital, but formed by in situ alteration. The high concentrations of Ti and Zr are typical of alkaline rocks. The presence of highly expandable illite – smectite in the claystones at a depth of around 2700 meters indicates incipient burial diagenesis. In the Zr/TiO_2 vs. SiO_2 diagram, the bentonite of the "3 dedos" marker falls within the trachite field. The high REE content (893 ppm) also suggests that the precursor rock was alkaline.

4. FINAL REMARKS

1. The in situ alteration of volcanic ashes (Grim and Güven, op cit.) lead to the formation of smectitic or kaolinitic bentonites. Tonsteins in the Candiota Mine (Brazil), is the product of alteration of volcanic ash to kaolinite in a palustrine environment.

2. Bentonites of Aceguá (Brazil) and Melo (Uruguay) were formed by the alteration of volcanic ash in a fluviat – lacustrine environment under semi-arid conditions, leading to the formation of smectite. Hein and Scholl (1978, in Weaver, 1989) described various stages in the early formation of smectite as a product of the alteration of acidic ashes in the Bearing Sea. Bentonites are generally montmorillonitic (not beidellitic). The later diagenetic reactions occur in a semi-arid environment. Diagenesis is restricted to the increase in tetrahedral charge and crystallinity of montmorillonite.

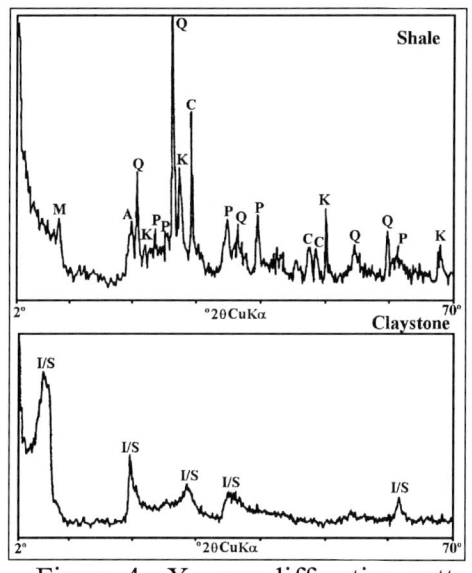
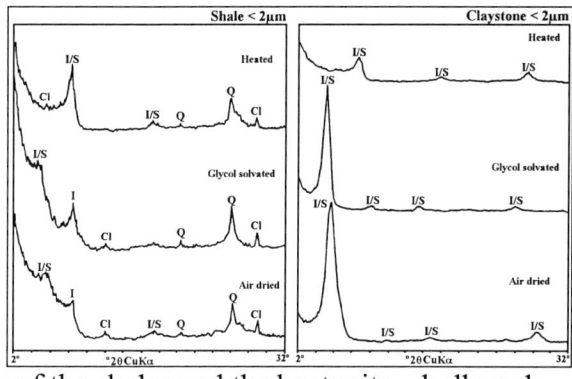

Figure 4 - X – ray diffraction patterns of the shales and the bentonite – bulk rock.
Claystone: I/S = Illite – smectite, M = muscovite, Q = quartz, P = plagioclase, K = kaolinite, C = calcite, A = anatase. Bentonite: Cl = chlorite, I/S = illite – smectite, Q = quartz, I = illite

3. K – bentonites represent the product of diagenesis or low-grade metamorphism of smectites. I/S shows 1.5 ordering and also R.3 ordering (Bergstrom et al., op cit) ; Cingolani et al. (op cit.) described R.1 ordering in K – bentonites of the Precordillera de San Juan (Lower Paleozoic K – bentonites). The Cretaceous bentonites of the Ubatuba Formation – Campos Basin (Brazil) are I – S with disordered inter-stratification with 85 – 90% of smectite layers, and therefore they are not K - bentonites.

4. The REE patterns of the tonsteins (Aceguá and Melo), with high La/Yb, negative Eu anomalies and Zr/TiO_2 vs. Nb/Y diagrams, indicate that they have rhyolitic and rhyodacitic to dacitic composition. The Zr/Y defines a calc – alkaline affiliation for the Aceguá bentonite. The bentonite of the Campos Basin ("3 dedos" marker) is derived from a trachytic rock (much higher La/Yb, very highREE, plotting in the trachyte field in the Zr vs. SiO_2 diagram) (Alves et al, 1995b).

5. The bentonites or bentonitic material (tonsteins, Aceguá and Melo, and Aba del Despeñadero) are of Permian age, but only the tonsteins have a radiometric age determined through U/Pb method (267.1 ± 3.4 Ma).

6. Biostratigraphic calibration may be achieved by using the radiometric age of bentonites and palynological data (Cazzulo – Klepzig et al., 2001). TheK – bentonites of the Arisaig beds in Canada and in the Precordillera de San Juan (Bergstrom op cit. and Cingolani, op cit.) had their radiometric ages used for biostratigraphic calibration of graptolite zones.

7. Tectonomagmatic studies may be done through the location of volcanic sources, aiming at the definition of the geotectonic environment. Andreis (op cit.) concluded that the bentonites in the Rio do Rastro and Yaguari Formations are related to volcanic events that spanned the Carboniferous through the Triassic – Lower Cretaceous in northern Argentinian Patagonia. Iñiguez Rodrigues et al. (op cit.) suggested that the pyroclastic activities responsible for the generation of tephra falls (bentonitic material) would be related to the volcanic arc located in SW of the Sierras Australes.

REFERENCES

1. W.C.Knight, Eng. Min. J., 66: 491 (1898).
2. D.F. Hewit, 1917, J. Wash. Acad. Sci., 7: 196-198.
3. R.E. Grim And N. Güven, 25, Elsevier Scient. Publishing Company, 249p (1978).
4. S.M. Bergström.; W.D. Huff; D.R. Kolata and M.J. Melchin, Can. J. Earth Sci., 34: 1630 – 1643, (1997).
5. W.E. Huff.; D. Davis; S.M. Bergström; M.P.S., Krekeler, D.R. Kolata And C. Cingolani, Episodes, 20, 1:29-33 (1997).
6. C.A Cingolani; W. Huff; S.M. Bergström And D. Kolata, Revista da Associacion Geologica Argentina, 52 (1): 47-55, (1997).
7. R. G. Alves, Porto Alegre, Pós-graduação em Geociências, UFRGS (1994).
8. L. M. Calarge, São Leopoldo, UNISINOS, Mestrado em Geologia Sedimentar, Brasil (1997).
9. M.L.L. Formoso; L.M. Calarge; J.V. Garcia; D.B. Alves; M.E.B. Gomes and A.M.P. Misusaki, In: Clay in our future. Proceedings of the 11th International clay conference. Kodama, Mermut & Torrance eds. Ottawa, Canada: 613-621 (1997).
10. D.A. Spears And P.C. Lyons, In: M.K.G. Whateley and D.A. Spears (eds), European Coal Geology, Geological Soc. Sp. Pub. 82:137-146, (1995).
11. D.A. Spears And R. Kanaris-Sotiriou, Sedimentology, 26:407-425, (1979).
12. S.L.F. Matos, Instituto de Geociências, Universidade de São Paulo, USP, Brazil. 158p. Master thesis (unpublished).
13. M.A.S. Basei; O. Siga Junior; K. Sato; W.M. Sproesser, Anais da Academia Brasileira de Ciências. 67(2): 221-236 (1995).
14. J.A. Winchester And P.A. Floyd, Chem. Geol., 20: 325-343 (1977).
15. M.C. Barberena; D.C. Araujo; E.L. Lavina; S.A. Azevedo, In: Congresso Brasileiro de Paleontologia, 8, Rio de Janeiro, 149-153. Sociedade Brasileira de Paleontologia(1992).
16. R.R Andreis & M.S. Japas, In: El Sistema Permico en la Republica Argentina y en la Republica Oriental del Uruguay. P. 45-64. Acedemia Nacional del Uruguay. Cordoba. Argentina (1996).
17 L.M. Calarge; A. Meunier;M.L.L. Formoso, (in preparation) (2001).
18. R. Greene-Kelly, J. Soil Sci., 4: 233-237 (1953).
19. A.M. Iñiguez-Rodriguez And P.E. Zalba, Proceedings of 10th International clay conference, Adelaide, Australia, 436-438 (1994).
20. C.R. Marconi., Proceedings Intern. Workshop of Activated Clays, 97-107 (1998).
21. D.B. Alves; A.M.P Misusaki.; L.F.G . Caddab; In: G.J. Churchman; R.W. Fitz Patrick and R.A Eggleton (ed.), Proceedings of 10th International Clay Conference, Adelaide, Australia (1995a).
22. D.B. Alves. Congresso Brasileiro de Geologia, Niterói (1995b)
22. C.E. Weave, in: Clays, Muds and Shales. Developments in sedimentology, 44: 372-382, 819p (1999).
23. M. Cazzulo-Klepzig; M. Guerra-Sommer; M.L.L. Formoso; L.M. Calarge; Journal of South American Earth Sciencs, in press (2001).

Clay mineralogy, illite crystallinity and polytypes in the Campana Mahuida Porphyry Cu Deposit, Neuquén, Argentina.

A. Impiccini[a], M. Franchini[a-b], G. H. Grathoff[c], A. Schalamuk[b-d] and L. Meinert[e]

[a]Universidad Nacional del Comahue, Facultad de Ingeniería, CIMAR. Buenos aires 1400, 8300, Neuquén, Argentina. aimpicc@uncoma.edu.ar
[b]CONICET – Consejo Nacional de Investigaciones Científicas y Técnicas, Argentina.
[c]Department of Geology, Portland State University, P.O. Box 751, Portland, OR, 97207-0751 USA
email: GrathoffG@pdx.edu
[d]INREMI, Facultad de Ciencias Naturales, Universidad Nacional de La Plata.
[e]Department of Geology, Washington State University, Pullman, Washington 99164-2812. USA

Clay mineralogy, illite crystallinity, and polytypes were determined for 67 whole-rock samples from 24 drill cores of the Campana Mahuida porphyry copper deposit to develop a general exploration model for porphyry copper deposits. In the potassic core, between 10 m and 228 m depth, Mg-rich chlorite is the dominant clay-fraction mineral followed by illite ± smectite or smectite ± illite. Illite crystallinity indices decrease from 0.3 to 0.1° 2θ with depth, and correspond to the high temperature, $2M_1$ polytype. The chlorite-rich zone may have resulted from the break down of secondary biotite at intermediate values of a(K+)/a(H+) as the system cooled. Clay-fraction minerals in the outermost propylitic halo are similar to the potassic zone except that chlorite is more abundant and consists of both Fe-rich and Mg-rich varieties.

In the phyllic halo that surrounds the potassic core, clay abundance between 5 and 149 m depth outlines two zones: 1- an illite-rich zone ($2M_1$ polytype) to the south, northeast-southwest and east of the potassic core, and 2- a less well-developed smectite-rich zone in the northwest section of the phyllic halo. Within the smectite (beidellite) zone, chlorite accompanies smectite in drill-holes near the potassic core, whereas farther northwest, illite is present with smectite. In both zones, the illite crystallinity index decreases with depth, except in a single drill hole that intercepts a zone of argillic alteration where smectite dominates in the hypogene ore zone and illite crystallinity indices remain low and constant. This zonation within the phyllic halo may reflect differences in cation removal due to differences in faulting intensity and water: rock interaction.

1. INTRODUCTION

Conventional mapping and petrography in porphyry copper exploration may not identify fine-grained minerals or define important compositional variations within them. Fine-grained minerals are commonly classified as argillic or phyllic without determination of their actual mineralogy, and this may loose valuable information. The objective of this investigation is to examine the clay-size fraction of alteration zones at different ore types of a porphyry copper system, and evaluate if the resulting patterns can be useful in exploration. For this purpose, the clay mineralogy, illite crystallinity, and polytypes were determined for 67 whole-rock samples from 24 drill cores of the Campana Mahuida porphyry copper deposit. These results correlate well with alteration and mineralization patterns in the porphyry copper hydrothermal system.

2. LOCATION AND GEOLOGICAL SETTING

The Campana Mahuida deposit is located at Loncopué in the Neuquén province of Argentina (Fig. 1). Previous investigations include: Plan Cordillerano 1968; Falconbridge 1972; CEGM I 1974-75; CORMINE, 1990-91; RAA, 1993 and GMASA 1996, 1997). Extensive drilling since 1968 has resulted in an estimated resource of 40 Mt of 0.49 % Cu at a cuttoff grade of 0.15 percent Cu equiv. (1). The porphyry copper deposit is hosted by fluvial sediments (fine-grained sandstone with siltstone and claystone intercalations) from the Tordillo Formation (upper Jurassic) and the El Sillero dioritic porphyry stock (upper Cretaceous) that intruded into a NNE trending structural zone.

Hydrothermal alteration is characterized by a potassic core and a phyllic halo surrounded by a large propylitic zone. Argillic alteration patches occur within the phyllic zone (Fig. 1). The main hypogene copper mineral is chalcopyrite that is spatially related to the potassic alteration zone. Supergene alteration resulted in a vertical zonation of: leached capping, oxide copper, supergene sulfide copper and hypogene sulfide copper zones. The thickness of the leached zone increases towards the edges. The oxide zone occupies the core and pinches out towards the periphery and SW of the core. The supergene zone forms a halo around a core free of secondary sulfides and extends outside the oxide zone, to the S and SW. Finally, the hypogene mineralization with more than 0.25% Cu is concentrated in the deeper parts of the potassic alteration zone (1).

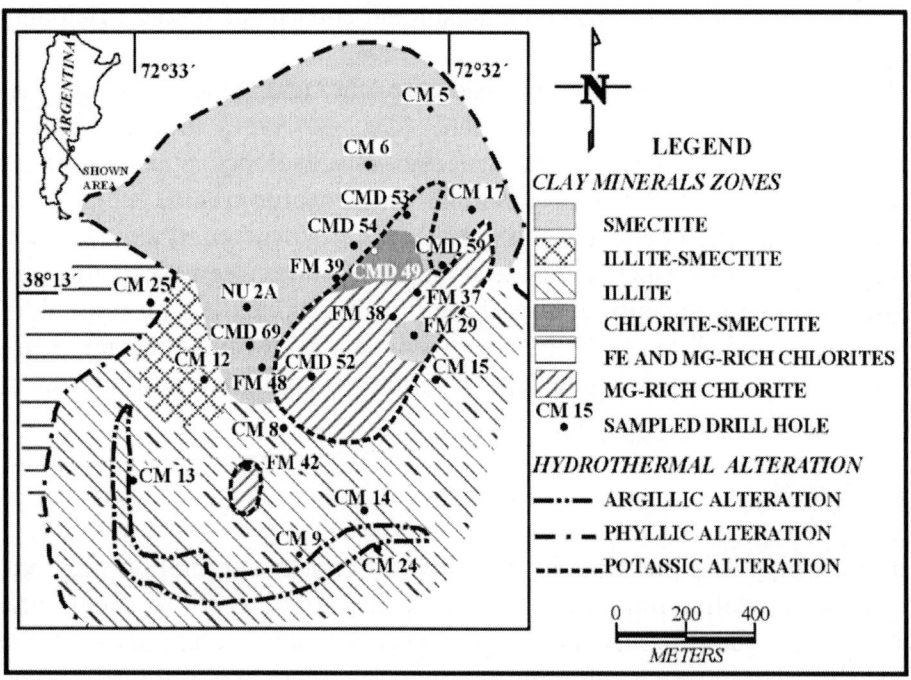

Fig. 1. Location map of the Campana Mahuida porphyry Cu deposit with the sampled drill holes, alteration zones (after Chabert and Zanettini, 1999) and clay mineral zones.

3. MATERIAL AND METHODS

Sixty seven samples of the hydrothermally altered Tordillo sediment and the El Sillero andesite were collected from 24 drill cores owned by CORMINE (Neuquén Mining Corporation). Fig. 1 shows drill hole locations within the porphyry Cu system that range from the potassic core to the propyllitic zone, extending more than 500 m outboard of the potassic core.

Whole-rock samples were crushed to a powder and dispersed in 30g/l distilled water with added Na-hexametaphosphate and then mixed for 5 minutes with an ultrasonic probe. The <2 µm fraction was separated by timed centrifugation. Oriented specimens for subsequent X-Ray diffraction analysis were then analyzed after drying, glycolation, and heating to 375°C and 550°C. For smectite and chlorite differentiation, disaggregated samples were saturated with Li^+ (2) and K^+ respectively.

X-ray diffraction. The clay minerals were identified using a Rigaku DMAX-2D diffractometer at the Centro de Investigaciones en Minerales Arcillosos de la Universidad Nacional del Comahue. Air-dried and glycol-solvated clay mineral separates (< 2µm) were analyzed from 2° to 40° 2θ, at 2° 2θ /min, with Cu-Kα radiation. The XRD reflections were evaluated with the Rigaku software, NEWMOD© (3) was used for the quantitative estimation of each clay mineral.

The illite crystallinity index was determined by measuring the width of the 001 illite peak at half its height above background, measured in degrees 2θ on tracing of the <2 µm fraction, using the method of (4). For polytype determination, the <1µm size fraction was analyzed using a Philips X'pert™ MPD X-ray diffractometer (XRD). The samples were top packed due to the small amount of sample. The powders were run in the step scanning mode (0.02 degrees per step) from 16-44 degrees 2-THETA and a long count time (20 seconds). The experimental patterns were afterwards compared to calculated XRD patterns using the computer model WILDFIRE© (5, 6).

4. RESULTS

4.1. Clay Mineralogy

X-ray diffractograms of the 67 samples indicate the presence of the following micaceous minerals: smectite, chlorite, illite and kaolinite. Based on their distribution and abundance, clay minerals were grouped in 4 well-defined zones and 2 mixed zones. The latter contain similar quantities of the dominant clay mineral in the proximal zones. Smectite, chlorite and illite are the most abundant clay-size fractions; kaolinite occurs only in small amounts in the illite zone or in traces in 2 drill cores from the smectite and chlorite zones.

The cluster of drill holes CMD 52, FM 38, FM 37 and CMD 49, between 10 and 228 m depth intercept a zone where Mg-rich chlorite is the dominant clay-fraction mineral followed by illite ± smectite or smectite ± illite. Mg-rich chlorite is also dominant up to 60 m depth in drill hole FM 42, located to the SW (Fig. 1); downwards illite becomes dominant followed by chlorite ± smectite (Fig. 3). Clay-fraction minerals in samples from drill core CM 25, from the deposit outermost halo, are similar to the Mg-rich chlorite zone except that chlorite is more abundant and consists of both Fe-rich and Mg-rich varieties.

A smectite-rich zone forms a halo surrounding the Mg-chlorite zone in the NW and NE quadrangle. It was intercepted by drill holes CM 5, CM 6, CM 17, CMD 53, CMD 59, CMD

64, FM 39, NU2A, CMD 69 and FM 48, between 5 and 149 m depth (Fig. 1). Smectite is the predominant clay size fraction with minor amounts of chlorite ± illite. Smectite was distinguished by the 001 peak shift from 12.1-13 Å in air-dried samples to 16.3-17.0 Å after ethylene glycol saturation. After treating sample with Li^+ the 001 basal reflections of CM 5-30m and CM 13-69m expanded to 17.7 and 17.8 Å on glycerol saturation indicating the presence of beidellite.

To the south and southeast of the smectite zone, illite becomes dominant followed by subordinate amounts of smectite, kaolinite and chlorite. This illite-rich zone was intercepted by drill holes CM 13, CM 8, CM 14, CM 24 and CM 15, between 10 and 131 m depth (Fig. 1). Illite was distinguished in air dried, glycolated and calcinated samples by the sharp reflection of the 001 peak at 10 Å, and by the successive reflections of 2nd, 3rd, and 4th order peaks. Exceptions are samples CM 14-100 m and CM13-50m -69 m, where chlorite and smectite, respectively, are more abundant than illite.

Samples from drillholes CMD 49 and CM 12 have similar proportions of the dominant clay minerals from the contiguous zones, thus, defining two mixed zones of chlorite -smectite and illite-smectite, respectively (Fig. 1).

4.2. Illite crystallinity and polytypes

Illite is present in most analyzed samples and was the dominant clay mineral in four samples. Illite with minor chlorite were also determined as the only clay-size fraction minerals in the fresh sandstone outcrops of the Tordillo formation outside the district. Illite crystallinity indices of the <2 μm fraction are shown with corresponding elevation and ore type in the schematic S-N and E-W sections of Fig. 3 A and B. The application of the illite

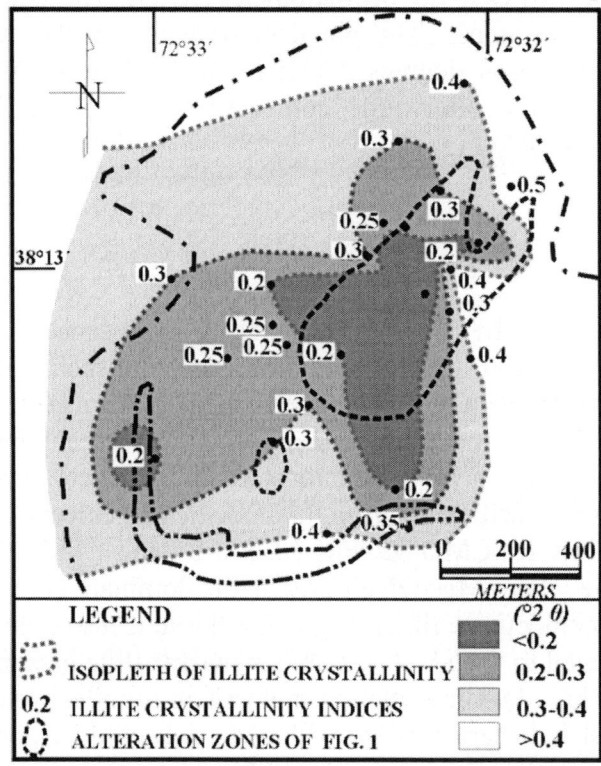

Fig. 2. Isopleth map showing projection of illite crystallinity indices variations in samples from 10 and 30 m depht.

crystallinity index is based on the assumption that temperature and fluid composition (K available) are the most important variables affecting the distribution of illite crystallinity indices, considering the relatively restricted lithological variations of the material sampled (6).

The illite crystallinity index in the illite fraction of the Tordillo sandstone is 0.7° 2θ. Within the deposit, in the samples that contain illite (63 samples) the indices show notable variation from 0.1° to 0.5° 2θ. The highest values (0.3 to 0.5 2θ; 11 values) were obtained from samples collected between 5 to 30 m depth in the leached capping and oxide zones of the system. Similar illite index (0.3-0.35 2θ) were measured in illite at 20 m depth in the hypogene sulfide and leached zones of drill core CM25 and CM 24, 512 m and 287 m from the potassic core, respectively. Fig. 2 is an isopleth map of illite crystallinity indices measured in samples between 5 to 30 m depth.

At different drill core elevations within the oxide copper, supergene sulfide copper and hypogene sulfide copper zones, the illite crystallinity indices drop to values of 0.3° to 0.1° 2θ, at 40 to 228 m depth. By contrast, illite crystallinity indices remain relatively low and constant at different elevations in drill core CM 13, located within a small zone characterized as argillic (Fig. 3).

Comparing the experimental with the calculated XRD patterns it was determined that the main illite polytype was the *$2M_1$* polytype, that no *$1M$* illite was present, and that only minor amounts of *$1M_d$* illite are present. The *$1M_d$* illite was interpreted from the elevated background between 23 and 33 degrees 2-THETA. Due to the small amount of fine fractions in the samples, it was difficult to obtain random orientation and quantification of *$1M_d$* illite was difficult due to preferred orientation and to the presence of smectite, chlorite, and kaolinite.

5. DISCUSSION

The main clay-size fractions from the porphyry copper deposit (67 samples) consist of smectite, chlorite and illite. Based on their abundance along the lateral and vertical porphyry copper extension, they were grouped into different clay zones. The Mg-rich chlorite zone is located in the potassic core where the hypogene sulfide mineralization occurs, both centered on the El Sillero diorite stock. This is the thermal center of the hydrothermal system. The chlorite-rich zone may have resulted from the hyrdothermal break down of secondary biotite at intermediate values of $(K^+)/a(H^+)$ as the system cooled. Argillic and phyllic alterations are superimposed on the potassic core, as indicated by the presence of smectite and illite accompanying chlorite. In the outermost propylitic halo chlorite is more abundant than in the potassic core and consists of both Fe-rich and Mg-rich varieties. This is consistent with the sample position relative to the hydrothermal system: low temperature chlorites tend to be ferrous and aluminous (7).

The smectite and illite zones are located in the phyllic halo that surrounds the potassic core (Fig. 1). Fluid inclusion evidence (8), suggests that the phyllic halo was caused by vapor phase separation from hypersaline fluids that expanded laterally and upwards. Progressive dilution and cooling may have promoted further leaching of most elements except Al and Si, resulting in the formation of beidellite and/or kaolinite. Thus, the clay mineral zonation within the phyllic halo may reflect differences in cation removal due to differences in faulting intensity or water: rock interaction.

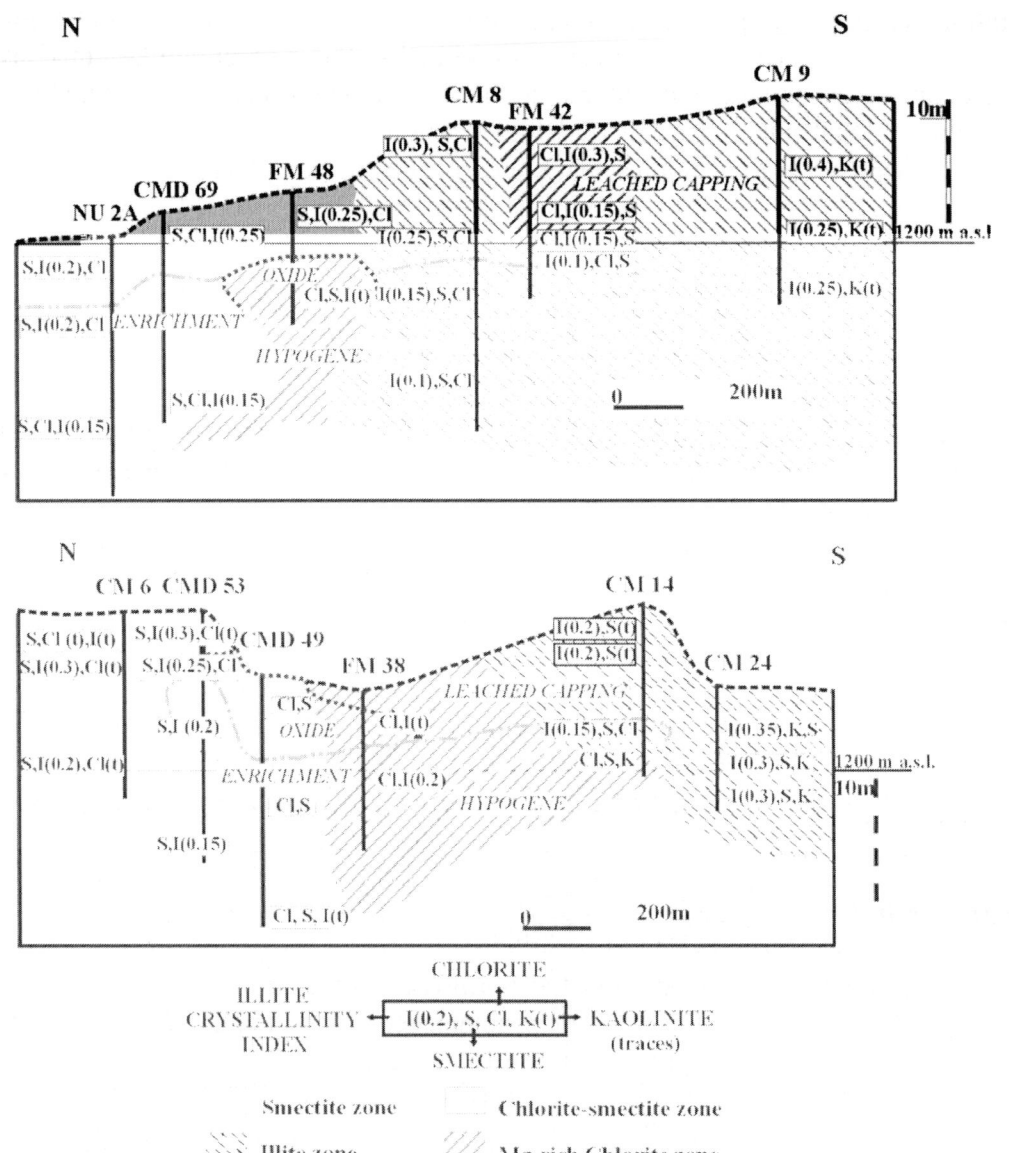

Fig. 3A. N-S cross sections through the deposit showing lateral and vertical distribution of clay minerals and illite crystallinity index.

The illite crystallinity index is another parameter that varies in the clay size fractions of fresh sandstone (0.7° 2θ), corresponding to the highest grade of diagenesis (9), and in the altered rocks in the uppermost level of the deposit, which range from 0.5 to 0.3°2θ, indicative of hydrothermal origin. Thus, crystallinity index differences between diagenetic and hydrothermal illite in areas with sedimentary rock outcrops may be useful in the porphyry copper exploration, as previously pointed out by (10).

The observed changes in the illite crystallinity within the leached capping are consistent with sample position relative to the porphyry copper hydrothermal system: the high grade illite crystallinity delineated by the 0.2° 2θ isopleth (Fig. 2) occurs in the potassic core, where the leached capping is weakly developed and higher temperature, hydrothermal alteration dominates. Fig 2 shows the extension of the lower illite index isopleth to the S and SW of the potassic core, indicating that these rocks were exposed to higher K availability, and probably higher paleotemperatures or less K lixiviation than those located to the N and NE section of the deposit.

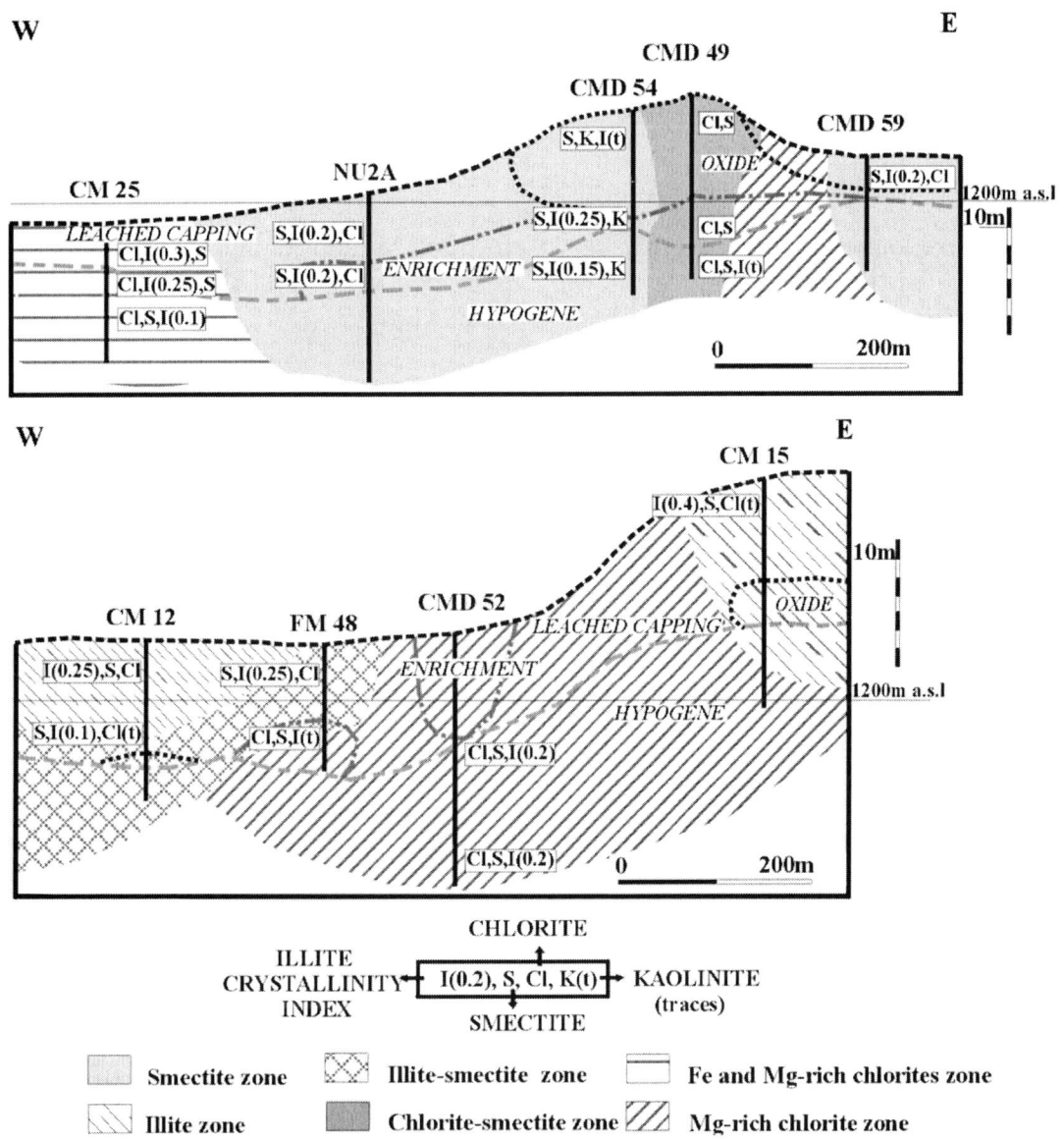

Fig. 3B. E-W cross sections through the deposit showing lateral and vertical distribution of clay minerals and illite crystallinity index.

Illite crystallinity indices also decline with depth, from the oxide copper to the hypogene sulfide copper zones and these changes are more drastic in samples from drill holes that intercept or are located nearby the potassic zone (FM 42, FM 39) than those from distal drill holes like CM14. Toward the periphery of the phyllic halo where the hypogene sulfide zone (without chalcopyrite) occurs proximal to the surface (20m; CM25), illite with intermediate crystallinity index (0.3° 2θ) is the dominant clay fraction.

6. CONCLUSION

In the Campana Mahuida deposit, variations in clay-size fractions abundance and distribution define zones that can be correlated with the alteration and mineralization zones of the porphyry copper system. Likewise, illite crystallinity varies laterally and vertically, suggesting thermal gradients and differences in K availability that also correlate with the alteration-mineralization halos in the deposit. Thus, these variations may be useful predicting altered-mineralized zones within a porphyry copper system.

AKNOWLEDGMENTS

We would like to thank Lic. Mario Chabert from Grupo Minero Aconcagua and Lic. Mario Dezza from CORMINE S.M.E.P, for allowing us to collect samples for this study and consult unpublished reports from the Campana Mahuida project.

REFERENCES

1. Chabert, M. y Zanettini, 1999. Pórfiro cuprífrero Campana Mahuida, Neuquén. En: Recursos Minerales de la República Argentina (Editor Eduardo O.Zappettini), Instituto de Geología y Recursos Minerales, SEGEMAR, Anales 35: 1279-1289. Buenos Aires.
2. Reynolds, R.C.Jr., 1985. Newmod©, a Computer Program for the Calculation of One-Dimensional Diffraction Patterns of Mixed-Layered Clays: R.C. Reynolds, Jr., 8 Brook Rd., Hanover, N.H.
3. Kubler, B.1964. Les argiles, indecateurs de metamorphisme. Institut Francais du Petrole Revue, 19: 1093-1112.
4. Reynolds, R.C., Jr. 1993. Three-dimensional x-ray powder diffraction from disordered illite: Simulation and interpretation of the diffraction patterns: Clay Minerals Society Workshop lectures, v. 5: Computer applications to x-ray powder diffraction analysis of clay minerals. p. 43-78.
5. Grathoff, G.H.; Moore, D.M. 1996. Illite Polytype Quantification using WILDFIRE* calculated XRD patterns. Clays & Clay Minerals, 44: 835-842.
6. Panno, S.; Moore, D.M. 1994, Mineralogy of the clay-sized fraction of the Davis Shale, Southwest Missouri: Alteration associated with a Mississippi Valley-type ore deposit. Economic Geology. Vol. 89, p. 333-340.
7. Velde, B., 1985. Clay Minerals. A physico-chemical explanation of their occurrence. En: Developments in Sedimentology, Elsevier, Amsterdam, 427 p.
8. Franchini, M., Curci, M., Schalamuk, A e Impiccini, A., Inclusiones fluidas en la cubierta lixiviada y oxidada del depósito Campana Mahuida, Neuquén. Implicancias en la exploración de pórfiros de Cu. VII Congreso Argentino de Geología Económica, in press.
9. Kisch, J. H. 1991. Illite crystallinity: Recommendations on sample preparation, X-ray diffraction settings and interlaboratory samples: Journal of Metamorphic Geology, v. 9p. 665-670.
10. Williams-Jones, A.E., 1986. Low-temperature metamorphism of the rocks surrounding les Mines Gaspé, Quebec. Implications for mineral exploration. Economic Geology, 81, p. 466-470.

A neoformed kaolinitic mineral in the upper Pleistocene of northeastern Argentina

M. Iriondo and D. Kröhling

CONICET – Facultad de Ingeniería y Ciencias Hídricas, Universidad Nacional del Litoral, C.C.217 (3000) Santa Fe, Argentina.

A micaceous neoformed mineral occurs in El Palmar Formation, a quaternary fluvial unit located in the Uruguay river basin. It forms macroscopic plates up to 4 mm long. X-ray analyses, microscopic observations and chemical composition indicate a mineral belonging to the Kaolinite Group. The sediment is very fine quartz sand containing up to 20% plates.

The deposit was TL dated at 80,670 ± 13,420 years Before Present. The maximal burial of the sediment is less than 10 m. The present climate of the region is humid subtropical. Several climatic changes occurred during the 80 kyr since the deposition of the sediment, but without a significant departure from subtropical environments. The soils in the region are characterized by pH 5, abundant exchangeable aluminium and amorphous silica. We propose the kaolinite has originated from the primary combination of silica and alumina.

The microscopic characteristics of the mineral are: perfect basal cleavage; elliptic general form with smooth curved fringes; surfaces frequently composed of 50-100 μm long scales which form a pseudo-rhombohedral pattern; transparent; RI low (η> Eugenol); birefringence weak; interference colour first-order gray; extinction angle <5°; scarcely visible parallel exfoliation traces; optically biaxial negative.

The X-ray diffractogram of the clay fraction shows kaolinite peaks (at *d:* 7.21 Å and 3.58 Å) and smectites and a lesser amount of quartz.

SEM observations showed the large relative size of plates in comparison with quartz grains. EDAX spectra indicate equivalent proportions of Si and Al, with significant presence of K.

Owing to the large relative sizes of the plates in comparison with the sand grains (up to 10:1) and the remoteness of potential sources of detrital plates, we deduce that the mineral is authigenic. A model of this mineral is proposed.

1. INTRODUCTION

A mineral belonging to the Kaolinite Group occurs in very large plates (up to 4 mm long) in Pleistocene quartz sands in northeastern Argentina. The deposit is located at Federación, NE of Entre Ríos province (31°00´ lat.S and 58°00´ long.W). The mineral has also been observed by the authors in the Ituzaingó Formation in Empedrado and in the Holocene Paraná delta, both in northeastern Argentina.

The mineral was erroneously cited as muscovite in former field surveys owing to the large size and general characteristics. However, a closer attention to its occurrence showed that the existence of a detrital mica is hard to explain in the context of the regional geology. Granitic rocks which can be considered as potential sources of mica are located at too long distances, hundreds of kilometers upstream in the fluvial basin.

Moreover, the quartz grains which form the bulk of the sediment are considerably smaller than most of the plates (five to ten times shorter in length) and originated in the destruction of an aeolian sandstone, a process which precluded the conservation of detrital micas. And (last, but not least) the edges of plates appear smooth and fresh, without indication of attrition or abrasion.

Hence, it was necessary to consider an authigenic origin for the unidentified mineral. And such condition excludes mica, because the host sediment lays at the surface and it has never been buried into the Lithosphere. Hence, the neoformed mineral was generated in a near-surface environment, at depth less than ten meters in an environment directly depending on the local climatic conditions. The climate in that region was humid subtropical with 19°C mean temperature and 1,500 mm/year precipitation during the Twentieth Century. The region underwent several climatic changes to warm/humid and dry/cool temperate conditions since the deposition of the host sediment [1; 2; 3], but all those changes belong in the general subtropical realm. Soils in the region have a high content of free alumina and pH of 5-5.5. In short, it is a climate wich favours the formation of kaolinite [4].

The search was then oriented to the clay minerals, in spite of the large size of the plates. The mineralogical analyses were performed by optical microscopy, X-ray diffractometry, scanning electron microscopy, microsonde and geochemical determinations. In all cases the results coherently indicate a mineral belonging to the Kaolinite Group.

The particular interest of this case is the extremely large size of the plates, and the fact that the maximal age and the environmental conditions of the mineral formation can be quite precisely identified.

2. GEOLOGICAL SETTING

The neoformed mineral is included in the El Palmar Formation [5], a sedimentary deposit generated by the Uruguay river. It is composed of mature well-sorted quartz sands coated by iron hydroxides, and scarce feldspars. The sand is composed of sub-rounded to well-rounded quartz grains, originated in the destruction of Cretaceous aeolian sandstones. Clayey sand and sandy silt, red in colour, and lenses formed by siliceous gravels and pebbles are intercalated with the sand. In some strata composed of particularly clean sand, the kaolinite occurs in the form of micaceous plates in large proportions (up to 20% of the total sediment).

The El Palmar Formation is an old flood plain of the Uruguay river, developed during the warm Oxygen Isotopic Stage 5a in the Upper Pleistocene. A thermoluminescence dating of 80,670 ± 13,420 years Before Present was obtained in the deposit; a second one sampled near Salto in Uruguay indicates a mean age of 88,000 years Before Present.

Sand strata and gravel lenses represent channel facies and fine sediments form facies of inundation. This formation, 3 to 12 m thick, lies at the surface and has not been buried since its deposition [2].

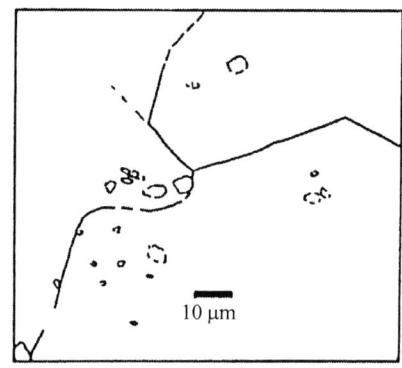

Figure 1. Drawing from a micrograph with 1000x magnification showing the 125° angles in individual layers.

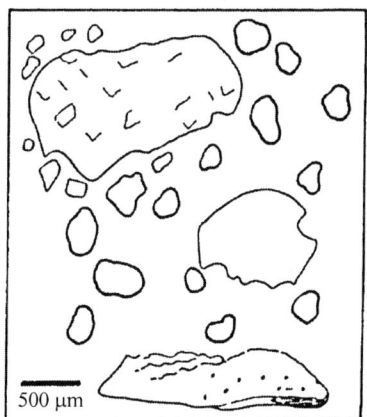

Figure 2. A composition of the kaolinite plates described under optical microscope.

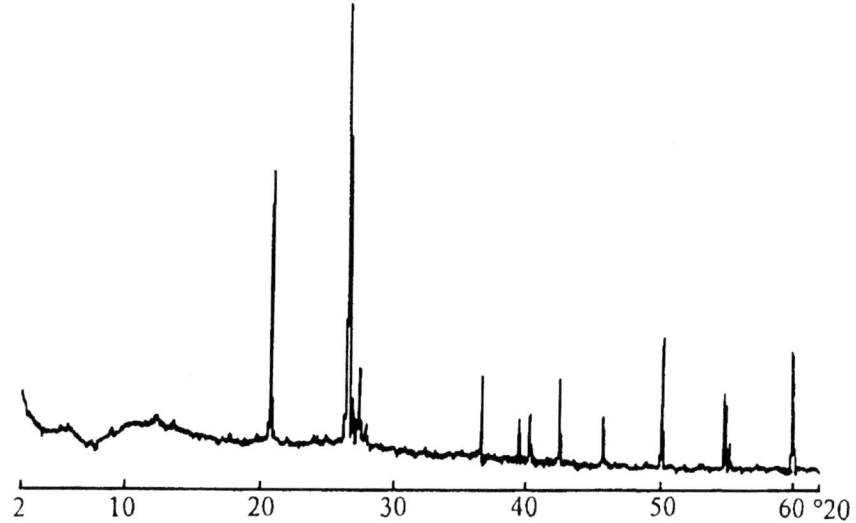

Figure 3. X-ray diffraction of the bulk sediment of El Palmar Fm.

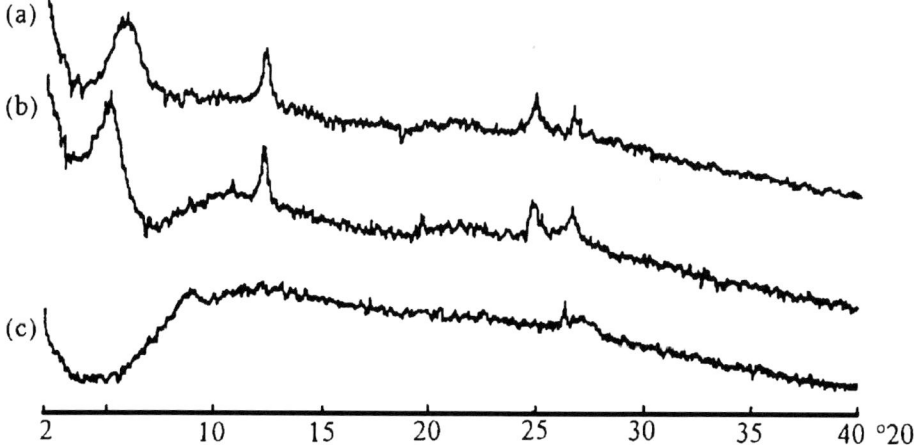

Figure 4. X-ray diffraction clay mineralogy of El Palmar Fm, indicating a kaolinitic nature. (a) air dried. (b) treated with ethylene glycol. (c) heated to 550°C.

3. INSTRUMENTS AND METHODS

The identification of the mineral species was made by:

Optical microscopy: Morphological and physical characteristics were determined under stereoscopic lens and by using orthoscopic and conoscopic methods with a polarizing microscope (Carl Zeiss Jena, model Amplival).

Scanning electron microscopy (SEM): Morphological features and photographs were performed by an electronic scanning microscope (JEOL model JSM-35C), equiped with EDAX for X-ray electron sonda (EPXMA) for determination of elemental composition.

X-ray diffractometry (XRD): X-ray patterns were obtained by using a X-ray diffractometer with CuKα-radiation and Ni filter (RIGAKU, model Dmax II-C).

Geochemical analyses (XRF): Chemical data on bulk samples were obtained by the X-ray fluorescence method in X-Ral Laboratories, Ontario (lower reported limit: 0.001%).

The geological scenario was performed through regional surveys made by the authors of the paper and published elsewhere during the last several years [6; 1; 7; 8; 3].

The different analytical methods employed in this study are not exactly equivalent, but it is clear that coherent results are necessary in this point.

4. MICROSCOPIC DESCRIPTION

The mineral occurs in transparent, very fragile plates, with typical sizes between 0.4 and 0.6 mm long and maximum length of 4 mm. They exhibit perfect basal cleavage and smooth rounded fringes. Pseudo-rhombohedral scales and growth features are present in some cases. Scales and layers have frequently angles of 125° (Figure 1). Some representative grains are described:

(a) Plate 1.8 x 1.2 mm (Figure 2) general elliptic shape with smooth rounded borders. Colourless, transparent, micaceous luster. Straight or curved 50 to 100 µm long pseudo-rhombohedral scales are conspicuous. Birefringence weak. Interference colour is first-order white. Refraction index low (η> Eugenol), extinction angle < 5° (with reference to exfoliation estriae), the elongation of the plate is approximately perpendicular to the extinction axis. This plate was found included in very fine quartz sand with a mean diameter of 100 µm, that is 18 times smaller.

(b) Plate 0.6 mm long x 0.4 mm wide (Figure 2). It represents the most frequent size in the deposit. Colourless, transparent, pearly luster. General elliptic contour, smooth fringes with large rounded irregularities. Birefringence weak. Interference colour is first-order gray. Poor interference figure (biaxic negative).

(c) Elongated plate 0.6 mm long x 0.15 mm wide (Figure 2). Similar to (a) and (b), with circular black holes, 10 to 20 µm in diameters, coalescent in some places. Such features occur in a small proportion of individuals and are probably originated in a corrosion process. No indication of transport can be noted.

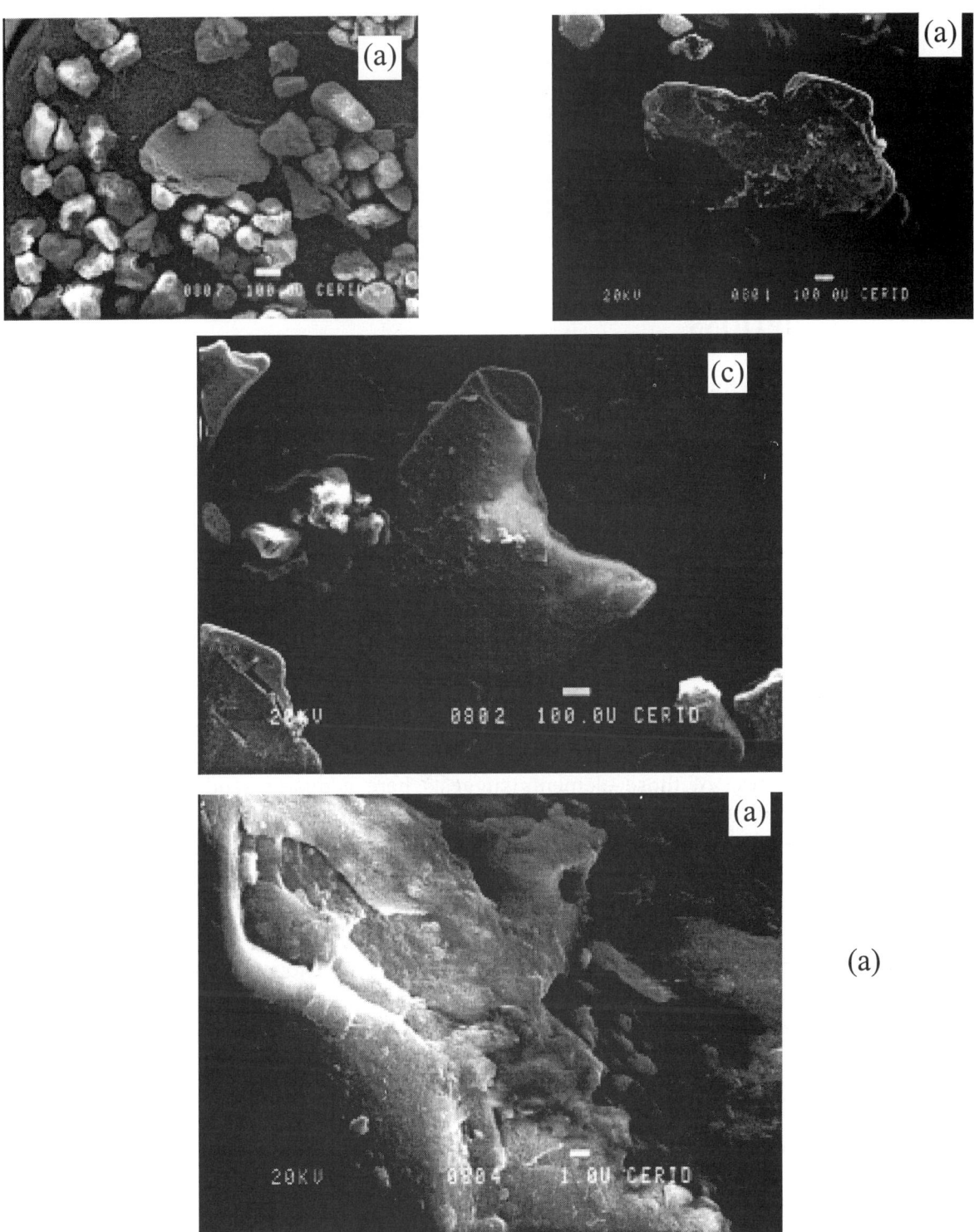

Figure 5. Electron micrographs. (a) 500 μm long plate inside very fine-fine sand. General elliptic form with perfect basal cleavage and smooth fresh fringes. (b) and (c) Very large 1.5 and 1.2 mm long plates. Observe the smooth minor forms in the fringes, indicating that the grains have not underwent transport neither attrition. (d) Grow features and angles of 125° are shown in a magnified sector (4400x) of the concave border of the plate in (c).

5. X-RAY AND LOOSE-GRAIN ANALYSES

The X-ray diffractogram of the bulk sediment shows almost pure quartz, with scarce feldspars and poorly defined clay peaks (powder sample; Figure 3). The scarce clay fraction (oriented sample) is composed of kaolinite (at d: 7.21 Å and 3.58 Å) and smectites, reduced amount of quartz. Kaolinite peaks follow as usual the classical treatments: no change at glycolation and disappear at 550°C (Figure 4).

The loose-grain count technique provided the following results: 74.4% monocrystalline quartz, 19% kaolinite, 3.2% policrystalline quartz, 2.7% feldspars and 0.7% alterites.

6. SEM DESCRIPTIONS

Observations made by SEM at different magnifications confirm and reinforce the optical description. The perfect basal cleavage and the smooth fringes of kaolinite plates, besides minor features such as neat angles of 125° and growth features appear very clearly (Figure 5).

7. MICROSONDA AND GEOCHEMISTRY

The microsonda analysis shows that (for elements heavier than Neon) the mineral is composed of similar amounts of Si and Al, with smaller but significant presence of K. Medium values of four samples resulted in: 45.3% Si, 38.7% Al and 15.4% K. Minor percentages of Fe and Mg appear in some samples (Figure 6).

The geochemical analysis of bulk samples is coherent with the results obtained by other methods: 87.3% SiO_2, 4.49% Al_2O_3, 0.40% CaO, 0.50% MgO, 0.19% Na_2O, 0.96% K_2O, 3.34% Fe_2O_3, 0.02% MnO, 0.707% TiO_2, 0.02% P_2O_5, 0.07% Cr_2O_3, 1.60% LOI.

A modal calculation of those elements suggests here a quartz sand with a significant percentage of kaolinite and iron coats. A direct calculation of the former analyses results in 13.9% kaolinite, a value which is coherent with the other analytical approaches.

8. PROPOSED MODEL

If one normalizes the composition data, a persistent proportion of about 15% potassium occurs in the different analyses of the mineral. It allows us to propose the crystalline structure of Figure 7, which slightly differs from the general kaolinite composition, sensu [9] and [10] for this particular mineral. The molecular weight in this case is 267, of which 39 ($\approx$15%) belongs to potassium.

9. CONCLUSIONS

The following conclusions result from the referred analyses:
(a) A neoformed mineral of the Kaolinite Group occurs in the El Palmar Formation.
(b) The mineral occurs in extremely large plates, with mean sizes around 0.5 mm, and contains 15% potassium.
(c) The mineral was generated in a surficial environment under subtropical climates during a maximum interval of 85,000 years.
(d) This kaolinite was generated by direct combination of alumina and silica.
(e) There is a reasonable possibility that this finding corresponds to a new mineral species.

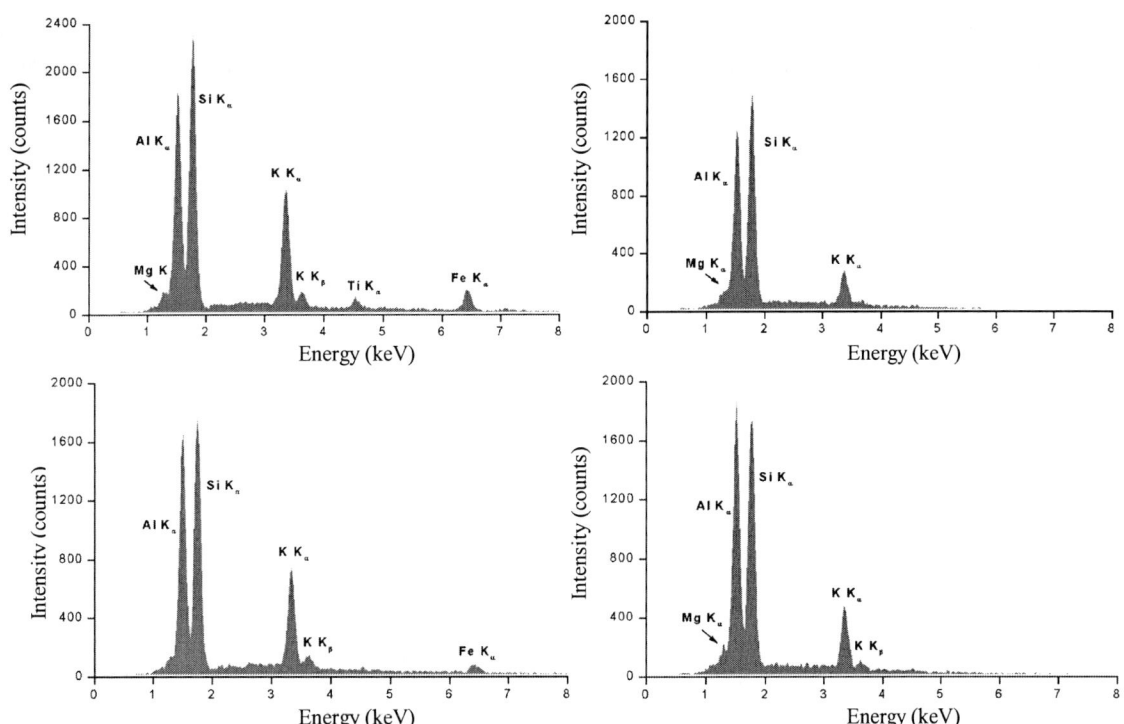

Figure 6. Elemental composition of representative samples. Note the persistence of values of Si, Al and K. (a) corresponds to the kaolinite in figure 5a. (b) corresponds to the kaolinite in figure 5b. (c) analysis of the particle in figure 1. (d) another representative sample.

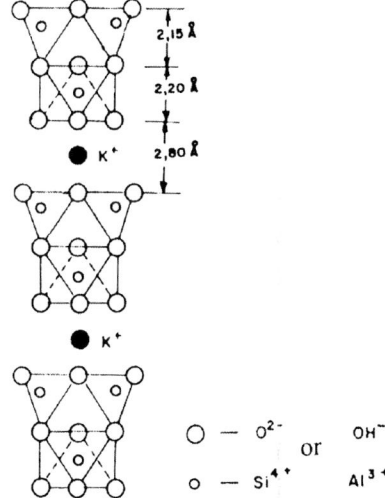

Figure 7. Proposed crystalline structure of the molecule. Note the regular intercalation of the K atoms (adapted from Souza Santos, 1975).

REFERENCES

1. M. Iriondo and N. García, Climatic variations in the Argentine plains during the last 18,000 years, Palaeogeography, Paleoclimatology, Palaeoecology, 101(1993) 209-220, Elsevier Sc. Publ. B.V, Amsterdam.

2. M. Iriondo and M. Santi, La Formación Salto Chico en el subsuelo de Entre Ríos, II Congreso Latinoamericano de Sedimentología, AAS, Abstracts (2000) 91. Mar del Plata.

3. M. Iriondo and D. Kröhling, Cambios ambientales en la cuenca del Uruguay (desde el Presente hasta dos millones de años atrás), Colección Ciencia y Técnica, Universidad Nacional del Litoral, Santa Fe, in press.

4. R. Hardy and M. Tucker, X-ray powder diffraction of sediments, In: M. Tucker (ed.): Techniques in Sedimentology, 191- 228 (1988), Blackwell Science Ltd, Oxford.

5. M. Iriondo, El Cuaternario de Entre Ríos, Revista de la Asociación de Ciencias Naturales del Litoral, 11(1980) 125-144, Santa Fe.

6. M. Iriondo, The Quaternary of Northeastern Argentina, In: J. Rabassa (ed.): Quaternary of South America and Antarctic Penninsula, 2 (1984) 51-78, A.A. Balkema, Rotterdam.

7. M. Iriondo and D. Kröhling, Los sedimentos eólicos del noreste de la llanura pampeana (Cuaternario superior), XIII Congreso Geológico Argentino, Actas IV (1996) 27-48, Buenos Aires.

8. D. Kröhling, Upper Quaternary of the Lower Carcarañá Basin, North Pampa, Argentina, In: T. Partridge; P. Kershaw and M. Iriondo (eds.): Paleoclimates of the Southern Hemisphere, Quaternary International, 57/58 (1999) 135 a 148, Pergamon Press, Oxford.

9. L. Pauling, The structure of micas and related minerals, Proceedings of the Natural Academy of Sciences, 16(1930) 123.

10. P. Souza Santos, Tecnologia de argilas, Vol 1. Fundamentos, Editora Edgard Blücher Ltda. 340 pp, São Paulo, 1975.

Surface microtopography of pyrophyllite from different modes of occurrence

Mayumi Jige [a], Ryuji Kitagawa [a],
Victor Zaykov [b], Irina Sinyakovskaya [b], Varely Udachin [b]
and Jin-Yeon Hwang [c]

[a] Earth and Planetary Systems Science, Graduate School of Science, Hiroshima University, Kagamiyama, Hisgashihiroshima, 739-8526 Japan

[b] Institute of Mineralogy, Urals Branch Academy of Sciences, Miass, Chelyabinsk District 456301, Russian federation

[c] Department of Geology, College of National Sciences, Pusan National University, Pusan 609-735, Korea

The surface microtopography of pyrophyllite crystals from various localities has been investigated using transmission electron microscopy (TEM) and the Au-decoration technique. The pyrophyllite specimens investigated were collected from the southern Urals (Russia), California (USA), the southwestern part of the Korean Peninsula, and from southwest Japan. Pyrophyllite deposits in the southern Urals are related to the western and eastern Devonian island arcs, and are associated with massive sulfide ores. The structure of the ore field is complicated by well-developed schistosity produced by late Paleozoic dynamometamorphism. In contrast, massive pyrophyllite ores in the southwestern part of the Korean Peninsula and in southwest Japan occur in hydrothermal metasomatites of Cretaceous acidic rocks such as rhyolites and dacites. Two specimens occurring in veins were also collected from the southern Urals and California.

Gold decoration successfully revealed growth steps ~1nm in height on the crystal surfaces. Pyrophyllite specimens from schistose and massive ores exhibit characteristic growth patterns with narrowly separated parallel steps, whereas polygonal spiral growth patterns with wider separation were found on specimens from veins. The morphology of pyrophyllite crystals seems to reflect the growth processes under different conditions (temperature and/or supersuturation), e.g., parallel steps on the irregular plates, and polygonal spiral steps on polygonal plates.

1. INTRODUCTION

The microtopography of surfaces of layered silicates such as kaolinite, mica, chlorite and rectolite have been studied by many investigators to elucidate crystal growth mechanisms (Baronnet, 1972, 1980, Sunagawa and Koshino, 1975, Tomura et al., 1979 and Kitagawa, 1997; 1998). Growth-induced step-patterns a few nanometers in height are revealed by the gold-decoration technique employed with transmission electron microscopy (TEM), as developed by Bassett (1958) and Gritsaenko and Samotoyin (1966). Such examination has revealed spiral and parallel-step patterns on the (001) crystal surfaces of various layer silicates, so that their growth mechanisms and conditions of growth could be inferred (Sunagawa and Koshino, 1975, Tomura et al., 1979).

Using the Au-decoration technique, Sunagawa and Koshino (1975) and Tomura et al. (1979) demonstrated that: 1) coalescence of crystals is common in hydrothermal solutions; 2) in metasomatic environments, crystals grow in a malformed or circular spiral pattern, 3) in regional metamorphic environments, Ostwald ripening plays an essential role in the growth of layered silicates, and the spiral-growth mechanism is absent. These results demonstrate that observation of surface microtopography provides important information on the genesis of layer silicates. The microtopography of pyrophyllite crystals has not been described to date. We have applied the decoration technique to the (001) surfaces of pyrophyllite crystals collected from genetically different localities in the southern Urals (Russia), California (USA), the southwestern part of the Korean Peninsula and in southwest Japan. We here describe the results of these observations, and interpret them in relation to the growth environment of the samples.

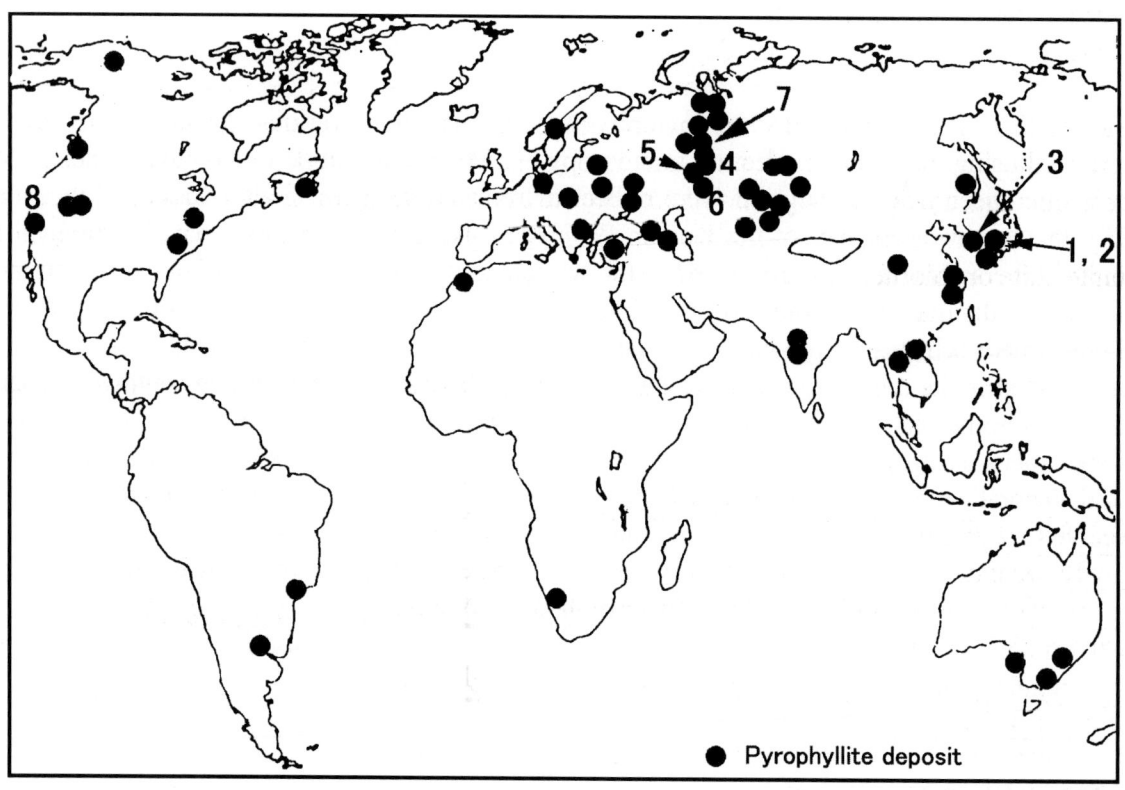

Fig.1 Location of pyrophyllite deposits and sampling points.
1. Shokozan (Japan), 2. Uku (Japan), 3. Gussi (Korean Peninsula),
4. Chistogor (Southern Urals), 5. Kul-Yurt-Tau (Southern Urals), 6. Gay (Southern Urals),
7. Berezovsk (Southern Urals), 8. California (USA).

2. MATERIALS AND METHODS

2.1. Specimens

Many sulfide deposits occur in the southern Urals, including porphyry-copper (Grabezhev and Borovikov, 1993), Kuroko-type and vein-type deposits. Pyrophyllite ores in the Urals are formed in hydrothermal metasomatic zones in Devonian palaeo-island arc rhyolites or dacites in association with massive sulfide ores, and in veins in gold deposits (Zaykov and Udachin, 1994). Pyrophyllite specimens were collected from the alteration zones of three sulfide deposits in the Urals, i.e. the Gay, Kul-Yurt-Tau and Chistogor deposits. The Gay massive sulfide deposit is the largest in the Urals. The schistose pyrophyllite specimen investigated in this study was collected from the central part of the ore field. In the Kul-Yurt-Tau deposit, the structure of the ore field is complicated by a well-developed schistosity which was produced by Late Palaeozoic dynamometamorphism (Zaykov and Udachin, 1994). An additional specimen was collected from a different occurrence in the Urals, from the hydrothermal veins of the Berezovsk gold deposit near Ekaterinburg (Fig. 1). In the Berezovsk deposit, pyrophyllite occurs as aggregates of pale-green radiating clusters (1~2cm in diameter) within veins. The pyrophyllite specimen from the deposit in California also occurs as light-brown radiating clusters within veins (Fig. 1).

Numerous massive pyrophyllite deposits occur in Cretaceous rhyolitic and dacitic rocks in southwest Japan and in the southern part of the Korean Peninsula (Kitagawa et al., 1988, Kim and Nagao, 1992). Pyrophyllite specimens were collected from the Shokozan deposit in southwest Japan and the Gussi deposit in the southwestern Korean Peninsula (Fig. 1). The general modes of occurrence of the pyrophyllite specimens are summarized in Table 1, along with the assemblages of associated clay minerals. As shown in Table 1, some specimens contain a small amount of kaolinite, illite or chlorite.

Table 1 Mode of occurrences of pyropyllite with associated clay mineral.

Specimen	Clay mineral	Color	Occurrence
Southwest Japan			
Shokozan	Pyrophyllite	green, white	massive
Uku	Pyrophyllite>>Kaolinite	white	massive
Korean Peninsula			
Gussi	Pyrophyllite	light green	massive
Southern Urals			
Chistogor	Pyrophyllite>>>Illite	gray	schistose
Kul-Yurt-Tau	Pyrophyllite>>>Chlorite	gray	schistose
Gay	Pyrophyllite	light green	schistose
Berezovsk	Pyrophyllite>>>Chlorite	pale green	vein (radiating aggregate)
USA			
California	Pyrophyllite	light brown	vein (radiating aggregate)

2.2. Observation Method

The decoration technique of electron microscopy was applied to observe the surface microtopography. In the Au-decoration technique, gold is flash-evaporated in a vacuum-evaporation apparatus onto the crystal surfaces. Minute grains of gold preferentially nucleate along the steps on the surfaces, thus revealing the surface microtopography of the crystal faces. Steps the height of a unit cell or less can be clearly revealed by this technique with electron microscopy, if the coating is correctly applied. The Au-decoration technique applied in this study followed that by Kitagawa (1997 and 1998), as outlined below.

The specimens were dispersed in distilled water and collected on a thin cover glass. After drying, these mounts were heated at 400~500°C in a vacuum of 10^{-4} torr for 2-3 hours. Heating the specimens gives clear surfaces lacking impurities and greater gold mobility, enabling selective nucleation of Au along the growth steps (Basset, 1958, Tomura et al., 1979). Gold was flash-evaporated from a tungsten coil heater. A carbon coat was then applied. Following this, the specimens were immersed in a solution of ~10% HF for 8-10 days (occasionally one week), to completely dissolve the specimens. After soaking the remains in distilled water, the thin carbon films and their gold grains were collected on copper mesh grids for observation with a TEM (JEM2000ES, 120kV).

3. RESULTS

The electron micrographs in Figs. 2 and 3 show that the grains of gold preferentially nucleated along steps on the crystal surfaces. However, these are not cleavage patterns (Sato, 1970, Kitagawa, 1997). The single steps are assumed to be the height of a unit-cell layer, based on the shadowing technique.

Single and paired parallel step patterns with narrow step separations are characteristically observed on the pyrophyllite crystals from massive and schistose-textured specimens associated with sulfide ores. In contrast, pyrophyllite from veins in the Berezovsk and Californian deposits shows polygonal spiral growth patterns with wider step separation, as shown in Table 2. However, circular spiral growth patterns, which are generally observed on illite, kaolinite and chlorite (Kitagawa, 1998), are not found on these pyrophyllite crystals.

Pyrophyllite is generally formed under the higher temperature than kaolinite and mica (Sverjensky et al. 1991). According to Sunagawa (1982), spiral growth mechanism is controlled under lower temperature and/or lower supersaturation conditions. On the other hand, parallel growth mechanism is assumed for the higher temperature and/or higher supersaturation. Therefore, these observations indicate that pyrophyllite in the massive and schistose-textured ores grew under higher hydrothermal temperatures and/or more supersaturated solution conditions, compared with other clay minerals such as kaolinite, illite and chlorite. The growth mechanism of pyrophyllite formed in veins may be controlled by much slower growth rate, compared with the schistose and massive ores. This slower growth rate allows the crystals in the veins to reach a greater size.

X-ray powder analysis of pyrophyllite shows that both monoclinic (2M) and triclinic (1T) polytypes occur. Single step and paired steps are commonly observed on the pyrophyllite crystals (Figs. 2 and 3). When a structure contains zigzag stacking, one can expect interlaced patterns or paired steps to appear by a spiral-growth process, as has been demonstrated with 2M illite (sericite) by Sunagawa and Koshino (1975) and Sunagawa (1984). Interlaced patterns and paired steps will

Table 2 Relationship between mode of occurrence and growth pattern.

Specimen	Morphology	Step pattern	Step separation	Growth mechanism
Southwest Japan				
Shokozan	irregular	malformed	5-50nm	parallel
Uku	irregular	malformed	10-50nm	parallel
Korean Peninsula				
Gussi	irregular	malformed	10-30nm	parallel
Southern Urals				
Chistogor	irregular	malformed	10-50nm	parallel
Kul-Yurt-Tau	irregular	malformed	20-50nm	parallel
Gay	irregular	malformed	8-40nm	parallel
Berezovsk	polygonal	polygonal	100-120nm	spiral
USA				
California	polygonal	polygonal	100-125nm	spiral

appear depending upon the symmetries of the elemental sheet, their mode of stacking, and the number of stacked sheets, owing to the difference in rate of advance in the same direction among the successive elemental sheets. Therefore, paired and single steps are inferred to correspond to 2M and 1T polytypes, respectively.

Morphologically, pyrophyllite crystals in massive and schistose specimens form irregular plates. However, crystals in veins occur as polygonal plates. Parallel steps are observed on the surfaces of pyrophyllite crystals from massive and schistose specimens, whereas polygonal spiral growth patterns are found on crystals from veins. Therefore, the morphology of the crystals reflects the growth mechanisms.

ACKNOWLEDGMENTS

We are grateful to Professor B. Roser, Professor A. E. Lalonde and Dr. J. B. Percival for valuable comments on a preliminary version.

REFERENCES

1. A. Baronnet, Am. Mineral., 57 (1972) 1272-1293.
2. A. Baronnet, In Current Topics in Material Sci., 5 (E. Kaldis, ed.), North-Holland, Amsterdam, (1980) 447-548.
3. I. Sunagawa and Y. Koshino, Am. Mineral., 60 (1975) 407-412.
4. S. Tomura, M. Kitamura and I. Sunagawa, Phys. Chem. Minerals., 5 (1979) 65-81.
5. R. Kitagawa, Clay Mineral., 32 (1997) 89-95.
6. R. Kitagawa, Canadian Mineral., 36 (1998) 1559-1567.
7. G. A. Bassett, Phil. Mag., 3 (1958) 1042-1045.
8. G. Gritsaenko and N. Samotoyin, Proc. Int. Clay Conf. (Jerusalem), 3 (1966) 91-400.
9. A. Grabezhev and Y. Borovikov, Resource Geol. Spec. Issue, No.15 (1993) 275-284.

10. V. V. Zaykov and V. N. Udachin, Applied Clay Sci., 8 (1994) 417-435.
11. R. Kitagawa, H. Nishido, and S.Takeno, Mining Geol., 38 (1988) 357-366.
12. I. J. Kim and K. Nagao, J. Petrol. Soc. Korea, 1 (1992) 58-70.
13. H. Sato, J. Japan, Assoc. Mineral. Petrol. Econ. Geol., 64 (1970) 192-198.
14. R. Kitagawa, Jour. Mineralogical Soc. Japan, 26 (1997) 215-220.
15. D. A. Sverjensky, J. J. Hemley and W. M. D'angelo, Geochim. Cosmochim. Acta., 55 (1991) 989-1004.
16. I. Sunagawa, Estudios geol., 38 (1982) 127-134.
17. I. Sunagawa, In Material Science of Earth's Interior (I. Sunagawa, ed.). Terra scientific publishing co., Tokyo, Japan, (1984) 63-105.

Geochemistry of hydrothermal kaolins in the SE area of Los Menucos, Province of Río Negro, Argentina

P. J. Maiza[1,2]; D. Pieroni[1] and S. A. Marfil[1,3]

1. Dpto. de Geología. Universidad Nacional del Sur. San Juan 670. 8000 Bahía Blanca. Argentina. Te: 054-0291-4595184. FAX: 054-02914595148. email: smarfil@criba.edu.ar
2. Principal Researcher at CONICET. Professor of Universidad Nacional del Sur.
3. Adjunct Researcher at CIC. Professor of Universidad Nacional del Sur.

Rhyolitic tuffs of the Sierra Colorada Fm (Triassic – Middle Jurassic) accommodate remarkable argillaceous mineralization with a zonation characterized by strong silicification at the upper levels, kaolinite, dickite and alunite at the intermediate levels, and sericitization at the bottom, where relicts of the original texture of the rock can be recognized.

The alteration process has wholly obliterated the original texture. Relicts of embayed quartz, phantoms of mafic minerals, defined by the precipitation of iron oxides, and feldspar pseudomorphically replaced by almost pure kaolinite and dickite with crystals that can be up to 70 μm in size, can be recognized.

Ore samples taken from a quarry about 15 meters in width were analyzed by XRD, TG-DTA, IR, SEM, EDAX and ICP. These methods allowed the identification of dickite, kaolinite and alunite occurrence.

Rare earth elements (REE) were analyzed showing a marked Eu negative anomaly, but no Ce anomaly, which would be typical of sedimentary deposits.

There is a close spatial relationship between kaolinite and dickite crystallization, which is concentrated at the intermediate levels. Alunite distribution is being concentrated at the most permeable sections.

The presence of dickite and alunite, the textures and values of REE support a hydrothermal origin. Ore-bearing fluids reached a quite acid pH, determined by the presence of alunite and temperatures of about 250°C, which allowed dickite crystallization.

1. INTRODUCTION AND GEOLOGIC SETTING

In this work kaolin mineralization, whose petrographic and mineralogical characteristics had been described by Maiza (1972) and Maiza et al. (1975), was studied. Its aim was to support the hydrothermal origin of the deposit based on a geochemical analysis.

An assemblage of vulcanites and tuffs and scarce clastic sedimentary rocks overlie the Mesozoic basement. The volcanic complex starts with a mesosiliceous lithology (mainly

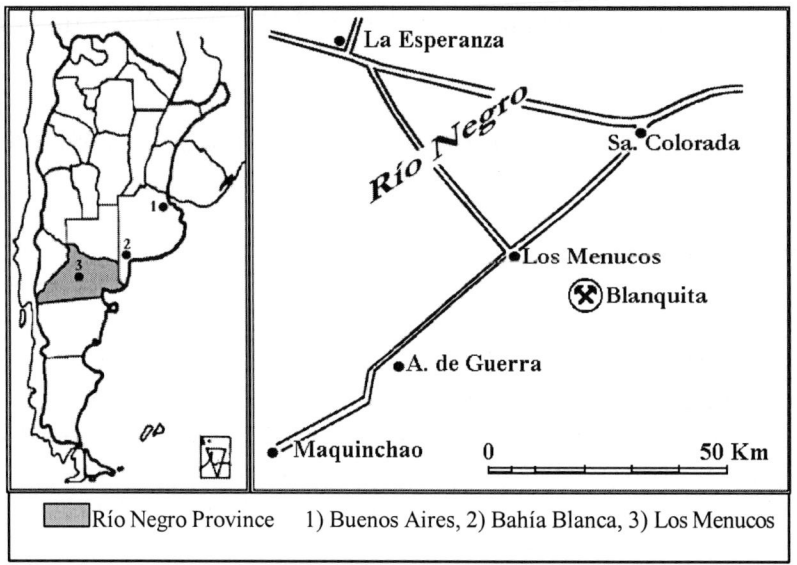

Figure 1: Location map.

andesites), defined as Los Menucos Formation. This formation is overlain by a sandy tuffaceous sequence rich in fossils, called Flora de Dicroidium Fm. It ends with a series of rhyolitic flows, ignimbrites and tuffs called Sierra Colorada Formation. This assemblage was subsequently termed Marifil Complex (Cortés 1981).

Blanquita is one of the main quarries of the so-called Blanquita - Don Sergio area. It is located at 30 km SE of the locality of Los Menucos (Province of Río Negro - Argentina). (Figure 1).

The kaolin mineralization occurs in rhyolitic tuffs from the Sierra Colorada Formation, of Triassic – Middle Jurassic age. The kaolinized zone lies between two impermeable rhyolitic flows that control the alteration process, with silicification in the upper rhyolite and weak argillization in the lower one.

The volcanic complex has undergone randomly oriented faults, although in Mina Blanquita it occurs in a predominantly NE – SW structure. This cracking developed a number of higher temperature centers in the area where kaolin enrichment occurs and are characterized by a mineralogy of dickite well crystallized kaolinite, alunite and pyrophyllyte.

In nearby regions it is possible to recognize fluorite veins, silicified structures with evidence of sulfides and dikes of andesitic-trachytic-rhyolitic lithologic composition, of uncertain age, though attributed to Late Mesozoic.

The mineralized area is about 5 to 8 km wide by approximately 20 km long and covers about 100 km^2. Furthermore, in the region of Los Menucos there are other two areas with similar characteristics but with lower temperature mineralogies. They are the so-called Adelita – Fortuna and Aguada de Guerra areas (5 km and 45 Km south of Los Menucos, respectively). In all the cases the rock that accommodates the mineralization is a rhyolitic tuff whose upper levels are silicified. These levels have protected the kaolinized areas.

Blanquita is an open pit quarry and its irregular shape indicates the presence of iron, the main impurity that has limited the exploitation of the kaolin ore.

Work was carried out at the upper part of the deposit, at a 15-meter high quarry. Marfil et al. (1999) identified dickite and alunite associated with kaolinite that had not been mentioned before.

2. MATERIALS AND METHODS

Eighteen (18) samples were collected from a quarry 15 meters in width. Vertical and horizontal sampling was performed.

Petrographic and mineralogical characteristics were studied using a petrographic microscope, X-ray diffraction, SEM, TG - DTA and IR. Determinations were performed on bulk samples on a Rigaku X-ray diffractometer D-Max III - C with Cu Kα radiation and a graphite monochromator at 35 kV and 15 mA; a scanning electron microscope JEOL JSM; a Rigaku simultaneous thermoanalyzer and a Perkin Elmer infrared spectrometer 599-B, with a scanning range between 4000 and 200 cm^{-1}. Chemical analysis of major and trace elements were performed by ICP in ACTLABS (Canada).

3. RESULTS

3.1 PETROGRAPHY AND MINERALOGY

Kaolinized bodies are emplaced within a rhyolitic tuff overlying porphyritic rhyolites with microcrystalline matrices, with abundant quartz, sanidine, plagioclase, hornblende and biotite (Figure 2a). Alteration is moderate, with sericitized and kaolinized feldspars and chloritized amphiboles and biotites. Zeolites are common in the matrices (Figure 2b). The tuffaceous level is of varying texture, from very fine-grained tuffs to conglomeratic types.

It is possible to observe strongly kaolinized zones obliterating textures where only original quartz and relict forms of tuffaceous components can be detected.

The upper level is a rhyolite with a felsic matrix, with relict volcanic glass, locally silicified with scarce phenocrysts. Cracked embayed quartz and scarce pseudomorphically replaced feldspar are identifiable (Figure 2c).

SEM observations showed large dickite crystal bundles in a well-crystallized kaolinite matrix (Figure 2d).

X-ray diffraction indicated that the samples are composed mainly of kaolin minerals (kaolinite - dickite), with varying amounts of quartz and scarce alunite. The latter was identified in the samples with higher kaolin minerals content. The richest zone is the upper-middle level of the profile.

The quarry mineralogy, determined by XRD, is shown in Figure 3. Samples 12, 13 and 14 correspond to a strongly silicified zone and sample 10 to the upper level of the sampled profile.

IR and TG -DTA were used to characterize the kaolin-group minerals. The infrared spectrum showed absorption bands at 3.698 cm^{-1}, 3.658 cm^{-1} and 3.620 cm^{-1} that are related to the H-O-H bond typical of kaolinite and dickite.

The differential thermal analysis showed an endothermic peak at 585°C related to kaolinite, with an inflection of the curve at 665°C due to the presence of dickite. The exothermic peak appears at 990°C.

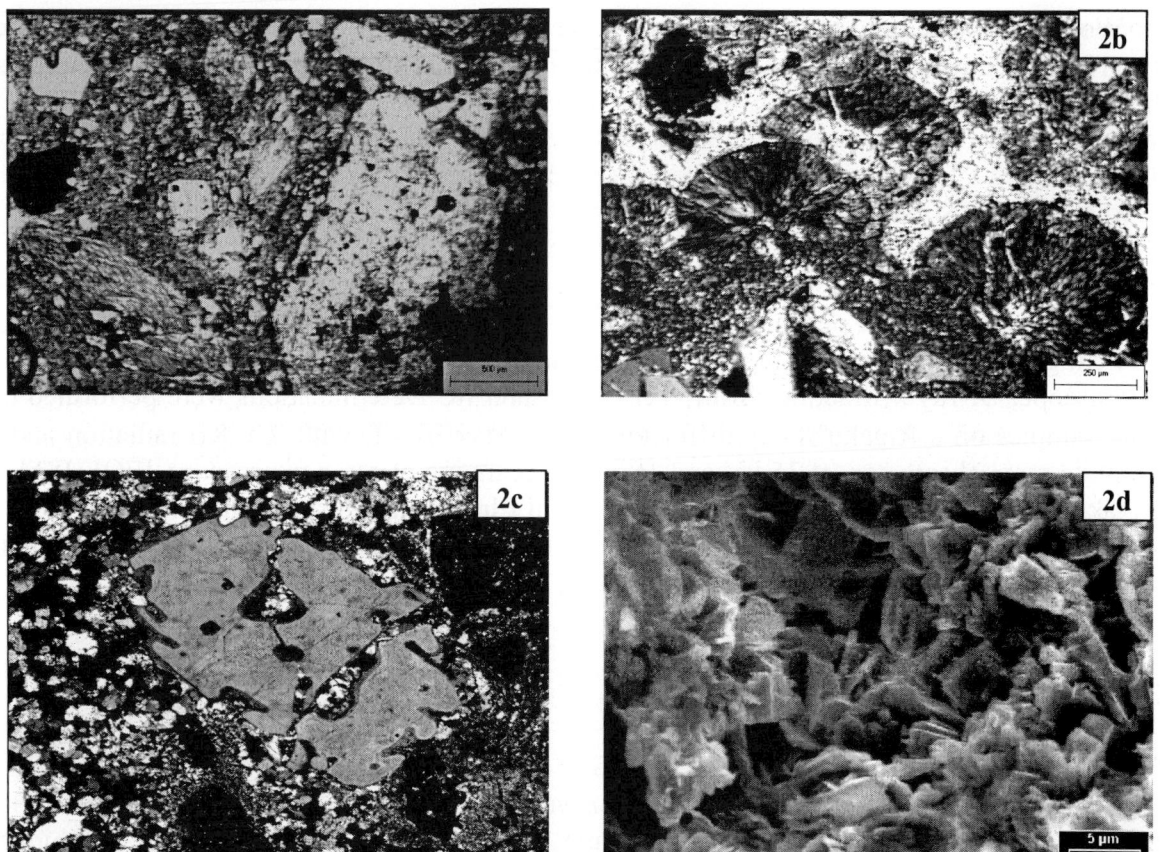

Figure 2. a) porphiritic rhyolite with moderate alteration, b) zeolites in the matrix, c) cracked embayed quartz, d) large dickite crystals in a kaolinite matrix.

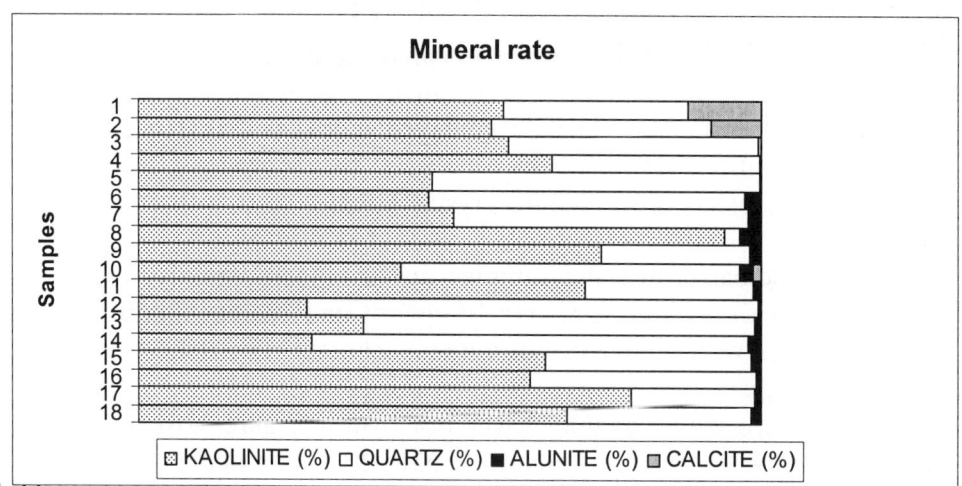

Figure 3: Quarry mineralogy.

3.2. CHEMICAL ANALYSIS

3.2.1 Major elements

From the analysis of major elements (Table 1) the kaolin minerals (kaolinite + dickite) content is calculated to range between 27 and 73 %. The samples with higher silica content correspond to the silicified zone. Sulfur allowed us to estimate an alunite content that varies between 0.3 and 2.1 %, which is closely linked to the most intense kaolinization.

Although major elements cannot be used as a guide to determine the source of mineralization, the increase in alumina content, which is close to theoretical values for pure kaolinite, and the decrease in alkalis are closely related to increased alteration.

Table 1. Chemical analysis of major elements on a whole rock (%).

Sample	SiO_2	Al_2O_3	Fe_2O_3	MnO	MgO	CaO	Na_2O	K_2O	TiO_2	P_2O_5	S	LOI
01	55.23	22.56	0.73	0.007	0.16	6.28	0.23	0.08	0.118	0.23	0.34	13.48
02	56.84	24.38	0.66	0.002	0.08	4.85	0.19	0.05	0.055	0.28	0.19	12.84
03	66.55	23.15	0.40	0.004	0.05	0.29	0.19	0.02	0.097	0.25	0.19	9.38
04	63.60	26.10	0.18	-	0.04	0.11	0.15	0.03	0.202	0.08	0.11	9.87
05	70.69	20.15	0.17	0.002	0.02	0.08	0.23	0.02	0.225	0.12	0.11	8.09
06	73.43	18.44	0.26	0.002	0.04	0.08	0.26	0.07	0.201	0.07	0.11	7.32
07	72.32	18.51	0.32	0.002	0.08	0.19	0.32	0.12	0.201	0.09	0.11	7.58
08	46.28	37.21	0.43	0.002	0.04	0.11	0.17	0.08	0.081	0.64	0.34	14.60
09	58.24	29.46	0.25	0.002	0.15	0.12	0.17	0.06	0.205	0.33	0.19	11.48
10	73.92	16.76	0.80	0.002	0.03	0.70	0.17	0.06	0.230	0.17	0.16	7.24
11	60.29	28.37	0.11	0.002	0.03	0.07	0.09	0.04	0.103	0.18	0.15	10.99
12	84.78	10.84	0.14	0.002	0.01	0.04	0.06	0.01	0.247	0.10	0.05	4.22
13	79.53	14.40	0.11	0.002	-	0.04	0.05	0.01	0.237	0.12	0.10	5.59
14	82.86	11.14	0.13	0.002	-	0.02	0.05	0.03	0.235	0.05	0.03	4.51
15	63.46	25.86	0.06	0.002	0.03	0.09	0.19	0.08	0.212	0.27	0.15	10.01
16	65.37	24.98	0.07	-	0.01	0.06	0.06	0.10	0.125	0.17	0.09	9.26
17	56.60	31.36	0.11	-	0.03	0.06	0.07	0.03	0.051	0.24	0.14	11.91
18	61.42	27.28	0.11	-	-	0.04	-	0.12	0.150	0.24	0.27	10.79

3.2.2 Trace elements

The results from trace elements for the analyzed samples are shown in Table 2 where their high content in Ba, Sr and As is apparent.

Table 2: Chemical analysis of trace elements (ppm).

Sample	Ba	Sr	Y	Sc	Zr	V	Cr	Ga	Ge	As	Nb	Mo	Ag
01	9041	1300	11	7	47	274	55	14	2	213	4	4	2
02	488	1509	10	7	33	126	58	12	2	166	3	5	1
03	1067	1479	3	17	58	82	170	13	3	295	4	17	6
04	372	1441	5	7	87	48	42	17	11	701	8	3	4
05	824	882	5	7	95	98	142	13	5	181	8	15	3
06	180	1390	5	8	88	71	74	14	6	624	8	6	1
07	214	1213	5	9	92	140	103	20	8	707	10	9	11
08	676	4457	6	10	52	249	56	34	4	650	3	0	0
09	1010	1929	7	7	79	76	88	28	5	585	9	4	3
10	714	1293	5	5	91	100	93	22	2	172	9	9	1
11	203	2315	2	12	47	75	64	26	9	686	2	3	21
12	90	1083	3	7	105	41	145	11	4	293	6	11	16
13	265	1716	4	8	101	57	93	12	4	503	6	7	16
14	100	335	3	5	110	28	85	10	3	215	6	6	12
15	432	1603	5	12	90	126	113	16	3	243	9	5	5
16	235	1023	3	9	65	67	32	14	3	140	3	0	12
17	140	3009	3	12	34	111	33	21	6	445	0	0	11
18	485	1898	2	11	65	164	0	27	4	218	5	0	5

3.2.3 Rare earths

The behavior of rare earth elements (REE) during the interaction of hydrothermal solutions with the country rock is not well known.

Michard (1989), concludes that hydrothermal solutions have low REE concentrations, about 5×10^2 to 10^6 times less than the country rock. Therefore such hydrothermal activity will not vary the REE content of the solid unless the hydrothermal solution-country rock relation is very high. Furthermore, she noted that REE concentrations in hydrothermal fluids increase with the decrease in pH, regardless of the rock type or temperature. In this work it is also noted that Eu anomalies are restricted to high temperature solutions, rich in chlorides and with acid pH.

Chondrite-normalized results from samples of a cross section, following Boynton, (1984) (Rollinson H., 1992), are plotted and compared with fresh rock data in Figure 4.

REE are impoverished with respect to the country rock, with marked parallelism in element distribution. There is a small Eu negative anomaly. Although Ce behavior in marine environments is well known, Eu positive anomalies have been identified in weathering profiles related to laterites, (Rankin et al., 1976). Cravero et al. (2001) have detected a Ce positive anomaly in kaolin deposits in the Provinces of Chubut and Santa Cruz (Argentina) of sedimentary origin. No Ce anomaly is present in the analyzed samples from Blanquita.

4. DISCUSSION

Dill et al. (1997) discriminated between hypogene and supergene kaolinization using the relations between several elements present in kaolin (P vs. S, Zr vs. Ti, Cr + Nb vs. Ti + Fe, and Ce + Y + La vs. Ba + Sr). S, Ba and Sr are appreciably enriched in the kaolins developed during hydrothermal alteration, whereas Cr, Nb, Ti and lanthanide elements are concentrated mainly during weathering.

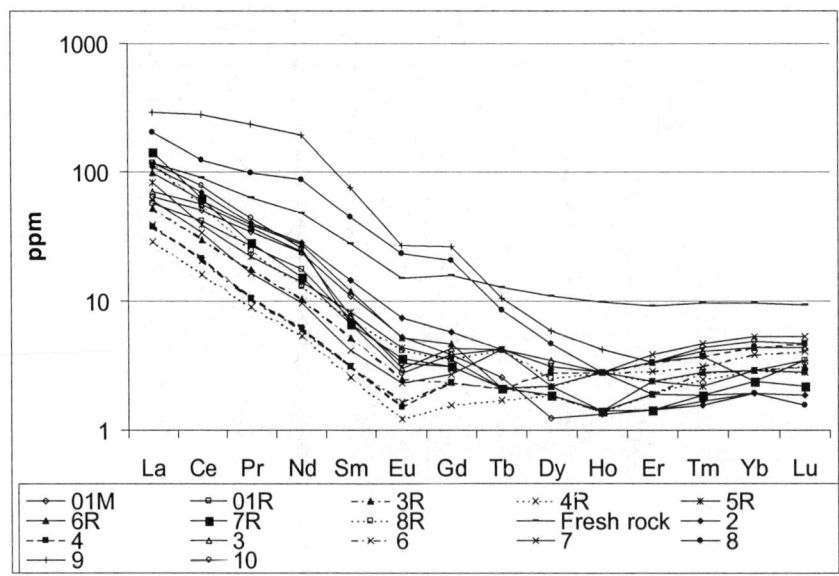

Figure 4: Chondrite-normalized REE profiles.

Dill et al. (2000) used binary diagrams to distinguish the origin of kaolin deposits for diagnostic major and trace elements. The results obtained for Zr versus Ti suggest that titanium is released from primary minerals during hypogene and supergene kaolinization. However, Ti concentration is much more effective during supergene kaolinization. By plotting Cr + Nb versus Ti + Fe contents it can be observed that kaolin deposits of supergene origin have high Cr and Nb values, while these values are low in hypogene type deposits. By plotting Ba + Sr versus Ce + Y + La contents, the authors observed that supergene deposits contain high amounts of Ce + Y + La, whereas hypogene deposits exhibit high Ba + Sr contents.

The results from Blanquita samples are plotted in the graphs shown in Figure 5. By a comparison with data published by Dill (2000), their hypogene origin becomes apparent.

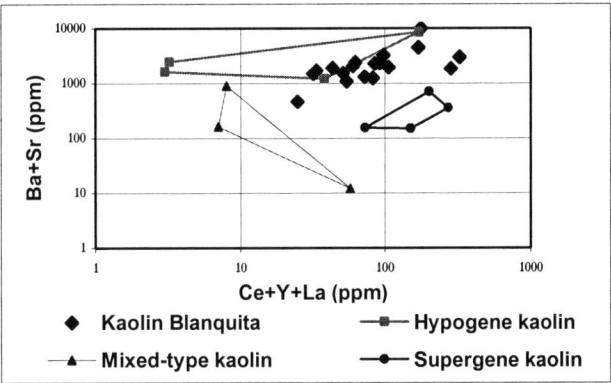

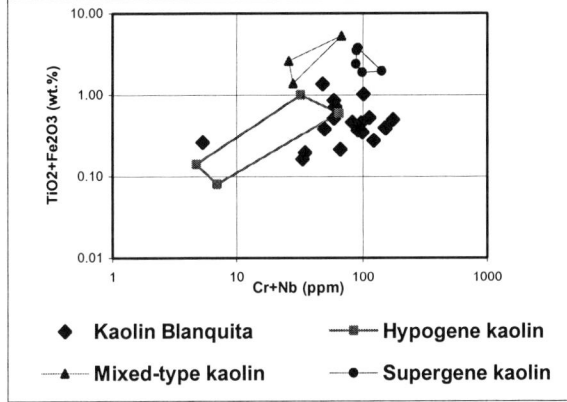
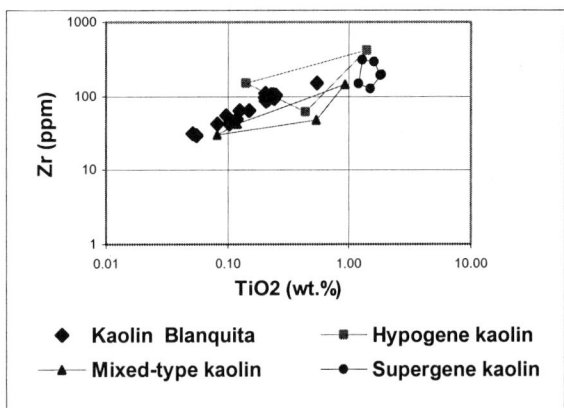

Figure 5: Cross plots to discriminate hypogene and supergene kaolin using major and trace element contents. a: Ba + Sr vs. Ce + Y + La. b: $TiO_2 + Fe_2O_3$ vs. Cr. c: Zr vs. TiO_2 (Dill et al. 2000).

CONCLUDING REMARKS

- The presence of well crystallized alunite, dickite and kaolinite, replacing feldspars, mafic minerals and rock matrix, and their arrangement in high purity veinlets give clear evidence for the development of hydrothermal activity.
- The ore shoot structure and heterogeneous mineralization relate the kaolinization process to the endogenous environment.
- Alunite-dickite association leads to an acid environment and temperatures above 270ºC (Maiza, 1972).
- The high Sr, Ba and As contents and the low Ce content are clear indicators of hypogene origin for the mineralization of Mina Blanquita.
- The low $TiO_2 + Fe_2O_3$ vs. $Cr + Nb$ relations and the $Zr - TiO_2$ relation contribute to support hydrothermal origin (Dill 2000).

ACKNOWLEDGEMENTS

The authors are grateful to CIC from the Province of Buenos Aires and to UNS for their support. They also thank Mr. Rodolfo Salomón for his cooperation in making graphs and photomicrographs.

REFERENCES

1. Maiza, P. Thesis. Universidad Nacional del Sur.(1972).
2. Maiza P. J. and K. Hayase. II Cong. Ibero-Americano de Geol. Ec., Bs. As. II. (1975).365-383.
2. Marfil, S.; D. Pieroni y P. Maiza. Mineralogía y Metalogenia. INREMI. (2000). 281-286.
3. Cortes, J. M. Asociación Geológica Argentina. XXXVI. Nº 3. (1981). 217-235.
4. Dill H.; R . Bosse; H. Henning and A. Fricke. Mineralum Deposita. 32 (1997). 149-163.
5. Dill H. and H. Bosse. Economic Geology. 95. (2000). 517-538.
6. A. Michard. Geochimica et Cosmochimica Acta., 53, (1989) 745-750.
7. Rankin, P. and C. Childs. Chemical Geology. 18. (1976). 54-64.
8. Cravero, F.; E. Dominguez and C. Iglesias. Applied clay science. 18. (2001). 157 - 172.
9. Rollinson H. Ed. University of Zimbabwe. (1992).

Chlorite-smectite geothermometry of two wells from the Copahue geothermal field, Argentina

G. R. Mas[a], L. Bengochea[a], L. C. Mas[b]

[a] Departamento de Geología. Universidad Nacional del Sur. San Juan 670. (8000) Bahía Blanca.
 Consejo Nacional de Investigaciones Científicas y Técnicas
[b] Ente Provincial de Energía del Neuquén. Rioja 385. (8300) Neuquén.

The mineralogy and paragenesis of phyllosilicate minerals in core samples from two exploration wells from the Copahue Geothermal Field (Neuquén, Argentina) were investigated by means of XRD and optical study, in order of evaluate their potential use as indicators of geothermometric conditions.

Clay minerals constitute one of the main hydrothermal products of the metamorphic alteration of andesites, and show a rather wide range of variation in their structural and optical characteristics, regarded with the depth of occurrence.

In this paper we use the data of a detailed study of alterations within the geothermal field of Copahue to evaluate the use of chlorite/smectite (C/S) mixed layer minerals at the view of their use as a quick and simple mean for the thermal characterization of the drilling zone.

1. INTRODUCTION

Clay minerals have received considerable attention in geothermal fields during the last twenty years. Being extensive products of water-rock interaction processes, they are reliable indicators of spatial organization of active or fossil geothermal fields. Smectite (S), illite/smectite (I/S) and chlorite/smectite (C/S) are by far the most widely extended clay minerals and both mixed layers series are of interest for geothermometry. The crystal-chemical properties of these mixed layer minerals (especially the smectite layers content and the ordering degree) change in response to thermal conditions, ranging from less than 50°C to about 230-240°C (Srodon and Eberl, in Bailey 1984; among others). Consequently, the compositional variations of I/S and C/S have been interpreted in terms of geothermometry. However, for a given alteration stage, the spatial variation of mixed layer clays is controlled by kinetics of mixed layer transformation and not only by temperature. In fact, the present crystal-chemical states of the mixed layer minerals is not their initial state, but it was acquired during the overall hydrothermal history which postdate the nucleation of smectitic clay materials at high temperature.

2. GEOLOGICAL SETTING

The Copahue Geothermal Field is a volcanic complex of Tertiary-Quaternary age, formed by a great caldera of about 15x 20 km, which would have originated from a big stratovolcano, in which several extrusive centers have developed associated to the major structures related to the former caldera. Some of these centers would have evolved into explosion craters.

It is located at latitude 37°50'S and longitude 71°05'W, some 1170 km WSW of Buenos Aires City, and adjoining the border with Chile. The area is connected with the city of Neuquén by national and provincial routes, covering nearly 360 km through Zapala, Las Lajas and Loncopue. This geothermal field is on the east side of Los Andes, in the ridge which forms the watershed separating the river basins of the Pacific and Atlantic sides, as typified by Volcan Copahue and Paso Copahue in the western part of the area. This area rises to about 2000 m above sea level.

The magmatism shows characteristics of a calcakaline series with predominance of pyroclastics and lava flows from dacitic to andesitic compositions. The present day thermal anomaly of Copahue is centered on an area delimited by interconnected active faults. Three exploration wells have been drilled in this field, in a sector about 6 km northeast of the Copahue volcano. The three wells form a triangle with 1 km side length, placed over a predominant fault WNW-ESE. They confirm the occurrence of a vapor-dominated reservoir below a depth of 800 m.

All the volcanic formations drilled in the geothermal field have been subject to strong hydrothermal alteration. A detailed study of the alteration petrography evidenced a vertical zoning of alteration that results from time-space super-impositions of at least three hydrothermal stages. The first stage of hydrothermal alteration affected the totality of the geothermal zone and led to a zoned distribution of mineralogical facies ranging from clay-zeolite to propylitic facies with increasing depth and temperature. Alteration paragenesis suggests that temperatures grades from less than 100°C at surface to 250°/300°C at 1200m depth (chlorite-epidote-prehnite assemblage). This stage represents a thermal event typical of the external part of aureoles during which zoned pervasive alteration developed in response to mainly conductive thermal gradient (Mas, 1993).

The second stage of hydrothermal alteration was initiated by a hydraulic fracturation of part of the system (between 800 m and 1200 m depth). Infiltration of the permeable newly fractured horizons by aqueous fluids of meteoric origin promoted the intense alteration of the wall rocks. Fluid inclusions data evidence boiling during this alteration stage. Quartz, wairakite, and even garnet, precipitated mainly in open fractures at temperatures of about 240°-280°C. Clay minerals formed also in open fractures and replaced preexisting igneous and hydrothermal minerals of the early stage in surrounding wall rocks. Clays are essentially non-expandable chlorites or chlorite rich C/S mixed layers. These newly formed clay minerals occur along the fractures of the permeable horizons.

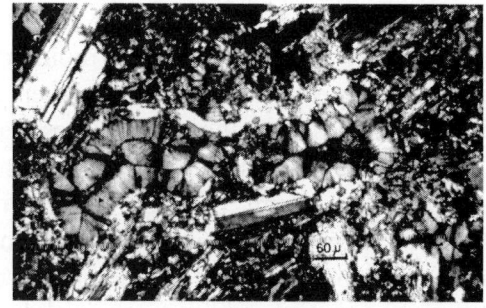

Nowadays, active circulation of hydrothermal fluids is channeled by active vertical faults. In permeable horizons, coating of very fine grain material overprint the coarser grain chlorite and C/S formed during the previous stage. Fig.1 shows a photomicrography of a vein of coarse chlorite coated by a film of C/S.

Fig. 1: chlorite coated by C/S

3. CLAY ALTERATION

3.1. Methods

X-ray diffraction (XRD) is the most useful method for identification of C/S mixed layer minerals in altered rocks. In order to eliminate contaminant-drilling mud, samples were carefully washed with distilled water. Clay minerals were mechanically separated in an ultrasonic bath and by repeated sedimentation. Oriented samples were produced by pipetting the clay slurry onto a glass slide. All samples were run in a Rigaku D-Max IIIC diffractometer with Ni-filtered copper radiation.

In spite of the fact that the basal reflections give very useful indications of the minerals present in a clay sample, there are considerable limitations to an analysis based only in this kind of data. The studied clays are really a mixture of clay minerals and a number of accessory minerals, and besides, reflections usually are broad because the crystals are thin and the layer stacking is frequently disordered. It was necessary, therefore, to examine clay specimens before and after various treatments that alter the diffraction patterns of the components.

The most common of these techniques are the treatment with polar liquids such as ethylene glycol or glycerol to aid the identification of smectites and various heat treatments that alter the diffraction patterns of the components by collapsing swelling minerals and destroying or transforming minerals such as kaolinite. In this study clay samples were saturated in an ethylene glycol atmosphere at 60°C, and to ensure clay identification same samples were heated at 200°C, 350°C and 500°C. After each treatment, the samples were X-rayed.

The recognition of kaolinite in presence of chlorite may be a problem for the 002 and 004 reflections of chlorites (d ≈ 7.1Å, 3.55Å) overlap the 001 and 002 reflections of kaolinite (d ≈ 7.15Å, 3.58Å). However, in this study the treating of samples with warm dilute HCl in which chlorites are soluble proved that there is not kaolinite in the analyzed material. Besides the samples were heated to 450°-500°C for 4 hours to analyze the behavior of the minerals, because kaolinite transforms in a poorly ordered metakaolin while chlorites survived this treatment and even increased the intensity of the 14 Å reflection.

To determine the number of layers (N) in a crystal by means of the Scherrer equation, the breath of the basal 00l reflections should be measured. Many basal reflections may be obtained by using a well-orientated sample. Then, after making the corrections for instrumental broadening, values of crystal thickness (L) can be found for the whole sequence of basal reflections. Scherrer equation allows determining the number of layers in a clay mineral crystal, however diffraction peaks may be broadened by still another causes than the number of ordered layers, for example random interstratification. As a whole, the result may be considered reliable if the same value of L is obtained from all the observed peaks, and the number of layers in the crystals is given by $L/d_{(001)}$.

3.2. Results

Smectite minerals are common in the upper 200m of the wells, but they are also present immediately below the intensively fractured and altered zone of reservoir between 800 and 1200m depth. Smectites have a basal reflection d_{001} between 14.3Å and 15.0Å under dry conditions, which shift to 16.6Å to 18.0Å in glycolated samples. Fig.2a shows the diffractogram of a smectite from 50 m depth and the same one after ethylene glycol treatment.

Swelling chlorites are the most common phyllosilicates in drill hole cores at depth greater than 150 m, with abundance only comparable with that of the chlorites. These minerals have d_{001} (dry) values between 14,0Å and 15,0Å, and expanded up to values of 18Å after the treatment with ethylene glycol. The degree of expansion increases with the increment of expandable components. Also d_{002} reflection suffers the glycolation effect, showing a decrease of intensity and a conspicuous increase of width. Fig. 2b, 2c and 2d shows the diffractograms of a smectite rich/, chlorite/smectite and a chlorite rich/ mixed layer minerals from 170 m, 280 m and 450 m of depth respectively.

Chlorite with non-expandable d_{001} spacing is also common in Copahue although it is less abundant than the swelling one and it is restricted to the higher temperature and more intensive alteration levels. With the increase of depth the percentage of chlorite layers in C/S mixed layer minerals increases and they change to chlorite rich clay.

Figs.2: XRD between 3° and 15° 2θ, of samples from 50 m, 170 m, 280 m, and 450 m, dry conditions and treated with ethylene glycol.

(2a): almost pure smectite;

(2b): Smectite rich C/S;

(2c): C/S.

(2d): chlorite rich C/S.

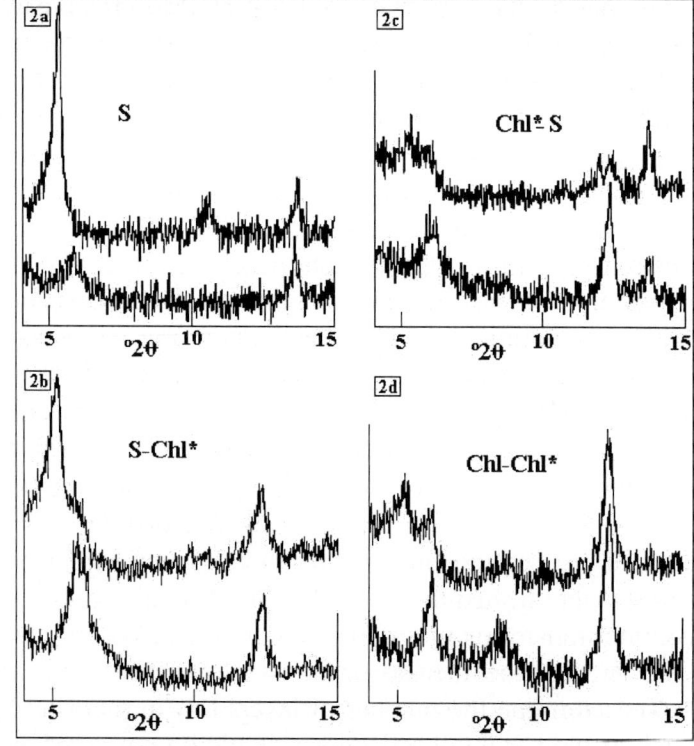

The relative intensities of the 14Å and 7Å in dry samples can be used to roughly estimate the relative proportion of expandable chlorites (Liou et. al.1985). With the increasing of chlorite contain the peak height ratio of 14Å / 7Å decreases from ≈1.5 to ≈0.4 and d_{001} vary from 15Å to 13.5Å. Fig.3 shows the plot I_{001}/I_{002} vs. d_{001} for 40 samples, where the mentioned tendency can be observed. The samples with d_{001} smaller than 14.5Å have I_{001}/I_{002} below 1, and this relation gradually diminishes with the reduction of the basal spacing, in coincidence with the increment in the chloritic phase. As a whole, this ratio decrease with increasing depth of bore holes.

Only few samples of C/S mixed layer clay show a reflection in 31Å (d_{001}), which point out their regular interbedded character. All the others are randomly interlayered.

Fig.4 illustrated the ratio between d_{001} values of dry clays and their glycolated equivalent. The diagonal straight-line joints points with the same values of d (without expansion). It is clearly observed that the expansion increases with the increment of d_{001}, and that there is a gap between smectites and C/S mixed layer clay while a more gradual transition exists between this last group and non-expandable chlorite.

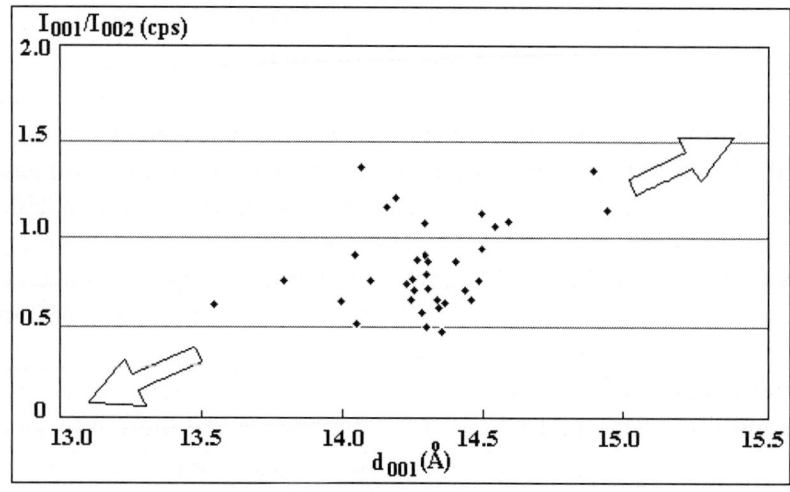

Fig. 3:

I_{001}/I_{002} vs. d_{001} diagram of C/S mixed layer clay of Copahue Geothermal Field.

The XR diffractogram of well-crystallized chlorite gives additional chemical and structural information (Brindley & Brown, 1984). For example, the b_o parameter of the unit cell varies between 9,294 Å and 9,252 Å, with a tendency to decrease with increasing depth. The main variable that can be estimated from this measurement is the substitution of Mg by Fe^{2+}. The ratio Fe/Fe+Mg changes in the chlorites of Copahue from 0,45 to 0,32. On the other hand, the Al^{IV} proportion is relatively constant, varying between 1,0 and 1,2 without an apparent relationship with depth. All the non-expandable chlorites analyzed are trioctahedrical IIb and Ib polytype with IIb type dominant.

The size of crystallites perpendicular to the basal spacing direction was estimated applying the equation of Scherrer (Brindley and Brown, 1984), which relates the width of the basal reflections halfway their height, with the number of basal planes. It is extremely variable, and it clearly increases with depth, as can be seen in the following example:

        120 m of depth→ randomly variable number of planes
        620 m   → regularly variable number of planes
        660 m   → 21 basal layers
        880 m   → 27 basal layers

The number of planes is different for each basal reflection in the clay materials of the upper levels and this difference may be regular or random according to the regular or disordered character of the interstratificación. Non-expandable chlorite, on the other hand, shows almost constant values for the whole series of basal spacing in each sample. This number grows up with depth, up to the most fractured and intensely altered level where it remains approximately constant between 27 and 30 basal layers.

The optic analysis of clay minerals from different levels shows that their optic characteristics also change according to depth and so do the associated mineral assemblages.

Swelling chlorite occurs as distinct, thin, platy and generally curved pale green crystals and as aggregates of curved plates in the form of rosettes. Texture of non-expandable chlorite is laminate parallel, with crystals of about 100 µm, in veinlets, radial aggregates and books. Although it is frequent that both varieties coexist in the samples of a given level, the relative

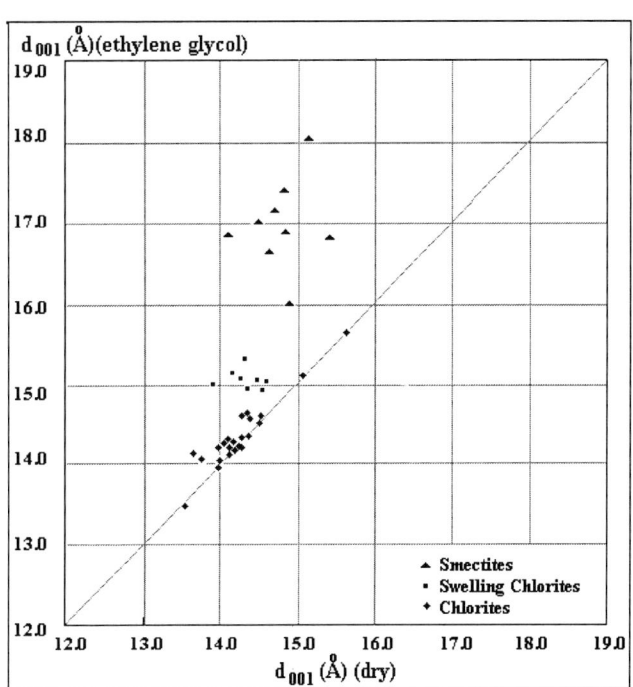

Fig. 4: d_{001}(dry) vs. d_{001}(ethylene glycol) plot of smectite, C/S mixed clay

and chlorite mineral series in Copahue Geothermal Field.
abundance of each one changes with depth. Besides, smectite and S/C mixed layer clay occurs as replacement of groundmass or plagioclase and mafic phenocrysts while chlorite occurs mainly as precipitation in veins and along fractures.

Chart 1 summarize the optic characteristics of expandable and non expandable chlorites:

Type	(001) Orientation	Grain size	Interference color	Pleochroism	Orientation/	Optic sign
C/S	Erratic, laminar forms poorly developed	< 20 µm	1° order gray	Weak or unknown	+	?
C	Parallel, regular	>100 µm	Gray to Anomalous blue	Greenish weak to colorless	+	-

4. DISCUSSION

When the distribution of clay minerals is examined as a function of depth in drill holes of Copahue geothermal area, three zones with distinctive characteristics can been distinguished:
- an upper zone, approximately up to 700 m of depth, in which the intense and low graded alteration has pervasive character and replacement minerals predominate. Smectite is common near the surface, besides very subordinate illite and zeolites. From 80m and up to 620m S/C and C/S mixed layer clays occur, with a progressive increasement of chlorite. In Cop 3 well, there is some subordinate illite between 450 m and 660 m of depth.
- a deeper zone, from about 700 and up to1020m, in which minerals show evidences of higher temperatures of formation. Secondary minerals formed by precipitation in vesicles and along fractures coexist with replacement mixed clays. Non expandable chlorite is the dominant clay mineral. This is a zone of total drillig mud losses, and corresponds to a fracture controlled permeable horizon submitted to active flow regime, i.e. a reservoir.
- Below 1022m a conspicuous change takes place, since the chlorite diminishes again and it is substituted gradually by swelling chlorite and smectite.

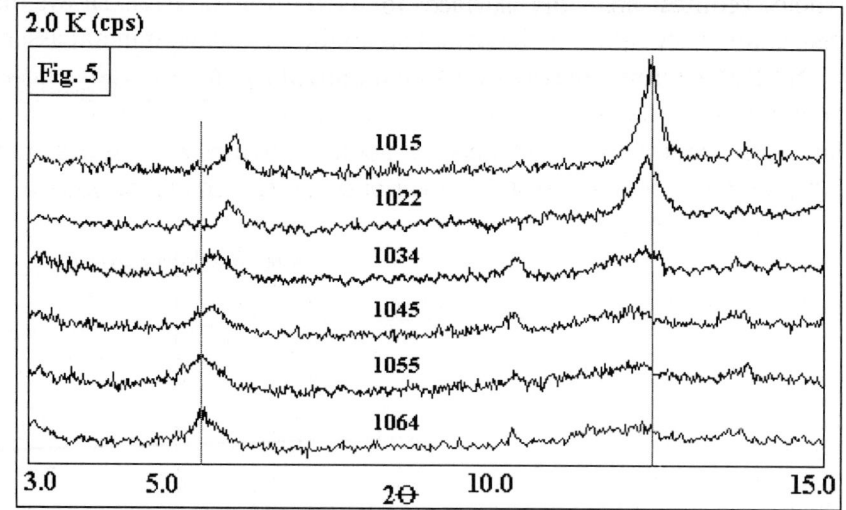

Fig.5: sequence of XRD of ethylene glycol treated clays from the deepest sector of well COP-3, between 1002 and 1065m.

Figure 6 summarized the results obtained in the XR diffractometric analysis of samples between 20 and 1064 m of depth from the well Cop3.
- Smectite occurs in the upper 100m of depth, at relatively low temperatures, 120°-130°C;.

- C/S mixed layer clays occur mainly between 50 and 650 m, at a temperature of about 200ºC. - Non expandable chlorite occurs in deeper levels, between 650 m and 1000 m depth, at temperatures of 250º and higher. It is associated with epidote, prehnite, wairakite, garnet and amphibole. Al^{IV} ratio (1,0-1,2) in chlorites corresponds to temperatures between 240º and 300ºC, according to the graphic proposed by Cathelineu and it Nievas (1985).

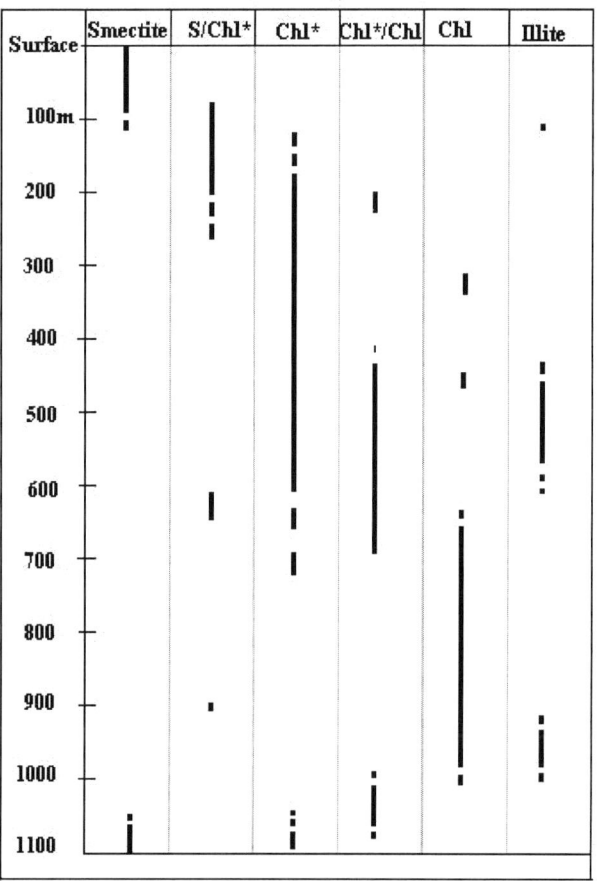

The thickness of crystallites increases with depth, indicating a better crystallization degree. The presence of IIb and Ib polytypes confirms this tendency. Hayes (in Bailey, 1988) pointed out that it is necessary a temperature higher of 200ºC to get the transition Ib to IIb, for the structural changes are very sensitive to temperature changes and relatively insensitive to changes in composition. Walker (1989), on the other hand, said that the chlorite IIb is the end product of thermal metamorphism and the transition would take place to a temperature between 150º and 250ºC.

The secondary minerals associated to clays, which vary between cristobalite and calcic zeolites near the surface to calcic secondary minerals (epidote, prehnite, tremolite, garnet, etc.), in deeper levels confirm the zoning. This association has been formed by the circulation of hot fluids through fractures. In the same way, Ca zeolites change from stilbite and mordenite near surface, through laumontite at about 500m depth, to wairakite at the bottom (Mas et al. 1995)

Kristmannsdottir (in Browne, 1990) describes a similar tendency in Reykjanes, Island, where smectite is present as a discreet phase where the temperature is lower than 200ºC, it becomes erratically interbedded with chlorite at temperatures between 200º and 270ºC. Above 270ºC chlorite is the only clay mineral present. Fujishima and Fan (1977) pointed out that in Keolu Hills the increment of temperature and pressure with the increasement of depth reduces the incidence of the expandable layers of chlorite.

Fluid inclusions in quartz contain low salinity waters and have homogenization temperatures of up to 280ºC in the deepest samples of the reservoir (Mas et al, 1993). The coexistence of liquid- and vapor-rich fluid inclusions suggest that boiling took place when quartz is forming. The static temperature for the bottom hole is about 239ºC for COP-1, 235ºC for COP-2 and 237ºC for COP-3. The sporadic presence of illite in certain levels, and the even scarcer pyrophyllite, seems to be more related with changes in pH of solutions that with temperature, probably because of boiling episodes.

The occurrence of smectite and C/S mixed layer clays in the deepest level of the wells, below the mineral assemblages of high temperature, suggests that they are consequence of a inverse zonation below the area of circulation.

5. CONCLUSIONS

Depth zonation of secondary minerals was delineated for phyllosilicates as:
smectite → C/S → swelling chorite → non expandable chlorite.
Based on the characteristics of the clay minerals and their relative proportions it is posible to define ranges of temperature. A vertical zonation has been determined for clay minerals as follows:

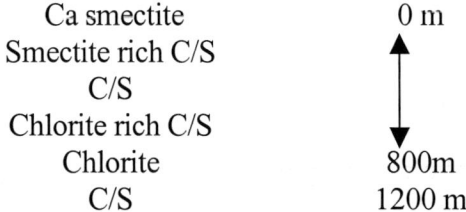

Ca smectite	0 m
Smectite rich C/S	
C/S	
Chlorite rich C/S	
Chlorite	800m
C/S	1200 m

Hydrothermal clay minerals, among other geothermal minerals, may be used as thermoindicators in geothermal fields because their structure and chemical composition are sensitive to thermal changes. Temperatures based on the proportion of clay minerals are in good agreement with those proposed for other geothermal systems as Reykjanes, Iceland or Los Azufres, Mexico. In the two well studied here microthermometric measurements on fluid inclusions are equivalent to the clay minerals temperatures, but yielded slightly higher temperatures than the measured values. Possibly hotter fluids occurred at the moment of hydrothermal mineral deposition. The difference among temperatures could indicate that a cooling has occurred in the reservoir.

REFERENCES

Bailey; S.W. ed.; 1988. Reviews in Mineralogy; P.H. Ribbe ed.; Vol.19; 725 pp.
Brindley, G.W. & G. Brown ed.; 1984. Mineralogical Society; Mon.N°5; 495 pp.
Browne, P.R.L.; 1990: Geology Lectures Course. N°86-102. Nueva Zelanda.
Cathelineu, M & D. Nieva; 1985: Contribution to Mineralogy and Petrology. 235-244.
Fujishima, K. & P. Fan; 1977: Vol.62; N° 5-6; 574-583.
JICA, 1992. The Feasibility Study on the Northern Neuquén Geothermal Development Project. Final Report. Japan International Cooperation Agency.
Liou J.G., Y. Seki, R. Guillemette & H. Sakai; 1985. Chemical Geology; Vol.49; 1-20, Elsevier Science Publishers. Amsterdan.
Mas, L.C.; 1993. Doctoral Thesis. Central Library Univ. Nac. del Sur. Bahía Blanca
Mas, G.; L.C.Mas & L.Bengochea; 1995. World Geothermal Congress 1995 Proceedings. Firenze, Italy. Vol.2: 1077-1082. Berkeley; Ca. USA.
Mas, G.R.; L.C. Mas & L. Bengochea; 1993. Actas del XII Congreso Geológico Argentino. V: 92-98. Mendoza.
Walker, J.R.; 1989. American Mineralogist. Vol.74; N°7-8; 738-743.

Clay Mineral diagenesis of a Pleistocene volcanogenic sequence, Mexican Basin

L. de Pablo-Galan[a], J. J. de Pablo[b] and M. L. Chavez-Garcia[c]

[a]Instituto de Geologia, Universidad Nacional Autonoma de Mexico,
Ciudad Universitaria, Mexico, D. F.*

[b]Department of Chemical Engineering, University of Wisconsin,
Madison, Wisconsin

[c]Facultad de Quimica, Universidad Nacional Autonoma de Mexico,
Ciudad Universitaria, Mexico, D. F.

Diagenesis of the Pleistocene volcanogenic depositional sequence of the Mexican Basin transformed volcanic ash to dioctahedral smectite, a 7Å-phyllite and mixed-layers of this 7Å-phyllite and smectite when high permeability and low retention of fluids prevailed. The predominant interstratifications are R1 and R3 7Å-phyllite(0.80)/2H$_2$O-smectite. In the mud underlying the coarse ash diagenetic alteration was to dioctahedral smectite. The 7Å-phyllite corresponds with $(Si_{1.87}Al_{0.13})(Al_{1.65}Fe^{3+}_{0.36}Mg_{0.03}Ti_{0.04})O_5(OH)_4$. The smectites are aluminian with layer charges from 0.43 to 0.57 when replacement was in the octahedral sheet and of 0.25-0.42 when replacement was in the tetrahedral and octahedral sheets. The hydroxyl complexes are characterized by $(Al_{1.57}Fe^{3+}_{0.26}Ti_{0.03}Mg_{0.16})(OH)_6$.

1. INTRODUCTION

Trioctahedral smectite and Mg-rich chlorite are widespread products of diagenesis and low-temperature metamorphism. They are formed in intermediate to mafic volcanic rocks and volcanogenic sediments (1, 2, 3), marine evaporites (4), lacustrine mudrocks (5, 6), evaporitic sequences (7) and mafic intrusions (8). Low-temperature smectite is transformed to higher-temperature chlorite through an intermediate sequence of interstratified chlorite/smectites and corrensite (3, 6, 9, and 10). The transformation of smectite to corrensite to chlorite has been described as continuous, characterized by random and regular interstratifications (5, 11) and as a series of discontinuous steps (6, 12, and 13).

The diagenesis of younger sediments has not been so completely documented although the transformation of volcanogenic sediments to dioctahedral smectite is ubiquitously accepted.

*The authors are indebted to the Consejo Nacional de Ciencia y Tecnologia – National Research Foundation, grant 400324-CO46A. They express their appreciation to the Dirección General de Construcción y Operación Hidraulica, Departamento del Distrito Federal, for supplying material for study and pertinent data and to A. Altamira, L. Cabrera, A. Maturano and M. Reyes for their contribution.

Associated with the smectite may occur a 7Å layer mineral which has been loosely referred to as 7Å-chlorite, serpentine-like 7Å layers, 7Å-layer or 7Å-clay (10, 14). Karpova (14) described it as a Fe-7Å Ib(β 90°) chlorite that transforms to Fe-14Å Ib(β 90°) chlorite to Mg-Fe-14Å IIb chlorite as a function of increasing burial. This Fe-rich 7Å chlorite could possibly correspond with a high-Fe-14Å chlorite or any of the 7Å high-Fe minerals (15, 16, and 17). It could be presumed that diagenesis of young volcanoclastic sediments to 7Å-phyllite or to dioctahedral smectite could be associated with mixed-layers of 7Å-phyllite/smectite distinct from those developed under more intense stages of diagenesis or low-grade metamorphism. A logical postulate would be to expect dissimilar mechanisms of diagenesis and mineral assemblages within the pores and channels of the coarse ash and sand and in the less permeable fine ash and mudstones.

This paper documents the occurrence, clay mineralogy and diagenesis in lacustrine sediments from the Pleistocene depositional sequence in the Mexican Basin of Central México.

2. GEOGRAPHIC AND GEOLOGIC SETTING

The Mexican Basin extends between 19°00' and 20°15'' N latitude and 98°15' and 99°33' W longitude, covering an area of 7160 km^2 in central México. The Basin was formed in the Middle Tertiary when active volcanism developed thick sequences of basaltic andesite, andesite, dacite and latite. Volcanic activity decreased towards the end of the Miocene. Tectonism associated with the Clarion fault disrupted (18) the crust along NNW-SSE fractures. In the Pliocene, intensive rains created the abrupt relief of the Middle and Upper Tertiary volcanic complexes and formed extensive fluvial deposits intercalated with strata of siltstone, lava flows and tuffs, with lenses of limestone. In the Pleistocene the climate was humid and cold (18,19). The area was covered with thick layers of basalt and pumice until intense voluminous lava flows closed the drainage at the south end of the Basin. This was followed by deposition of air-borne and water transported ashes that settled in the low areas (Tacubaya and Becerrra Formations). The formations are characterized by strata of pyroclastic sands and ashes that in the lacustrine environment were transformed into the high swelling clays common to the Basin (18, 19, 20).

A simplified stratigraphic column of the Mexican Basin shows a randomly distributed layered sequence of silt, mud, sand and gravel. The thickness of the mud strata varies from 40 to 100 m. the mud retains large amounts of water, is mechanically unstable, highly compressible and swells substantially (21). Other studies have described the hydrogeology (22) and the stratigraphy and paleoenvironment of the southeastern part of the Basin (23).

3. METHODOLOGY

Samples were obtained from more than 20 drillholes distributed across the Basin down to depths of 324 m. They are black to brown mudstones with variable contents of sand and ash, and sandstone with volcanoclastic gravel-size fragments, lapilli, sand and ash. Samples were prepared by dispersion in deionized water, screening, and gravity settling and centrifuging. The separation of very fine ash required addition of sodium hexametaphosphate, dilution and long settling periods but it could not be complete due to the strong jellying of the clay and the fine particle size of the ash. Analyses were made by X-ray diffraction; in some cases, the diffraction pattern of ash was subtracted from those of the clay fraction to improve the characterization of the clay minerals. Interstratified 7Å-phyllite/smectite minerals were compared with those calculated using NEWMOD (27) assuming serpentine with 0.5 Fe atoms as equivalent to 7Å-phyllite and montmorillonite with 0.5 Fe atoms. Smectite and vermiculite were differentiated by saturation with Mg and glycerol. Morphology and chemical composition were studied by scanning electron microscopy coupled with energy dispersive X-ray fluorescence spectrometry. Structural formulae were calculated respectively for the $O_{10}(OH)_2$ half-cell of montmorillonite and the $O_5(OH)_4$ and $(OH)_6$ cells of the 7Å-phyllite and the hydroxyl complexes (16, 24, 25, 26).

4. RESULTS

4.1. X-ray diffraction

The fine ash shows broad diffraction bands typical of glass extending between 6-14 and 20-22° 2θ, with associated plagioclase, cristobalite, quartz and K-feldspar. Smectite is $1H_2O$- and $2H_2O$-smectite, dioctahedral ($d(060)$ 1.506 Å). Vermiculite does not occur. The 7Å-phyllite is characterized by intense 001 and 002 reflections at 7.246 and 3.533 Å. The interstratified 7Å-phyllite/smectite are of types Reichweite 1 and 3, with characteristic reflections at 11.62-11.15 Å, 9.30-10.49 Å, 7.78-7.32 Å and 3.72-3.16 Å.

In the clay fraction dioctahedral $1H_2O$- and $2H_2O$-smectite are the principal clay minerals. They predominate from the surface to 26 m depth, associated with minor R1 7Å-phyllite(0.60)/$2H_2O$-smectite and R3 7Å-phyllite(0.80)/$2H_2O$-smectite. At 26 m depth the 7Å-phyllite decreases and from 86 m down to 324 m smectite is essentially the single mineral with hardly any traces of 7Å-phyllite. This mineralogical sequence is reversed in the south area (Tezonco) of the Basin where a 46 m section of mudstones formed essentially by $2H_2O$-smectite overlies the gravel containing ash, smectite and the 7Å-phyllite.

In the clay fraction dioctahedral $1H_2O$- and $2H_2O$-smectite are the principal clay minerals. They predominate from the surface to 26 m depth, associated with minor R1 7Å-phyllite(0.60)/$2H_2O$-smectite and R3 7Å-phyllite(0.80)/$2H_2O$-smectite. At 26 m depth the 7Å-phyllite decreases and from 86 m down to 324 m smectite is essentially the single mineral with hardly any traces of 7Å-phyllite. This mineralogical sequence is reversed in the south area (Tezonco) of the Basin where a 46 m section of mudstones formed essentially by $2H_2O$-smectite overlies the gravel containing ash, smectite and the 7Å-phyllite.

In the clay fraction dioctahedral $1H_2O$- and $2H_2O$-smectite are the principal clay minerals. They predominate from the surface to 26 m depth, associated with minor R1 7Å-phyllite(0.60)/$2H_2O$-smectite and R3 7Å-phyllite(0.80)/$2H_2O$-smectite. At 26 m depth the 7Å-phyllite decreases and from 86 m down to 324 m smectite is essentially the single mineral with hardly any traces of 7Å-phyllite. This mineralogical sequence is reversed in the south area (Tezonco) of the Basin where a 46 m section of mudstones formed essentially by $2H_2O$-smectite overlies the gravel containing ash, smectite and the 7Å-phyllite.

Table 1. Mineralogy of the clay fraction.

Depth (m)	1H-S	2H-S	7Å-P	R1 P(0.60)/2H-S	R1P(0.50)/1H-S	R3 P(0.25)/2H-S	R3 P(0.80)/2H-S	R1 P(0.75)/G-S	R3 P(0.80)/G-S	R3 P(0.25)/G-S
8		xx	xx	x			x	x	x	
10		xx	xx	x			x	x	x	x
16	xxx							x	x	
18	xx		xx		x	x				x
26		xx	x						x	x
86		xx	x	x	x		x	x		
90	x	xx	x				x		x	x
148		xxx								
218		x								
254		xxx								
262		xxx				x				
324	xx	xx								

1H-S, 1H$_2$O-smectite; 2H-S, 2H$_2$O-smectite; 7Å-P, 7Å-phyllite; R1 P(0.6)/2H-S, R1 7Å-phyllite(0.6)/2H$_2$O-smectite; R1 P(0.75)/2H-S, R1 7Å-phyllite(0.75)/2H$_2$O-smectite; R1 P(0.50)/1H-S, R1 7Å-phyllite(0.50)/1H$_2$O-smectite; R1 P(0.25)/S, 7Å-phyllite(0.25)/ 1H$_2$O-smectite; R3 P(0.75)/2H-S, R3 7Å-phyllite(0.75)/2H$_2$O-smectite; R3 P(0.80)/2H-S, R3 7Å-phyllite(0.80)/2H$_2$O-smectite; R1 P(0.75)/G-S, R1 7Å-phyllite(0.75)/ EG-smectite; R3 P(0.80)/G-S, R3 7Å-phyllite(0.80)/ EG-smectite; R3 P(0.25)/G-S, R3 7Å-phyllite(0.25)/EG-smectite. The relative abundance of minerals estimated from the intensity of their principal diffraction peaks is indicated by xxx when predominant and x when least abundant. Under the same heading and depths are indicated the minerals in the bulk and in the glycolated samples.

4.2. Scanning electron microscopy

The typical morphology of the <2 μm fraction (10 m depth) shows glass and glass transforming to pseudohexagonal forms. At 86 m, glass, diatoms, plagioclase, glass transforming to hexagonal forms, pseudohexagonal plates and opal are recognized. At 92 m, glass, pseudohexagonal platelets, smectite and minor plagioclase are identified. At 188 m, glass partially transformed to smectite and smectite are present, and at 254 m smectite occurs associated with glass.

The energy dispersive X-ray fluorescence analysis of selected fragments from the upper gravel, 10, 18 and 26 m deep, indicates fragments of high-SiO$_2$ contents corresponding to glass or contaminated by it. Clay lamellae have compositions that fit the O$_{10}$(OH)$_2$ half-cell of smectite with contents falling within the limits Si 3.67-4.00, ^{IV}Al 0.33-0, ^{VI}Al 0.95-1.54, Mg 0.05-0.66, Fe^{2+} 0.22-0.71, Ti 0.02-0.17, Ca 0.04-0.26, Na 0.05-0.39, K 0-0.17. The 7Å-phyllite is (Si$_{1.87}$Al$_{0.13}$)(Al$_{1.65}$Fe$^{3+}_{0.36}$Mg$_{0.03}$Ti$_{0.04}$)O$_5$(OH)$_4$. The hydroxyl complex, (Al$_{1.57}$Fe$^{3+}_{0.26}$Ti$_{0.03}$Mg$_{0.16}$)(OH)$_6$, was recognized at depths of 90 m. The compositions confirm the dioctahedral character of smectites with 0 to 0.33 Al atoms per half-cell in the tetrahedral sheet; their octahedral cation is largely Al^{3+}, replaced partially by Fe^{2+} and Mg^{2+}. The layer charge varies from 0.25 to 0.67 units per O$_{10}$(OH)$_2$ which are in part within the 0.5-0.7 range reported for montmorillonite of high hydration and swelling characteristics comparable to those of vermiculite (24, 28).

5. DISCUSSION

The 7Å-phyllite is more common in the coarse gravel. Its composition, (Si$_{1.87}$Al$_{0.13}$)(Al$_{1.66}$Fe$^{3+}_{0.36}$Mg$_{0.03}$Ti$_{0.04}$)O$_5$(OH)$_4$, does not correspond with kaolinite which has minimal ionic replacement (29,30), berthierine (R$^{2+}_a$R$^{3+}_b$□$_c$)(Si$_{2-x}$Al$_x$)O$_5$(OH)$_4$, odinite or chlorite (14, 15, 16, 17). The mineral here referred as the 7Å-phyllite has 0.13 tetrahedral Al atoms per cell, well below the range of 0.4-1.8 Al atoms per four tetrahedral positions accepted for the chlorite half-cell (26, 31); the tetrahedral charge is insufficient to develop the 14 Å chlorite layer. The octahedral 0.36 Fe^{3+} atoms are less than the 6 atoms per half-cell considered as evidence for the oxidation to Fe^{3+} chlorite (26). Octahedral $Al^{3+}+Fe^{3+}$ is higher than in the associated smectites, which are richer in Mg. The mineral is dioctahedral and from this standpoint it may be considered as kaolinite with tetra- and octahedral substitution.

The smectites in the Basin are dioctahedral. They have between 3.67 and 4 ^{IV}Si atoms per half-cell; on the average, compositions with less Si and more ^{IV}Al are from lamellae in the overlying coarse ash whereas those higher in Si and lower in ^{IV}Al correspond to lamellae from the underlying mudstones. Tetrahedral ^{IV}Al ranges between 0 and 0.33 atoms per $O_{10}(OH)_2$ half-cell. Smectites without ^{IV}Al have layer charges from 0.43 to 0.57 whereas those with tetrahedral substitution have lower charges of 0.25-0.42. Some of these layer charges are close to the range of high-swelling vermiculite (0.6-0.9); at these high values there is no difference in the stability of the layers (32, 33). Their octahedral sheet has a uniform predominantly aluminian composition; the octahedral cations are between 1.83 and 1.96 for the dioctahedral types and from 2.01 to 2.27 for those smectites with substitution by Mg. The smectites have in their octahedral sheet more Mg and less Fe^{2+} than the 7Å-phyllite and their Fe/(Fe+Mg) ratios are lower. The Fe/(Fe+Mg) ratio varies between 0.45 and 0.92, average 0.68, nearly the same ratios that characterize basalt (34). Octahedral Al+2Ti is essentially constant whereas the tetrahedral Al and the octahedral Mg/(Mg+Fe) ratio change within wide limits.

Interstratified 7Å-phyllite/smectite minerals occur at various depths. They may suggest transformations between the 7Å-phyllite and smectite, comparable to that reported for trioctahedral smectite-chlorite. Comparable transformations between trioctahedral chlorite and smectite have been reported in other systems (1, 2, 5, 9, 10, 34, 35, 36, 37, 38, 39, 40, 41).

The 7Å-phyllite that prevails in the upper gravel formed directly from the volcanic glass. It is associated with mixed-layers of 7Å-phyllite/smectite. They both decrease in abundance with depth, as permeability and flow become increasingly limited and smectite increases. In the smectite mud they do not occur. Hydroxyl complexes were recorded; opal was found to be rare or absent. The data suggest that diagenesis may have followed the reaction path: volcanic glass → 7Å-phyllite → 7Å-phyllite/smectite → smectite + hydroxyl complex. A second diagenetic reaction path is represented by the direct transformation of fine ash to smectite in a less permeable environment higher in fluids.

The progressive transformation of volcanic ash to 7Å-phyllite to smectite, and of ash directly to smectite of high layer charge would be an expected outcome in a system of heterogeneous behavior and of increasing instability. A uniform set of properties across the depositional sequence should not be expected. The effect would be opposed to the decreasing expandability associated with the trioctahedral smectite-chlorite transformation (2, 10).

6. CONCLUSIONS

Low-grade diagenesis of the Pleistocene volcanogenic depositional sequence of the Mexican Basin developed a 7Å-phyllite, interstratified 7Å-phyllite/smectite and smectite. This 7Å-phyllite is dioctahedral, of composition different from that of kaolinite or other 7Å minerals. It predominates in the upper stratums of the overlying coarse ash where high permeability and low retention of fluids prevailed. With increasing depth, decreasing permeability and more availability of fluid of augmenting ionic strength transforms to smectite. A possible diagenetic reaction path is: volcanic glass → 7Å-phyllite → 7Å-phyllite/smectite → smectite + hydroxyl complex.

REFERENCES
1. W.R. Almon, L.B. Fullerton and D.K. Davies, J. Sed. Petrol., 46(1976), 89.
2. H.K. Chang, F.T. Mackenzie and J. Schoonmaker, Clays and Clay Minerals, 34(1986), 407.
3. W.T. Jiang, D.R. Peacor and P.R Buseck, Clays and Clay Minerals, 42(1994) 593.
4. M.W. Bodine. and B.M. Madsen, In: L.G. Schultz, H. van Olphen and F.A. Mumpton (eds.), Proceedings of the International Conference Denver 1985, The Clay Minerals Society, 1987.
5. R.H. April, Clays and Clay Minerals, 29(1981) 311981).
6. S. Hillier, (1993) Clays and Clay Minerals, 41(1993) 240.
7. O.C. Kopp and S.M. Fallis, Amer. Miner., 59(1974) 623.
9. I. Hutcheon, A. Oldershaw and E.D. Ghent, Geochim. Cosmochim. Acta, 44(1980) 1425.
10. W.T. Jiang and D.R. Peacor, Clays and Clay Minerals, 42(1994) 497.
11. L. Bettison-Varga and I.D.R. Mackinnon, Clays and Clay Minerals, 45(1997) 506.
12. Y.H. Shau and D.R. Peacor, Contrib. Mineral. Petr., 112(1992) 119.
13. P. Schiffman and H. Staudigel, J. Metam. Petr., 13(1995) 487.
14. G.V. Karpova, Sedimentology, 13(1969) 5.
15. G.W. Brindley, Clays and Clay Minerals, 30(1982) 153.
16. S.W. Bailey, In: S. W. Bailey (ed.), Hydrous Phyllosilicates, Reviews in Mineralogy v. 19, Mineralogical Society of America, Washington, 1988.
17. G.S. Odin, S.W. Bailey, M. Amouric, F. Frohlich and G.S. Waychunas. In: G. S. Odin (ed.), Green Marine Clays, Developments in Sedimentology, Elsevier, 1988.
18. F. Mooser, Informe sobre la Geología de la Cuenca del Valle de México, Secretaria de Recursos Hidraulicos, Comision Hidrologica del Valle de México, México, 1956.
19. D.A. Gasca and C.M.Reyes, La cuenca lacustre Plio-Pleistocenica de Tula-Zumpango. Instituto Nacional Antropologia e Historia, Informe 2, 1977.
19. E. Lopez-Ramos, Geología de México, 1979.
20. Sistema Hidraulico del Distrito Federal (1994) Cronologia. Memoria de las Obras del Sistema de Drenaje Profundo del Dustrito Federal, 1994.
21. R.J. Marsal and M. Mazari, El Subsuelo de la Ciudad de México. Facultad de Ingenieria, Universidad Nacional A. de México, México, 1962.
22. J. Durazo and R.N. Farvolden, J. Hydrology, 112(1989) 171.
23. J. Urrutia-Fucugauchi, S. Lozano, O. Ortega, M. Caballero, R. Hansen, H. Bohnel and J.F.W. Negendank, (1994) Geofisica Internacional, 33(1994) 421.
24. N. Güven, (1988) In S. W. Bailey (ed.), Hydrous Phyllosilicates, Reviews in Mineralogy, Mineralogical Society of America, Washington, 1988.
25. D.M. Moore and R.C. Reynolds, X-ray diffraction and the identification and analysis of clay minerals. Oxford University Press, Oxford, 1997.
26. M.D. Foster, US Geological Survey Prof. Paper 414-A, 1962.

27. R.C.Jr Reynolds and R.C.III Reynolds, NEWMOD: The calculation of one dimensional x-ray diffraction patterns of mixed-layered clay minerals. Computer program, 8 Brook Road, Hanover, New Hampshire 03755, USA, 1996.
28. H. Suquet, J.T. Iiyama, H. Kodama and H. Pezerat, Clays and Clay Minerals, 25(1977) 231.
29. R.F. Giese, In: S. W. Bailey (ed.), Hydrous Phyllosilicates, Reviews in Mineralogy, Mineralogical Society of America, Washington, 1988.
30. W.A. Deer, R.A. Howie and J. ZusssmanRock Forming Minerals, Longmans, Great Britain, 1962.
31. S.W. Bailey, In: S. W. Bailey (ed.), Hydrous Phyllosilicates, Reviews in Mineralogy v. 19, Mineralogical Society of America, Washington, 1988.
32. J. Mering and G. Pedro, Bull. Groupe Franc. Argiles, 21(1969) 1.
33. H. Suquet and H. Pezerat, H. Clays and Clay Minerals, 36(1988) 184.
34. T. Sudo and H. Kodama, Zeitsch.Kristall., 109(1957) 379.
35. W.E. Seyfried, W.C. Shanks and D.E. White, Earth and Planetary Sci., 41(1978) 265.
36. J. Hoffman and J. Hower, (1979) Clay mineral assemblages as low grade metamorphic geothermometers: application to the thrust faulted disturbed belt of Montana, USA: Soc. Econ. Paleo. Mineral. Spec. Publ., 26(1979) 55.
37. B. Roberts and R.J. Merriman, (1990) Geol. Magaz., 127(1990) 31.
38. N. Vergo and R.H. April, Clays and Clay Minerals, 30(1982) 237.
39. C.E. Weaver, Clays, Muds and Shales, Elsevier, Amsterdam, 1989.
40. A. Inoue and M. Utada, Amer. Mineral., 76(1991) 628.
41. J.H. Ahn and D.R. Peacor, Clays and Clay Minerals, 33(1985) 228.

Analysis of colour rhythmites in sensitive marine clays (Leda Clay) from eastern Canada

J.B. Percival[a], J.M. Aylsworth[a], and A. Fritz[b]

[a] Geological Survey of Canada, 601 Booth Street, Ottawa, Ontario, Canada K1A 0E8
[b] 645 Waterloo Row, Fredericton, New Brunswick, Canada E3B 1Z6

In eastern Canada, Leda clay is a highly sensitive sediment associated with catastrophic earthflows. Leda clay is composed of clay- and silt-sized particles of glacially-ground material derived from the Precambrian Shield and deposited in the Champlain Sea. Within the marine deposits is a thick sequence (~20-25 m) of colour-banded rhythmites formed of couplets of alternating blue-grey and reddish-grey silty-clay. Formation of the rhythmites has been under speculation for many years. There are no significant differences between the grey and red bands with respect to major mineral or total Fe content. However, Mössbauer results show that the colour is related to iron mineralogy. The red layers contain hematite whereas the grey layers contain only traces of magnetite. In the non-magnetic species, the ferric/ferrous ratio is dramatically different, being much larger in the red bands. Based on Mössbauer results, these cores represent extreme redox banding in marine sediments. The rhythmites may be varves, however, age-dating is necessary for confirmation. The availability of oxides, the fine-grain size, cool climatic conditions, low organic content, and rapid sedimentation all contributed to their formation and preservation.

1. INTRODUCTION

Highly sensitive marine sediments are found throughout the Ottawa-St. Lawrence Lowlands of eastern Canada. Although informally known as Leda Clay, these sediments are actually neither composed of clay minerals nor necessarily clay-sized particles. It is composed of clay- and silt-sized particles of glacially-ground material largely derived from the Precambrian Shield and deposited in the Champlain Sea, a temporary sea at the end of the last glaciation (Fig. 1). These sediments, upon disturbance, may undergo a significant loss of shear strength, liquefy and flow as a viscous mud, rapidly devastating vast areas (up to 45 ha) of flat land behind the initial slope. Their sensitivity is a constant threat to local communities, resulting in millions of dollars in damage in the event of a landslide (Fig. 2).

An unusual characteristic of Leda clay is the occurrence of a thick sequence (~20-25 m) of colour-banded rhythmites formed of couplets of alternating blue-grey and reddish-grey silty-clay. Several theories attempt to explain the origin of this coloured banding. These include size sorting of red clays, relation to seasonal melt cycles, and oxidation events based on exposure of the sea bottom and the availability of oxides [1]. None of these, however, have been proven explicitly. The development of red clay is usually associated with tropical environments, and is typically a function of iron content [2]. As tropical conditions were not present in this region at the time of deposition, or since then, the colour must be derived from the *in situ* primary materials or from their oxidation products [1]. This paper reports on the mineralogical and geochemical properties of the rhythmites and speculates on their origin.

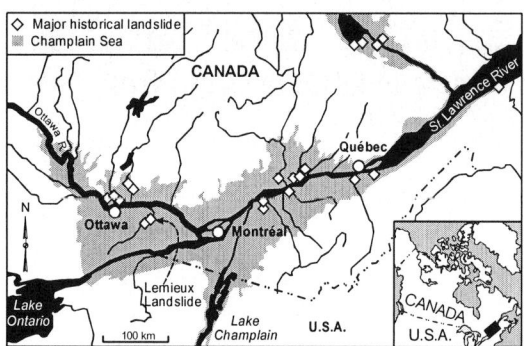

Figure 1. Maximum extent of Champlain Sea and location of major historical landslides.

Figure 2. Lemieux landslide: debris has impounded the river, which has just started to overtop the dam.

2. CHAMPLAIN SEA SEDIMENTS

At the end of the last ice age, the Champlain Sea inundated approximately 55, 000 km^2 of the St. Lawrence Lowlands (Fig. 1) while the land was still isostatically depressed from the weight of the ice sheet. The Ottawa Valley inundation commenced ca. 11.4 ka B.P. [3]. As the ice sheet withdrew from the vicinity, isostatic rebound of the land caused the Sea to regress. A large prograding delta followed the retreating shore, forming an extensive, and locally thick, cap of sand and silt over much of the fine marine deposits. By ca. 10 ka B.P. the Champlain Sea had receded from the Ottawa basin and by 9 ka B.P. the Champlain Sea ceased to exist [4]. As uplift continued, rivers eroded the deltaic and marine sediments, cutting terraces and forming deep valleys. To this day, these slopes remain a landslide hazard where underlain by thick deposits of sensitive marine clays, and, particularly, where the clays are overlain by the sand unit.

The marine sequence records an upward change from a deep-water, high-salinity, marine environment to estuarine conditions (Fig. 3). Frequency of marine shells and salinity decrease upwards in the sequence. The stratigraphically lowest marine sediments are grey, massive to weakly-colour-stratified clay and silty clay [1], deposited at the time of deglaciation in a salinity-stratified sea [5]. These are overlain by a coarsening-upward sequence of rhythmites of grey- and red-coloured, fine-grained sediment with occasional silt or sand layers and represent the distal deposits of the prograding delta. The rhythmites are overlain by coarser deltaic deposits.

In the boreholes studied in the Ottawa region, a maximum of 270 colour-pairs were counted. The red bands are characterised by occasional black mottles which disappear after a few hours exposure to air. The grey bands are characterised by occasional, small blebs or thin partings of silt or fine sand. The rhythmites are seen first, at depth, as narrow clay bands (< 0.2 - 1 cm thick) with very subtle colour differences. The thickness of individual bands increases with elevation in the section, approaching 10 cm or more per band nearer the top. With the exception of the lowermost part of the sequence, a silt band is associated with each colour pair. Thickness of this silt band also increases with elevation, varying from narrow partings at depth to thicknesses of several centimetres. Within each rhythmite, red clay grades up into grey clay which in turn grades up into the thin silt parting or band.

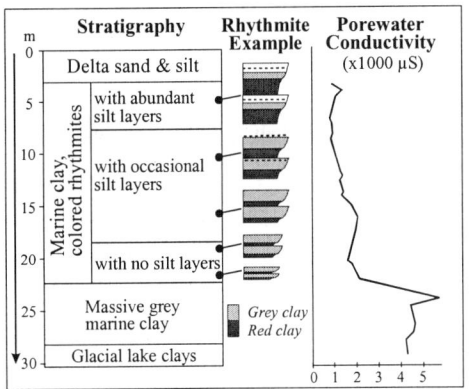

Figure 3. Stratigraphic log from borehole 97-1, with schematic representation of changing thickness of colored rhythmites. Higher conductivity indicates higher porewater salinity.

Figure 4. Scanning electron photomicrograph (SEM) of grey banded sediment sample, 97-1-19 (15.86 m) showing typical texture (1=amphibole; 2=plagioclase; 3=quartz; 4=K-feldspar; 5=biotite; 6=chlorite; 7=pyrite;

There is a sharp contact between the silt and the overlying rhythmite. In the lower part of the section, the thickness of the grey band is generally greater than the red band. Near the top of the section the thickness of the red band is greater, the thickness of the silt band increases, and the thickness of the grey band diminishes until it finally disappears, ending the rhythmite sequence.

3. MATERIALS AND METHODS

Following the 1993 Lemieux landslide (Fig. 2), much interest was generated in the study of Leda clay. Nine geotechnical boreholes were drilled in the vicinity of the landslide. For the rhythmite study, samples were collected for detailed analyses from core recovered from several of the boreholes. Pairs of samples were collected throughout the core from adjacent red and grey bands. Samples were selected based on the brightness of the red band [6].

Total iron as Fe_2O_3 and FeO was determined using XRF and chemical methods, respectively. Semi-quantitative mineralogy of clay-sized (<2 μm) separates was determined by XRD analysis using a Philips PW1710 automated powder diffractometer equipped with a graphite monochromator, Co Kα radiation set at 40 kV and 30 mA. The XRD patterns were captured digitally and processed using JADE™ (v. 3.1) software [7]. Polished thin sections and grain mounts (dilute suspension pipetted onto a carbon impregnated tape) were examined under a Leica Cambridge Stereoscan S360 SEM equipped with an Oxford/Link eXL-II energy-dispersion X-ray analyser, Oxford/Link Pentafet Be window/light element detector, and an Oxford/Link Tetra backscattered electron detector. The SEM is operated at an accelerating voltage of 20kV and a working distance of 14 to 25 mm.

Four samples were selected for Mössbauer analysis. Room temperature (RT = 22 °C) ^{57}Fe Mössbauer spectra were collected in transmission mode using random orientation powder absorbers having 85.5(5) mg/cm^2 of freeze-dried sample in holders with 1/2 inch diameter windows. Calibration was obtained using an enriched ^{57}Fe foil at RT and all centre shifts (CSs) are reported with respect to the CS of metallic iron at RT. The transducer was operated in constant acceleration mode and folding was performed to achieve a flat background. All raw folded Mössbauer spectra were fit with the Voigt-based fitting method of Rancourt and Ping [8] for QSDs and simultaneous hyperfine field distributions (HFDs) for the bulk Fe-oxide phases that were present, using the Recoil software of Lagarec and Rancourt [9].

4. RESULTS AND DISCUSSION

4.1. Physical properties

The samples are silty-clay with an average silt content of 77 wt% containing a prominent peak at 4-8 μm. The clay size (< 2 μm) content ranges from about 6 to 45% and averages 21% and sand averages about 2%. Sand and coarser silt found in a few samples may represent thin partings or small lenses within the finer-grained sediments, although attempts were made to avoid sampling these lenses. These were most noticeable in the grey bands, particularly higher in the section. The grey bands are somewhat coarser than the red bands although this difference decreases with depth as the grey bands become finer-grained. Typical moist Munsell colours for the grey bands are 10Y 4/1 to 10Y 5/1 and for the red bands 5YR 5/3 to 2.5YR 4/2.

There are slight differences in the physical properties between colour bands. There is also a relationship with depth in the sections, which is likely a function of grain size, moisture content, and pore water salinity. Atterberg Limit tests indicate that liquid and plastic limits range from 27.3 to 75.3, and 13.6 to 37.1, respectively. Plasticity indexes are 10.3-44.13; liquidity indexes are 0.40-2.75. The limits of the red bands are higher, decreasing slightly with depth. The limits of the grey bands increase with depth as the sediment fines, becoming closer to those of the red bands. The moisture content varies greatly, from 30 to 80+ wt%. Red bands have a greater moisture content than the grey, reflecting the finer texture of the red sediment. Moisture content of the red bands decreases significantly with depth, becoming less than the grey near the bottom. The moisture content commonly exceeds the liquid limit. The liquidity index is highest in the grey bands.

Leda clay is very sensitive to shear stress. In general, undrained shear strengths increase with depth, reflecting the increased salinity of the lower units (salt in the pore water offers a bonding strength to these sediments). There is little difference between the two colours. Remolded shear strengths are extremely low (commonly 0.03-0.6 t/m^2) with the lowest values associated with grey bands.

4.2. Chemical and mineralogical properties

The iron content of the grey and red bands are similar (Table 1). Average total Fe_2O_3 content is 7.56 and 7.26 wt% for grey and red bands, respectively. The FeO content averages 3.8 wt% for grey and 3.1 wt% for red. Although it appears that there is a slight difference between the red and grey bands, it may not be significant given the small number of samples analyzed.

Semi-quantitative mineralogy of the clay-size fraction is also shown in Table 1. The samples are dominated by quartz, plagioclase and K-feldspar with subordinate illite/mica, chlorite and amphibole. Illite is partly biotitic. Carbonate content (calcite + dolomite) averages less than 5 wt%. Trace amounts of smectite and/or a mixed-layer clay mineral (illite/smectite) were detectable in some samples. Hematite was observed in trace amounts in a few of the red samples and magnetite was not detectable by XRD in any of the samples. There appear to be no significant differences between the red and grey bands.

Table 1: Comparison of ferrous iron (measured as Fe_2O_3T) with ferric iron and semi-quantitative mineralogy of red and grey banded samples as determined by XRD.

Sample	Col.	Fe_2O_3T	FeO	Qtz	Pfs	Kfs	Amp	Ill	Chl	Sm	ML	Cal	Dol	Hem
96-1-5C	grey	8.21	4.1	14	30	21	5	11	8			7	4	
96-1-6	red	6.90	3.5	19	27	19	6	9	17		tr	tr	3	
96-1-7B	grey	8.34	4.2	19	36	17	9	5	8		tr	6		
96-1-8B	red	7.47		17	26	23	5	9	8	tr	tr	8	4	tr
96-1-11	grey	7.25	4.2	21	34	11	7	12	13	tr			2	
96-1-12	red	7.92	3.3	18	27	22	6	8	11	tr	tr	4	4	tr
96-1-22	grey	8.42	4.2	16	31	20	6	8	12			5	2	
96-1-23	red	6.90	3.1	15	26	26	5	8	9	tr	tr	8	3	tr
96-2-20	grey	6.88	3.0	18	26	24	6	6	12		tr	5	3	
96-2-21	red	8.08		15	31	19	8	10	12	tr	tr	3	2	
96-2-23	grey	8.17		19	31	17	6	10	13		tr	2	2	tr?
96-2-22	red	5.96	2.6	23	26	27	4	10	8		tr		2	tr
97-1-M1	grey	6.30	3.3	20	29	14	6	12	15			2	2	
97-1-M2	red	7.59	3.1	12	21	17	5	17	22	tr		4	2	
97-1-M3	grey	6.94	3.6	15	31	12	8	16	16			1	1	
97-1-M4	red	7.27	2.8	18	20	16	6	13	21	tr		4	2	

Qtz = quartz; Pfs = plagioclase feldspar; Kfs = K-feldspar; Amp = amphibole; Ill = illite; Chl = chlorite; Sm = smectite; ML = mixed-layer clay mineral; Cal = calcite; Dol = dolomite; Hem = hematite.

A host of trace and heavy minerals were detected during SEM analyses of polished thin sections. Pyroxene, epidote, rutile, zircon, monazite, apatite, ilmenite, sulphides such as pyrite, chalcopyrite and sphalerite all occur in trace amounts in both the red and grey samples. In addition, trace amounts of iron oxides occur in both types of bands and include hematite and titano-magnetite. Figure 4 illustrates the texture typical of the Leda clay materials. Note that the grains tend to be angular and very poorly sorted. Although detrital pyrite appears in both types of bands, framboids were only observed in the grey bands as shell replacements (Figs. 5a and b). The presence of the framboids indicates reducing conditions.

Mössbauer results are summarized in Table 2 and examples of the spectra are shown in Figures 6a and b. The Fe^{3+}/Fe^{2+} ratio is the ferric/ferrous ratio of species that are para-magnetic and superparamagnetic at room temperature. Note that the ratios are different between the red and grey bands, with the red bands having a higher ferric iron content. This may indicate the presence of ferrihydrite or that ferrous-iron bearing silicates such as biotite

Table 2: Mössbauer parameters for Leda clay samples.

Sample No.	Colour	Fe^{3+}/Fe^{2+} P/SP	% Fe Hematite	% Fe Magnetite	% Fe Amorphous
97-1-M1	grey	0.531	2	2	13
97-1-M2	red	0.884	7	0	13
97-1-M3	grey	0.514	1	2	16
97-1-M4	red	0.864	9	0	12

P = paramagentic; SP = superparamagnetic

have become oxidized. The remaining columns show the percentage of Fe in hematite, magnetite and "amorphous Fe oxides" which would include poorly crystalline maghemite and hematite. Although there is a minor amount of Fe tied up in hematite for the grey bands, there is no Fe present in magnetite for the red bands. The amount in "amorphous Fe-oxides" is similar for both red and grey bands. Based on these results, it is evident that hematite is contained in the red bands and the grey bands contain only traces of magnetite. The spectra for 97-1-M3 (grey) and 97-1-M4 (red) show distinct differences illustrating the data presented in Table 2. These results indicate that these materials represent extreme redox banding in marine sediments.

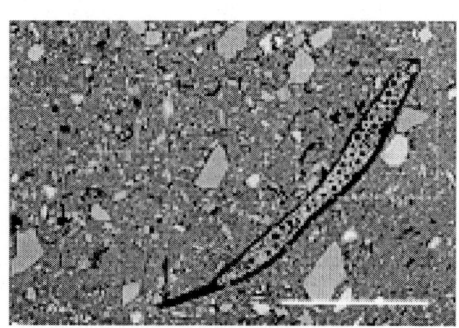

Figure 5. SEM photomicrograph of grey- banded sediment sample 97-1-37 (25.64 m).
(a) pyrite framboids replacing a shell (1=K-feldspar; 2=plagioclase; 3=dolomite; 4=quartz; 5=amphibole; 6=calcite); (b) close-up of framboidal pyrite.

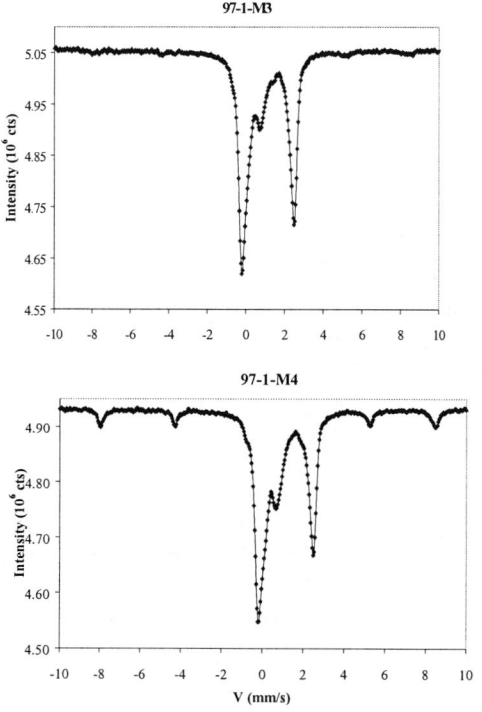

Figure 6. Mössbauer spectra of 97-1-M3 (grey) and 97-1-M4 (red).

4.3. Origin of rhythmites

Several theories to explain the origin of these rhythmites have been suggested [1], including grain size effects, relation to seasonal melt cycles (i.e., formation as varves), or syn- to post-depositional oxidation/reduction effects.

Fritz [6] conducted an experiment to test if the red colour was strictly a result of the presence of clay-sized material. The clay fraction was completely removed (by sedimentation) from a bulk sample composed of equal amounts of grey and red material. The clay-size material was then added to the "de-clayed" material in increments of 5% to give a range from 20% to 95% clay-size. The samples became more reddish in colour with the increase in clay content, from dark grayish brown (2.5Y 4/2) at 20% to brown (10YR 5/3) at 100% clay-sized content. A drastic colour change occurred with each 10% increase in clay content up to 60%. At higher clay contents, the sample colour remained the same. These results indicated that size sorting played an important role in the colour difference found in the red and grey bands. Fritz [6] suggested that the differences in grain size were related to the depositional environment and were probably seasonal. Fritz [6] also suggested that the colour difference was probably not just due to grain size but may be due to an increase in iron content within the red bands. However, differences in iron content were later determined not to be significant (see Table 1).

Could these rhythmites be varves? Varves formed in a glacial environment typically are composed of a lighter-coloured, coarser-grained, summer layer that grades gradually into a darker-coloured, finer-grained, winter layer. Normally there is a sharp boundary between the varves, at the base of the coarser-grained layer, which represents arrival of the spring freshet. The rhythmites in the Leda clay have a slightly coarser grey to blue-grey band (spring/summer?) and a slightly finer-grained red layer (winter?). There is a sharp boundary, but it occurs at the bottom of the red band. The red band grades upwards into the grey band and there is a fine sand- to silt-rich parting at the top of the grey (see Fig. 3). This parting increases in thickness towards the top of the sequence as does the thickness of the red bands. Although the rhythmites do not possess the common characteristics of varves as described above, their consistent cyclicity throughout a thick sediment sequence, and occurrence here as well as in other areas of the Champlain Sea suggest that they could be annual. Regional glacial history implies that the entire deep-water marine, pro-delta and deltaic sequences represent about a 1500-year period from initial transgression to final regression of the Champlain Sea from the Ottawa valley. This package of rhythmites may indeed represent a 270-year period of cyclical sedimentation distal to a prograding delta and the underlying massive grey marine unit and overlying deltaic sands account for the remaining period of time. No age dates are available at this time.

Red and grey rhythmites/varves have been found elsewhere in Canada. They have been described in glaciolacustrine sediments in the Dryden area of northwestern Ontario where they have been attributed to sediment from the Proterozoic Sibley Group red beds and derived tills of the Thunder Bay area [10, 11, 12]. Red and grey rhythmites in Holocene sediments in Georgian Bay of Lake Huron, Ontario, have been attributed to flow from glacial Lake Agassiz through the Thunder Bay area into the Great Lakes [13]. Similar colour rhythmites have also been observed in glaciomarine sediments in northern British Columbia [14]. Colour banding in marine sediments is common throughout the Ottawa valley. Although the environment of deposition differs (e.g., marine vs. freshwater), the relationship to a temporary deglacial waterbody is constant. These environments would comprise cold waters and rapid deposition consistent with glacial retreat. Note, although the Champlain Sea was a marine environment, the amount of meltwater entering into this estuary ensured that water salinity was low (see Fig. 2). The red and grey rhythmites in all these areas may indeed be related to provenance, however, at present, in some areas, sources may be unknown.

It seems unlikely that oxidation of the layers during or after deposition caused the colour banding. The iron oxide grains (hematite and (titano-)magnetite) observed under the SEM seem to be detrital rather than authigenic. Mössbauer results indicate that the difference between the red and grey bands is mainly due to the type of iron oxide present or absent (i.e., in the case of no magnetite in the red band). The formation of framboidal pyrite in the grey bands indicates that reducing conditions persisted during incipient diagenesis. Weathering (except in the immediate near surface zone) has been limited by the present-day temperate climatic conditions, pore-water salinity, low hydraulic conductivities and low hydraulic gradients [15]. The absence of sufficient organic matter for appropriate microbial activity may have prevented reduction of the ferric oxides in the red bands [16]. Rapid sedimentation during this time could quickly bury the layers thus preserving their inherited redox characteristics.

5. SUMMARY AND CONCLUSIONS

Within the highly sensitive marine sediments of the Champlain Sea is a thick sequence of colour-banded rhythmites formed of couplets of alternating blue-grey and red silty-clay. Grain size, physical properties, iron chemistry and major mineralogy are not significantly different between the bands. Mössbauer results do show that hematite is more abundant in the red band and only traces of magnetite occur in the grey band; there is no magnetite in the red band. These rhythmites possess characteristics similar to varves and may represent a 270-year period in the regressive phase of the Champlain Sea. The availability of oxides, the fine-grain size, cool climatic conditions, low organic content, and rapid sedimentation all contributed to the formation and preservation of these rhythmites/varves.

ACKNOWLEDGEMENTS

The authors are grateful for reviews by H. Thorliefson, W. Huff and Th. De Putter and discussions with H. Thorleifson (GSC) and J.K. Torrance (Carleton Univ.). Special thanks to D. Rancourt (Univ. Ottawa) for Mössbauer analyses, A. Tsai and P. Hunt for SEM analyses, P. Belanger for chemical analyses, and R. Lacroix for digital cartographic assistance.

REFERENCES

1. N.R. Gadd, Geol. Surv. Can. Paper 85-21, (1986).
2. U. Schwertmann and R.W. Fitzpatrick, in H.C.W. Skinner and R.W. Fitzpatrick (eds.), Biomineralization, Processes of Iron and Manganese - Modern and Ancient Environments. Catena Suppl. 21, Catena Verlag, Cremlingen-Destedt, Germany, (1992) 7.
3. C.G. Rodrigues, in N.R. Gadd (ed.), The Late Quaternary Development of the Champlain Sea Basin, Geol. Assoc. Can., Spec. Paper 35 (1988) 155.
4. S. Occhietti, in R.J. Fulton (ed), Quaternary Geology of Canada and Greenland, Geol. Surv. Can. v. 1 (1989) 350.
5. C.G. Rodrigues and S.H. Richard, in Current Research, Part B, Geol. Surv. Can., Paper 85-1B (1985) 401.
6. A. Fritz, B.Sc. Thesis, Environmental Science, Carleton Univ., Ottawa, 1998.
7. J.B. Percival, P. Hunt and M. Wygergangs, in Current Research Part E, Geol. Surv. Can. Paper 2001-E9 (2001).
8. D.G. Rancourt and J.Y. Ping, Nucl. Instr. Meth. Phys. Res. B (NIMB), 58 (1991) 85.
9. K. Lagarec and D.G. Rancourt, Recoil: Spectral analysis and data treatment software for Mössbauer spectroscopy (http://www.science.uottawa.ca/phy/~recoil/) (1998).
10. G. Rittenhouse, Amer. J. Sci., 28 (1934) 110.
11. S.C. Zoltai, Proc. Geol. Assoc. Can., 13 (1961) 61.
12. T.A. Warman, M.Sc. Thesis, Geology, Univ. of Manitoba (1991).
13. M. Lewis, Personal Comm., 1999.
14. M. Geertsema, Personal Comm., 1998.
15. J.K. Torrance and J.B. Percival, 12th ICC Proceedings, Argentina (this volume).
16. J.K. Torrance, Personal Comm., 2001.

Berthierine formation in spotted slates

M.D. Ruiz Cruz[a] & E. Galán[b]

[a]Departamento de Química Inorgánica, Cristalografía y Mineralogía. Facultad de Ciencias. Campus de Teatinos. 29071 Málaga (Spain)

[b]Departamento de Cristalografía, Mineralogía y Química Agrícola. Facultad de Químicas. 41071 Sevilla (Spain)

FAX: 3452201300
e-mail: mdruiz @ uma. Es

Berthierine has been identified in spots developed in slates from the Maláguide Complex (Betic Cordillera, Southern Spain) and has been ascribed to a thermal metamorphism subsequent to the Alpine-regional one. Two assemblages have been found: a) berthierine+muscovite±chlorite assemblage, characterized by the development of Al-poor oxidized berthierine, which preferentially develops in Fe-rich spots; and b) muscovite+chlorite+biotite association, developed in micaceous spots. Textural features indicate that berthierine formation occurred from both chlorite and muscovite. In the first case berthierine packets frequently contain 14-Å interstratified layers. This transformation involves a notable Fe-enrichment relative to the original chlorite. Formation of berthierine from muscovite occurs, on the contrary, through a complex set of muscovite-berthierine interstratifications and intergrowths. AEM data for biotite and berthierine suggest that the Fe availability may be an important factor controlling the berthierine formation in the Maláguide spotted slates.

1. INTRODUCTION

Spots formed by muscovite, biotite and/or chlorite are frequent in contact aureoles in pelitic rocks. These textures have been interpreted classically as the result of cordierite poikiloblasts alteration [1]. Nevertheless, these phyllosilicate aggregates can also represent the first product of the metamorphic reactions at low temperature [2], mainly in the outermost zone of the contact aureoles.

The assemblage chlorite-muscovite-biotite is common in both contact and regional low-grade rocks but the association muscovite-berthierine has not been previously reported in these contexts. In fact, berthierine has been preferentially described in sedimentary iron formations [3-4], and has been commonly considered as a low-temperature variety of chlorite, which evolution during diagenesis and progressive metamorphism leads to Mg-chamosite [5-

9]. +Nevertheless, Jiang *et al.* and Slack *et al.* [10-11] found hydrothermal and metamorphic berthierine partially replacing chlorite, this replacement implying that berthierine is more stable than chlorite under certain (undetermined) conditions. Similarly, the occurrence of berthierine in rocks of the prehnite-pumpellyite facies suggests the possibility of a stability field for berthierine, common to chlorite with similar but non identical composition [12].

This study supplies new data on the relationships between chlorite and berthierine, found in contact aureolas, which has not been previously investigated.

2. GEOLOGICAL SETTING

The Maláguide Complex, a part of the Betic Cordillera, Southern of Spain (Fig. 1). displays a low metamorphic grade. Detailed descriptions of this Complex in the Málaga area, are given by Azèma and Mäkel [13-14]. Both tectonical and petrological data, indicate that two main metamorphic events, the Variscan and the Alpine, affected this Complex [15-16]. Temperatures deduced from variations on chlorite composition were in the order of 300 ºC during the Variscan event whereas the range 150 ºC - >300 ºC, was identified in Alpine chlorites [15]. On the basis of the illite crystallinity (IC) and chlorite crystallinity (CC) indices evolution, three zones, the diagenetic, the anchizone and the epizone, have been identified in the studied area, which approximately correspond to the Permo-Triassic, the Carboniferous-Devonian and the Silurian-Ordovician sequences, respectively. The later sequence, also known as the phyllite member is also characterized by the presence of numerous dikes of intermediate rocks. Because the scarce thickness of the dikes (< 1m), the contact metamorphism on the enclosing slates cannot be identified in most of the Maláguide rocks, but spotted slates, with textural features characteristic of the external aureole zone, have been identified in a restricted zone near Málaga.

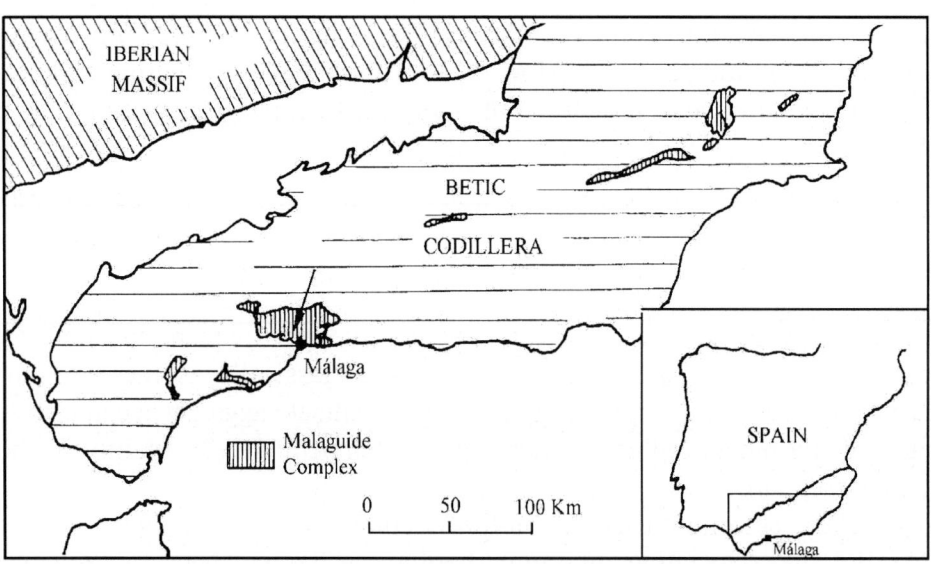

Figure 1. Geological setting of the Maláguide Complex and location of the studied samples (arrow).

3. SAMPLES AND EXPERIMENTAL

The studied samples have been collected between 10 and 20 km Northern Málaga (Fig. 1), in a zone with upper epizone characteristics [17]. Samples were mineralogically investigated by X-ray diffraction and optical microscopy. Modal analyses were used in the determination of the spot content. From this previous study, three samples showing different grades of spots development were selected for a detailed study by transmission and analytical electron microscopy (TEM/AEM).

X-ray patterns were recorded using a Siemens D-5000 diffractometer with CuKa radiation and graphite monochromator, operated at 40 mA and 40 kV; 0.01° step size and 1 s counting time. Thin sections prepared with orientation perpendicular to the main schistosity were used for the transmission and analytical electron microscopic (TEM-AEM) study. Washers were attached to selected areas and later separated from the glass backing. These areas were ion-thinned and carbon-coated and examined in a 200 kV Philips CM-20 scanning-transmission electron microscope (STEM) fitted with a solid-state detector for energy dispersive analysis (EDX). Standards were natural albite for Na, natural muscovite and biotite for K, natural albite, spessartite and muscovite for Al, natural biotite and olivine for Mg and Fe, natural spessartite for Mn and natural titanite for Ca.

Although numerous analyses were carried out, only those obtained from areas well characterized by lattice-fringe images and electron diffraction patterns were used in tables and figures. Normalization of muscovite and biotite formulae was based on charge balance relative to 11 oxygens. Normalization of berthierine and chlorite analyses was carried out following Jiang *et al.* [10], based on charge balance relative to 14 oxygens and a total of ≤10 tetrahedral and octahedral cations. The relative number of Fe^{2+} and Fe^{3+} ions was determined by charge balance.

4. RESULTS

The characteristic mineral assemblage of the spot-free slates, deduced from the XRD patterns is quartz-albite-muscovite-chlorite, although minor mica/chlorite mixed-layers are present in some of the samples. Slates show a well defined schistosity, and frequent crenulation. Mineralogical differences among the selected samples mainly affect the ratio mica/chlorite, the presence of chlorite-bearing mixed-layers and the spot content, which represents about 18% in volume in samples with the highest spot development.

Most of the spots are ovoid in form and show sizes in the order of 0.5-2 mm. Spots preferentially develop in chlorite- Fe-oxides- and mica-rich bands, and show some concentrical zones, with variable development. Spots formed from chlorite- and/or ore-rich bands are characterized by the presence of an external yellow to reddish zone, which surrounds an internal lighter zone. Spots formed in mica-rich bands are colorless, although they also appear frequently surrounded by an external yellow rim with very variable thickness. These spots appear is some cases formed by aggregates of phyllosilicates and quartz but more frequently they show an inner, isotropic zone.

4.1. TEM/AEM data from slates

TEM observation of the spot-free areas from spotted slates reveals that muscovite and chlorite crystals form thick packets, which are parallely oriented following the schistosity planes, and show variable deformation degree. Muscovite crystals show lattice fringe images displaying periodicities of both 10 and 20 Å. The electron diffraction (SAED) patterns reveal an ordered two-layer polytype. The high-resolution images of the largest non-deformed chlorite crystals show a high regularity and scarce structural defects and the SAED patterns correspond to ordered one-layer or two-layer polytypes.

AEM data of muscovite indicates that Si content ranges between 3.10 and 3.25 (± 0.02) *apfu*. The (Fe+Mg) content is variable, reaching up 0.50 (± 0.04) *apfu*. AEM data of chlorite show high Al(IV) content (from 1.28 to 1.50 (± 0.05) *apfu*) and high Fe content. The structural formulae show slightly lower Si content than most of the chlorite in spots-free slates with similar stratigraphical position. In some of the analyzed chlorites Al(IV)>Al(VI), which requires, according to the criteria by Foster [18] the presence of some Fe^{3+}. These chlorites may be classified as chamosite (following the recommendations of the AIPEA Nomenclature Committee, Bailey, 1980) or as ripidolite (following the Hey's (1954) classification).

4.2. TEM/AEM data from the spots

TEM observation and specially the AEM data obtained in differently textured spots reveal that they contain two characteristic mineral associations, which are present, although with a very different development, in most of the studied spots. The assemblage muscovite+chlorite+biotite preferentially develops in micaceous spots, and in a lesser extent, in the external rim of the other spots. The assemblage muscovite-berthierine (with occasional chlorite) is the most abundant in yellow to reddish spots, where it appears in both the external zone and the inner part of the spots, in contact with amorphous phases.

In micaceous spots, muscovite packets appears in contact with both chlorite and biotite ones. Chlorite shows in these areas variable structural and chemical features. So disordered to very ordered polytypes appear either as alternating packets or within a single packet. The AEM data of both types of chlorites (Table 1) reveal that disorder is accompanied by slight changes in the Si/Al ratio, which leads to high Al(IV) content in the disordered packets. Since most of the obtained structural formulae show Al(IV)>Al(VI), the obtaining of octahedral occupancies = 6 also requires that some Fe to be as ferric iron. AEM data of muscovite (Table 1) reveal a relatively high Na content (about 25%), as well as very low contents of both Fe and Mg.

Biotite packets frequently contain structural defects, mainly interstratifications of thin packets with 7-Å and 14-Å periodicities. The 14 Å chlorite layers appear regularly spaced within the biotite packets, and suggest that biotite mainly formed from transformation of chlorite. AEM data of biotite (Table 1) reveal an uniform composition, with Si content near 2.5 (±0.02) *apfu*, and Al content in the order of 1.8 (±0.05) *apfu*. Analyses of biotite also contain Na in a proportion similar to muscovite. The low but constant Ti content also characterize biotite.

In Fe-rich spots, the assemblage muscovite+berthierine±chlorite dominates. In the external zones of the spots, muscovite and berthierine form either crystals containing a single phase or fine intergrowths of both phases. The high-resolution images of the discrete berthierine packets (Fig. 2) frequently show large areas with regular 7-Å periodicity in which 14-Å fringes appear regularly spaced, indicating that berthierine formation involved chlorite. The

SAED patterns only show reflections with 7-Å spacing, which confirm that these packets have a serpentine-like structure.

The analyses of these packets correspond to a Fe-Al-Mg phyllosilicate, similar to some Fe-rich oxidized chlorites. The structural formulae, calculated on the basis of 14 oxygens indicates that tetrahedral Al is in the order of 1.8 *apfu*. In the octahedral positions Al content is frequently lower, the charge deficit being compensated, in appearance by the presence of Fe^{3+}. Some of the obtained formulae are within the compositional field of berthierine, according to the data by Brindley [19]. Nevertheless, most of these show total Al content clearly lower than that characteristic of berthierine. In fact the obtained formulae indicates some grade of solid solution between berthierine and other 7-Å phase, probably Fe-amesite [20].

Table 1.- Selected AEM analyses of muscovite (Ms), chlorite (Chl), berthierine (Ber), biotite (Bt) and amorphous phases (A.Ph)

Element (wt.%)	Ms	Ms	Chl	Chl	Bt	Bt	Ber	Ber	A.Ph	A.Ph
Si	45.17	42.29	26.55	24.52	32.11	31.28	21.85	21.52	41.58	46.96
Al	41.12	41.34	24.49	27.41	23.27	22.08	22.61	20.97	41.95	39.26
Fe	0.63	0.68	35.18	32.41	19.30	20.08	41.46	47.17	7.97	2.84
Mg	1.40	1.46	13.26	15.15	11.06	13.76	12.96	9.05	2.98	--
Mn	--	--	0.52	0.52	0.20	0.43	1.12	1.28	--	--
Ti	--	--	--	-	1.10	2.30	--	--	--	--
K	8.75	10.52	--	-	10.56	9.17	--	--	1.04	--
Na	2.93	3.71	--	-	2.41	1.10	--	--	--	--
Structural formulae										
Si(IV)	3.08	3.05	2.66	2.45	2.55	2.47	2.18	2.16		
Al(IV)	0.92	0.95	1.34	1.55	1.45	1.53	1.82	1.84		
Al(VI)	1.88	2.03	1.11	1.19	0.40	0.21	0.44	0.26		
Fe(+3)	--	--	0.21	0.36	--	--	1.38	1.58		
Fe(+2)	0.04	0.05	3.31	2.88	1.53	1.55	2.76	3.14		
Mg	0.19	0.11	1.33	1.52	0.88	1.08	1.30	0.90		
Mn	--	--	0.05	0.05	0.02	0.03	0.12	0.12		
Ti		--	--	--	0.09	1.17	--	-		
Total oct.	2.11	2.19	6.00	6.00	2.92	3.04	6.00	6.00		
K	0.60	0.76	--	--	0.84	0.73	--	--		
Na	0.20	0.27	--	--	0.19	0.08	--	--		
O	11	11	14	14	11	11	14	14		

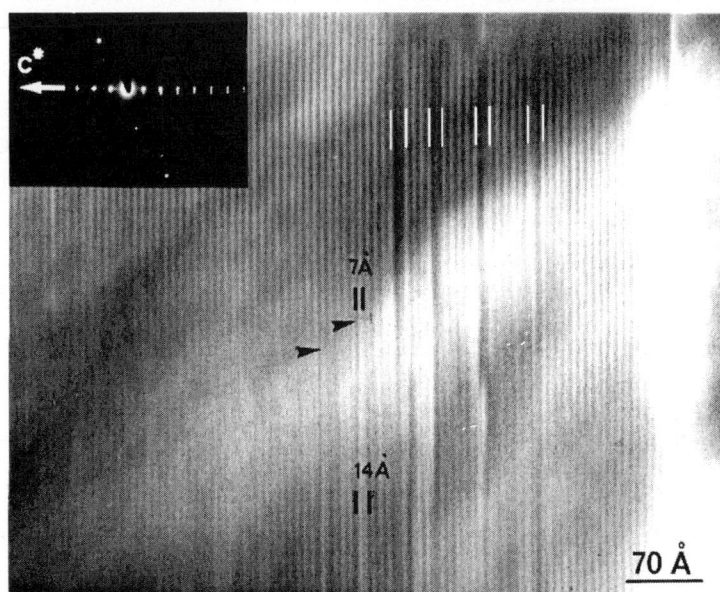

Figure 2. High-resolution image of a berthierine packet showing 7-Å periodicity. Some fringes with 14-Å periodicity appear interstratified. The SAED pattern (inset) shows intense reflections of berthierine.

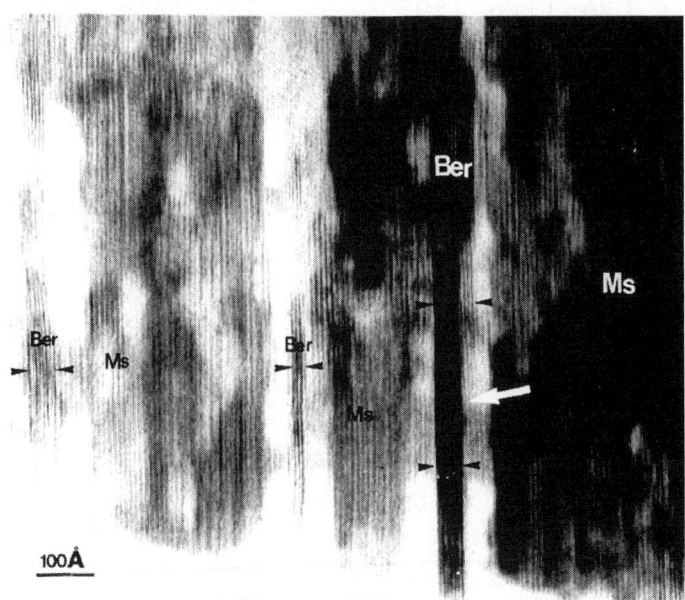

Figure 3. High-resolution image of a muscovite-berthierine packet, showing intergrown 10-Å and 7-Å groups of layers. Black arrows mark the limits of the berthierine packets. White arrow marks the transition muscovite-berthierine layers. Ms: Muscovite; Ber: Berthierine.

Muscovite crystals frequently contain discrete packets of berthierine in the order of 50-100 Å thick. The transition between 10-Å and 7-Å packets frequently occurs through areas where narrow packets showing 10-Å and 7-Å fringes alternate. Figure 3 shows a complex intergrowth of muscovite and berthierine and mixed-layers muscovite/berthierine, with mineral boundaries preferentially parallel or subparallel to (001). These types the textural relationships between muscovite and berthierine suggest that berthierine formed from a previous phase with 10-Å periodicity, which has been, in a great part, preserved.

In the innermost part of the spots, the assemblage muscovite-berthierine±chlorite frequently appears in contact with the optically isotropic areas. The "amorphous" phases appear, at the TEM scale as short lamellae, frequently bent, and originating rounded aggregates. The AEM data collected in the isotropic areas reveal that the "amorphous" phases, mainly contain Si and Al, in a proportion similar to kaolinite, variable amount of Fe and minor Mg (Table 1).

5. DISCUSSION AND CONCLUSIONS

The present results show that Fe-rich berthierine is an important constituent of spots developed during a thermal metamorphic process. This occurrence again posses the problem of the stability field of berthierine, which has been classically considered as a low-temperature phase. Presence of berthierine in metamorphic terrains is seldom interpreted as being a relic of its incomplete transformation into chlorite, although Coombs *et al.* [12] point out the possibility that chlorite and berthierine may share a field of stable coexistence. In the spotted slates berthierine only has been identified in the spots, whereas spot-free slates only contain chlorite. Thus, the formation of berthierine during the thermal metamorphic process is clearly demonstrated.

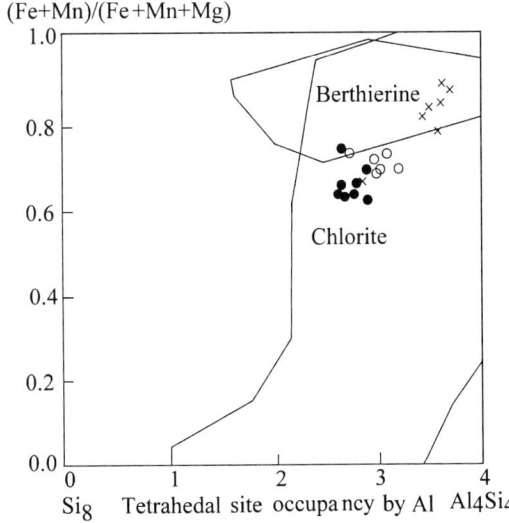

Figure 4. Fields of berthierine and chlorite after Jiang et al. [10] and plot of some chlorite and berthierine analyses from the spotted slates. Full circles: Chlorite from the spot-free areas of the spotted slates. Open circles: Chlorite from spots. Crosses: Berthierine.

Some berthierine packets show abundant deformation features and contain regularly spaced 14-Å chlorite layers, suggesting that berthierine formed from chlorite throughout a topotactic reaction. Chemical modifications of berthierine relative to chlorite include the decrease of Si and the Fe enrichment (Table 1). These modifications induce a higher Al(IV) content, which requires that some Fe to be as ferric iron in the octahedral positions. So, the structural change from the chlorite to the berthierine is accompanied by important chemical variations and suggests that berthierine was formed in microdomains where Fe content increased.

Representation of the composition of berthierine and chlorite from the studied samples in a such plot diagram similar to that used by *Jiang et al.* (Fig. 4) reveals that only one of the obtained analyses of berthierine is similar to chlorite, the other occupying a clearly different field. Thus, chlorite and berthierine cannot be regarded in this cases as polymorphs, and it is possible that berthierine represents a stable phase.

ACKNOWLEDGMENTS

This study has received financial support from the Project BTE-2000-1150 (Ministerio de Educación y Cultura, Spain). The authors are grateful to Dr. F. Nieto (University of Granada) for help in discussion concerning interpretation of data and to Dr. M. M. Abad for help in obtaining the TEM/AEM data.

REFERENCES

1. D.R.M. Pattison and R.J. Tracy. Rev. In Mineral., 26 (1991) 105.
2. A. Miyashiro. Metamorphic Petrology. UCL Press, London (1994)
3. A. Iijima and R. Matsumoto. Clays Clay Miner., 30 (1982) 264.
4. T.A. Toth, and S.J. Fritz. Clays Clay Miner., 45 (1997) 564.
5. B. Velde. Mineral. Mag., 39 (1973) 297.
6. H.J. Kisch. In: *Diagenesis in sediments and sedimentary rocks,* 2. (G. Larsen & G. V. Chilinger eds.), Elsevier, New York (1983).
7. C.D. Curtis; C.R. Highes; J.A. Whiteman and C.K. White. Mineral. Mag., 49 (1985) 375.
8. C.E.Weaver. Clays, Muds and Shales. Elsevier, Amsterdam (1989).
9. J.R. Walker and G.R. Thompson. Clays Clay Miner., 38 (1990) 315.
10. W-T. Jiang; D.R. Peacor and J. Slack. Clays Clay Miner., 40 (1992) 501.
11. J.F. Slack; W-T. Jiang; D.R. Peacor and P.M. Okita. Can. Mineral., 30 (1992) 1127.
12. D.S. Coombs; G. Zhao and D.R. Peacor. Mineral. Mag., 64 (2000) 1037.
13. J. Azèma. Estudios Geol., 17 (1961) 131.
14. G.H. Mäkel. GUA Papers of Geology, 22 (1985).
15. M.D. Ruiz Cruz. Can. Mineral., 35 (1997) 923.
16. M.D. Ruiz Cruz and B. Andreo. Clay Miner., 31 (1996) 133.
17. M.D. Ruiz Cruz and P. Rodríguez Jiménez. Clay Miner. (in press)
18. M.D. Foster. U.S. Geol. Surv. Prof. Pap., 414-A (1962) 1.
19. G.W. Brindley. Clays Clay Miner., 30 (1982) 153.
20. S.W. Bailey. Rev. in Mineral., 19 (1988) 169.

The Application of Clay Mineralogical Analysis to the Reconstruction of a Greek Bronze Age Coastal Environment

C.M. Shriner[a] and H.H. Murray[b]

[a]Departments of Classical Studies and Geological Sciences, Indiana University, Bloomington, IN 47405

[b]Department of Geological Sciences, Indiana University, Bloomington, IN 47405

The Argive plain is a subsiding coastal basin in the Peloponnese, Greece. Although there are varying explanations for the rise of a complex society on the Argive plain in the Greek Early Bronze Age (3000-2000 B.C.), archaeologists believe that there is a close link between changes in the environment and human activity.

Previous geoarchaeological analysis of the Pleistocene and Holocene stratigraphy of the Argive plain has delineated cycles of marine transgression/regression (Zangger 1991, 1993). On the west side of the bay Zangger's Holocene reconstruction for the plain has a freshwater lake, which experienced a major alluviation event in the Early Helladic II period (2500-2150 B.C.). He considers this event to be the result of soil erosion caused by increased human activity. The east side of the bay was a marine environment.

Certain aspects of Zangger's Early Bronze Age landscape reconstruction could equally apply to Holocene deltaic formation as suggested by Stanley and Warne (1997). Clay-rich sediments from outcrops on the Argive plain were compared with Zangger's Holocene stratigraphy. These sediments were examined using X-ray diffraction and particle size analysis. This was done in order to more fully understand the transition from marine to fluvial conditions.

The results of particle size analysis and determination of clay mineral content and quantity strongly suggest that the depositional environment is better explained as being deltaic. The variability over short intervals of the sand-silt-clay ratios and the differences in the smectite-illite-kaolinite ratios fit well with a prograding delta.

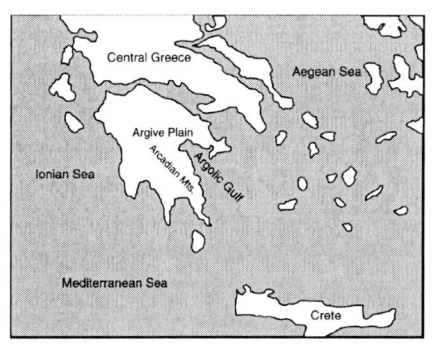

Figure 1. Generalized map of Greece and the surrounding area showing the location of the Argive Plain.

1. INTRODUCTION

The Argive plain lies in the northeast Peloponnese, Greece (Fig. 1). The plain is bounded on the west by the ridges of the Argolic-Arcadian mountain range, dominated by the Ktenias (1599 m) and Artemision (1772 m) moutain peaks. The area is locally broken by narrow valleys and small mountains. The Argive plain itself is the subaerial section of a Plio-Pleistocene sediment wedge built up by the rivers draining the surrounding mountain ranges (van Andel et al. 1993).

The drainage system runs east, either directly into the Argolic Gulf or over the Xerias and Inachos rivers, and occupies ca. 1200 sq. km^2 (Zangger 1993).

Reconstructions of the Holocene landscape for this archaeologically important area are needed in order to establish an environmental context for its significant Late Bronze Age sites, e.g. Mycenae, Tiryns and Argos. More specifically, archaeologists are interested in understanding why settlement patterns on the Argive plain changed in the later Early Bronze Age (2300-2000 B.C.). A major stumbling block for this type of geoarchaeological research is the lack of a methodology that can distinguish and verify climatological effects on geomorphic systems from those of an anthropogenic nature (Wagstaff 1981:261).

Clay mineralogical analysis offers a method to differentiate soil profiles from depositional sequences. In the present study clay-rich sediments, dated to the Early Bronze Age, from the central Argive plain were analyzed with X-ray diffraction and particle size analysis. We studied the clay mineralogy, clay mineral ratios, and particle size distribution of these sediments in order to more fully understand the differences between fluvial and marine environments on the plain itself. Subsequently, the results were used to verify and evaluate existing geomorphological and sedimentological hypotheses for physical changes in the local environment (Kraft 1972:38; Zangger 1991:13, 1993:83; Stanley and Warne 1997:2).

2. ARCHAEOLOGICAL PROBLEM

2.1 Nature of the Problem

The Argive plain witnessed over the course of the Third Millennium B.C. the emergence of a complex society. Wiencke, utilizing evidence from recent surface surveys, pointed out a significant increase in the number of sites and perhaps of population for the Argive plain in the Early Bronze Age (1989). This change in settlement patterns was gradual, beginning in the later Neolithic (ca. 5300-4500 B.C.) and continuing through the Early Helladic (EH) II (2650-2150 B.C.) period. A pivotal period in this process of urban development was the later part of EH II, i.e. Lerna III: C-D (2300-2150 B.C.). The development of central and important sites in this period coincided with the abandonment of smaller sites. She suggested a process of site nucleation that appeared to accelerate in the EH III (2150-2000 B.C.) period (1989:499).

Kraft (1972) constructed a hypothetical Holocene vertical sedimentary environmental sequence for the Argive plain. Assuming a stable or slowly rising relative sea level, he suggested a marginal marine embayment as a paleogeographic reconstruction. Almost two decades later Zangger suggested a series of transgressive/regressive (alluviation) cycles with no formation of a delta for the Argive plain in the Holocene. He used a soil science facies model for his geoarchaeological reconstruction (1993).

Zangger postulated that intensified land use due to an increased population caused soil erosion and deposition on the Argive plain. He suggested that this soil erosion resulted from a shortened fallow or a rapid clearance of steep, marginal soils for agricultural purposes. Evidence for this human-induced aggradation was the reconstruction of a specific alluviation "event" (the Pikrodafni alluvium) on the Argive plain. This alluviation "event" was securely dated to the later phases of the EH II period (Zangger 1991:7-13; Pullen 1992:47-48). Zangger's results were consistent with previous descriptive analysis for the Southern Argolid. Using a soil science facies model, Pope and van Andel had gathered a great deal of evidence from the Southern Argolid for a major episode of alluviation (the Pikrodafni alluvium) in the later Early Bronze Age (1984). Archaeologists were aware that this alluviation "event" indicated important changes in the environment may have occurred (Rutter 1993:766).

Zangger's reconstruction was never compared with Kraft's hypothetical sedimentary facies data or verified with another analytical technique. However, his hypothesis of human-induced aggradation was widely accepted by archaeologists. It has frequently been used as evidence for a cultural collapse in the EH II/III transition caused by human activity, e.g. warfare, overpopulation, deforestation, and the consequences of poor agricultural techniques (van Andel, Runnels and Pope 1986; Wiencke 1989; Pullen 1992; Maran 1998).

2.2 Context for Present Study

In order to construct a complete cross section clay-rich sediments were collected from the Keromotechniki Argos clay pit in the central part of the Argive plain and described using standard geologic sedimentary facies analysis. All relevant data, i.e. sediments, sherds, bone/tooth and shell collected from the pit were compiled. The objective was to compare the data from the central clay pit profile with that of Zangger's construction trench south of Argos (AP-SP-14) (Fig. 2). Zangger did not sample the center of the plain during his fieldwork. The Argive Plain clay pit is roughly comparable to auger core 52 and drill hole I of his cross-section II (1993:28, fig. 17). AP-SP-14 intersects cross-section II between auger cores 67 and 78, ca. one-half kilometer behind our profile. Zangger used this construction trench (AP-SP-14) to verify "the contact between Pleistocene red beds and floodplain or subaqueous deposits found in cross-section II" (1993:37). The trench trends east-west, is 1 km long and 5.5 - 6.5 m deep. For this study the sequences in Zangger's trench and our Argive Plain clay pit were well matched.

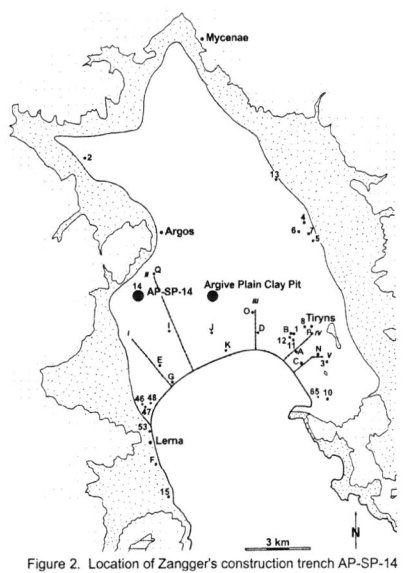

Figure 2. Location of Zangger's construction trench AP-SP-14 and the Argive Plain Clay Pit (after Zangger 1993).

The results of our analysis suggested a physical rather than anthropogenic cause for the Pikrodafni alluviation "event". There appeared to be little or no paleosol formations. Instead the sediments were interpreted as high cyclical discharge of fluvial deposits or annual seasonal fluctuation of climate patterns. Perhaps there was an earlier presence of another river channel or lateral migration of the channel. The clay formations seen in the distinctive A horizons of Zangger's model were considered depositional clays rather than soil formation (1993:25-34). Zangger's fining-upward sequences were seen as part of a larger context.

Moreover, the results of the sedimentary facies analysis for the central plain correlated with Kraft's earlier hypothetical sequence for the plain (1972) and a recent model for global Holocene delta formation (Stanley and Warne 1997). It appears that Zangger's EH II alluvium can be interpreted equally well as Kraft's original early Holocene transgression to progradation sequence and as the base of Stanley and Warne's deltaic unit III. Depending on the choice of facies model for morphological description, the sediments on the Argive plain can be explained as either clay-rich sediments or pedogenesis.

In order to verify the results of our comparative facies analysis, the present clay mineralogy study was undertaken. Based on the results presented below, our data questions Zangger's descriptive hypothesis for a human-induced environmental change on the Argive plain in the EH II/III transition. Alternatively, we propose that there was a physically-induced early

Holocene environmental change on the Argive plain. Within that hypothesis changes in settlement patterns and population density on the plain in the later Neolithic and EH periods can be explained as a human response to a new natural environment.

3. ANALYTICAL TECHNIQUES AND SAMPLE PREPARATION

Samples for the clay mineralogy analysis were taken from levels in our clay pit profile which were dated to the Early Bronze Age or earlier by accompanying sherd material (Fig. 3, levels Lowest Sand, Mottled Zone, LLG, LMG, LUG, Upper and Lower Posthole). The samples were fractioned into > 62 μm for sand, 62 - 4 μm for silt, and - 4 μm for clay. The silt and clay slurries were dried and all samples were weighed. Silt and clay fractions were micronized for X-ray diffraction to determine their bulk mineral content. Oriented slides were prepared for all seven levels from the - 4 μm fraction and analyzed for clay mineralogy. Slides were analyzed on a Phillips X-ray diffractometer with CuK radiation and settings of 45 kilovolts and 20 milliamperes. A scanning rate of 2 degrees 2θ/min over an angular range of 2 - 60 degrees was used.

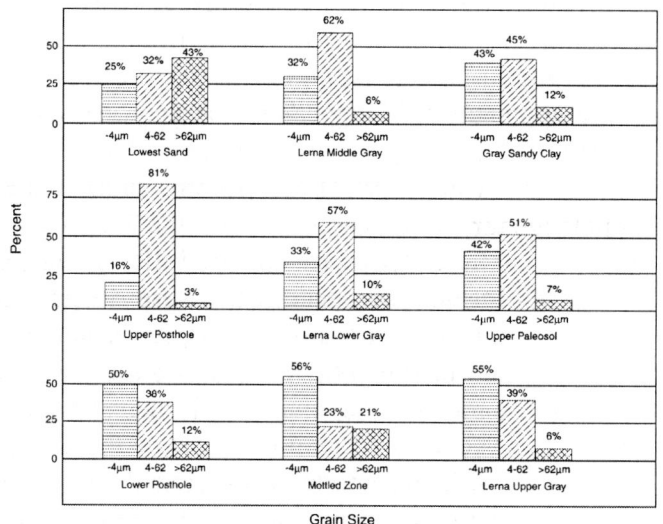

Figure 3. Sand-silt-clay percentages for the sedimentary zones in the Argive Plain Clay Pit profile.

4. RESULTS

The sand-silt-clay percentages are shown graphically on Fig. 3. The highest sand percentage is in the Lowest Sand, the highest silt percentage is in the Upper Posthole, and the highest clay percentage is in the Mottled Zone. A cross section of the clay pit is shown on Figure 4. The sand-silt-clay percentages indicate in general the energy present in the depositional environment. The coarsest sediment is the Lowest Sand and as the percentages indicate the overall depositional sequence indicates that it is fining upward.

X-ray diffraction of the clay fraction of the samples that were analyzed for their particle size showed that each of the sediment zones contained smectite, illite, mixed-layer illite-smectite, kaolinite, quartz and calcite in varying proportion. The Mottled Zone and the Upper and

Middle Gray (LUG and LMG) clay contained the highest amount of smectite. In order to be sure that the diffraction peak at about 14.9 Å was smectite, the slide was heated. The peak shifted to a 10 Å spacing therefore indicating that it was a calcium smectite. All the samples contained some illite and kaolinite, indicating that the deposition was in a fresh or brackish environment, typical of a prograding delta. The quartz and calcite quantity in the clay fraction was very consistent. Studies of the sediments in the Mississippi River delta indicate a similar mineralogical composition (Johns and Grim 1958).

5. DISCUSSION

Two major schools of research into the causes of late Quaternary alluviation in Greece emerged in the 1970's and 80's: (1) Kraft et al. and (2) van Andel et al. Both defined their approaches in a series of articles and reconnaisance studies. Both schools rejected the simplistic model of two universal phases of erosion, which produced the "Older Fill" and "Younger Fill" as initially proposed by Vita-Finzi for the entire Mediterranean Basin (1969) and subsequently supported and applied in detail to Southern Greece by Bintliff (1977).

While both schools of research would argue that they have interrelated surficial processes with stratigraphic sequences, the use of core drilling to give the third-dimension was developed by Kraft (1972) and sets his original methodology apart. Van Andel et al. always concentrated on the headwaters of alluvial valley systems, mapping vertical sequences in accessible locations along the drainage systems which were later developed into lateral river sequences (Facies A occurs upstream, B in the middle, and C nearest the coast). Kraft et al., on the other hand, concentrated on the interface between fluvial and marine. It is the relationship of the lower alluvial valleys and coastal plains to the marine area that is most critical in reconstructing paleogeographic settings (Kraft et al. 1975:1195).

For reasons which are difficult to understand fully from the literature in question, these conflicting schools of thought are combined in Zangger's Argive Plain research project. Van Andel was anxious "to test whether the models of landscape destabilization developed in the Southern Argolid possessed a wider applicability" (van Andel and Zangger 1990:145). Thus, Zangger, his student, undertook a parallel study "in a region closer to the heartland of Greek history, where human exploitation began sooner and on a larger scale" (145). Kraft offered a methodology which van Andel felt suited the differences in depositional environment between the Argive plain and the Southern Argolid. Zangger's description of trench AP-SP-14 agrees in gross detail with our observations for the Argive Plain clay pit (Fig. 4). In general, this is the same sedimentary sequence that
Zangger identified throughout his cross-sections (Zangger 1993:25-35). There are a few differences, however. Overall carbonate nodule sizes do not appear as large as those identified by Zangger and are, therefore, not adequate evidence for the soil development he infers. Subsequent evidence from electron microprobe analysis (Shriner and Dorais 1999) makes a strong case for deposition from clay-rich sediments carried by the Xerias River system behind Argos to the plain rather than from formation as part of pedogenesis. The earlier force of the sediment load from the Xerias River is substantiated by the identification

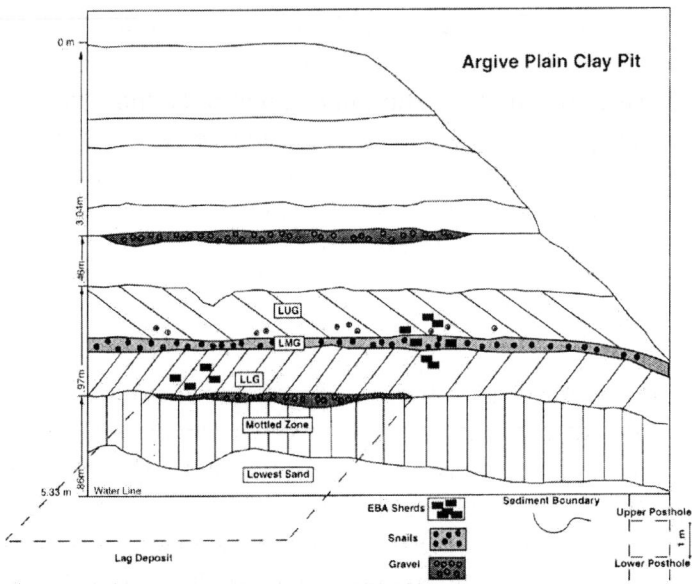

Figure 4. Section Drawing of the Argive Plain Clay Pit.

in the Upper Gray level deposits (Fig. 4, level LUG) of chloritoid in a 50 micron size, derived from the metamorphic window behind Argos, some 20 kms to the west.

Zangger's paleocology report for the Argive Plain addressed only ostracod populations, and did not include a discussion of the specific species of gastropods and cockles observed in his cores and outcrops (Zangger 1993:59-61). The snail species (*Oxychilus and Helicella*), which are found in our profile as well, are terrestial land snails, not a freshwater species. They live by a shore and feed off rotting vegetation. Because they appear in a thin broad horizon in the Middle Gray level (Fig. 4, level LMG) of our clay pit and then are dispersed throughout the level above, they may be an indication of a periodically inundated land surface. The *Cerastoderma lamarkii*, which was found at the same level in our pit, is a lagoonal cockle. Unlike its cousin the *edule*, whose environment is marine, the *lamarkii*'s habitat is either: (1) brackish water fed by freshwater (2) an area of high ground water (3) a river emptying into a sea or (4) most often a bay or lagoonal environment. The Middle Gray level contains very little sand size particles and is dominantly silt and clay. The mineralogy as determined by X-ray diffraction is quartz, calcite, kaolinite, illite and smectite. This composition indicates a probable lacustrine delta environment. Kaolinite does not generally occur in marine sediments.

Zangger made no mention of the lag deposit composed of sand, shells, gravels, mud and wood fragments at the base of the Holocene sequence nor of the prominent upper gravel zone (stone line) which must imply a river channel or fluvial floodplain. The lag deposit at the base of the Holocene sequence and an upper gravel zone are conspicuous in our sedimentary sequence and signify a transition from a low energy environment to a high energy environment.

In fairness to Kraft, it would have seemed natural for Zangger to reconstruct his data collected from the fieldwork on the Argive plain using Kraft's sediment sequence approach and his hypothetical vertical sedimentary sequence for the Argive Plain in his reconnaissance studies (1972:36). In that sequence Kraft assumed a slowly rising relative sea level in alluvial valleyfill-delta complex-coastal plains at the head of deep water embayments of variable widths. Our clay pit profile data substantiates that the Argive plain is in all probability a deltaic sequence with alternating fresh to brackish water deposits along with channel gravels which is typical of a birdsfoot delta. The clay mineralogy indicates that the clays vary in abundance from a mixture of kaolinite, illite, and smectite to dominantly smectite which is indicative of a more marine environment. Kaolinite is found in fresh and brackish environments. Illite and smectite are found in fresh, brackish and marine environments.

Stanley and Warne's recent model for global delta formation in the early Holocene confirms Kraft's earlier observations. They propose that "delta development, inextricably linked to deceleration in rate of sea-level rise, provided newly formed, resource-rich environments conducive to occupation and subsequent development of sedentary human cultures worldwide" (Stanley and Warne 1997:2). Their model was developed through a preliminary study of 34 representative sites, identified worldwide by archaeologists. Zangger offered his data from Dimini Bay-Seskolitis (1991) for this project but not the data from the Argive plain, even though he had the required radiocarbon dates for material older than 5,000 yr B.P.

They postulate that the threshold from marine transgression and coastal erosion to sediment accretion and progradation at the mouth of rivers took place on a worldwide basis within a span of ~2,000 years (8,500 - 6,500 B.P.). As river gradients decreased, there was a change from ephemeral braided to more stable meandering river systems (1997:5). These localities shared the following features: a generally prograding shoreline, permanent freshwater sources, high water table, aquatic habitats (fresh, brackish and marine), well-developed systems of distributary channels and fertile silt-rich soil (1997:5). All of these physical characteristics, as well as the appropriate clay mineralogy and ratios, are present on the Argive plain.

6. CONCLUSIONS

Our clay mineralogical data contradicts Zangger's descriptive hypothesis for human-induced environmental change on the Argive plain in the EH II/III transition. A more plausible conclusion is that this is a deltaic environment with all of its micro-environments. This new environment with its abundant freshwater resources, spawning grounds for fish, access to inland settlements via meandering river systems and quickly changing wetland areas would have had a dramatic effect on human behavior and settlement patterns. More importantly for the archaeological community this suggestion allows for a discussion of settlement pattern shifts and site density within the context of multiple working hypotheses for environmental change.

ACKNOWLEDGEMENTS

Funding for travel for C. Shriner to the 12th ICC was provided by the Schrader Endowment Fund for Archaeological Fieldwork at Indiana University. The authors would also like to express their gratitude to Chris Kraft for his valuable discussions.

REFERENCES

Bintliff, J. L. 1977. *Natural Environment and Human Settlement in Prehistoric Greece*. BAR Supplementary Series 28 (i). Oxford: British Archaeological Reports.

Johns, W.D. and R.E. Grim. 1958. Clay Mineral Composition of Recent Sediments from the Mississippi River Delta. *Journal of Sedimentary Petrology* 28: 186-199.

Kraft, J. 1972. *A Reconnaisance of the Geology of the Sandy Coastal Areas of Eastern Greece and the Peloponnese.* Research in the Coastal and Oceanic Environment Geography Branch Office of Naval Research, Technical Report Number 9, University of Delaware, Newark, Delaware.

Kraft, J., G. Rapp, Jr. and S. Aschenbrenner. 1975. Late Holocene Paleogeography of the Coastal Plain of the Gulf of Messenia, Greece, and its Relationship to Archaeological Settings and Coastal Change. *Geological Society of America Bulletin* 86:1191-1208.

Maran, J. 1998. *Kulturwandel auf dem griechischen Festland und den Kykladen im späten 3. Jahrtausend v. Chr.* Universitätsforschungen zur prähistorischen Archäologie aus dem Institut für Ur- und Frühgeschichte der Universität Heidelberg. Band 53, Teilen I und II. Bonn: Rudolf Habelt GmbH.

Pope, K. O. and van Andel, T. H. 1984. Late Quaternary Alluviation and Soil Formation in the Southern Argolid: its Causes and Archaeological Implications. *Journal of Archaeological Science* 11:281-306.

Pullen, D. J. 1992. Ox and Plow in the Early Bronze Age Aegean. *American Journal of Archaeology* 96:45-54.

Rutter, J. B. 1993. Review of Aegean Prehistory II: The Prepalatial Bronze Age of the Southern and Central Greek Mainland. *American Journal of Archaeology* 97:745-797.

Shriner, C. and M. J. Dorais. 1999. A Comparative Electron Microprobe Study of Lerna III and IV Ceramics and Local Clay-Rich Sediments. *Archaeometry* 41:25-49.

Stanley, D. J. and A. G. Warne. 1997. Holocene Sea-Level Change and Early Human Utilization of Deltas. *GSA Today* 7:1-7.

van Andel, T. H. and E. Zangger. 1990. Landscape Stability and Destabilizations in the Prehistory of Greece. In *Man's Role in the Shaping of the Eastern Mediterranean Landscape*, eds. S. Bottema, G. Entjes-Nieborg and W. van Zeist, 139-157. Rotterdam.

van Andel, T. H., C. N. Runnels, and K. Pope. 1986. Five Thousand Years of Land Use and Abuse in the Southern Argolid, Greece. *Hesperia* 55:103-128.

van Andel, T. H., E. Zangger, and A. Demitrack. 1990. Land Use and Soil Erosion in Prehistoric and Historic Greece. *Journal of Field Archaeology* 17:379-396.

van Andel, T. H., C. Perissoratis and R. Rondoyanni. 1993. Quaternary Tectonics of the Argolikos Gulf and Adjacent Basins, Greece. *Journal of the Geological Society* 150: 529-539.

Vita-Finzi, C. 1969. *The Mediterranean Valleys - Geological Changes in Historical Times.* Cambridge: Cambridge University Press.

Wagstaff, J. M. 1981. Buried Assumptions: Some Problems in the Interpretation of the 'Younger Fill' Raised by Recent Data from Greece. *Journal of Archaeological Science* 8:247-264.

Wiencke, M. H. 1989. Change in Early Helladic II. *American Journal of Archaeology* 93:495-509.

Zangger, E. 1991. Prehistoric Coastal Environments in Greece: The Vanished Landscapes of Dimini Bay and Lake Lerna. *Journal of Field Archaeology* 18:1-13.

Zangger, E. 1993. *The Geoarchaeology of the Argolid.* Deutsches Archaologisches Institut Athen. Argolis, Zweiter Band. Berlin: Gebr. Mann Verlag

Facies distribution and pedogenic evolution of clayey deposits in Caxambu Hill, Quadrilátero Ferrífero, Minas Gerais, Brazil.

A. F. D. C. Varajão[a] and M. C. Santos[b]

[a]DEGEO/EM/UFOP, Campus Universitário/Morro do Cruzeiro, 35400-000
Ouro Preto, MG, Brazil. angelica@degeo.ufop.br

[b]DEGEO/EM/UFOP, Campus Universitário/Morro do Cruzeiro, 35400-000
Ouro Preto, MG, Brazil. du@degeo.ufop.br

Based on stratigraphic relationships, micromorphological and mineralogical studies with X-ray diffraction (XRD), differential thermal and gravimetric analysis (DTA-DGA), optical and scanning electronic microscopy (SEM), four lithofacies were identified in the clayey deposits of Caxambu Hill: fragmentary, nodular, massive and friable. The fragmentary facies is composed of extraclast fragments (3-30 cm) of the local Paleoproterozoic basement and sand size quartz-grains dispersed in a kaolinite-muscovite-goethite-hematite matrix. The nodular facies comprises nodules with recognizable rock and soil fabrics (2-5 cm) dispersed in a matrix similar to the fragmentary facies. The massive facies is characterized by quartz grains dispersed in a matrix of kaolinite, hematite and goethite with minor amounts of muscovite. The friable facies differs from the massive facies by its channel morphology, higher quartz and kaolinite content and the presence of clay-balls millimeter size.

These clayey deposits originated by syntectonic sedimentation in a NW/SE oriented graben during the Cenozoic. The differentiation of the facies is related to gravity mass-flow processes. The fragmentary facies was deposited during the early stage of opening of the basin. After that, under sub-arid conditions, the slumping of lateritic materials from the surrounding regolith led to the formation of the nodular and massive facies. The friable facies originated by the action of unidirectional flow that reworked the clayey sediments. The increasing kaolinite content towards the top is related to the chemical weathering action after the deposition of the sediments.

1. INTRODUCTION

Deposits generated by gravity mass-flow processes are texturally immature and comprise poorly sorted sediments dispersed in a mud matrix (1). In spite of the widespread occurrence of these processes in depositional systems, their characterization is not always obvious in highly weathered terrains in which iron mobilization and pedogenic mechanisms can homogenize the materials, leading to the formation of apparently massive clayey facies.

In intertropical regions the weathered mantle is relatively thick and chemically highly evolved with the formation of laterites. A typical lateritic profile shows a vertical succession of horizons or group of horizons (2, 3) that comprises: 1) coarse grained saprolite or saprock; 2) fine grained saprolite; 3) pedolith comprising mottled clay zone and nodular horizons with Fe- and Al accumulation. The coarse grained saprolite forms the basal horizon in which relict fragments of all sizes and some primary minerals remain unweathered. The fine grained saprolite is the horizon in which most primary minerals have been altered to secondary minerals, such as kaolinite and iron oxyhydroxides. In the pedolith, the pre-existing macrostructure of the parent rock is destroyed, with the formation of mottled clay zones and ferruginous nodules. So the lateritic profile results of the progressive development of each horizon by the transformation of the immediately underlied horizon.

The present study is a contribution to understanding the origin of the clayey deposit from Caxambu Hill. The investigations were based on sedimentological and pedological field observations and, mineralogical and micromorphological analysis.

2. LOCALIZATION AND GEOLOGICAL SETTING

The Quadrilátero Ferrífero is a mountainous Precambrian region located in the center-southeast of Minas Gerais State, southeastern Brazil. The geomorphologic features of the Quadrilátero Ferrífero are related to structural and lithologic controls (4). The geologic architecture of the Quadrilátero Ferrífero defines mainly dome-and-keel structures formed by domes of Archean basement surrounded by keels (troughs) containing Archean and the Paleoproterozoic rocks (5).

The clayey deposit of Caxambu Hill is inserted in the Dom Bosco syncline trough which is composed by Paleoproterozoic rocks comprising quartzites, ferruginous quartzites, phyllites, carbonatic phyllites, dolomites and schists. The Dom Bosco syncline is a structure with W-E orientation and has an extension of 80 km and a width of nearly 10 km. Santos (6) and Santos *et al.* (7) observed that during the Tertiary small grabens were generated in this syncline and sediments were deposited by gravitational process forming several clayey deposits as that of Caxambu hill.

3. MATERIALS AND METHODS

The study of the Caxambu deposit was initiated by field investigations to identify the different facies in which representative samples were collected for mineralogical and micromorphological studies. Samples were previously crushed in an agate mortar and ultrasonically dispersed in distilled water. The < 2 µm fraction was separated from suspension using sedimentation procedures. A mineralogical identification was performed on whole samples and on the clay fraction, as both random and oriented samples, using a Rigaku D/MAX-2B diffractometer with a monochromatic CuKα radiation. The Hughes and Brown index, HB index (8), was measured from the XRD patterns of samples of randomly oriented < 2 µm fractions, because it is not possible to determine the Hinckley index (9) for high-defect kaolinites due to the absence of discrete reflections at the range 20°-24° 2θ Cu. Differential and gravimetric thermal analysis (DTA-DGA) were performed using a SDT 2960 TA DTA-GTA equipment. Morphological aspects of the clay particles were investigated using a scanning electron microscope (SEM - Jeol JXA-50A). For micromorphological studies, thin sections were described under an optical microscope.

4. RESULTS

4.1. Distribution, mineralogy and petrology of the deposit facies

The studied clay deposits (Figure 1), originated by syntectonic sedimentation during the Tertiary, occur in a NW/SE to N/S oriented graben, according Santos (6) and Santos *et al.* (7).

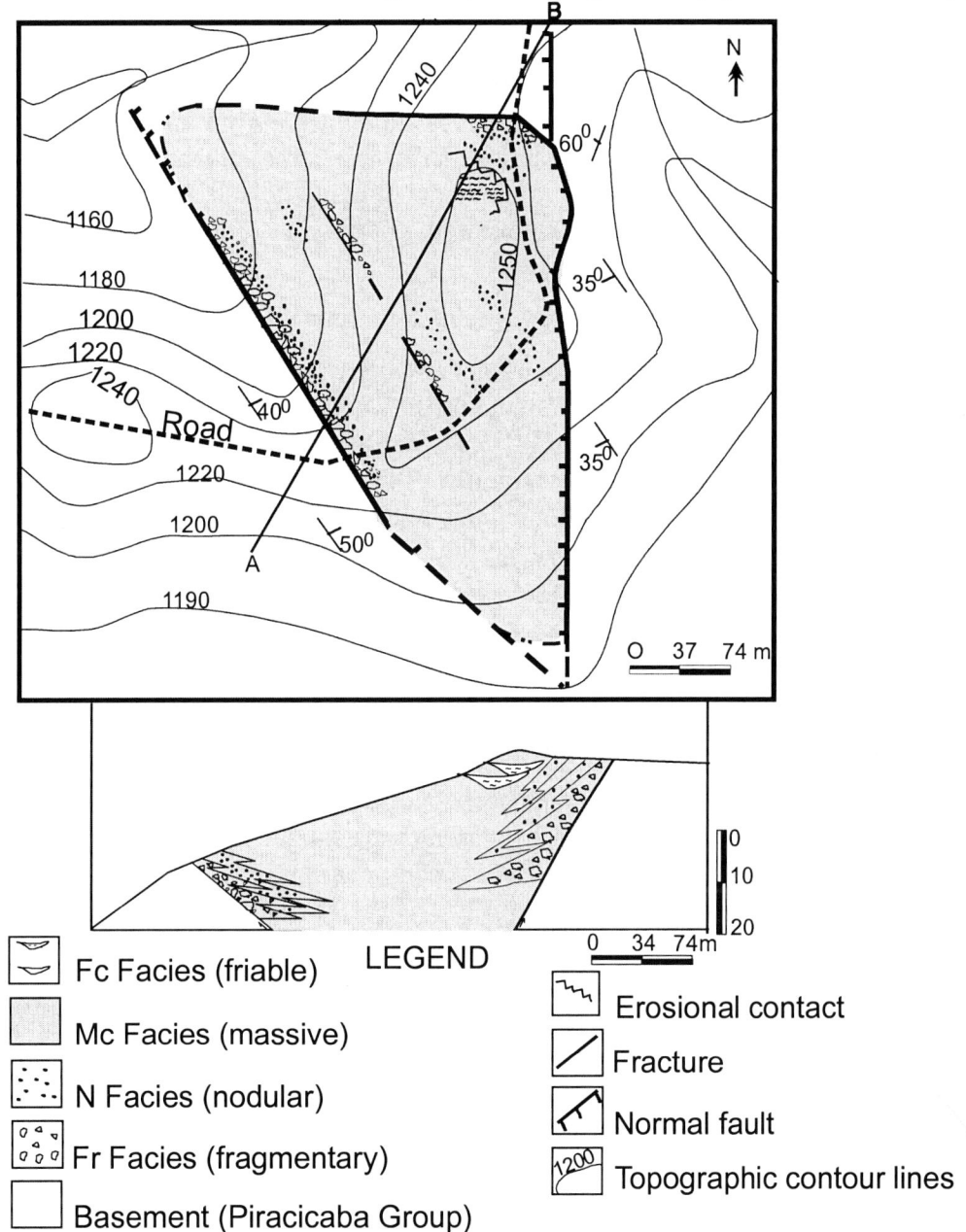

Figure 2. a) Geologic map of the Caxambu hill deposit and b) the schematic section AB showing the distribution of the facies.

Four facies were identified in the Caxambu deposit: fragmentary (Fr); nodular (N); massive clayey (Mc) and friable clayey (Fc) facies (Figure 1):

4.1.1. Fr - Fragmentary Facies

The fragmentary facies (Fr) occurs in the lower part and at the western and northern borders of the deposit, in contact with the local basement (Piracicaba Group, Minas Supergroup,). It has well defined contacts with the nodular (N) and massive clayey (Mc) facies and displays a thickness ranging from 50 cm to 1 m.

This facies is composed of polymictic massive diamictite, comprising lithic fragments of different sizes (3,0 to 30,0 cm), associated to some ferruginous nodules (5%) with recognizable rock fabric with sizes nearly to 2,0 cm. The fragments are angular to sub-rounded and randomly oriented. The matrix is clayey with a brown–reddish (Color Munsell - 5YR 4/4) to yellow-reddish (Color Munsell 5YR 4/6) color and shows the dominant presence of muscovite and kaolinite, with minor amounts of hematite and goethite. The matrix contains dispersed quartz-grains, with size ranging from coarse silt (0,06 mm) to coarse sand (0,6 mm) and some granules.

4.1.2. N - Nodular facies

The nodular facies (N) with a mean thickness of 1.5 m occurs adjacent to the fragmentary facies (Fr) facies and grades basinwards to the massive clayey facies (Mc).

This facies consists of nodules of different sizes (0,30 to 2cm, locally 5cm) immersed in a clayey matrix with dispersed quartz-grains (0,5 mm to 4mm, occasionally, 5mm). The nodules are: ferruginous, clayey and nodules with recognizable rock fabric. The color matrix is brown-reddish (2.5YR 4/4), with yellow-white stains, suggesting a deferruginization process. Several fault and fracture planes form a regular lattice, in which higher concentration of white stains is observed.

Microscopic examination (Figure 2a) reveals a skeleton formed by quartz (25%) and muscovite (15% a 20%), floating in a clayey matrix and forming a porphyroskelic structure (10). Several types of nodules and lithic fragments are also disseminated within the matrix. In addition, there are heavy minerals (1%), such as: kyanite, leucoxene, rutile, tourmaline and zircon. The matrix has kaolinitic-hematitic-goethitic composition with traces of muscovite (Figure 2b) forming silasepic and argillasepic structures. It contains irregular dark-brown ferruginous nodules of goethite and hematite composition.

The fabric composition and relationships with the matrix permitted the separation the glaebules into two groups. The first group presents round-shaped nodules with undifferentiated fabric, black and dark-red color, with hematite composition and different sizes (0.30 cm to 5 cm). These nodules with recognizable soil fabric have an irregular distribution and can possess a yellow cortex of goethite composition with a sharp external boundary with the matrix (Figure 2c). They also can present internal fractures with goethite crystallization in abrupt contact with the matrix. Together with these nodules, glaebules with concentric fabric occur with rather sharp external boundaries whose surfaces consist of smooth curves related to deferruginisation mechanisms. In addition, nodules with recognizable rock fabric of phyllite (Figure 2a), quartzite, with or without red/yellow halos and fragments of ferruginous cortex are present.

The second group is characterized by nodules with the same undifferentiated internal fabric, but showing only the enclosing soil fabric (Figure 2d) .They have heterogeneous size (0,064 to 3mm) and the transition to the matrix is generally gradual.

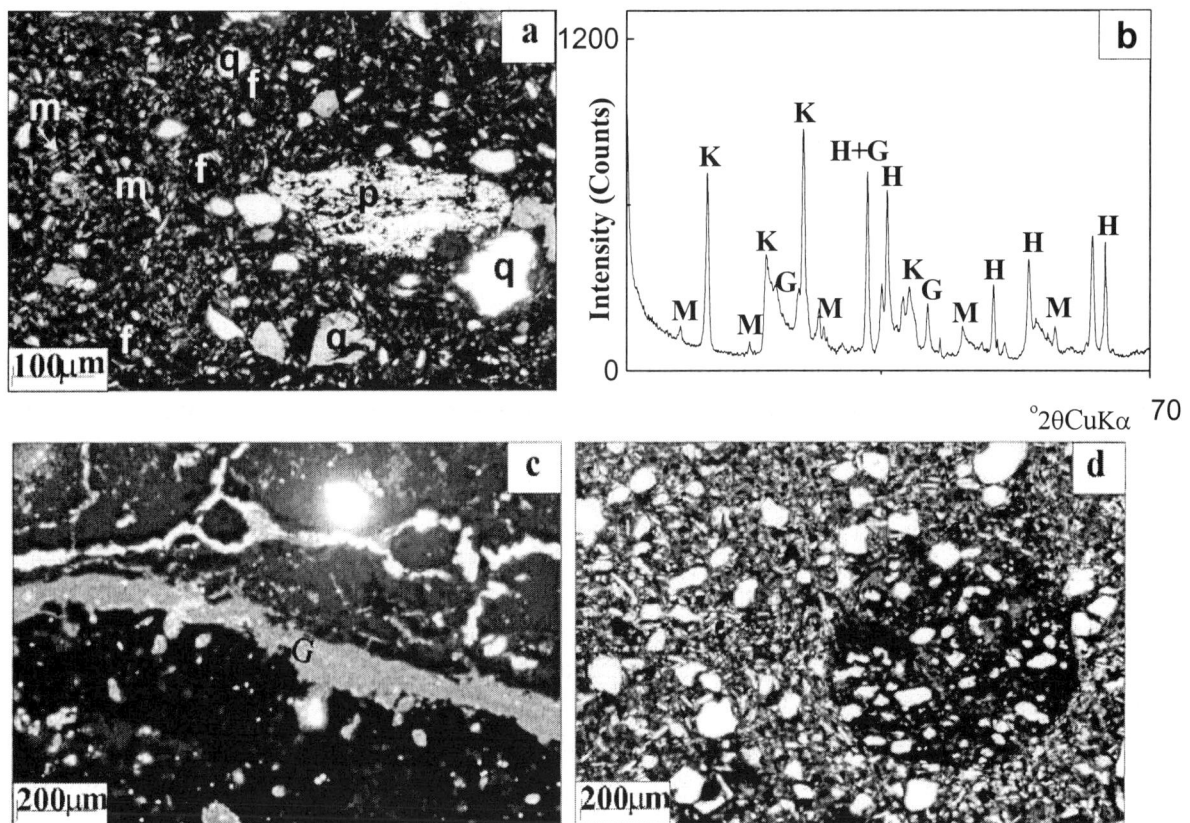

Figure 2. Nodular facies (N). a) Optical micrograph (plane-polarized light) of the nodular facies showing the porphyroskelic structure formed by quartz (q) and muscovite (m) skeleton together with nodules with recognizable rock fabric of phyllite (p) and ferruginous nodules with recognizable soil fabrics (f). b) X-ray diffraction pattern (Cu kα radiation) showing the composition of the matrix of the nodular facies; K= kaolinite, M = muscovite, G = goethite, H = hematite. c) Optical micrograph from the border of a ferruginous nodules showing a cortex of goethite defining a sharp external boundary with the matrix. d) Optical micrograph of a nodule with undifferentiated internal fabric showing the enclosing soil fabric.

4.1.3. Mc - Massive clayey facies

The massive clayey facies occurs in the most basinwards part of the deposit and generally it has a thickness of ~3 m. The intercalation of nodule facies within the massive clay facies is common, suggesting differentiation of the source of sediments

This facies is characterized by a dark-red (2.5YR 4/4) to red (2.5YR 4/6) color with dispersed quartz-grains (0,25 to 0,80 mm, locally, 4 mm). A lattice of fracture planes, forming blocks up to 1m large surrounded by white stains, occurs as well. Locally, some planes of fractures are filled with ferruginous material, originating a duricrust.

Optical microscope observation revealed that this facies is homogeneous and analogous to the other facies with the same skeletons, without any preferential orientation. Likewise to the nodular facies, several types of nodules are present but in minor proportion (1%) and in smaller sizes (0,20 a 1,00mm). The clayey matrix has kaolinite-muscovite-hematite and goethite composition and a silasepic to argillasepic structure (10). The DTA-GTA analyses of the clay fraction show the presence of kaolinite with an endothermic peak at 500 °C, suggesting high-defect concentration or small crystal size (11, 12).

4.1.4. Fc - Friable clayey facies

The Friable clayey facies is characterized by a dark-red (2.5YR 4/4) clay with white to yellowish stains surrounding the fracture/fault surfaces similar to the massive clayey facies. The contact of these stains with the dark-red clayey matrix is gradual. These stains become abundant near the top, coalescing to form the entire upper part of the clayey body. The friable clayey facies has dispersed quartz grains (0,26 to 4,00 mm), sub-rounded to sub-angular. Clay balls with diameters of ~0,5 cm are also associated with the quartz-granules (3,00 to 4,0 mm) at the base. A lattice of fractures/faults also cuts this facies.

Microscopically this facies is mainly formed by a quartzose skeleton (45%) and, secondarily by particles of muscovite (2%), heavy minerals (2%) as kyanite, leucoxene, rutile, tourmaline, zircon, and, some lithic fragments (1%) and nodules (2%) with high degree of degradation, all dispersed in a kaolinitic matrix. The quartz grains have similar optical characteristics to the underlying facies, but they are more frequent and their surfaces show more dissolution pits. The matrix has an argillasepic structure, beige-white color with stains of different sizes and different shades of red. These stains have a gradual contact with the beige matrix and contain black to dark-red ferruginous glaebules. The diffuse external boundary with the matrix suggests degradation process due to the deferruginisation mechanisms.

Mineralogical analyses by XRD of the white-beige matrix show the dominant presence of kaolinite with minor amounts of muscovite and goethite. The HB index (8) was calculated and the values ranged from 3-4, typical of kaolinitic soil. These results are consistent with DTA-GTA analysis that show a peak temperature of kaolinite desydroxylation at ~500 °C, confirming a high-defect concentration or small crystal size of the kaolinite. In agreement, SEM analyses of the beige-white matrix show island of well defined hexagons of kaolinite, typical of muscovite alteration, surrounded by poorly-defined and small crystals and flakes of kaolinite typical of pedogenic kaolinite (13, 14).

5. DISCUSSION AND CONCLUSION

The studies performed at the clayey deposit of Caxambu Hill showed that several factors influenced its formation and comprise syntectonic deposition, climate, lithology of the source-area and depositional systems. In addition, pre and post-depositional weathering action is evident in all the facies defined in this deposit.

The deposition of clayey strata occurred in small grabens whose opening proceeded in the NW-SE direction, probably reflecting the direction of the faults that affected the crystalline basement and the Proterozoic rocks (6, 7). In this context, the fragmentary (Fr) facies, constituted by polymictic fragments of the underlying rocks, can be related to the processes of rock falls by faulting during the formation of the basin, generating their accumulation at the border of the deposit.

The high iron content of the Caxambu Hill clayey deposits are evidence of subaerial environment conditions ("red beds") and the mineralogical/textural immature characteristics are indicative of gravity mass-flow process, relating these sediment-filled grabens to colluvial deposits under arid/sub-arid climate. This deposition probably occurred in the Tertiary during filling of other basins in the Quadrilátero Ferrífero (15).

In this scenario, the climate becomes a fundamental factor during the evolution of all the phases: pre, sin and post-depositional. As the Caxambu deposit is constituted, predominantly, of clayey sediments with subordinate portions of lithic fragments and pedogenic nodules, this association indicates that the rocks of the source area already presented a developed weathered cover, formed under humid climate and originated from the local basement rocks. This lateritic cover was probably composed of a vertical succession of coarse saprolite and fine saprolite horizons overlain by a pedolith horizon rich in nodules and kaolinitic, goethitic and hematitic materials. The features observed in the nodular facies of the Caxambu deposit suggest that the most of the nodules were remobilized pedolith and saprolite materials already available in the source area, in the lateritic profile from higher landscape. The presence of a second group of nodules in the nodular facies, with the enclosing soil fabric, suggests an *in situ* accretion of iron, which accumulated by diffusion or crystallization in the small pores in the matrix due to suitable local chemical conditions (16, 17). The iron may have originated from the saprolite basement and was accumulated by iluviation in clayey domains with nodules with recognizable rock and soil fabrics of the nodular facies.

In the same way, the sediments that now constitute the massive clayey facies would be related to the mottled clayey zones and inter-nodular portions of the pedolith horizon and to the deeper portions of the alteration profile in the source area. The similarity of this massive clayey facies and the nodular facies, relative to the skeleton composition and to the different kinds of glaebules (nodules with recognizable rock and soil fabrics), suggests for both an allocthonous origin by gravitational process and a common supply.

The similar mineralogical skeletal association between the friable facies and the previous facies, together with the analogous micromorphological and pedogenic features, suggest that all the facies have the same allochtonous common origin associated with the gravitational process. The presence of channeled features, clay balls, relict nodules with a high degree of degradation and the higher content in quartz skeleton suggests that the friable clayey facies (Fc) was generated by the action of unidirectional flow on the materials of the massive clayey and nodular facies. Another fact that corroborates the allocthonous features of this facies is the presence grouping of kaolinite pseudomorphs, typical of muscovite alteration from lithorelictal nodules of phyllite, surrounded by poorly-defined and small crystals typical of pedogenic kaolinite.

The action of secondary processes due to post-depositional climate changes was probably responsible for the presence of ferruginous duricrusts occurring in zones of preferential percolation of water in the fractures, faults and concave surfaces of the channels as for the nodules formed by *in situ* accretion of iron. The presence of white stains, typical of deferruginization mechanisms, also suggests weathering action related to a less arid climate. The pedogenic action in all the facies has obliterated the pre-existing sedimentary structures, leading to the formation of mixture of muddy sediments with a increasing of kaolinite content towards the surface of the clayey deposit.

ACKNOWLEDGEMENTS

The authors thank FAPEMIG, CAPES and CNPq for financial support. We are grateful for reviews of the manuscript by Frank U. H. Falkenhein.

REFERENCES

1. D.R. Lowe. Sediment gravity flow: II. Depositional models with special reference to the deposits of high-desensity turbidity currents. Journal of Sedimentary Petrology, 52:0279-0297 (1982).
2. Y. Tardy. Géochimie des Altérations. Etude des Arènes et des Eaux de Quelques Massif Cristallins d'Europe et d'Afrique. Ph.D. Thesis, University of Strasbourg, 199p. (also Mémoire Sérvice Carte Géologique, 31: (1969).
3. D. Nahon. Evolution of iron crusts in tropical landscapes. In: Rates of Geochemical Weathering of Rocks and Minerals, Academic Press, London (1986).
4. C.A.C. Varajão. A questão da ocorrência das superfícies de erosão do Quadrilátero Ferrífero, Minas Gerais. Revista Brasileira de Geociências., 21:131-145 (1991).
5. F.F. Alkmim and S. Marshak. Transamazonian Orogeny in the Southern São Francisco Craton region, Minas Gerais, Brazil: evidence for Paleoproterozoic collision and collapse in the Quadrilátero Ferrífero. Precambrian Research, 90: 29-58 (1998).
6. M.C. Santos. Gênese dos Corpos Argilosos do Morro do Caxambu e da Mina do Vemelhão, Sinclinal Dom Bosco, Quadrilátero Ferrífero, Minas Gerais, Brasil. DEGEO/EM/UFOP, Ouro Preto, Dissertação de Mestrado, 176p. (1998).
7. M.C. Santos, A.F.D. Varajão, P.T.A Castro, A.P.A. Moreira. Gênese dos "depósitos argilosos" da Mina do Vermelhão e do Morro do Caxambu, Sinclinal Dom Bosco, M.G. In: SBG/Núcleo Brasília, Simpósio de Geologia do Centro-Oeste e Simpósio de Geologia de Minas Gerais, Resumos Expandidos, p 40 (1999).
8. J.C. Hughes and G. Brown A crystallinity index for soil kaolins and its relation to parent rock, climate and soil maturity. Journal of Soil Science, 30, 557-563 (1979).
9. D.N. Hinckley. Variability in "crystallinity" values among the kaolin deposit of the coastal plain of Georgia and South Carolina. Clays and Clay Minerals, 11, 229-235 (1963).
10. R. Brewer. Fabric and mineral of soils. New York, John Wiley & Sons. 1964.
11. R.C. Mackenzie. The Differential Thermal Investigation of Clays. London, Mineralogical Society, 1957.
12. R.E. Grim. Clay Mineralogy, 2nd edition. McGraw-hill, New York, 1968.
13. A.F.D.C. Varajão, R..J. Gilkes and R..D. Hart, The relationships between kaolinite crystal properties and the origin of materials for a Brazilian kaolin deposit. Clays and Clay Minerals, 49/1, 45-59 (2001).
14. B.Singh and R.J. Gilkes, An electron optical investigation of the alteration of kaolinite to halloysite. Clays and Clay Minerals, 40, 212-229 (1992).
15. J. R. Maizatto and P.T.A. Castro. Origem e evolução da bacia do Gandarela-Quadrilátero Ferrífero, Minas Gerais. In: SBG, Simpósio Nacional de Estudos Tectônicos, Belo Horizonte, 12:325-329 (1993).
16. B. Boulangé. Les formations bauxitiques latéritiques de Côte d'Ivoire. Les facies, leur transformation, leur distribution et l'évolution du modelé. Trav. et Docum., ORSTOM, 175, 363p. (1984).
17. Y. Tardy. Pétrologie des Laterites et des Sols Tropicaux. Paris, Masson. 416p. (1993).

The effect of deformation on illite crystallite sizes

K. Wagner (née Hetfeld)[a]

[a]Department of Geology, Trinity College, Dublin 2, Ireland

Samples used in this study were taken from uppermost Devonian rocks in the Munster Basin, SW Ireland, Mississippian (Upper Carboniferous) rocks within the Clare Basin, W Ireland, and Jurassic rocks from the Rockall Trough margin, North Atlantic. The rocks sampled from the Munster Basin and the Clare Basin have all experienced tectonic deformation during the Variscan orogeny, but to a different degree. Isoclinal folding and the development of cleavage in the Munster Basin contrast with the open folding and lack of cleavage in the Clare Basin. Vitrinite reflectance measurements have been carried out for all samples. 'MudMaster', computer modelling software was used to calculate fundamental particle size distributions from XRD traces after PVP-10 (polyvinylpyrrolidone) treatment of the samples. MacEwan crystallite sizes were determined using the Scherrer equation. In the Munster Basin section uncleaved samples yield larger MacEwan crystallite sizes than the cleaved samples. Even though the Munster Basin samples have experienced a similar palaeotemperature according to vitrinite reflectance as the Clare Basin samples, MacEwan crystallite sizes and fundamental particle sizes far exceed those of the Munster Basin. The Jurassic samples comprise undeformed mudstones plus one obviously tectonically deformed mudstone. All samples yielded similar illite fundamental particle sizes around 5nm except for the tectonically deformed sample, which has a 'best mean' of around 8nm. Taking all these results into account, it seems that deformation has a major impact on illite crystallite sizes.

1. INTRODUCTION

Reaction progress of clay minerals has been widely used as a metamorphic indicator (Jaboyedoff *et al.* 2001) with crystallinity measurements the most commonly used. To express the crystallinity of illite, several indices have been proposed. The Kübler Index (the full width of half the maximum peak height) is probably the most widely used. Although several factors influence the Kübler Index, temperature change is considered to be the main controlling factor (Frey 1987; Merriman & Peacor 1999). The effect of stress on clay crystallinity is less well known, but there is some evidence that it promotes crystal thickening (Merriman & Peacor 1999).

The main effect of increasing metamorphism is an increase in illite crystallite size, particularly thickness. Based on the assumption that the individual crystallites grow in thickness along their c-axis with increasing temperature, an attempt has been made to relate size distributions to different maturation levels. To achieve this in a reasonable time at low cost a software package called 'MudMaster' (Eberl *et al.* 1996) was used to model crystallite size distributions from XRD traces. The resulting mean sizes were compared to sizes calculated using the Scherrer equation. The Scherrer equation is believed to represent an averaging of the MacEwan crystallite as described by Moore & Reynolds, Jr. (1997) and Eberl & Środoń (1988). By treating the sample with PVP (cf. Methods) a disarticulation of those MacEwan crystallites is believed to be achieved which allows the coherent scattering domains to consist mainly of so called fundamental particles (Moore & Reynolds, Jr. 1997; Eberl *et al.* 1998). Dislocations separate coherent scattering domains in either model.

2. GEOLOGICAL SETTING

The study area includes two sedimentary basins in the west of Ireland, the Clare and the Munster basins (Figure 1). Both basins underwent tectonic deformation during the Variscan Orogeny in the Upper Carboniferous and Early Permian. The rocks within the Munster Basin experienced deformation, which resulted in isoclinal folding and the generation of cleavage. Sediments within the Clare Basin were only gently folded and generally display no cleavage (Figure 2).

Ten samples (TH1-TH10) were collected from Uppermost Devonian Rocks of the Munster Basin. They were taken from the Old Head Sandstone Formation at Toe Head, Co. Cork (National Grid 114400 027100). The section is about 14m in thickness. It can be assumed that the rocks in this section have undergone the same thermal history. Two sets of samples can be distinguished. One set was taken from lateral extensive mudrock beds, which have developed cleavage. The other set comes from mudrock lenses within the sandstone layers with no apparent cleavage. (Figure 3)

Sixteen samples were chosen from the Slievecallan borehole, Co. Clare (National Grid 117900 176050). They are part of the sedimentary filling of the Clare Basin and consist of grey shales of undifferentiated Namurian age (Goodhue & Clayton 1999). (Figure 1)

Six samples from a recently drilled shallow borehole in the Rockall Trough (Figure 1) are being used as an example to underline the conclusions drawn from the data obtained from the Munster Basin and Clare Basin samples.

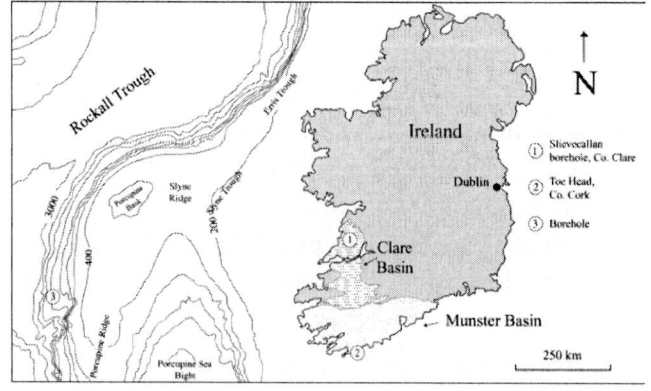

Figure 1. Map of Ireland indicating the Clare and Munster basins in the Southwest and main offshore basins to the West of Ireland.

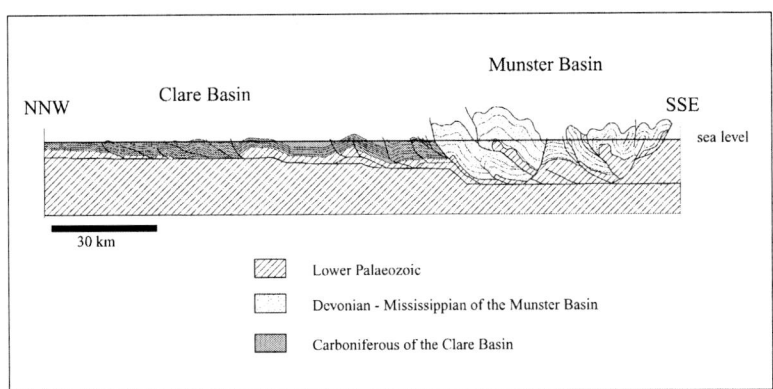

Figure 2. Schematic cross-section through the Clare Basin and the Munster Basin showing the different degrees of deformation (after Bresser 2000).

3. METHODS

3.1 Vitrinite Reflectance

Organic residue was extracted by dissolving the samples with 40% hydrofluoric acid. The dried residue was mounted onto a block using an epoxy resin. Mean random vitrinite reflectance (R_r) was measured from polished blocks under a Leitz MPV-1 compact microscope/photometer. The standards used were glass (Leitz No. 998) with a reflectance of

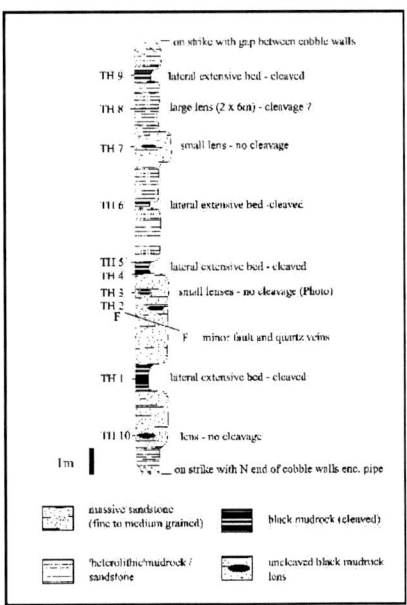

Figure 3. Profile of the sampled section at Toe Head, Munster Basin. TH1-TH10 indicate sample locations.

1.23%R_r in oil, cubic zirconia (McCrone No. 319) with a %R_r of 3.28 in oil, and silicon carbide (McCrone No. 339) with a reflectance of 7.54%R_r in oil. The actual measurements were made in non-polarised light. Mean and standard deviation were calculated and the data were plotted in histograms.

3.2 Illite crystallinity and Crystallite Size Distributions (CSD)

For each sample four slides were run analysed by XRD: air-dried, glycolated, Na-saturated and PVP-treated sample. Oriented samples were mounted by simply letting the <2μm fraction in suspension dry on the slide. After scanning them in the XRD the first time, they were glycolated with ethylene glycol vapour at 60°C in a desiccator and run in the XRD again. A separate set of material was saturated with 1 M NaCl solution. Part of it was mounted as air-dried, oriented slides while the rest was treated with PVP-10 (Polyvinylpyrrolindone-10) before it was mounted on Si-slides. Both sets were consequently run in the XRD. Scans were usually carried out between 2°2θ and 35°2θ. The step-size was 0.02 at a time constant of 1. The X-ray diffractometer used was a Philips PW 1050. The machine uses CuK$_\alpha$ radiation at a wavelength of 1.5418Å. It is run at 20 mA and 40 kV. The divergence slit is set at 1°. A curved graphite crystal monochromator is used as a filter.

Illite crystallinity (IC) was calculated using the Kübler Index. The Kübler Index measures the full width of half the maximum peak height (FWHM) of the illite 001 peak at 10Å. The data were standardised following the crystallinity index standard (CIS) using interlaboratory standards provided by L. N. Warr (Warr & Rice 1994). The Scherrer equation was used to calculate the thickness of MacEwan crystallites. Finally, the scan data of the PVP treated samples were run in MudMaster (Eberl *et al.* 1996) to calculate the size distributions for illite fundamental particles.

4. DATA

4.1 Munster Basin samples

The mean random vitrinite reflectance ranges from 3.6%R_r to 4.54%R_r. The main clay mineral components are an illite phase with a small amount of swelling interlayers plus a mixed-layered phase, probably illite and smecite. Illite crystallinity values for air-dried samples vary between 0.21 and 0.28 Δ°2θ. They decrease when glycolated. Na-saturated and PVP-treated slides show even lower crystallinity values. Illite fundamental particle sizes range round 18nm as their best mean whereas MacEwan crystallites range between 84 and 228 nm. (Table 1)

4.2 Clare Basin samples

The vitrinite reflectance values measured by Goodhue & Clayton (1999) range from 5.1%R_r to 7.56%R_r for the studied samples. The only clearly distinguishable illite phase shows a small amount of swelling interlayers. Air-dried samples yield illite crystallinity values of around 0.5 Δ°2θ whereas the values drop for glycolated samples. Na-saturated samples still show fairly high illite crystallinity values, but the PVP-treated samples again yielded very low illite crystallinity values. Illite fundamental particle sizes vary between 5 and 7 nm. MacEwan crystallites show thicknesses between 15 and 39 nm. (Table 1)

Table 1

IC (air-dried) values correspond to CIS (crystallinity index standard) values. Fundamental particle sizes for illite were modelled using MudMaster (Eberl *et al.* 1996) after PVP-treatment of the samples. The Scherrer equation was used to determine the sizes of MacEwan crystallites from air-dried sample scans.

Locality	Sample	IC ($\Delta°2\theta$)	Fundamental particle sizes (nm)	MacEwan sizes (nm)	VR (%R_r)
Munster	TH 1	0.23	18	145	4.16
Munster	TH 2	0.24	18	123	4.4
Munster	TH 3	0.28	15	84	4.26
Munster	TH 4	0.23	15	145	3.84
Munster	TH 5	0.23	21	145	3.6
Munster	TH 6	0.24	15	123	3.85
Munster	TH 7	0.21	16	228	4.36
Munster	TH 8	0.24	17	123	3.64
Munster	TH 9	0.22	8	177	4.54
Munster	TH 10	0.25	20	123	4.16
Clare	14	0.73	5	15	5.1
Clare	26	0.50	6	27	5.88
Clare	35	0.54	5	24	5.75
Clare	38	0.53	6	25	5.89
Clare	40	0.52	7	25	5.34
Clare	41	0.54	6	24	6.11
Clare	43	0.40	6	39	5.94
Clare	44	0.51	7	26	5.88
Clare	45	0.51	7	26	6.15
Clare	46	0.53	6	25	5.56
Clare	47	0.52	5	25	7.56
Clare	48	0.52	5	25	6.29
Clare	50	0.49	5	27	5.85
Clare	53	0.51	6	26	5.59
Clare	54	0.55	5	23	6.07
Clare	57	0.57	6	22	5.48
Rockall	12105	0.96	5	14	-
Rockall	12104	0.71	8	20	-
Rockall	12103	0.61	5	25	-
Rockall	12101	0.70	6	21	-
Rockall	12102	0.79	6	18	-

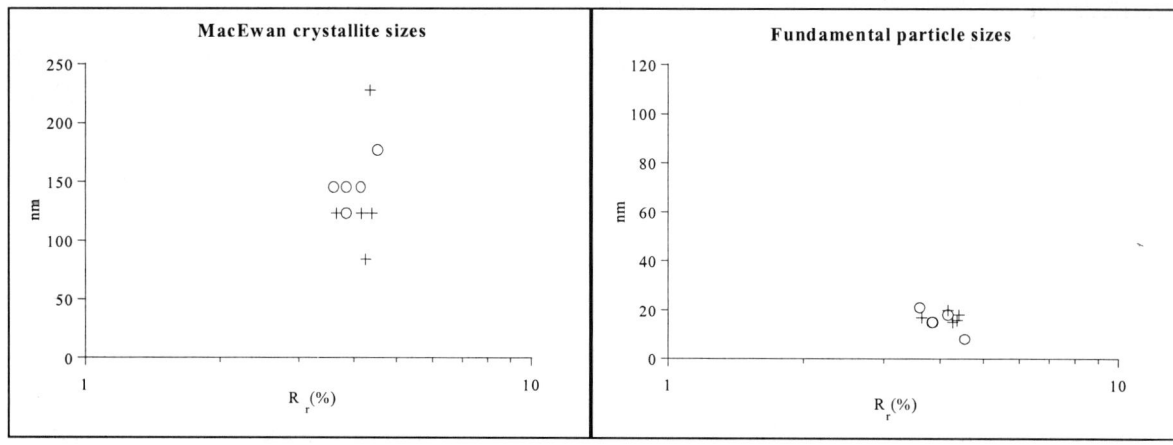

Figure 4. Illite crystallite sizes of the Munster Basin samples. Circles represent uncleaved samples, crosses indicate cleaved samples. Vitrinite reflectance is plotted on the x-axis as mean random reflectance on a log scale. MacEwan sizes were calculated using the Scherrer equation. Fundamental particle sizes were determined by using MudMaster (Eberl *et al.* 1996).

4.3 Rockall Trough samples

The illite phase present in these samples contains a moderate amount of swelling interlayers. Illite crystallinity in the Rockall Trough samples ranges in the diagenetic zone with values around 0.5 $\Delta°2\theta$. Fundamental particle sizes of around 5 to 6 nm have been measured for four samples. Sample no. 12104 shows a fundamental particle size of 8 nm. MacEwan crystallite thicknesses range between 14 to 25 nm. (Table 1)

5. DISCUSSION AND CONCLUSIONS

Vitrinite reflectance measurements for the Munster Basin suggest a uniform peak palaeotemperature throughout the section. There seems to be no apparent trend for the two groups of samples, uncleaved and cleaved, in the vitrinite reflectance. However, the illite crystallinity values of the air-dried and the glycolated samples show a separation of data related to the development of cleavage in the samples which can also be seen in the MacEwan crystallite sizes (Figure 4). The separation seems to disappear when the samples are saturated with NaCl and treated with PVP. Fundamental particle sizes derived from MudMaster (Eberl *et al.* 1996) cluster and show no separation of cleaved and uncleaved samples (Figure 4).

Separation of illite crystallite sizes can also be observed on a much larger scale for both the Clare Basin and the Munster Basin. Vitrinite reflectance for both basins is very similar. However, both the MacEwan crystallite sizes and the fundamental particle sizes differ considerably in the two basins. Generally the Clare Basin samples show much smaller sizes than the Munster Basin. As differences due to lithology, mineralogy and age can be excluded it is likely that differences in the geological history, particularly, the degree of deformation, is the controlling factor causing the size differences in the two basins. (Figure 5)

Amongst the five Rockall Trough samples, which were studied, one sample was macroscopically deformed. This shows clearly enhanced illite fundamental particle sizes in comparison with the other samples (Figure 6).

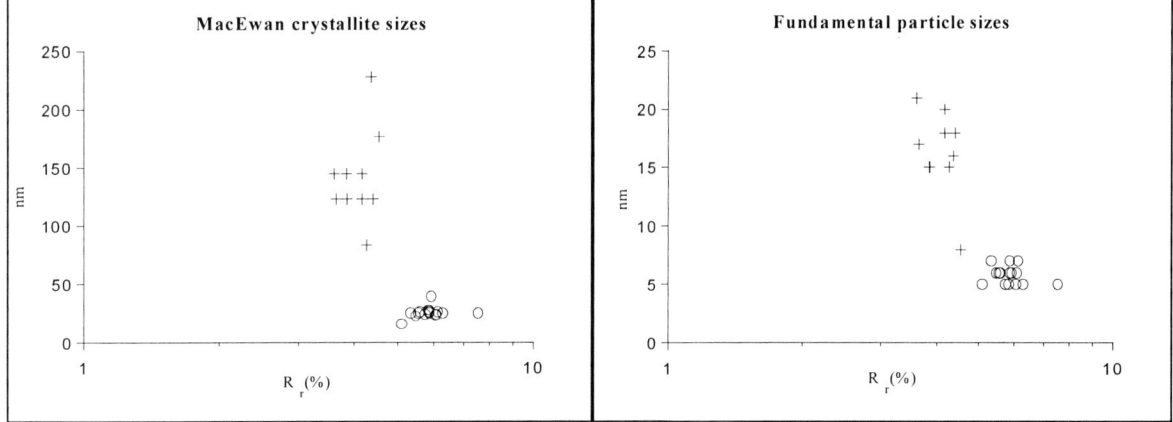

Figure 5. Crosses indicate Munster Basin samples, circles represent Clare Basin samples. Vitrinite reflectance is plotted on the x-axis as mean random reflectance on a log scale. MacEwan sizes were calculated using the Scherrer equation. Fundamental particle sizes were determined by using MudMaster (Eberl *et al.* 1996).

Thus, the results of deformation affecting illite crystallite sizes observed in this study can be summarised as followed. It was found that
1. increases in MacEwan crystallite sizes, but not in fundamental particle sizes,
2. increase in MacEwan crystallite sizes <u>and</u> fundamental particle sizes, and
3. increases in fundamental particle size, but not in MacEwan crystallite size

occur. A satisfactory explanation for these phenomena has yet to be found. As a first attempt it may be proposed that increases in MacEwan crystallites might be due to "deformation-enhanced syntectonic recrystallisation" (Burkhard & Badertscher 2001) whereas increases in fundamental particle sizes may be caused by "dislocations..[that]..migrated...and thus annealed subgrains developed in mica crystals" (Merriman *et al.*1995).

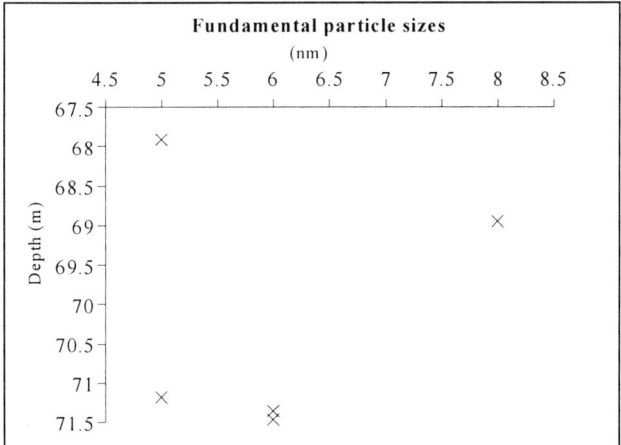

Figure 6. Fundamental particle sizes of the Rockall Trough samples were determined by using MudMaster (Eberl *et al.* 1996).

6. ACKNOWLEDGMENTS

I like to thank my supervisor Prof. Geoff Clayton (TCD) for making this study possible in the first place and for all the support given throughout. A thank you to Dr Richard Merriman (BGS) for his help with the clay side of my project. I also thank BG International Ltd, Anardarko Ireland and Enterprise Energy Ireland for funding this research studentship. The Rockall Studies Group (RSG) is thanked for access to the Rockall shallow borehole material and the permission of usage of the data herein. I would also like to thank Pete Wagner and Jacqueline Connolly (TCD) for their encouragement and helpful discussions.

The ownership of and copyright to all the Rockall data and interpretations contained herein resides with PIPCo RSG Ltd.

REFERENCES

1. Bresser, G. (2000). An integrated structural analysis of the SW Irish Variscides. AGB Band 35, Wissenschaftsverlag Mainz.
2. Burkhard, M. & Badertscher, N. (2001). Finite strain has no influence on the illite crystallinity of tectonized Eocene limestone breccias of the Morcles nappe, Swiss Alps. Clay Minerals, **36**: 171-180.
3. Eberl, D.D. & Środoń, J. (1988). Ostwald ripening and interparticle-diffraction effects for illite crystals. American Mineralogist, **73**: 1335-1345.
4. Eberl, D. D., Drits, V.A., Środoń, J., Nüesch, R. (1996). MudMaster: a program for calculating crystallite size distributions and strain from the shapes of x-ray diffraction peaks. Boulder, Colorado, U.S. Geological Survey.
5. Eberl, D.D., Nüesch, R., Sucha, V. & Tsipursky, S. (1998). Measurement of fundamental illite particle thicknesses by X-ray diffraction using PVP-10 intercalation. Clays and Clay Minerals, **46**: 89-97.
6. Frey, M. (1987). Low Temperature Metamorphism. Glasgow, Blackie & Sons Ltd.
7. Goodhue, R. & Clayton, G. (1999). Organic maturation levels, thermal history and hydrocarbon source rock potential of the Namurian rocks of the Clare Basin, Ireland. Marine and Petroleum Geology, **16**: 667-675.
8. Jaboyedoff, M., Bussy, F., Kübler, B. & Thélin, Ph. (2001). Illite "crystallinity" revisited. Clays and Clay Minerals, **49** (2): 156-167.
9. Merriman, R.J. & Peacor, D.R. (1999) Very low-grade metapelites; mineralogy, microfabrics and measuring reaction progress. In: Frey, M. & Robinson, D. (eds.) Low-Grade Metamorphism. Blackwell Sciences Ltd., Oxford: 10-60.
10. Merriman, R.J., Roberts, B., Peacor, D.R. & Hirons, S.R. (1995). Strain-related differences in the crystal growth of white mica and chlorite: a TEM and XRD study of the development of metapelitic microfabrics in the Southern Uplands thrust terrane, Scotland. Journal of metamorphic Geology, **13**: 559-576.
11. Warr, L. N. & Rice, A.H.N. (1994). Interlaboratory standardisation and calibration of clay mineral crystallinity and crystallite size data. Journal of Metamorphic Geology, **12**: 141-152.

III
Soil Mineralogy

III

Soil Mineralogy

X-Ray analysis of clay and silt fractions of soils developed in allochthonous materials of aeolian (loess) and alluvial origin in the southwestern Pampa, Argentina

M. del C. Blanco [a], I. M. Natale [a,b], N. Amiotti [a], E. A. Ferreiro [a,c] and M. E. Mandolesi [a]

[a] Departamento de Agronomía, Universidad Nacional del Sur. Altos de Palihue, (8000) Bahía Blanca, Argentina

[b] CIC - Comisión de Investigaciones Científicas, Prov. de Buenos Aires, Argentina

[c] CONICET - Consejo Nacional de Investigaciones Científicas y Técnicas, Argentina

Mineral analyses were performed on the <1μm, <2 μm, and 2-50 μm fractions to study clay mineral composition of soils developed on loess (zonal soil) and alluvial (azonal soils) parent materials in the southwestern Pampean region. The phyllosilicates are primarily inherited from the soil parent materials, although moderate weathering conditions have transformed illite and chlorite into irregularly interstratified illite-smectite and irregularly stratified chlorite-smectite. Weathering has produced identical secondary clay minerals without translocation in both soils depending on water percolation and the moderately aggressive chemical characteristics of soil solution. Despite of changes in the climate-paleoclimate during the Holocene epoch and differences in type, origin, texture, age and landscape position of allochthonous soil parent materials in the southwestern semiarid Pampean region, clay and silt mineralogy is quite similar.

1. INTRODUCTION

The geo-pedogenetic history of soils and the changes that have occurred in parent materials during soil formation are recorded in the soil minerals. In particular, the secondary minerals of the clay fraction reflect the imprint of the evolutionary tendency of weathering and pedogenetic processes [1]. The objective of this paper is: 1) to describe the minerals of the clay and silt fractions of two soils developed in Holocene age loess and alluvium in the southwestern Pampean region of Argentina, and 2) to explain weathering and the evolution of soil forming processes through the interpretation of soil fine fractions. Very few studies of soil mineral analyses on the fine fractions of soils of the SW of the Pampean region are known at present [2, 3, 4].

2. METHODS AND MATERIALS

Two soils, one representative of the zonal soil developed on stable upland positions and the other developed in alluvium, were studied in different geomorphological positions in the distal extreme of the Llanura Subventánica Occidental Bonaerense [5] at the southwestern extreme of the Pampean region (Fig. 1). Soil descriptions and classification followed Soil Survey Manual [6] and Soil Taxonomy [7]. Mineral analyses were performed in the <1 μm, <2 μm and 2-50 μm fractions. Organic matter was first removed with hydrogen peroxide, after which the fine particles were separated by successive sedimentation according to the Stokes's law and removed by siphoning. The following determinations were carried out on the <2 μm fraction: pH (1:2.5 in water and 1 N KCl, potentiometric), total organic carbon (Walkley and Black) and cationic exchange capacity (ammonium acetate 1 N).

Separates of the <1 μm, <2 μm and 2-50 μm fractions without organic matter were then treated with sodium citrate-sodium bicarbonate-sodium dithionite to remove Fe-oxides and hydrated oxides of Al and Fe, followed by treatment with 0.5 M NaOH to remove poorly crystalline aluminosilicates [8]. After each treatment, residues were suspended in water and prepared for X-ray analyses (total of 150 samples). Diffractograms were obtained using a Rigaku Denki Geiger Flex D Model Max 3C diffractometer equipped with graphite monochromator. The Cu-target x-ray tube was operated at 35 kV and 15 mA. The diffractograms were smoothed with a 25-point smoothing routine but the background was not reduced.

3. SOIL SITES AND ENVIRONMENTAL FACTORS

The climate is temperate semiarid with a mean annual rainfall of 615 mm, potential evapotranspiration of 770 mm and mean annual temperature of 15.5 °C. There is evidence of a humid paleoclimate in the buried geosol developed in the Agua Blanca Formation, that has been correlated to the Puesto Berrondo paleosoil identified in other parts of the Buenos Aires province. The soil moisture regime is ustic and soil temperate regime is thermic. Natural vegetation was grassland (dominant *Stipa sp.*) but the soil is currently cultivated.

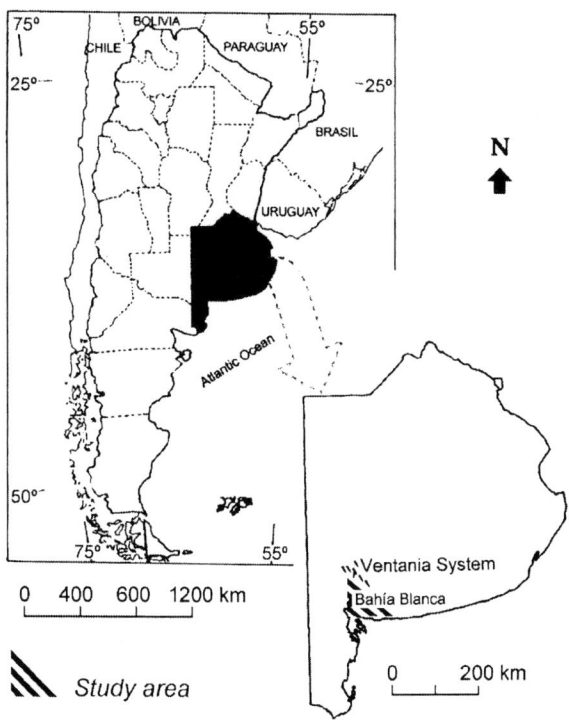

Figure 1. Geographical location of the study area.

The Muzzi soil is in the distal sector of the Llanura Subventánica Plain [5], a landscape position characterised as a level and stable interfluve. The Canesa soil is on a level terrace of Napostá Grande River. Erosion is not expected to occur at either site as both sites have little relief.

Geologically, the area can be described as follows: sediments of the Agua Blanca Formation-Arenoso medio member (<10.000 BP) were deposited on a substratum of Plio-Pleistocene age materials under a cold and semiarid climate [9]. The sequence continues towards the silty and silty clay deposits of the Agua Blanca Formation that culminates with a paleosoil developed under a humid temperate paleoclimate. A new aridization cycle lead to deposition of coarse texture sediments that correspond to the Chacra La Blanqueada Formation, which is not identified in this study. The stratigraphic sequence at Puente Canesa ends with aeolian sandy sediments belonging to the Matadero Saldungaray Formation deposited during the most recent semiarid pulse (Late Holocene up to Present) that buried the underlying alluvial deposits. The plain interfluve is mantled by calcareous loess sediments of aeolian origin, which unconformably covered a pre-existing undulated paleotopography marked by a thick tosca layer (Petrocalcic horizon) of Plio-Pleistocene age. Loess sediments are generally refered to Pampeano Formation, Pampeano, Pampiano or Postpampiano [10, 11, 12].

4. RESULTS AND DISCUSSION

4.1. Profile description and classification

The parent materials of the zonal soils (Muzzi soils) on the interfluve plain are calcareous loess of Holocene age correlated to the Saavedra Formation.

The azonal soils (Canesa soils) on the terrace are developed in sandy aeolian sediments of the Matadero Saldungaray, juxtaposed to alluvial sediments of the Agua Blanca Formation. The horizon sequence of the Muzzi soil is Ap-ACd-C-2Ck-3Ckm. The Canesa soil contains a buried soil and has a more complex horizon sequence of Ap-C_1-$2C_2$-3Ab. The Muzzi soil is classified as a Petrocalcic Paleustoll and the Canesa soil is a Typic Ustifluvent characterised by an irregular distribution of organic matter, alternating layers of different textures, and pedofeatures developed by salinization and gleying at depth.

4.2. Silt (2-50 µm) and clay mineralogy (< 1 µm and < 2 µm)

Illite is the dominant mineral in the parent material and in the solum, which has been transformed into illite/smectite throughout Muzzi soil (Table 2). The production of organic acids in the Ap horizon and alternating wetting and drying of the surface horizon should contribute to weathering reactions. The Ck horizon receives less water than the overlying horizons, but since continued downward movement is limited by the Petrocalcic horizon, the contact time between mineral surfaces and water is longer than in the soil solum. Chlorite inherited from soil parent material, occurs in the Ck and C horizons. Irregularly interstratified chlorite-smectite occurs as an intermediate stage of transformation into smectite in the horizons overlying the Petrocalcic horizon as a result of a loss of potassium, probably due to crop (wheat) consumption, since these mixed layers are identified in the Ap and ACd horizons, which are the most highly rooted horizons. Wheat consumes 57 kg/ha of K and 14 kg/ha are stored in the grain [13]. Other minerals such as quartz, microcline and orthoclase as K-feldspars, plagioclase, quartz, tridymite, cristobalite, probably pyroxene or amphibole, ilmenite, and apatite were identified as accessory minerals. Loess sediments are initially calcareous, and calcium carbonate solubilized in upper horizons moves with the soil solution to lower horizons and reprecipitates as pseudomycelium, mottles, matrix impregnations, coatings and nodules. Dolomite is also present in the Ck horizon (Fig. 2).

Few changes in the clay mineralogy were detected in this loess-derived soil and differences within the profile caused by weathering are minimal. Native grassland does not greatly acidify the soil. The surface horizons have a neutral pH, the subsurface horizons are slightly alkaline and the cation exchange complex is dominated by Ca^{2+} over other bases (Table 1). Consequently, mineral degradation resulting from depotasication only results in 2:1 clay minerals as secondary weathering products. The combined effect of climate-pedoclimate, mineral composition of the parent material, type of natural vegetation and time controlled the moderate pedogenetic weathering and soil clay minerals evolution. Hence, the Muzzi soil clay mineral composition is interpreted as a property inherited primarily from the loess parent material and reflect pedogenetic processes of moderate intensity resulting in clay transformation without neoformation.

Pedogenetic pathways on each sedimentary mantle encompass Holocene climatic changes giving rise to a sequence of juxtaposed soils. Nevertheless, clay composition of the <1 µm and <2 µm fractions of the Canesa soil (Table 3) are quite homogeneous vertically, independently of lithologic discontinuities, typic of river basins, and marked texture and age differences, given the superposition of sediments during different periods of alluviation. X-ray patterns of the clay fractions of the 2C₂ and 3Ab horizons are quite similar (Fig. 3). They mostly differ in the higher content of interstratified clays in the buried A horizon (Table 3).

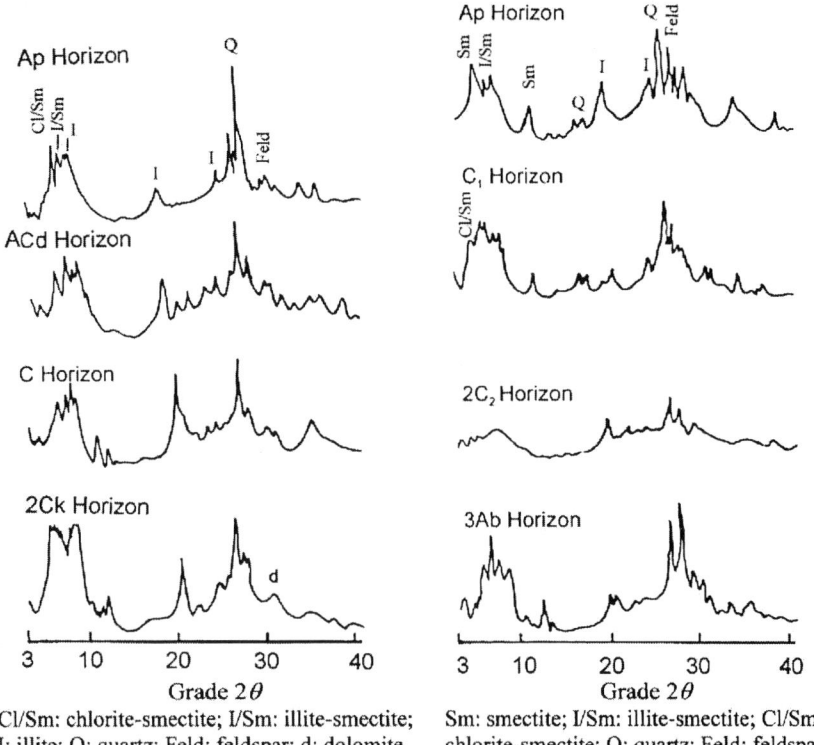

Cl/Sm: chlorite-smectite; I/Sm: illite-smectite; I: illite; Q: quartz; Feld: feldspar; d: dolomite.

Figure 2. X-ray diffraction pattern of Muzzi soil.

Sm: smectite; I/Sm: illite-smectite; Cl/Sm: chlorite-smectite; Q: quartz; Feld: feldspar.

Figure 3. X-ray diffraction pattern of Naposté soil.

Table 1: Physic-chemical characteristics of Muzzi and Canesa soils

Soil	Horizon	pH H₂0/ KCl		OC gr.kg⁻¹	CEC cmol.kg⁻¹	Clay < 2 μm
Muzzi	Ap	6.7	5.5	15.7	25.5	31.9
	ACd	7.1	6.0	10.0	22.8	27.3
	C	7.5	6.4	5.5	20.7	25.4
	2Ck	8.4	7.0	2.1	19.1	23.0
	3Ckm	Petrocalcic Horizon				
Canesa	Ap	7.4	6.5	16.0	16.3	14.5
	C₁	7.7	6.8	8.0	13.8	12.3
	2C₂	7.5	6.3	2.0	6.7	6.8
	3Ab	7.8	6.7	29.0	28.0	30.8

OC: Total organic carbon.

This is consistent with an older soil age for the alluvial soil parent materials, a humid paleoclimate and a longer period of stability subjected to weathering and pedogenetic processes transforming secondary clays.

Chlorite and interstratified chlorite-smectite were not detected in the buried soil. Clay mineralogy in Ap, C₁ and even 2C₂ horizons is comparable to the Muzzi soil, since the upper section (Ap-C_1) of the Canesa soil is developed in sediments of the Matadero Saldungaray Formation of historic age, which in turn resulted from erosion of the adjacent zonal soils. The underlying aeolian sediments of the $2C_2$ horizon that buried the pre-existing alluvial soil, represents a brief period of stability and pedogenesis. Moreover, the silt fraction of the Canesa soil indicates intrapedon similarity. However, smectite that is lacking in the silt fraction of uppermost horizons appears in the 3Ab horizon. Calcite is irregularly distributed with depth and is identified in the Ap, C_1 and 3Ab horizon but is lacking in the $2C_2$ horizon.

Table 2: Clay mineralogy of Muzzi soil

Horizon	Fraction	I	I/S	Ch	Ch/Sm	Sm	V	Accessory Minerals
Ap	< 1 µm	x	x		x	x		Qz ; Tr ; F ; A ; a ; He ; Ap
	< 2 µm	x	x		x			F ; A ; a ; Ap
	20-50 µm	x	x					Qz ; F ; Px/Am
ACd	< 1 µm		x		x	x		F ; A ; a
	< 2 µm	x	x		x	x		Qz ; F ; Ca ; A ; a
	20-50 µm	x						Qz ; F ; A ; a
C	< 1 µm	x	x		x	x		Qz ; F ; Ca ; Ap ; Il
	< 2 µm	x	x			x	x?	Qz ; F ; A ; a ; Ca ; Ap
	20-50 µm	x						Qz ; F
2Ck	< 1 µm		x	x	x	x		Qz ; F ; A ; a ; Ca ; Ap ; Cr ; Px
	< 2 µm	x	x			x		Qz ; F ; A ; a; Do ; Ca ; Ap ; He
	20-50 µm	x	x					Qz ; F ; Ca

Clay minerals: I: illite; I/S: illite/smectite; Ch: chlorite; Ch/Sm: chlorite/smectite; Sm: smectite; V: vermiculite. ***Accessory minerals***: Qz: quartz; Px: pyroxene; Am: amphibole (hornblende); A: anorthite; a: albite; Mi: mica; Ca: calcite; Go: goethite; He: hematite; F: K-feldspar; Tr: tridymite; Ap: apatite; Do: dolomite; He: hematite; Il: ilmenite; Cr: cristobalite; An: anatase; Y: gypsum.

Table 3: Clay Mineralogy of Canesa soil

Horizon	Fraction	I	I/S	Ch	Ch/Sm	Sm	Accessory minerals
Ap	< 1 µm		x			x	F ; Qz ; A ; a ; Ca;
	< 2 µm	x	x	x		x	F ; Qz ; a ; A
	20-50 µm	x					Tr ; Qz ; F ; a ; A ; He
C_1	< 1 µm	x	x				F ; a ; A ; Qz ; Ca
	< 2 µm	x	x		x	x	F ; Qz ; a ; A ; Ca ; Am
	20-50 µm	x					F ; Qz ; a ; A ; Ca ; Am
$2C_2$	< 1 µm	x	x		x		F ; Qz ; a ; A
	< 2 µm		x			x	F ; Qz ; a ; A
	20-50 µm	x					F ; Qz ; a ; A ; Am
3Ab	< 1 µm	x	x				Mi ; Qz ; Ca ; Ap ; Cr
	< 2 µm	x	x			x	F ; Qz ; a ; A ; Ca ; Y? ; Px ; Go
	20-50 µm	x				x?	Qz; F; a; A; Ca

Clay minerals: I: illite; I/S: illite/smectite; Ch: chlorite; Ch/Sm: chlorite/smectite; Sm: smectite. ***Accessory minerals***: Qz: quartz; Px: pyroxene; Am: amphibole (hornblende); A: anorthite; a: albite; Mi: mica; Ca: calcite; Go: goethite; He: hematite; F: K-feldspar; Tr: tridymite; Ap: apatite; Do: dolomite; He: hematite; Il: ilmenite; Cr: cristobalite; Y: gypsum.

5. CONCLUSIONS

We conclude that phyllosilicates are primarily inherited from the parent materials and that transformation processes of moderate intensity change illite and chlorite into irregularly interstratified illite-smectite and chlorite-smectite. Weathering has produced identical secondary clay minerals without translocation in both soils, depending on water percolation and moderate aggressive chemical characteristics of the soil solution. The more humid paleoclimatic conditions of the Middle Holocene have not influenced neoformation of clay minerals in the buried soil, probably due to insufficient time and a mild climate. Despite of changes in climate-paleoclimate during the Holocene epoch and differences in type, origin, texture, age and landscape position of allochthonous soil parent materials in the southwestern part of the semiarid Pampean region, clay and silt mineralogy is quite similar.

REFERENCES

1. R. E. Hughes, D. M. Moore and H. D. Glass, Quantitative Methods in Soil Mineralogy, SSSA Miscellaneous Publication (1994) 330.
2. M. del C. Blanco and L. Sanchez, Turrialba, 45 (1995) 78.
3. M. del C. Blanco, L. Sánchez, M. Vera and J. Aguilar Ruiz, Edafología, Boletín de la Asociación Española de la Ciencia del Suelo (1997) 120.
4. N. Peinemann, N.M. Amiotti and M. B. Villamil, Ciencia del Suelo, 18 (2000) 69.
5. M. Gonzalez Uriarte, IX Congreso Geológico Argentino, Actas (1984) 557.
6. Soil Survey Staff, USDA, Soil Survey Manual Handbook, 18 (1991) 437.
7. Soil Survey Staff, USDA, Soil Taxonomy, Agriculture Handbook, 436 (1999) 863.
8. M. L. Jackson, H. L. Chin and L. W. Zelasny, A. Klute (ed.), Methods of Soil Analysis I, Agronomy 9 (1986) 101.
9. Ch. J. Rabassa, I Jornadas Geológicas Bonaerenses, Actas (1985) 80.
10. C. Darwin, Geological Observations in South America, London (1846) 120.
11. J. Frenguelli, (ed. Universidad Nacional de La Plata), Loess y Limos Pampeanos, Serie Técnica y Didáctica (1925) 88.
12. M. Zárate and A. Blassi, Symp. Int. sobre el Holoceno en América del Sur (1988) 15.
13. D. Fanning and M. Fanning, Soil Morphology, Genesis and Classification, John Wiley and Sons (eds.), New York (1989) 387.

Coastal dune soils in Oregon, USA, forming allophane, imogolite and gibbsite

Georg H. Grathoff[a], Curt D. Peterson[a], and Darren L. Beckstrand[a],

[a]Applied Mineralogy Laboratory, Dept. of Geology, Portland State University, Portland, OR 97207-0751, USA email: GrathoffG@pdx.edu

Three diagenetic Al phases: allophane, imogolite, and gibbsite formed in dunal soils on the central coast of Oregon, USA, within the last 70,000 years. The allophane and imogolite completely replaces large pieces of wood and roots up to 10 cm in diameter. Only the outside of the roots is replaced by allophane and imogolite. Random powder X-ray diffraction (XRD) patterns show broad reflections at 3.3 and 2.25 Å typical of allophane with additional smaller broad reflections typical of imogolite at ~15, 7.7, 5.7, and 4.2 Å. The samples lose more than 75 % by volume on drying. Microprobe and SEM analyses showed the imogolite has a Al to Si ratio of about 1.8 and the allophane has a Al to Si ratio of about 1.4. The gibbsite occurs in the coarser sections of the sand as apparent cavity fillings and is not replacing preexisting minerals (e.g., feldspars).

Our preliminary conclusions are that the formation of the Al bearing phases in these dunes are caused by i) high Al concentrations, supplied by weathering of Al-silicates, ii) varying permeabilities within the dunes iii) change in pH of the groundwater changing the stability of the Al phases and iv) interactions (inorganic and/or microbial) with woody material.

The information from these soils is used to interpret aquifer geochemistry, age of paleosols to date dune emplacement, and potential preservation of artifacts.

1. INTRODUCTION

Weathering is primarily a function of precipitation, time, and source mineralogy. The coastal dunes in Oregon are excellent examples of young, highly permeable soils that develop in climates characterized by high precipitation (averaging 1600 to 2200 mm per year), wet winters, relatively dry summers, and mild temperatures throughout the year (Taylor and Hannon, 1999). The weathering in this environment is quite extreme due to the high precipitation. The high permeability of the soils limits the surface run off and reduces the residence time of the groundwater.

This study is part of the first regional compilation of the age and extent of emplacement of coastal dune-sheets in Western North America since Cooper (1958 & 1967) (Figure 1). Most of the samples for this study were taken from the dunes of the Oregon Dunes National Recreation Area (ODNRA), which include the Florence and Coos Bay dune sheets. For more detailed information on sample sites and regional geology see the fieldtrip guide of Peterson et al. (2002). These dune sheets are the largest dune fields in Oregon comprising about 70 km of coastline and characterized by moderately high levels of unstable minerals and high permeability (Twenhofel, 1946; Scheidegger et al., 1971).

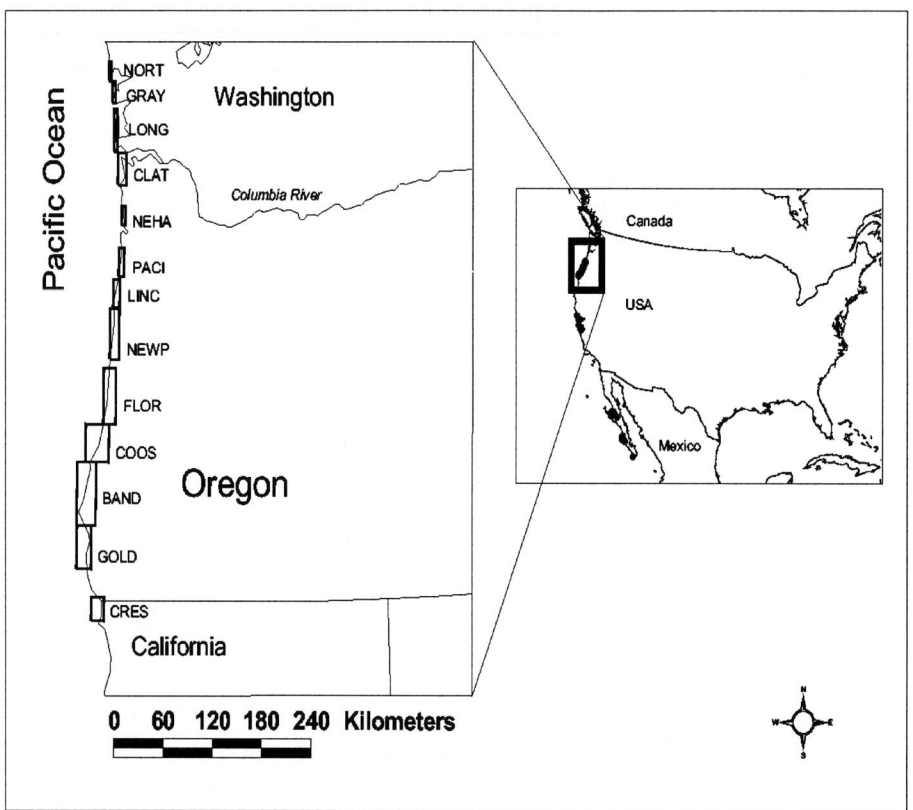

Figure 1: Map of the study area. The rectangles show the different dune sheets. Samples for this publication were all taken from the FLOR (Florence) and COOS (Coos Bay) dunesheets with more detail in Figure 2.

These dune plains are covered by basal (Pleistocene) dune sheets, deflation deposits, and widespread paleosols (Beckstrand, 2001). The dune sheets themselves are between 3 to 20 km wide and 30 to 100 km long (Cooper, 1958; Dupre et al., 1980, Murillo et al., 1999). These dune sheets migrated onshore from the exposed continental shelf during late-Pleistocene periods (24-70 + thousand years ago TLYBP (thermoluminescence years before present) (Beckstrand, 2001). By comparison, very-small dune advances occurred during the late-Holocene (Orme, 1990; Hart and Peterson, 1997). The TL ages of these dune field deposits in the central coast range from 70 to 1k (Beckstrand, 2001) (Figure 2).

Within these dune sheets we find a number of paleosols, which appear to control the groundwater flow and chemistry as well as the stability of the dunes. These low permeable paleosols, which stratify the dune deposits, were formed during episodes of dune sheet stabilization, vegetative colonization, and surface leaching. Secondary mineral precipitation from groundwater flow along paleosol redox-boundaries further cemented these horizons.

The occurrence of gibbsite is often attributed to old and highly weathered soils because Al is assumed to be immobile and silica and iron need to be removed before gibbsite forms. Gibbsite is usually found in bauxites and other highly weathered soils such as Ultisols (Hsu 1989). Two methods of gibbsite formation in soils have been suggested in the literature (e.g. Sherman et al., 1967; Young and Stephen, 1965). Either Al-silicates can weather directly to gibbsite, often preserving the original cleavage and crystal form, or gibbsite can form in the porespace. In this study we find gibbsite precipitating in the porespace of young highly permeable soils that have seen high precipitation.

Allophane and imogolite are usually known as metastable weathering products formed in volcanic ash and pumice soils within the B horizons. The processes that cause the formation of allophane and imogolite are usually accompanied by the formation and transformation of minerals (e.g., opal, clay minerals, Fe-oxides) and the accumulation of humus (Wada, 1989). The only alteration that we find associated with the allophane in the dune sheets are Fe-oxides.

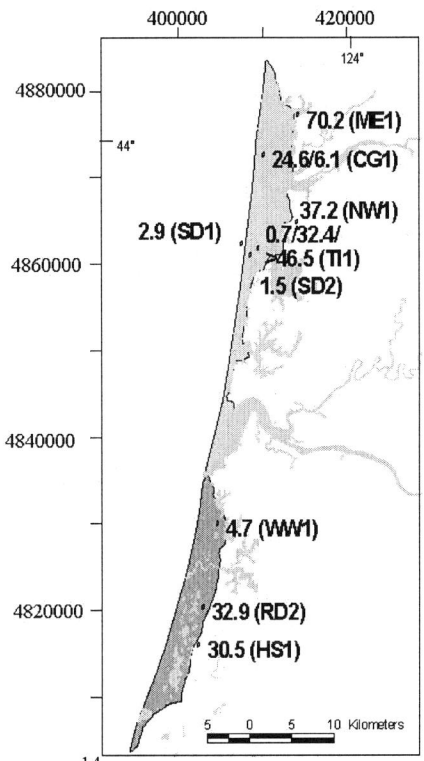

Figure 2: Study area with TL and C^{14} dates in 1000s of years. Radiocarbon dates are the 6.1 (CG1), 0.7 & >46.5 (TI1), other dates are TL (from Beckstrand, 2001).

2. METHODS

The mineralogy was determined using X-ray diffraction (XRD) with a Philips theta-theta X-pert™ system, using Cu K-alpha radiation (40kV and 30mA), and a Peltier™ cooled energy dispersive detector. Side packed samples were run in the step-scanning mode, from 2.5-75° 2-Theta, at 0.025°/step and 40 sec count time. Thermal Gravimetric Analysis was done using a Perkin Elmer™ TGA 7 with heating from 20 to 900°C at 10 deg/min. The chemistry of the allophane and imogolite was determined using the microprobe at Oregon State University's College of Oceanography as well as the SEM from Portland State University's Geology Department. The microprobe used was a Camica™ SX50 four-spectrometer electron microprobe operated by Dr. Roger Nielsen.

3. RESULTS

3.1. Allophane and imogolite

The samples that contained allophane and imogolite were collected at the sea cliff dune exposure north of Haceda Head, Oregon, USA (UTM 410330E; 4888745N) and the partially mineralized (replaced) tree roots from Seal Rock State Park, Oregon, USA. The allophane and imogolite completely replaces large pieces of wood up to 10 cm in diameter (figure 3) and replaces roots within paleosols. Only the outside of the roots is replaced by allophane and imogolite. Random powder X-ray diffraction (XRD) patterns of the mineral replacing the wood (figure 4) show broad reflections at 3.3 and 2.25 Å, which are typical for allophane with additional smaller broad reflections at ~15, 7.7, 5.7, and 4.2 Å, which are typical for imogolite (Wada, 1989). The samples lose more than 75 % by volume on drying. Microprobe and SEM analyses showed the imogolite has a Al to Si ratio of about 1.8 and the allophane has a Al to Si ratio of about 1.4.

The differentiated TGA data shows peaks at 100 and 350ºC reflecting highest change in weight loss at these temperatures. TGA and DTA data from Wada (1989) show peaks at 100ºC for both allophane and imogolite with an additional peak at about 400ºC for imogolite.

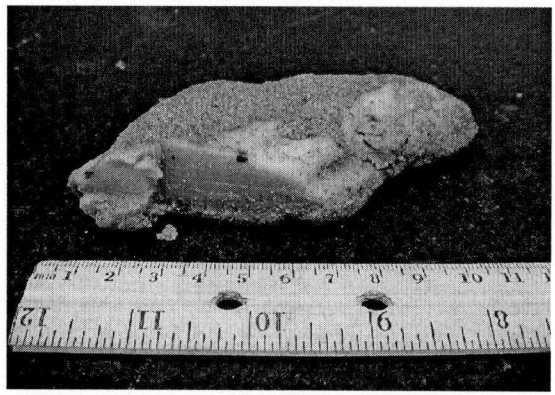

Figure 3: 10 cm piece of wood completely replaced by allophane and imogolite collected north of Haceda Head, Oregon, USA. Sand covers outside of sample, interior of sample shows preservation of original wood laminae (rings).

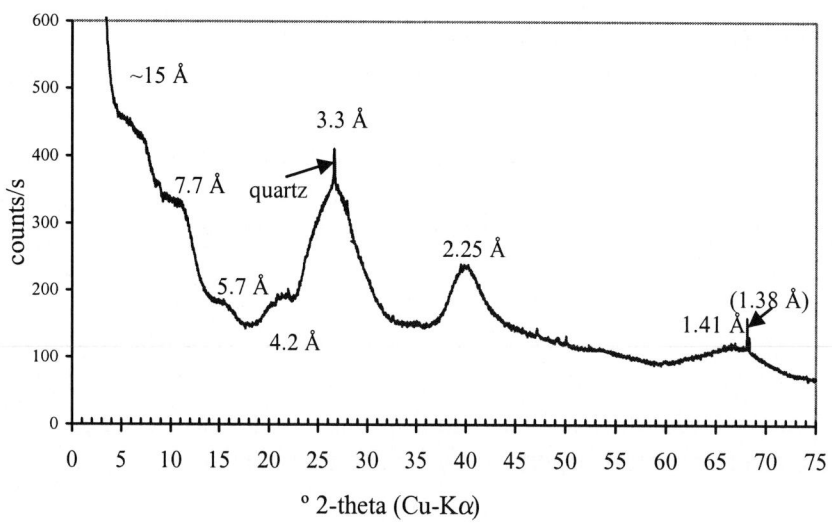

Figure 4: Random Powder XRD pattern of wood replacement

3.2. Gibbsite

Gibbsite was found within the Pleistocene dune sheets at Mercer Lake (ME1), the oldest thermoluminescence (TL) dated site at 70.2 ka, and also at the Houser Slough site (HS1), TL dated at 30.5 ka (Beckstrand, 2001). Gibbsite occurs in the coarser sections of the sand as apparent cavity fillings (Figure 5) and is not replacing preexisting minerals (e.g., feldspars).

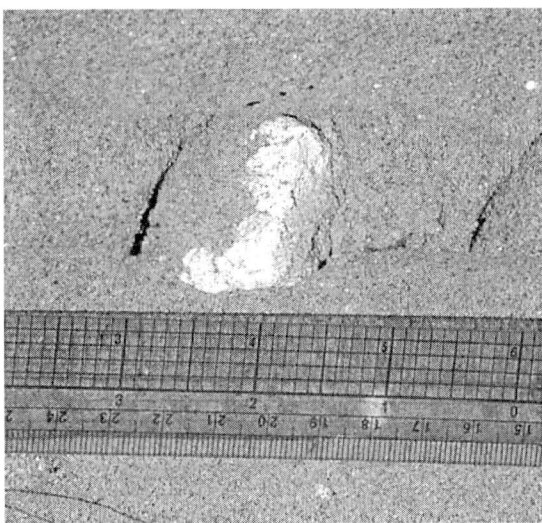

Figure 5: 2 cm gibbsite nodule within the Pleistocene sand dunes at Mercer Lake (ML1), Oregon, USA

4. DISCUSSION

We found two different areas where three different Al rich phases are formed, allophane, imogolite, and gibbsite. Often in sandy soils the groundwater is supersaturated with allophane, imogolite, and gibbsite but neither actually precipitate in the soils (e.g. Arocena et al. 1994). In the soils we studied all three minerals form. In our case high permeability apparently fosters gibbsite precipitation while the interaction with the woody material fosters allophane and imogolite formation.

4.1. Allophane and imogolite replacing woody materials

Petrification of wood by replacement with silica is common and the mechanism well understood. The silica enters the wood either in solution or as a colloid with subsequent penetration of the cell walls and progressive dissolution of the cell walls (Scarfield and Segnit, 1984). A similar method could be responsible for the allophane and imogolite replacing the woody material. The only difference would be that the groundwater would have to be enriched in Al. However it is unlikely that the groundwater would have the same Si/Al ratio as the allophane. Usually the groundwater concentration of silica is at least 10 times higher than Al (Drever, 1988). Therefore there has to be a mechanism that increases the Al/Si ratio in the pore water. A possibililty would be that the woody material could be used as a semipermeable membrane to increase the Al and Si concentration to supersaturate the solution with respect to allophane and imogolite. The woody material could also simply cause an increase in the residence time of the water, eliminating the kinetic restraints common to Al-silicates at surface temperatures. The groundwater would need to be undersaturated with respect to amorphous silica and the Al concentration would need to be high enough to make allophane and imogolite stable. Another likely possibility is that microbes are associated with the precipitation of the allophane and imogolite. Wada (1989) points out that allophane is often associated with humus. Humus is likely to contain wood and root material and therefore wood may be used as a substrate.

4.2. Gibbsite formation

Gibbsite can either precipitate directly from solution or replace Al-silicates (e.g. Sherman et al., 1967; Young and Stephen, 1965). In our case gibbsite is precipitating in the coarser sand size fractions. But why does gibbsite form and not an Al-silicate (e.g. kaolinite or halloysite)? The formation of kaolinite or halloysite versus gibbsite is controlled by the Si activity in solution. Apparently gibbsite formation is very rapid once the Al is separated from the Si. If the movement of the solution slows down Al may recombine with Si and kaolinite will precipitate (Keller, 1958 and Kittrick, 1970). This is consistent with our observation that the gibbsite forms in the coarser and more permeable sands.

Hsu (1989) concluded that gibbsite only precipitated directly from solution under tropical and subtropical climates, in basic and intermediate rocks, and upland topography. Gibbsite in our samples precipitated in a different climate and rock type. The ODNRA dunes climate is not tropical or subtropical it is characterized by high rainfall and moderate temperatures. The sands are quartz and feldspar rich with moderately high levels (1-5 weight percent) of unstable minerals (Twenhofel, 1946; Scheidegger et al., 1971).

Gibbsite can precipitate due to a change of pH. The solubility of gibbsite increases two orders of magnitude with a change from pH 6 to pH 4.5 (Paterson et al. 1991; Drever, 1988). For gibbsite to be stable and not kaolinite the Si activity has to be 10^{-5} or less (H_4SiO_4). Groundwater analyses of the sand dunes near Coos Bay, in the southern part of the ODNRA, have been reported between 5 and 37 mg/L silica (SiO_2) (Magaritz and Luzier, 1985; Dobberpuhl et al. 1985). The groundwater is undersaturated with respect to amorphous silica, supersaturated with respect to quartz. Why does gibbsite form? An explanation could be that in these highly permeable soils the groundwater drains too quickly to increase the silica concentrations. A reason that no gibbsite precipitates in the areas of lower permeability may be that the gibbsite becomes unstable due to the increase in silica concentration and kaolinite or another Al-silicate becomes stable, but due to the slow kinetics of this reaction no Al-silicate precipitates. However in the highly permeable sands there has to be a mechanism that supplies the Al.

The pH in dunal soils was commonly between 4.5 and 5.0 in the upper portions of the soils. The pH of core ME1 containing the most abundant gibbsite in the Cox horizon varied between 5.0 and 5.3. Although the pH in the upper portion of the soil profile might have changed over time, the current pH should not mobilize Al. One proposed scenario is that when the pH of this Al-enriched solution became more basic lower in the section, as is typical in the dunes, the Al precipitated out of solution, forming gibbsite in the more porous material of the crossbeds.

Another possibility is that gibbsite forms due to the salinity of the groundwater. Gibbsite is produced on an industrial scale by the Bayer process, where Na-aluminate solutions are cooled in the presence of gibbsite seed crystals. Adding Na(OH) to $AlCl_3$, gibbsite precipitates in a matter of hours at a pH of higher than 12 (Hsu, 1989). The pH and the temperatures are quite different in the dunes but the presence of NaCl could speed up the reaction of gibbsite precipitation.

5. CONCLUSION

Gibbsite is forming in coarser sands and fills large cavities in Oregon dunal soils. The controlling factors for the precipitation of gibbsite may be high Al concentrations, change in pH, low silica concentration, and the interaction with the Na and Cl.

Allophane replaces woody materials in the dunal soils. The controlling factors may be the interaction of organic matter and microbes with the pore water, the woody material may cause an increase in the residence time, or the woody material may serve as a semipermeable membrane.

ACKNOWLEDGEMENTS

This study was supported by SeaGrant # R/SD-04 Index NA650K and NSF grant EAR 9807085. We would like to thank Rodger Hart for supplying some samples and helpful discussions, and D.M. Moore for comments and review of the manuscript. Errol Stock assisted with site mapping and preliminary interpretations of gibbsite nodules.

REFERENCES

Arocena, J.M., Pawluk, Dudas, M.J., 1994, Mineral transformation in some sandy soils from Alberta, Canada. Geoderma, v.61, p.17-38.

Beckstrand, D.L., 2001, Origin of the Coos Bay and Florence dune sheets, south central coast, Oregon. [Unpublished M.S. Thesis]: Portland State University, Portland, Oregon, 192 p.

Cooper, W.S., 1958, Coastal sand dunes of Oregon and Washington: Geol. Soc. Am. Mem. v. 72, 169 p.

Cooper, W.S., 1967, Coastal dunes of California. Geol. Soc. Am. Mem. v. 104, 131 p.

Dobberpuhl, R.A., Luzier, J.E., Collins, C.A., 1985, Selected water-quality data for a coastal dunes aquifer near Coos Bay, Oregon – 1971 to 1983. USGS Open File report 84-858, 192 p.

Drever, J.I.,1988, The Geochemistry of Natural Waters. Prentice Hall, Engelwood Cliffs, New Jersey, USA, 2nd ed. 437p.

Dupre, W.R., Clifton, H.E., and Hunter, R.E., 1980, Modern sedimentary facies of the open Pacific Coast and Pleistocene analogs from Monterey Bay, California, in Field, M.F., Bouma, A.H., Colburn, I.P., Douglas,R.C., and Ingle, J.C., eds., Quaternary Depositional Environments of the Pacific Coast: Pacific Coast Paleogeography Symposium, Society of Economic Paleontologists and Mineralogists, p. 105-120.

Hart, R., and C. Peterson, 1997, Episodically buried forests in the Oregon surf zone: Oregon Geology, v. 59, p. 131-144.

Hsu, P.H., 1989, Aluminum Oxides and Oxyhydroxides, in Dixon, J.B. and Weed, S.B. (eds). Minerals in Soil Environments, 2nd Ed.: Soil Science Society of America, Madison, Wisconsin, USA, p. 331-371.

Keller, W.D., 1958, Argillation and direct bauxitization in terms of concentrations of hydrogen and metal cations at surface of hydrolyzing aluminum silicates: Bulletin of the American Association of Petroleum Geologists, v. 42, p. 233-245.

Kittrick, J.A., 1970, Precipitation of kaolinite at 25 C and 1atm. Clays Clay Minerals, v. 18, p. 261-267.

Magaritz, M., and Luzier, J.E., 1985, Water-rock interaction and seawater-freshwater mixing effects in the coastal dunes aquifer, Coos Bay. Oregon: Geoch. Cosmochim. Acta, v. 49, p. 2515-2525.

Murillo De Nava, J.M., Gorsline, D.S., Goodfriend, G.A., Vlasov, V.K., and Cruz Orozco, R., 1999, Evidence of Holocene climatic changes from aeolian deposits in Baja California Sur, Mexico: Quaternary International, v. 56, p. 141-154.

Orme, A.R., 1990, The instability of Holocene coastal dunes: the case of the Morro dunes, California, in Nordstrom K.F., Psuty, N., and Carter, B., eds., Coastal Dunes, Form and Process, Wiley & Sons, New York, USA, p. 315-336.

Paterson, E., Goodman, B.A., Farmer, V.C., 1991, The Chemistry of Aluminum, Iron and Manganese Oxide in Acid Soils, in Ulrich, B., and Summer, M.E., eds., Soil Acidity: New York, USA, Springer Verlag, p. 97-124.

Peterson, C.D., Beckstrand, D.L., Clough, C.M., Cloyd, J.C., Erlandson, J.M., Grathoff, G.H., Hart, R.M., Jol, H.M., Percy, D.C., Reckendorf, F.F., Rosenfeld, C.L. Steeves, P., and Stock, E.C. 2002 Pleistocene and Holocene landscapes of the Central Oregon Coast: Newport to Florence, in Moore, G.W. ed., Field Guide to Geologic Processes in Cascadia, Oregon Department of Geology and Mineral Industries Special Paper 36, p. 201-222.

Scarfield, G. and Segnit, E.R., 1984, Petrification of Wood by Silica Minerals: Sedimentary Geology, v. 39, p.149-167.

Scheidegger, K.F., Kulm, L.D., and Runge, E.J., 1971, Sediment sources and dispersal patterns of Oregon continental shelf sands: Journal of Sedimentary Petrology, v. 41, p. 1112-1120.

Sherman, G.D., Cady, J.D., Ikawa, H., Blumsburg, N.E., 1967, Genesis of bauxitic Hailu soils: Hawaii Agricutural Experimental Station Technical Bulletin, 56.

Taylor, G.H. and Hannon, C., 1999, The Climate of Oregon – from Rain Forest to Desert. Oregon State University Press, Corvallis, USA; 211p.

Turner, R.V. and Bryden, J.E., 1967, Effect of length of time of reaction on some properties of suspensions of Arizona bentonite, illite and kaolinite in which aluminum hydroxide is precipitated. Soil Sci., v. 103, p.111-117.

Twenhofel, W.H., 1946, Mineralogical and physical composition of the sands of Oregon Coast from Coos Bay to the mouth of the Columbia River, in Bulletin – Oregon, Department of Geology and Mineral Industries, p. 64.

Wada, K., 1989, Allophane and Imogolite, in Dixon, J.B. and Weed, S.B. (eds). Minerals in Soil Environments, 2nd Ed.: Soil Science Society of America, Madison, Wisconsin, USA, p. 1051-1081.

Young, A., and Stephan, I., 1965, Rock weathering and soil formation of high-altitude plateau of Malawi: Journal of Soil Science, v. 16, p. 803-809.

Deep and actively forming illuvial clay in the regolith and on bedrock

D.L. Johnson[a], D.N. Johnson[b], and D.M. Moore[c]

[a]Geography Department, University of Illinois, Urbana, IL 61801, USA

[b]Chemistry Department, University of Illinois, Urbana, IL 61801, USA

[c]Illinois State Geological Survey, University of Illinois, Urbana, IL 61801, USA

Recent observations in the Santa Barbara area of California suggest that clay eluviated in upper soil profiles does not stop entirely in subsoil B and C horizons, but can bypass them and be transported completely through the soil and regolith to deeper geologic materials, even to subjacent bedrock where it can be illuviated (deposited). Data are presented to support these observations and interpretations. They suggest an explanation for commonly observed deep and often thick illuvial clay films in C horizons of some soils, for clay films and channel-fillings of clay in the regolith below C horizons, and for deposited clay found at regolith-bedrock interfaces, and even in or on bedrock. They also suggest an explanation for thick (± 1m) clay bodies that occur as solutional vug-fillings below residual limestone soils as viewed in some limestone quarries, and for water-sedimented clay bodies seen in many caves, so called 'cave clays', which probably originate at least partly as soil derived translocated clay.

Key Words: Eluvial-illuvial clay; clay films; regolith processes; soil formation; translocations

1. INTRODUCTION

The usual concept of illuvial clay is that of clay translocated in soils, either vertically downward from upper to lower soil horizons, or horizontally downslope as throughflow (interflow) within soil horizons, or both, then deposited as clay films (or clay skins, clay coatings, etc.).

Evidence for illuviation is the presence of clay films on subsoil structural aggregates (clay-coated peds), as blebs within peds, as bridges between grains, as envelopes on clasts, and as increases of clay in subsoils (Bt, argillic horizons) over that of topsoils and E horizons (from which the clay is presumed to have been derived). Downward- and lateral-moving soil water flowing as 'wetting fronts' from rainfall and/or snowmelt events are the assumed energy vectors of clay translocation. Very thick clay films (> 0.5-1cm) not uncommonly occur in soil C horizons, sometimes deeply into what some soil scientists consider 'non-soil' geologic materials. For example, thick (> 1cm) clay films are formed in basal (>5-10m deep) soil C horizons on the Channel Islands of California (Johnson, 1972).

1.1. Study site location, and geologic-pedologic contexts

On the adjacent California mainland, on the north wall of a deep and steep railroad cut several km west of and very near Santa Barbara, clay films measuring 1-2 mm thick were observed in April, 1999 to be actively depositing directly onto Miocene (Monterey Shale) bedrock from milky vadose water issuing from the regolith-bedrock interface (Fig. 1). The illuviating bedrock surfaces were exposed in the railroad cut at the base of the regolith, well below the soil C horizon.

The study site consists of an uplifted marine bedrock platform cut across Monterey Shale bedrock, and overlain by thin (< 1m) marine gravelly sands. This marine unit is in turn overlain by a thick (> 8 m) deposit of stratified gravelly continental deposits (regolith). A well developed Concepcion-like soil (Argialboll), with shallow though well expressed A and E horizons and deep Bt and Ct horizons, is formed in the upper and middle parts of the continental regolith pile.

1.2. Observed clay suspensions and active illuviations

Clay suspensions, evident at the base of the regolith as milky vadose (meteoric) water, were observed issuing from numerous seeps along the interface between the bedrock marine platform and the overlying regolith-soil pile, as shown in Figure 1. As the thin clay suspensions seeped down across the vertical bedrock face they settled (deposited) as evaporation set in, and clay thus was observed to be actively illuviating, leaving a moist and shiny clay surface that coated the exposed bedrock. The uplifted wave-beveled marine platform, which defines the bedrock-regolith interface, is immediately above the overhanging vegetation in the upper photo of Figure 1, and slopes gently upward and inland (away) from the wall of the railroad cut.

Downslope from the actively clay-illuviating bedrock surface, freshly deposited clay had commenced drying, and in places had completely dried. Where dried the clay had cracked and curled, and in places had flaked off and had accumulated as miniature clay talus piles at the base of the bedrock face. It was also clearly evident that the clay suspensions had been much more extensive only days before when vadose flow was more copious, with flow entirely over the bedrock surface to lower lying areas near the railroad bed where it had accumulated in several small shallow pools as a sedimented clay sheet. Dried and drying, curled and partly curled illuviated clay could be traced continuously from the actively drying illuviated bedrock surface to small shallow pools below where the more copious earlier suspensions had ponded and sedimented (not shown in figures).

Rainfall events during preceding days and weeks had obviously initiated the gravity-driven wetting fronts. The vertical throughflow of soil-vadose water had apparently entrained clay from the upper horizons of the surface soil profile. Wetting fronts carrying suspended clay were apparently passing through the A, E, Bt (argillic), and Ct horizons of the surficial soil, and through the underlying sediment pile (regolith), to the marine-beveled Monterey bedrock below. (The bedrock obviously functions as an effective aquiclude. Flow was thence downslope along the bedrock aquiclude-regolith interface (± 1% slope) and out to the exposure in the railroad cut.

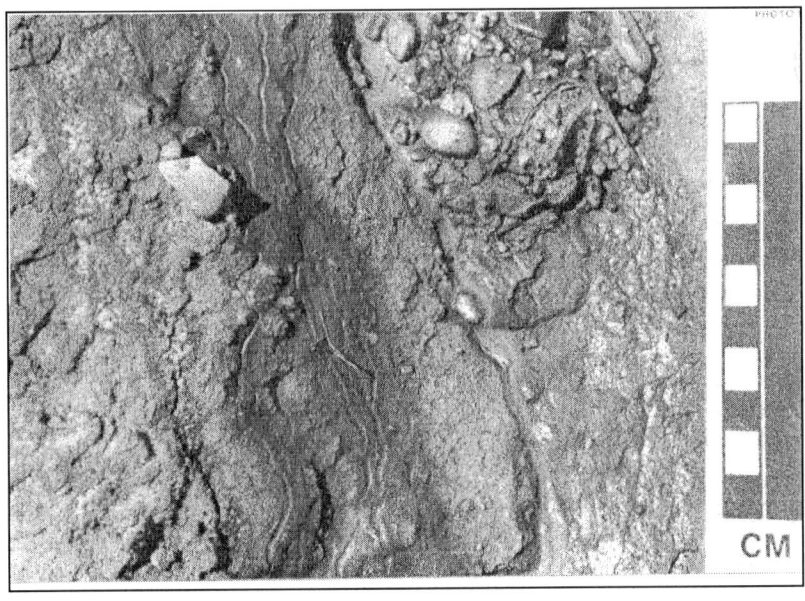

Figure 1. Above, coauthor D.N. Johnson pointing to where clay was actively illuviating onto Monterey Shale bedrock (Miocene), along north wall of railroad cut just west of Haskell Beach, Santa Barbara area, California in April 1999. Below, close up showing almost micro-detail of actively illuviating clay from slowly downward-moving soil-vadose suspensions.

2. FIELD AND LABORATORY METHODS

Beyond photographing in detail our fortuitous observations (Figure 1, plus other photos not included due to page limitations) – photos that essentially speak for themselves regarding deep illuviations, field work consisted of describing the profile and regolith, and sampling the soil by horizon (and by 15cm depth increments within most horizons), with select sampling of the regolith, down to the bedrock platform. Clay samples were also collected by scraping clay from the actively illuviating Monterey shale bedrock (Fig. 1). Laboratory methods included particle analysis of the soil profile and regolith (Soil Survey Staff, 1992), and clay mineral analysis of the clay fraction (Scintag© X-ray diffraction (XRD) of clay-sedimented slides using a theta-theta goniometer, CuKa radiation, a germanium N_2 cooled detector, and DMS© software). We used standard sample preparation techniques (Moore and Reynolds, 1997), for air-dried and ethylene solvated oriented aggregates.

3. RESULTS AND DISCUSSION

The soil profile description and particle size data (neither included -- page limitations) indicate that the soil is close to the Conception Series, and might be classified as a Xeric Argialboll (or Mollic Abruptic Haploxeralf), or some such closely related taxon (Shipman, 1981). It has a highly invertebrate-bioturbated dark colored A horizon (0-36cm thick) and a somewhat bleached and leached E horizon (36-48cm), also highly bioturbated, that contains small shot-sized ironstone pisoliths. The E horizon has an abrupt lower boundary with a clay-rich and deep Bt (argillic) horizon (48-432cm) whose clay content generally decreases with depth, though it sometimes contains thick and very thick clay films with depth. The combined Ct horizon and regolith (4.3-8.5m) also contain clay films, sometimes as thick joint and channel fillings, and also as coatings around clasts, though clay through the Ct and regolith also generally decreases in depth, until the Miocene bedrock is reached wherein it abruptly increases.

Select X-ray tracings of the clay mineral suite from the A horizon down to the clay-illuviated bedrock are shown in Figure 2. The traces, together with the observations made (Fig. 1) and particle size data suggest, if not collectively indicate, that the clay distributed throughout the soil profile and the regolith, and the clay illuviated onto the subjacent bedrock surface, is genetically linked. (Though not presented, the soil descriptions augment this interpretation.)

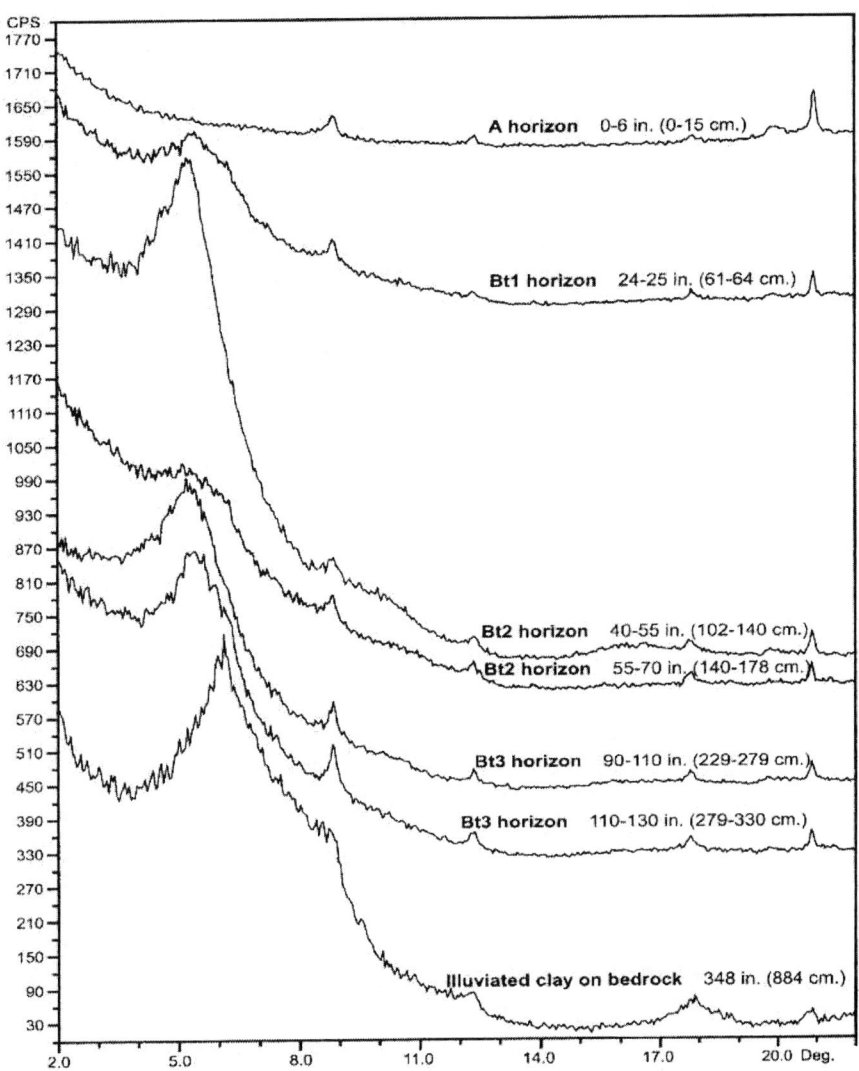

Figure 2. Select X-ray diffractograms of sedimented slides of clay fraction from horizons and sub-horizons of Concepcion-like soil, from the regolith, and from bedrock-illuviated clay west of Haskell Beach, Santa Barbara area, California. (The E horizon XRD trace, not shown, is virtually identical to the A horizon trace.) A genetic linkage between the clay in the soil, in the regolith, and that illuviated onto bedrock, is suggested by coincident positions of the expandable mixed-layered peaks, illite and kaolinite peaks, and the shift to the right with depth of the expandable mixed-layered peaks.

4. CONCLUSIONS

The observations presented here (Fig. 1) were extremely fortuitous, and were at the right time and place, and under the right circumstances. It is noteworthy that *all* the abundant and thick clay coats observed and described in this report were produced historically, after the railroad cut into the Monterey Shale bedrock was made in the late 19th Century. The observations together with the clay mineral, particle size data, and profile descriptions confirm what we have observed on other continents -- that clay films are commonly much thicker in soil C horizons, and often in the regolith, than in soil Bt/argillic horizons. These observations and data suggest that clay illuviation, usually thought to be largely a surficial and upper profile phenomenon, may be common in deep "non-soil" geologic materials well below soil profiles. Clay translocations and illuviations may in fact be common C horizon/regolith processes in general. If so, our concept of soil illuvial clay and it's depth domain should be expanded to include the C horizon and regolith, and wherever vadose waters deposit suspended clay. Further, 'cave clay', often observed as thick underground accumulations in karst landscapes, and thought to represent surface clay washed during rainstorms from soils into sinkholes and other karst channels by direct fluvial action (e.g., Gillieson, 1987), may be partly derived by the clay eluviational-illuviational processes described here. By the same token, large clay bodies commonly observed as solutional vug and cavity fillings in limestone terrains (Frolking et al., 1983, Webster, 1888, Nicholson, 1882, and pers. obs.) probably also owe their origin to such soil-regolith illuvial processes.

ACKNOWLEDGEMENTS

We thank Jane Domier for cartography and Barbara Bonnell for typing, and the Geography Department and LAS Dean's office at the University of Illinois for supporting travel to Santa Barbara (for another purpose, but which enabled us to work on the site).

REFERENCES

Frolking, T.A., M.L. Jackson, and J.C. Knox, 1983. Origin of red clay over dolomite in the loess-covered Wisconsin driftless uplands. Soil Sci. Soc. Am. J. 47: 817-820.

Gillieson, D., 1987. Tropical caves in retrospect and prospect. Prog. Phys. Geog. 11: 511-532.

Johnson, D.L., 1972. Landscape evolution on San Miguel Island, California. Ph.D. thesis, Univ. Kan., Lawrence (Univ. Microfilms, Univ. Mich., Cat. No. 73-11902).

Moore, D.M.. and R.C. Reynolds, 1997. X-ray diffraction and the Identification and Analysis of Clay Minerals (2nd. Ed.). Oxford Univ. Press, 387 p.

Nicholson, F. 1882. A review of the Ste. Genevieve copper deposit. Trans. Am. Inst. Min. Eng. 10: 444-456.

Shipman, G.E., 1981. Soil Surv. Santa Barbara Co., So. Coastal Part. U.S. Dept. Agric., Wash., D.C.

Soil Surv. Lab. Staff, 1992. Soil Survey Laboratory Methods Manual. Soil Surv. Invest. Rept No. 42, version 2.0 (method 3A),. U.S. Dept. Agric., Wash., D.C.

Webster, C.L. 1888. Notes on the geology of Johnson County, Iowa. Am. Nat. 22: 408-419.

Pedogenic mineral formation as environmental indicator in a paleosol sequence from the Miocene to the Holocene at Mergelstetten, Southwest-Germany

K. Stahr, P. Kallis, M. Zarei and K. E. Bleich

Institute of Soil Science and Land Evaluation, Hohenheim University, D-70593 Stuttgart, Germany

The karstic limestone plateau of the Swabian Alb is a marine deposit which has risen above sea level at the end of the Jurassic period. Since then it has undergone terrestrial land development for more than 100 Million years. Old soil formations of Cretacous and Tertiary period have been mainly removed by erosion because of epirogenic uplift of the area by 1,000 m. However, in the south of this limestone plateau during the development of the alpine orogeny, there was a sedimentary basin formed between the Alps and the Swabian Alb. This molasse basin was filled with the debris of the rising Alps, under fresh water and partly under marine conditions. The Miocene Upper Marine Molasse formed a cliff coast line from southwest to northeast crossing the whole limestone area.

Outside the marine development, the erosion of older soils and soil sediments was restricted through the rising sediment and groundwater level. Therefore in these places soils and soil sediments reflecting Tertiary and Holocene environmental development are partially preserved. The authors have been able to analyse such a sequence at Mergelstetten near Heidenheim. The sequence shows five separated periods of development. At the bottom are found limestones and dolomites of the Upper Jurassic as relicts of a strong karstic dissolution. The dissolution residue under tropical ferralitic soil development left a mineral assemblage of hematite, kaolinite and gibbsite behind. This sequence was in the depression later transformed through reductomorphic processes. The weathering under a strong loss of silica is reflecting the period earlier than 60 Million years B.P. Subsequently, under the environment of the Upper Marine Molasse a few kilometre south, the hollow was sedimented with clays and marls from the neighbouring hills. This has undergone a weathering in a depression of the landscape with reductomorphic stagnant features and a smectitic mineral assemblage. The next hiatus follows before the Kaiserstuhl volcanism, which is 20 Million years B.P. Then a sedimentation without very strong silicate weathering left behind carbonatic illitic materials. This after the Kaiserstuhl volcanic events from 17 Million years B.P. onwards showed a soil formation under strongly alkaline conditions within a swamp or gleyic environment. The main feature is the formation of thick columnar and pisolitic calcrete (limestone) reflecting this alkaline development under mild, mediterranean subtropical, environment. The main clay mineral which was probably formed with silica addition is smectite, but also a clear tendency of palygorskite formation can be observed. The youngest Tertiary period of the Pliocene is only ill documented. However, Pliocene and Holocene weathering in the first meter of the 10 meter deep profile show influence of alpine - derived dust deposition, the solution of limestone and the formation of a Chromic Cambisol again with smectitic mineralogy towards a Terra Fusca (Chromic Luvisol). This modern development showed a leaching environment which produced beside dominant illitic mixed layer and smectitic minerals also chlorite. The sequence is manifested through changes in the heavy mineral assemblage, in geochemical observations, in the carbonate petrography as well as with oxygen isotopy.

1. INTRODUCTION AND PROBLEM DEFINITION

The Swabian Alb is a karstic limestone plateau which was under marine influence until Upper Jurassic period. For more than 100 Mill. years it has undergone terrestrial influence. However, during the Miocene period a perialpine marine influence from the south was observed. This marine influence can still be observed at the major landscape divide of the Miocene cliff line (ancient shore line). In the vicinity of this old shore line already the early geologists observed rather irregular sediments including secondary carbonates, which have been found on both sides of the shore line. Fraas (1868) already supposed that this sediments, especially the carbonates, are beach formations of the molasse sea. Later Berz (1924) argued for a fresh water formation of the same material. The relevant period for the formation of this thin and very variable deposits is the Middle Miocene. After geological mapping, Hüttner (1961) suggested that the carbonates found there have been formed under lacustrine conditions in a relatively flat landscape covered with small lakes and drowned valleys. Later Müller and Reiff (1993) suggested that these limestones or calcrete are of pedogenic origin. The authors had the privilege to study a sequence of soils and sediments covering the Tertiary until Holocene development of the landscape at Mergelstetten close to the city of Heidenheim.

We tried to use methods of soil micromorphology and mineralogy as well as geochemistry to reconstruct the development of this sequence including the history of the landscape. In this work we try to explain the sequence of soils and sediments through analysing the mineral fractions.

2. ENVIRONMENTAL SETTING

The Miocene cliff line is a major landscape divide. To the west, the karstic area has been affected by terrestrial weathering since the end of the Jurassic period. Remnants of Cretaceous and Paleogene soils are still preserved in protected relief positions. East of the cliff line we have a sequence of molasse sediments namely the Lower Freshwater Molasse, the Upper Marine Molasse and the Upper Freshwater Molasse, documenting a Neogene sedimentation in the molasse basin (Kallis et al., 2000). The sequence of soils and sediments was investigated at a limestone quarry (Company E. Schwendt) at Heidenheim,

Mergelstetten, 2 km north of the Miocene cliff line (figure 1).

The northern wall of the quarry exposed a 200 m wide and up to 20 m thick sequence of soils and sediments including two massive tertiary carbonate formations (Kallis et al, 2000).

The whole form is a hollow which was formed by Cretaceous and Paleogene karstic weathering. A riff formation imbedded in the upper Jurassic limestone and marls forms the bottom of the sequence. This karstic limestones are covered by several meters of clay soil residue. This soil material shows strong stagnic properties (red, ochreous and white mottles). Its upper part shows features of preworking such as accumulation of broken iron and silica concretions. At the deepest point of the hollow this early paleosol layer is overlain by an up to 1 m thick bank of fluvial gravel derived from Jurassic limestone (Jura Nagelfluh), which has been secondarily cemented by carbonate. This is embedded in a paleosol horizon derived from the weathering of resedimented Jurassic marls, which show olive and ochreous mottling.

A subsequent layer of fine textured sediment was transformed into soil horizons with gleyic and stagnic properties intercalated with carbonates. Above, white chaeky columnar carbonates within a gleyic matrix have grown upwards to form a massive horizon of columnar limestone, which grades upwards into an embedded and mottled pisolitic limestone. The top of this sequence is overlain by a gleyic soil horizon rich in carbonate concretions. Fossil shells of Cepaea sp., a terrestrial gastropode, dates the sequence to the Carpatian period within the middle Miocene (Müller und Reiff, 1993). This is grey clay material with carbonate concretions and red zone which show again gleyic properties. At top of the sequence a recent terrestrial soil with a mollic epipedon and a clay subsoil with blocky aggregation was observed. The methods used in this work have been described in detail by Kallis (2001).

3. CLAY MINERALS

The profile at Mergelstetten according to its clay mineralogy can be subdivided into four general units (compare figure 1):

The bottom (unit I) is dominated by kaolinite (70 - 90 %). The X-ray-diffraction diagram in this part shows mainly the kaolinite peaks with a very weak illitic peak at 1,0 nm and some reflections of quartz and goethite. Samples from other locations in the vicinity also showed the occurrence of gibbsite in this part of the profiles. This kaolinitic sequence at the bottom of the profile is up to 4 m thick. It is followed by an only 1 m thick layer (unit II), which is dominated by smectite (up to 60 %). Here also kaolinite, illite and irregular illite-smectite interlayers are observed. This part has some similarity to the clay mineral association, which is observed in the Upper Jurassic cement marls (Upper Jurassic Zeta 2). With a sharp hiatus a 1 to 2 m thick layer with the predominance of illitic clay minerals is observed (unit III). Only about 10 % of kaolinite and mixed layers occur here. Therefore the illitic dominance seems to be strange in the whole sequence. The upper 4 to 5 m (unit IV) again returns to a smectitic dominance, but having generally a much more mixed mineralogy with about 10 % of kaolinite, 10 to 30 % of illite, 5 - 30 % of mixed layers and even some secondary chlorites. Within this sequence from the residue of the carbonates, palygorskite was clearly detected (Kallis et al., 2000).

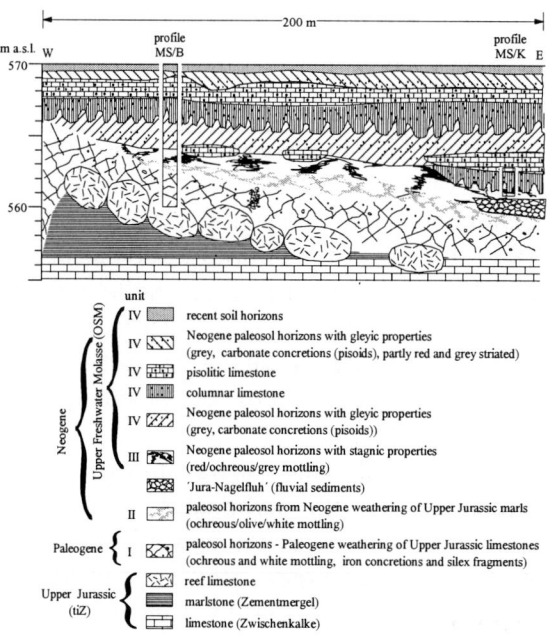

Fig. 1: Generalized Tertiary soil-sediment sequence of an exposure at Heidenheim-Mergelstetten (SW-Germany)

4. SECONDARY CARBONATES

The whole profile sequence is resting on the Tithonian limestones of the Jurassic period. Samples of these limestones showed calcite contents of 96 - 97 % (Tab. 1). This is pure calcite. No dolomite or magnesium-calcite was found. However, the clay of this residual bottom zone (unit I) is almost free of calcium carbonate. That means it had been decalcified and there was no recalcification at all during the later periods. In the second zone (II) there are some clay parts, which are free of calcium carbonate, but there was also calcium carbonate precipitation in the hollow which resulted in this horizon in high carbonate contents of 6 % in concretions of stagnic zones and up to 89 % in the soft limestone. Part III of the profile was again almost free of lime. In the more complex part IV, the calcium carbonate content in the columnar and pisolitic limestones rise from 16 % in concretions to 83 - 92 % in the limestone parts. In the top part of the profile again a calcium carbonate free soil matrix can be observed. The recent profile is only acidified down to 50 cm. All the other horizons have pH-values above 7, up to 8.3. That means, modern silicate weathering could not affect heavily the mineralogy of the sequence. All the calcium carbonates, which have been found in the filling of the hollow could by isotopic analysis of carbon and oxygen be related to fresh water situations. The mean values are $\delta^{13}C = -6,9$ and $\delta^{18}O = -$

4,33, which is typical for lacustrine environments. However, the lighter isotopic composition shows some pedogenic influence especially in sparitic pore fillings, while the micritic centers of pisoids seem to be developed in a lacustrine environment.

Tab. 1: Carbonate content of the different horizons and sediment layers of profile at Mergelstetten

Layer	Depth cm	$CaCO_3$ % w/w
Modern soil	0 - 80	0.05 - 1.1
Hard pisolitic limestone Unit IV	100 - 400	82.0 - 92.0
Soft limestone Unit IV	100 - 400	83.0 - 88.0
Relic gleyic prop. Unit IV	400 - 500	0.1 - 1.2
Clay, stagnic prop. Unit III	500 - 620	0.0 - 0.3
Soft limestone Unit III	680 - 720	83.3
Neogen, mottled clay Unit II	620 - 700	0.03 - 6.6
Paleogene clay, stagnic prop. Unit I	700 - 1070	0.0 - 0.1
Upper Jurassic Limestone	> 100 m	95.5 - 98.9

5. HEAVY MINERALS

The heavy minerals from the sand fraction show a clear sequence of development through the whole profile. The basal clay horizons are dominated by a group of very stable minerals, which are characterized by zircon and rutile. The zircon content is more than 40 %, the rutile generally more than 30 %. As a third important component tourmaline was determined. All the others are of minor importance. In the second sequence, the dominance of zircon and rutile is still maintained in the same percentage as before, however, few minerals of the metamorphic group, like epidote, have been found. Only in the third part of the sequence, which is again clayey is evidence for transition from the stable to the metamorphic and vulcanic heavy minerals observed. In the upper part of this section III melanite, a garnet mineral, is above 30 %. The upper section (IV) is clearly characterized by the metamorphic alpine minerals of the garnets, almandine and melanite, together with the epidote group minerals. In this part the stable minerals together represent less than 10 % of thesum. To the top of the profile, the variability increases significantly in the uppermost horizons. It may be also affected by very modern loess minerals, which has led to a great variability in the heavy mineralogy.

Single zircons from different horizons have been picked and by means of fission track dating method we have determined the age of the formation of soils or sediments. From this method we got a very clear sequence of dating: In all samples there have been 15 - 20 % of the zircons, dated back to 140 Mill. years and earlier. This means they are derived from the Jurassic limestone directly without alteration. However, the zircon of the basal unit I had their maximum of distribution at 50 - 60 Mill. years with the range from 30 - 80 Mill. In contrast the zircons of the unit IV had the maximum at 20 Mill. years with a range from 10

- 30, but with a secondary maximum from 40 - 70. Thus, the youngest sediments have contain mainly zircons of a young Tertiary age, but they also retranslocated old Tertiary minerals to the extend of 20 - 30 %.

6. RECONSTRUCTION OF THE DEVELOPMENT PHASES

The beginning of the sequence started when the retreat of the Jurassic sea gave the possibility for terrestrial development. The period of the Cretaceous is ill documented. However, it can be expected that already during the Cretaceous we found a leaching environment producing clayey limestone residues. The zircon of Cretaceous age strengthen this estimate. However, the better documented phase is the older Tertiary (Paleogene) which started with a warm humid phase of intensive decalcification, development of a karstic geomorphology and mineral newformation. The mineral newformation is characterized by decalification and leaching of alkaline and earth alkaline elements and a respective Al and Fe residual accumulation as oxides. Therefore we find gibbsite, goethitic and hematitic concretionary iron oxides in addition to Kaolinite. The gradual development of the clayey bottom of the sequence in the unit I ended with a transport of broken concretions of iron and also silica within a kaolinitic matrix, which contained only stable heavy minerals. The period of leaching and desilification is the longest period which lasted for almost 100 Mill. years. With the beginning alpine orogeny and the development of the molasse basin, the regional groundwater table rose and karstic development was stopped for a first time (Hüttner, 1961 and Reiff, 1989). The vertical water transport was replaced by a partially horizontal one, which filled small lakes and rivers. During this period, the bottom of our karstic depression was filled with marls and even with limestone gravels by fluvial activity. This new sediments, with a changing environment, brought back alkalinity to the landscape. The tendency towards the smectitic minerals in this time was rather strong and the first lacustrine lime precipitation took place. The hiatus towards the younger Tertiary (still molasse influenced) showed a clay and calcium carbonate sedimentation together with influence of some alpine minerals, which brought strange mineralogy from metamorphic origin. At least in a temporary lake there was a formation of soft and pisolitic limestone in two phases. This took place around 20 - 15 Mill. years before now. The period was after the Kaiserstuhl event (20 Mill.). Limestones are intercalated with red partially illitic soil sediments infilled from the terrestrial environment. In this sequence we found the clay mineral palygorskite. Therefore with the red clays the palygorskite in the depression and the result of the oxygen isotopes we expect a mediterranean climate during formation of this soil sequence. The uppermost part of the profile is developed in the last 17 Mill. years in the Pliocene, Pleistocene and Holocene again under a leaching environment. However, this has not affected the sequence deeply enough to transform the well documented old and young Tertiary paleosols.

REFERENCES

Berz, K.L., 1924: Kreide und Tertiär-Jüngere Bildungen - Das Werden der Landschaft-

Begleitworte zur geognostischen Spezialkarte von Württemberg, Atlasblatt Heidenheim, 2. Aufl., Stuttgart, pp. 105-115.

Fraas, O. 1868: Begleitworte zur geognostischen Spezialkarte von Württemberg, Atlasblatt Heidenheim mit Umgebungen von Weißenstein und Steinheim, Stuttgart.

Hüttner, R., 1961: Geologischer Bau und Landschaftsgeschichte des östlichen Härtsfeldes, Schwäbische Alb. Jh. Geol. Landesamt Baden-Württemberg **4**, 49-125.

Kallis, P., 2001: Tertiary soil formation along the northern border of the Southwest German Molasse Basin, Swabian Alb. Hohenheimer Bodenkundliche. Hefte, Band **60**, 280 S. in German.

Kallis, P., K. E. Bleich, K. Stahr, 2000: Micromorphological and geochemical characterization of Tertiary 'freshwater carbonates' locally preserved north of the edge of the Miocene Molasse Basin, SW Germany. Catena, **41**, S. 19-42.

Müller, S., W. Reiff., 1993: Ein Vorkommen von Oberer Süßwassermolasse bei Heidenheim und seine Bedeutung für die Landschaftsgeschichte. N. Jahrb. Geol. Palaeontol. Abh. **189**, 255-274.

Reiff, W., 1989: Das Kliff in Heldenfinden und die Klifflinie auf der Heidenheimer Alb. Jber. Mitt. Oberrhein. Geol. Ver., N.F. **71**, 467-482.

Experiences with selective extraction procedures for iron oxides

J. Kenneth Torrance[a] and Jeanne B. Percival[b]

[a]Department of Geography and Environmental Studies, Carleton University, Ottawa, Ontario K1S 5B6, Canada

[b]Geological Survey of Canada, 601 Booth Street, Ottawa, Ontario, K1A 0E8, Canada

Four extraction procedures - acid ammonium oxalate, hydroxylamine hydrochloride, Tiron and dithionite-citrate - designed to extract ferrihydrite and specific crystalline iron oxides, were assessed for relatively unweathered, sensitive, post-glacial marine clays and for intensively altered, argillaceous material surrounding a uranium deposit. The procedures were more aggressive in extracting oxides other than the target oxide, as well as Fe-bearing and non-Fe-bearing aluminosilicates, than the literature suggests. Reassessment of geochemical investigations where these and other extraction procedures have been used may be needed.

1. INTRODUCTION

Iron oxides (e.g. goethite, hematite, magnetite, and amorphous ferrihydrite) may occur as surface coatings on other minerals or as fine, discrete, colloidal particles [1]. They may influence surface charge, adsorption characteristics, rheological behaviour and strength. The problems reported occurred: a) during studies of the landslide-prone Leda clays of eastern Canada [2,3], where the concerns were the dominant iron oxide forms present and their short to medium range variations in concentration, with particular interest in amorphous iron oxide, as prior claims [4] of 2-6% lacked credibility [5]; and b) during the development and application of sequential extraction procedures to assess the role of large amounts of iron oxides in restricting metallic element migration in the alteration halo of a hydrothermal uranium deposit in northern Saskatchewan, Canada [6,7,8]. In both investigations, major problems were encountered based on the standard interpretations of the selectivity of the extraction procedures used.

2. EXTRACTION PROCEDURES

Extraction procedures have been developed for which claims of selectivity for various forms of iron oxides have been made. The amounts of iron extracted are commonly called "free iron", "amorphous iron oxides", etc. Despite criticisms of each procedure's individual selectivity, a hierarchy of aggressiveness is generally accepted, and new methods continue to be developed.

Four extraction procedures were tested, using variations of the original procedures in which the sample-to-extractant ratio was increased - a change which would decrease the aggressiveness of the extractant: **a) Acid ammonium oxalate (AAO)** is claimed to selectively dissolve amorphous Fe-oxides when used under dark conditions [9], only slightly attacks silicates (hornblende, montmorillonite, illite, and Al-chloritized clays), hematite and goethite [10,11,12], and selectively dissolves magnetite and finely-divided, easily-weathered silicates [12,13]. Ten mL of AAO (pH 3) were added to 0.5 g of sample and shaken in the dark for 4 hours [12]. **b) Hydroxylamine hydrochloride (HAH)** is claimed, when acidified by variable concentrations of acetic, nitric or hydrochloric acids, to be selective for Mn-, amorphous Fe- or crystalline Fe-oxides [14,15,16,17]. It is reported to extract less Al and Fe from Spodosols and from pedogenic amorphous material, more from ground material, and to dissolve less magnetite than AAO [18]. Samples weighing 0.5 g were extracted with 20 mL of 0.25 M $NH_2OH.HCl$ in 0.25 M HCl (pH 2) for 30 minutes in a 50°C oven [11]. **c) Tiron** (catecholdisulphonic acid di-sodium salt ($C_6H_4Na_2O_8S_2$)) is claimed to selectively extract amorphous Fe-, Al-, and Si-oxides from soils under alkaline conditions (pH 10.5) [19,20,21], and extract small amounts of Fe from goethite, hematite and magnetite [20]. Samples weighing 0.1 g were extracted with 30 mL of Tiron in an 80°C water bath for 1 hour [19], followed by high-speed centrifugation, decantation and analysis. **d) Dithionite-citrate (DC)** extracts 'free iron oxides' by reduction of Fe^{+3} to Fe^{+2}, and chelation [14,22]. It is claimed to extract crystalline and amorphous Fe-oxides, although large crystals of goethite, hematite and magnetite are only partly dissolved. Slight attack on Fe-bearing silicates occurs and Al-chloritized clays are only slightly affected [10]. Twenty mL of 0.68 M Na-citrate and 0.4 g of Na-dithionite were added to 1.0 g samples followed by shaking overnight [12].

With each procedure claimed to preferentially extract individual or combinations of iron oxides, they have been used to assess the nature and abundance of Fe-oxides in earth materials. Sequential extractions, tailored to specific applications, have been used to extract increasingly resistant mineral forms and determine the elements associated with each form [23,24,7].

In this investigation, the supernatants were monitored for Al, Fe, K and Mg, to assess the extent of dissolution of the aluminosilicate minerals. X-ray powder diffraction of the alteration halo materials, before and after the initial and final (10th) extractions, was used to monitor the effects of the reagents on clays and other minerals.

3. LEDA CLAYS

The Leda clays, dominated by quartz and feldspar, have moderate amounts of illite and chlorite, traces of amphibole, vermiculite, smectite, oxide and amorphous minerals, variable amounts of carbonate, and traces of other primary minerals [25,26], are silty-clay/clayey-silt marine sediments derived from glacially-ground rocks of the Canadian Shield.

The Leda clay from Gatineau, Québec (in the Ottawa Valley, an Aquept) and from St. Barnabé, Québec (in the St. Lawrence Lowlands) which were examined have similar mineral suites (Table 1) with tectosilicates comprising 65-80% of the total, and illite and chlorite dominating the remainder [26]. Magnetite was identified by Mössbauer spectroscopy at St. Barnabé but not at Gatineau [27]. Hematite is assumed to be present at Gatineau on the basis of its detection at the nearby South Nation River landslide site [28]. Total iron oxide content is generally <1 wt % in unweathered material, but may exceed 1 wt % in the A and B horizons.

3.1. Gatineau

AAO, HAH and DC extraction results for samples from the strongly oxidized surface, through the oxidation-reduction transition at 2.5 m, to a strongly reduced environment at 4 m [29], are presented in Figure 1a. DC extracted 0.79 to 3.00 wt % Fe_2O_3, following the pattern of

Table 1. Qualitative mineralogy by XRD and Mössbauer. (A = abundant, M = minor, Tr = trace, no symbol = not detected)

	St. Barnabé	Gatineau	57-02B	57-02C	103-05B	103-05S
Quartz	A	A-M				
Plagioclase	A	A				
K-feldspar	A-M	A-M				
Amphibole	M-Tr	M-Tr				
Calcite	Tr?					
Siderite					A	
Illite	M	M	A	A	M	A
Chlorite	M-Tr	M-Tr		M (siderite)		
Kaolinite			Tr		M	A-M
Smectite	M-Tr	Tr?				
Mixed layer clay				Tr		
Hematite	Tr?	Tr	M	M-Tr	A	A-M
Magnetite	Tr					

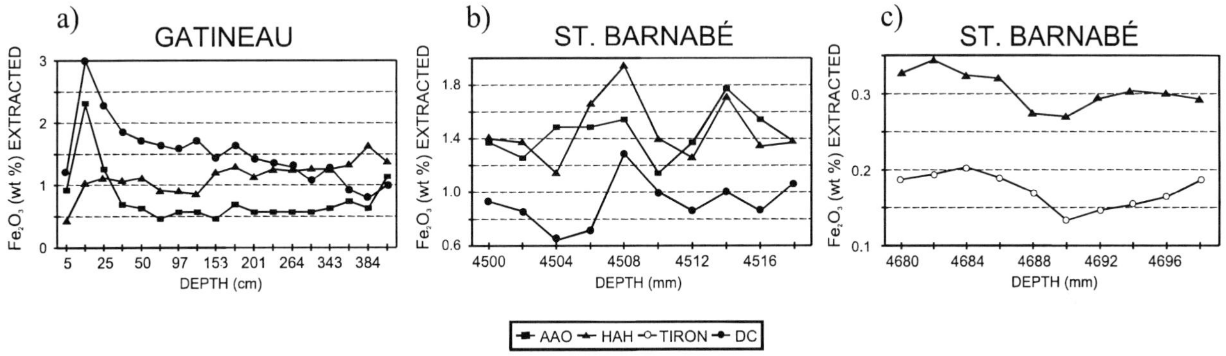

Figure 1. Weight % Fe_2O_3 extracted from Leda clay: a) Gatineau, Québec; b), c) St. Barnabé, Québec.

a less than average amount from 0-5 cm, the greatest amount from 10-15 cm (upper B horizon), and decreasing amounts with increasing depth below 20 cm. This is the expected trend where near-surface oxidation produces free iron oxides, which eluviate and accumulate in the B-horizon, and where reduced conditions prevail at depth. In the absence of magnetite, DC should extract all the crystalline and amorphous iron oxides present, and should be (and was) the most aggressive treatment. The trend was similar for AAO extraction, but with lesser quantities removed (0.46 to 2.29 wt % Fe_2O_3), except at 4.1 m (reason unknown). HAH-extractable Fe ranged from 0.43 to 1.63 wt % Fe_2O_3; HAH was least aggressive above 25 cm, intermediate between AAO and DC from 10 to 300 cm, and most aggressive below 3 m, in the reduced zone.

3.2. St. Barnabé

The extractions were made on sets of 2-mm thick slices of air-dried soil from 450-452 cm depth (AAO, HAH, DC, Figure 1b) or 468-470 cm depth (HAH, Tiron, Figure 1c), from the zone of reduced conditions (below 2 m [30]). AAO (1.14 to 1.77 wt % Fe_2O_3) was more aggressive than DC (0.64 to 1.28 wt % Fe_2O_3), as is consistent with magnetite presence (detected by Mössbauer), rather than hematite, as the dominant crystalline iron oxide. HAH and AAO extracted similar amounts of Fe suggesting similar ability to extract finely-divided magnetite. The Tiron and HAH comparison, on the samples from 468-470 cm (Figure 1c), indicates that Tiron (0.13-0.20 wt % Fe_2O_3) was notably the least aggressive reagent in extracting iron oxides.

3.3. Comparisons

Problems with quantitative estimation of various iron oxides by selective extraction may arise from extraction of iron oxide forms other than that (those) targeted, or from extraction of iron from other iron-bearing minerals. Extracted Al and K were monitored to assess the severity of phyllosilicate attack. Al is believed to derive mainly from chlorite and biotitic octahedral sheets and K from illite interlayers. HAH is the most aggressive for Al in all cases, followed by AAO and DC, and the most aggressive for K followed by DC and AAO. The Tiron/HAH comparison indicated HAH is more aggressive than Tiron for Al, but that Tiron is the most aggressive of all reagents in extracting interlayer K. No consistent relationships occur between Fe extraction and Al and K extraction by any of the reagents.

Results consistent with claims are that DC extracted more iron where hematite was the crystalline iron oxide (Gatineau), and AAO more where magnetite was the oxide present (St. Barnabé). Nonetheless, DC decreased the magnetite content at St. Barnabé below the detection limit by Mössbauer spectroscopy [28]. The amount of iron extracted by both DC and AAO exceeded the amount of amorphous iron that could be present in these sediments (ferrihydrite was undetected by Mössbauer spectroscopy [28]). HAH was unexpectedly aggressive in Leda clay, being comparable to AAO in magnetite-bearing samples and as aggressive as DC in some of the hematite-bearing samples. Al and K results indicate consistent attack on both octahedral sheets and K interlayers of clay minerals.

Are the problems inherent to the procedures, or are specific characteristics of the Leda clays involved? A possible explanation relates to the origin of Leda clays as silt- and clay-sized, glacially-ground, predominantly-igneous, rock material from the Canadian Shield which was not subjected to subaerial weathering before deposition in the Champlain Sea ca. 12,000 years ago. Since uplift above sea level ca. 10,000 ago, weathering (except in the immediate surface) has been severely inhibited by: reducing conditions, low hydraulic conductivities, low hydraulic gradients, cool climate and, for a substantial portion of their existence, high pore-water salinity. Very sharp XRD peaks for the silicates confirm that little weathering has occurred [28]. In contrast, the extraction procedures were developed for, and tested with, soils that had experienced more intense weathering, and little or no grinding. In addition, the grain size of the crystalline hematite and magnetite (estimated at 20 nm for hematite at the South Nation site [28]) is much smaller than that of hand-ground samples of pure oxides on which the procedures were tested. It appears probable that two factors, lack of prior chemical weathering and the fine grain size of the iron oxides are the major explanations for the discrepancy between expected (claimed) and observed effects of the various extractants in the Leda clays.

4. ALTERATION HALO CLAYS

The procedures were also tested on four clay-rich materials from the alteration halo surrounding the unconformity-type uranium deposit at Cigar Lake, Saskatchewan (mineralogy - Table 1): an illite-, kaolinite- and siderite ($FeCO_3$)-bearing bulk sample (103-05), with 58 wt %

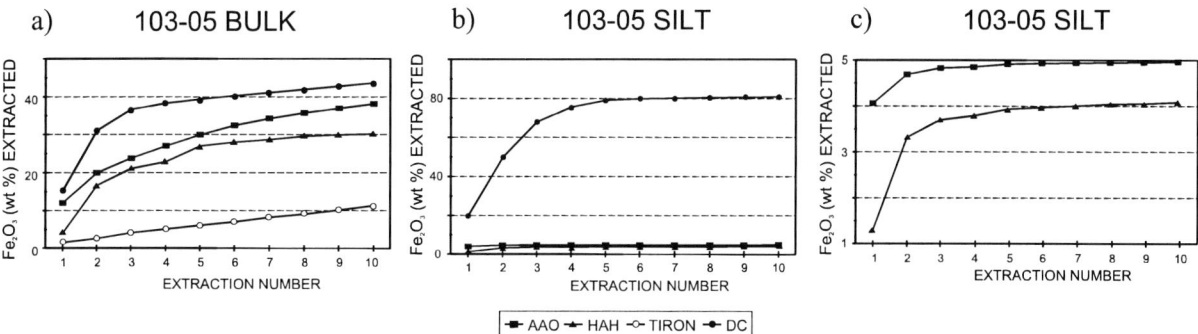

Figure 2. Cumulative weight % Fe$_2$O$_3$ extracted from: a) 103-05 bulk sample through 10 extractions; b) 103-05 silt size (2-5 μm) separate through 10 extractions; and c) scale expanded for AAO and HAH.

Fe$_2$O$_3$; a fine silt (2-5 μm) separate of 103-05, with 70 wt % Fe$_2$O$_3$; an illite- and chlorite (variety sudoite)-bearing bulk sample (57-02), with 10 wt % Fe$_2$O$_3$; and a clay (<2 μm) separate of sample 57-02, with 30 wt % Fe$_2$O$_3$. Bulk samples were obtained by mill-crushing to <50 μm. Fine silt (2-5 μm) and clay (<2 μm) fractions were obtained by hand-crushing core samples, followed by ultrasonic dispersion in demineralized water, sedimentation and centrifugation. The goal was to develop a sequential extraction procedure to determine the partitioning of uranium within the halo [6,7,8]. Each material was extracted 10 times with each reagent to test both the robustness of the procedure and reagent selectivity.

For the bulk 103-05 sample (Figure 2a), HAH extracted significant Fe up to the seventh cycle, whereas AAO, DC and Tiron extracted Fe through all 10 cycles. Aggressiveness was DC>AAO>HAH>Tiron after each step. XRD analysis revealed: decrease in the (104) hematite XRD peak intensity after one extraction by all extractants; DC had removed most of the hematite after 10 cycles, whereas AAO, HAH and Tiron were less effective, as expected; HAH and AAO completely dissolved siderite in 10 cycles and Tiron decreased it, whereas DC left it relatively untouched; and HAH and AAO reduced the 7.1 and 3.58 Å X-ray peak intensities of kaolinite. The 2-5 μm silt fraction of 103-05 (Figure 2b) behaved somewhat differently: DC effectively ceased extracting iron after five cycles, and AAO and HAH after three cycles. The more rapid extraction is presumed attributable to the fine grain size. Aggressiveness was DC>>AAO>HAH after each step. XRD revealed complete removal of hematite by DC, but only slight reduction with AAO and HAH. All treatments enhanced the clay mineral XRD peak intensities.

For the bulk 57-02 sample (Figure 3a), Fe removal beyond 3 extractions was insignificant for all extractants. XRD indicated that hematite was removed by 10 cycles with DC, but appeared unaffected by the other procedures. A trace of mixed-layer clay mineral (illite-smectite?) was detectable after 10 extractions by Tiron. Sudoite was slightly reduced by HAH. For the clay-size fraction of 57-02, DC ceased extracting Fe after two cycles, HAH after three cycles, whereas AAO and Tiron removed small amounts throughout (Figure 3b). The hematite XRD peak disappeared after one extraction by DC and after 10 by Tiron. Aggressiveness for both bulk and clay-size samples was DC>>AAO≈HAH>Tiron. Extracted Al and K results (Table 2) indicate that, in all these high-Fe materials, the Fe extracted from non-oxide minerals is small compared to the total extracted (Figure 2).

Extraction of the alteration-halo clays yielded some inconsistent results. Hematite was the only crystalline iron oxide identifiable by XRD and no evidence was found for ferrihydrite. With this oxide mineralogy, it is unclear why AAO, HAH and Tiron were so effective in extracting iron from the 103-05 bulk sample, when their performance with the other samples, especially the finer-grained separates, was essentially as predicted. In the 103-05 bulk sample, siderite is the main

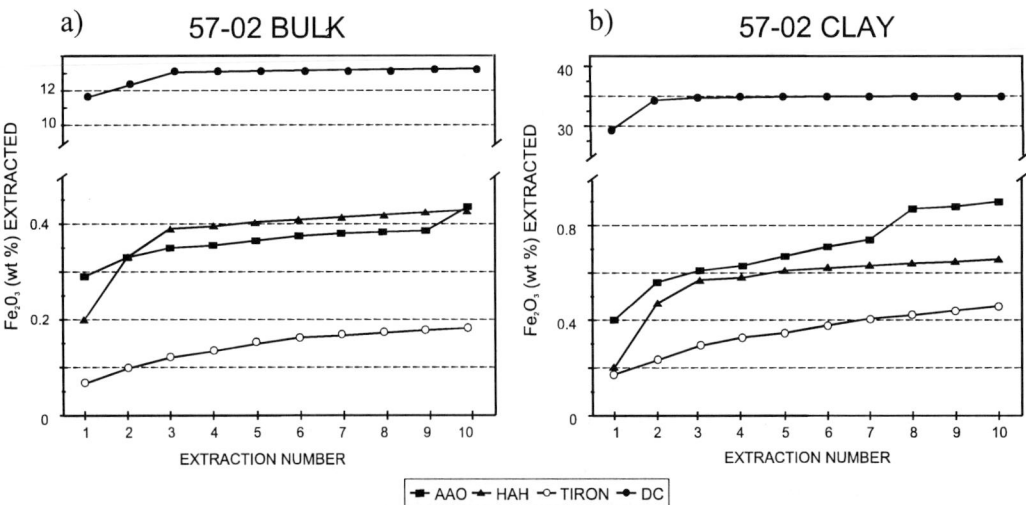

Figure 3. Cumulative weight % Fe_2O_3 extracted from a) 57-02 bulk sample through 10 extractions; b) 57-02 clay size (< 2 μm) separate through 10 extractions. Note the vertical scale change.

non-oxide Fe-bearing mineral, although illite probably contains traces of iron. K, Al and Mg extraction indicate limited attack on non-oxide minerals by all methods. Overall, XRD traces after 10 cycles of extraction provided: no evidence of attack on non-target silicate minerals by DC or AAO; limited attack on siderite by Tiron; substantial attack on siderite by HAH and AAO; and slight attack of sudoite by HAH. Although non-target silicate mineral extraction was not a serious problem in these relatively mineralogically simple rock materials, only DC consistently behaved in the claimed manner.

5. CONCLUSIONS AND RECOMMENDATIONS

The goals of applying "selective" Fe-extraction techniques to the Leda clays and the alteration halo samples were frustrated by the problems elaborated above.

With the Leda clays, the relative effectiveness of DC and AAO was consistent with whether hematite or magnetite dominated, but DC unexpectedly extracted very fine-grained magnetite. Extraction of Al and K from non-target, Fe-containing phyllosilicates, by all the extractants, also suggests selectivity problems.

With the alteration halo materials, relative extractant aggressiveness was generally as expected, but there are troubling features. Most apparent was that HAH, AAO and Tiron attacked siderite, and hence cannot be used to assess ferrihydrite when siderite is present. With hematite as the only XRD-identifiable iron oxide, DC should be and was consistently the most effective, but all other extractants removed unexpectedly high amounts of iron. Also, the selectivities of other extractants are in question because, through many cycles, they extracted Fe from the alteration-halo clays, especially bulk sample 103-05. Dithionite reduction limits DC extraction of iron and, with abundant hematite present, multiple extractions were expected to be necessary. For the other extractants, mineral surface area and contact time control or partially control the extent of reaction.

How can one judge, for a specific situation, the appropriate contact time for extraction to be complete? Should extraction continue until successive extractions yield equal amounts of Fe? Does this equal extraction represent Fe from non-target minerals, allowing a correction factor to

Table 2: Al_2O_3 and K_2O (wt %) extracted after one cycle and cumulative after ten cycles by AAO, HAH, Tiron and DC.

SAMPLE NO.	AAO		HAH		TIRON		DC	
	Al_2O_3	K_2O	Al_2O_3	K_2O	Al_2O_3	K_2O	Al_2O_3	K_2O
57-02 BULK - 1st ext.	0.106	0.03	0.076	0.036	0.091	0.036	0.095	0.049
10th ext.	0.174	0.042	0.187	0.058	0.418	0.096	0.095	0169
57-02 CLAY - 1st ext.	0.134	0.048	0.095	0.048	0.334	0	0.289	0.067
- 10th ext.	0.225	0.063	0.285	0.058	0.554	0.260	0.386	0.180
103-05 BULK - 1st ext.	0.679	0.019	0.321	0.024	0.076	0.021	0.183	0.023
- 10th ext.	1.004	0.220	1.247	0.046	0.336	0.059	0.274	0.122
103-05 SILT - 1st ext.	0.673	0.005	0.302	0.007			0.278	0.005
-10th ext.	0.852	0.012	0.982	0.014			0.556	0.109

be applied? Might some non-target mineral be completely consumed before successive extractions yielded the same amount of Fe?

The results indicate that selectivity for specific iron oxides is neither good nor consistent; the problems differ with the material being extracted. No procedure tested was unambiguously selective for ferrihydrite: presence of siderite negates ferrihydrite selectivity by both HAH and Tiron; HAH is sufficiently aggressive against very fine-grained magnetite and probably the octahedral sheet of phyllosilicates (Leda clays) to negate its selectivity for ferrihydrite in relatively unweathered glacially-ground material; and Tiron, the least aggressive extractant in the Leda clays, appears to extract some hematite from the alteration halo, and thus probably overestimates ferrihydrite in the Leda clays. The amount of iron extracted by the least effective extractant provides, at best, the upper limit to the amount of ferrihydrite present in the Leda clay.

Recommendations for using these or other "selective" extraction procedures are: 1) determine the sample mineralogy; 2) test the procedures individually on your samples to determine relative aggressiveness; 3) monitor the concentration, in the extracts, of elements which may signal attack on non-target minerals; 4) check the mineralogy of the samples after extraction; 5) even with these precautions, treat the results with caution; 6) because the extractions are operationally-defined, report the results as DC-extractable iron, Tiron-extractable iron, etc., unless there is independent confirmation of the mineral(s) removed.

For sequential extractions, comparison of results may only be valid when the same extractant sequence is applied to mineralogically similar materials. Individual, rather than sequential, application of the procedures to multiple replicate samples and comparison of the results would determine their relative effectiveness in each case, and relative effectiveness with different materials would presumably be indicative of differences in the materials. Sequential extraction results need to be re-evaluated in the light of the evidence presented here, and the approach to achieving the goals of sequential extraction procedures needs to be re-examined.

7. ACKNOWLEDGEMENTS

The authors acknowledge: laboratory assistance by J.-L. Bouvier, A. Brereton, J. Cliff, R. Conlon, R.A. Meeds, D. Omond, S. Phaneuf and K. Stoker; computer-aided drafting by C. Earl, D. Viner and R. Lacroix; support from NSERC Grant No. A8503 and Atomic Energy of Canada

Limited. Thanks to H. Kodama, J.A. Percival and J.N. Ryan for reviewing a draft manuscript.

REFERENCES

1. E.A. Jenne, Adv. in Chem. Ser., 73 (1968) 337.
2. J.K. Torrance, Applied Clay Sci., 5 (1990) 307.
3. J.K. Torrance, Applied Clay Sci., 14 (1999) 199.
4. R. Yong, A.J. Sethi, and P. La Rochelle, Can. Geotech. J., 16 (1979) 511.
5. J.K. Torrance, Can. Geotech. J., 32 (1995) 535.
6. J.B. Percival, Ph.D. thesis, Ottawa-Carleton Geoscience Centre, Carleton Univ., Ottawa, Canada, 1989.
7. J.B. Percival, J.K. Torrance, and K. Bell, in J.W. Gadsby, J.A. Malick and S.J. Day, (eds.), Acid Mine Drainage: Designing for Closure, BiTech Publ. Ltd., Vancouver, B.C., (1990) 51.
8. J.B. Percival, K. Bell, and J.K. Torrance, Can. J. Earth Sci., 30 (1993) 689.
9. O.K. Borggaard, in J.W. Stucki, B.A. Goodman and U. Schwertmann, (eds.), Iron in Soils and Clay Minerals, D. Reidel Publ. Co., Dordrecht, Holland, (1988) 83.
10. J.A. McKeague and J.H. Day, Can. J. Soil Sci., 46 (1966) 13.
11. T.T. Chao and L. Zhou, Soil Sci. Soc. Am. J., 47 (1983) 225.
12. B.H. Sheldrick, (ed.), Analytical Methods Manual 1984, Land Resource Res. Inst., Cont. 8430, Agric. Canada, Ottawa, 1984.
13. F.E. Rhoton, J.M. Bigham, L.D. Norton, and N.E. Smeck, Soil Sci. Soc. Am. J., 45 (1981) 645.
14. T.T. Chao and P.K. Theobald, Jr., Econ. Geol., 71 (1976) 1560.
15. S. Gatehouse, D.W. Russell, and J.C. Van Moort, J. Geochem. Explor., 8 (1977) 483.
16. E.W. Bogle, and I. Nichol, J. Geochem. Explor., 15 (1981) 405.
17. L.H. Filipek and P.K. Theobald, Jr., J. Geochem. Explor., 14 (1981) 155.
18. C. Wang, P.A. Schuppli, and G.J. Ross, Geoderma, 40 (1987) 345.
19. V. Biermans and L. Baert, Clay Miner., 12 (1977) 127.
20. H. Kodama and G.J. Ross, Soil Sci. Soc. Am. J., 55 (1991) 1180.
21. G.J. Ross, in M.R. Carter (ed.), Soil Sampling and Methods of Analysis. Lewis Publ., Boca Raton, FL, (1993) 745.
22. M.L. Jackson, Soil Chemical Analysis - Advanced Course. 2nd ed., Madison, WI., 1979.
23. A. Tessier, P.G.C. Campbell and M. Bisson, Analyt. Chem., 51 (1979) 844.
24. T.T. Chao, J. Geochem. Explor., 20 (1984) 101.
25. J.K. Torrance, in N.R. Gadd, (ed.), The Late Quaternary Development of the Champlain Sea Basin, Geol. Assoc. Can. Spec. Paper 35, (1988) 259.
26. R.W. Berry and J.K. Torrance, Can. Miner., 36 (1998) 1625.
27. L. Bowen, Personal comm., 1992.
28. J.K. Torrance, S.W. Hedges, and L.H. Bowen, Clays and Clay Miner., 34 (1986) 314.
29. J. Cliff, Hon. Proj., Geography, Carleton Univ., Ottawa, 1986.
30. N. Tomas, Hon. Proj., Geography, Carleton Univ., Ottawa, 1987.

Neoformed Halloysite in Podzols developed on the Bärhalde granite, Southern Black Forest, Germany

M. Zarei[a], M. Sommer[b] and K. Stahr[a]

[a]Institut für Bodenkunde und Standortslehre, Emil-Wolff-Straße 27, 70593 Stuttgart, Germany
E-mail address: zareimeh@uni-hohenheim.de

[b]Institut für Biomathematik und Biometrie, GSF-Forschunszentrum für Umwelt und Gesundheit, Ingolstädter Landstr. 1, 85764 Neuherberg, Germany, E-mail address: sommer@gsf.de

In Podzols of the Southern Black Forest developed on granite the presence of kaolinite, illite and hydroxy-interlayered vermiculite (HIV) were detected using X-ray diffraction analyses. The different minerals indicate different microenvironments. The occurrence of different kandite group minerals (kaolinite, halloysite, dickite and nacrite) may give specific information about the development. In the clay-size fraction of two Podzols kaolinite was first identified, however, morphological studies using SEM and TEM analyses distinguished kaolinite from halloysite. The horizon sequence offered a clear separation of mineral associations. The eluvial horizon showed kaolinite and illite with some hydrated illite. The primary minerals in the very acid topsoil are strongly corroded. Orthoclase surfaces shows deep etch pits whereas plagioclase shows straight surfaces implying direct dissolution. In the spodic horizon, a 14 Å peak of HIV was clearly detected. The underlying C horizon showed a similar pattern but with better developed X-ray peaks. The 7 Å peaks are slightly asymmetric and show a broad basis. Where micas occur next to halloysite, they can hardly be separated by XRD. This dehydrated halloysite was expanded to 10.8 Å through N-methylformamide saturation. Halloysite was concentrated by ultrasonic treatment from feldspar in silt and sand fractions. Plagioclase and orthoclase weathering was observed along cleavage planes. We observed a transformation of feldspars into halloysite tubes. Kaolinite plates are situated above the halloysite. Halloysite later is transformed into kaolinite.

1. INTRODUCTION AND PROBLEM DEFINITION

Generally in soil mineralogy 7 Å group minerals (kaolinite, halloysite, dickite and nacrite) are undifferentiated as kandite group minerals. Kaolin minerals are difficult to separate from each other analytically because they are similar in chemistry and structure. The halloysite contain an interlayer of water in contrary to other 7 Å group minerals. At about 60 °C the water escapes and halloysite dehydrates to form metahalloysite.
Churchman and Gilkes (1989) described that halloysite tubes could be altered to kaolinite through dehydration in solid state. This was confirmed by Jeong (1998) Robertson and Eggleton (1991) as well as Singh and Gilkes (1992), however, in their TEM studies reported a conversion vice versa from kaolinite to halloysite, but they could not find an explanation for

the hydration process. From earlier soil mineralogical work on Podzols of the Black Forest the clay mineral assemblage of kaolinite, illite and hydroxy-interlayered vermiculite was reported (Stahr, 1979). Although the 7Å X-ray peaks were asymmetric. Stahr (1979) suggested it was kaolinite. In the separated clay fraction from the fine earth (i.e., <2 mm fraction) halloysite was not observed, however, through SEM-observations halloysite was observed along cleavage planes of feldspars as well as in soil aggregates (Gudmundsson and Stahr, 1981). Based on these initial findings the following questions were addressed. How can halloysite and kaolinite be distinguished morphologically and by X-ray diffraction (XRD) after N-methylformamide treatment in these soils?
Do these minerals occur at different locations of origin?
Do morphological or mineralogical observations help explain the development of the two minerals at the same site?

2. MATERIAL AND METHODS

The study area is located in the southern Black Forest, Germany, about 1000-1300 m above sea level. A catena of two Podzols developed on granite in the cool-perhumid climate was studied. In the area the climate, vegetation and time of soil formation are similar. The granite is the Bärhalde two-mica-granite intruded by another porphyritic granite. The Bärhalde-granite is composed of 34% quartz, 33% orthoclase, 25% plagioclase (albite, An 2-10) and 4% biotite as well as 3% muscovite (Retief, 1970). The granite-porphyry shows a mineralogical composition of quartz, orthoclase and albite. The mineralogical properties are reflected in the chemical composition with much higher contents of Ca, Mg, K, Fe and Ti:Zr ratios, respectively, in the granite-porphyry compared to the Bärhalde granite. Soil development occured in physically pre-weathered periglacial debris. All soils are of Holocene age.
Soil samples were selected from the eluvial, illuvial and C horizons of both soil profiles. The pH of the soils was determined in 0.01 M $CaCl_2$ (1:2.5) and measured electrometrically.
Pre-treatment for particle-size analysis included destruction of humus using 10% H_2O_2, with ultrasonic treatment. Fine-earth (<2 mm) has been obtained by wet sieving, silt and clay have been separated by sedimentation. The preparation of oriented specimens included K and Mg saturation, heat treatment of the K-saturated mounts to 400 and 600°C and glycerol solvation of the Mg- saturated samples. A Siemens Instrument (D-500) with Cu Kα radiation was used for XRD analysis. Semiquantitative analysis of clay fraction was calculated with the software package Diffrac AT 3.3 from Siemens Company. Scanning electron microscopy (SEM) was carried out on a Leo 420 instrument and energy dispersive X-ray (EDX) spectrometry (EDAX, Pv 9800) was performed. TEM information was gained with a Zeiss EM 10.
The clay fraction was oven-dried for 24 hours at 60 °C. All samples were treated for two hours with N-methylformamide followed by XRD analysis.

3. RESULTS

3.1 Clay minerals in the Podzols of the Bärhalde Catchment

Mineralogical analyses of the clay-size fractions of the eluvial, illuvial and C horizons of the catena are presented in Figure 1 (Sommer, et al., 2000). In the C horizon beside illite (10 Å) and kaolinite (7.2 Å) occur aluminous interlayered minerals (1.43 Å). The latter peak remained stable under treatment by glycerol and K-saturation. This peak disappeared only by heating at 400 °C and showed an increase of the 10 Å peak (Fig 1c). In the eluvial and illuvial horizons there are distinct 10 and 7 Å peaks (with narrow width) present. In the eluvial horizon, a shoulder between 10 and 14 Å occurs in the Mg-saturated samples, but did not expand with glycerol. This is interpreted as an illite-vermiculite interlayered mineral (Fig 1a). The 14 Å peak in the illuvial horizon did not react with glycerol, partly disappeared after K-saturation, and completely disappeared after heat treatment at 400 °C. This is an indication of the presence of the hydroxy-interlayered-vermiculite (HIV).

The clay fraction of both eluvial soil horizons contained illite, illite-vermiculite and kaolinite. In the illuvial (spodic horizon) and C horizons hydroxy-interlayered-vermiculite, illite and kaolinite were detected.

3.2 SEM observation of weathering on sand and silt minerals

In the clay fraction of the two Podzols only kaolinite was first identified. Morphological studies using the SEM and TEM clearly allowed the distinction of kaolinite and halloysite as two separate mineral phases (Figure 2-9).

Churchman and Gilkes (1989) reported that halloysite tubes could be altered to kaolinite through dehydration in the solid state. Jeong (1998) observed a solid state transformation of halloysite tubes into layered kaolinite plates by coalescence. The low crystallinity and broad basal peaks of kaolinite supports the hypothesis, that they are derived from halloysite through solid state transformation. Jeong (1998) suggested that the curved layers of halloysite are flattened by partial rupture or weak dissolution rather than by complete recrystallization.

In the very acid topsoil, primary silicate minerals, in general are strongly weathered. Along the cleavage planes formation of new minerals (clay minerals) are observed in protected hollows. Nevertheless, in A and E horizons conditions to form new minerals are no longer favourable, but destruction and dissolution of the primary minerals do occur. A heavily weathered feldspar is shown in Figure 2. Within the cleavage planes and (dissolution) hollows well-formed pseudohexagonal kaolinite plates occur. Aggregation of kaolinite as well as halloysite covering the primary minerals in the A and E horizons are rare, but frequent in the subsoil. Platy kaolinite and the tubular halloysite structures can cover the primary mineral surfaces (Figure 3). Morphological observations show that in the range below pH 3.5 halloysite seems to disappear (Figure 4). The weathering of plagioclase as well as orthoclase starts generally along cleavage planes. Cations like Ca, Na and K together with some Si are dissolved and halloysite and/or kaolinite can crystallise from the residue. Kaolinite is observed on plagioclases as well as on orthoclases, whereas halloysite is mainly detected along the cleavage planes of plagioclases. The newly formed minerals are of clay size. Often the whole matrix seems to be consisting of kaolinite and/or halloysite. The long tubes of the halloysite can be discrete but often they appear to grow in bundles or are interwoven. The connected particles of halloysite form kaolinite-like morphology (Figures 4 and 5). The halloysite from the illuvial and eluvial horizons exhibit defects. Only in the subsoil (Cw or Rw horizon) are newly formed kaolinite-group minerals accumulated. Here they have a well-developed structure, are aggregated on weathered primary minerals or form small clay-size clusters (Figure 5). Often halloysite and kaolinite occur in voids of weathered orthoclase and plagioclase (Figures 6 and 7). During the study spheroidal halloysite was not observed. They form on easily weatherable minerals and may have Al, Fe and Si oxyhydroxides as precursors (Eggleton, 1987).

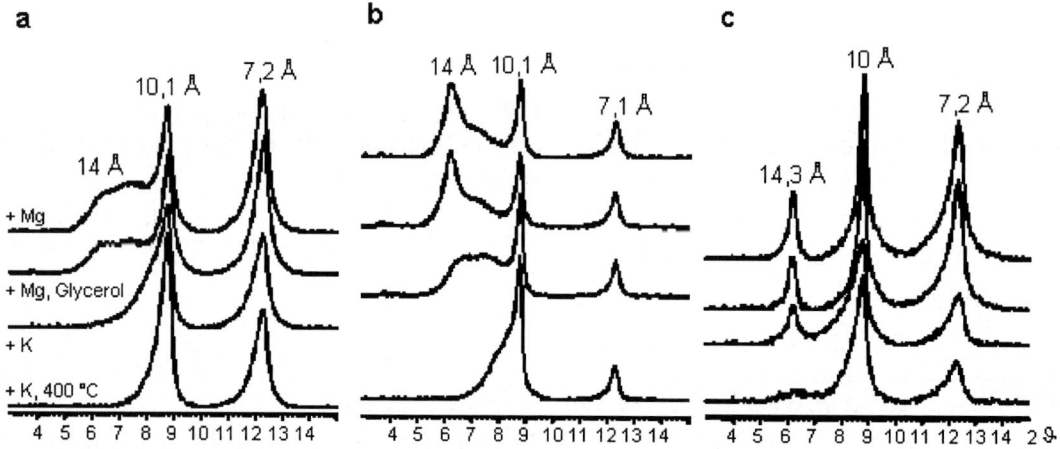

Figure 1. XRD patterns of the representative samples from the a) eluvial, b) illuvial and the (c) C horizons of two Podzol profiles in the southern Black Forest.

3.3 Clay fraction separated from sand and silt

With the oriented specimens of the clay fraction as shown before it is difficult to separate different minerals of the kandite group because the crystal structure is not very different. The basal reflection at 7 Å is broad and relatively low for all horizons. Therefore, this mineral is probably dehydrated halloysite. The identification of halloysite with kaolinite and illite is difficult to be proven only using XRD. Halloysite can be expanded by solvation in formamide (Churchman & Theng, 1984 and Churchman, et al., 1984). Best results are achieved using, N-methylformamide which expands-halloysite from 7.2 Å to 10.1 Å. All the samples have been treated with N-methylformamide (Figure 8). Thus in all samples the 7.2 reflection was reduced in favour of a secondary reflection at 10.8 Å. However in the illuvial horizon samples this effect was less distinct possibly because of existing iron-organic coatings, which in general weakens the X-Ray signals. In these horizons the ratio between Fe_o and Fe_d is from 0.5 to 0.8, indicating a high proportion of Fe-oxides with low crystallinity.

4. DISCUSSION

Keller (1977) described feldspar weathering in granites and concluded that kaolinite formed from solution. In a similar study, Banfield and Eggleton (1990) observed halloysite and kaolinite as weathering products of feldspars and biotite. The halloysite occured in tubes and spheroids, formed by precipitation in voids from solution. Singh and Gilkes (1992) suggested solid state transformations of micas into halloysite bundles can occur. The parallel oriented bundles later could be transformed into kaolinite plates. Jeong (1998) also postulated the solid state transformation of halloysite tubes into kaolinite plates. All this work was done on samples from the tropics or subtropics, partly in deeply weathered, leached profiles.
This study, however, deals with a perhumid cool temperate site where the weathering of Bärhalde granite occurs. Stahr et al. (1995) reported a significant weathering and leaching of bases, up to 2 Mol · ha^{-1} · a^{-1}. In the subsoil, tubular and platy pseudohexagonal forms of

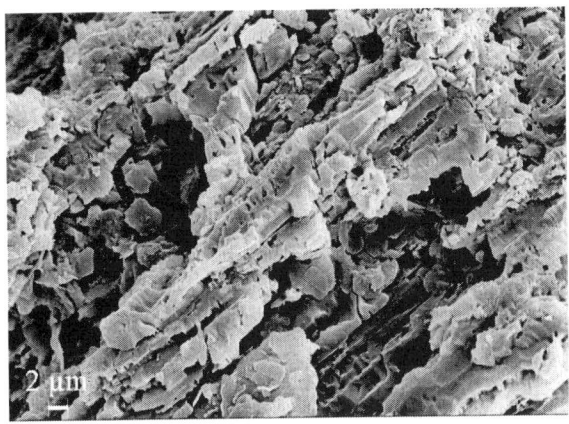
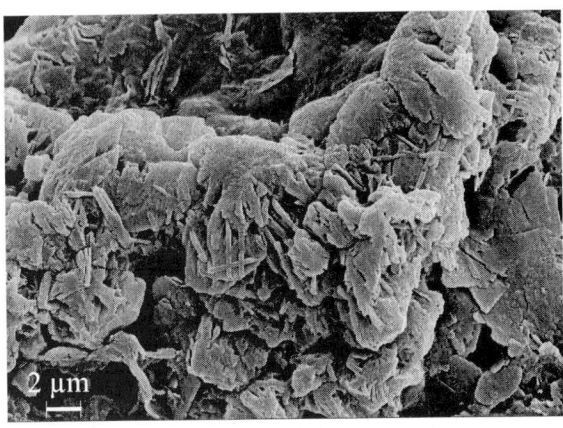

Figure2. Scanning electron photomicrograph of weathered feldspar, kaolinite with pseudo-hexagonal plates form in the cleavage pit, eluvial horizon, Podzol Bärhalde.

Figure3. Scanning electron photomicrograph of vermicular hal-loysite and platy kaolinite eluvial horizon, Podzol Bärhalde.

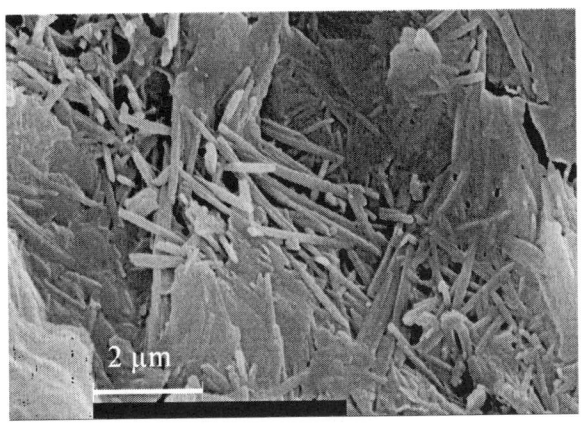
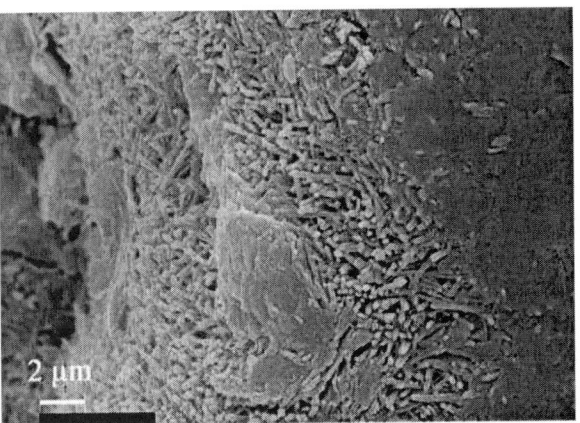

Figure 4. Scanning electron photomicrograph of halloysite with tubular structure grading into plates, Cw horizon, Podzol Bärhalde.

Figure5. Scanning electron photomicrograph of halloysite tubes in a clay aggregate, Cw horizon, Podzol Bärhalde.

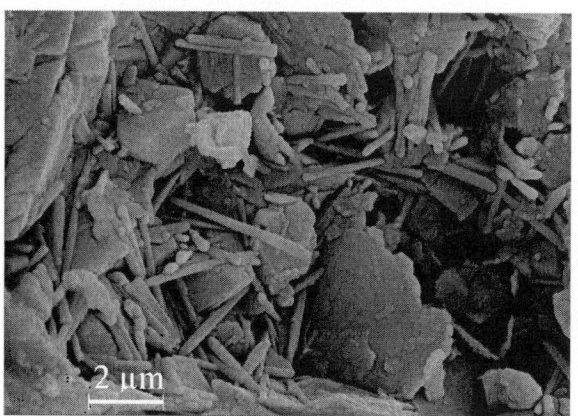
Figure 6. Scanning electron photomicrograph of halloysite tubes and kaolinites plates in an orthoclase void, Cw horizon, Podzol Bärhalde.
ärhalde.

Figure 7. Scanning electron photomicrograph of weathered plagioclase and neoformed minerals along the cleavages, Cw horizon, Podzol, Bärhalde.

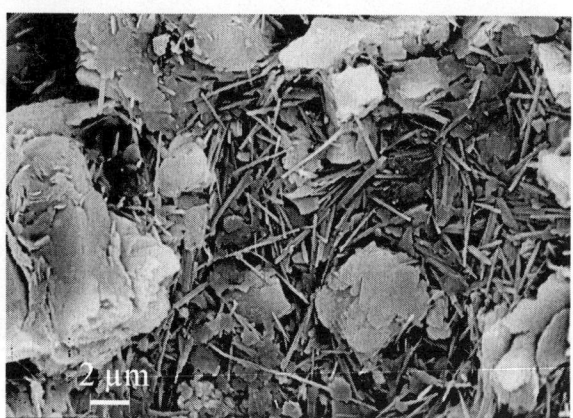
Figure 8. Scanning electron photomicrograph showing irregular halloysite tubes and several kaolinite platelets, Cw horizon, Podzol Bärhalde.

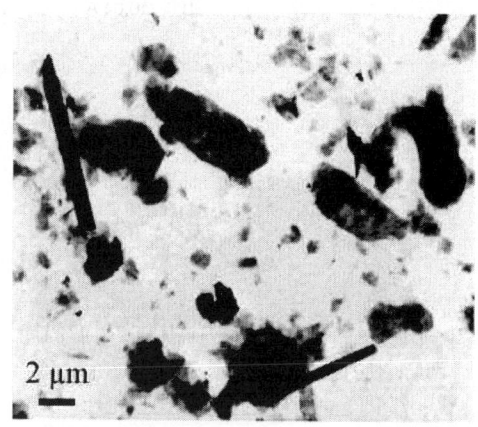
Figure 9. Transmission electron photomicrograph of halloysite and kaolinite in the clay fraction, E horizon, Podzol Bärhalde.

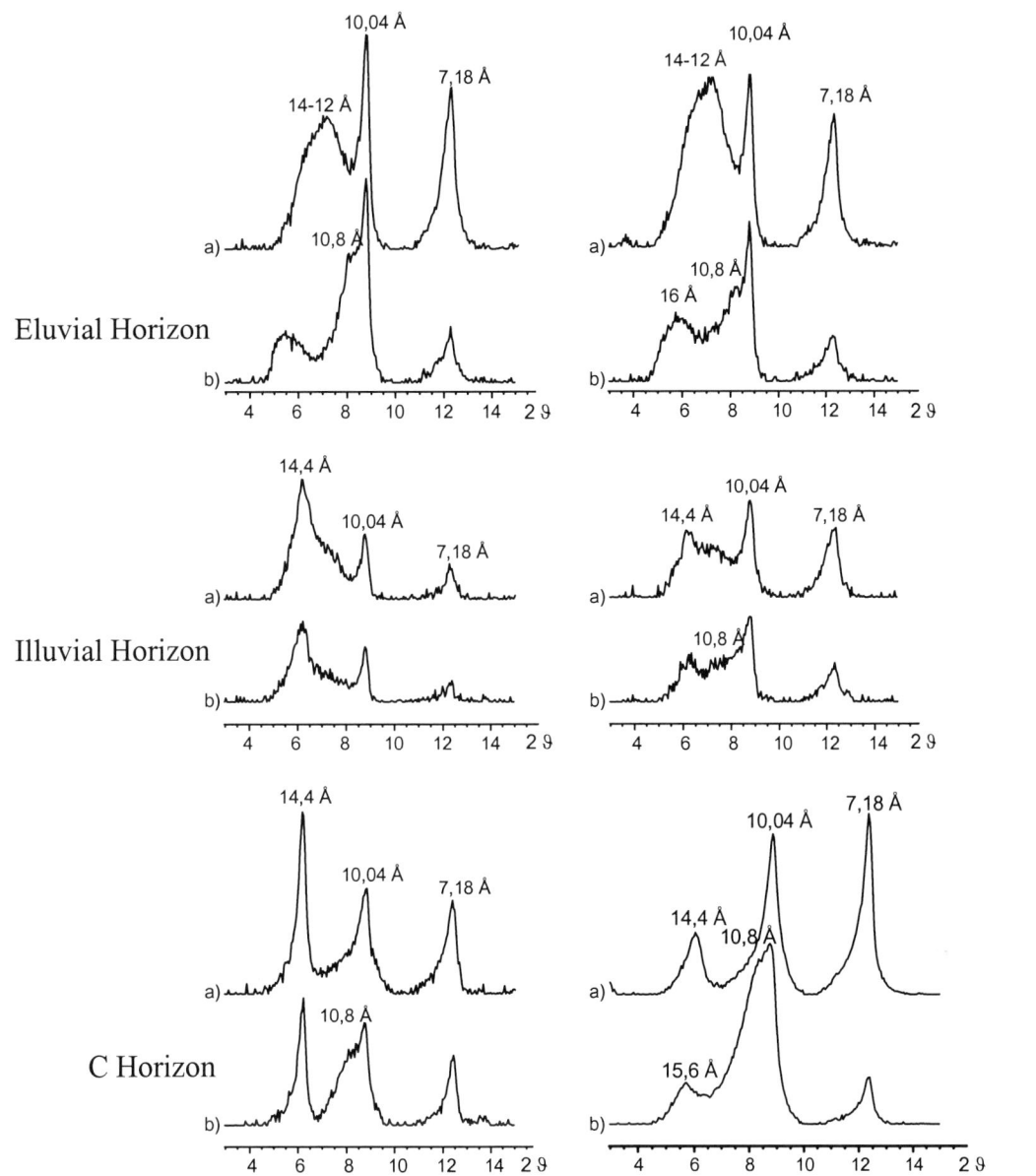

Figure 10. XRD of two Podzols in a catena (a) Mg-saturated, dried 24 hours at 60 °C (b) saturated with N-methylformamide.

clay-size grains have been identified mainly covering both plagioclase and orthoclase. EDX-analysis proved the existence of Al and Si as the only metal constituents. The separation of clay from silt and sand fractions and consecutive treatment with N-methylformamide confirmed the existence of both halloysite and kaolinite. The minerals generally grow along cleavage planes or in dissolution pits. Therefore a solid state transformation is most probable. Kaolinite is almost exclusively formed on top of halloysite tubes or bundles. Therefore, a solid state transformation is suggested such as noted by Churchman and Gilkes (1989) and Jeong (1998).

The fact, that in the illuvial horizon the secondary mineral crystallinity is low, is explained by the presence of amorphous Fe-Al-organic complexes. In the topsoil the very acid condition (pH 2.8-3.5) results in extremely strong weathering of the primary minerals. The newly formed halloysite and kaolinite mineral grains show more irregular morphology and crystallinity. Therefore, the new formation of kandite group minerals is restricted to the mild weathering in the subsoil, whereas in the topsoil these minerals are relicts of earlier weathering stages. Spheroidal halloysite was not detected in any of the Podzols of the Bärhalde area.

REFERENCES

Banfield, J. F., and Eggleton, R. A. (1990): Analytical tranmission electron microscope studies of plagioclase, muscovite, and K-feldspar weathering. Clays and Clay Minerals **38**, 77-89.

Churchman, G.J., and Gilkes, R.J. (1989): Recognition of intermediates in the possible transformation of halloysite to kaolinite in the weathering profiles. Clay Minerals **24**, 579-590.

Churchman, G. J., and Theng, B. K. G. (1984): Interactions of halloysites with amides: Mineralogical Factors affecting complex formation. Clay Minerals **19**, 161-175.

Churchman, G. J., Whitton, J. S., Claridge, G. G. C., and Theng, B. K. G. (1984): Intercalation Method Using Formamide for Differentiating Halloysite from Kaolinite. Clays and Clay Minerals **32**, 241-248.

Eggleton, R. A. (1987): Noncrystalline Fe-Si-Al-oxyhydroxides. Clays and Clay Minerals **35**, 29-37

Gudmundsson, Th., and Stahr, K. (1981): Mineralogical and geochemical alterations of the "Podsol Bärhalde". Catena, **8**, 49-69.

Jeong, G. Y. (1998): Formation of vermicular kaolinite from halloysite aggregates in the weathering of plagioclase. Clays and Clay Minerals **46**, 270-279.

Keller, W. D. (1977): Scan electron micrographs of kaolins colected from diverse environments of origin. IV. Georgia kaolin and kaolinizing source rocks. Clays and Clay Minerals **25**, 311-354.

Retief, E. (1970): petrology of the Schluchsee and Bärhalde granite plutons, Southern Schwarzwald . Ber. Natur. Ges. Freiburg I.Br. **60**, 139-172.

Robertson, I. D. M., and Eggleton, R.A. (1991): Weathering of granite muscovite to kaolinite and halloysite and of plagioclase-derived kaolinite to halloysite. Clays and Clay Minerals **39**, 113-126.

Singh, B., and Gilkes, R.J. (1992): An electron optical investigation of the alteration of kaolinite to halloysite. Clays and Clay Minerals **40**, 212-229.

Sommer, M., Halm, D., Weller, U., Zarei, M., and Stahr, K. (2000): Lateral Podzolization in a Granite Landscape. Soil Science Society of America Journal, **64**, 1434-1442.

Stahr, K. (1979): Die Bedeutung periglazialer Deckschichten für Bodenbildung und Standortseigenschaften im Südschwarzwald. Freiburger Bodenkundl. Abh. **9**, p. 273.

Stahr, K., Feger, K.-H., Zarei, M., and Papenfuß, K.-H. (1995): Estimation of actual weathering rates in small catchments on Bärhalde granite, Black Forest, SW Germany. In: Clays Controlling the Environment. Proceedings of the 10th Inter. Clay Conference, Adelaide, Australia, July 18 to 23, 1993, Published by CSIRO Publishing, Melbourne, Australia, 494-498.

IV
Applied Clay Science

Clay mineralogy of raw materials and ancient pottery from archaeological sites in the Ambato valley, Catamarca, Argentina.

S. R. Bertolino[a] and M. Fabra[b].

[a] CONICET. Facultad de Matemáticas, Astronomía y Física. Ciudad Universitaria, (5016) Córdoba, Argentina. silvana@quechua.fis.uncor.edu

[b] Museo de Antropología. Facultad de Filosofía y Humanidades. Hipólito Irigoyen 174, (5000) Córdoba, Argentina. <mfabra00@tutopia.com>

Pottery sherds from El Altillo (1900±70 BP radiocarbon age) and Piedras Blancas (ca.600-1000 AC) archaeological sites, in the Ambato valley, and two local clay sources, were characterized for their mineralogy and chemistry. Raw materials used were predominantly the red clays with minor proportions of the off-white clays. Firing rages of <700ºC, 700-800ºC and 800->900ºC were inferred from the mineralogy and SEM textures. No significant changes were observed in the ceramic technology. Thus, it does not reflect the socio-political changes that occurred in the IV century confirming the continuity of the ceramic tradition, a good usage of natural resources and a knowledge of their properties within a period of 1000 years.

1. INTRODUCTION

Clay minerals and provenance study of ancient pottery are applied to better understand some archaeological problems, such as the reconstruction of manufacturing technology, management of natural resources, organization of goods production, goods circulation, in cultures of the southern Andes. A socio-political and ideological change occurred in the Ambato valley [1] in the IV century AC, promoting more specialized goods production and standardized work; it was important to verify if the ceramic technology reflected those changes.

This study focuses on defining the provenance and firing temperatures of pottery sherds from two archaeological sites, El Altillo (1900±70 BP radiocarbon age) and Piedras Blancas (ca.600-1000 AC), within the Ambato valley, Catamarca province, Argentina, where cultural and chronological sequences allow the study of the effects of social changes. The provenance was defined comparing the mineralogical and chemical data of sherds and two possible local clay sources.

Between the Yungas (rainforests) and the pre-Puna (arid to semi-arid) provinces, the valley is an strategic area with access to valuable natural resources, favoring the settlement of human groups [2] since the Formative Period (600 BC-300 AC). El Altillo (EA) is considered the first settlement in the valley. Villages of the Formative period had a relatively equal social organization. Two art styles are found here: Ciénaga, [3] and Condorhuasi. Piedras Blancas (PB), Regional Integration Period (300-1000 AC), represents a well settled, more complex and unequal society [1]. A new style appears, Aguada or "draconian style" with designs of characters with power attributes expressing the lordliness power configuration of this period, coexisting with Condorhuasi and Ciénaga.

2. MATERIALS AND METHODS

Highly representative and unique technological classes of pottery sherds were selected for both sites. Two outcrops of clay-rich rocks found at 2 km west of El Altillo, were also sampled; they consist of a red sediment (RC3, 9) and an off-white altered tuff (OWC11, 13), both probably of Tertiary age. Experimental methods applied for mineralogical and chemical determinations were: X-ray diffraction (XRD) on the bulk and the <2μm fractions (air dried, glycolated and heated at 500ºC, oriented) using a PW1710 diffractometer, Cu-Kα radiation at 36kV and 20mA, with monochromator. A Jeol 6400 and a Leica 360 scanning electron microscopes (SEM), both with Oxford Mod. Inca energy dispersive system (EDS) were used to analyze chips on fresh fractures; individual <2μm particles were analyzed in an Hitachi H-800 electron microscope (TEM-STEM at 200 kV) with a Kevex Delta Class EDS.

3. RESULTS AND DISCUSSION

The mineralogy (Fig.1 A) of clay sources is characterized by: quartz, calcite, albite, microcline, micas, kaolinite and hematite with a <2μm assemblage of kaolinite, I/S (35% I, R=0) and illite in the red one; and in the off-white by: quartz, albite, smectite, micas, amphiboles and cristobalite, with pure smectite (traces of I/S 15%I) in the clay fraction. All pottery pieces (Fig.1 B) are composed of quartz, with varying proportions of feldspars (usually low albite), micas, occasional hematite, cristobalite and hornblende or calcite; some sherds contain one or two spinels (Mg-Al and/or Mn-Fe-Ti oxides). The antiplastic is usually quartz, feldspars, micas or ground ceramic.

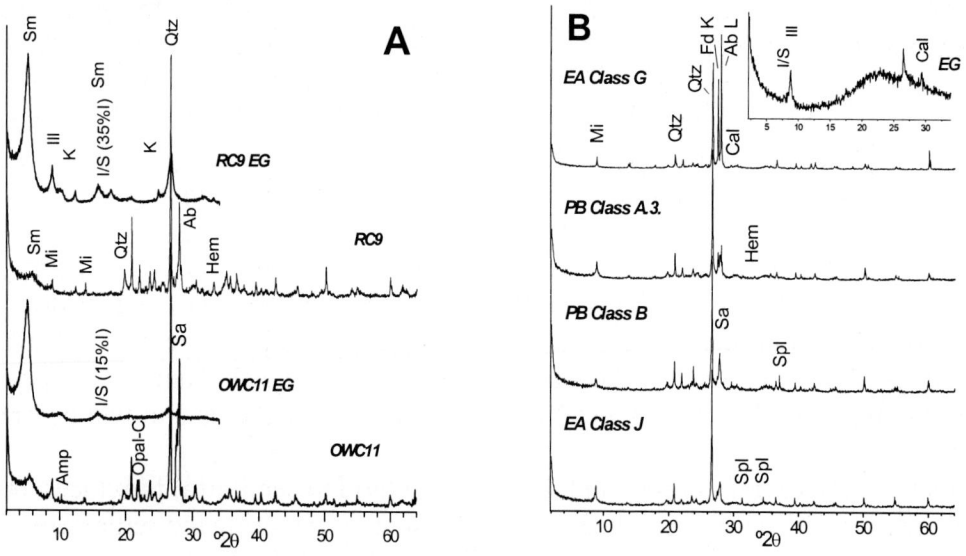

Figure 1. XRD patterns of bulk and glicolated (EG) samples, clay sources (A) and sherds (B).

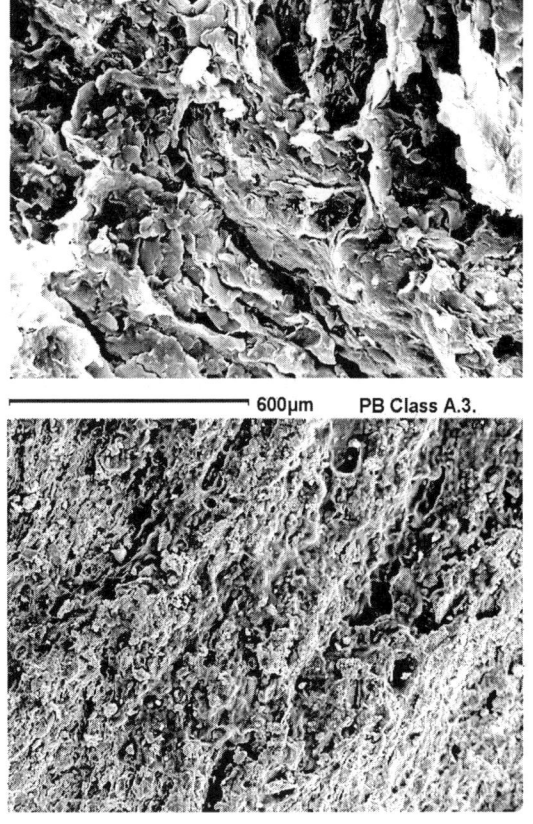

Figure 2. SEM images of red clay (RC9) and sherds showing different degrees of sintering.

Red clays (Fig.3, RC9) are iron rich, contain K, Mg and minor Na as interlayer cations and aprox. 0.8w% of Ti, while the off-white clays (Fig.3, OW11) have no Ti, minor Fe and dominant Mg, Na and Ca with scarce K in the interlayer. The clay matrix of sherds (Fig.3, PB Classes A.3) has a composition very close to that of red clays; typically contain aprox. 0.4 to 1.3w% of Ti, is iron rich and have dominant K, Mg and minor Na and Ca.

Comparing the mineralogy and clay chemistry of sherds and clay sources, it can be clearly seen that the raw material used was predominantly the red clay sometimes mixed with some proportions of the off-white clay (amphiboles). This is logical because the mineral composition of the red clay reflects their good properties for ceramic usage while the other clay, dominated by smectite is suitable, only if added in small proportions to improve the plasticity.

Inferred burning temperature [4], based on bulk mineralogy and SEM studies (Fig.2) of sherds define three ranges: 1) <700°C, sherds EA classes G, D, C and PB class A.1. with no signs of sintering and absence of new faces. 2) 700-800°C, sherds EA Class A and PB Class A.3 exhibiting

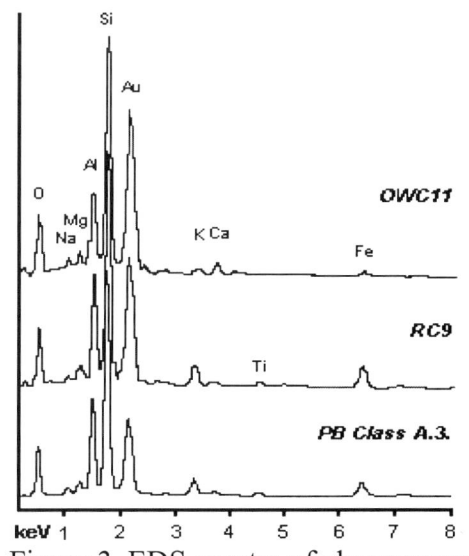

Figure 3. EDS spectra of clay sources and a representative sherd.

some evidences of sintering and few mineral changes. 3) 800-900ºC (probably over 900ºC), sherds EA Classes F and J and PB Class B, major sintering, new faces: spinels formation.

Table 1. Mineralogy and estimated burning temperatures of pottery sherds and clay sources. Mineral names after Kretz (1983). [a] Condorhuasi; [b] Ciénaga; [c] Aguada.

Site	Class	T ºC	Mineralogy
Red Clay	RC 3		Qtz Cal>Ab L>Mc>Mi>K, Hem; Clay: K>Ill>I/S(35%I)
Red Clay	RC 9		Qtz>>Ab>>Hem>Mi I/S K; Clay: K, I/S(15 y 35%I)
Off-white Clay	OWC 11		Qtz>Alk Fd >>Mi Sm Opal-C>>>Amp; Clay: Sm
Off-white Clay	OWC 13		Qtz>>Alk Fd>>Mi Sm Opal-C; Clay: Sm, tr. I/S(15%I) Fd
El Altillo	EA Class G	<700ºC	Ab L>Qtz>Mc>>Mi: Clay: Ill>>>Fd>Cal, tr. I/S
El Altillo[a]	EA Class D	<700ºC	Qtz>Ab L>Mi, Hem; Clay: Ill, tr. Cal y Fd
El Altillo[b]	EA Class C	<700ºC	Qtz>>>Ab L>Mi>>Mc
Piedras Blanca[b,c]	PB Class A.1	<700ºC	Qtz>>Alk Fd>Mi, tr. Hem
El Altillo[b,c]	EA Class A	700-800ºC	Qtz>>Ab>Mi
Piedras Blancas[a]	PB Class A.3	700-800ºC	Qtz>>>Alk Fd>Mi, tr. Hem
El Altillo[a]	EA Class F	800-900ºC	Qtz>>>Mi>>Ab L, San, Spl
El Altillo	EA Class J	800-900ºC	Qtz>>An Mi, Spl
Piedras Blancas	PB Class B	800-900ºC	Qtz>>Alk Fd>>Mi, Spl (Mg-Al; Mn-Fe-Ti)

4. CONCLUSIONS

1- The use of local clay sources, in particular the red clays, is strongly suggested by the mineralogy and the chemical compositions of the clay matrix in all sherds.

2- Pottery sherds from both sites were fired at different temperature ranges with similar technologies and using the same clay sources; the correlation of these results with previous information on the ceramic production [6] suggests that no significant changes in the treatment of the ceramic body (raw materials, antiplastics), decoration or burning temperatures occurred within a period of 1000 years. They do not reflect the social-political changes observed in the iconography of the Aguada style.

3- The results show that these cultures (at El Altillo and Piedras Blancas) have a continuity in the ceramic tradition and a well-known usage of their natural resources for over 1000 years. They also developed knowledge on clay properties to select the raw materials creating a specialized pottery production.

REFERENCES

1. J. A. Perez Gollán, Public. 46 Arqueología. CIFFyH., Córdoba (1992).
2. S. Assandri, A. Avila, R. Herrero and S. Juez, Public. 46 Arqueología, CIFFyH (1992).
3. A. Serrano, Public. Inst. Antrop., I, Univ.Nac.Lit., (1953).
4. J. Schomburg, App. Clay Sci., 6 (1991) 215-220.
5. R. Kretz, Am. Miner., 68 (1983) 277-279.
6. M. Fabra, Rev. Estudios Atacameños, Chile (2001), in press.

Transformation of Chlorinated Aliphatic Compounds by Ferruginous Smectite

Javiera Cervini-Silva[a], Richard A. Larson[b], Jun Wu[b], and Joseph W. Stucki[b]

[a] Department of Environmental Science, Policy, and Management, University of California, Hilgard Hall #3110, Berkeley, CA 94720-3110,
[b] Department of Natural Resources and Environmental Sciences, University of Illinois, W-521 Turner Hall, 1102 South Goodwin Avenue, Urbana, IL 61801

A series of chlorinated aliphatic compounds (RCl, including carbon tetrachloride, 1,1,1-trichloroethane, 1,1,2,2-tetrachloroethane, pentachloroethane, hexachloroethane, trichloroethene and tetrachloroethene, trichloronitromethane (chloropicrin), and trichloroacetonitrile was reacted with ferruginuous smectite, SWa, in aqueous suspension under anoxic conditions. Incubations with SWa-R promoted RCl reduction (chloropicrin, trichloroacetonitrile) or dehydrochlorination (1,1,2,2-tetra and pentachloroethane). The reduction of structural Fe catalyzes the transformation of RCl *via* Brønsted and Lewis-basic promoted pathways. This study indicates that oxidation state of the structural Fe in SWa greatly alters surface chemistry and has a large impact on clay-organic interactions.

1. INTRODUCTION

The swelling nature of smectite clays permits H_2O and other molecules into the interlayer where numerous chemical processes may occur, including the transformation of organic compounds. Recent studies show that the oxidation state of structural Fe in smectite clays alters the Brønsted (1-2) and Lewis (3-4) basicity and (1, 4) provide evidence that indicates a strong correlation between the Fe(II) content in clay minerals (nontronite, montmorillonite, smectite), {Fe(II)}, and the rate of transformation of chlorinated aliphatic compounds *via* dehydrochlorination (e.g, 1,1,2,2-tetra and pentachloroethane; 1-2) and dechlorination (e.g, chloropicrin; 4). Because most clay minerals contain some Fe that is susceptible to redox cycling in their crystal structure, RCl could be transformed by interacting with oxidized and reduced clay minerals.The aim of this study is to understand the effect of molecular structure in the fate of chlorinated aliphatic compounds (carbon tetrachloride, chloropicrin, trichloroacetonitrile, tri-and tetrachloroethene; 1,1,1-, 1,1,2,2-tetra-, penta-, and hexachloroethane) with redox manipulated Fe-bearing smectites and to determine the extent of structural Fe participation. The outcome of this study will be of particular importance in predicting the fate of RCl in natural soils and sediments during burial, submersion, wetting, drying, and other events.

2. METHODS

A 0.1 mM solution of RCl (98% Mallinckrodt) was prepared in 9:1 water-ethanol under an argon atmosphere. The concentration of RCl was confirmed by gas chromatography with a standard curve. The initial pH in solution was 8.

A sample of the < 2-μm particle-size fraction of SWa-1 (Source Clays Repository of The Clay Minerals Society; $(Na_{0.82}(Si_{7.84}Al_{0.16})(Al_{3.10}Fe^{3+}_{0.30}Mg_{0.66})O_{20}(OH)_4$; 5) was saturated with Na$^+$, dialyzed, and freeze-dried. Thirty mg portions of the freeze-dried SWa-1 were dispersed in 50-mL polycarbonate centrifuge tubes with 20 mL of NaCl (5 mM) by shaking gently overnight. Reduction was performed according to the method of Stucki et al. (6) (using dithionite and a citrate-bicarbonate buffer solution at pH 8.4). An aliquot of the clay suspension was taken to determine {Fe(II)}, and total Fe, {Fe$_T$}, using a colorimetric method (7).

Two ml of chlorinated aliphatics (RCl) and clay suspension, in deoxygenated NaCl, was added to a 4.5-ml vessel. Each vessel was septum-sealed; the headspace was saturated with argon. The final concentration of RCl was 0.05 mM; total clay content was 2.5 mg /ml; NaCl concentration was 2.5 mM; pH was 7.5. The vessels were placed horizontally in a wrist-action shaker, and removed periodically to collect samples.

The identification of reactants and products was confirmed using a Hewlett-Packard 5890 quadrupole GC/MS, equipped with an ECD and a Hewlett-Packard column (HP-5, 12 m x 0.2 mm ID, and 0.33-μm film thickness). Prior analysis, aliquots of the reaction mixture were concentrated by solid-phase extraction using Varian cartridges (C-18, 1cc/100mg).

3. RESULTS AND DISCUSSION

We identified four different courses of reaction between RCl and SWa-R. Profiles I and II illustrate the case where RCl are reduced (chloropicrin, trichloroacetonitrile; Figures 1 and 2) and dehydrochlorinated (1, 1, 2, 2-tetra and pentachloroethane; Figures 3 and 4). Profile III refers to the case when the concentration of RCl decreases over time (tri- and hexachloroethane, and carbon tetrachloride; not shown) but no reaction products were identified after 24h of contact time. Profile IV is a particular case of profile III where RCl loss is extremely slow.

3.1. Adsorption

The faster adsorption of chlorinated alkanes that are highly polarizable (e.g., 1,1,2,2-tetra- ($CHCl_2CHCl_2$) and pentachloroethane (CCl_3CHCl_2)) or that share substituents which facilitate charge delocalization [chloropicrin (CCl_3NO_2) or trichloroacetonitrile (CCl_3CN); e.g, eq. 1] may be attributed to dipole-dipole interactions between the substituents and the clay surface and/or hydrating water molecules:

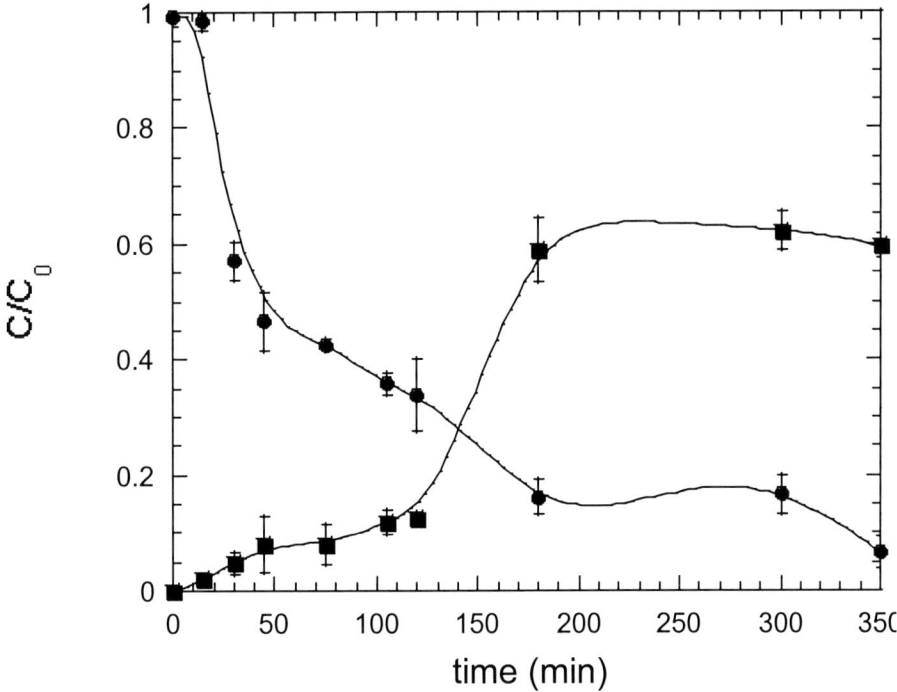

Figure 1. The reaction of trichloroacetonitrile (●) with reduced SWa. The production of dichloroacetonitrile (■) is normalized to the initial concentration of trichloroacetonitrile, C_0.

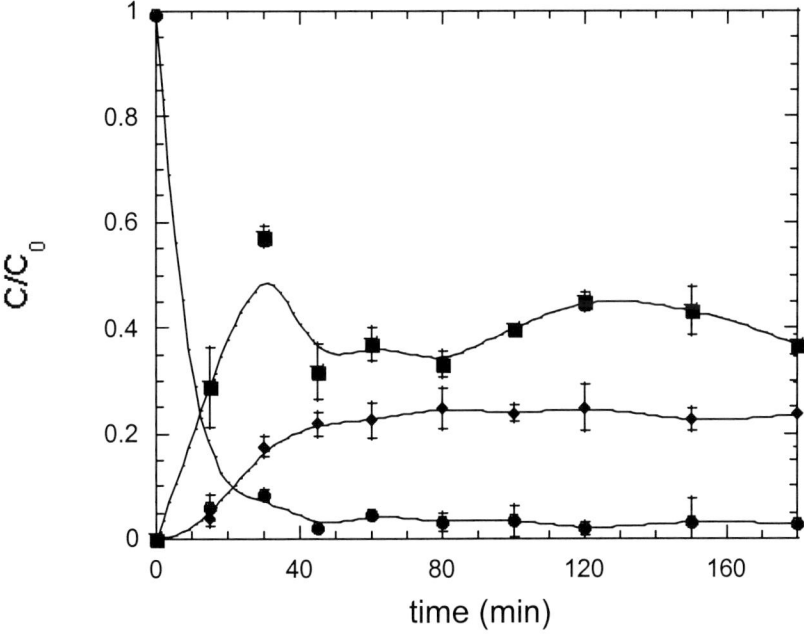

Figure 2. The reaction of chloropicrin (●) with reduced SWa. The production of dichloroacetonitrile (■) and chloroacetonitrile (◆) are normalized to the initial concentration of chloropicrin, C_0.

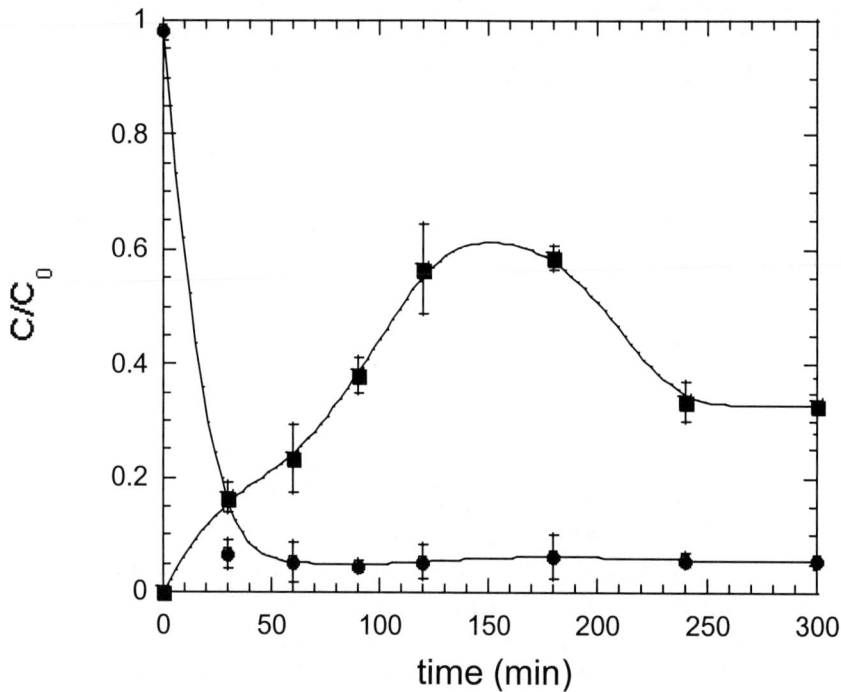

Figure 3. The reaction of pentachloroethane (■) with reduced SWa. The production of tetrachloroethene (●) is normalized to the initial concentration of pentachloroethane, C_0

$$C_0. (Cl)_3C-C\equiv \ddot{N} \leftrightarrow (Cl)_3-C-C=\ddot{N}{:}^- + H_2O \rightarrow (Cl)_3C-C=\ddot{N}{:}^{\delta-}\cdots^{\delta+}H-O-H \quad (1).$$

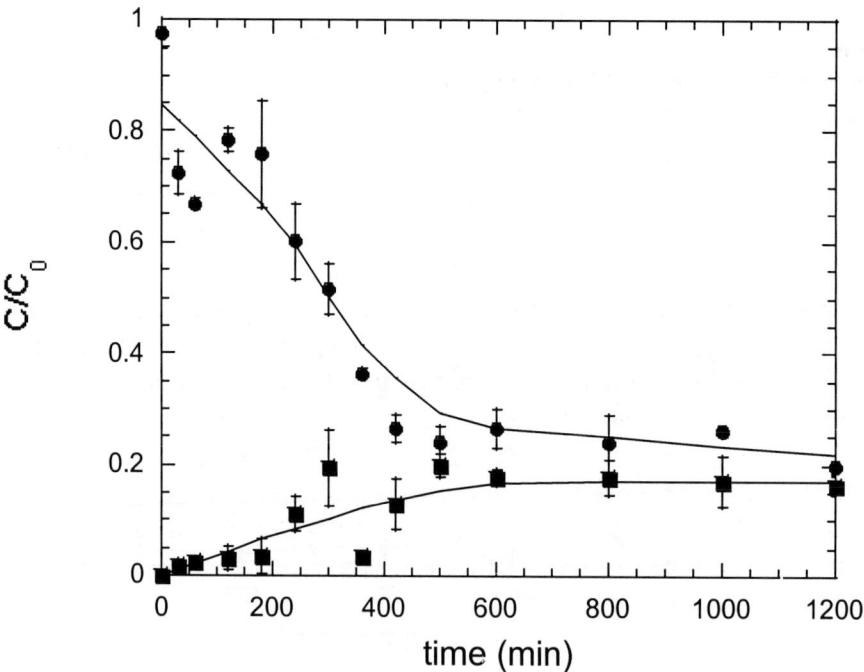

Figure 4. The reaction of 1,1,2,2-tetrachloroethane (●) with reduced SWa. The production of trichloroethene (■) is normalized to the initial concentration of 1,1,2,2-tetrachloroethane,

On the other hand, the small susceptibility of HCA and PCM adsorption to the reduction of structural Fe can be in part explained because the lack of polarity in the structure and of substituents to participate in electrostatic interactions with interlayer water molecules and/or the clay surface.

3.2. Dehydrochlorination.

PCA and TeCA are dehydrochlorinated to PCE (58.6 ± 8.3 % conversion) and TCE (19.4 ± 6.8 %), respectively, which has been attributed to the abstraction of structural hydrogen atom and intramolecular rearrangement (eq. 2).

$$CCl_3CHCl_2 \rightarrow CCl_2 = CCl_2 + H^+ + Cl^- \qquad (2)$$

The susceptibility of 1,1,1-tri-, 1,1,2,2-tetra-, and pentachloroethane to adsorb and undergo dehydrochlorination in the presence of SWa-R was similar to their trend in reactivity to basic hydrolysis in pure water (k_b=7.86x10^{-3} h^{-1}, 5.82 x10^{-7}, and 0.00, respectively; k_b= base-catalyzed hydrolysis rate constant; 8-9). Likewise, the slow adsorption observed for tri- and tetrachloroethene in the presence of SWa-R was consistent with their tendency to undergo slow basic hydrolysis (k_b=6.42x10^{-11}, 8.22x10^{-14} h^{-1}; 8-9). On the other hand however, the reduction of structural Fe accelerated the adsorption 1,1,1-trichloroethane (k_b=0; k_n =7.44 x10^{-5} h^{-1}, where k_n=neutral-catalyzed hydrolysis rate constant, and k_{hyd}= k_b + k_n, where k_{hyd} = hydrolysis rate constant). These results suggest that the interactions of RCl with redox-manipulated smectites are not limited by the hydrolysis mechanism of RCl in bulk water, but on the ability of RCl to hydrate.

3.3. Reduction.

Trichloroacetonitrile and chloropicrin underwent single- and double-dechlorination (Figures 1 and 2). The dechlorination of both trichloroacetonitrile and chloropicrin was fast to produce dichloroacetonitrile (62 ± 3.3 %), and dichloronitromethane (57 ± 1.9 %) and chloronitromethane (24.8 ± 4.3 %), respectively. The loss of trichloroacetonitrile is coupled with the formation of dichloroacetonitrile, which reacts much more slowly.

The reaction of chloropicrin with SWa-R led the simultaneous formation of dichloronitromethane and chloronitromethane (Figure 2) and indicates that the SWa surface served as a bulk reductant and participated in two independent electron transfer pathways with chloropicrin. The dechlorinations of trichloroacetonitrile and chloropicrin are presumably initiated by the transfer of an electron from the SWa surface, with the odd electron delocalized to the C-Cl bond to form a three-electron bond prior to homolysis (eq. 3). The dechlorination of trichloroacetonitrile (Figure 1) and chloropicrin (Figure 2) but not carbon tetrachloride shows that the simultaneous resonance and inductive contributions from the substituents to the carbon center facilitate charge delocalization during clay-organic interactions.

$$C-Cl(Cl_2)NO_2 + e^- \rightarrow C\dot{-}Cl(Cl)_2NO_2^- \rightarrow {}^{\bullet}CCl_2NO_2 + Cl^- \qquad (3)$$

On going research shows that dechlorination of chlorinated aliphatics occurs also in the presence of smectites reduced by microorganisms that are commonly present in natural soils and sediments (10). In that study, structural Fe in ferruginous smectite (SWa) was reduced by incubations with two cultures of microorganisms, *Shewanella oneidensis* strain MR-1 and an enrichment culture from rice paddy soils, then the clay sample was reacted with pentachloroethane, 1,1,1-trichloroethane, and carbon tetrachloride in aqueous suspension under anoxic conditions. Microbial reduction of structural Fe in the clay accelerated the removal of these compounds from suspension, whereas reoxidation of microbially reduced clay samples decreased the removal of these compounds by 50 to 70%. Microbially-reduced SWa promoted the dechlorination of pentachloroethane, 1,1,1-trichloroethane, and carbon tetrachloride to tri- and tetrachloroethene (80% conversion), 1,1-dichloroethene (54%), and chloroform (trace amounts), respectively. The ability of microbially-reduced SWa to promote dechlorination reactions was also attributed to increases in the Brønsted basicity of near-surface water molecules, resulting from increases in electron density at the clay surface. These data indicate that clay-bound Fe(II) produced by microbial Fe(III) reduction has the potential to be an important reductant controlling the fate of organic chemicals in contaminated sediments.

REFERENCES

1. J. Cervini-Silva, J. Wu, J. W. Stucki, and R.A. Larson. Clays Clay Miner. 48 (2000) 132.
2. J. Cervini-Silva, R.A. Larson, J. Wu, and J. W. Stucki. Environ. Sci. Technol. 35 (2001) 405.
3. J. E. Amonette, J. S. Fruchter, Y. A. Gorby, C. R. Cole, K. J. Cantrell, and D. I. Kaplan. U.S. Patent 5 783 088, 1998.
4. J. Cervini-Silva, J. Wu, R.A. Larson, and J. W. Stucki. Environ. Sci. Technol. 34 (2000) 915.
5. A. Manceau, B. Lanson, V. A. Drits, D. Chateigner, W. P. Gates, J. Wu, D. F. Huo, and J. W. Stucki. Am. Minerol. 85 (2000) 133.
6. J. W. Stucki, D. C. Golden, and C. B. Roth. Clays Clay Miner. 32 (1984) 350.
7. P. Komadel and J.W. Stucki. Clays Clay Miner. 36 (1988) 379.
8. P. M. Jeffers, L. M. Ward, L. M .Woytowitch, and N. L. Wolfe Environ. Sci. Technol. 23 (1989) 965.
9. R. A. Larson and E. J. Weber. Reaction Mechanisms in Environmental Organic Chemistry. Lewis Publishers: Chelsea, MI, 1994.
10. J. Cervini-Silva, R.A. Larson, J.E. Kostka, and J.W. Stucki. (in review)

Porous alumina-nanoclay functionally gradient membrane structures

K. Darcovich, F.N. Toll and L.S. Kotlyar

Institute for Chemical Process and Environmental Chemistry
National Research Council of Canada
Ottawa, Ontario, Canada, K1A 0R6

A processing method has been established where metastable colloidal suspensions of alumina powder of a broad particle size distribution sediment under gravity in a drain casting mold to form a continuously gradient porous structure. By controlling the colloidal phase state of these suspensions, an association of the finest particles with all other size classes contributes to forming a consolidated structure which may be sintered at relatively low temperatures without warping or cracking because of the spatial gradient in sintering rate that is also achieved by neighbouring particles with large curvature differences.

The present project advances this theme by investigating effects of adding clay nanoparticles to these suspensions. The particles are ultra-thin plate-like kaolinite and mica alumino-silicates recovered from tar sands processing, hexagonal in shape and averaging about 200 nm^2 in area on one side. Control samples free of the nano-clays as well as samples containing these nanoparticles in the matrix of the structure, and/or added secondarily as a surface coating will be prepared. The samples are then sintered to form a fused contiguous body.

Structural characterization of the samples included data from bulk porosimetry and microscopy for top layer pores, as well as some preliminary permeation and separation data from a pressurized cross-flow filtration module. The effects of adding small amounts of nano-clays to a porous ceramic membrane structure will be discussed in view of potential improvements in permeation and separation performance.

1. INTRODUCTION

Ceramic filtration media are of interest because of their high chemical and thermal resistance. In order to produce such a material, normally an ultrafine coating or film layer is fused to a porous substrate that imparts mechanical integrity to the unit. The coating or film layer is typically referred to as a "membrane".

An on-going project has investigated fabricating highly porous asymmetric ceramic structures designed to serve as filtration media [1-2]. A polydisperse colloidal sedimentation method was conceived and implemented. In a single processing step, samples of alumina with surface pore sizes in the ultrafiltration regime were obtained, with diameters in the 50 to 100 nm range. Analysis by AFM has given surface roughness data in the range of 25 to 75 nm. Such structures without any subsequent modification would be suitable for coarser types of filtration, such as beverage clarification or soot removal. These structures would also be suitable materials to act as substrates for sol-gel coatings, or film deposition by any number of techniques.

A continuing objective in ceramic filter research is to be able to produce materials suitable for ever finer types of separations, and ultimately gas separations. In view of this, a number of projects are currently underway to use these structures as substrates for applying ultrafine coatings and/or for film deposition. It has always been an advantage however, to minimize the number of processing steps involved in preparing these structures, since their porous morphology is highly dependent on any thermal treatments they undergo, and even ultrafine coatings must be fused to the substrate with the application of heat. Thus, the rationale for the present work is to incorporate a means to achieve even finer surface pore properties into the single processing step by the addition of co-stable nanoclays to the suspension used to cast the entire structure.

To date, there have been a number of projects that have examined the preparation of composite structures of clays and conventional metal oxide ceramics. Isayama and Kunitake [3] did some early work on the preparation of self-supporting films of clay minerals and metal oxides. These structures were made from ultrafine clays mixed with alumina sols. Calcination at 1000 °C was performed on samples self-assembled into a bilayer template, and the layered structure remained intact. These samples were prepared with catalytic applications in mind. Lao et al. [4] attempted membrane preparation using alumina-pillared montmorillonite and γ-alumina. This was an initial study and only confirmed that a porous pillared structure was achieved. Vercauteren et al. [5] researched the preparation of pillared clay structures as membranes on top of alumina substrates, investigating various aspects such as interlayer cations, processing time, coating thicknesses on pore properties and gas separation performance. The clay pillaring was applied as a coating atop a γ-alumina interlayer, as it was found that direct adhesion to the α-alumina substrate was ineffective. A top layer thickness of 0.15 μm was suitable to achieve a permselectivity similar to a γ-alumina membrane for the H_2/O_2 system, but at a very low flux. Brinza et al. [6] have reported on the preparation and testing of sepiolite clay used as a membrane making material. A clay sample with a specific surface area of about 300 m^2/g, and sintered at 700 °C was able to produce ultrafiltration membranes with a molecular weight cut off around 20000. These particular samples were found to have a high degree of mechanical flexibility. Membranes only 250 μm thick were able to withstand 90 psi pressures, and the same material was successfully made into mechanically sound hollow fibers.

The present work is an attempt to provide a kind of intermediate state of material that might lend itself to easier subsequent coating operations and/or an improved porous asymmetric structure with finer ultrafiltration properties.

2. MATERIALS AND SAMPLE PREPARATION

The separation of the clay ultrafines (<300 nm) from oil sands was carried out using the Cold Water Agitation Test described in detail elsewhere [7]. Typically, distilled water, or 0.1 weight percent sodium pyrophosphate solution (60 g), was added to an oil sands sample (15 g) and the mixture agitated vigorously before mild centrifugation at 200xG for 1 hour. Both ultrafines (<300 nm) and the coarser solids (>300 nm), remaining after ultrafines removal, were separated into hydrophilic and biwetted components using emulsion flotation with toluene.

A range of analyses of these particulates revealed that they are platelets with a major dimension between 50 and 400 nm across, and whose composition is primarily kaolinite and mica [8]. Representative transmission electron micrographs of the nanoclay are shown in Fig. 1. The particles are thin plates and somewhat hexagonal in form, but with a fairly irregular overall range of shapes.

Aqueous alumina suspensions were prepared at 5 volume percent solids. A mixture of two powders (Sumitomo AKP-30 (25%) and Ceralox APA-0.2 (75%)) was used to provide a broad particle size distribution, with grain sizes between 0.05 μm and 5 μm with a mode at about 0.2 μm [9]. The nanoclay particles were provided as a suspension, and quantities were added to achieve the desired loadings. Stabilization of these suspensions was provided electrostatically, by adjusting the pH to around 11. This was a suitable pH to disperse both the alumina and the clay nanoparticles, and with non-minimum viscosities at the selected solids concentration.

The ceramic structures were then produced by sedimentation. They were drain cast into the form of flat discs in 45 mm diameter tube sections over milled gypsum slabs. The discs were cast to a thickness of 3.3 mm. Prior to sintering, the green bodies were dried in an oven at 50 °C, and then they were sintered over a 15 hour period, attaining a maximum temperature of 1200 °C at a heating rate of 175 °C/hr and a soak time of one hour.

2.1 Porosimetry

To investigate the pore size distribution of the sintered ceramic structures, mercury porosimetry measurements were done with a Quantachrome Pore Master 60 porosimeter. This is a bulk property, so a contact angle of 140 degrees was assumed, corresponding to an alumina surface. A pore size distribution representing the entire cross-section was obtained for each sample.

2.2 Surface Roughness

The root mean square roughness of some sintered samples was measured with a Dektak 3 Surface profiler magnetic force tip instrument. This provided a two-dimensional integration along a single line covering a 1 mm length.

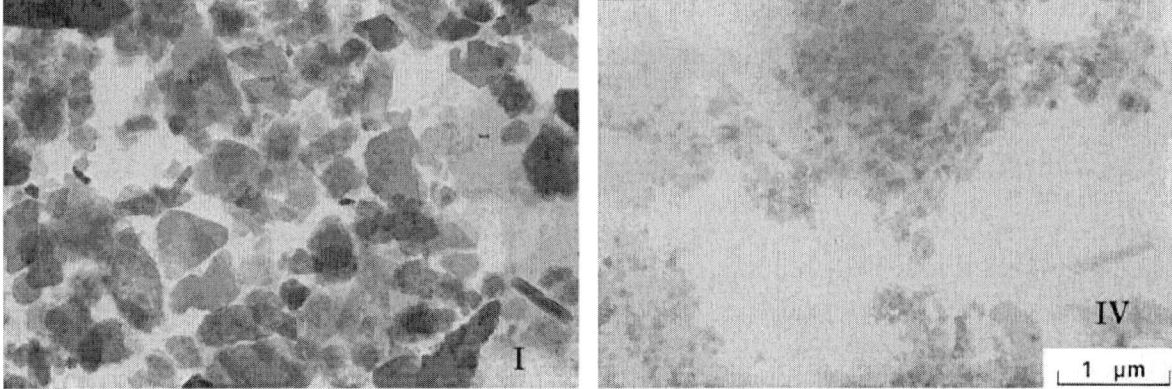

Figure 1. TEM of the nanoclay sample. I (400 nm fraction) and IV (50 nm) fraction of platelets.

2.3 Surface Pore Properties

SEM imaging was performed with a Hitachi S4700 Field Emission Scanning Electron Microscope instrument, with images obtained for at 50000 times magnification. These images were then imported into Image-Pro Plus software, where digital image analysis was run to determine a surface pore size distribution. Two levels of gray level discrimination were used since at the first level to determine a number count of the pores, a visible amount of the pore area was excluded. A second level was used to visually account for all the surface pore area, which caused individual pores to run together. The area ratio was then used as a factor to correct the individual pore sizes identified at the first gray level. The second level provided a value for the two-dimensional surface porosity. Typical area differences between the two gray levels were between 10 and 15%, which can be considered as an estimate of the experimental error for this technique.

2.4 Project Experimental Overview

A number of different sintered samples were prepared to compare some of the resulting properties. One series of samples was prepared with the nanoclay integrated into the alumina suspension prior to casting (i-series), and another with the nanoclay suspension coated over a previously sedimented alumina compact (c-series). Quantities of nanoclay to provide the equivalent of 10, 100 and 1000 layers over the discs were added.

For each sample, SEMs were taken and then used for determining the surface pore properties. Each sample was also analyzed by mercury porosimetry for bulk pore properties. A few preliminary tests with a permeation and separation cell were also conducted to situate the samples in the filtration spectrum. A filter disc 36 mm in diameter was mounted in a cross-flow test cell. Tests were run with pure water as well as aqueous polyethylene oxide (PEO) solutions made from solutes of 100K, 300K and 900K molecular weight to assess separation performance. Tests were done at pressures of 25 and 50 psi (1.7 and 3.4 bar).

3. RESULTS AND DISCUSSION

The SEM images in Fig. 2 show that the presence of increased amounts of the nanoclay reduces the overall surface porosity and the mean surface pore diameter. The bonding of the nanoclays to the alumina substrate appears to be thorough and stable. The grain sizes in the SEM images appeared to enlarge with nanoclay addition level, suggesting that the clays sintered and amalgamated neighboring alumina grains into smoothed over clusters. X-ray diffraction work by Sainz et al. [10] was able to demonstrate that sintering alumina and kaolinite mixtures between 1150 and 1300 °C caused primary mullite formation of rod-like growth by a solid state process. This would be consistent with the observations of the present samples and would help explain the reduced image sharpness in the c-series SEM images in Fig. 2.

According to the image analysis procedure described earlier, the plots in Fig. 3 show the resulting surface pore size distributions. These distributions were obtained by summing pore areas of each progressively larger pore to first make a numerical cumulative distribution, which was then curve fit with a third-order polynomial and differentiated to give the frequency distribution curves. In general, the mode of the distribution moves to finer diameters with increasing nanoclay content. This happened to a larger extent with the coated samples, as presumably some of the nanoclay content in the integrated samples became associated with the material forming the bulk of the structure and could not contribute to surface coating. A visible difference is evident when comparing the 10i to the 10c case. In the 10i case, the surface pore size distribution is very close to the distribution obtained for the alumina without nanoclay, while with the 10c case, the position of the distribution peak clearly shows a significantly reduced mean diameter. At a 10 layer addition level, a slight association of the nanoclay with the bulk of the structure could significantly reduce any influence on surface modification from the nanoclay. With the nanoclay added as a coating, it apparently remains at the surface and makes more of a change on the surface pore size distribution.

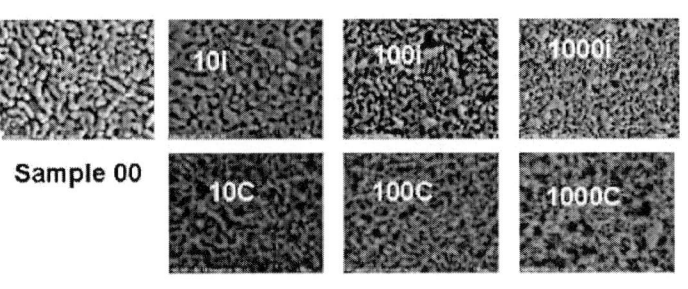

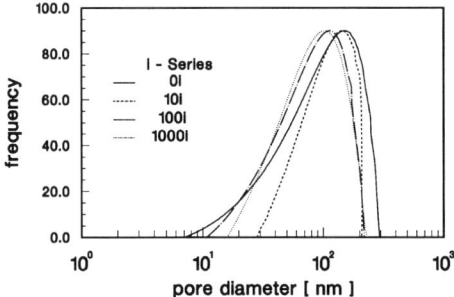

Figure 2. SEM images of top surfaces of i and c-series samples at 50000 X.

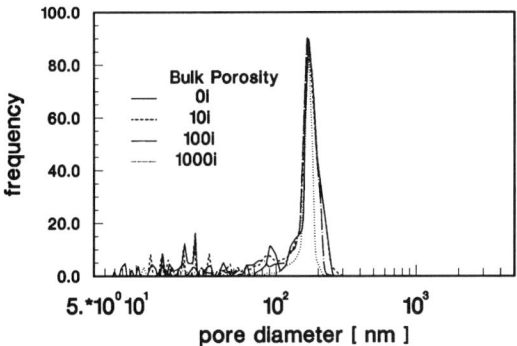

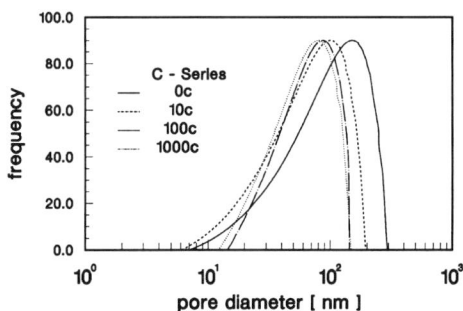

Figure 4. Bulk pore size distributions for the i-series samples obtained by mercury porosimetry.

Figure 3. Surface pore size distributions for the i-series (top) and c-series (lower) samples obtained by image analysis.

Fig. 4 shows bulk pore size distributions for the i-series samples done by mercury porosimetry. All the pore size distributions in this case are essentially the same, and that the addition of the nanoclays would mainly affect the surface of the structures. The mean pore size from the bulk measurements is around 168 nm, compared to a surface pore size of 149 nm. This is consistent with our structure preparation method where there is an increasing mean pore size from the top to the bottom of the sample. Additionally, mention should be made that the surface and bulk pore sizes are calculated assuming geometrical symmetry. In view of this, any deviation from purely round (in two-dimensions) or spherical (in three dimensions) pores will underestimate the bulk pore sizes to a greater extent. Porosimetry also provided structural porosity data, ranging from 48 to 51%, indicating that the nanoclay does not change the bulk structure of the samples.

Fig. 5 shows the total surface porosity plotted as a function of the amount of nanoclay added to the system, expressed in total potential layers of coverage. The samples made from the integrated clay-ceramic suspensions showed a larger surface porosity, likely attributable to nanoclay being associated with other particulates and being incorporated throughout the structure. The fact that the i-series does show a decreasing surface porosity with nanoclay level at least indicates that the sedimentation of these clays was slow enough so that their main effect occurred at the structure surface.

The RMS surface roughness of the samples is shown in Fig. 6. The coated samples contribute surface roughness to a greater extent at equal additions. This corresponds to information from the SEM images that show the nanoclays tending to adhere to alumina particulates, rather than filling pores. Additionally, differences in surface energy between species would promote the cohesion of the nanoclays upon sintering such that they would

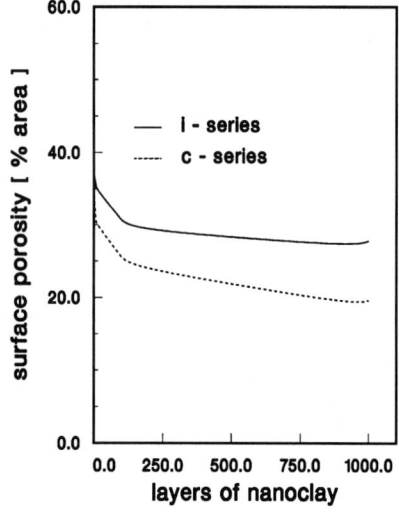

Figure 5. Surface porosity versus quantity of nanoclay added to the system.

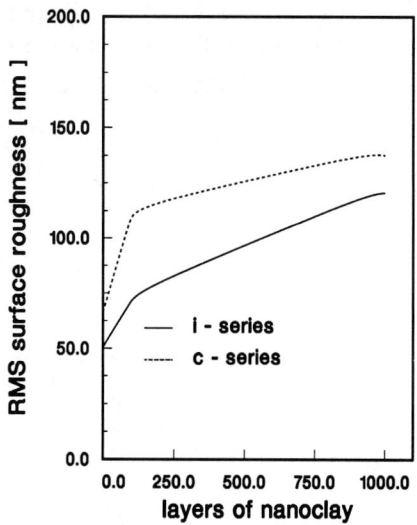

Figure 6. RMS surface roughness versus quantity of nanoclay added to the system.

be liable to coalesce and contribute to additional external structure on the surface. Surface roughnesses upwards of 100 nm can be problematic for subsequent sol-gel coating operations to make ionic or gas separation membranes. Some preliminary permeation and separation tests were carried out with sample 10c. The results from these tests are shown in Fig. 7. The standard near linear flux dependence on pressure was observed, with a normalized flux of about 5.9 L/m2·h·bar. The presence of the nanoclays on the surface reduced the flux compared to samples prepared with only particulate alumina [2]. Examining the separation data shows a molecular weight cut-off (near 90%) around 900K. A 900K molecular weight polyethylene oxide has a radius of 24.83 nm, thereby giving an effective separation pore radius of about 32 nm. This corresponds suitably with the optically determined surface pore radius of about 50 nm. The difference between these two figures suggests that the optical method could overestimate pore size by virtue of considering the entire exposed pore opening in view as opposed to the narrowest gap in the pore which would effectively dictate the separation performance. Coupling SEM with AFM could give more information about pore morphology at the surface and provide a means to improve the optical pore size determination.

The flux values obtained for the 10c sample compares to about 9.0 L/m2·h·bar for filtering 35K polyethylene glycol with sol-gel coated Kerafil membranes with a 25K molecular weight cut-off [11]. The lower flux in the present case can be attributed to the sample thickness of over 3 mm. It was not the present objective to optimize a filtration device, so the fact that the performance data lie in the same order of magnitude is promising.

4. CONCLUSIONS

From the preliminary investigation carried out it can be said that the addition of the nanoclay material extracted from tar sands tailings is of potential interest to improving the single step processing method for preparing asymmetrical inorganic membrane subtrates. The data have shown that the top surface pore sizes are reduced with increasing addition levels of the nanoclays. The nanoclays had a sufficiently impeded sedimentation such that

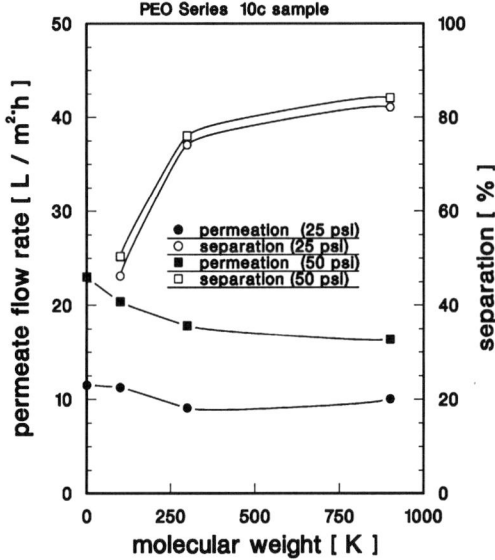

Figure 7. Permeation and separation data for sample 10c, tested with pure water and a number of PEO solute weights.

the vast majority of this material ended up in the surface region of the samples. Bulk porosimetry tests revealed that the bulk pore properties of the samples were not modified by the addition of the nanoclays. Preliminary permeation and separation tests show that the addition of the nanoclays help to achieve finer levels of separation, and that the global structure achieved shows data comparable to other similar ultrafiltration materials.

The effect of the nanoclays was localized to the surface where they acted as precursors for primary mullite formation after sintering to 1200 °C. Despite having a chemical affinity that allows a co-sintering of the nanoclays and alumina to take place, the increase in surface roughness with addition level of nanoclay suggests that there is a mullite phase emergence taking place with the growth of rod-like crystals. In general, contrary to what was observed here, sintering finer homogeneous material at a surface results in less surface roughness. To improve on this problem, the present results suggest processing with a separate step to drain cast a nanoclay layer followed by a much lower temperature sintering to avoid mullite formation. In this way, the nanoclays could be bonded to the alumina without causing them to degrade surface smoothness.

Acknowledgements

The authors would like to express their appreciation to Mr. Jeff Fraser and Mr. John Phillips of NRC-IMS, for providing respectively the SEM images and the image analysis.

REFERENCES

1. K. Darcovich and M.E. Price, J. Can. Cer. Soc., 66 [2] (1997) 146.
2. K. Darcovich, D. Roussel and F.N. Toll, J. Membrane Sci., 183 (2001) 293.
3. M. Isayama and T. Kunitake, Adv. Mater., 6 [1], (1994) 77.
4. H. Lao, C. Detellier, T. Matsuura and A.Y. Tremblay, J. Mater. Sci. Lett., 13 (1994) 895.
5. S. Vercauteren, M. Vayer, H. Van Damme, J. Luyten, R. Leysen and E.F. Vansant, Colloids and Surfaces. A: Physicochem. Eng. Aspects, 138 (1998) 367.
6. D.L. Brinza, T. Matsuura, C. Detellier and E. Rutinduka, Proc. 5th Intl. Conf. Inorganic Membranes, Nagoya, Japan, June 22-26, 1998, p. 46.
7. L.S. Kotlyar, B.D. Sparks, R. Schutte and C.E. Capes, AOSTRA J., 8 (1992) 55.
8. L.S. Kotlyar, Y. Deslandes, B.D. Sparks, H. Kodama and R. Schutte, Clays and Clay Minerals, 41 [3] (1993) 341.
9. K. Darcovich and C.R. Cloutier, J. Am. Cer. Soc., 82 [8] (1999) 2073.
10. M.A. Sainz, F.J. Serrano, J.M. Amigo, J. Bastida and A. Caballero, J. Eur. Cer. Soc., 20 (2000) 403.
11. K. Pflanz, N. Stroh and R. Riedel, Key Eng. Mater., 150 (1998) 135.

Effect of Chemical Modification of *Stilbite* Zeolite on Removing Lead from Wastewaters

A. C. P. Duarte; M. B. M. Monte; A. A. Neto and A. B. da Luz.

Centre for Mineral Technology – CETEM. Coordination of Mineral Processing. Av. do Ipê 900, Cidade Universitária, Ilha do Fundão, CEP 21941-590 Rio de Janeiro / RJ Brazil.
E-mail: acduarte@cetem.gov.br

Stilbite is a natural zeolite that allows both reversible ion exchange and ion adsorption. These properties are due to the size and the charge of the hydrated cation and to the distribution of exchange sites in the zeolite. The suitability of Brazilian natural zeolite for wastewater treatment is still not known. Two kind of treatments were carried out with original *Stilbite* to convert it into a homoionic state: in treatment 1, the zeolite was mixed with a 2N NaCl solution; and in treatment 2, the solution was exchanged for a 0.5N NaOH. These procedures improved significantly the zeolite cationic exchange capacity (CEC). Comparatively, CEC values increase from 0.92 meq/g (original *Stilbite*) to 2.97 and 2.56 meq/g (treatment 1 and 2, respectively). The increase on lead removal efficiency ranges from 45.94% for natural zeolite (original) to 79.59 and to 99.97% for treatment 1 and 2, respectively.

1. INTRODUCTION

Natural zeolites belong to a group of aluminosilicates of alkaline and alkaline-earth metals with cationic exchange and ion adsorption capabilities. These phenomena can occur due to their large inner area and net negative residual charge as a result of more extensive isomorphism in the tetrahedral sites on the outer surface (1, 2). Based on these properties, natural zeolites can be largely used for wastewater treatment, soil amendment, among other applications (3).

Stilbite is a calcium rich zeolite that can be described as a fundamental polyhedral configuration containing four and five membered rings of tetrahedra (4). In order to be used on wastewater treatment, it is important to investigate its chemical composition and cationic exchange capacity. For an accurate determination of this latter parameter, it is necessary to convert the zeolite to the homoionic form to minimize cation competition (5,6).

In this report, it is presented the characterization of the natural zeolite from Parnaíba Basin / Brazil which has a detrital sedimentary origin, and to investigate the influence of zeolite sodium treatment, under different conditions, for lead removal applications.

2. MATERIALS AND METHODS

2.1. Technological Characterization

X-ray diffraction (XRD) and scanning electron microscope (SEM) coupled with energy dispersive X-ray detector (EDS) were applied to confirm the mineral identity of the sample and its qualitative composition. In addition, chemical composition was also determined by atomic absorption spectrometry (AAS).

2.2. Experimental Methods

Two types of treatments were studied aiming to convert *Stilbite* into a homoionic state in sodium form. In the treatment 1, zeolite (25 g) were put in contact with 250 mL of 2N NaCl solution under continuous stirring at room temperature and for 24h. The suspension was then filtered and the filtrate dried at 100 °C. For treatment 2, about 25 g of zeolite were mixed with 250mL of 0.5N NaOH solution, i.e., the same procedure presented for the first treatment.

Cationic exchange capacities (CEC) were measured by sodium displacement by potassium cations. This was accomplished by mixing 1g of each sample (original and treated) with 40 mL of a solution of 1M KCl. The suspensions were filtered and the filtrates were dried at 100 °C. All determinations were carried out on the initial and final solutions using atomic absorption spectrometry (AAS).

Lead removal performance from a synthetic effluent was carried out using *Stilbite* before and after treatments, 1 (NaCl) and 2 (NaOH), respectively. Zeolite (1g) was mixed with 40 mL $Pb(NO_3)_2$ solution at concentrations ranging from 100 to 1,500 mg/L. The suspension was stirred up at constant speed (16 rpm/min), for a period of 3h, at room temperature. The amount of removed metal was determined by comparison with a control zeolite that had not received any treatment. All metals concentrations were determined by atomic absorption spectrometry (AAS) at the initials and finals solutions.

3. RESULTS AND DISCUSSION

3.1. Technological Characterization

SEM examination of the sample (Figure 1) revealed non-defined crystals that could be related to the *Stilbite* structure. It can also be seen in the Figure 1 that the original zeolite has small quantities of impurities, which were identified by EDS as magnesium silicate and quartz. The qualitative chemical composition determination by SEM/EDS identified Si, Al, Ca, Mg, Na, K and other trace elements in *Stilbite*. The analysis of XRD data confirmed that the original zeolite studied was *Stilbite*. XRD using the Rietveld technique shows that the magnesium silicate (saponite) is, in fact, present in trace quantities and that the sample is mainly composed by *Stilbite* (90.5%) and quartz (9.5%).

Figure 1. Micrograph of virgin *Stilbite*[*] sample.

The chemical composition of the original zeolite was determined by AAS, volumetric and gravimetric techniques. The results are shown in table 1.

Table 1
Stilbite chemical composition

Compound	%	Compound	%
MnO	0.02	K_2O	0.65
Na_2O	0.65	CaO	9.50
Fe_2O_3	0.92	SiO_2	55.95
MgO	1.20	Al_2O_3	16.60

Traces of P, Cr, Ni, Zn and Co.

This analysis revealed that *Stilbite* from Parnaíba Basin is mainly compose of silicon oxide (55.95%), aluminum oxide (16.60%), iron oxide (0.92%), magnesium oxide (1.20%) and sodium oxide; potassium oxide and manganese oxide are present with smaller concentrations. These results showed that *Stilbite* had a Si/Al ratio of 3.37.

Based on these results about the molecular formula for this zeolite is:

$$(CaO)_{3,4}(Na_2O)_{0,23}(K_2O)_{0,14}(MgO)_{0,62}(Fe_2O_3)_{0,12}(Al_2O_3)_{3,08}(SiO_2)_{18,9}$$

The molecular weight of *Stilbite* is 1,711.24 g/mol. Based on the aluminum content and this molecular weight, the theoretical exchange capacity is 3.6 meq/g zeolite.

3.2. Experimental Methods

Table 2 shows the sodium content incorporated by the zeolite and the CEC after treatment 1 (NaCl) and 2 (NaOH). When compared with the original *Stilbite*, these results show an increase by factors of 7.1 and 6.4, in terms of sodium content, for treatment 1 and 2, respectively. CEC in original *stilbite* did not exceed 1 meq/g but with treatment 1 it increases to 2.97 and 2.56 with treatment 2. This is an improvement of 2.7-3.0 times when comparing with the original samples.

Table 2
Sodium content determined before and after treatments.

	Sodium (mg/g zeolite)	Ratio of Sodium content	CEC (meq/g zeolite)
Original	3.01	1	0.92
NaCl	21.35	7.1	2.97
NaOH	19.14	6.4	2.56

Figure 2 shows the effect of the chemical treatment of the zeolite on the Pb removal ability. These results show that for lead initial concentrations lower than 312 mg/L, treated and original *Stilbite* have similar removal performances (approximately 6 mg of Pb per gram of zeolite). Lead uptake is enhanced from about 6 to approximately 39 mg of Pb per gram of zeolite, when Pb initial concentrations increased from 312 to 625 mg/L for treatments 1 and 2, respectively. On the other hand, the lead uptake for the original *Stilbite* did not exceed 17 mg of Pb per gram of zeolite, even in the system where the initial Pb concentration was 625 mg/L. Treatment 2 presented the best Pb removal capacity exceeding 39 mg of Pb per gram of zeolite.

As illustrated by Figure 3, *Stilbite* treated with sodium hydroxide solution has a removal efficiency exceeding 99% at all concentration levels ranging from 62 to 625 mg Pb/L. Original samples performed poorly compared to the treated samples, with removal efficiency decreasing drastically from 95% at 62 mg Pb/L to 45% at 625 mg Pb/L. It can also be observed that the removal efficiency for zeolite treated with NaCl slightly decreased up to 79% at 625 mg/L of initial lead concentration.

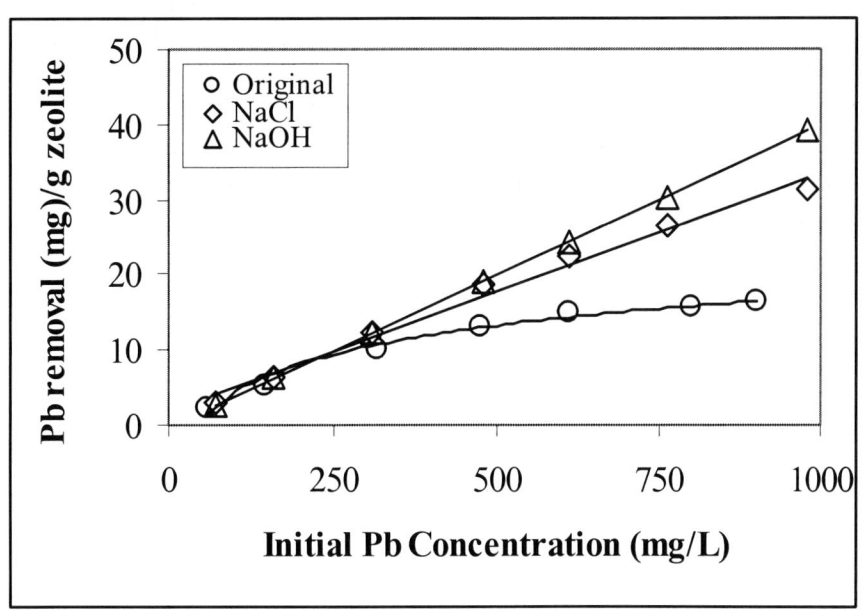

Figure 2. Effects of treatment on Pb removal by *Stilbite*.

Figure 3 shows the effect of chemical treatments on the Pb removal by *stilbite*.

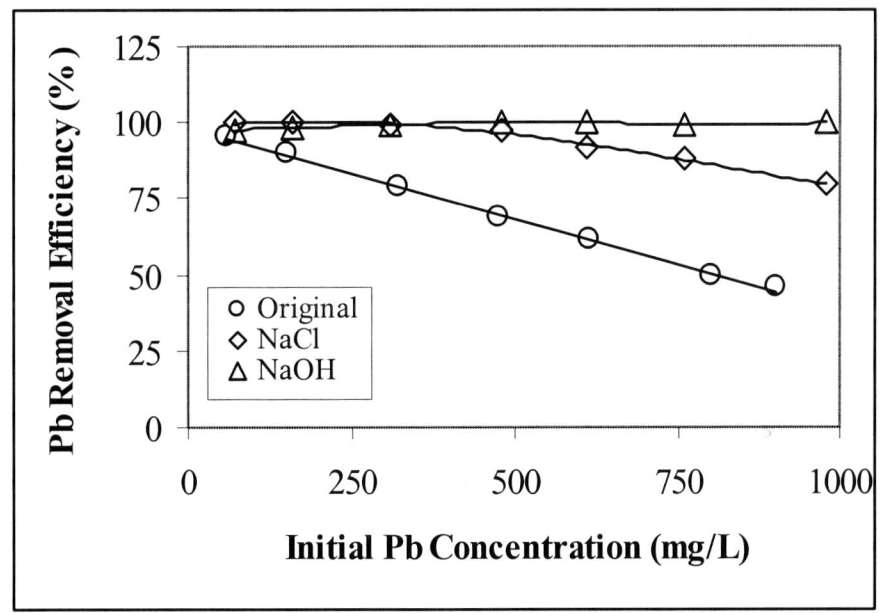

Figure 3. Pb removal efficiency by *stilbite* in three different forms.

4. CONCLUSIONS

This preliminary study revealed that the adopted treatment improves both CEC and lead removal performances when operating at metal initial concentration higher than 312 mg/L. For Pb removal efficiency, it was found that sodium hydroxide treatment presents the best results when compared to original samples. For initial lead concentrations of 625 mg/L, it yields an increase on removal efficiency of 45.94% for original *stilbite* and of 79.59 and 99.97% for treatment1 and 2, respectively.

REFERENCES

1. M. Kithome, J. W. Paul, L. M. Lavkulich and A. A. Bomke. Soil Science Soc. Am. Journal. 62:622-629 (1998).
2. M. F. Brigatti, C. Lugli and L. Poppi. Applied Clay Science. 16:45-57 (2000).
3. A. Hanson. Industrial Minerals. p.40-53. (1995).
4. J. Li, J. Quin, Y. Sun and Y. Long. Microporous and Mesoporous Materials. 37:365-378 (2000).
5. E. L. Cooney, N. A. Booker *et al*. Separations Science and Technology. 34(12): 2307-2327 (1999).
6. S. Kesraoul-Ouki, C. Cheeseman and R. Perry. Environmental Science and Technology. 27: 1108-1116 (1993).

Pore lining chlorites in hydrocarbon reservoirs : structure and composition related to their origin

Claudine Durand, Bernadette Rebours, Elisabeth Rosenberg and Etienne Brosse
Institut Français du Pétrole, 1 et 4 avenue de Bois-Préau ; 92852- Rueil-Malmaison, France

Pore lining chlorites are authigenic clays common in deep hydrocarbon reservoirs. Preventing silicification, they contribute to the preservation of an economically useful porosity. The structure and composition of about forty samples, coming from six hydrocarbon fields, have been thoroughly studied either on size separated fractions, for XRD, or on polished sections, for EDS-SEM analyses. Most of the samples are Ib chamosites, but some plot in the IIb clinochlore area. Equations proposed in the literature provide temperature values higher than those known from field data, so that the use of chlorites as geothermometers cannot be recommended based on these case studies. On the other hand, the polytypes and the composition of tetrahedral versus octahedral layers appear more closely related to the nature of the possible precursors, 10 Å or 7Å phases, than to temperature conditions.

INTRODUCTION

Pore lining chlorites are authigenic clays common in deep hydrocarbon reservoirs. Preventing silicification, they contribute in many cases to preserve an economically useful porosity. The availability of data from different hydrocarbons fields having undergone various geological histories has provided a unique occasion to compare the compositions versus structure and texture of the chlorites. This was expected to shed light on the nature of the precursors and the conditions of formation. Since more than a century, chlorite composition and structure have indeed been investigated, trying to relate ranges of composition and polytypism to the occurrences and the conditions of formation, mainly availability of chemical elements and temperature.

The purpose of this paper is i) to characterise the composition of chlorite, ii) to provide data about the species (chlorite, and/or 7 Å phase) and the polytypes (I or II) present in different hydrocarbon fields. These data allow the classification of chlorites by comparison with other works. They give arguments for a discussion about their origin and formation conditions.

EXPERIMENTAL
Sampling

More than 40 polished sections of rocks coming from several wells in 6 hydrocarbon fields were analysed for chlorite composition by Scanning Electron Microscopy (SEM) fitted with an Energy Dispersive Spectrometer (EDS). More than 30 fine fractions were separated from whole rocks and analysed by X-ray Diffraction. The wells are referred to as follows:
LS1, for Lower Silurian of Libya; S1, for Strunian (Upper Devonian) of Algeria; D1 and D2, for Lower Devonian of Algeria (these four wells will be referred to as "Saharian"); J1, for Jurassic of the North Sea; T1, for Tertiary of the North Sea; and C1 to C5, for Cretaceous of Argentina. Cca-1 reference of the Clay Mineral Society (CMS) repository was analysed for comparison.

Chemical formulae of chlorites

Hundreds of microanalyses were performed by a FEG (Field Emission Gun) SEM. JSM6300F fitted with an EDS detector IMIX PGT. Samples were imbedded in an epoxy resin, polished and coated with carbon. Operating conditions were 15 kV accelerating voltage, approximately 1 nA current and a counting time of 60 seconds per analysis of the whole spectrum. Based on calibration results, the values obtained in this work can be compared with confidence to results in the literature, mostly obtained under similar conditions (Electron Probe Micro Analysis, 15 kV).

Following the rules given by Foster (1962) and Brown and Brindley (1980) the structural formula was calculated by normalisation on $O_{10}(OH)_8$, i.e. 14 O (or 28 charges) and 4 H_2O molecules for the half cell, for the cations Si, Al, Mg, Fe, without differentiating the brucite and talc-like sheets, which are shown between square brackets in the following formula: $[(Mg_{2+}, Fe_{2+}, Al_{3+})_3(Si, Al_{IV})_4O_{10},(OH)_2] [(Mg_{2+}, Fe_{2+}, Al_{3+})_3(OH)_6]$ All Fe was considered as divalent. The results are given in Table 1. Results of chlorite analyses were not correct for well J1, owing to abundant illite. Other chlorite values for the same well (Hillier et al., 1992) are shown for comparison.

	LS1 well average n=277	σ	S1 well average n=50	σ	D1-D2 wells average n=161	σ	J1 *	T1 well average n=19	σ	C1-C5 wells average n=132	σ
Si	2.78	0.08	2.85	0.05	2.65	0.03	2.75	3.04	0.03	3.02	0.11
Al IV	1.22	0.08	1.15	0.05	1.35	0.03	1.26	0.96	0.03	0.98	0.11
Al VI	1.72	0.08	1.68	0.09	1.59	0.09	1.73	1.56	0.06	1.90	0.12
Mg	0.68	0.09	0.67	0.08	0.55	0.03	0.88	1.42	0.04	1.95	0.06
Fe	3.36	0.21	3.38	0.16	3.73	0.17	3.09	2.72	0.12	1.68	0.22
Ca	0.01	0.01	0.04	0.01	0.02	0.00	0.02	0.04	0.01	0.02	0.03
Ti	0.01	0.00	0.01	0.00	0.00	0.00	0.01	0.00	0.00	0.01	0.01
K	0.02	0.01	0.00	0.00	0.02	0.00	0.02	0.00	0.00	0.02	0.02
Na	0.02	0.02	0.00	0.00	0.01	0.00	0.02	0.00	0.00	0.00	0.00
Σ octa	5.75	0.07	5.73	0.03	5.88	0.10	5.71	5.70	0.04	5.54	0.11
Si/Al	0.94	0.04	1.01	0.06	0.90	0.01	0.92	1.20	0.02	1.04	0.04
Al_{tot}	2.94	0.05	2.86	0.13	2.94	0.06	2.98	2.53	0.04	2.89	0.05
$Al^{VI} - Al^{IV}$	0.49	0.15	0.53	0.07	0.24	0.11	0.47	0.61	0.08	0.92	0.23
Fe/Fe+Mg	0.83	0.03	0.83	0.02	0.87	0.04	0.78	0.66	0.02	0.46	0.04

Table I. Structural formulae for chlorites averaged on all the samples analysed in the wells of the same field. n is the number of analysed points in each field (*) see text

Very homogeneous compositions were found within each sample, even on chlorites with different textures, so that no argument could be found for differentiating textures from the chemical point of view.

Structural data from X-ray Diffraction

Identification of chlorite polytypes and associated mineral phases was based on comparison of powder patterns obtained from disoriented samples of the purest size separated fractions (mostly < 2 µm) with those compiled in the ICDD (International Committee for Diffraction Data) database. Main results are shown in Figure 1.

Type I polytype was identified in all the Saharan samples, as well as in the North Sea fields. Type II was identified in Argentinian samples, and in mixture with type I in J1 samples. Berthierine was identified in LS1, S1 and D1-D2 samples which show the presence of ooids on thin sections. In these wells, whole rock powder diffraction patterns also clearly show the presence of berthierine.

Further calculations were performed according to the methods given by Brown and Brindley (1980). To determine the Al^{IV} content from the position of the 00l lines, the Bailey (1972) equation was used. The Fe content was determined from the position of the 060 line, and as well from the intensities (areas) of the 00_ lines. In many cases of Fe-rich chlorite, the odd lines 001, 003, 005 were substantially wider than the lines of even order 002, 004. The amount of 7 Å interstratified phase (S for serpentine) was calculated according to Reynolds et al. (1992). Results are given in Table 2.

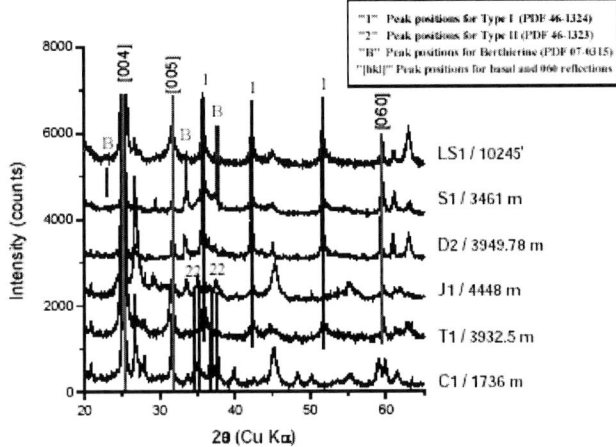

Figure 1: Identification of mineral phases and chlorite polytypes on typical samples of each field.

	Method	LS1	S1	D1-D2	T1	C1-C5	Cca-1
		av	av (σ)	av (σ)	av (σ)	av (σ)	
Al^{IV}	EDS	1.22	1.17 (0.05)	1.33 (0.13)	0.96 (0.03)	0.97 (0.14)	1.32
	XRD	1.60	1.60 (0.07)	1.61 (0.08)	1.22 (0.04)	1.56 (0.11)	1.62
Fe	EDS	3.33	3.33 (0.16)	3.75 (0.49)	2.72 (0.12)	1.60 (0.25)	2.13
	XRD 060	3.21	3.61 (0.45)	4.01 (0.04)	2.64 (0.18)	0.60 (0.32)	1.77
	XRD (A)	4.53	4.09 (0.83)	4.06 (0.42)	2.18 (0.16)	0.34 (0.24)	1.4
% S		14.23	5.44 (1.79)	0.31 (0.66)	2.79 (0.56)	0.27 (0.93)	-0.25

Table 2: Comparison between the chlorite chemical parameters determined from XRD and EDS

DISCUSSION

Classification

Both EDS and XRD provide data on the amounts of Al$_{IV}$ and Fe, which are the main parameters used in the classification of chlorites. Comparison between the values obtained on the same samples is shown in Table 2. Clearly, Al$_{IV}$ content from EDS is in all cases lower than from XRD, by about 0.2-0.3 atom per half cell. On the other hand, Fe estimations by the different manners are relatively close, particularly from EDS and those from the position of the 060 line.

The discrepancy between EDS and XRD for Al$_{IV}$ is obviously the highest for the

Argentinian samples, where Al_{VI} is the highest, and the lowest for the D1-D2 samples, where

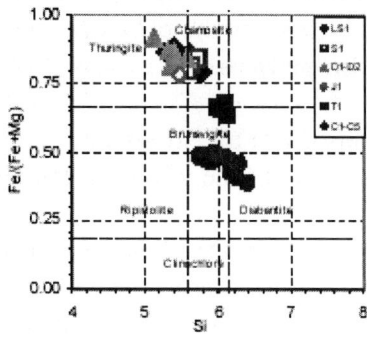

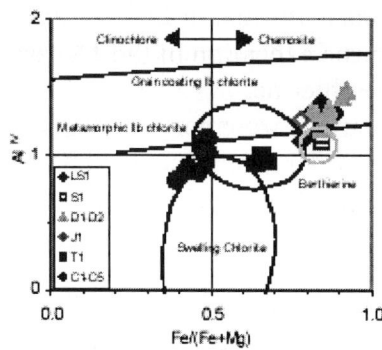

Figure 2: Chlorites classification after Foster (1952), disclosing the composition as a function of Si, Fe and Mg contents

Figure 3: Chlorites classification as to the occurrence domains defined by Brown an Bailey (1962), modified by Curtis (1985), using Al_{IV} from EDS.

Al^{VI} is the lowest. Bailey (1972) already remarked that Al_{IV} from XRD was found too high in the dioctahedral chlorites. Later, Whittle (1986), and Spötl et al. (1994) noticed that XRD values were higher than direct ones for Al_{IV}, and lower for Fe. Whittle concluded that these relationships may work for IIb polytypes, but are not relevant for Ib.

Whatever the "true" value of tetrahedral Al, samples cluster in the same three groups. In the Foster's classification, Si vs Fe/Fe+Mg (Fig. 2) the Saharian samples, as well as the J1 ones, fall in the Thuringite-Chamosite area, while the T1 are at the limit Brunsvigite-Diabantite, and the C1-C5 as well, but with lower values of Fe/Fe+Mg. In the Al_{IV} vs Fe/Fe+Mg diagram of Brown and Bailey (1962) completed by Curtis(1985), three groups are found as well (Fig. 3). If we take the Al_{IV} from EDS, the three groups fall below the separation lines already drawn for metamorphic IIb chlorite, grain coating Ib chlorite, and berthierine, although almost none of the samples fall in the swelling chlorite area. If, instead of the EDS value, the XRD derived values be taken, all points would be shifted to higher Al_{IV} values. From this classification, Argentinian samples would be "metamorphic", Saharian samples and J1 would be at the limit between Ib, IIb and berthierine, while T1 samples only would be in the "grain coating Ib" area.

Another way of plotting chlorite data is by the Si vs R2+ = Fe+Mg diagram that was derived from Wiewióra and Weiss (1990), and used by Hillier and Velde (1992) to analyse the mixed-layering. In this diagram (Fig. 4), all the samples plot along and close to the Altotal= 3 in the half cell, with the exception of the T1 samples, which have lower values.

The D1-D2 samples are the closest to the amesite-serpentine line, while S1 and LS1 shift towards more abundant 7 Å layers. Although the nature of this 7 Å phase is uncertain, this spreading is consistent with the evaluation of the amount of serpentine layers by the Reynolds method: very little interstratification in D1-D2, substantial values in the other Saharian wells as well as the North Sea T1 and J1. The location of the Argentinian samples in this diagram cannot be interpreted as an amount of mixed-layering. Indeed an amount of 20 to 30 % of 7 Å layers is not consistent with the XRD diagram. These samples hold a higher amount of Si, and a lower amount of R_{2+} than the Fe-rich samples. Fe determinations set Argentinian samples totally apart from the others, with values in all cases, except one, lower than 1 Fe out of 6 octahedral positions. On the other hand, a tetrahedral sheet with a formula Si_3Al is the same as in muscovite. Therefore the most likely explanation to this position in diagram relies on a possible relationship with a 10 Å phase either smectite or illite, which have a Si/Al ratio>1. However, the difficulties encountered to get "clean" chlorite analyses in the Argentina series suggest to be cautious on interpretation. Values higher than 1 for Al_{IV}, like the ones in Fe-rich chlorites do not compare correctly with values found in 10 Å phases.

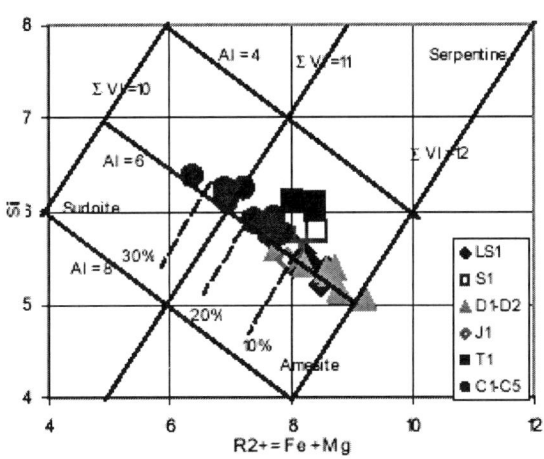

Figure 4: Chlorites classification afterWiewióra (1990) and Hillier (1992), showing the spreading of octahedral occupancy interpreted as the amount of mixed layering

Identification of phases and polytypes

Mainly two types of chlorites were identified in this work: Ib ferriferous, and IIb magnesian. As shown in Fig. 3, Ib is frequently assumed to be a low temperature form, while IIb would be a high temperature form. The Saharian samples were all identified as Ib type, with some of them including berthierine (Fig.1). The abundance of berthierine, which is supposed to be a mineral unstable with increasing depth, implies that the mineral assemblage is not in equilibrium. The most striking variation in this set of Saharian samples is the decrease both of the line widths, which corresponds to an increase in crystallinity, and of the differences of the line width between odd and even 00_ lines, which corresponds to a decrease in the amount of 7 Å interstratification. These variations can be related to an increase of burial over a medium which retained a good porosity. This, albeit not a proof, would also agree with the hypothesis that the actual chlorite comes from a 7 Å precursor, whose abundance decreases with evolution.

The two North Sea wells have a priori no geological relationship. The chlorites are both ferriferous, but less than in the Saharian samples, and present some interstratification. However, they differ from one another by all the other parameters: Al, Fe, and interstratification amount. The presence of IIb polytype in mixture with Ib in the J1 samples

(Fig. 1) can be a sign of evolution related to the burial depth, or related to a different precursor structure.

Argentinian samples are quite different from Fe-rich samples: the polytype is IIb, there is little Fe, there is no 7/14 Å interstratification. These characteristics, together with the distribution of cations in octahedral/tetrahedral layers, may be consistent with a magnesian 2:1 precursor. The absence of swelling means that, if a 2:1 precursor was admitted, and if an analogy with the smectite/illite transformation was taken, the reaction had proceeded fairly to its end. Assuming that, the polytype may be consistent with a relatively high temperature, in agreement with the assumption of IIb as a stable phase at high temperatures.

About geothermometry and thermodynamical stability

For the studied wells, the in situ present day temperature can be considered as the maximum burial temperatures. The exception is for C1-C5, where a burial value taking into account about 2000 m sediments more should be calculated, placing the value at about 125°C. A first observation (Table 3) is that polytype Iib do not appear to be related to higher field temperatures than polytype Ib. Calculation of temperature using the published formulas based on Al$_{IV}$ content (Cathelineau, 1988), or corrected with Fe/Fe+Mg (Kranidiotis and MacLean, 1987, Jowett, 1991) provides higher values than those measured on the field (Table 3).

	LS1		S1		D1-D2		T1		C1-C5	
	T°C	σ	T°C	σ	T°C	σ	T°C	σ	T°C	σ
In situ Well T°C	90		125		160		140		125	
T°Cath	332	26	313	16	372	29	248	8	259	35
T°Kr&McL	236	10	229	3	253	13	190	4	172	15
T°Jowett	348	26	329	15	389	29	259	9	263	35

Table 3: Comparison between temperatures got from direct measurements in the wells and those calculated from different equations (see text).

Using Al$_{IV}$ estimated from XRD should give even higher values. If the field temperatures are considered as sound, estimations based on these literature equations cannot be accepted. This adds arguments to several papers (Caritat et al., 1993, Walker, 1993, Jiang et al., 1994) estimating that the use of chlorite composition as a geothermometer lacks reliability outside the cases for which they have been designed. However, the interval depth in each field is relatively narrow, so that evolution related to depth is not expected to be significant within a well.

About precursor phases

If the polytypes and the composition cannot be directly related to formation temperature, another line of investigation can be followed: relating the composition, through the layer occupancy, to the nature of the precursor. Following this idea, the C1-C5 samples have tetrahedral layers with a composition close to that of muscovite, and octahedral layers relatively rich in Al. This suggests a transformation: muscovite (or illite or beidellite) + Fe => chlorite, where a possible segregation of trivalent cations in the mica layer and divalent ones in the brucite layer cannot be rejected. Such a cation segregation would be consistent with a decrease in the d(001) which is indeed observed, inducing a large discrepancy between the determination of Al$_{IV}$ from EDS and XRD, the latter determination being based on the assumption that the distribution of charges is random. The D1-D2 samples have a low Si content, thus a tetrahedral charge very high, if interpreted in a 2 :1 formula. This would be better interpreted as close to twice a 1 :1 formula, like berthierine. In this mineral the charge compensation is within the 1 :1 cell, between one tetrahedral and one octahedral layer. A « polymerization » of the berthierine would lead to tetrahedral or octahedral layers having the same composition either in the talc-like or in the brucitic layer. The presence of berthierine ooids in some samples of the Saharian area, even if they cannot be considered as a unique source for chlorite formation, is a witness that all the elements constitutive of the chlorite are, or have been present. Similarly the presence of siderite shows that Fe and Mg are still present. On the contrary, it is difficult to guess what were the elements available in the Argentinian neighbourhood. Assuming that the abundant feldspars (albite and orthoclase) proceed from volcanic material, it may be assumed that Mg and Fe have been leached from the framework minerals in former evolution states.

CONCLUSIONS

Two main types of chlorites have been found:
Fe-rich, polytype Ib, proceeding from mixed layered with a 7 Å phase to well crystallized prismatic particles Fe-poor, polytype IIb, with crystals more flexuous, and no mixed layering. The first case encompasses the Saharian samples. The second case is found in the Argentinian samples. Special situations are found for the T1 samples, which are Ib, with a composition in between the two main classes, and the ones from the J1 well, which are highly ferriferous, but may contain some IIb polytype. Further studies should be needed to clarify these late cases.
In all the studied samples, the estimations of Fe are quite consistent whatever the method, and the estimations of possible interstratification can be compared to other data in the literature. On the contrary, estimation of Al_{IV} gives results substantially different, whether they are measured from EDS, or estimated from XRD, through literature relationships. This allows us to assume that reconciliation of these estimations needs further studies in order to be used in a general purpose.
As a consequence, the geothermometer relationships based on Al_{IV} content are also inconsistent depending on the data used. However, they give in any case values higher than those based on field measurements, which are considered to be more reliable. These relationships cannot be used in any of the studied field. Based on the Argentinian samples, polytype IIb does not appear as linked to high temperatures.
Another line of interpretation is proposed: relating the composition of the to the nature of the precursor phases. Although this is risky, because most of the precursor phases have disappeared, it seems likely, from their composition and texture, to assume that the Fe-rich chlorite may proceed from berthierine, and the Fe-poor from a 2:1 phase. Further studies should help documenting this interpretation. Evolution of the composition and crystallinity state with burial, thus temperature, based on samples encompassing large depth intervals of homogeneous series should give arguments on geothermometry.

Acknowledgments: The authors thank AGIP, Anadarko, Elf, Sonatrach, TotalFina and IFP for authorization to publish this work. They thank P. Gueroult and S. Belin for participation with SEM work, and D. .Alves Barbosa and an anonymous reviewer for their help to clarify the paper.

References

Bailey S.W. (1972) – Determination of chlorite composition by X-ray spacings and intensities. Clays and Clay Min., vol 20, 381-388.
Brown B.E., and Bailey, S.W., (1962) – Chlorite Polytypism : I. Regular and semi-random one-layer structures. Amer. Min. vol 47, 819-850.

Brown, G. and Brindley, G.W. (1980) X-ray Diffraction procedures for clay minerals identification. In Crystal structures of clay minerals and their X-ray identification, G.W. Brindley and G. Brown Eds. Mineralogical Society, London.

Caritat, P. de, Hutcheon, I., and Walshe, J.L., (1993) Chlorite geothermometry : a review. Clays and Clay Min., vol 41, No 2, 219-239.

Cathelineau, M. (1988) Cation site occupancy in chlorites and illites as a function of temperature. Clay Min. 23, 471-485.

Curtis, C.D., Hughes, C.R., Whiteman, J.A., and whittle, C.K. (1985) Compositional variation within some sedimentary chlorites and some comments on their origin. Mineralogical Magazine, vol 49, 375-386.

Foster, M. (1962) Interpretation of the composition and a classifiacation of the chlorites. USGS Prof. Paper 414A.

Hillier S. and Velde, B. (1992) Chlorite interstratified with a 7 Å mineral: an example from offshore Norway, and possible implications for the interpretation of the composition of diagenetic chlorites. Clay Min., 27, 475-486.

Jiang, W-T, Peacor, D.R., and Buseck, P.R. (1994) – Chlorite geothermometry ? – Contamination and apparent octahedral vacancies. Clays and Clay Min. vol 42, n° 5, 593-605.

Jowett, E.C. (1991) Fitting iron and magnesium into the hydrothermal chlorite geothermometer. GAC/MAC/SEG Joint Annual Meeting, (Toronto, May 27-29) Program with Abstracts, 16, A62.

Kranidiotis P., and MacLean, W.H. (1987) Systematics of chlorite alteration at the Phelps Dodge massive sulfide deposit, Matagami, Quebec. Econ. Geol. 82, 1898-1911.

Reynolds, R.C. Jr, DiStefano, M.P.,and Lahann, R.W. (1992) Randomly interstratified serpentine/chlorite : its detection and quantification by powder X-ray diffraction methods. Clays and Clay Min., vol 40, No 3, 262-267.

Spötl, Ch., Houseknecht, D.W., and Longstaffe, F.J. (1994) Authigenic chlorites in sandstones as indicators of high temperature diagenesis, Arkoma Foreland Basin, USA. J. Sediment . Res., vol A4, n° 3, 553-566.

Walker, J. R. (1993) - Chlorite polytype geothermometry. Clays and Clay Min., vol 41, 260-267.

Whittle, C. K. (1986) Comparison of sedimentary chlorite compositions by X-ray diffraction and analytical TEM. Clay Min., 21, 937-947.

Wiewióra A. and Weiss Z. (1990) Crystallochemical classifications of phyllosilicates based on the unified system of projection of chemical composition . II. The chlorite group. Clay Min. 25, 83-92.

Mechanochemical activation of kaolinite surfaces

R. L. Frost*[a], É. Makó,[b] J. Kristóf,[c] Z. Ding[a] and J.T. Kloprogge[a]

[a]Centre for Instrumental and Developmental Chemistry, Queensland University of Technology, 2 George Street, GPO Box 2434, Brisbane, Queensland 4001, Australia.

[b]Department of Silicate and Materials Engineering, University of Veszprém, H-8201 Veszprém, P.O.Box 158, Hungary.

[c]Department of Analytical Chemistry, University of Veszprém, H8201 Veszprém, PO Box 158, Hungary.

Kaolinite surfaces were modified by grinding kaolinite/quartz mixtures for periods of time up to 10 hours. X-ray diffraction shows the loss of intensity of the d(001) spacing with mechanical treatment resulting in the delamination of the kaolinite. Thermogravimetric analyses show the kaolinite surface is significantly modified and surface hydroxyls are replaced with both coordinated and adsorbed water molecules. Changes in the molecular structure of the surface hydroxyls of the kaolinite/quartz mixtures were followed by DRIFT spectroscopy. Kaolinite hydroxyls were lost after two hours of grinding as evidenced by the decrease in intensity of the OH stretching vibrations at 3695 and 3619 cm^{-1} and the deformation modes at 937 and 915 cm^{-1}. Changes in the surface structure of the OSiO units were reflected in the SiO stretching and OSiO bending vibrations. The decrease in intensity of the 1056 and 1034 cm^{-1} bands attributed to kaolinite SiO stretching vibrations were concomitantly matched by the increase in intensity of additional bands at 1113 and 520 cm^{-1} ascribed to the new mechanically synthesized kaolinite surface.

1. INTRODUCTION

Kaolin has a widespread industrial application as raw material for the production of paper, ceramics, plastics, etc. Normally the first step in kaolin processing is to mill, blunge or grind the kaolinite with impurities. The industrial application of kaolinite depends on its physical, chemical, structural and surface properties. Many minerals are modified by mechanochemical treatment trough grinding, milling, blunging and other mechanical means. Such treatment often results in the production of small particles with high surface areas. This is mechanochemical activation. Mechanochemical activation causes significant changes in the kaolinite structure through increasing the number of lattice defects, in surface energy and in chemical reactivity. Schrader observed that the crystal structure was deformed mainly along the *c*-axis during mechanical treatment and proved to be more resistant along the *b*-axis [1]. The rupture of O-H, Al-OH, Al-O-Si and Si-O bonds during grinding was reported earlier [2-5]. Some researchers suggested that OH groups were displaced irreversibly during milling [6-

8]. The structural changes of kaolinite -especially with respect to its hydroxyl groups- was mainly investigated by means of infrared spectroscopy [6,8].

Kaolinite hydroxyl groups have four infrared stretching bands centred on 3695, 3670, 3650 (v_1, v_2, v_3) and 3620 cm^{-1} (v_5) [9-13]. The intensity of these bands depends on the defect structure of kaolinite and the degree of substitution into the octahedral layers. Two types of hydroxyl groups are present in the kaolinite structure: inner surface hydroxyls and inner hydroxyls. The three higher frequency vibrations (v_1, v_2, v_3) are assigned to the three inner surface hydroxyls of the gibbsite-like octahedral sheet hydrogen bonded to the oxygens of the adjacent tetrahedral layer [12,13]. The v_5 band is assigned to the inner hydroxyls located in the plane shared with the apical oxygen atoms of the tetrahedral sheet (13). The disintegration of kaolinite crystals is amorphization of the structure characterised by the cleavage of bonding forces between clusters of atoms and the irreversible displacement of atomic groups, which do not revert on release of the external load. The product of such mechanochemical treatment is a water-containing xerogel with random structure, which develops in several stages via intermediates [7].

2. EXPERIMENTAL

2.1 Materials

The kaolin used in experiments was the high-grade natural kaolin from Sedlec (Zettlitz) in Slovakia. Its chemical composition in wt. % is MgO, 0.26; CaO, 0.54; SiO_2, 46.97; Fe_2O_3, 0.37; K_2O, 1.21; Al_2O_3, 36.32; TiO_2, 0.05; loss on ignition, 13.38. The major mineral constituent is low-defect kaolinite (92 wt. %) with a Hinckley index of around 0.7. Some minor amounts of quartz (4 wt. %) and illite (4 wt. %) are also present. This kaolin was selected for this experiment because of its low quartz content. The specific surface area is 18.5 m^2/g. The arithmetic mean diameter determined by a Fritsch Laser-Particle-Sizer "analysette 22" is 4.4 µm. The kaolin contains 10.1 % of particles less than 1 µm and 9.3 % of particles greater than 10 µm in size. The natural sand from Fehérvárcsurgó in Hungary was used to produce samples with different quartz content. Its chemical composition in wt. % is CaO, 0.2; SiO_2, 91.97; Fe_2O_3, 0.14; K_2O, 1.11; Na_2O, 0.12; Al_2O_3, 5.65; TiO_2, 0.02; loss on ignition, 0.35. This natural sand consists mainly of quartz (90 wt. %) with some impurities of feldspar (6 wt. %) and corundum (4 wt. %). The specific surface area is 3.1 m^2/g. The arithmetic mean diameter determined by a Fritsch Laser-Particle-Sizer "analysette 22" is 3.6 µm. The sand consists of 9.7 % of particles less than 1 µm and 3.2 % of particles greater than 10 µm in size.

2.2 Milling procedure

A Fritsch pulverisette 5/2 type laboratory planetary mill was used to grind the kaolinite. Samples were ground for different periods of time up to 24 hours. Each milling was carried out with a 10 g air-dried sample in an 80 cm^3 capacity stainless steel (18 % Cr +8 % Ni) pot using 8 (31.6 g) stainless steel balls (10 mm diameter). The applied rotation speed was 374 r.p.m.

2.3 X-ray Diffraction

The X-ray diffraction (XRD) analyses were carried out on a Philips PW 3710 type diffractometer equipped with a PW 3020 vertical goniometer and curved graphite-diffracted beam monochromator. The radiation applied was CuKα_1 from a long fine focus Cu tube, operating at 40 kV and 40 mA. The samples were measured in step scan mode with steps of 0.02° 2θ and a counting time of 1 s. Data collection and evaluation were performed with PC-APD 3.6 software. Profile fitting was applied to extract information on the microstructure and structural defects of kaolinite and its alteration products. The Profile Fitting option of the software uses a model that employs twelve intrinsic parameters to describe the profile, the instrumental aberration and wavelength dependent contributions to the profile.

2.4 Thermoanalytical methods

Thermogravimetric analyses (TG) were carried out in a Netzsch (Germany) TG 209 type thermobalance under dynamic heating conditions (10°C/min heating rate) in flowing argon atmosphere of 99.995 % purity (Messer Griesheim, Hungary). A ceramic crucible was used for the experiments filled with approx. 20 mg sample in each case.

2.5 Diffuse Reflectance Spectroscopy

Diffuse Reflectance Fourier Transform Infrared spectroscopic (DRIFT) analyses were undertaken using a Bio-Rad 60A spectrometer. 512 scans were obtained at a resolution of 2 cm^{-1} with a mirror velocity of 0.3 cm/sec. Spectra were co-added to improve the signal to noise ratio. Approximately 3 wt. % of samples were dispersed in 100 mg oven dried spectroscopic grade KBr with a refractive index of 1.559 and a particle size of 5-20 μm. Reflected radiation was collected at ~50% efficiency. Background KBr spectra were obtained and spectra rationed to the background. The sample cup (3 mm deep, 6 mm in diameter) accommodates powdery samples mixed with KBr using an agate mortar and pestle in 1-3 % concentration. The reflectance spectra expressed as Kubelka-Munk unit versus wavenumber curves are very similar to absorbance spectra and can be evaluated accordingly.

3.0 RESULTS and DISCUSSION
3.1 X-ray Diffraction

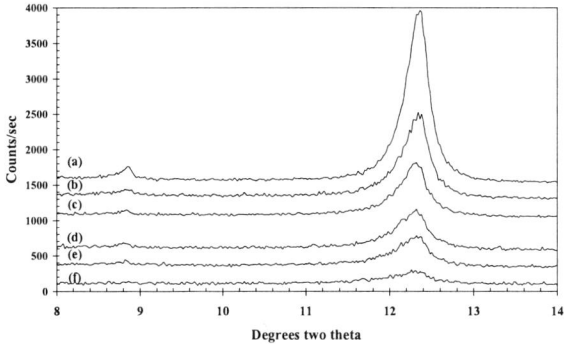

Figure 1 X-ray diffraction patterns of kaolinite ground for (a) 1 (b) 2 (c) 3 (d) 4 (e) 6 (f) 10 hours

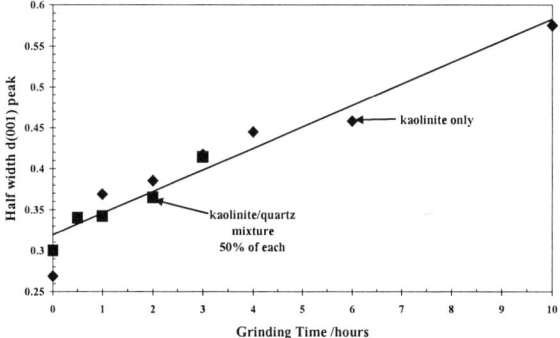

Figure 2 Variation in the crystallite size with time of grinding for kaolinite and kaolinite/quartz mixtures.

The XRD patterns of the mechanically ground kaolinite show the rapid changes in the kaolinite structure during the grinding process (Figure 1). The effect of grinding causes the diminution of the d(001) spacing and after some 10 hours of grinding almost no intensity remains in this peak. The additional peak at around 9° (2θ) is due to illite impurity. The significance of the loss of intensity of the d(001) peak means the stacking between the layers is disrupted and lost. It is suggested that mechanochemical treatment has broken the hydrogen bonding between adjacent kaolinite layers. Thus, the kaolinite has been completely delaminated through the mechanical grinding process. After 10 hours of grinding no XRD pattern of the kaolinite is present. This means that the long-range ordering in the layers is disturbed so that there is no regular pattern of atoms, which can cause the diffraction. The XRD pattern only shows a broad peak centred on around 25° (2θ). The significance of this pattern remains with the production of a poorly diffracting material through grinding. This material has had its long-range structure destroyed. The modified kaolinite is it is poorly diffracting but still retains the 7Å peak. The decrease in basal spacings is results from the increased packing disorders generating a loss of periodicity. The rapid structural degradation of kaolinite is connected with an increase of the mean lattice distortion and is not the consequence of the reduction of the particle size. Figure 2 displays the variation of the halfwidth of the d(001) peak (the coherent scattering thickness) as a function of grinding time. The width of the peak is related to the crystallite size according to the Scherrer equation:

$$L = \frac{\lambda K}{\beta \cos \theta}$$

L is the mean crystallite dimension in Ångstroms along a line normal to the reflecting plane, K is a constant close to unity, and β is the peak broadening expressed in radians of 2θ. Figure 2 indicates that the structural degradation of kaolinite increases to 10 hours. The effect of the mechanochemical grinding is to cause the delamination of the kaolinite. The d(001) peak almost disappears after 10 hours. The mechanochemical treatment changes the morphology of the kaolinite particles from hexagonal-platy to small spherical particles which can be observed by scanning electron microscopy. SEM shows only an agglomeration of small spherical particles with no surface morphology.

3.2 Thermal Analysis

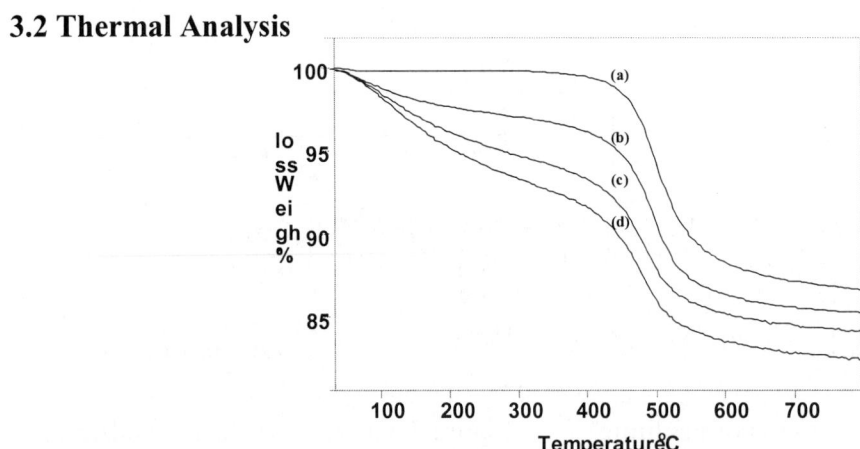

Figure 3a

Figure 3a, 3b shows the thermogravimetric and differential thermogravimetric analyses of the ground kaolinite. Pure kaolinite shows a steep TGA curve centred upon 475°C with a weight loss approaching 22%. The TG curve for the pure kaolinite shows a single weight loss at around 450 ^{0}C. Curves b, c and d show a two step weight loss. With increased grinding time, the curves shift to lower temperatures. Further it appears that the weight loss in the first step (>100 and < 400 ^{0}C) increases as the weight loss in the second step decreases (>400 ^{0}C). Such observations are confirmed by the differential thermogravimetric curves. Two weight losses are observed. The first weight loss is attributed to the liberation of water formed as a result of mechanochemical dehydroxylation of the kaolinite, while in second stage (above 450 °C) water is lost in the thermal dehydroxylation process. This dehydroxylation temperature decreases with grinding time. The area of this peak decreases concomitantly as the area of the first weight loss peak centred at approximately 150 ^{0}C increases. This weight loss profile ascribed to a water weight loss is rather complex, because of different types of water present in the mechanochemically-synthesized material.

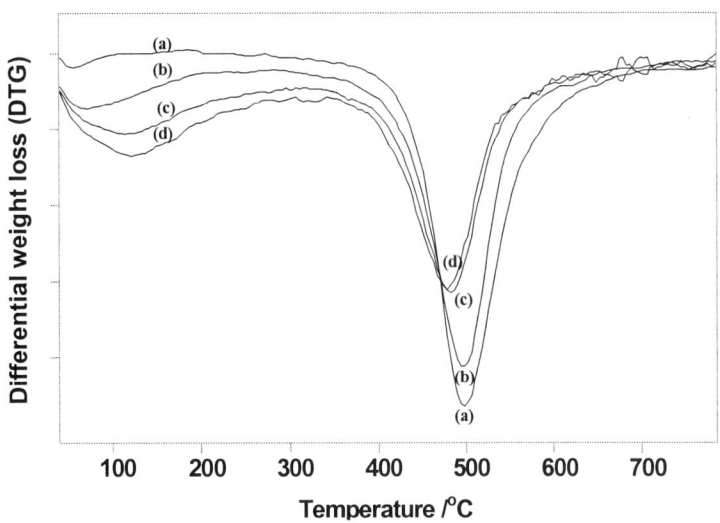

Figure 3b Thermogravimetric (upper figure) and differential thermogravimetric analysis (lower figure) of kaolinite ground for 1, 3, 5, 6 hours

3.3 Drift spectroscopy

The destruction of the crystalline structure of the ground kaolinite as determined by DRIFT spectroscopy is illustrated in Figures 4 and 5. Figure 4 displays the decrease in intensity of the hydroxyl stretching vibrations as a function of grinding time and the concomitant increase in intensity of OH stretching vibrations, attributable to water. The bands at 3695 (ν_1) and 3685 (ν_4) cm^{-1} are attributed to the longitudinal and transverse optic vibrations. This latter band is intense in Raman spectra of low defect kaolinites but is of low intensity in the infrared spectra and is only determined as a component in the overall band profile. The bands observed at 3668 (ν_2) and 3652 (ν_3) cm^{-1} also show a decrease in band position with the length of grinding, and after 10 hours of mechanochemical treatment no intensity is observed in these bands. These bands result from the out-of phase vibrations of the inner surface hydroxyls corresponding to the in-phase vibrations observed at 3695 and 3685 cm^{-1}. This means that the inner surface hydroxyls are no longer behaving in a cooperative vibrational pattern. The mechanochemical treatment causes significant changes in the surface structure at the molecular level. The position of the band attributed to the stretching vibration of the inner hydroxyl at 3619 cm^{-1} (ν_5) is not affected by the grinding process, even though the intensity decreases. An increase in bandwidth is noted. This may result from the intense localised heating of the kaolinite surfaces during mechanochemical treatment. Alternatively the effect

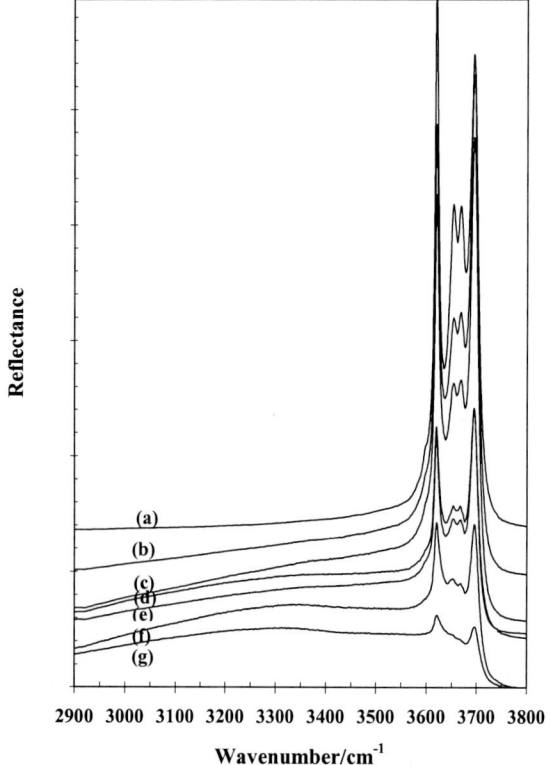

Figure 4 DRIFT spectra of the hydroxyl stretching region of mechanochemically activated kaolinite after grinding for (a) 1 hour (b) 2 (c) 3 (d) 4 (e) 6 (f) 10 (g) 20 hrs.

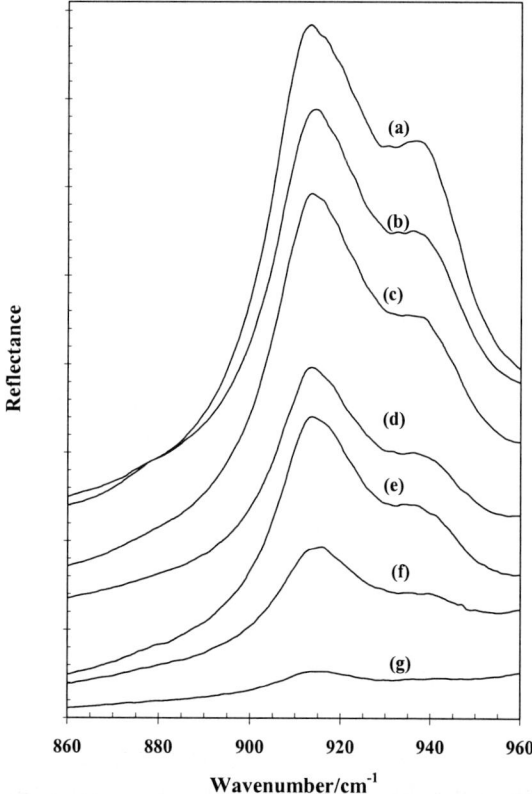

Figure 5 DRIFT spectra of the hydroxyl deformation region of mechanochemically activated kaolinite after grinding for (a) 1 hour (b) 2 (c) 3 (d) 4 (e) 6 (f) 10 (g) 20 hrs.

could result from a distribution of local environments generated by mechanical effects. Figure 5 shows the hydroxyl deformation modes observed at 937 and 914 cm^{-1} attributed to the inner surface and inner hydroxyls, respectively. In harmony with the decrease in intensity of the hydroxyl stretching vibrations, is the decrease in intensity of the kaolinite hydroxyl deformation vibrations with.

Figure 6 DRIFT Spectra of the SiO stretching region of mechanochemically activated kaolinites for (a) 1 hour (b) 2 (c) 3 (d) 4 (e) 6 (f) 10 (g) 20 hrs.

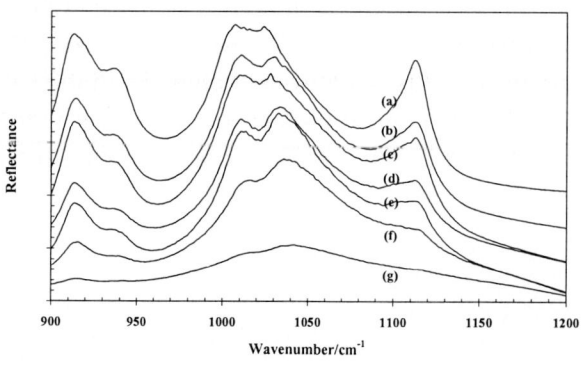

grinding time. There is an apparent increase in the intensity of the 914 cm^{-1} band but this results from the 937 cm^{-1} band intensity reaching zero. This result is significant as it means that the grinding process results in the loss of the inner surface hydroxyls before the inner hydroxyls

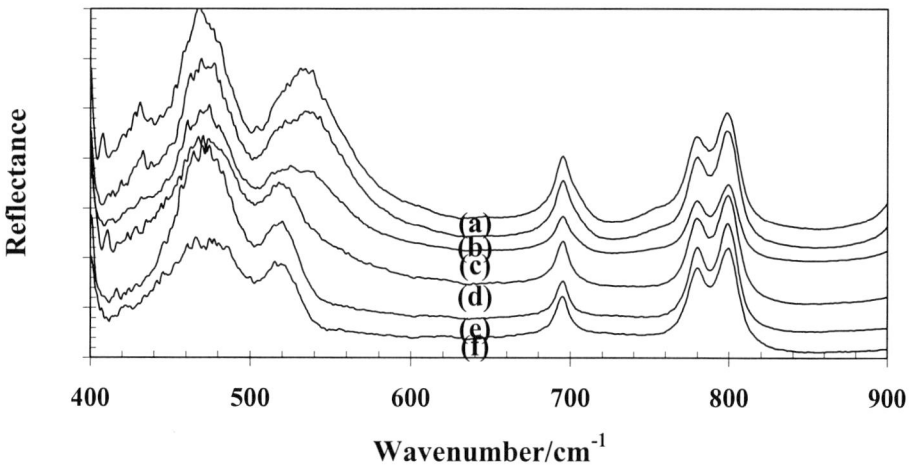

Figure 7 DRIFT Spectra of the low wavenumber region of mechanochemically activated kaolinites for (a) 1 hour (b) 2 (c) 3 (d) 4 (e) 6 (f) 10 (g) 20 hrs.

Two regions are identified for silicon-oxygen vibrations (a) the 980 to 1200 cm^{-1} region ascribed to SiO stretching vibrations and (b) the low wavenumber vibrations between 400 and 850 cm^{-1} ascribed to lattice vibrations. Figure 6 displays the SiO stretching region for the kaolinite, the quartz and the products of the mechanochemical treatment. The bands at 1056 and 1034 cm^{-1} are attributed to the SiO stretching vibrations of the silicate layer of the kaolinite. The bands observed at 1196, 1159 and 1103 cm^{-1} are associated with the SiO stretching vibrations of quartz. The relative intensities remain constant upon grinding. Figure 6 also shows the increase in intensity of an additional band at 1113 cm^{-1}, which appears upon the mechanochemical treatment. This band must be associated with the new surface phase produced upon grinding. The bands observed at 799, 780 and 696 cm^{-1} are vibrations associated with the sheet-lattice structure of the kaolinite (Figure 7). The major changes in the low wavenumber region are associated with the bending vibrations of the OSiO units. Thus, the band at 431 cm^{-1} diminishes in intensity with grinding time as does the bands at 537 cm^{-1}. Significant decrease in intensity occurs in this band. In fact after 2 hours of grinding no intensity remains in this band. This band is assigned to the OSiO bending vibration of the hexagonal ring structure of the silicate layer. Concomitantly a significant increase in relative intensity occurs in the 520 cm^{-1} band. This band is assigned to the OSiO vibrations of a tetrahedral Si. Thus, this suggests that some opening of the ditrigonal units of the silicate layer occurs.

4. CONCLUSIONS

Kaolinite was ground for periods of time up to 10 hours and a poorly diffracting silicate phase was produced. The kaolinite surface hydroxyls were lost after 2 hours of grinding and were replaced with water both coordinated and adsorbed on the surface of the aluminium containing layer. Upon mechanochemical treatment of the kaolinite with quartz, significant structural alteration occurred rapidly to form a new material with a significantly modified kaolinite surface of reduced crystal size with a somewhat higher surface area.

Of the techniques used to characterise the new mechanochemically synthesized material, only infrared spectroscopy and thermal analysis proved worthwhile. Thermal analysis appears to show the loss of hydroxyls from the kaolinite surface and the replacement with water. Infrared spectroscopy shows that the hydroxyl stretching vibration intensity is lost after two hours of grinding. This suggests the optimal time for the production of the delaminated kaolinite. The appearance of water bands upon grinding supports the concept of the replacement of the kaolinite hydroxyl units with water. Such observations are in harmony with the conclusions drawn from thermal analysis. The study of the hydroxyl deformation mode indicates the inner surface hydroxyls are lost before the inner hydroxyls. The observation of new SiO stretching and bending vibrations at 1113 and 520 cm^{-1} suggests that the new material has a different molecular structure from that of kaolinite. The concomitant decrease in the SiO stretching and bending modes of kaolinite support the concept of the synthesis of a new material which has a very different surface structure from that of the untreated kaolinite.

ACKNOWLEDGMENTS

Financial support from the Hungarian Scientific Research Fund under grant OTKA T25171 and F023531 are also acknowledged. Mr. Tamás Szilágyi is thanked for carrying out the thermoanalytical measurements. The financial and infra-structural support of the Queensland University of Technology, Centre for Instrumental and Developmental Chemistry is gratefully acknowledged.

REFERENCES

1. R. Schrader, Silikattechnik, 21 (1970) 196.
2. J. G. Miller and T. D.Oulton, Clays Clay Miner., 18 (1970) 313.
3. J. Hlavay, K. Jónás, S. Elek and J. Inczédy, Clays Clay Miner., 25, 451 (1977).
4. E. F. Aglietti, J. M. Porto Lopez and E. Pereira, Int. J. Miner. Proc., 16 (1986) 125.
5. A. Z. Juhász, and L. Opoczky, Mechanical Activation of Minerals by Grinding: Pulverising and Morphology of Particles. Academic Press, Budapest, 1990.
6. É. Kristóf, A. Z. Juhász, and I. Vassányi, Clays Clay Miner., 41 (1993) 608.
7. F. Gonzalez Garcia, M. T. Ruiz Abrio, and M.Gonzalez Rodriguez, Clay Miner., 26 (1991) 549.
8. R. L. Frost, É. Makó, J. Kristóf, E. Horváth and J.T. Kloprogge ,J. Colloid Interface Sci. (in press).
9. R. L. Frost and A. M. Vassallo, Clays Clay Miner., 44 (1996) 635.
10. R. L. Frost Clay Miner., 32 (1997) 73.
11. G. N. Paroz and R. L. Frost, The Analyst, 123 (1998) 2813.
12. R.L. Frost, and J.T. Kloprogge, J. Raman Spec., 31 (2000) 415.
13. R.L. Frost and J.T. Kloprogge, Spectrochimica Acta, 57 (2001) 163.

The role of hydrophobic solids in the separation and upgrading of bitumen from Athabasca oilsands

L. S. Kotlyar, B. D. Sparks, J. Woods, J. Kung and K. H. Chung[a]

National Research Council of Canada, Montreal Road Campus,
Ottawa, ON, Canada, K1A 0R6.

[a] Syncrude Canada Ltd. Edmonton Research Centre 9421 – 17th
Avenue, Edmonton, AB, Canada, T6N 1H4.

A reduction in the supply of Canadian light crudes in recent years is forcing refineries to deal with bitumen from oilsands as an inexpensive substitute. In water based bitumen separation processes, the transfer of mineral particles from the oilsands into an aqueous phase is the critical step. When the mineral solids are both hydrophobic and finely divided they are not easily removed and may remain with the bitumen during the upgrading process. During refining, some solids may be carried over to the downstream units where they are a possible source for the solids found in the coker gas oil fraction. In plant upsets these solids may overload filters and contaminate catalyst in hydro-treating units. In this work we separate and characterize solids from different oil sands and process streams in order to determine wheather the solids encountered in the gas oil fraction originate from bitumen or are an artifact of the upgrading process itself. To this end samples from different process streams were subjected to bulk and surface analysis and compared to fractions separated from oilsands in earlier work.

1. INTRODUCTION

The Canadian oilsands deposits in northern Alberta occur in layers up to 100 m thick, spread over an area totaling more than 40,000 km^2. These reserves contain about 1.3 trillion barrels of crude oil equivalent. The largest of the four major formations, containing nearly three quarters of Alberta's bitumen reserves, is found in the Athabasca region. About one tenth of the oil sands in this deposit lies within 50 m of the surface and is economically recoverable by conventional surface mining techniques [1, 2]. The Hot Water Extraction process (HWEP) is the commercial process used to recover bitumen from mined oilsands. Oilsands are slurried with slightly alkaline hot water to release bitumen from the largely sand matrix. The resulting bitumen froth is recovered by a combination of gravity separation and air flotation. The critical step in this process is the transfer of mineral particles from the oilsands matrix into the aqueous phase. Typically, the bulk of the solids in oilsands are 'inert', i.e., they participate in the process only through mechanical entrainment. However, the behavior of an 'active' component of the mineral fraction, comprising about 2-w/w% of the total solids, affects both the separation of bitumen and its quality. These solids encompass a mineralogically diverse group of particles. The common factor in each case is the presence of significant amounts of toluene insoluble, organic matter (TIOM), hence the generic description 'organic rich solids' or ORS [3-6]. Two types of ORS have received particular attention. The first is a coarser fraction, up to 100 µm in diameter, typically present as aggregates of smaller particles bound together by humic matter and precipitated minerals. The second is comprised of very thin, ultra-fine clay particles with a major dimension of <0.3µm. This fraction (ORS-BS) remains with the bitumen and may be carried over to the upgrading process. Because of potential process problems associated with organic rich solids in oilsands, this present work evaluates these components from a number of diverse oilsands samples and discusses their role in determining bitumen recovery and quality.

2. EXPERIMENTAL

Syncrude Canada Ltd supplied three marine and two estuarine oilsands samples from their Athabasca mine; they also provided a sample of middle range distillate, coker gas oil, produced during the fluid coking of bitumen. Bitumen separation was carried out in a Batch Extraction Unit (BEU) based on a standard design supplied by Syncrude. Recovered bitumen samples were fractionated into asphaltenes and maltenes by the drop-wise addition of n-pentane to bitumen solutions in toluene [7]. Organic rich solids co-precipitate with the asphaltenes because of chemically similar external surface characteristics. These solids (ORS-BS) were removed from the asphaltenes fractions by centrifugation of 5-w/w% toluene solutions of the precipitates for 1 h at 366,000 gravities. The ORS-AGG fraction is separated directly from oilsands by a cold water agitation test described in detail elsewhere [6]. In brief, samples of oilsands (~10 g) are mixed for 10 min on a high intensity Spex shaker with 0.1w/w% sodium pyrophosphate solution (~60 g) and toluene (~20 g). After settling under gravity, for about 15 min, the top layer, comprising toluene, ORS-AGG and dissolved bitumen is skimmed off, or removed by a pipette. The rest, Hydrophilic Bulk Solids (HBS), comprising water wet sand and clay, remains as a sediment. This treatment is repeated with successive aliquots of fresh toluene until no more ORS-AGG is separated. The ORS-AGG fraction is washed with fresh aliquots of toluene until all bitumen is removed. The samples are then dried to remove residual toluene. Samples of this fraction from each oil sands ore were treated with hydrogen peroxide to oxidize the TIOM and release the primary particles [8]. Centrifugation of a 30-w/w% toluene solution of the coker gas oil, for 1 hour at 366,000 gravities, removed 200 ppm of solids. This fraction was designated ORS-CGO. The sample was washed repeatedly with fresh toluene to remove soluble organics. A LECO CHNS-932 analyser determined total carbon (organic and inorganic), hydrogen, nitrogen and sulfur. Carbonate carbon was determined by coulometric titration after acid digestion, using a Carbon Dioxide Coulometer, Model 5010. Induced coupled plasma (ICP) atomic adsorption provided metals contents. Surface spectra, by X-Ray Photo-Electron Spectroscopy (XPS) were recorded using a PHI 5500 Instrument (PHI Electronics, Eden Prairie, Minnesota, USA) with an Al Kα source of X-rays. High-resolution spectra were obtained at a pass energy of 29.6 eV; survey spectra were recorded at a pass energy of 156 eV. Transmission Electron Microscopy (TEM) provided information on particle morphology and mineralogy. The Philips CM20 200 kV electron microscope is equipped with an Oxford Instruments LINK EDX (energy dispersive X-ray spectroscopy) detector and a Charge-Coupled Device (CCD) camera. The CCD camera produces electronically enhanced image contrast, allowing extremely thin objects to be viewed at high resolution. The EDX detector allows quantitation of elements heavier than nitrogen. Samples were deposited on a 50Å thick carbon film on a copper grid, then dried at 35C° for an hour and degassed in the instrument. XRD cannot be used because the particles are too thin and not properly orinted.

3. RESULTS AND DISCUSSION

Table 1 shows that the estuarine sample E-2 and the marine sample M-2 have the highest and lowest bitumen contents respectively. The amounts of asphaltene and ORS-BS present in the separated bitumens were quite similar for all five oilsands. However, the marine oilsands samples exhibited a significantly greater contribution from the ORS-AGG fraction than the estuarine samples. Figure 1 is a SEM micrograph comparing an ORS-AGG sample with similarly sized sand grains from an HBS fraction. The primary particle size distributions for this fraction, from each oilsands, are shown on Figure 2. Size distributions are similar in each case.

Table 1
Oil sands properties

Depositional Environment	Sample	Component			
		(w/w% of oil sands)		(w/w% of bitumen)	
		Bitumen	ORS-AGG	Asphaltenes	ORS-BS
Estuarine	E-1	11.2	0.6	16.7	0.9
	E-2	12.7	0.5	17.1	1.0
Marine	M-1	10.9	1.1	17.6	0.8
	M-2	7.4	1.0	17.0	0.9
	M-3	8.9	1.8	16.6	1.0

TEM is required to resolve the ultra-fine particles in the ORS-BS fractions; the micrograph presented on Figure 3 shows thin, 10 nm at the most, ultra-fine clay crystallites with a lateral extension of less than 200 nm.

The bulk elemental analyses of the various solids fractions are listed in Tables 2 and 3. Regardless of their source, the ORS type solids have a similar composition. Typically, the fraction is associated with large amounts of organic carbon and hetero-atoms in combination with heavy metals, particularly Fe and Ti. Comparatively, the hydrophillic bulk solids (HBS), remaining in the sediment after removal of the ORS fractions, are low in carbon, hetero-atoms, Fe and Ti. The surface characteristics of oil sands solids are the major factor governing their distribution during processing. X-ray photoelectron spectroscopy (XPS) provides an excellent tool for probing the first 7 nm of the surface layer. Table 4 presents a summary of the major elements detected in the surface layers of ORS-AGG solids from different oilsands and compares the results with HBS material. The most important factor to be noted is that organic carbon is strongly concentrated at the surface of ORS-AGG particles, with typical values falling between 38 to 45 atomic percent. By comparison, surface carbon is less than 10 atomic percent for the HBS particles.

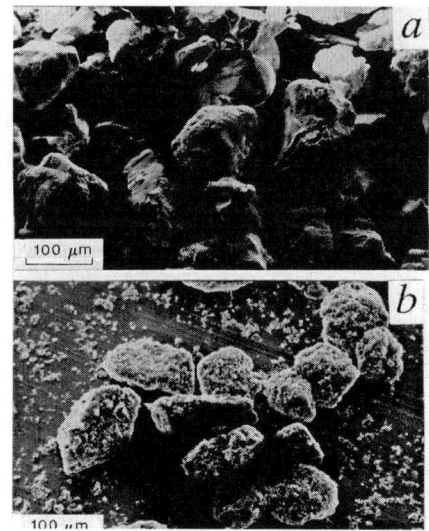

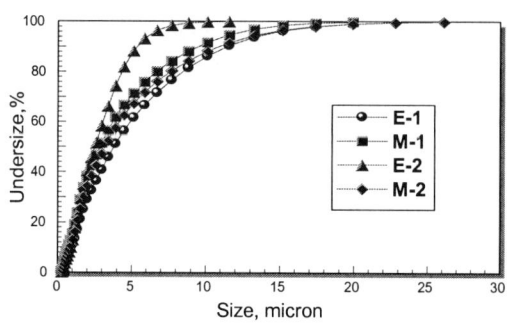

Figure 1. SEM of ORS-AGG (b) and sand grains from HBS fraction (a).

Figure 2. Size distribution of primary particles from ORS-AGG samples.

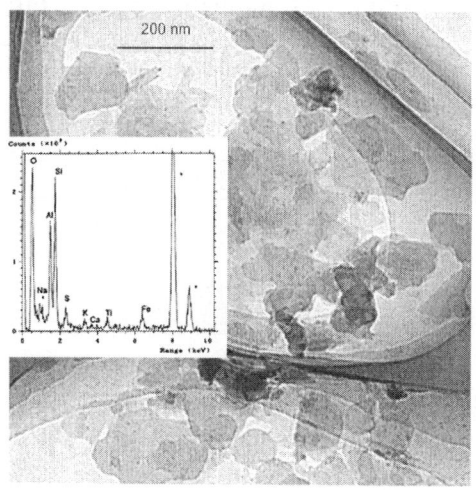

Figure 3: TEM and EDX of ORS-BS solids (*Cu from grid).

These values are close to the amount of carbon contamination which is usually difficult to avoid with this techniques. The fact that inorganic elements are detected by XPS in each ORS-AGG case is an indication that the surface layer of organic material is either thin or patchy.

Significantly the HBS solids have much higher Al, Si and oxygen contents than the comparable ORS-AGG fractions. In earlier work [5], similar ORS fractions were subjected to time-of–flight secondary ion mass spectroscopy, which probes the first 1 nm of the surface layer. Those results showed that Al, Si, Fe and Ti dominate parts of the very top surface layers, confirming that the organic coverage is definitely patchy rather than continuous. This determination provided a molecular basis for the observed biwettable characteristics of the ORS fractions [3, 4].

Table 2.
Bulk analysis of ORS-AGG and HBS solids separated from oilsands

Solids Type	Sample	Elemental Composition (w/w%)					
		C	H	N	S	Fe	Ti
ORS-AGG	E-1	29.5	2.5	0.4	1.8	5.0	1.0
	E-2	30.9	2.6	0.2	3.6	3.9	1.5
	M-1	32.1	3.0	0.5	2.6	4.2	0.9
	M-2	28.3	2.8	0.4	3.0	6.8	2.5
	M-3	33.0	3.1	0.6	2.4	5.5	1.8
HBS	E-1	0.3	0.1	0	0	0.2	0.1
	M-1	0.5	0.2	0	0	0.3	0.2

Table 3.
Bulk analysis of ORS-BS and ORS-CGO solids separated from asphaltenes and coker gas oil.

Solids Type	Sample	Elemental Composition (w/w%)					
		C	H	N	S	Fe	Ti
ORS-BS	E-1	27.7	2.4	0.4	1.5	3.3	1.0
	E-2	30.8	2.4	0.2	1.9	2.4	1.3
	M-1	26.6	3.1	0.3	2.0	2.2	0.9
	M-2	28.0	2.8	0.4	3.0	5.2	0.9
	M-3	28.3	3.1	0.5	2.2	4.7	1.9
ORS-CGO	Coker gas oil	38.6	3.9	0.6	2.0	1.5	0.5

Table 4.
Surface analysis of ORS-AGG and HBS solids separated from oilsands.

Solids Type	Sample	Elemental Composition (atomic %)					
		C	O	Al	Si	Fe	S
ORS-AGG	E-1	39.6	43.9	5.4	7.9	1.5	0.6
	E-2	42.8	41.6	5.1	7.8	0.9	0.4
	M-1	45.0	38.3	4.9	7.9	0.2	1.6
	M-2	41.8	36.2	4.8	7.8	0.6	1.4
	M-3	37.7	44.1	6.7	10.6	0.6	0.3
HBS	E-1	7.0	60.0	16.9	14.0	0	0
	E-2	9.8	60.0	17.3	12.9	0	0

Table 5 compares the surface analysis of ORS-BS solids, i.e., that fraction remaining with bitumen during upgrading, with the solids separated from coker gas oil (ORS-CGO). The latter fraction shows higher concentrations of surface carbon but lower values for the elements associated with clay minerals. This is an indication that the ORS-CGO may be a more hydrophobic fraction than the ORS-BS material. A comparison between the EDX spectra of ORS-BS and ORS-CGO, Figures 4 a,b, show the same elements present in both samples but in different concentrations. Alluminosilicate clays are more predominant in the ORS-BS sample while zinc and chlorine dominate the ORS-CGO material. However, the data is not quantitative and further work will be required to yield more definitive results.

4. PROCESS IMPLICATIONS

In bitumen separation processes the organic matter associated with various ORS fractions facilitates their interaction with bitumen. As a result of these interactions it is not surprising that the transfer of ORS-AGG solids from the ore matrix into the aqueous, slurry phase is extremely difficult. Smaller (<30 µm), low-density ORS-AGG particles may be preferentially collected with the bitumen froth during flotation. After diluting the froth with naphtha, centrifugation removes most of the particulates larger than about 1 µm. These solids are

Table 5.
Surface analysis of ORS-BS and ORS-CGO solids separated from aphaltenes and coker gas oil.

Solids Type	Sample	Elemental Composition (atomic %)					
		C	O	Al	Si	Fe	S
ORS-BS	E-1	54.7	28.0	6.0	7.4	0.4	0.9
	E-2	57.0	30.0	4.7	3.4	0.2	0.6
	M-1	60.5	29.0	2.8	5.4	0.7	0.8
	M-2	61.1	29.0	2.4	4.4	0.2	0.3
	M-3	55.9	31.0	3.7	6.3	0.9	1.1
ORS-CGO	Coker gas oil	69.9	25.0	2.0	2.9	n.d	n.d

rejected to the centrifuge tailings, together with any associated bitumen. Coarser, higher density ORS-AGG particles may be too heavy to float in the froth and can remain with the flotation middlings stream or settle into the primary tailings with any attached bitumen. In all cases the result is a reduction in bitumen recovery proportional to the amount of ORS-AGG solids originally present in the oilsands.

The diluted bitumen from froth clean up still contains ultra-fine clay and heavy mineral fractions, i.e., the ORS-BS material. The nano-sized ORS-BS fraction may be responsible for

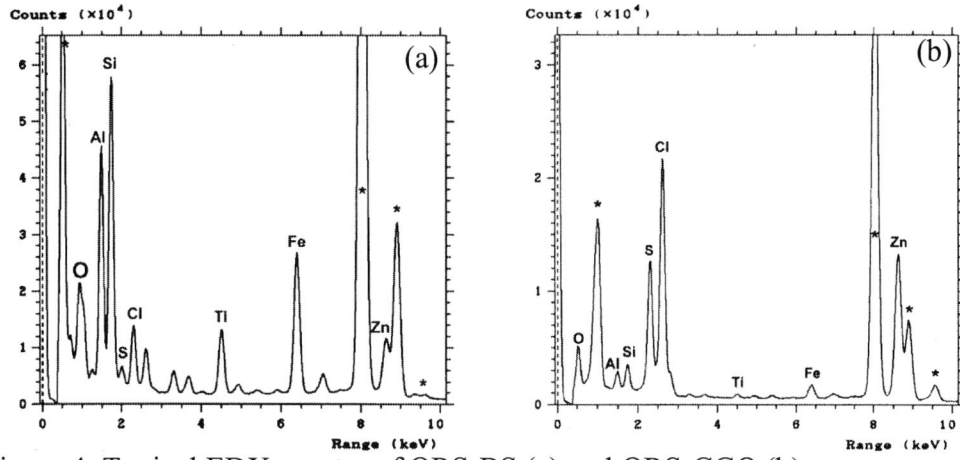

Figure 4: Typical EDX spectra of ORS-BS (a) and ORS-CGO (b).

the retention of salty connate water by bitumen during its separation from oilsands. Because of their biwettable character ORS-BS material can form clay-water clusters, or stabilize emulsions, by developing a rigid barrier between water droplets and the continuous oil phase. The extremely small particle size allows stabilisation of miniscule emulsion droplets.

The ORS-BS materials remaining in the bitumen are transported to the bitumen coking process and may become entrained with volatile overheads and carried-over to cause problems in down-stream units [9,10]. Although the ORS-BS and ORS-CGO are similar they are not the same. The latter fraction appears to contain less clay and more of a stoichemetric compound of zinc, oxygen and chlorine. At the moment it cannot be determined whether the zinc compound is present in the original oilsands ore or whether it is introduced during processing.

NOMENCLATURE

EDX	Energy Dispersive X-ray Diffraction
HBS	Hydrophilic Bulk Solids
HWEP	Hot water extraction process
ORS	Organic Rich Solids
ORS-AGG	Organic Rich Solids separated from oilsands
ORS-BS	Organic Rich Solids separated asphaltenes
ORS-CGO	Organic rich solids from coker gas oil
SEM	Scanning Electron Microscopy
TEM	Transmission Electron Microscopy
TIOM	Toluene Insoluble Organic Matter
XPS	X-ray Photo Electron Spectroscopy

ACKNOWLEDGEMENTS

The authors express their appreciation to Y. LePage for TEM and G. Pleizier for XPS analysis.

REFERENCES

1. Hocking, M.B., "The chemistry of oil recovery from bituminous sands"; J. Chem. Educ., 54 (1977) 725.
2. Berkowitz, N. and Speight, J.G., "The oilsands of Alberta"; Fuel, 54 (1975) 138.
3. Kotlyar, L.S., Sparks, B.D., Woods, J., Raymond, S., Le Page, Y. and Shelfantook, W.; Petroleum Science and Technology, 16 (1998) 1.
4. Kotlyar, L.S., Sparks, B.D., Woods, J.R., and Chung, K.H.; Energy & fuels, 13 (1999) 346.
5. Bensebaa, F., Kotlyar, L.S., Sparks, B.D. and Chung, K.H.; J. Chem. Eng., 78 (2000) 610.
6. Kotlyar, L.S.; Sparks, B.D.; Chung, K.H.; Rev. Process Chemistry and Eng. 1 (1998) 81.
7. Mitchell, D.L. and Speight, J.G.; Fuel, 52 (1973) 149.
8. Van Langeveld, A.D., Van der Gaast, S.J. and Eisma, D.; Clays and Clay Minerals, 26 (1978) 361.
9. Chung, K., Xu, C., Hu, Y. and Wang. R.; Oil and Gas Journal, 1997.
10. Chung, K.H., Xu, C., Gray, M., Zhao, Y., Kotlyar, L.S. and Sparks, B.D.; Reviews in Process Chemistry and Engineering, 1 (1998) 41.

The physico-chemical and rheological characterizations of two industrial bentonites.

C. Malfoy[a], A. Besq[b], A. Pantet[a], P. Monnet[b].

[a]HydrASA-ESIP, UMR 6532, 40 avenue du recteur Pineau, 86022 Poitiers France.
[b]LEA, UMR 6609, boulevard Marie et Pierre Curie, 86962 Futuroscope cedex France.
Groupe Géomécanique et Génie Civil de Poitiers.

This paper describes the contribution of rheology to obtaining a better knowledge of structural network established by smectites in water. Physico-chemical and rheological properties of two industrial bentonitic fluids (Volclay and Wyoming) are explored with two criteria: concentration (20g/l to 60g/l), and granulometric fraction (total powder, < 2µm).

Suspensions made with distilled water generally exhibit the properties of non-newtonian fluid which are measured by a high-resolution rheometer (Stresstech). Thixotropy, shear shinning, and yield stress are the main characteristics of these viscoelastic fluids. Flow tests are undertaken with very accurate procedures to guaranty good reproducibility.

As a preliminary, several analytical techniques were used to determine the mineralogical composition including X ray diffraction, infrared spectrometry, and chemical analysis.

Rheological data show the creation of a macroscopic network which is very influenced by each of the previous criteria. Results allow the determination of rheological parameters of a Herschel Bulkley law which have permitted the prediction of steady flow in circular pipes for Wyoming bentonite (Besq et al., 2000).

1 INTRODUCTION

Industrial bentonite (i.e. mainly activated montmorillonite clay type) mixed with water are commonly used in civil engineering technologies such as slurry shield tunneling or horizontal directional drilling. Bentonite mud is useful to maintain cutting face stability and convey cuttings (pipe flow). A better knowledge of mechanical behavior of theses systems is necessary in order to improve drilling fluid performances, but it also provides useful information about clay structural networks.

In contact with water, clays are transformed (hydration, cationic exchange, adsorption), and several physical states are observed depending on the percentage of available water, which controls the state as a solid, gel or liquid. In every case, the mechanical behavior relies on the structural organization. At rest, a three dimensional structure is developed. Under flow, this structure is broken with specific kinetics.

Further, it is widely reported that 2:1 phyllosilicates and clay water systems exhibit rheological properties that depends on interlayer cation nature, layer charge value, concentration, pH... Thus, studies of such mixtures are complex. Given the diverse industrial treatments of bentonite : a huge range in rheological behaviors can be observed.

We propose an original approach which consist in studying physico-chemical and rheological properties of two industrial bentonites (Volclay and Wyoming) according to mineral powder proportion (20 to 60 g/l i.e. 2 to 6%) and powder fraction (raw powder, <2µ) in distilled water. Rheological investigations are realized with a high-resolution rheometer for all suspensions

2 MATERIALS AND METHODS

In this paragraph, mineralogical techniques are briefly discribed whereas rheological methods are widely described in detail.

In this study, no mineralogical pre-treatment were carried out on Volclay, a poorly activated bentonite (addition of Na_2CO_3). The Wyoming powder was Na saturated with 1M NaCl solution (washed 3 times and rinsed with alcohol 5 times, and tested with $AgNO_3$), in order to be able to compare with Volclay in which Na is dominant.

2.1 Powder identification procedure

The <2µ particles were extracted by centrifugation (Jouan GR 422). They were then softly dried at 60°C (to keep swelling properties) and pulverized with pestle and mortar.

Oxides composition was established by chemical analysis and minerals were analyzed on random powder and oriented film of clays (natural and ethylene glycol solvated) on a Philips X-Ray diffractometer using $CuK\alpha$ radiation. Infrared spectrometry was used in order to determine exchangeable cationic capacity and layer charge, according to Petit et al methods (1998). Natural, NH_4, Li-NH_4 saturated powders (in any granulometric fraction) were prepared with continuously air-dried purged Nicolet 510 FTIR spectrometer at 4 cm^{-1} resolution. IR spectra were measured using KBR pellets (1mg of sample mixing with 150mg KBR), in the 200-6400 cm^{-1} spectral range

Total and external surfaces were measured using a B.E.T (ASAP 2010 system after degassing for 6 hours at 60°C) and by methylene blue adsorption methods. The swelling index in static state was achieved by sprinkling 2 g of clay (<80µm pulverized and 52°C dried) in 100ml of distilled water and measuring the volume of expanded clay developed after 24 hours in water.

2.2 Suspensions characterization procedure

Suspensions were made by mixing clay powder in distilled water with an ultrasonic probe (Sonic Materials, Vibra Cell mod VC 500 500watts, 50/60 Hz, frequency 20 Hz), during 2min at power 7. It is important to note that level of energy must be controlled in order to ensure reproducibility and allow comparisons. For all experiments, suspensions were left at rest for 24 hrs. Their pH fluctuated between 9 and 10, which is far from IEP for Na smectites (pH $_{IEP}$= 7)(Benna et al., 1999). Fractional sedimentometry effect was observed by testing both the bulk sample and the <2µ powders, for both the Volclay and Wyoming samples. Raising of concentration from 2% to 6% was tested only on Volclay for each powder fraction.

A Rheologica-Stresstech HR control stress rheometer was used with coaxial cylinders. The outer cylinder, filled with suspension, is fixed whereas the inner one is shearing the fluid. Torque applied on cylinder and its rotating speed relies on the specific rheological properties of the material. Analysis software allows controlling torque and measure rotating speed. Shear stress (τ) and shear rate ($\dot{\gamma}$) are then calculated. The ratio $\tau/\dot{\gamma}$ defines the apparent viscosity of the suspension. It's well known that clay-water suspensions are yield stress and thixotropic materials (Mollet (1996), Pignon et al., (1996), Ramsay et al., (1993)). Specific rheometric procedures must be performed to obtain reliable and reproducible measurements. One of them consists in establishing a "quick" flow curve. First, experiments start with a shearing stage ($\dot{\gamma} = 500\,s^{-1}$ for 240 s), followed by a rest period (600 s), in order to define a structural reference state. This one is characterized by a yield stress value which depend on particle association kinetics at rest. Then, the material is subjected to increasing and decreasing shear stress gradients for 30 min. The difference between the increase and decrease values gives indications about thixotropic properties (i.e. structural evolution) of the fluid. In order to avoid water evaporation, a film of poorly viscous silicon oil is deposed on the upper free surface of the suspension (T=20°C). Many authors have reported that behavior of clay-water suspensions could be described with a Herschel-Bulkley law (Pignon et al.,1996;Coussot, 1997). It can also predict pressure drop for industrial steady flow in pipe (Besq et al., 2000). To account for structural evolution, we applied a modified Herschel-Bulkley model for rheological data processing. This choice was motivated by the pertinence of fitting and by the simplicity and efficacy of this model.

$\tau = \tau_0(\lambda) + k\dot{\gamma}^n$ is the model expression. (1)

τ_0 is the yield stress value depending on λ, k and n are specific values (n<1 for shear shinning fluids). In this equation λ is a normalized structural parameter, which characterize the fluid evolution according to time, history of the flow. λ=0 correspond to a fully broken structure, and λ=1 is associated with a structure strongly connected. A least squared fitting procedure has been used to determine these parameters of experimental flow curves.

Even if it is not submitted here, a complete rheological program was realized on these suspensions. Four complementary tests were performed to determine λ parameter.

A more absolute state of reference was obtained by breaking down the fluid at 500 s^{-1} during 3h. The fixed rate gradient was then associated at the beginning with a level of stress which decrease as far as the fluid break down. It was possible to calculate characteristics times of breaking down kinetics.

Then recovery was link to measure the structure evolution at rest with an oscillatory shear. As stress involve distortion, storage (G') and viscous (G") modulus were quantified.

When build up was achieved, creep recovery tests were realized with progressive growth stress in order to determine the yield stress τ (1). Finally, a complete flow test (named pseudo-established diagram) was achieved during 3 days in order to respect structural evolution between each stress.

3 RESULTS AND INTERPRETATIONS

3.1 Powder characterization

3.1.1 Chemical analysis.

The calculated chemical composition (raw powders) and cationic exchange capacities (for <2μ particles) are presented in table 1.

sample	Structural formula base 11O	charge IV	Charge VI	total charge	C.E.C meq/100g
Volclay	$(Si_{3.98}Al_{0.02})(Fe^{3+}_{0.17}Al_{1.53}Mg_{0.27}Ti_{0.01})O_{10}(OH)_2$ $(Ca_{0.08}Na_{0.16}K_{0.02})$	-0.02	-0.32	0.34	68
Wyoming	$(Si_{3.95}Al_{0.05})(Fe^{3+}_{0.17}Al_{1.54}Mg_{0.29}Ti_{0.01})O_{10}(OH)_2$ $(Ca_{0.08}Na_{0.12}K_{0.02})$	-0.05	-0.25	0.30	67.5

Table (1): Structural formula and cationic exchange capacity.

3.1.2 X Ray diffractometer

- Volclay.

Identified minerals in powder diagram are: montmorillonite, micas (biotite, muscovite), quartz (diminution as far as the particle size decrease), gypsum, feldspar, plagioclase. The 060 peack $d_{060} = 1.496$ Å indicate that it is a dioctahedral smectite.

- Wyoming

Wyoming contain dioctahedral montmorillonite ($d_{060} = 1.491$ Å), quartz, albite, muscovite, calcite.

All of non-clay minerals decrease in concentration with decreasing particle size.

3.1.3 Infrared spectroscopy.

Thanks to this method we can verify that there is no powder contamination by polymers: spectrum are not disturbed. NH_4 vibration band for calibration, areas of peacks before and after Li saturation, with respective percentages of tetrahedral and octahedral charge are reported on table 2.
This confirm that the smectites are indeed montmorillonites (main part of charge in octahedral site). In series variations in peak area between samples are negligible. One notes that relative proportion of tetrahedral and octahedral charge correlates well with the structural formulae.

Powder	fraction	normalization pick (cm^{-1})	δNH_4 NH_4	δNH_4 Li-NH_4	IV + variable Charge %	VI Charge %
Volclay	total	525	8.13	0.40	5	95
	<2μ	525	8.27	0.54	6.6	93.4
Wyoming	total	525	9.90	2.29	23.1	76.9
	<2μ	525	9.59	1.913	20	80

Table (2): Charge quantification by F.T.I.R spectrometry

3.1.4 Specific surfaces and swelling index.

Specific surface areas (m^2/g) are presented in table 3. All B.E.T experimental measurements were made in triplicates using the same heating and degassing procedures.

Powder	Volclay		Wyoming	
Fraction	total	≤ 2µ	total	≤ 2µ
BET surfaces	19.93	16.89	31.33	29.49
Value of blue / 100g	28.03	28.39	30.55	31.28
Active surface Sa (m2/g)	586.73	594.27	639.27	654.53
Swelling index	31	44	30	40

Table (3) : External (B.E.T) and total (methylene blue chemisorption) specific surfaces

Total and external surfaces values are different from theoretical data for pure smectites (800m^2/g, 60m^2/g). This is due to non-swelling minerals present in our powders. It needs to remark the logical augmentation of specific surface, when powder becomes finer. Swelling is more important for fine fraction. This has already been described by Kotlyar (1998), who concluded that aggregates made from small clay particles are more voluminous than those made for coarser particles.

Based on this mineralogical analysis one note that the Volclay and Wyoming bentonites have similar properties.

3.1.5 Rheological characterization

All data are presented in Figures 1, 2, 3, 4.

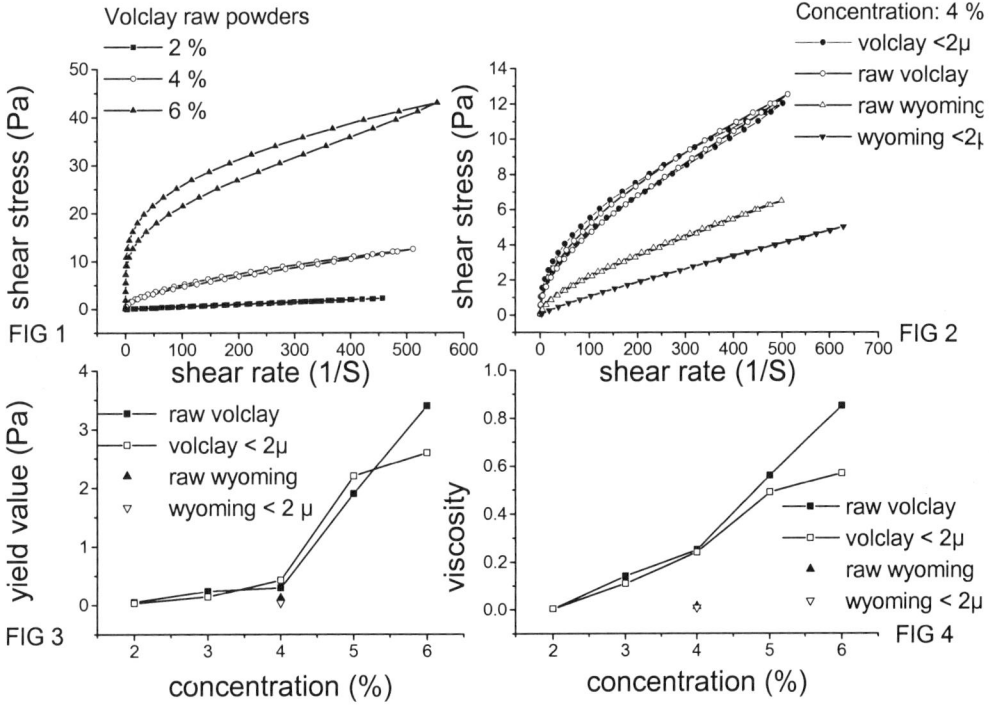

Comments on trends:

FIG 1. One notes that fluctuations in clay concentration generate a radical change in rheological parameters. Flow curves exhibit similar trends but yield stress, thixotropy and viscosity increase with concentration.

FIG 2. The flow curves for the 2 Na-montmorillonites are of the Herschel-Bulkley type. Nevertheless, intensity of interparticle interactions and thixotropy are weaker for Wyoming than for Volclay at same concentration. For Wyoming, this phenomenon is accentuated with a lower granulometry.

FIG 3 and 4. Yield stress and viscosity strongly increase with concentration. A slope modification appears for around 4%, suggesting it is a critical concentration. This concentration defines two domains associated with two different structural organizations. Raw suspensions and <2 µ samples seem to differentiate from each other above this value. Such properties have already been observed by Pignon et al (1997) for laponite (synthesis hectorite) suspensions thanks to fractal analysis.

4 DISCUSSION AND CONCLUSION.

Previous studies have determined that, in hyper diluted suspension case (e=3900, where e is the void index), dispersion of Na montmorillonites is extreme, layers are almost independent and form little tactoides, while Ca smectites make quasicrystal. In case of saturated clays (e=30 to 40), Touret (1988) defined 4 types of structures, thanks to SEM explorations:
- layers,
- particles made with layers in turbostratic association,
- aggregates of particles, and
- association of aggregates.

Four types of water are associated:
- inter-layer water,
- lenticular water internal of particles,
- inter-particles water, and
- inter-aggregates water.

Each time, it suppose that there is different types of bonding, inter-layer forces, inter-particles forces and inter-aggregates strength. Nature and intensity of those forces depend on elements which are in interaction.

In this study, suspensions made with Na montmorillonite have intermediate concentration (e = 45 to 70). All rheological tests show that all of them present an apparently continuous and constant structure (no fissure), during our experiments. This macrostructure can offer more or less resistance to flow. We therefore think that elementary structures build up in suspension, which explains this apparent stability. Rheological techniques allow a macroscopic approach to those structures. Resistance measured under flow conditions give the sum of bonding strain, without possibility to distinguish them. The program of rheological tests was carried out in order to determine parameters of Herschel Bulkley model which is particularly suit for circular pipe flow fitting. The mineralogical analysis confirmed that the Na Wyoming and Volclay samples are both dioctaedral sodium smectites with equivalent

C.E.C, charge, swelling index), but Na Wyoming doesn't contain significant quantities of non-clay minerals in the <2μ fraction. A similar pH was obtain for both samples.
The overall conclusions of flow curves are that:
- For one powder, there are differences which depend on concentration (Volclay 2% to 6%). In all cases, suspensions made with raw powders present higher yield stress values and are more thixotropic.
- Two suspensions with the same concentration, made from the powdered forms (Volclay and Wyoming), have different behaviors.
- There is a critical concentration (4%) value around which structural changes are obvious for Volclay bentonite.

Theses established facts lead us to the following conclusions:

At low concentration, Brownian and hydrodynamic effects dominate (Coussot, 1997), and are most dominant in finest fractions. Structure exists, in such suspensions but it is very fragile. Interactions between particles in a total clay suspension are very small and due to dilution, and the network is very slack. A unitary volume may be composed with flexible aggregates which form micro-cavities full of water or tense aggregates which enveloped some macro-cavities.

At higher concentration, there is an increase in the probability of encounter for small platelets of clay. The possibilities of collision, and of course aggregation are more important. Aggregates are less fragile and the structure is less alveolar, as shown in the diagram of Tessier (1984) and Touret (1988).

Our experiments show that there are two different types of structures below and above 4% for Volclay bentonite. This 4 % concentration seems to be the critical one as shown by slope variations in FIG 3 and 4. But one notices that it points to the fact that if structures seems to be equivalent for raw and < 2μ powders below 4 %, it look as if each granulometric fraction develop its own type of network above this value as the curves move away. This critical concentration must be associate with the evaluation of attractive and repulsive forces which induce an equilibrium and stability for each suspension. Many authors evoke a critical volume fraction, that depend on powder mineralogy and fraction. This poses the problem of secondary minerals effect.

XRD and chemical analysis have showed that secondary, non-smectite minerals represent 12% of total powder which is an important proportion in Volclay <2μm fraction. The strength of the raw powders is higher than that of the < 2μ fractions, and it become worse when concentration increase. For the <2μ fractions, the best resistance of Volclay 4% with regard to Wyoming at same concentration could probably be correlated with the bigger percentage of secondary minerals that are present in the fine fraction.

For this reasons we conclude that suspensions developed a very strong network in case of sufficient volume concentration and / or when they contain secondary minerals (especially in case of low concentration).

It seem necessary to take in account the structural parameter λ, which is why we carried out complementary tests such as creep-recovery and oscillation tests, which provide useful information about structural networks and the kinetics of break down under flow. It appears, thanks to this study, that a complete rheological program could give macroscopic information about smectites crystallochemistry in water. We broach investigations about variations of one smectite macroscopic behavior, in response to crystallochemistry disparities (by changing interlayer cation nature, charge value…). It appears that a rheological investigation, in conjunction with a mineralogical study could be very instructive for mineralogist, specialists in fluid mechanic, and industrials.

It is therefore essential to appreciate that the simple notion of concentration, is not sufficient to predict bentonite mud performances. The performance of Wyoming clay is not equal with Volclay at 4% solids. The bentonite granulometric fraction is also important as impurities often cause damage to pumps. A systematic study and the fitting of behaviors with Herschel Bulkley model could also be useful for predicting bentonite fluids behaviour during drilling.

REFERENCES

M. Benna, N. Kbir-Ariguib, A. Magnin, F. Bergaya. Effect of pH on rheological properties of purified sodium bentonite suspensions. Journal of Colloïd and Interface Science 218,p 442-455 (1999).

A.Besq, Ph. Monnet, A. Pantet. Flow situations of drilling muds. Effects of thixotropic property Actes du VIème congrès international flucome 2000 (8 / 2000).

P. Coussot. Mudflow ,Mudflow Rhéology and Dynamics. Balkema IAHR AIRH 1997 255p

L.S.Kotlyar, B.D Sparks, Y.LePage, J.R Woods. Effect of particle size on the flocculation behaviour of ultra-fine Clays in salt solutions. Clay minerals vol 33, 103-107 (1998).

F.Mollet. Contribution à l'étude des fluides thixotropes en écoulement. PHD thésis, University of Nancy I, (1996).

S. Petit, D.Righi, J. Madejová, A. Decarreau. Layer charge estimation of smectites using infrared spectroscopy. Clay minerals 33,579-591 (1998).

F. Pignon, A. Magnin, J.M Piau. Processus de désagregation dans les gels d'argile thixotropes sous écoulement de cisaillement. Les Cahiers de la Rhéologie. XV, n°12 p294-300 (1996).

F. Pignon, A. Magnin, J.M Piau, B. Cabane, P. Linder, O. Diat. Yield stress thixotropic clay suspension : Investigation of structure by light, neutron, and X-ray scattering. Physical Rewiew E volume 56 number 3. September p 3281-3289 (1997).

J.D.F Ramsay, P. Linder. Small-angle neutron scattering investigation of the structure of thixotropic dispersions of smectite clay colloïds , J.Chem.soc.Faraday Trans , 89(23), p 4207-4214 (1993).

D. Tessier. Etude expérimentale de l'organisation des matériaux argileux.hydratation, gonflement et structuration au cours de la dessication et de la réhumectation. INRA PHD theses Paris VII 361p (1984).

O. Touret. Structure des argiles hydratées thermodynamique de la déshydratation et de la compaction des smectites. PHD thesis of Strasbourg university geology department, january (1988).

Application of bentonite and organoclay in stabilization/solidification of tannery waste

C.A.Pinto[a]; S.M.Toffoli[b]; L.T.Hamassaki[c]; F.R.Valenzuela-Diaz[b] P.M. Büchler[d]

[a]Ph D student at Chemical Engineering Department, Polytechnic School, University of São Paulo, São Paulo, Brazil.
[b]Mettalurgical and Materials Engineering Department, Polytechnic School, University of São Paulo, São Paulo, Brazil.
[c]Technologic Research Institute of São Paulo State, São Paulo, Brazil.
[d]Chemical Engineering Department, Polytechnic School, University of São Paulo, São Paulo, Brazil.

Stabilization/solidification is applied to immobilize hazardous wastes for final disposal in industrial landfills. Hazardous wastes are usually toxic chemicals or heavy metals. The final solidified material is a cement matrix. Although cement is only effective for inorganic wastes, organic wastes are not well fixed in the cement matrix. Furthermore the organoclay can be used to adsorb the organic molecules. Chromium, both organic and inorganic, is found in the waste of the tannery industry. The leather treatment consists in the formation of a chemical complex of animal protein and chromium. This paper is related to previous experimental studies of bentonite and organophilic clays that are used to adsorb both inorganic and organic chromium. The Brazilian sodium bentonites used were extracted and processed in Campina Grande located at State of Paraíba, and organoclays were obtained from the replacement of the sodium cation by a quaternary ammonium cation. This salt was got from animal fat. The properties measured were compressive strength, porosity and leachate concentration. High compressive strength allows for an effective pilling up of the cement block in the industrial landfill. Low porosity difficult the leachability of wastes by the rainwater or groundwater when the cement block is buried. The cement probes were allowed to hydrate for 28 days in a wet chamber for compressive strength and porosity tests. After that, the measured compressive strength was in the range of 28 to 37MPa and the porosity range varied from 28 to 35%. The leachate test presented a range of 2,07ppm to 1,06ppm. These values show that the cement blocks are not going to brake after pilling up and the water penetration is satisfactorily restricted according to literature.

1. INTRODUCTION

The waste management industry is facing increasing pressure from legislation. The necessity of new treatments for industry wastes has been taken to the development of several researches mainly related to heavy metals that are hazardous wastes, as in the tannery industry.

The treatments for hazardous wastes are incineration, recycling, agriculture and landfill disposal A study in U.K. showed that landfill is currently the cheapest disposal option with the lowest cost (Harries-Rees, 1993).

Chemical fixation of toxic and hazardous waste using cement has been practiced for many years as a form to stabilize the waste, which is placed in landfill. Cement-based stabilization/solidification, according to Wiles and Pollard et al. (1991) is a chemical treatment process, which aims to either bind or complex the components of a hazardous waste within a solid cement matrix.

---

* This paper is part of Miss C.A. Pinto Master's Dissertation sponsored by The Brazilian Research Council.

This paper presents a proposal of treatment and final disposal of tannery waste with chromium, which corresponds to stabilization/solidification process. The method consists in mixing of cement and water with the waste such that, when the system is solidified, a rigid block is obtained with low hazardous of contamination of soil and groundwater after its disposal into landfill (Pinto et al. 2000).

Cement, water, sodium bentonite, organoclay and waste constitute the elaborated system. The addition of clay has the property to adsorb the waste and to be fixed at the cement, decreasing the ability for leachate of the waste by the water to the land. This paper addresses the viability of this process aiming its application at the waste with chromium. Some experiments focused the behavior of each component added to the cement. Compressive strength and porosity tests were carried out to analyze the mechanical performance of the solidified matrix formed.

Studies of the effect of heavy metals oxides (Cr, Cu, Zn, As, Cd, Hg, Pb) on the physical properties of cement (Tashiro and Kawaguchi 1977) have shown that metals interact with the hydration and microstructure of the hydrated cement in the early stages of hardening affecting seriously strength development. Some metals have been found to promote the growth of ettringite crystals and induce significant changes in the microstructure of the hydrated product of the tricalcium aluminate (C_3A) phase (Poon et al. 1985).

According to Lo and Liljestrand (1996) clays have high sorption capacity and low hydraulic conductivity in the study of a liner system applied to minimize the rate of contaminant transport through the liner by retarding the advective and diffuse fluxes of leachate of pollutants.

Sodium bentonite has been used in a liner system in order to reduce the waste permeability through the soil and groundwater (Büchler 1987). It adsorbs inorganic pollutants. The adsorption by organoclay and sodium bentonite is obtained through Van de Waals forces.

The use of adsorbents such as organophilic clay to solidify wastes containing up to 15% organic compounds has been shown to be effective in overcoming adverse organic-cement interactions. The presence of clay in cement paste has been known by affecting adversely its strength (Montgomery et al. 1991).

The replacement of exchangeable inorganic cations with quaternary alkylammonium cations can produce Organoclays. This treatment reduces the hydration of clay, increases the interlamellar spacing and simultaneously decreases the exposed aluminosilicate mineral surface area. As the inorganic cations are progressively replaced by the organic cations, the surface properties of clay may change considerably from highly hydrophilic to increasingly hydrophobic (Nzengung et al. 1997).

2. MATERIALS AND METHODS

2.1. Materials
The materials used in this study are sodium bentonite, organoclay, Portland cement and tannery waste.

Sodium bentonite
Commercial sample, beige in powder. It was industrially submitted at exchangeable cation for sodium, proceeded from State of Paraíba, Brazil. The clay was supplied by Bentonit União Nordeste S.A.

Valenzuela Diaz (1994) prepared some organophilics clays in laboratory which were characterized, as sodium bentonite, and the results are shown as follows.

The exchangeable cation capacity is 51meq/100g of dried clay at 110°C, with 32meq corresponding to cation sodium. The sodium bentonite presents Foster's swelling of 12mL/g, meaning that it is a bentonite that swells in water. The chemical analyses of sodium bentonite are described at the Table 1. Sodium bentonite was characterized by X-Ray Diffraction. The curve is represented in the Figure 1.

The X-Ray Diffraction curve of the received sodium bentonite was made in X-Ray Diffractometer Philips, 1880 model. The radiation was K-α of copper, with step of 0,040° (2theta) and time for step of 0,500s. The samples were prepared initially with dispersion and mixed mechanically for 20minutes and in repose for 48hours. The not sedimental material was transferred to a glass plate and it was dried at room temperature.

Analyzing the X-Ray Diffraction is observed the esmectitic clay peak at 14,5Å.

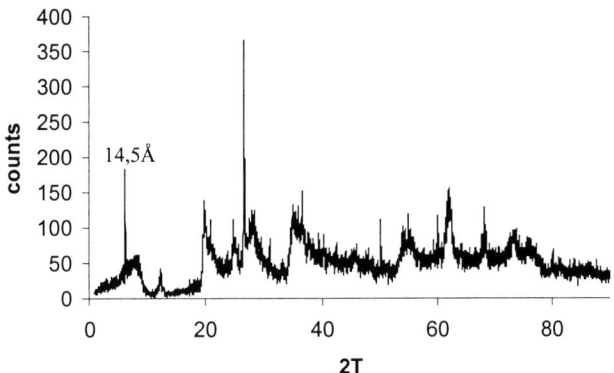

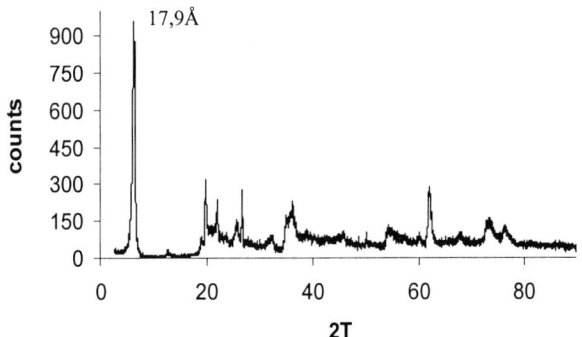

Figure 1. X-Ray Diffraction curve to sodium bentonite.
Figure 2. X-Ray Diffraction curve to organophilic clay.

Table 1
Chemical composition of sodium bentonite.

Chemical Analyze	Quantity present (%)
Fire lose	10,80
SiO_2	64,40
Al_2O_3	12,50
Fe_2O_3	7,85
CaO	0,87
MgO	1,30
Na_2O	2,29
K_2O	1,91
TiO_2	not determinated

Font: Valenzuela Diaz, 1994.

Organoclay

Commercial clay, cream color. The clay produced in Brazil, is commercially known as Tixogel-A, produced from an Argentinean sodium bentonite and a Brazilian quaternarium amonium salt (alquil-benzil-dimetil amonium chlorite, with babassu nut oil radical and C_{12} saturated fraction) which was obtained from Staucel - Produtos Químicos Ltda.

The clay was characterized by Foster's swelling and X-Ray Diffraction by Valenzuela Díaz (1994). The Foster's swelling analysis to organophilic clay presented low swelling in kerosene, soy-bean oil and "Varsol" (about 5mL/g); medium swelling in hydrated ethanol and absolute ethanol (9mLg) and high swelling in toluene (20ml/g without agitation and 27mL/g with agitation). The X-Ray Diffraction curve of the organophilic clay was studied by powder method. The Figura 2 presents the X-ray diffraction to the clay, which observes the interlamellar spacing of 17,9Å.

Tannery waste

The waste was collected dried, in tannery industry located at City of Franca, in São Paulo State. Most of the chromium obtained was in forms of $Cr(OH)_3$, $Cr(OH)SO_4$ and proteins complexes composts. It was grinned, obtaining a powder with retained fraction of 96% ABNT 200 mesh.

It was characterized by X-Ray Diffraction. The proceeding followed the analyses described previously and the curve is represented in the Figure 3. The analyze presented the peaks: $Cr(OH)_3$ at 2,81 Å, $Cr(OH)SO_4$ at 1,98 Å, $CaSO_4$ at 4,25Å, Na_2SO_4 at 1,62 Å.

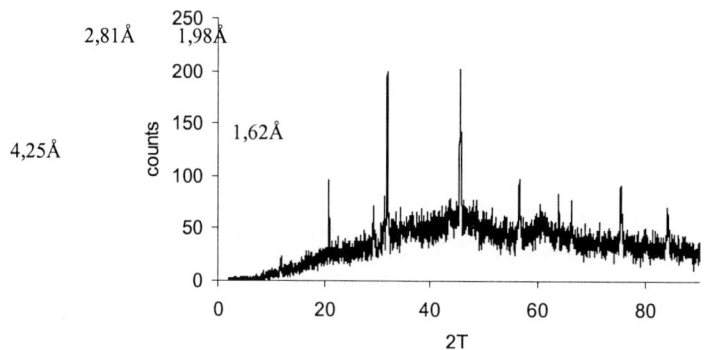

Figure 3. X-Ray Diffraction curve to tannery waste containing chromium.

Portland cement

The Portland cement used was a Brazilian commercial type II E-32 obtained from Votorantim. It contains retained fraction of 14% ABNT 325 mesh. The compressive strength for water/cement relation of 0,5 obtained in laboratory after 28 days was 33,2MPa (test realized according to ABNT 7215/96 norm - Cement Portland - Compressive strength determination) and density of $1,84 g/cm^3$.

Distillated water

2.2 Methods

This study analyzed the mechanical behavior and leachability of the block formed after the addition of the clay and waste in cement.

2.2.1 Compressive strength and porosity tests
The formulations used for those tests are represented in the Table 2.
Each system was homogenized in Becker, initially adding dried components and after

adding water, mixing manually until getting a homogeneous paste.

The paste was placed in testing cylindrical samples with 100mm of high and 50mm of diameter. The samples remained into a pattern for 24hours and they were taken to a wet chamber at 22°C (99% of humidity) for 28 days. After this, the tests were done.

Table 2
Formulations used for compressive strength and porosity tests.

Formulation	Components
CW	Commercial Portland cement; Water (50%)
CWB	Commercial Portland cement; Water (50%) Sodium Bentonite (5%)
CWO	Commercial Portland cement; Water (50%) Organophilic Clay (5%)
CWBO	Commercial Portland cement; Water (50%) Sodium Bentonite (5%); Organophilic Clay (5%)
CWBOT	Commercial Portland cement; Water (50%) Sodium Bentonite (5%) Organophilic Clay (5%);Tannery waste (10%)
CWT	Commercial Portland cement; Water (50%) Tannery waste (10%)

Compressive strength

The strength can be defined as the tension measure to break the material. It is measured applying a charge tension with velocity of 0,25MPa/s until the rupture of the sample. The rupture charge is measured and the individual strength (IS) is calculated as Equation 1.

$$IS = \frac{P}{S} \qquad (1)$$

where: P - rupture charge (kgf); S - sample sectional area

Porosity

The material porosity is defined by the presence of pores or empties.

The test was carried out according to ABNT 9778/87 norm - Mortar and Cement hardened- Water absorption determination by immersion; empties index and specific mass.

After the test, the water absorption (*Abs*) and the empties index (EI) were calculated respectively as Equations 2 and 3.

$$Abs = \frac{M_{sat} - M_s}{M_s} \times 100 \qquad (2)$$

where: M_{sat} - sample mass with dried surface after water saturation in water and boiling (g)
M_s - sample mass dried (g)

$$EI = \frac{M_{sat} - M_s}{M_{sat} - M_i} \times 100 \tag{3}$$

where: M_i - mass sample immerses in water after saturation and boiling.

2.2.2 Leachate test

The formulations used for this test are represented in the Table 3.

Table 3
Formulations used for leachate test.

Formulation	Components
CWT	Commercial Portland cement; Water (60%) Tannery waste (10%)
CWBT	Commercial Portland cement; Water (60%) Tannery waste (10%); Sodium Bentonite (5%)
CWOT	Commercial Portland cement; Water (60%) Tannery waste (10%); Organophilic Clay (5%)
CWBOT	Commercial Portland cement; Water (60%) Tannery waste (10%); Sodium Bentonite (5%); Organophilic Clay (5%)

3. RESULTS AND DISCUSSION

The results obtained in this study are represented in Tables 4 and 5.

Table 4
Results for compressive strength and porosity tests*.

Formulation	Compressive strength (MPa)	Porosity (%)	
		Water Absorption	Empties index
CW	33,2	29,8	43,8
CWB	37,0	28,6	42,2
CWO	31,6	29,9	42,8
CWBO	35,7	28,1	41,3
CWBOT	28,9	32,2	44,9
CWT	17,1	37,5	50,0

*All results were compared to the CW formulation.

3.1 Compressive strength

The results showed that the presence of sodium bentonite increases the compressive strength in 11,5%. The organophilic clay decreases it in 4,8%. The presence of both clays increases the compressive strength in 7,5%. The formulations containing the waste presented a decrease for this tests. The formulation CWBOT reduced in 16% and the formulation CWT reduced in 48,8%.

The study of stabilization/solidification of phenol using cement and organoclay, by Oliveira and Büchler (1991) showed that the presence of phenol decreases the compressive strength. The increase of phenol solution concentration decreases the compressive strength measurements for the same hydration time.

This discussion can be confirmed through the study of Wan and Vipulanandan (2000) which focused the behavior of chromium in K_2CrO_4 form added to cement with water/cement relation of 0,5. The compressive strength analysis showed that the presence of chromium in the cement matrix reduces this value because the chromium delays the hydration of the cement resulting in a low compressive strength.

3.2 Porosity

The results (in terms of water absorption) obtained for porosity test showed that the sodium bentonite reduces the porosity in 3,5% compared to cement/water system. The presence of the waste increases the porosity in 8,0% to CWBOT formulation and in 26% to cement/water/tannery waste system.

According to Tango (1983), when the water volume exceeds the standard necessary to cement hydration, the proportion of empties becomes bigger, and the compressive strength is shorter.

Table 5
Results for leachate test*.

Formulation	Leachated Chromium (ppm)
CWT	2,07
CWBT	1,85
CWOT	1,47
CWBOT	1,06

* All results were compared to the CWT formulation.

This analysis was performed according to NBR-10005/87 norm. The formulations were prepared mixing the components and after 24 hours they were grinded and placed in acetic acid with pH 5,5 in a ratio of 100g of waste solidified/liter of acetic acid.

The systems were mixed for 24 hours and the leachate was filtered. The analysis was measure by atomic absorption spectrophotometry.

3.3 Leachate

Analyzing the results, the presence of bentonite and organoclay decrease in 49% the quantity of leachate chromium related to behavior of each clay. The system containing sodium bentonite, decreased in 11% the leachate and for the system with organoclay this value decreased 30%.

CONCLUSIONS

The results obtained in these tests showed that the sodium bentonite presence increases the compressive strength and consequently decreases the porosity. On the other hand the presence of organophilic clay doesn't change the mechanical performance of the block containing cement and water.

The tannery wastes containing chromium promotes a reduction of the compressive strength, which delays the hydration of the cement and increases the porosity regarding water absorption and empties index. The formulation containing clays and waste presented good results for compressive strength test. The building construction values for this test are in a range of 32 to 40MPa, which demonstrate the viability of this process.

The leachate test showed that the presence of organoclay and sodium bentonite to adsorb chromium waste, reduces the chromium leachate.

ACKNOWLEDGMENTS

The acknowledgments to Brazilian Research Council (CNPq) and Brazilian Ministry of Education (CAPES) for financial support and Tematic Project sponsored by São Paulo State Foundation for Endowment of Research (FAPESP), Process n° 1995/0544-0 for the purchase of equipment.

REFERENCES

Harries-Rees, K. Minerals in waste and effluent treatment. Industrial Minerals, n.308, p.29-39, 1993.

Pinto, C.A.; Kozievitch, V.F.J.; Hamassaki, L.T.; Vieira-Coelho, A.C.; Xavier, C.; Büchler, P.M.; Valenzuela-Diaz, F.R.. Clays influence in the properties of Portland cement pastes. Proceedings.14th Brazilian Engineering and Materials Science Congress, 2000.

Pollard, S.J.T. ; Montgomery, D.M. ; Sollars, C.J. ; Perry, R. Organic compounds in the cement-based stabilization / solidification of hazardous mixed wastes-mechanistic and process considerations. Journal of Hazardous Materials, v.28, n.3, p.313-327, 1991.

Poon, C.S. ; Peters, C.J. ; Perry, R. ; Barnes, P. ; Barker, A.P. Mechanisms of metal stabilization by cement based fixation processes. The Science of the Total Environment, v.41, n.1, p.55-71, 1985.

Tashiro, C.; Kawaguchi, K. Cem. Concr. Res., v.7,n.69, 1977.

Büchler, P.M. Sorption of vinasse in organoclays. São Paulo, 1987. 78p. Dissertation (in Portuguese).

Lo, I.M.C.; Liljestrand, H.M. Laboratory sorption and hydraulic conductivity tests: evaluation of modified-clay materials. Waste Management and Research, v.14, n. 3, p. 297-310, 1996.

Montgomery, D.M. ; Sollars, C.J. ; Perry, R. ; Tarling, S.E. ; Barnes, P. ; Henderson, E. Treatment of Organic: contaminated industrial wastes using cement-based stabilization/ solidification I. Microstructural analysis of cement – organic interactions. Waste Management & Research, v.9, n.2, p.113-125, 1991.

Nzengung, V.A., et al. Organic cosolvent effects on sorption kinetics of hydrophobic organic chemicals by organoclays. Environmental Science Technology, v.31, n.5, p.1470-1475, 1997.

Valenzuela Diaz, F.R. Synthesis of organoclays. São Paulo, 1994. 256p. Ph D Dissertation (in Portuguese).

Oliveira, K.D.; Büchler,P.M. The stabilization/solidification of phenol. In: INTERNATIONAL ENVIRONMENTAL CHEMISTRY CONGRESS, 3., Salvador, 1991. Proceedings. [Yverdon], Switzerland, Gordon and Breach Science Publishers,1993.

Wan, S. ; Vipulanandan, C. Solidification/stabilization of Cr(VI) with cement. Leachability and XRD analyses. Cement and Concrete Research, v.30, n.3, p.385- 389, 2000.

Tango, C.E.S. Study of cements and concrete compressive strength. São Paulo, 1983. 462p. Dissertation (in Portuguese).

Plasticity of brick clays: comparison of several empirical tests and correlation with mineralogical composition and particle size distribution

M. Raimondo[a], M. Cocchi[b], G. Dircetti[c], M. Dondi[a], A. Ulrici[b], P. Zannini[b]

[a]CNR-IRTEC, Faenza (Italy)

[b]Chemistry Department, University of Modena (Italy)

[c]CERTI-LATER, Monte di Malo (Italy)

The aim of this study was to assess the most suitable analytical methods which give a direct or indirect measurement of plasticity of brick clays. For this purpose, eleven clays, with different physico-chemical characteristics, were selected. Correlations for results obtained with the different analytical methods are discussed, with the influence of the clays basic properties on their plastic behaviour being evaluated as well

1. INTRODUCTION

The term "plasticity" is usually referred to the characteristic behaviour of a ceramic material to become permanently deformed under the action of an external force and to retain its shape when the force is removed. This property, which is one of the most important to be considered in the working of clays, is closely related to the rheological characteristics of the clay-water system. On this basis, it is very difficult to provide an unambiguous meaning of the term "plasticity"; in fact, it is better related to the working behaviour rather than to the basic properties of clays.

The influence of clays properties, such as chemical and mineralogical composition, specific surface area and grain size distribution, is quite general and it is difficult to achieve a reliable statistical forecast of plasticity. Due to these factors, the measurements of such a parameter is usually performed by empyrical tests. The most widely used methods in the clay industry for measuring plasticity are the Atterberg, Pfefferkorn and Blue Methylene Index which don't define rheological parameters, but denote a degree of workability. Results from these methods are often difficult to interpret due to a large number of factors [1].

The collection of plasticity data through a bibliographic search seems to confirm these difficulties. Although a good agreement is generally found, the choice of one analytical method is related to the aim of the work (characterization or quality control) and to the length of the determination. The correlation between the analytical results and the mineralogical composition of clays has also been investigated through the definition of a "plasticity chart" which, on the basis of the amount of the mineralogical components, allows a forecast to be made of their plastic behaviour [3]. The use of alternative procedure for measuring plasticity are usually considered adequate for the quality control of specific materials, as the methylene blue index in the quality control of bentonites [4, 5].

From this survey, it is difficult to define an elected method for the measurement of the plasticity. The present study presents a survey of the most widely used analytical procedures, with the objective to identify the most suitable methods to characterize the plastic behaviour of clays.

2. MATERIALS AND METHODS

Eleven clays from different areas utilised in the Italian brick industry were chosen. The chemical, mineralogical composition and particle size distribution are reported in Table 1. The samples were selected on the basis of their wide range of plasticity demonstrated during the manufacturing process.

Table 1 Chemical, mineralogical composition and particle size distribution of clays

% wt	FID	FIG	COM	COG	VEM	VIM	VIG	TOS	SEC	PST	POS
SiO_2	58.6	51.0	49.5	58.2	52.4	53.1	59.3	68.4	54.5	39.0	56.0
TiO_2	0.5	0.8	0.6	0.7	0.6	0.9	0.8	0.6	0.6	0.6	0.8
Al_2O_3	11.2	15.9	11.9	14.6	12.6	13.9	17.2	12.6	11.4	10.1	12.8
Fe_2O_3	4.1	6.3	4.4	5.6	4.4	5.3	6.8	5.3	4.4	4.6	5.1
MgO	3.2	2.8	2.9	3.3	3.8	4.5	2.2	1.9	2.4	4.3	2.5
CaO	7.9	7.2	12.7	3.1	9.2	6.1	0.7	1.0	10.9	18.3	8.8
Na_2O	1.3	0.8	1.2	0.5	1.6	0.9	0.5	1.1	1.3	1.0	0.9
K_2O	1.9	2.9	2.2	3.2	2.8	2.8	4.3	2.3	1.9	2.1	2.1
L.o.I.	10.5	12.4	14.8	10.3	12.6	12.5	7.6	6.2	12.4	20.0	10.8
C org.	0.5	0.9	0.8	0.3	0.5	0.4	0.6	0.3	0.7	1.1	0.7
Illite	25	39	24	40	23	23	38	31	20	28	28
Kaolinite	5	9	7	6	3	9	11	5	8	2	9
Chlorite	8	10	5	6	2	9	7	8	8	9	9
Smectite	0	0	0	11	0	0	0	0	0	0	0
Fe-oxides	2	4	3	3	4	1	5	3	2	2	3
Quartz	32	17	22	24	31	30	27	39	28	14	28
Plagioclase	11	6	10	4	13	7	4	9	11	8	8
Calcite	10	12	18	3	8	0	0	2	18	27	15
Dolomite	7	2	8	3	16	20	2	0	3	11	2
Fraction > 10 μm	52.6	7.2	29.5	6.5	63.5	65.0	16.2	33.0	30.3	29.5	29.9
Fraction 2-10 μm	15.4	24.3	27.7	18.5	17.2	17.6	27.0	18.9	25.3	35.4	28.5
Fraction < 2 μm	32.0	68.5	42.8	75.0	19.3	17.4	56.8	48.1	44.4	35.1	41.6
Mean Particle size (μm)	12.6	0.7	3.1	0.5	18.2	22.1	1.3	2.4	2.8	3.8	3.3

The quantitative chemical composition was determined by a Philips PW 1480 X-ray fluorescence spectrometer on powders pressed on a boric acid support. The loss on ignition represents the weight percentage difference between 100°C and 1000°C. The organic substance content was determined with the Walkley-Peech method [6].

The mineralogical analysis was carried out with a Rigaku Miniflex X-ray powder diffractometer with CuKα radiation and Ni filter in the 4 – 64° 2θ range and the quantitative calculations were performed by comparison with PDF.

The grain size distribution was determined by X-ray photosedimentation with a Micromeritics Sedigraph 5100 (ASTM C958). The methylene blue index (MBI) was determined according to the ASTM C837, without pH adjustment, while the specific surface was measured by nitrogen absorption through a "single point" procedure (BET).

The plastic behaviour was investigated determining the Atterberg consistency limits (CNR-UNI 10014), Pfefferkorn index [7] and Barna index, this latter measured from the drying curve of Bigot, calculating the ratio between the processing (W_u) and residual (W_s) water content: BI=W_s/W_u.

The technological characterization was performed measuring the modulus of rupture (ASTM C 326), the drying shrinkage (ASTM C 326), the hygroscopic index (HYG), the Keeling coefficient (KEE) and the drying time of a clay thin film in acetone (ACE).

The hygroscopic index was calculated by the exposure of the clay sample, previously dried at 120 °C, to a humid atmosphere (75% of relative humidity, maintained with a saturated solution of NaCl) at 25°C for 48 hours. The Keeling coefficient of each material was calculated as the ratio between the loss on ignition and the hygroscopic index [9].

3. RESULTS AND DISCUSSION

The compositional differences among the raw materials considered are linked in a complex way to their plastic behaviour, so that this latter parameter is unpredictable. Generally the higher the Al_2O_3/SiO_2 ratio, the plasticity of the body improves with an increase in the clay minerals/quartz ratio. In addition, with all the others parameters being constant, there is a clear positive correlation between the percentage of organic matter and the workability of the body.

The mineralogical analyses indicate that the raw materials consist mainly of illite, quartz and some plagioclase, with the total carbonate amount (calcite and dolomite) being variable. COG is the only sample with an appreciable smectite content (11%). At this purpose, a statistical elaboration, performed by analysis of the main components, highlighted the role of the coarser grain fraction in depressing the plasticity, rather than a clear positive influence of the clayey fraction. A further aid to the identification of clays, in term of their different workability, is offered by the particle size analysis (Table 1): the lower the mean particle size, the higher the plasticity.

Table 2
Specific surface (BET), methylene blue index (MBI) and cation exchange capacity (CEC).

	FID	FIG	COM	COG	VEM	VIM	VIG	TOS	SEC	PST	POS
BET (m^2/g)	12.9	40.9	23.2	46.5	6.3	10.9	32.4	19.4	24.9	27.9	22.7
MBI (meq/100 g)	87	72	40	97	16	18	80	34	55	54	45
CEC (meq/100 g)	10.7	24.7	18.7	31.3	13.0	14.0	35.1	17.2	10.2	22.2	18.6

The BET and MBI values are well correlated, with the higher the specific surface, the higher the amount of methylene blue adsorbed, the only exception being the FID sample, whose BET value is quite low when compared with the MBI index (Table 2). The high specific surface of FIG, COG, VIG and PST clays is explained on the basis of their fine particle size distribution, often up to 75wt.% < 2 microns. On the contrary, the relation between the CEC values and BET ord MBI is not clear, since the amount of exchangeble cations is strictly identified with the clay mineralogical composition .

The determination of the plastic behaviour, in terms of plastic indices (Atterberg, Pfefferkorn and Barna), mechanical and technological properties of clays led to the results reported in table 3. On the basis of these latter, the following can be summarised:
- a good correlation exists between IB and W_P (Fig. 1c).
- there is little correlation of the Pfefferkorn index with Atterberg and Barna indices. For example, samples having very close values of I_P, show a wide range of PFK values and viceversa. In the correlation with W_P, PFK seems able just to discriminate among different classes of products, but not inside the same group of materials.

Table 3
Plastic limit (W_P), liquid limit (W_L), plastic index (I_P), Pfefferkorn index (PFK), Barna index (BI), modulus of rupture (MOR), drying shrinkage (RLE), hygroscopicity (HYG), Keeling coefficient (KEE) and acetone evaporation time (ACE).

	FID	FIG	COM	COG	VEM	VIM	VIG	TOS	SEC	PST	POS
W_P (wt %)	24.2	34.5	21.4	36.8	24.6	24.9	35.1	21.4	22.7	28.5	20.7
W_L (wt %)	30.6	66.3	42.1	78.9	27.8	30.0	58.7	40.1	48.0	51.7	45.1
I_P (wt %)	6.4	31.8	20.7	42.1	3.2	5.1	23.6	18.7	25.3	23.2	24.4
PFK (wt %)	31.8	30.4	22.6	29.3	20.8	19.8	30.9	21.3	20.3	28.2	20.0
BI (adim.)	0.28	0.62	0.47	0.70	0.19	0.24	0.60	0.52	0.57	0.34	0.49
MOR (MPa)	5.4	11.0	7.8	15.2	3.3	4.6	12.9	8.2	11.9	10.1	10.2
RLE (cm/m)	1.8	9.3	7.0	9.4	1.3	2.6	9.5	6.7	7.1	7.1	6.7
HYG (wt %)	5.2	4.3	2.7	5.3	1.2	1.7	5.1	2.4	2.7	3.0	2.5
KEE (adim.)	0.3	0.9	1.2	1.2	5.5	6.2	1.7	2.4	0.7	0.5	1.4
ACE (sec)	98	246	128	243	85	83	183	148	210	172	198

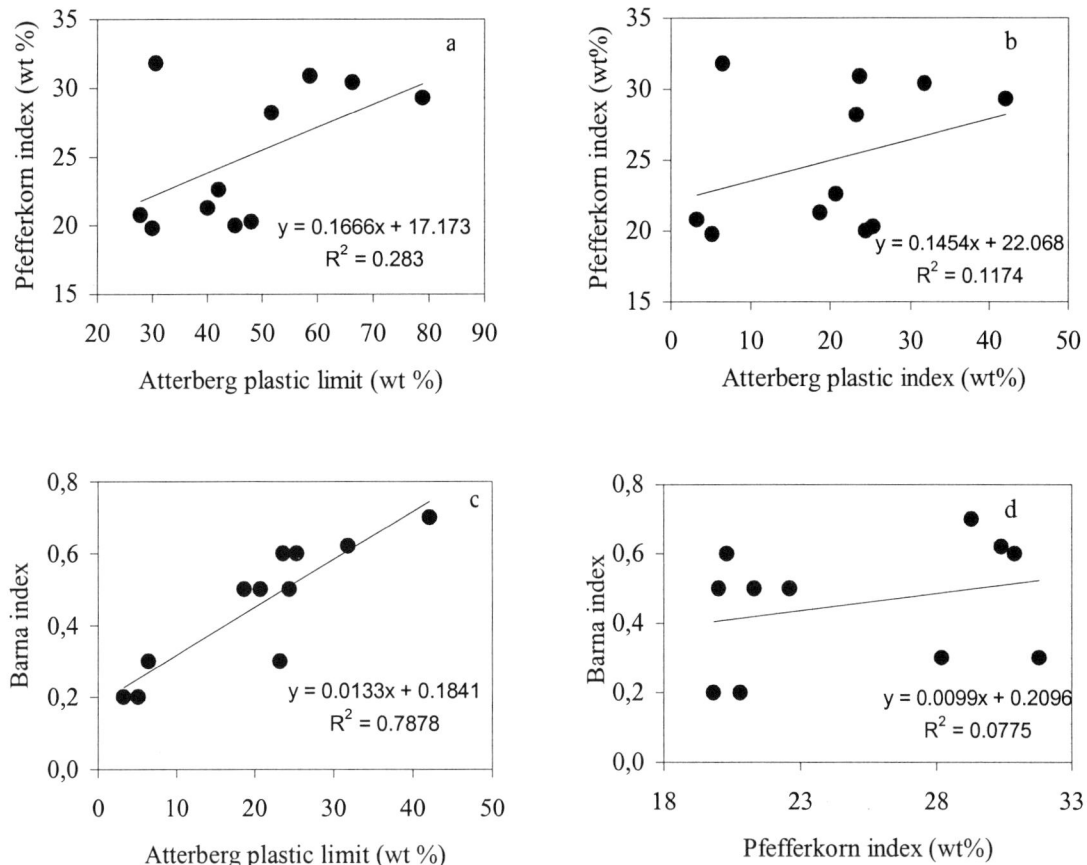

Figure 1. Plot between: a) Pfefferkorn index and Atterberg plastic limit; b) Pfefferkorn index and Atterberg plastic index; c) Barna index and Atterberg plastic limit; d) Barna index and Pfefferkorn index.

Both the liquid limit and the plastic index correlate well with the specific surface, with r^2 = 0.917 and 0.981, respectively (Fig. 2). The more plastic clays (FIG, COG and VIG) show values higher than 30 m^2/g, the less plastic ones (VEM, VIM and FID) below 10 m^2/g, with the others having intermediate values (18 – 25 m^2/g). Obviously, this trend can be also explained on the basis of the relative amount of the finer fraction, and, hence, of clay minerals. Moreover, the correlation between the liquid limit and the specific surface, from which the general trend can be established, highlights, within the same group of materials, the relative differences of plasticity.

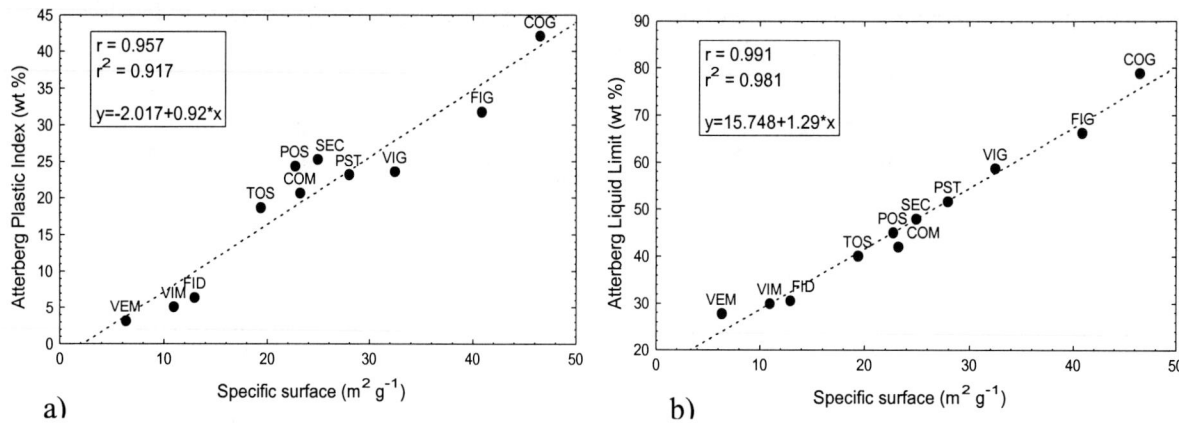

Fig. 2 - Correlation between the specific surface and the Atterberg plastic index (a) and liquid limit (b).

The hygroscopicity (HYG), being related to the nature and proportion of the clay fraction, allows indirectly to quantify the amount of this latter, while the Keeling coefficient, which is defined as the ratio between the loss of ignition and the hygroscopicity, is characteristics of the clay mineral [8]. On this basis, the greater the HYG, which implies a high clay mineral content, the higher the specific surface and, hence, the plasticity (Fig. 3a). On the other hand, since the ignition loss varies in the opposite way compared to hygroscopicity, the Keeling coefficient is able to distinguish between different types of clay minerals. The prevailing mixture of kaolinite and illite usually gives Keeeling values between 1 and 2, as happens for most of the materials here considered (Fig 3b). The unusual very high values of Keeling coefficient (5.5 and 6.2), which belong to VEM and VIM samples respectively, are due to their greater content of the fraction >10 µm.

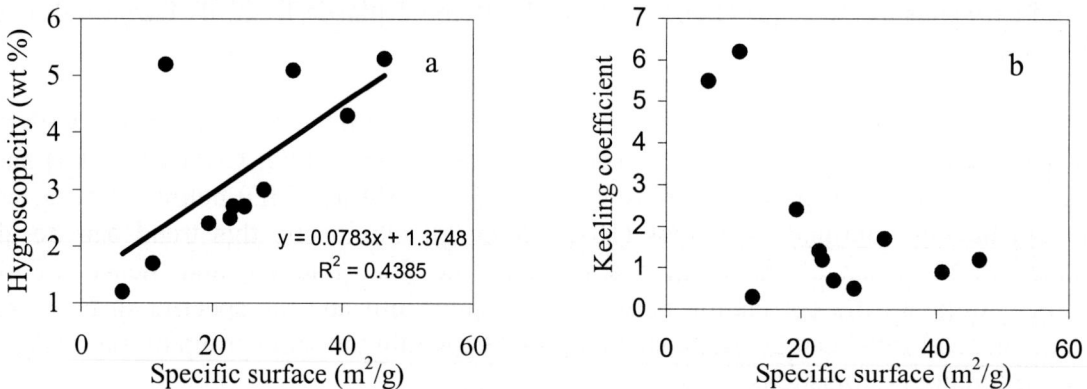

Figure 3. Plot between the specific surface and hygroscopicity (a) and the specific surface and Keeling coefficient (b).

A positive linear correlation is also found between the ACE values and the specific surface (Fig 4a). The longer the evaporation time of acetone, the finer the particle size and then the higher the plasticity. On the other hand, the rather good correlation between the specific surface and the drying shrinkage seems not able to distinguish among samples having very different BET values, and hence, different plasticity (Fig. 4b).

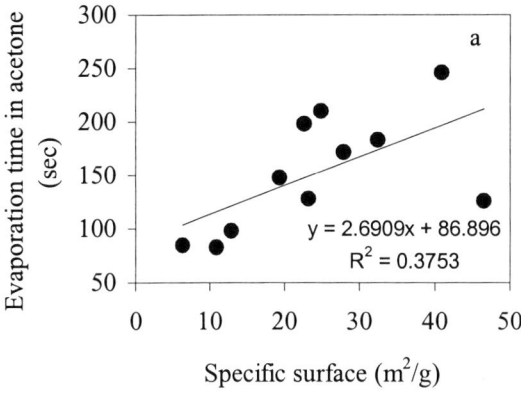 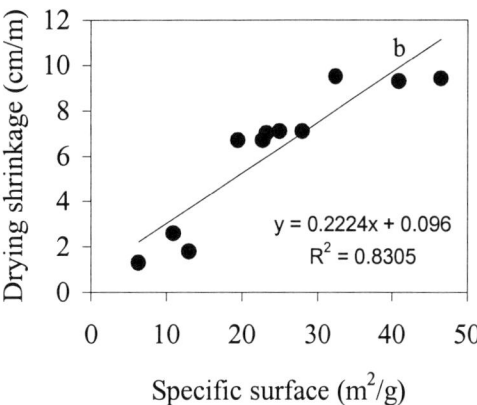

Figure 4. Correlation between the specific surface and the evaporation time in acetone (a) and the drying shrinkage (b).

4. CONCLUSIONS

The linear correlation coefficient and equation among different analytical methods for measuring plasticity were evaluated. A rather good correlation between IB and I_P was found, with little correlation of the Pfefferkorn index with the Atterberg and Barna indices. The mineralogical composition of clays is not a key component in plasticity which is considered mainly a function of the particle size distribution – the finer the clay the more likely it is to be plastic. This result is also confirmed from the importance of the specific surface values in predicting the plastic behaviour. This latter, in fact, is very well linearly correlated with the Atterberg plastic index and liquid limit, as well as with hygroscopicity, modulus of rupture, dryng shrinkage and evaporation time of clays in acetone.

REFERENCES

1. F. Ginès, C. Feliu, J Garcia-Ten, V. Sanz, Bol. Soc. Esp. Cerám. Vidrio, 36 [1] (1997) 25.
2. B. Chassefière and A. Monaco, Geologia Tecnica, 33 (1976) 122.
3. J. A. Bain, Clay Minerals, 9 (1971) 1.
4. A. Amarante J. and F. A. Boutros, Cerâmica, 27 (1981) 117.
5. J. A. Ferguson, P. J. Banks, Proc. 10th Int. Clay Conf., Adelaide, Australia 1993.
6. Methods of Soil Analysis (Part 2), Soil Science Society of America, 1982.
7. K. Pfefferkorn, Sprechs, 57 (1924) 297.
8. P. S. Keeling, Clay Minerals Bullettin, 3 (1958) 271.

Adsorption of phenol by organo-clays*

M.M.G. Ramos Vianna[a]; C.L. Vieira José[a]; F.R. Valenzuela Díaz[b]; P.M. Büchler[c]

[a]Ph D' students at Chemical Engineering Department, Polytechnic School, University of São Paulo, São Paulo, Brazil.
[b]Mettalurgical and Materials Engineering Department, Polytechnic School, University of São Paulo, São Paulo, Brazil.
[c]Chemical Engineering Department, Polytechnic School, University of São Paulo, São Paulo, Brazil.

Nowadays, increasing concern about pollution of groundwater by organic chemicals led to research on the use of various adsorbents. They can be applied to provide a barrier to the escape of organic contaminants from storage tanks and stabilization lagoons.

This study investigated, experimentally, the adsorption of phenol by three organo-clays, using phenol in aqueous solution. This is the way that wastewater is discharged.

The organo-clays used were a commercial smectite-organo-clay (ABDM-STiA), purchased in Brazil. The others were prepared with two different clays using a bentonite from the Brazilian State of Paraíba (ABDM-SVC), sodium exchanged in laboratory and smectite from Wyoming bentonite (ABDM-SWy) with one Brazilian quaternary ammonium salt. The cation exchange was alkyl benzyl dimethyl with dodecyl as the alkyl group. The hydrophilic character is then transformed into hydrophobic and organophilic.

Adsorption of phenol followed the order of ABDM-SVC > ABDM-SWy > ABDM-STiA. Their isotherms followed a convex up pattern.

The equilibrium curves that were obtained are well represented by the Freundlich isotherm model. The adsorption data showed that the prepared materials were effective to adsorb the phenol, the Brazilian clay is the most efficient of the three materials.

1. INTRODUCTION

Due to the isomorphous substitutions in the aluminosilicate layers, natural clay minerals usually have a net negative charge, which is balanced by alkali metal and alkaline-earth-metal cations such as Na^+ and Ca^{2+}. The strong hydration of these inorganic cations creates a hydrophilic environment on the surface and in the interlayer region of natural smectite clay (Souza Santos, 1992). The substitution of Na^+ or Ca^{2+} with quaternary ammonium cations of the form $[(CH_3)_3NR]^+$ or $[(CH_3)_2NRR']^+$ at the exchangeable sites of natural clays result in organo-clays derivative with organophilic properties that can act as sorbents contaminants hydrocarbons (Büchler, 1986a, 1986b, 1987; Jaynes and Boyd, 1990; Xu and Boyd, 1995).

---

* Tematic Project sponsored by São Paulo Foundation for Endowment of Research Process n° 1995/0544-0.

Organo-clays are effective sorbents for a variety of aqueous organic compounds including many common groundwater contaminants (Boyd, et al., 1988 a; Boyd et al, 1988 b; Lee et al., 1989, 1990; Sharmasarkar et al., 2000).

Organo-clays formed with quaternary ammonium salts such as trimethylphenylammonium (TMPA) or hexadecyltrimethylammonium (HDTMA), from smectite were effective sorbents for phenol and chlorophenols compared to unmodified smectite. Sorption of phenol was maximized when TMPA-clay was utilized as the sorbent (Jaynes and Boyd, 1990).

As phenols are considered as priority pollutants since they are harmful to organisms at low concentrations, three organo-clays were evaluated as adsorbents of water phenol. One organo-clay was commercial and the others were prepared using a bentonite from the Brazilian State of Paraíba, sodium exchanged in laboratory and the Wyoming bentonite with one Brazilian quaternary ammonium salt. The salt used was alkyl benzyl dimethyl ammonium chloride (ABDMA). We used this salt because it had the dodecyl and the benzyl groups, and they are used to make organo-clays (Jaynes and Vance, 1996; Sheng et al., 1997; Sheng and Boyd, 1998; Lo et al., 1998; Ceyhan et al., 1999; Sharmasarkar et al., 2000).

2. MATERIALS AND METHODS

The commercial organo-clay Tixogel A (ABDM-STiA) was obtained from Staucel-Chemical Products. The smectite from Wyoming (SWy) was purchased from Sigma Chemicals. The smectite clays from the Brazilian State of Paraíba (SVC) sodium exchanged in laboratory were prepared in the manner previously described by Valenzuela Díaz et al. (1996). The salt was obtained from Clariant Divison Surfactants. The clays had Na^+ cation exchangeable cation and a cation-exchange capacity (CEC) of ~ 100 cmol.kg^{-1} (centimoles of charge kg^{-1}). The Na-clays were dispersed by mixing with water, and these clays were settled overnight. The clays supernatant were removed and mixed with an aqueous solution of ABDMA chloride salt in the amount of 100 meq ABDMA per 100 g of the clays and stirred for 20 minutes. The ABDMA-smectite complexes were vacuum filtered and washed with distilled water until free of Cl$^-$, and stored at room temperature.

Basal spacings were determined by X-ray diffraction (XRD) analysis. X-ray diffraction patterns were recorded using CuKα radiation (λ= 1,5416 Å) and a Philips XPERT-MPD X-ray diffractometer, in steps of 0,02°2θ, at 1,0s/step.

Adsorption isotherms were determined using the batch equilibration technique. The initial concentrations for phenol were prepared in the range 200 to 800 mg/L. Glass tubes (35 mL) containing 500 mg of organo-clays and 25 mL of phenol aqueous solutions were closed and mixed. Every batch stood for 24 hours at 33 °C. Preliminary kinetic investigations indicated that sorption equilibrium was reached in less than 20 hr. After reaching equilibrium, the aqueous phase was separated by centrifugation at 3000 rpm for 60 min at room temperature using a BIO-ENG, type BE-4004. A 10 mL portion of the supernatant was analyzed by total organic carbon method (TOC) with Shimadzu type TOC-5000 A, in order to measure concentration of phenol equilibrium. The amounts of phenol adsorbed were calculated by the differences between the initial and equilibrium concentrations. Isotherms were obtained by plotting the amounts adsorbed against the equilibrium concentration in solution. All experiments were carried out in duplicate and each point on the isotherm is the average of two experiments.

The swelling capacity of organophilic clays in toluene, ethanol and soybean oil and Diesel were measured, as described before by Valenzuela Díaz (1994a, 1994b, 1996, 1999).

3. RESULTS AND DISCUSSION

Table-1 shows the swelling and basal spacings for the organoclays. High values (greater than 21 mL/g) for the swelling in toluene and Diesel were observed, showing the organophilicity of the clays.

Table 1: Swelling and basal spacings of organoclays.

Organo-clays	d-Spacing $d_{(001)}$ (Å)	Swelling without mixing (mL/g)				Swelling with mixing (mL/g)			
		Diesel	toluene	ethanol	soybean oil	Diesel	toluene	ethanol	soybean oil
ABDM-SWy	32,1	9	14	6	6	47	34	7	10
ABDM-SVC	30,2	9	12	5	7	14	16	5	7
ABDM-STiA	17,9	14	20	9	f*	17	27	10	5

*f = float

All the organoclays had low (below 8 mL/g) or middle (between 9 to 20 mL/g) swelling in ethanol and soybean oil (without and with mixing) and in Diesel and toluene (both without mixing).

The ABDM-STiA had relatively high swelling in toluene (with mixing).

The ABDM-SWy had high swelling in Diesel and toluene (both with mixing).

The interlayer spacing followed the order of ABDM-SWy > ABDM-SVC > ABDM-STiA.

Adsorption of phenol (figure 1) followed the order of ABDM-SVC > ABDM-SWy > ABDM-STiA.

The samples ABDM-SWy and ABDM-SVCL (d_{001} greater than 30 Å) adsorbed more phenol than ABDM-STiA (d_{001} = 17,9 Å).

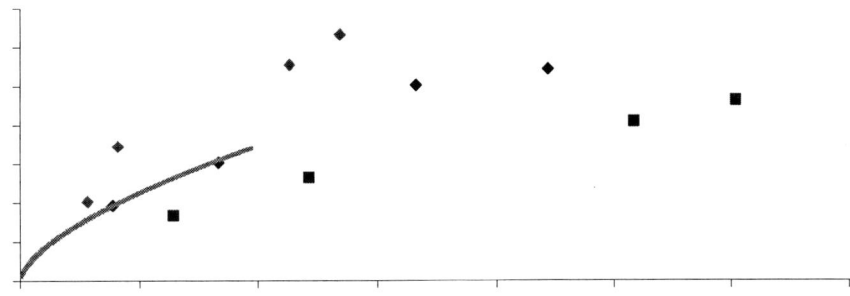

The data were fitted to Freundlich model resulting in greater correlation of the data (R^2). The Freundlich equation can be expressed as follows:

$$y = K_f C_e^{n_f^{-1}}$$ Freundlich equation

where K_f (L.mg^{-1}) and n_f (dimensionless) are constants related to sorbate binding capacity and a conditional index that describes the shape of the isotherm, respectively. The results of fitting models to phenol sorption on organo-clays are listed in table 2. The n_f values were greater than 1 for the three organo-clays. Freundlich fitting with $n_f > 1$ would indicate conformity of the data to multilayer formation at the adsorbent surfaces. Such observation could also be attributed to the molecular interactions between the sorbate species and subsequent aggregation in the pattern of surface monolayer. Such mechanistic paths could be the possible attributes to the conformity of our data with Freundlich fitting. The n_f values that are greater than 1 represent sorption results having convex up curvatures (Sharmasarkar et al. 2000). Isotherms for ABDM-SVC, ABDM-SWy and ABDM-STiA followed a convex up pattern. Banat et al. (2000) used natural bentonite to adsorb phenol. He concluded that its adsorptive capacity was limited (of an order of magnitude of 1 mg/g).

Validation of the models for adsorption of phenol by organo-clays was confirmed by testing the R^2 (correlation coefficient).

Table 2: Model parameters for phenol adsorption by organo-clays.

Organo-clays	d-Spacing $d_{(001)}$ (Å)	Freundlich		
		n_f	K_f (L/mg)	R^2_f
ABDM-SWy	32,1	1,5260	1,2893	0,9472
ABDM-SVC	30,2	1,6152	1,0002	0,9924
ABDM-STiA	17,9	1,5432	0,5717	0,9973

4. CONCLUSIONS

The ABDM-SVC, ABDM-SWy and ABDM-STia adsorption curves for phenol indicate that these organoclays were effective to adsorb the phenol and can be used as liners and waste disposal reservoirs. The adsorption data showed that the Brazilian clay is the most efficient of the three studied materials.

The Freundlich isotherms were found to be applicable for the adsorption equilibrium data.

ACKNOWLEDGMENTS

The acknowledgments go to Brazilian Research Council and Brazilian Ministry of Education for financial support and for the purchase of equipment.

REFERENCES

BANAT, F.A.; AL-BASHIR, B.; AL-ASHEB, S.; HAYAJNEH, O. Adsorption of phenol by bentonite. Environmental Pollution, v. 107, n. 3, p. 391-398, 2000.

BOYD, S.A.; MORTLAND, M.M.; CHIOU, C.T. Sorption characteristics of organic compounds on hexadecyltrimethylammoniun-esmectite. Soil Science Society American Journal, v. 52, n. 3, p. 652-657, 1988.

BOYD, S.A.; SUN, S.; LEE, J.F.; MORTLAND, M.M. Pentachlorophenol sorption by organo-clays. Clays and Clay Minerals, v. 36, n. 2, p. 125-130, 1988.

BÜCHLER, P.M.. Sorption of vinasse in organoclays. Brazilian Journal of Chemical Engineering, v. 9, n.1, p. 28-31, 1986 a (in Portuguese).

BÜCHLER, P.M.; PERRY, R. The use of clay liners in the attenuation of the organic load of vinasse in developing countries. In: THE INTERNATIONAL CONFERENCE ON

CHEMICALS IN THE ENVIRONMENT, Lisboa, 1986. Proceedings. London, Selper, 1986 b. p. 715-724.

BÜCHLER, P.M. Sorption of vinasse in organoclays. São Paulo, 1987. 78p. Dissertation (in Portuguese) – Escola Politécnica, Universidade de São Paulo.

CEYHAN, O. ; GULER, H. ; GULER, R. Adsorption of phenol and methylphenols on organo-clays. Adsorption Science & Technology, v. 17, n. 6, p. 469-477, 1999.

JAYNES, W.F. ; BOYD, S.A.. Trimethylphenylamonium-smectite as an effective adsorbent of water soluble aromatic hydrocarbons. Journal of the Air Waste Management Association, v. 40, n. 12, p. 1649-1653, 1990.

JAYNES, W.F. ; VANCE, G.F. BTEX sorption by organo-clays: cosorptive enhancement and equivalence of interlayer complexes. Soil Science American Journal, v. 60, n. 6, p. 1742-1749, 1996.

LEE, J.-F. ; MORTLAND, M.M. ; BOYD, S.A ; CHIOU, C.T. Shape selective adsorption of aromatic compounds from water by tetramethylammonium-smectite. Journal of the Chemical Society Faraday Transactions 1, v. 85, n. 9, p. 2953-2962, 1989.

LEE, J.-F. ; MORTLAND, M.M. ; CHIOU, C.T. , KILE, D.E. ; BOYD, S.A Adsorption of benzene, toluene, and xylene by two tetramethylammonium-smectites having different charge densities. Clays and Clay Minerals, v. 38, n. 2, p. 113-120, 1990.

LO, I.M.-C. ; LEE, S.C.-H. ; MAK, R.K.-M. Sorption of nonpolar and polar organics on dicetyldimethylammonium-bentonite. Waste Management & Research, v. 16, n. 2, p. 129-138, 1998.

SHARMASARKAR, S. ; JAYNES, W.F. ; VANCE, G.F. BTEX sorption by montmorillonite organo-clays: TMPA, ADAM, HDTMA. Water Air and Soil Pollution, v. 119, n. 1-4, p. 257-273, 2000.

SHENG, G.Y. ; XU, S.H. ; BOYD, S.A. Surface heterogeneity of trimethylphenylammonium-smectite as revealed by adsorption of aromatic hydrocarbons from water. Clays and Clay Minerals, v. 45, n. 5, p. 659-669, 1997.

SHENG, G.Y. ; BOYD, S.A. Relation of water and neutral organic compounds in the interlayers of mixed Ca/trimethylphenylammonium-smectites. Clays and Clay Minerals, v. 46, n. 1, p.10-17, 1998.

SOUZA SANTOS, P. Science and Technology of Clays. 2.ed. São Paulo, Edgar Blucher, 1992. 3 v. (in Portuguese).

VALENZUELA DÍAZ, F.R. Synthesis of organoclays. São Paulo, 1994a. 256p. PhD Dissertation – Escola Politécnica, Universidade de São Paulo (in Portuguese).

VALENZUELA DÍAZ, F.R. Recent studies of Brazilian smectites. In: CONGRESSO ÍTALO-BRASILEIRO DE ENGENHARIA DE MINAS, 3., São Paulo, 1994b. Anais. São Paulo, EPUSP,1994. p. 124-130, (in Portuguese).

VALENZUELA DÍAZ, F.R. ; ABREU, L.D.V. ; SOUZA SANTOS, P. Cation exchange in Brazilian smectites. Cerâmica, v. 42, n. 275, p. 290-293, 1996, (in Portuguese).

VALENZUELA DÍAZ, F.R. Synthesis of organoclays. In: Brazilian Ceramic Conference, 43, Florianópolis, 1999. Anais. São Paulo, Associação Brasileira de Cerâmica, 1999. p. 43201-43213 (in Portuguese).

XU, S. ; BOYD, S.A. Cationic surfactant sorption to a vermiculitic subsoil via hydrophobic bonding. Environmental Science & Technology, v. 29, n. 2, p. 312-320, 1995.

Portuguese clays used in geomedicine: a study of their relevant properties

J. Silva, C. Gomes and F. Rocha

Departamento de Geociências, Universidade de Aveiro, 3810-193 Aveiro, Portugal, cgomes@geo.ua.pt

Mud / clay used in two Portuguese spas (Vale dos Cucos and Vale das Furnas) for therapeutic treatments, particularly of arthro-rheumatic and skin affections, were studied in order to find out, on a scientific basis, the assets of those geomaterials. Relevant textural (particle size distribution), mineralogical (main and accessory minerals), chemical (chemical analysis of major, minor and trace elements, cation exchange capacity, exchangeable cations), specific surface area, expandability, abrasivity, liquid and plastic limits and thermal properties (specific heat, cooling rate) were assessed. Similar studies was carried out on a Portuguese bentonite intended to be used in a Portuguese Geomedicine Centre, in Porto Santo island, Madeira archipelago.

1. INTRODUCTION

Pelitic geomaterials such as clays have been used for a long time in geomedicine. In fact, particular types of clays are still used worldwide, either in thalassotherapy centres of some spas or in geologic sites, for therapeutic treatments generally related with arthro-rheumatic, and skin diseases.

Both sedimentary and hydrothermal clays, with or without little previous preparation, are being used as healing materials. In recent years the number of scientific publications concerned with the use of clays in geomedicine has increased, some of them deserving to be emphasised: Ferrand & Yvon (1991), Barbieri (1996), Yvon & Ferrand (1996), Novelli (1996), Veniale & Setti (1996) and Cara et al. (1996, 2000a; 2000b). The publications referred and 1996, and some others were issued in the Proceedings of an important scientific meeting held in Salice Terme, Italy, whose main topic was "*Argille Curative*" or Healing Clays. It is also well known that clays are used as well in cosmetics and in pharmaceutical formulations, particularly as excipients, and the following publications dealing with these subjects deserve to be emphasised: Browne et al. (1980), Hermosin et al. (1981), Galan et al.(1985), Ueda & Hamayoshi (1992) and Lopez-Galindo & Iborra (1996).

On an empirical basis, pre-historical man also utilised clay for therapeutic purposes. The clay plates of Nippur, in Mesopotamia, written down about 2500 yr. BC had a reference to clays suitable for the treatment of wounds and to stop haemorrhages. In ancient Egypt, Pharaos´s doctors used yellow ochre - mixture of clay and iron oxides/hydroxides - to cure skin wounds and internal affections, due to its antiseptic, absorbent and purifying properties. Clay was used as well in the mummification process.

The oldest book known so far, the famous Papyrus Ebers, written down in 1550 BC, reports some diseases treated with mineral based medicines consisting of clays (Reinbacher, 1999). In the famous medicinal clays *"Bolus Armenus"*, a red clay that occurs inside caves in Cappadocio mountains, in old Armenia and actual Turkey, is worthy of note as are the so-called *"terras"* of the Greek islands of Lemnos, Chios, Samos, Milos and Kimolos. The *"terra sigillata"* of Lemnos island deserves particular mention, because this white astringent and absorbent clay was prepared and shaped under the form of disks, which were marked or stamped with the goat stamp, the mark of the goddess Diana or Artemis. The clay from Kimolos island was identified as a Ca-smectite (Robertson, 1986).

The clays actually used in many medicinal applications come particularly from deposits related with thermal springs and are applied either in form of mud-baths or in form of warm cataplasms or patches. There is a growing interest on treatments using natural means as an alternative to conventional medicine. Properties of natural clays, such as: mineral and chemical composition, particle size distribution, cation exchange capacity and exchangeable cations, specific surface area, specific heat and heat diffusiveness, have been considered relevant for the use of clays in geomedicine. Smectitic clays, such as bentonites, are the ones more widely used in pelotherapy or mudtherapy applications due to their high swelling, high plasticity, high specific surface area and high ion exchange properties.

In Portugal there are several Thalassotherapy Centres which use particular types of clay in therapeutic treatments generally related with arthro-rheumatic, respiration and skin diseases. Both sedimentary and hydrothermal clays, without or with little previous preparation, are used as healing materials, under the form of both cataplasms or patches and mud-baths.

2. MATERIALS AND METHODS

This study discloses some of the relevant assets of two hydrothermal clays or muds, one used in the Thermal Centre of Vale das Furnas, São Miguel island, Azores archipelago, and the other used in the Thermal Centre of Vale dos Cucos, Torres Vedras, as well as of one clay, classified as bentonite, from Porto Santo island, Madeira archipelago.

This last clay referred to is not used yet in therapeutic treatments, however the studies already carried out did reveal its great potentialities in order to be used in complementary treatments to those which are actually used at the Porto Santo Hotel Geomedicine Centre. This Centre provides sand-baths recommended for rheumatic and orthopaedic affections, using the unique biogenic carbonate rich sands which occur naturally in the magnificent beaches and frontal dunes of Porto Santo island. The therapeutic assets of these sands are based upon their specific thermal and microchemical properties.

Properties of Vale das Furnas and Vale dos Cucos clays or muds, such as: mineral and chemical composition, particle size distribution, cation exchange capacity, specific surface area, specific heat and heat diffusiveness have been assessed.

As a matter of fact, the most relevant assets of peloids (warm patches or cataplasms of clay or mud) are as follows: high specific surface area, high cation exchange capacity, high specific heat and low rate of heat diffusiveness. All the properties referred to are dependent upon both granularity (particle size distribution) and mineral and chemical composition of clay or mud. Clay or mud particles can be smaller than bacteria and these whenever coated or encapsulated by clay particles lose activity and can be eliminated.

Certain clays containing smectite and palygorskite as major clay mineral components are the most currently applied as healing clays because they display more notoriously the properties referred to. According to Reinbacher (1999) in USA pharmacies it is sold an anti-diarrhoea pharmac made of about 75 % palygorskite and in Germany pharmacies for the same effect a pharmac is sold made of about 75 % kaolin. Geophagy was a current use of several populations who used clay for therapeutic and religious purposes or even to relieve famine.

It is known that when kaolinite, smectite, palygorskite and sepiolite are incorporated in pharmaceutical and animal feed formulations, due to either the adsorbent or absorbent properties of certain chemical compounds as well as of toxines, bacteria and virus, or due to the protective coatings formed in stomach, bowels and skin. In this last function, clays are utilised in dermatological and cosmetics applications.

Thermal Centres or Spas have less and less natural peloids available, deposited around their own thermal springs. Therefore, they make use of clay pastes brought about from other places, or they reutilize their own natural peloids previously utilised after submitting them to appropriate treatments.

Unfortunately, thermalism in Portugal is at a lower level of development compared to other European countries, such as, Italy, Spain and Germany, despite having great resources, particularly concerning thermal waters. It matters to reanimate Portuguese thermal centres, strategy that has to consider new products and treatments, to be applied on the basis of scientific and technical criteria.

Relative to treatments with peloids, mud-baths are applied at the Thermal Centre of Vale dos Cucos (presently under works of restoration), Torres Vedras, located on the right bank of Sizandro river. Mud/clay being used in the treatments, occurs along a fault associated with a spring, fault that is assumed to affect deep seated geologic formations. Water rises through a fault and bears in suspension a clayey material that is subsequentely deposited. About 6-7 tonnes of this dark grey clayey material or mud, extremely fine grained and exhaling sulphurous gas, are collected every year in a tunnel 30-40m long built on top of the fault. After being collected the mud/clay goes into maturation tanks where it ages over six months, becoming ready for the applications. The tectonics of the region where Vale dos Cucos Thermal Centre is located is very much influenced by the intrusion of evaporites. Salt, used in chemical industries, is extracted in the nearby Matacães diapir.

In Azores, more precisely in Caldeiras das Furnas, Vale das Furnas, São Miguel island, there are geysers of boiling water and sulphurous dark grey mud, one and the other being applied in the Furnas Thermal Centre in the treatment of both rheumatic and breathing affections.

This study discloses as well some of the relevant assets of a bentonite clay which occurs in Serra de Dentro, Porto Santo island, Madeira archipelago. Assessed on a scientific basis those assets make these geomaterials very interesting to be used separately or in engineered blends with local fine grained carbonate sand in therapeutic treatments to be carried out in a local Geomedicine Centre. Porto Santo is a volcanic island starting being formed at about 19 million years ago, in the Upper Miocene, stage Burdigalian. Most part of the geological materials consists of volcanic rocks, basalt, trachyte, trachyandesite, andesite, rhyolite and pyroclastic tuff. Sedimentary rocks, coralligenous limestone, beach and dune carbonate rich sands also occur in Porto Santo. The carbonate sands referred to are being used for many years in sand-baths for therapeutic purposes. Small bentonite deposits are widespread in the eastern part of the island, which did derive from the alteration, first under sea water and later outside sea water, of hyaloclastites.

3. RESULTS AND DISCUSSION

Vale dos Cucos mud represents a very rare mineral resource, because it is renewed every year. Within the short time corresponding to a month a clay layer around 20cm thick can be deposited. It is assumed that Vale dos Cucos mud/clay originates in the Hetangian clayey and calcareous formations. In the Thermal Centre of Vale dos Cucos mud baths at 38-39 °C and mud cataplasms at 43-45 °C are applied to patients.

Ingestion of thermal water does take place as well. These applications are used in the treatment of diseases such as: arthritis rheumatoid, arthritis psoriasis, extra-arthicular rheumatism, several other types of arthritis, anchilosant spondilite, etc. Vale dos Cucos mud/clay, utilised in the thermal treatments, has a complex composition made up of: kaolinite, illite, chlorite, smectite, interstratified chlorite-smectite, calcite, quartz, feldspar and sulphides (not identified). Table I shows the chemical composition data of Vale dos Cucos mud/clay as well as other relevant chemical and physical data.

Table I - Relevant chemical and physical properties of Vale dos Cucos mud

Chemical composition (<63 μm)	
SiO_2	38.49 %
Al_2O_3	17.21
TiO_2	0.75
Fe_2O_3	5.15
MnO	0.08
MgO	2.61
CaO	17.21
Na_2O	0.91
K_2O	2.57
P_2O_5	0.13
CO_2	13.0
SO_3	0.58
I.L.	5.04
Grain size distribution (<63 μm)	
< 10 μm	35%
< 2 μm	20%
Cation Exchange Capacity (<63 μm)	11 cmol.kg^{-1}
Main Exchangeable Cations (<63 μm)	$Ca^{2+}, Mg^{2+}, K^+, Na^+$
Specific Surface Area (<63 μm)	18 m^2.g^{-1}
Liquid Limit (< 63 μm)	36 %
Plastic Limit (<63 μm)	28 %
Expandability (<63 μm)	12 %
Abrasivity (<63 μm)	0,35 mg.m^{-2}
Apparent Specific Heat (<63 μm)	3.45 J.g °C
Cooling kinetics	30min., from 55°C down to 30°C

Table II - Relevant chemical and physical properties of Caldeira das Furnas mud

Chemical composition (< 63 μm)	
SiO_2	48.56 %
Al_2O_3	28.14
TiO_2	1.18
Fe_2O_3	2.38
MnO	0.05
MgO	1.02
CaO	0.24
Na_2O	0.84
K_2O	2.41
P_2O_5	0.19
SO_3	1.86
I.L.	10.56
Cation Exchange Capacity (<63 μm)	25 cmol.kg^{-1}
Main Exchangeable Cations (<63 μm)	Mg^{2+}, Ca^{2+}, K^+, Na^+
Specific Surface Area (< 63 μm)	50 m^2.g^{-1}
Grain Size Distribution (< 63 μm)	
< 10μm	80 %
< 2μm	30 %
Liquid Limit (< 63 μm)	85 %
Plastic Limit (< 63 μm)	25 %
Expandability (< 63 μm)	-
Abrasivity (< 63 μm)	0,15 mg.m^{-2}
Apparent Specific Heat (< 63 μm)	3.15 J.g°C
Cooling kinetics	25min. from 52 °C down to 30 °C

In Vale das Furnas there are 22 thermal springs, some with hot water, others with cold water. This fact is a good evidence of the great potential of this small area in hydrological terms.

The dark grey clay used in the mud-baths and mud-cataplasms being successfully applied in the Furnas Thermal Centre (São Miguel island, Azores archipelago) is very fine grained, smells to sulphur and is composed, essentially, by smectite and kaolinite. K-feldspar is an accessory component.

Chemical composition of the Caldeira das Furnas mud is shown in Table II together with other relevant chemical and physical properties.

Smectite content in the less than 63 μm fraction of Porto Santo bentonite from a site in Serra de Dentro (located in the eastern sector of the island) represents around 70%, whereas the smectite content in the less than 2μm fraction represents around 90%. Na-Ca feldspar, calcite, magnetite-maghemite and opal CT are the main accessory minerals identified in the less than 63 μm fraction.

Serra de Dentro smectite displays dioctahedral character and belongs to the series montmorillonite - nontronite.

Bentonite (<63μm) chemical composition is shown in Table III that contains as well data corresponding to other relevant chemical and physical properties.

Table III - Relevant chemical and physical properties of Serra de Dentro bentonite

Chemical composition (<63μm)	
SiO_2	45.01 %
Al_2O_3	18.57
Fe_2O_3	10.85
TiO_2	2.77
MnO	0.86
MgO	3.86
CaO	5.97
Na_2O	2.37
K_2O	0.45
P_2O_5	1.76
I.L.	7.54
Cation Exchange Capacity (<63 μm)	89 cmol.kg^{-1}
Exchangeable Cations (< 63μm)	Ca^{2+}, Na^+, Mg^{2+}
Specific Surface Area (<63 μm)	119 m^2.g^{-1}
Expandability (< 63 μm)	72 %
Particle Size	
< 10 μm	90 %
< 2 μm	71 %
Liquid Limit (<63 μm)	203 %
Plastic Limit (<63 μm)	42 %
Abrasivity (<63 μm)	0.11g/m^2
Apparent Specific Heat (<63 μm)	3.55 J. g °C
Cooling kinetics	38 min. from 58 °C down to 30 °C

Figure 1 shows the cooling kinetics of Vale dos Cucos mud, Vale das Furnas mud and Serra de Dentro bentonite, after being dried at 110ºC and cooled down to room temperature. In comparative terms, Serra de Dentro bentonite displays the lower cooling rate.

Experiences carried out with artificial mixtures of Porto Santo bentonite from Serra de Dentro and of Porto Santo carbonate beach sand show that the addition of carbonate sand, made up essentially of calcite (~80%), aragonite (~5%) and fine fragments of volcanic rocks (~15%), decreases the cooling rate of the bentonite pastes (containing 20% of moisture), i.e., its heat diffusiveness slows down. Both smectite and carbonate contents appear to influence very much the pastes thermal behaviour, lowering their cooling rate. In comparative terms, calcite particular crystallochemical features, including crystal structure and chemical bonds (type and contribution), make more difficult heat conduction than in common tectosilicates (such as quartz and feldspars) and phyllosilicates (such as clay minerals).

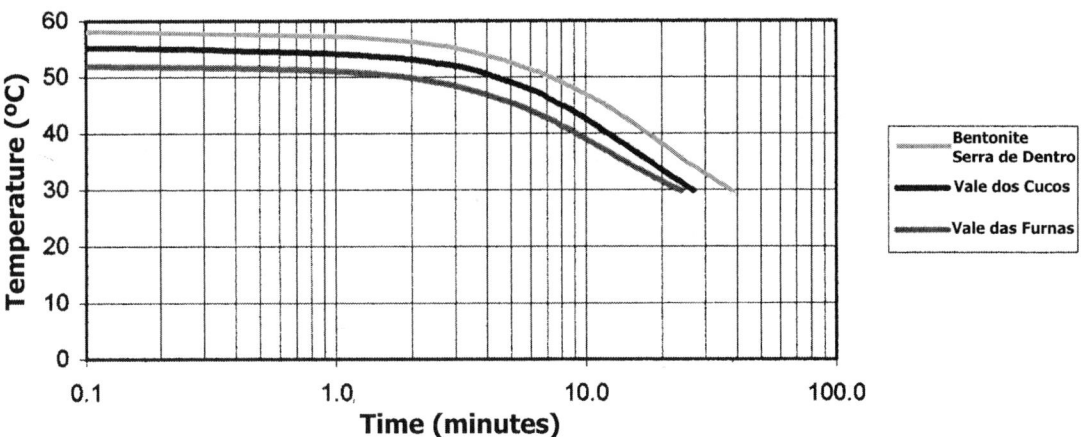

Figure 1- Cooling kinetics of Vale dos Cucos mud, Vale das Furnas mud and Serra de Dentro bentonite.

4. CONCLUSIONS

The three Portuguese muds/clays being studied show distinct physical and chemical properties. In fact they differ in textural parameters, in mineral and chemical parameters and in thermal behaviour. Smectite is present in all of them, but its content is very variable, discrete in Vale dos Cucos mud, medium in Vale das Furnas mud and high in Serra de Dentro bentonite. Concerning particle size distribution, Serra de Dentro bentonite is the finest one followed by Vale das Furnas mud and Vale dos Cucos mud.

It is considered that the therapeutic assets depend particularly on muds/clays thermal (heat diffusion and thermal shock) and microchemical (ion exchange) properties. Since skin acts as a semipermeable membrane, the existence of a gradient for some chemical exchangeable elements at the interface skin (Na, K) / mud (Ca, Mg, Sr, Zn), cation exchange reactions will take place.

REFERENCES

1. P. Barbieri, Atti. Conv. "Argille Curative" (Veniale, F., ed.), Salice Terme PV (Italy), Tipografia Trabella, Milano, 13-15 (1996).
2. J. Browne et al., J. Pharm. Sci., 69(1980), 816-823.
3. S. Cara et al.. Atti Conv. "Argille Curative" (Veniale, F., ed.), Salice Terme PV (Italy), Tipografia Trabella, Milano, 103-117 (1996).
4. S. Cara et al., Applied Clay Science, 16 (2000a), 117-124.
5. S. Cara et al., Applied Clay Science, 16 (2000b), 125-132.
6. T. Ferrand and J. Yvon, Applied Clay Science, 6 (1991), 21-38.
7. E. Galán et al., Bol. Soc. Esp. Min. (1985), 369-378.
8. M. Hermosin et al., J. Pharm. Sci., 70 (1981), 189-192.

9. A. Lopez-Galindo and C. Iborra, Atti Conv. "Argille Curative", (Veniale, F. ed.), Salice Terme (PV), Italy. Tipografia Trabella, Milano, 45-53 (1996).
10. G. Novelli, Atti Conv. "Argille Curative", (Veniale, F. ed.), Salice Terme (PV), Italy. Tipografia Trabella, Milano, 25-43 (1996).
11. G. Novelli, Incontri Scientifici. V Corso di Formazione "Metodi di Analisi di Materiali Argillosi". Gruppo Italiano aipea, 263-304 (2000).
12. W.R. Reinbacher, Clay Minerals Society News, 11, 1 (1999), 22-23.
13. R.II.S. Robertson, "Fuller's Earth: A history of calcium montmorillonite. Volturna press, Hythe, Kent,U.K., 421p. (1986)
14. H. Ueda, and M. Hamayoshi, J. Materials Sci., 27 (1992), 4997-5002.
15. F. Veniale, and M. Setti, Atti Conv. "Argille Curative", (Veniale, F. ed.), Salice Terme (PV), 139-145 (1996).
16. J. Yvon, and T. Ferrand, Atti Conv. "Argille Curative", (Veniale, F. ed.), Salice Terme (PV), Tipografia Trabella, Milano, 67-78 (1996).

Industrial clays of Brazil: a review

Pérsio de Souza Santos

Dept. of Metallurgical and Materials Engineering - Polytechnic School, University of São Paulo, P.O. Box 61.548, São Paulo, SP, Brazil

Brazilian kaolins, residual and sedimentary, range as 6th in world production. Deposits are in Jari and Capim rivers, Pará. Ball clays used for whiteware industry are partly supplied by imports from U.K. and Argentina. Refractory industry is supplied by fireclays and high alumina (gibbsitic) clays from several Brazilian deposits. Sodium-bentonite is produced for petroleum and metallurgical industries from Ca-bentonite from Northeast; there is an import of Na-bentonite from Argentina. Acid activated smectite is produced for bleaching of vegetable oils and there are imports from Mexico and USA. Organophilic bentonite is also produced with Brazilian raw materials. Palygorskite clays occur in the State of Piauí. Brazil is the 5th world producer of ceramic tiles; illite/smectite rich clays are widespread in the country and are used for that purpose and also for lightweight aggregates. Vermiculite deposits are exploited for production of exfoliated materials for thermal and acoustical insulation. Nickel clays, mostly smectitic, are used for production of nickel. Extensive deposit of oil shale is exploited for pilot-plant production of oil shale. The shale and the retorted shale can be used for self-firing production of structural clay products. Common clay is widespread and is used for Portland cement production, pozzolan, and structural clay products. There is small production of ground pyrophyllite for refractories and Al/Si alloy. There is a large production of different types of talc from massive and from clay-like rocks. Chrysotile asbestos is produced for internal consumption and for exportation.

1. INTRODUCTION

Brazil's economy is 8th in the world's economy. Clay production in Brazil is extensive and the country is almost self sufficient in the consumption of industrial clays, names given following the U.S. Bureau of Mines classification: kaolins; ball clays; refractory clays; bentonites; fuller's earths; common clays. The Brazilian chemical and metallurgical industries are the major users of their domestic industrial clays; there is significant exportation of kaolin for paper. Recent data[1] on clay production in Brazil are given in Table 1. In the Table, included under "common clay", is the production of ball clays and refractory clays. Detailed information on the industrial applications of Brazilian clays up to 2000 was published by Souza Santos[2] and Motta et al[3].

Table 1 - Production of refined processed clay in Brazil.

Year/Clay	1998	1999	2000
Kaolin[*]	1374	1517	1735
Bentonite[*]	220	275	274
Vermiculite[*]	24	23	23
Talc[*]+Pyrophyllite	452	454	450
(Agalmatolite)[*]	-	-	-
Bauxite[*]	11961	13839	13846
Al_2O_3[*]+Al	1266	1.362	1.987
Nickel Clays[*]	1080	1320	1280
Nickel[*]	27	33	32
Muscovite Mica[**]	4	3	4
Oil Shale[*]	8500	8500	8500
Chrysotile Asbestos[*]	198	188	209
Common Clay[*]	21190	21.490	238,40
(Diatomite)[***]	10	8	7
(Feldspar)[***]	59	65	61
Cement[*]	39942	40270	39208

*Thousands of metric tons; **Metric tons; *** Non-metallic minerals with industrial uses in parallel with clays

2. KAOLINS

Kaolin deposits in Brazil are either residual from pegmatites or sedimentary[4]. Measurements of four (crystallinity indices) of Brazilian kaolins are presented at this Conference[5]. The residual kaolins from the Center-South deposits (States of Minas Gerais; São Paulo; Paraná; Santa Catarina; Rio Grande do Sul) usually are composed of natural mixtures of kaolinite and halloysite in different proportions. The kaolinite plates are normally of irregular profile with medium to high degree of structural order. These kaolins are used for whitewares (sanitaryware, dinnerware, technical porcelain and wall tiles) and as filler for paper and plastics. Brazil is a large exporter of sanitaryware and dinnerware products, as well as of kraft paper. The homogeneous pegmatites of Northeast Brazil produce extremely white kaolins, composed of over 90% by weight of kaolinite, with euedral crystals, with very high degree of structural order and having a regular hexagonal profile. They are used for whitewares in the Northeastern ceramic industries and as coating kaolins for paper in the State of São Paulo[6]. The sedimentary kaolin from Jari River in the State of Amapá is exported as coating kaolin and is also used in the paper industry in the State of São Paulo[7]. Also in the Amazon region, in Capim River, São Domingos do Capim, State of Pará, there are large deposits of very white kaolins. They are now being processed for export as paper coating clay by two companies[8]. The processed kaolin production in 2000 was 1,735,000 metric tons. The apparent consumption of processed kaolin was 350,000 metric tons in 2000.

3. BALL CLAYS

Highly plastic clays showing white color, after firing in oxidizing atmospheres at the 1200°C-1250°C range, are called "ball clays". The most important examples of "true ball clays" are those from the Devon-Dorset counties in U.K. and from the States of Tennessee, Texas, Kentucky and Mississippi in USA. They are the major producers and exporters worldwide[9,10]; other example is from Ukraine. Owing to the relative scarcity of "true ball clays" other types of plastic kaolinitic clays, generally with lower white-firing characteristics, are utilized for much ceramic production. The whiteware industry in Brazil, specially sanitaryware, uses such ball clays from deposits from São Simão county, State of São Paulo, Tijucas do Sul, State of Paraná and Oeiras county, State of Piauí. The São Simão ball clay is used also as a binder in several types of refractories. Other ball clays from the State of Rio de Janeiro and São Paulo are used for the manufacture of wall tiles and dinnerware. Plastic refractory clays from Suzano and Ribeirão Pires counties, State of São Paulo containing significant amounts of the Al-hydroxide gibbsite are used by substituting up to 1/3 of the ball clay from São Simão. Higher percentages produce a significative thixotropic increase in the apparent viscosity of casting slips for sanitaryware manufacture. They can be made suitable for that purpose; by firing at 300°C, in neutral or reducing atmospheres: the gibbsite is dehydroxilated. The thixotropic behavior is then destroyed, making the clays usable for casting slips for the sanitaryware.

Some ball clays are imported from Argentina and from U.K.; however, better ball clays and feldspars, very white-firing, are very much needed in Brazil for the fabrication of porcelain tiles (gres porcelanato)[11].

4. REFRACTORY CLAYS AND FIRECLAYS

Brazil produces all types of refractory products in different sizes and shapes and as well the country is an exporter of these products. Refractory plastic kaolinitic clays, low in iron and fluxes, are widespread, in several States (São Paulo, Minas Gerais, Paraná), as well as plastic gibbsitic fireclays, containing up to 80% Al_2O_3 after firing at 1300°C. Unfortunately deposits of diaspore and flint fireclays are not known until the present; also, no commercial deposit of emery is known. Iron rich bauxites have been known for long time in the States of Minas Gerais, São Paulo, Pará and Maranhão. These bauxites are transformed industrially into alumina by the Bayer Process for Al production and export; also into electro-fused Ti-rich corundum for abrasives, synthetic mullite for refractories, and into gamma-Al_2O_3 for adsorbents and catalysts for sulfur production in the Claus Process, using H_2S from the Brazilian petroleum refineries. White bauxites from Poços de Caldas county are also used for fabrication of refractories. In 1995, 8103 thousand metric tons of clay refractories were produced in Brazil.

5. BENTONITES

Until today, smectitic clays for industrial uses are extracted from the State of Paraiba, county of Campina Grande, district of Boa Vista. In 2000, 274 thousand metric tons were produced[1] in Brazil; with 96% of Brazilian production coming from Paraíba. The green and red smectitic clays have CEC of 84/95 miliequivalents/100 grams of clay; the major exchangeable cation is Ca^{2+}; and Na^+ is responsible for 10% of the exchangeable cations. Ca-smectites, of green and red colors, are used as foundry bonding sands. Brown clays, exchanged with sodium carbonate, are used for foundry as well as binders for iron oxide pellets produced for export in the city of Vitória, capital of the State of Espírito Santo. Green and red smectite clays (localities of Bravo, Lages and Juá) are transformed into Na-smectites by chemical reaction with sodium carbonate in adequate humid environment. These "synthetic" Na-smectite clays meet the specifications for bentonite for drilling fluids made by Petrobrás, the larger state chemical petroleum and petrochemical industry in Brazil, based on API's specification 13-A. Small smectite deposits exist in several regions of Brazil, but are not exploited. In the southern part of State of Minas Gerais, there are enormous deposits of green nontronite formed from weathering of ultrabasic rocks. The nontronitic clays exchange with Na_2CO_3, but the Na-nontronite does not swell in water (Forster's swelling), and as a consequence, cannot satisfy the Petrobrás (and/or API's) specifications for Na-bentonite, for the minimun value for the plastic viscosity at 6% weight concentration. In the Southern States of Brazil, closer to Argentina than Paraíba, the foundry industries prefer to import Argentinean bentonites. The "synthetic Na-bentonite" from the State of Paraíba is also used for water treatment and for impermeabilization of foundations in buildings in the city of São Paulo as well as an adsorbent for animals litters. Weathering of granitic pegmatites may produce local concentrations of smectites, sometimes inside tunnels made in granite for water transportation in the cities of São Paulo and Rio de Janeiro; these dry smectites, embedded in granite, when wet by water, swell and develop very high pressures on the rocks, which produce collapse of the walls and, as a consequence cloggi the water tunnel. Similar damaging effect occurs when basalt weathered into nontronite is used as aggregate for the concrete employed in the dams on the rivers in the South of Brazil. Organophilic bentonites for drilling purposes are being produced in the city of São Paulo using synthetic Na-smectite from State of Paraíba.

6. FULLER'S EARTHS AND ACID ACTIVATED CLAYS

Fuller's earth is a traditional and biblical name given to clays which, naturally, have a high bleaching or decolorizing power for oils, fats and waxes, from animal, vegetal, mineral or synthetic origins[12,13] . Following to Robertson[14], the fuller's earths can be divided into two main groups: the English fuller's earths, constituted by calcium montmorillonite and the American fuller's earths, constituted by palygorskite. Sepiolite and some mixed-layers illite-smectite clays also have that natural decolorizing power. Ca-bentonites, treated by HCl or H_2SO_4, are transformed into acid activated clays with high bleaching power for oils and fats. Brazil is one of the bigger producers and exporters of animal fats and vegetables oils, specially of corn, soybean, cotton, peanut, castor and sunflower. Twelve thousands metric tons of acid activated clay were imported from Mexico in 1996. Natural smectite green from large clays, deposits over oil-shale in the Paraíba River valley in the State of São Paulo are extensively used for bleaching of vegetable oils. They are considered fuller's earths. Acid activated smectites, produced by reaction of H_2SO_4 with Ca-smectites from the Boa Vista district, Paraíba State, are now being produced in the city of Maceió, in the State of Alagoas. Successful developments were made by HCl activation of smectite clays from the region of the oil shale deposits of the Paraíba River valley for use in corn oil decolorization. The smectite used accounts for 4% of the Brazilian production. A review of the studies on the acid activation of Brazilian smectites was presented to the 11th International Symposium on Activated Clays in La Plata, Argentina[15].

Only small occurrences of no commercial value of palygorskite-sepiolite clays were known in Brazil until recently. Significant deposit of palygorskite mixed with white smectite was found in the bentonite deposit of Bravo, in the State of Paraíba. Also large deposits of good quality palygorskite were discovered in the States of Piauí and Maranhão. They are effective as fuller's earths for bleaching of soya bean oil [16]; and studies are in progress for use of these clays for offshore oil drilling muds in Southeastern Brazil.

7. COMMON CLAYS

This name was proposed by the U.S. Bureau of Mines to designate a group of clays, with varied mineralogical and chemical compositions, which are used in large tonnage in industry, The price is inexpensive in spite of their social importance, especially for producing building materials for houses. Reference [17] is an overview of the present Brazilian Ceramics Industries.

Structural Clay Products: - This name covers bricks; roofing tiles; paving tiles and all the "red ceramic products". Clays containing clays minerals rich in fluxes (Na; K; Ca; Mg; Fe), but with different geological origins, are used for this purpose. There is a widespread regional distribution, so each town or city has several ceramic factories for fabrication of these building products in the States of Brazil. Firing is mainly done by burning wood or fuel oil produced by Petrobrás. The green clays over oil shale and the retorted oil-shale from the Paraíba River Valley are excellent raw-materials for structural clay products.

Portland Cement: - Brazil is a large producer of Portland cement; about 70 factories exist spread around all regions of the country. Kaolinitic, illite and illite-smectite clays are used with argillaceous limestones or calcites. The present Portland cement production is adequate for Brazil's buildings needs and projects. 39.208 thousand metric tons were produced in the country in 2000.

Pozzolans: - Eleven factories in Brazil produce pozzolanic cement. Six of them use, as pozzolans, kaolinitic and/or illite clays, fired at 700°C-800°C in rotary kilns with oxidizing atmospheres. 20% to 30% of the Portland cement is clay pozzolan. The other pozzolans are clayey flyashes from coal in the States of Rio Grande do Sul and Paraná. These pozzolanic cements are used to minimize the alkali-aggregate reaction in concrete, specially that used in the dams built for production of electric energy for the country, as well as to economize thermal energy. The production of clay pozzolan in 1987 was about 270.000 metric tons.

Clays for Lightweight Aggregate: Illitic shale from the locality of Jarinu, near the city of Jundiaí, State of São Paulo is used as raw material for production of lightweight aggregate for concrete. The high quality coated aggregate pellets have very high mechanical strength. The most important buildings in São Paulo and Rio de Janeiro in the last 35 years have used the expanded clay lightweight aggregate. Oil shale retorted at 550°C, from the Paraíba River valley in the State of São Paulo and from the locality of São Mateus do Sul, State of Paraná (Petrosix Project from the Petrobrás Company) can be used to produce excellent coated lightweight aggregate for concrete. Activated sludge from the city of São Paulo, rich in kaolinite and illite as well as in organic matter, is now pyroexpanded for commercial production of non-coated lightweight aggregate for building purposes.

Floor Tiles: Brazil is the 5th largest producer and exporter of ceramic "coverings", that is, of ceramic tiles, wall and specially floor tiles. Sedimentary iron-potassium-rich clays are a major component of illite-smectite ceramic bodies used in factories in several States, specially in the Southern States of Brazil. These clays may also be used for production of coated lightweight aggregate and high quality structural clay products.

Stoneware Clays: Friable rocks constituted by finely divided quartz (40%), kaolinite and halloysite (30%) and sericite (finely divided muscovite mica) are used in sanitaryware and wall tile bodies as partial substitutes of feldspar, kaolinite and mica. Iron content is less than 1% and they vitrify after firing between 1200°C and 1250°C. The mines are in the city of Itapeva, in the State of São Paulo and Santo Antonio do Norte, State of Minas Gerais. The local name is "filito", which has no scientific meaning. It is also commercialized as an inexpensive filler for different products, sometimes under the wrong name of "industrial talc".

Partially weathered feldspar from pegmatites in the Southern part of the State of Minas Gerais and from the city of Mogi das Cruzes, State of São Paulo are fired at 700°C and ground to be sold as an inexpensive white pigment and filler; the local name is "feldspatito".

8. VERMICULITE

Exfoliated macrocrystals of vermiculite from deposits in the States of Piauí and Goiás are commercially used for thermal and acoustic insulation, either as blocks or in bulk or loose form. There is some exportat of crude vermiculite. Exfoliated vermiculite is used in pet litters. Delaminated vermiculite, after swelling in water, can be used in cement products in place of asbestos and muscovite mica[18].

9. MUSCOVITE MICA

Large crystals of moscovite mica are extracted from homogenous pegmatites for use as an electric insulator. The processing of residual kaolins from pegmatites produces large quantities of good quality mica in small crystals. These crystals are finely ground for use as filler for paints and plastics to give a pearls luster or processed for production of resin and glass bonded products for electrical use. The production of these micas comes from kaolin deposits from pegmatites in the States of Minas Gerais and São Paulo. The sericite (muscovite mica) is a fluxing component of the rock named "filito" very much used as substitution for feldspar in bodies for sanitaryware and wall tiles.

10. TALC

High quality talc is mined and processed from the magnesite deposits in Serra das Éguas, State of Bahia for use in cosmetics, filler for plastics, paper and other uses. Talc is also mined from the county of Ponta Grossa, State of Paraná. In the majority of deposits, the talc occurs as a clay, not as a massive rock. These talcs are formed by silicification of dolomitc rocks; their iron content ranges from very low values up to 6% to 8%. The low-iron talcs are used in cosmetics, as filler for plastics and for the fabrication of cordierite refractories used as kiln furniture and porcelain insulators. The iron rich talcs are used in the fabrication of ceramics bodies for vitreous floor and wall tiles[19].

11. PYROPHYLLITE AND AGALMATOLITE

Pyrophylite deposits in the city of Diamantina, State of Minas Gerais are exploited for the production of Al-Si alloys. Mines in Pitangui and Onça do Pitangui, Minas Gerais produce clays rich in pyrophylite which are used in ceramic bodies for wall tiles and dinnerware, and as fillers for plastics and other materials. In Ibitiara, Bahia, an extensive deposit of pyrophylite exists either as macrocrystals or as microcrystals; the large crystals, fired rapidly at 700°C to 800°C, exfoliate like vermiculite; unfortunately that property is lost by cominution of the macrocrystals.

Agalmatolite is a name of oriental origin which is used, in the regions of Pará de Minas, State of Minas Gerais, to hard greenish-bluish massive rock composed essentially of pyrophylite, with variable amounts of sericite (muscovite mica) and small contents of kyanite and diaspore. The massive rock is extensively used as a component in alumina-silica refractories to increase their mechanical strength. Finely ground (minus n° 325 sieve), agalmatolite is sold with the name "industrial talc".

12. NICKEL CLAYS

The production of nickel metal and Fe-Ni alloys in Brazil uses nickel clays (locally called garnierites) which contain Ni^{2+} in the crystalline structure of serpentines, smectites, vermiculites and chlorites. The major deposits of nickel clays are the localities of Niquelândia, State of Goiás and of Morro do Níquel, Pratapolis, State of Minas Gerais. Minor deposits occur in Jacupiranga, State of São Paulo and in Paulistana, State of Piauí.

13. CHRYSOTILE ASBESTOS AND ANTIGORITE

In the county of Mineraçu, State of Goiás, the Uruaçu mine produces chrysotile asbestos for consumption in the country of asbestos-cement products and also for exportat. Other serpentine clay mineral is antigorite, which is produced from a small deposit in the locality of Castro, State of Paraná. It occurs as a white clay; and sold for the same uses of the low-iron talcs of Ponta Grossa, Paraná.

14. OIL-SHALE

In the Paraíba River valley in the State of São Paulo, there was extraction of peat and oil shale for use as fuel during World War II. Now, Petrobrás, the largest chemical industry in Brazil, has built a big pilot plant for retorting, at 550°C, oil shale from the Irati formation in the locality of São Mateus do Sul, State of Paraná. The distilled shale-oil is now sold to ceramic industries in the State of Paraná as a substitute for fuel oil for the kilns. This pilot unit processes 25 thousand metric tons of shale per day, producing one thousand barrels of shale oil, $36,500 m^3$ of combustible gas and 17 tons of sulfur. The plan is to have several units as alternative source of oil, Brazil imports significant quantities of oil from several countries, especially from Argentina.

15. BAUXITE

Brazil is a major producer of alumina and aluminium from bauxite deposits formed by desilicatization of kaolinitc and halloysitic clays. These deposits are in the counties of Poços de Caldas and Ouro Preto, in the State of Minas Gerais and Paragominas, in the State of Pará. Gibbsite and kaolinite are the major minerals of these bauxites. Thermally activated bauxite is used as adsorbent and bleaching agents for mineral oils; also it is now extensively used as a catalyst for sulfur production by the Claus process by the Petrobrás units.

16. CONCLUSIONS

Brazil is a major producer of all types of industrial clays for the consumption of a population of 170 million people and also for a significant amount for export, especially of kaolin for paper. A total of 11,177 metric tons of several clays were imported in the year of 2000. No ball clay or fireclay were specially listed as imported in 2000. There was importation of 2,133 metric tons of kaolin clay for coating paper; 9,042 tons of bentonite; of which the majority is activated smectites for bleaching of vegetable oils for food and export and of palygorskite clays for diluent of pesticides and for oil drilling in saltybrine environments (listed as bentonite) and some vermiculite (2 thousand metric tons). 7,892 metric tons of diatomite were imported in 2000. The clay importation has decreased significantly since 1988, from 23,588 to 11,177 metric tons in 2000.

18. ACKNOWLEDGEMENTS

This paper is part of Projeto Temático FAPESP 1995/0544-0
Presented at the Symposium 6: Clay Resources in the Mercosur

REFERENCES

[1] DNPM - Sumário Mineral 2001, Ministério de Minas e Energia, Brasília, 2001.
[2] P. Souza Santos, Ciência e Tecnologia de Argilas, $2^aEd.$, 3v.; Edgar Blücher, São Paulo, 1992.
[3] J.F. Motta, M. Cabral Jr., J.V. Valarelli, P. Souza Santos, 31st IGC Special Symposium I, Rio de Janeiro, 2000.
[4] I.R. Wilson, H. Souza Santos and P. Souza Santos, Cerâmica 44 (1998) 118.
[5] A.C. Vieira Coelho and P. Souza Santos, This Conference, Abtracts, p. 163, Bahia Blanca, 2001.
[6] A.P. Ribeiro Filho, Geologia e Metalurgia 40 (1976) 353.
[7] A. Halward, Das Papier 33 (1979) 207.
[8] M.L. Costa and E.L. Moraes, Mineralium Deposita 33 (1998) 283.
[9] R.L. Virta, Clay Shale 1999 Annual Survey, US Geological Survey, Reston, 2000.
[10] M. O'Driscoll, Industrial Minerals, October 1998, p.21.
[11] J.F. Motta, Polo Cerâmico n°48 (2000) 2.
[12] R.B. Ladoo and W.M. Myers, Nonmetallic Minerals, 2nd Ed., McGraw-Hill, N.Y., 1953.
[13] R.E. Grim and N.Güven, Bentonites; Geology, Mineralogy, Properties and Uses, Elsevier, Amsterdam, 1978.
[14] R.H.S. Robertson, Fuller's Earth: a History of Calcium Montmorillonite, Volterna Press, Hythe, 1986.
[15] P. Souza Santos, Proceed. 1st Intl. Worksp. Activ. Clays, p.21. La Plata, 1998.
[16] J. Pereira Neto and S.L.M. Almeida, Atapulgita do Piauí, CETEM - Tecnologia Mineral n° 64; DNPM; Rio de Janeiro, 1993.
[17] M.A.P.J ordão and U.P. Santos, Bol. Soc. Esp. Cerámica Vídrio 40 (2001) 59.
[18] A.C. Coelho and P. Souza Santos, Cerâmica, 30 (1984) 307.
[19] P. Souza Santos and H. Souza Santos, Cerâmica 36 (1990) 127.

White bentonite from Patagonia, Rio Negro Province, Argentina

J.M.Vallés[a,b] and A.Impiccini[a]

[a]Universidad Nacional del Comahue, Facultad de Ingeniería, CIMAR. Buenos Aires 1400, 8300, Neuquén, Argentina. jvalles@uncoma.edu.ar
[b]CONICET – Consejo Nacional de Investigaciones Científicas y Técnicas, Argentina.

The exploitation of white and non-swelling bentonite whose geological and mineralogical features were not widely known, has recently begun in Río Negro Province, in South of Argentine. The deposits are alteration products of pyroclastic materials (volcanic ash) that originated in late Cretaceous volcanic events. These deposits are composed mainly of montmorillonite and disordered cristobalite-tridymite (opal-CT). Other impurities and accessory minerals are glass shards, quartz, albite, calcite, K-feldspars, biotite, pyrolusite, pyroclastic rock fragments and diatom shells. The whiteness index reaches 78 %. Their CEC varies between 83 and 98 meq/100g. Chemical composition shows a low Fe_2O_3 content that varies between 0.80 and 0.85% and a moderately high magnesium content: 3.5 and 3.9%. The main octahedral cation is aluminum. Calcium plus magnesium predominate over sodium as natural interlayer cations. The replacement of Si by Al is very low and the main part of the layer charge comes from substitutions in the octahedral sheet. White bentonites like these, are relatively rare in occurrence, compared to other bentonites, particularly in Argentina. Technologically, this bentonite has rheological properties that do not favor its use as a component in drilling muds and gels. Nevertheless, it shows interesting properties for it use as plasticisers in ceramic and refractories, as a filler and strengthener in plastics, as a filler in rubber, as a decolorizing (bleaching) and refining agent for oils, and many other applications.

Key words: White bentonite, smectite, technological properties

1. INTRODUCTION

In Argentina, non-swelling white bentonites, have only been exploited in the North of the country, in San Juan and Mendoza provinces. On the contrary, bentonites that have been mined up to now in the South of the country, are yellowish and green colored, swelling, sodium bentonites. The exploitation of a group of deposits of this type of white bentonites has recently begun in the Patagonia region, in the South of the country. The geological, mineralogical, genetical and technological aspects of these bentonites had not been studied in depth up to the moment. The mentioned deposits were included in a program that leads to improve the knowledge of the Southern bentonites, in order to study them, to evaluate their applications and to optimize their use.

2. LOCATION AND GEOLOGICAL SETTING

The bentonite deposits are situated in the South of Río Negro province, 5 Km Southeastward from Teniente Maza railroad station near route 23.

They are located in the Northpatagonian Massif and included in the Neuquén Group, an Upper Cretaceous-Paleocene continental sedimentary sequence. In North Patagonia, that red-beds sequence is generally more than 500 m thick, but in this place it is less than 50 meters thick. Lithology consists of coarse sandstones, conglomerates, siltstones, claystones, mudstones, tuffaceous sandstones and cinder tuffs, which were deposited in a fluvial environment (Dirección de Minería, 1999). Sediments lie in near horizontal position. In some outcrops of this formation, remains of dinosaur bones and petrified trees were found.

This is the first report of bentonite occurrence in this kind of sedimentary environment.

A geological map and a generalized stratigraphic section of the area with the position of bearing-beds, is shown in Figure 1a.

3. MATERIALS, SAMPLES AND METHODS

To describe geological features, 8 vertical sections of open quarries and trenches have been studied. Representative samples from the mines El Cerro, Catalina and Cateo 3 were taken.

X-ray diffraction (XRD) tracings were obtained using a Rigaku Dmax-2D diffractometer with CuKα radiation. Sample preparations on glass slides for the air-dried, ethylene glycol saturated, and heated conditions on <1-μm clay fraction were conventional as well as bulk powdered random condition (Moore and Reynolds,1997). XRD and optical microscopy identification of non-clay components were performed on coarse fractions retained in mesh # 270. Greene-Kelly (1951) tests after Li^+ saturation and dropping onto ceramic plate were performed. Scanning electron microscopy observations and EDX semiquantitative analyses were done with a Philips 515. Bentonite was chemically analyzed by ICP after $LiBO_2$ fusion. Cation exchange capacity (CEC) was determined after Jackson's (1956) method adapted by Giaveno and Chiachiarini (1995), after saturation with sodium acetate at pH 8.2 and extraction with ammonium acetate at pH 7. Specific surface area was measured by water adsorption, using $NaNO_3$, NH_4Cl and E.G.M.E. as absorbates.

Determination of foundry properties of samples were carried out on samples in their natural state. Swelling capacity of bentonites were tested on 2 g of clay. Green and heated compression and wet tensile strength, were measured on testing sands prepared with 5 % of bentonite and silica sand. Thermal stability was tested comparing methylene blue adsorption in natural and 550ºC heated conditions. Measurements of whiteness and yellowness were made by means of a Photovolt photocolorimeter.

Rheological properties were determined according to API (1993) specifications, with a Fann viscometer, filter-press Baroid type, on a 6.42 % solid/liquid dispersion.

4. RESULTS

4.1. Geology of deposits

Outcrops of bentonite bodies occur in an area of 35 square km. Clay deposits consist of extensive, subhorizontal lenticular shaped bodies with thickness usually thinner than 2.90 meters (average 2.25 m) interlayered between shaly sand, tuff, and mudstone beds. Clay beds

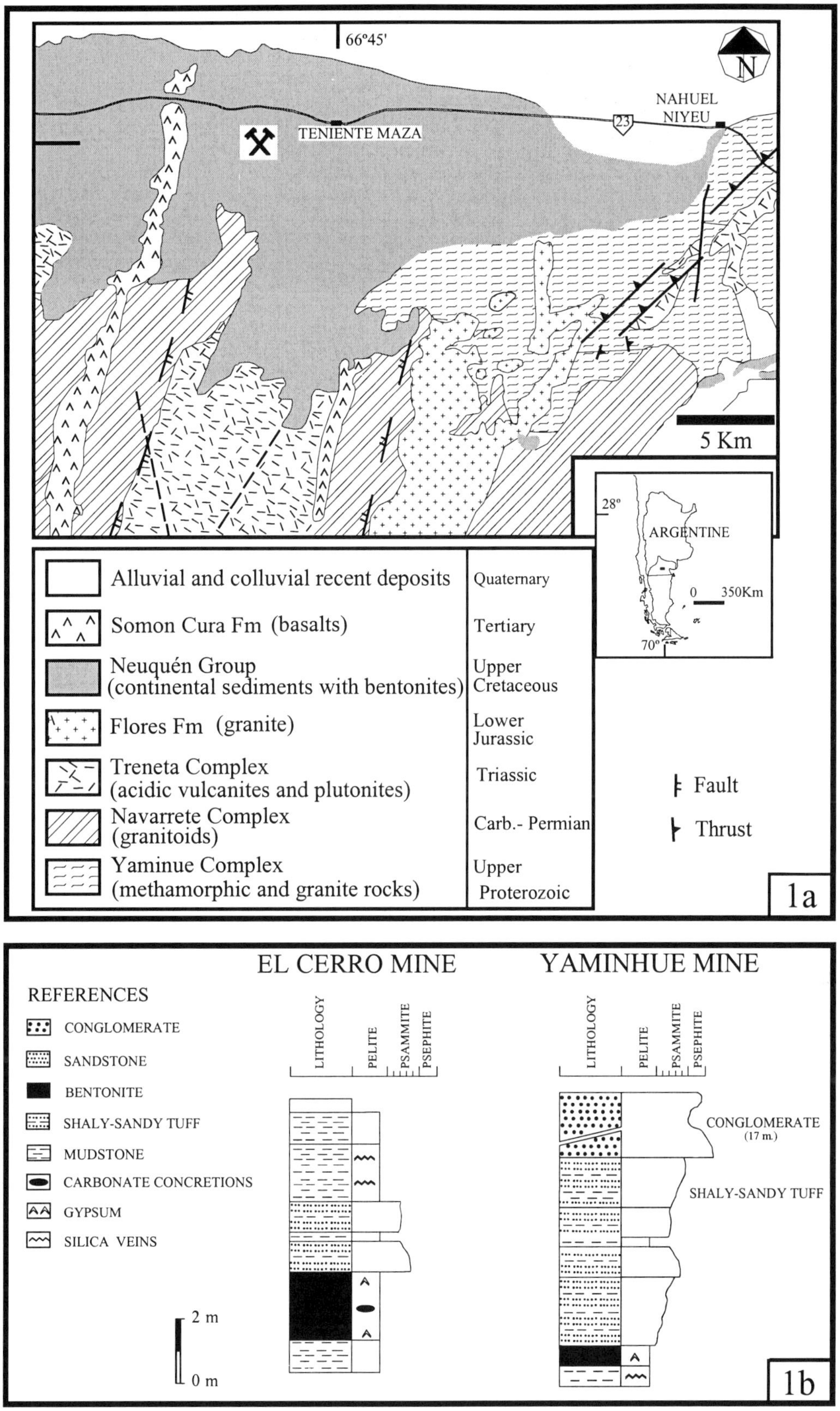

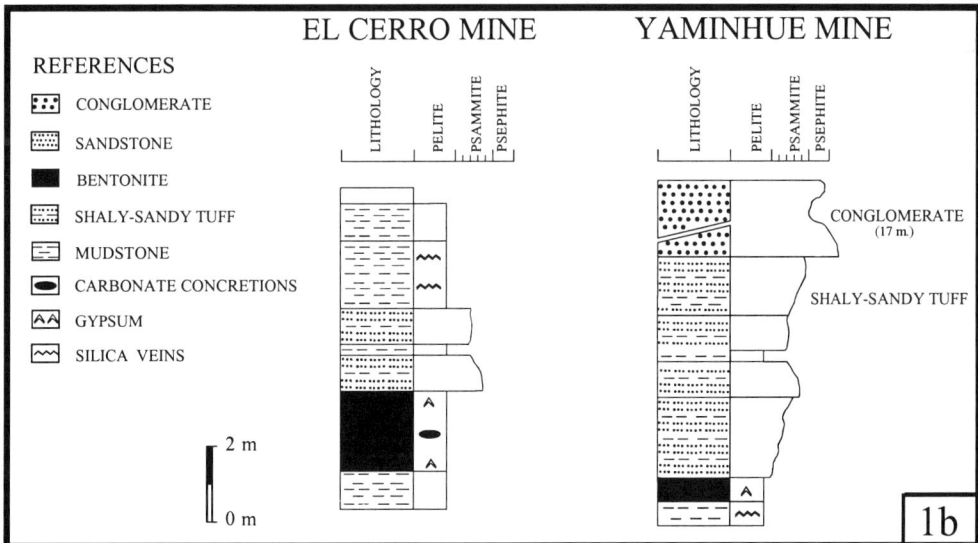

Figure 1: a) Geological map of the area of bentonite deposits, Río Negro Province, Argentine.
b) Schematic sections of El Cerro mine and Yaminue mine.

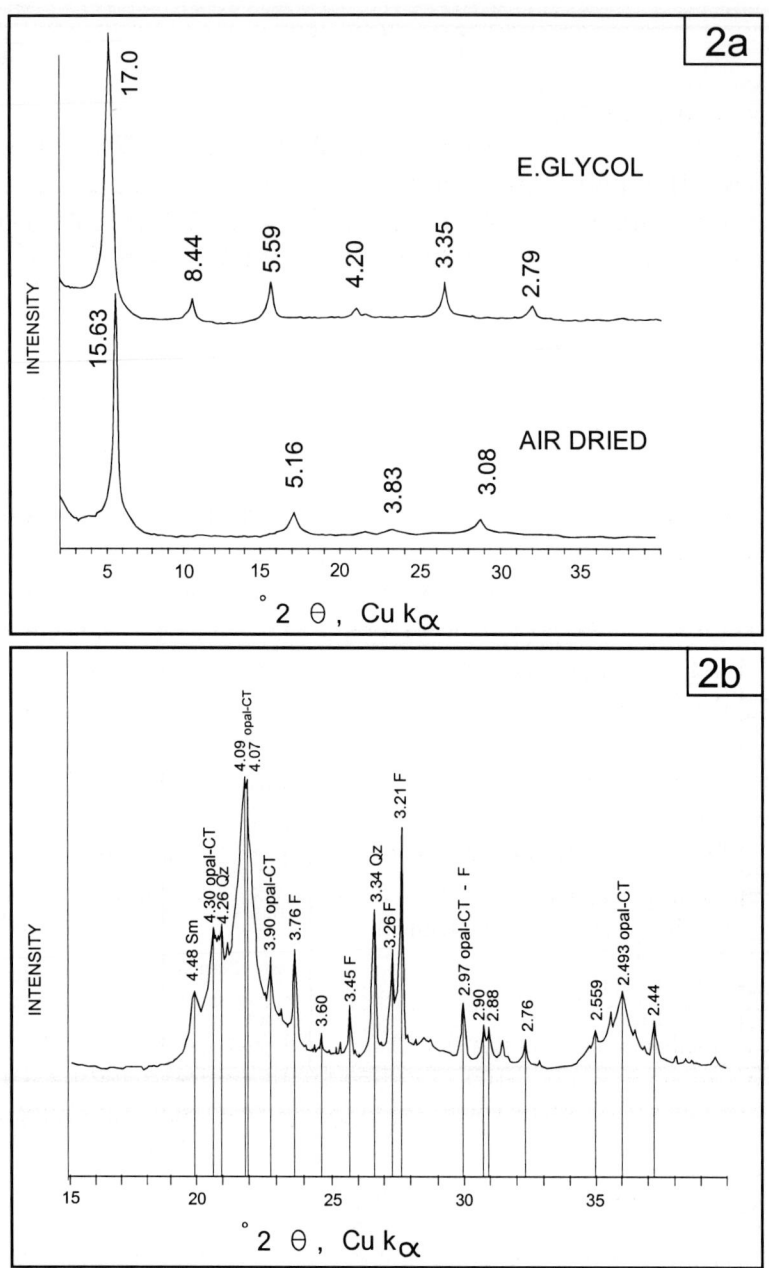

Figure 2: XRD powder diffraction patterns of samples of bentonite, a) oriented preparation, air dried and glycolated, b) unoriented preparation, Sm (smectite), Qz (quartz), F (feldspar)

have sharp contacts with the underlying layers and diffuse with overlying. In El Cerro mine, a thin calcite layer of 0,13 m thick constitutes the floor of the bentonite bed. This layer separates bentonite from the underlying red clay. Schematic profiles are shown in Figure 1b. Bentonite is massive, white colored, it has an opaline appearance and a conchoidal fracture. Layering is not evident.

In the clayey matrix, abundant calcite nodules of spheroidal shape and sizes between 5 and 25 cm (average 15 cm) can be found. They contain more than 60 % of calcite, 30 % of smectite, opal-CT, feldspar and quartz. Included into the bentonite beds frequent veinlets and nodules of hard microcristalline silica may be found. To a lesser extent, there are thin veinlets of gypsum 1 to 2 cm in thickness.

4.2. Mineralogical composition

The mineralogy of the bentonite beds is typically dominated by smectite and also includes non-clay minerals. The only clay mineral found is an aluminian dioctahedral smectite characterized as montmorillonite. XRD traces can be seen in Figure 2a. The montmorillonite occurs with a cornflake morphology. In SEM images it can be seen formed at expenses of the volcanic glass shards.

The mineralogical assemblage consists, besides montmorillonite, of disordered cristobalite-trydimite (opal-CT) as major component. Although glass shards, albite, quartz and calcite are common, appearing as minor phases. There are also pyroclastic rock fragments, biotite, K-feldspars and pyrolusite but in lesser amounts. Gypsum and diatom shells may also be present, but in trace amounts.

Opal-CT occurs either as silica veinlets like thin partition walls into the mass of smectite flakes or as massive overgrowth of microcrystals 0.3 to 0.5-µm in size. It is the prevailing mineral in coarsest fraction (> 64 µm) observed by means of optical microscope. Figure 2b shows a representative XRD pattern of a coarse fraction of a sample showing non-clay minerals, rich in opal-CT (Jones and Segnit, 1971) with broad peaks at 4.30-4.32 Å, 4.06-4.09 Å, 3.90 Å and 2.49 Å. The reflections are broad enough that the first three are never fully resolved.

Glass shards are colorless, with angular faces, fragments with a half-moon shape, with grooved textures, from less than 1 µm to 70 µm in size, refraction index 1,48. A semiquantitative SEM/EDX analysis of several glass particles gives as average result: SiO_2 71.4 %; Al_2O_3 14.6 %; Fe_2O_3 12.4 %; CaO 1.3 %. Plagioclase composition is An 0-10.

4.3. Chemistry

Table 1 shows bulk chemical analyses of characteristic samples from mines El Cerro (0612 and EC1), Catalina (0614) and Cateo 3 (0914). The relatively high content in Mg and low content in Fe and Ti stands out as well as a high content in SiO_2 coming from opal-CT. CEC measured in natural conditions is over 80 meq/100 g in all samples (Table 2). Calcium plus magnesium predominate over sodium as natural interlayer cations.

4.4. Surface area

Measurements of surface area by water adsorption using the three absorbates mentioned above, have given: 777, 627 and 720 m^2/g on sample EC1.

4.5. Technological properties

Rheological tests were performed on several samples of natural bentonites. This type of clay does not develop the gel properties suitable for drilling muds. Yield is extremely low: between 1.5 and 2.5 p/100 i^2, filtrate loss volume very high (between 14 and 26 ml) and plastic viscosity is less than 2, tested on a suspension 6.42 wt %.
In Table 3 the results of diverse tests are shown. They are applied mainly to the use as binders in the foundry industry. The adsorption of MB is appropriately high, although in some samples, the adsoptive capacity falls abruptly before heating at 550ºC. The swelling volumes are low, typical of calcium bentonites more than to sodium ones. The pH is slightly alkaline. The sand sized impurities included in the samples are low. Nevertheless, the high silica content reflected in chemical analyses concentrates in finer fractions.

Table 1. Chemical analyses of bulk bentonites

Oxides (%)	0612	0614	0914	EC1	EC1 < 1-μm
SiO_2	60.83	61.82	63.18	65.07	61.29
Al_2O_3	12.17	11.72	11.88	12.04	13.92
Fe_2O_3	1.00	1.17	1.00	0.85	1.04
CaO	0.98	1.29	0.82	0.79	0.86
MgO	3.89	3.69	3.49	3.46	4.14
Na_2O	1.26	1.24	1.40	1.15	1.52
K_2O	0.16	0.17	0.18	0.24	0.08
TiO_2	0.15	0.19	0.16	0.13	0.18
LOI	19.3	18.5	17.8	16.2	16.9

Table 2. Cationic exchange capacity (CEC) and exchangeable cations of bulk bentonites

	Meq/100 g sample					
SAMPLE	Na^+	Mg^{2+}	Ca^{2+}	K^+	Total cations	CEC
0612 El Cerro	43.62	25.43	41.08	0	110.12	97.93
0614 Catalina	41.47	26.92	54.87	0	123.26	91.61
0914 Cateo 3	47.67	16.34	33.55	6.20	103.75	83.12

Results of compressive and tensile strength tests, conclude that some of the samples collected are suitable for foundry industry, compared with commercial products.

The high whiteness showed is a prominent property of these bentonites, particularly to be used in paints, enamels and ceramic.

5. DISCUSSION AND CONCLUSIONS

The calculation of the structural formula, after correction for free silica, was based on the chemical analyses, the values of CEC and the exchangeable cation determination (Ross and Hendricks, 1945). Only non-exchangeable Mg was allocated to the octahedral layer (Foster, 1951). It is as follow:

$$(Si_{7.92}Al_{0.08})^{-0.08}(Al_{2.69}Mg_{0.92}Fe_{0.15}Ti_{0.02})^{-1.56} O_{20}(OH)_4 (Na,Ca,Mg,K)^{1.65}$$

Their main octahedral cation is aluminum. Mg was assigned partially in octahedral site and partially as exchangeable cation. Layer charge arising from octahedral substitutions predominate. Calcium plus magnesium predominate over sodium as natural adsorbed cations, with lesser amount of potassium. According to Schultz(1969) classification, the aluminous smectites from this deposits can be classified as Otay-type montmorillonites.

These bentonite beds were derived from alteration of fine-grained pyroclastic materials (volcanic ash) originated in late Cretaceous volcanic events of rhyolitic-rhyodacitic composition. They were deposited in shallow continental fresh-water bodies, in which very few retransported detrital particles were introduced. Later, in a continental wet environment with moderate drainage that provide alkaline media, glassy shards were transformed into smectite with segregation of silica that remained in the system as opal-CT. Calcium that forms calcite, was introduced in the system but not everywhere. It possibly comes from upper carbonate marine sedimentary layers that overlie the Cretaceous sediments.

Based on SEM/EDX analysis, low refraction index, shape, and color of relict shards, a rhyolitic-rhyodacitic composition for the parental volcanic ash is postulated. This acidic or intermediate composition agrees with the low content of Fe and Ti in bentonite, but not with the relative high content in Mg which could be attributed to an external source, similar to that of calcium. Also, the high tenor in silica as opal-CT, suggests an acidic composition for the parental rock.

As regards its uses, this clay can be classified as a low-swelling calcium-magnesium white bentonite. White bentonites with this characteristics, are relatively rare in occurrence and commercial exploitation, compared to commodity bentonites. The presence of abundant opal-CT diminishes the possibility of uses in various applications.

Table 3: Physical properties of bentonites and foundry molding sands using these bentonitas

	TEST	SAMPLE		
		6121099	6141099	9141099
Raw Bentonites	Original moisture (%)	12,6	13,0	10,8
	Methylene blue adsorption (ml)	63.5	57.2	57.7
	Methylene blue adsorption at 550°C (ml)	21	16.6	34.3
	Thermal stability (%)	33	29	59.5
	Swelling 2 g (ml)	15	13	12
	pH	7.72	8.10	7.63
	Whiteness (%)	74,5	73,6	78,5
	Yellowness (%)	5,0	6,8	4,1
	Grit retained # 200 mesh	1.1	1.8	1.2
Foundry molding sands probe	Green compressive strength (N/cm^2)	16.7	17.0	15.2
	Wet tensile strength (N/cm^2)	0.23	0.24	0.30
	Hot compressive strength (psi) (950°C)	110	117	123
	Compactability (%)	47	47	48

Hydroclassification to obtain the purest and highest priced products can be carried out but this process must be well studied because of the presence of very fine silica. This kind of bentonite does not develop rheological properties to be used as drilling fluids. Neither it is the best to be used as a binder in the foundry industry. Nevertheless, due to their whiteness and brightness, chemical and mineralogical composition, high surface area, cationic exchange capacity and low swelling, these bentonites are suitable for the following applications, among others: plasticisers in ceramic and refractories, filler and strengthener in plastics, filler in rubber and in others products, decolorizing (bleaching) and refining of vegetable, animal, and petroleum based oils, low temperature foundry sand binders, industrial absorbents in granules, personal care, liquid abrasive cleaners, soap and detergents.

AKNOWLEDGMENTS

These works were carried out with the financial support of the Universidad Nacional del Comahue. We thank to: María A. Giaveno for the CEC analysis, Luis Muldon for surface area measurements, and Juan Altamira, for whiteness determinations. The authors aknowledge also to Raúl Hernández and Minera José Cholino e Hijos S.R.L. (the company owner of the mines) who gave us information about the deposits and helped us in the fieldwork.

REFERENCES

API (1993) Specification 13ª, Specification for drilling fluid materials.

Dirección de Minería, Provincia de Río Negro-SEGEMAR (1999). Geología y Recursos Minerales de la Hoja 4166-I, Valcheta.

Foster, M.D.(1951). The importance of exchangeable magnesium and cation exchange capacity in the study of montmorillonitic clays. Am.Miner.(36): 717-730.

Giaveno, M. y Chiachiarini,P.(1995). Estudio de los cationes de intercambio en muestras de bentonitas de la región Norpatagónica. Argentina. Rev. Información Tecnológica (6), 77-81.

Greene-Kelly, R. 1953. The identification of montmorillonoides in clays. Jour. Soil Science (4): 233-237.

Jackson,M.L. (1956). Soil Chemical Analysis. Advanced Course. Dept. of Soils, Univ. Of Wisconsin, Madison. USA.

Jones, J.B., Segnit, E.R. 1971. The nature of opal I. Nomenclature and constituent phases. J.Geol.Soc.Aust., 18 (1): pp57-68.

Moore,D.M. and Reynolds,R.C. (1997). X-Ray Diffraction and the Identification and Analysis of Clay Minerals, 2nd Ed. Oxford University Press. NY.

Ross, C.S. and Hendricks, S.B. (1945). Minerals of the montmorillonite group, their origin and relation to soils and clays. Prof. Paper US Geol.Surv. (205-B): 23-79.

Schultz, L.G.(1969). Lithium and pottasium absorption, dehydroxilation temperature and water content of aluminous smectites. Clay Clay Miner. 17, 115-149.

Corrensite, a stratigraphic marker in the Quintuco Formation, Neuquén Basin, west-central Argentina, and Mississipian carbonates of the Illinois Basin, Illinois, USA

[1]J.M.Vallés, [1]G. R. Pettinari and [2]D.M. Moore

[1] CIMAR – Departamento de Geología y Petróleo, Facultad de Ingeniería, Universidad Nacional del Comahue, Buenos Aires 1400 – 8300 Neuquén, Argentina.
jvalles@uncoma.edu.ar

[2] Illinois State Geological Survey, 615 Peabody Dr., Champaign, IL 61820, USA (now at University of New Mexico, Albuquerque, NM 87131).

We have identified corrensite and corrensite-like mixed-layer clay minerals in two basins, one in Argentina and one in USA. In both instances, these minerals are associated with a depositional environment of carbonates deposited in warm, shallow, marine seas. They are a potential correlation tool and, because no two of the corrensites that we studied were the same, may be (1) indicators of depositional conditions and diagenesis, and (2) site specific. Individual characteristics were discerned using experimental and calculated X-ray diffraction tracings. Some samples met the AIPEA nomenclature committees requirements for the mineral corrensite, but several were best modeled at ratios slightly different than 50:50 corrensite: smectite. Small amounts in excess of 50% of either component can be detected in samples with an apparently rational series of basal reflections.

Key words: corrensite, chlorite-smectite, mixed-layer.

1. INTRODUCTION

Corrensite and corrensite-like minerals were identified in samples of the Jurassic-Cretaceous age Quintuco Formation from the Entre Lomas, Centenario and Río Neuquén oil fields, 100 km apart in the Neuquén Basin. The Quintuco-Loma Montosa Formation is in the largest and most diversified oil and gas-producing carbonate depositional system in this basin, which is in west-central Argentina in the provinces of Neuquén and Río Negro. The corrensite minerals occur at the same stratigraphic level in four wells in three different oil fields. From an unpublished manuscript (Fraser, et al., 1975) we were aware that corrensite had been identified in samples from carbonates in the Illinois Basin representing approximately the same depositional environments as those of the Quintuco Formation, and where they also seem to serve as a correlation tool.

These samples are from the southern edge of the Illinois Basin and are Meramecan (Middle Mississippian or Lower Carboniferous) carbonates. Our goal for this study was to refine our identification techniques for corrensite and evaluate their use as a correlation tool. (Hereinafter, when we use the word corrensite, we imply that corrensite-like minerals are included. This is to cover those intergrowths of chlorite and an expandable component that may be slightly different than 50:50).

2. GEOLOGIC SETTING

The foreland backarc Neuquén Basin of west-central Argentina is located in the southernmost segment of the sub-Andean orogen of South America, and is a major oil and gas producer. Digregorio and Uliana (1980) described the lithostratigraphic setting, Carozzi et al (1993) reported on the Jurassic to Eocene part of the column. The Quintuco-Loma Montosa Formation is a shallow water carbonate sequence deposited over platform during earliest Cretaceous Berriasian Stage in an environment that was subtidal, of moderate energy, and associated with siliciclastic sediments representing a shoreface with lateral variation to semi-restricted conditions, probably lagoonal. It grades upward and laterally into the Loma Montosa Formation, a marginal marine, oolitic limestone interbedded with evaporates and fine-grained clastics. Inmediately below the Quintuco is the Vaca Muerta Formation, the chief source rock for the Neuquén Basin, consisting of dark, bituminous shales and siltstones.

The Illinois Basin, a mature, petroleum-producing, interior cratonic basin, covers the southern three-quarters of Illinois. It has produced oil since the early 1860s. Its deepest part, which contains approximately 4600 m of Cambrian to Pennsylvanian sedimentary rocks, is in southeastern-most Illinois (Nelson, 1995). Cretaceous marine sediments unconformably overlap the southern tip of the Illinois Basin (Leighton et al 1990).

The upper St. Louis Limestone in the area from which the samples came is characterized by bioclastic wackestone to packstone, lime mudstone, and some bioclastic-peloidal grainstone. The overlaying Ste. Genevieve Limestone is light gray, mainly oolitic, cross-bedded, and crinodal, but includes some darker lithographic beds similar to those of the St. Louis Limestone (Lasemi et al.,1999). Baxter et al. (1986) reported corrensite at the same stratigraphic level in two cores, one in Massac County and one in Pope County, both 25 to 30 km east of our two sample sites. The oolitic beds of the Ste. Genevieve Limestone form some of the most prolific hydrocarbon producers in the Illinois Basin.

3. METHODS

We analyzed 108 samples of cuttings from four wells in three oil fields in the northeastern and southeastern parts of the Neuquén Basin, 23 of them from the Quintuco formation. For the Illinois Basin, our samples were taken from two quarries, 15 from the lower Ste. Genevieve Limestone in the Anna Quarry in Union County (Sec. 17, T12S, R1W, Anna 7.5-minute quadrangle), and 8 from the lower Ste. Genevieve and upper St. Louis Limestones in the Cypress Quarry, in Johnson Co. (SW Sec. 5, T14S, R2E, Cypress 7.5-minute quadrangle).

For X-ray diffraction (XRD), carbonates were dissolved in 6N acetic acid, washed, Sr-saturated, dispersed, and pipetted onto glass slides to make oriented aggregates. Slides were run in the air-dried and ethylene glycol solvated states, and after heated to 375° C for at least

Table 1. Calculated Peak Positions for Chlorite/Smectite-Ethylene Glycol Mixed-Layered Clay Mineral

%Chlorite	C001/S001 2q/d	C001/S001 2q/d	C002/S002 2q/d	C002/S003 2q/d	C003/S003 2q/d	C003/S004 2q/d	C004/S005 2q/d	D/2q
0.999		6.19/14.27	12.52/7.07				25.28/3.52	
0.975		6.16/14.35	12.50/7.08			18.81/4.72	25.29/3.52	
0.95		6.10/14.49	12.45/7.11			18.82/4.72 asym	25.31/3.52	
0.9		6.06/14.58	12.38/7.15			18.83/4.71 asym	25.38/3.51	
0.85		6.01/14.72	12.25/7.23			18.85/4.71 asym	25.44/3.50	
0.8		5.94/14.85	12.10/7.32		18.00/4.93	18.93/4.69 asym	25.51/3.49	7.51/1.44
0.75		5.91/15.01	11.94/7.41	shl	17.59/5.04	19.10/4.65	25.58/3.48	7.99/1.56
0.7	no shl*	5.85/15.10	11.77/7.52	shl 13.79/6.43	17.50/5.07	19.36/4.58	25.65/3.47	8.15/1.6
0.65	bkgrd rising	5.80/15.25	11.66/7.59	shl 13.79/6.42	17.39/5.10	19.52/4.55	25.70/3.47	8.31/1.63
0.6	shl 2.58/34.2	5.74/15.39	11.51/7.69	13.90/6.37	17.30/5.13	19.67/4.51	25.77/3.46	8.47/1.67
0.575	2.63/33.57	5.70/15.50	11.49/7.70	13.99/6.33	17.26/5.14	19.80/4.48	25.79/3.45	8.53/1.69
0.55	2.73/32.37	5.68/15.55	11.45/7.73	14.06/6.30	17.21/5.15	19.84/4.47	25.82/3.45	8.61/1.7
0.525	2.78/31.81	5.67/15.60	11.40/7.76	14.15/6.26	17.17/5.16	19.90/4.46	25.85/3.45	8.68/1.71
0.5	2.87/30.73	5.65/15.64	11.34/7.80	14.26/6.21	17.13/5.18	19.94/4.45	25.87/3.44	8.74/1.74
0.45	2.83/31.26	5.61/15.75	11.26/7.86	14.25/6.22	17.03/5.21	20.00/4.44	25.92/3.44	8.89/1.77
0.4	shl 2.68/32.96	5.56/15.88	11.20/7.91	14.15/6.26	16.92/5.24	20.06/4.43	25.97/3.43	9.05/1.81
0.35	bkgrd convex	5.53/15.99	11.09/7.98	slt shl	16.78/5.28	20.14/4.41	26.02/3.42	9.24/1.86
0.3	bkgrd rising	5.46/16.19	10.99/8.05		16.60/5.34	20.20/4.39	26.07/3.41	9.47/1.93
0.25		5.41/16.33	10.89/8.12		16.40/5.40	20.31/4.37	26.13/3.41	9.73/1.93
0.2		5.38/16.44	10.76/8.22		16.15/5.49	20.40/4.35	26.18/3.40	10.03/2.09
0.15		5.33/16.59	10.61/8.34		15.92/5.57	20.49/4.33	26.23/3.40	10.31/2.17
0.1		5.28/16.74	10.51/8.42		15.80/5.61	20.61/4.31	26.28/3.39	10.48/2.22
0.05		5.24/16.86	10.38/8.52		15.74/5.63	20.74/4.29	26.33/3.38	10.59/2.25
0.001		5.22/16.92	10.32/8.57		15.70/5.64	20.84/4.26	26.37/3.38	10.67/2.26
						20.92/4.25		

Conditions: Chl d_{001}=14.1Å, Sm d_{001}=16.9 Å, R1; Fe^+ 0.2; For Chl, Sil Fe=1.8, Hyd Fe=1.8, hydroxyl sheet=1; defect free distance=3, HiN015; CuKα radiation assumed. *shl=shoulder; bkgrd=background; slt=slight; C1=001; S1=S001, Δ2θ =(C003/S003-C004/S005). Original tracings at 0.2; re-runs for .575, .550, .525, & .500 at .1 increment; β*= peak width.

2 hrs. The heating samples were transferred quickly from the oven to an environmental chamber on the X-ray diffraction machine. Dry N_2 was streamed through the chamber. We used a Rigaku DII-Max© and a Scintag© diffractometer, both with Cu tubes. XRD tracings, reference intensity ratios, and tables of 2θ for identification were calculated using NEWMOD© (Reynolds, 1985; Moore and Reynolds, 1997). Measurement of experimental peak positions for the basal series were adjusted to the first quartz peak at 20.85° 2θ (CuKα). For comparison, we made tables of calculated peak positions as an aid to analysis of these minerals (Tables 1 and 2).

4. RESULTS

We identified corrensite in cuttings of the Quintuco Fm in the Entre Lomas, Centenario and Río Neuquén oil fields, in the Neuquén Basin. In the Illinois Basin, it is present at or below the boundary between the Ste. Genevieve and St. Louis Limestones at the Anna Quarry and the Cypress Quarry, in addition to the occurrences reported by Baxter et al. (1986).

No two corrensites were exactly alike. We used the calculated values from Tables 1 and 2 to find Δ 2θ for our corrensites in the ethylene glycol solvated and heated states. By this standard, only the Clay Mineral Society's standard for corrensite, CorWa-1, came close to 50:50 chlorite/smectite. Comparison of experimental and calculated X-ray diffraction tracings yielded percent standard deviations for the corrensites for two samples below the 0.75% limit set by the AIPEA nomenclature committee for regularly interstratified, 1:1 mixed-layered minerals (Bailey, 1982). Several samples are best modeled at ratios slightly off of the 50:50 ratio. Variations, based on values in Tables 1 and 2, of the corrensites that we studied are shown in Table 3. Note that the group apparently averages excess chlorite. Table 4 gives examples of rational sequences of five of the corrensites that we studied.

Table 2. Calculated Peak Positions for Chlorite/Smectite-Heated Mixed-Layered Clay Mineral

%Chl	C1+S1 2θ/d	C1/S1 2θ	C1/S1 D	C1/S1 Area	C1/S1 β*	C2/S1 2θ	C2/S1 D	C2/S1 β*	C2/S2 2θ/d	Δ2θ
62		6.93	12.75	5602	0.95	11.45	7.72	1.153		4.52
58		7.1	12.44	5812	0.906	11.33	7.81	1.049		4.23
54	slt shld	7.18	12.32	5783	0.832	11.21	7.89	0.952	14.83/5.97	4.03
50	3.8/23.2	7.27	12.15	5926	0.757	11.09	7.98	0.861	14.93/5.93	3.82
46	slt shld	7.35	12.03	5216	0.839	11.02	8.03	0.959	14.95/5.92	3.67
42		7.42	11.91	4466	0.891	10.92	8.1	1.06	14.95/5.92	3.5
38		7.44	11.87	4042	0.969	10.8	8.19	1.15	14.91/5.94	3.36

Refer to Table 1 for key abbreviations.

5. DISCUSSION

We have found corrensite in quite similar environments in two basins. Clearly in the Neuquén Basin, the corrensite is found at the same stratigraphic level in four different wells from different fields separated by 80-100 km. The case in the Illinois Basin is similar in that

Table 3. Δ2θ values for experimental tracings of C/S

Sample	Heated					EG				
	C1+S	C2/S1	C1/S1	Δ°2θ	%Chl	C1+S	C3/S3	C4/S5	Δ°2θ	%Chl
425		10.15	5.94	4.21	57.5	2.72	25.71	17.13	8.58	56
		10.84	6.7	4.14	56					
Lippmann		?	6.49			2.81	25.86	17.23	8.63	54
CorWa-1	3.81	10.5	6.72	3.78	49	2.9	26.11	17.34	8.77	49
		11.2	7.39	3.81	50					
3850F	3.39	10.1	5.74	4.36	60	2.82	25.85	17.19	8.66	53
		10.77	6.73	4.04	54					
41772		10.97	7.18	3.79	49	2.71	25.81	17.22	8.59	56
41805	3.83	10.8	7.03	3.77	49	2.64	25.76	17.15	8.61	55
		10.8	6.67	4.13	56					
41835		10.85	7	3.85	50.5	2.72	25.82	17.16	8.66	53
		10.85	6.33	4.52	62					
3853B		10.94	7.07	3.87	51	2.79	25.95	17.24	8.71	51
		10.94	6.4	4.54	62					
3853C	3.33	11.14	7.28	3.86	51	2.86	26.04	17.39	8.65	54
		11.14	6.87	4.27	58.5					
3853E		10.88	7	3.88	51	2.77	25.87	17.27	8.6	55
3853F		10.94	6.74	4.2	57.5	2.79	25.91	17.25	8.66	53
3853G	2.95	10.94	7.01	3.93	52	2.82	26.02	17.3	8.72	51
				Ave.=	54.2±4.5%			Ave. =		53.3±2.2%

Table 4. Examples of rational sequences for experimental corrensites, EG state

	3850F		3853B		CorWa-1		41835x2G		41772x2G	
	32.16		31.66		31.00	31.00	32.48		32.60	
	15.66	31.32	15.51	31.02	15.08	30.16	15.66	31.31	15.61	31.23
	10.07	30.21*	10.54	31.62*	10.43	31.28	10.43	31.28	10.43	31.28
	7.80	31.18	7.72	30.88	7.58	30.30	7.76	31.05	7.70	30.78
	6.19	30.93*	6.17	30.84*	6.13	30.65			6.13	30.65
	5.18	31.09	5.14	30.84	5.08	30.49	5.15	30.91	5.15	30.90
	4.45	31.17	4.43	31.03	4.38	30.63	4.46	31.21	4.46	31.23
	3.86	30.90	3.83	30.66	1.11	1.11			3.78	30.22
	3.47	31.20?	3.46	31.17	3.41	30.73	3.44	30.97	3.45	31.05
	3.45	31.06?								
Ave.		31.01		31.01		30.49		31.12		30.90
Std dev.		0.33		0.29		0.22		0.17		0.38
CV**		1.06		0.94		0.72		0.55		1.23

*Peaks very small, very broad. **as defined in Bailey (1982)

corrensite seems to be confined to the upper St. Louis Limestone, but is not as well documented. In both basins, the environment of deposition was carbonate banks in marine waters in which there were oolitic shoal systems in very shallow, tidally-dominated shelf environments (Carozzi et al., 1995; Lasemi et al., 1999). However, because no two corrensites that we examined were exactly alike, there is an opportunity to more tightly tie corrensite characteristics to environment of formation, if we can more fully understand these minerals.

Understanding corrensite and corrensite-like minerals is not a simple task. The apparent components have been well studied, chlorite (Bailey, 1988) and smectite (Środoń, 1980; Güven, 1988) or vermiculite (de la Calle and Suquet, 1988; Douglas, 1989). When these components are joined to form a mixed-layered mineral, characteristics of individual components are compounded and not necessarily recognizable (Reynolds, 1988). Overlaying this, there is the controversy about whether corrensite is a discrete phase formed without a continuous sequence from 100% smectite or vermiculite to 50:50 chlorite/smectite or vermiculite. Or, starting from the other end, from 100% chlorite (Roberson et al. 1999, and references therein).

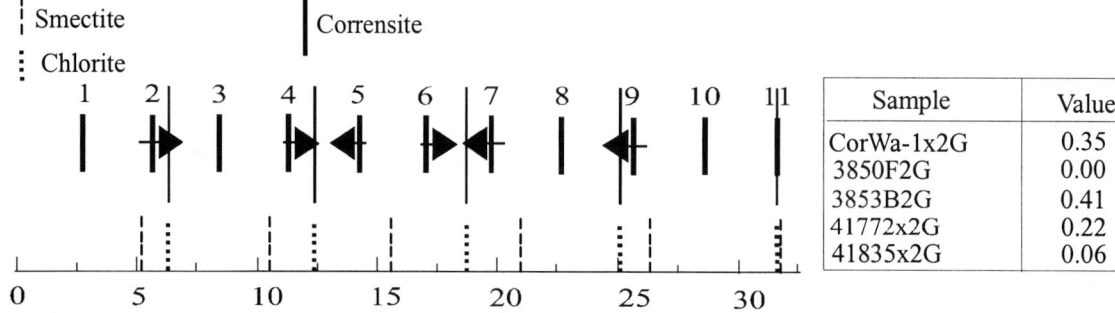

Value = (Ave. d001 odd-order peaks) - (Ave. d001 even-order peaks)
Figure 1. Méring´s rules. Three examples with positive values indicating small amounts of randomly intergrown chlorite, and two examples indicating no, or almost no, excess chlorite or smectite.

Therefore, we were not surprised that each corrensite was slightly different from all the others we examined resulting in a lack of detailed agreement. Corrensite 3853B shows the largest difference in shift of peak positions between even -and odd- order peaks (Fig. 1), but shows little excess chlorite in Table 3. CorWa-1x2G shows excess chlorite in the table in Fig. 1, but shows a slight excess of smectite in Table 3. Two observations emphasize these differences. First, a comparison of Tables 1 and 2 with Tables 3 and 4 indicates that peak positions are only close to 50:50 proportions of the two components. Second, some of the variation may be explained by using Méring´s principle (1949) to test the possibility that chlorite, as a discrete phase, is an intergrown component in corrensite. We point out two observations that emphasize these differences. First, note that from a comparison of Tables 1, 2, and 3, peak positions are only close to that which would indicate exactly 50:50 proportions of the two components. Second, as a way to explain some of the variation, we used Méring's principle (1949) to test the possibility that chlorite was an intergrown component in corrensite as a discrete phase. Small amounts of chlorite, randomly intergrown, should cause slight shifts in the peak positions toward the position of the nearest chlorite basal peak. The odd-order basal peaks of corrensite would shift toward the low-angle end of the tracing and the even-order ones would shift toward the high-angle end of the tracing (Fig. 1). By averaging the d001 from the odd-order peaks, as one would do to test for a rational series (Table 4), and then subtracting the average d001 from the even-order peaks, we extracted a number that should be zero if there was no shift in the peaks, but some positive value if there had. The small table included in Fig. 1 shows three examples with positive values indicating small amounts of randomly intergrown chlorite, however, not enough to keep the basal series from being rational. If small amounts of smectite were intergrown, all of the shift directions would be the opposite and the result would be a negative number.

The source of variations of corrensite may be site specific, or may be an artifact of the way in which peak positions were modeled. The calculated peak positions in Tables 1 and 2 were made for a specific set of conditions, i.e., specified Fe content, size of coherent diffracting domain, sample length, specified d001 values for chlorite and smectite. Our experience with calculating corrensite tracings, and comparing them to experimental tracings, suggests that one needs to match the conditions used for calculating patterns to the minerals of each project and the conditions used in each particular laboratory. However, if we can satisfy ourselves that we have modeled tracings correctly, then the possibility of additional information about the environment of formation of corrensite and corrensite-like minerals may be achieved.

6. CONCLUSIONS

The corrensite in samples from carbonates of the Quintuco Formation of the Neuquén Basin, and carbonates of the St. Louis and Ste. Genevieve Limestones of the Illinois Basin are apparently the product of burial diagenesis. Because the corrensite is found at the same stratigraphic level in each basin, it is useful as a correlation tool and an indicator of the conditions of deposition and diagenesis. However, the details of the variable physical and chemical characteristics of this mineral, which are almost certainly indicators of conditions in the environment of deposition and diagenesis, have yet to be elucidated. The critical next step is to discover the relations between the variations of the corrensites and the conditions of their formation, especially given the hint in Tables 3 and 4 that corrensite characteristics may be site specific. These minerals contain a wealth of information, but in a language we are only beginning to be able to translate.

ACKNOWLEDGMENTS

This works was carried out with the financial support of the Universidad Nacional del Comahue, of the Illinois State Geological Survey, USA, and the FOMEC, Ministery of Education, Argentina. The authors acknowledge C. Arregui, J.Quinteros and O.Carbone and Perez Companc S.A. We gratefully acknowledge the use of an unpublished manuscript by G.S. Fraser, J.W. Baxter, and R.D. Harvey. We used some of the samples they collected at the Anna, Cypress, and Mermet Quarries. Z. Lasemi and R. Norby also brought us samples from the Anna and Cypress Quarries.

REFERENCES

Bailey, S.W., (1982). Nomenclature for regular interstratifications. Amer. Min. 67, 394-398.
Bailey, S.W., (1988). Chlorites: Structures and crystal chemistry. In Bailey, S.W., ed., Hydrous Phyllosilicates. Reviews in Mineralogy, vol. 19, MSA, 347-403.
Baxter, J.W., Hughes, R.E., and Glass, H.D., (1986). Corrensite in the Paleozoic strata of Illinois. Annual Mtg of The Clay Minerals Soc., Jackson, MS, 38.

Carozzi, A.V., Orchuela, I.A., and Rodriguez Schelotto, M.L. (1993). Depositional models of the Lower Cretaceous Quintuco-Loma Montosa Formation, Neuquén Basin, Argentina. Jour. Petrol. Geology, v.16, 421-450.

De la Calle, C., and Suquet, H., (1988) Vermiculite. in Bailey, S.W., ed., Hydrous Phyllosilicates (exclusive of micas). Reviews in Mineralogy, vol. 19, MSA, 455-496.

Digregorio, J.H., and Uliana, M.A., (1980). Cuenca Neuquina, in J.C.M. Turner, coord., Simposio Geología Regional Argentina, Córdoba, v.2, 985 – 1032.

Douglas, L.A., (1989). Vermiculites. In Dixon, J.B., and Weed, S.B., eds. Minerals in the Soil Environment, 2nd ed., Soil Science Society of America, 635-674.

Fraser, G.S., Baxter, J.W., and Harvey, R.D., (1975). Occurrence and paleoenvironmental significance of corrensite on the southern margin of the Illinois Basin. Unpublished manuscript in the files of the Illinois State Geological Survey, Champaign, Illinois.

Güven, N., (1988) Smectites. in Bailey, S.W., ed., Hydrous Phyllosilicates (exclusive of micas). Reviews in Mineralogy, vol. 19, Mineralogical Society of America, 497-559.

Lasemi, Z., Norby, R.D., Devera, J.A., Fouke, B.W., Leetaru, F.E., and Denny, F.B., (1999). Mississippian Carbonates and Siliciclastics in Western Illinois. Guidebook 31 for 33rd Annual Meeting, April 1999, North-Central Section, Geological Society of America, Illinois State Geological Survey, 60.

Leighton, M.W., Kolata, D.R., Oltz, D.F., and Eidel, J.J., eds., (1990). Interior Cratonic Basins. AAPG, Mem. 51 (World Petroleum Basins), Tulsa, OK., 819.

Moore, D.M., and Hughes, R.E., (1994). Systematic analysis of chlorites and illites: Annual Mtg of the Clay Minerals Soc., Saskatoon, Saskatchewan, 29.

Moore, D.M., and Reynolds, R.C., Jr., (1997). X-Ray Diffraction and the Identification and analysis of Clay minerals, 2nd Ed., Oxford Univ. Press, 384.

Nelson, W. J., (1995). Structural Features in Illinois. Bulletin 100, Illinois State Geological Survey, Champaign, IL, 144.

Reynolds, R.C., Jr., (1985). NEWMOD $^{©}$ a Computer Program for the Calculation of One-Dimensional Diffraction Patterns of Mixed-Layered Clays. 8 Brook Rd., Hanover, NH.

Reynolds, R.C., Jr., Di Stefano, M.P., and Lahann, R.W., (1992). Randomly interstratified serpentine/chlorite: Its detection and quantification by powder X-ray diffraction methods. Clays and Clay Minerals, v.40, 262-267.

Roberson, H.E., Reynolds, R.C., Jr., and Jenkins, D.M., (1999). Hydrothermal synthesis of corrensite: A study of the transformation of saponite to corrensite. Clays and Clay Minerals, v.47, 212-218.

Środoń, J., (1980). Precise identification of illite/smectite interstratifications by X-ray powder diffractions: Clays and Clay Minerals, v.28, 401-411.

Reynolds, R.C., Jr. (1988) Mixed layer chlorite minerals. in Bailey, S.W., ed., Hydrous Phyllosilicates (exclusive of micas). Reviews in Mineralogy, vol. 19, Mineralogical Society of America, 601-629.

Study of the structural order degree of Brazilian kaolinites by X-ray diffraction

A.C. Vieira Coelho, P. Souza Santos

Laboratório de Matérias-Primas Particuladas e Sólidos Não-Metálicos – Depto. de Engenharia Metalúrgica e de Materiais - Escola Politécnica da USP - Av. Prof. Luciano Gualberto, Travessa 3, 380 CEP 05508-970 - São Paulo, SP, Brasil – Email : acvcoelh@usp.br

This paper presents the results of the evaluation of the structural order degree in Brazilian kaolinitic clays by indices determined by X-ray diffraction, specifically those by Hinckley, Lièrard, Amigó and the Plançon and Zacharie's "expert system", and the results from the calculation of linear correlation coefficients (r) between these indices. Values of (r) higher than 0.9 were found between some indices when the samples were grouped according to their origin, residual or sedimentary; the same was observed for %ldp (from Plançon and Zacharie's expert system) vs Hinckley index for all the samples of this work.

1. INTRODUCTION

Since 1950 the low/high-defect characteristics of the crystalline structure of kaolinite (nature and evaluation of crystal defects) and the possible correlations between these characteristics and the properties of industrial interest have been studied [1-13]. The first studies were made on kaolins for the paper industry [1,2] and lead to the proposal of a "crystallinity index", called Hinckley Index (HI), evaluated from their XRD patterns. Several other methods to evaluate the degree of structural order/disorder of kaolinites [5,14] ("the crystallinity indices"), based on the profiles of the XRD patterns have been presented in the literature. However, difficulties are observed because of the great complexity of the kaolinite crystalline structure, which allows a large diversity of stackings arrangements that may occur in its formation along its geological history [11]. Furthermore, the presence of associated minerals can confuse the calculation of the indices: the effects of mineralogical interference of those minerals on kaolinite crystallinity index measurements were reported by Aparicio and Galán [12]. Those difficulties are shown up even in the terminology. The first expression was "crystallinity index", used by Hinckley in 1963 to quantify "crystallinity" of samples of kaolinite [2]. Some years later, the expression "degree of structural order/disorder" (of kaolinite crystals) was set in and it was widely used until the last years [5,14-17]. Recently, the Nomenclature Committee of the AIPEA, based on Prof. R.F. Giese's suggestion in 1998, recommended the substitution of the former terms by the expressions "low-defect kaolinite" and "high-defect kaolinite" for samples with "low" or "high" amounts of defects. For the several indices, the appropriate terminology is to refer to a "Hinckley index" or "Lièrard index" ("name of author index"). The term "crystallinity index" is not recommended to be used any more [18].

This paper is part of an extensive research program on characterization of Brazilian kaolins. Most of Brazilian kaolins are produced in the Amazon basin; they have sedimentary origin and are intensively studied due to their industrial applications in the paper industry. Other kaolin deposits exist in Brazil, formed in different geological environments : the majority is residual, formed usually by the weathering of pegmatites, volcanic rocks, granites and anorthosites [19,20]. They are used for paper, ceramics and as fillers in Southern Brazil chemical process industries; they have a wide range of values of their technological properties. However, no systematic information exists in literature on the degree of structural order of the kaolinite from all these clays, as measured by any "crystallinity indices". This paper presents the results of the calculation of four of these indices and the tentative results from the calculation of the linear correlation coefficients between those four indices in 30 samples of Brazilian kaolinitic clays [19-30], formed from several different geological environments.

2. MATERIALS

2.1. North Region
Sedimentary kaolins from four different localities of the Amazon Region.
Jari (Amapá State): commercial samples (CADAM) from the kaolin mine from Morro do Felipe; samples Ja19 and Ja23 (different levels of the profile). The area is occupied by the Alter do Chão sediments; they were submitted to an intensive lateritic weathering processes [21]. Manaus (Amazonas State): non exploited sedimentary clays from three places distant 50km from the city of Manaus, named RM-15, RM-35 and C samples. The samples are taken in different levels of the profiles. The geology of the area is described in reference [22]. Rio Negro (Amazonas State): non exploited clay samples from the margins of the Negro river (Rio Negro); place distant 20km from Manaus; three samples named AC (from Alter do Chão), taken in different levels of the profile. The locality is mentioned in [23]. Rio Capim (Pará State) : commercial samples (Pará Pigmentos) of sedimentary kaolin, named Capim 1 and Capim 2. The geology of the deposit (Alter do Chão sediments) is described in [19,20,22].

2.2. Northeast Region
These kaolins are residual and formed in the decomposition of Pre-Cambrian pegmatites (Seridó Group rocks); they are constituted by euhedral platy hexagonal kaolinite crystals; halloysite tubes are absent; they contain small amounts of muscovite mica [19,20]. The samples are not commercial kaolins: Equador (Rio Grande do Norte State): localities Tanquinho, Alto da Favela, Alto do Giz and Serra Redonda; Junco do Seridó (Paraíba State) : localities Noruega, Olho d'Água de Baixo, Cajazeiras and Alto da Aldeia II; Recife (Pernambuco State).

2.3. Southeast / South Region
Horii kaolin (São Paulo State): commercial sample; residual; formed from the Pre-Cambrian granitic rock; produced in Jundiapeba County, near the city of Mogi das Cruzes (State of São Paulo); used as filler for paper and in ceramics; rare halloysite tubes [19,20]; MM kaolin (São Paulo State): commercial sample; residual; formed from granitic gneiss; used in ceramics; 7Å-halloysite, as thick and long tubes [19,20]. Mar de Espanha kaolin (Minas Gerais State): commercial sample; residual; formed from pegmatites; used in ceramics; 7Å-halloysite tubes

are a significant component. [19,20,24,25]. Floresta kaolin (Santa Catarina State): commercial sample; residual; formed from acid volcanic tuffs; white color; used in ceramics; 7Å-halloysite tubes and very fine grained quartz are significant components [19,20,26].

1. Jari, Amapá State
2. Manaus and Rio Negro, Amazonas State
3. Capim, Pará State
4. Equador, Rio Grande do Norte State
5. Junco do Seridó, Paraíba State
6. Recife, Pernambuco State
7. Horii and MM kaolins, São Paulo State
8. Mar de Espanha kaolin, Minas Gerais State
9. Floresta kaolin, Santa Catarina State

Figure 1 : Sample localities in States of Brazil

3. EXPERIMENTAL METHODS

The clay powdered samples ($\phi < 0,210$ mm, #65 ASTM sieve), without preferential orientation, were analysed in a XRD Philips instrument, X'PERT-MPD model, with automatic slit and a rotary holder operating a 30 rpm, necessary to minimize any preferential orientation of the platy crystals of kaolinite and tubes of halloysite. Kα-Cu radiation (λ = 0.15416 nm) was used. Step scanning was used, each step of $0,02°(2\theta)$. The accumulation time for each step was 10 seconds.

Eventual minor amounts of associated minerals present in the sieved samples were maintained prior the XRD analysis. When necessary, what sometimes happened with samples containing quartz or anatase, a mathematical treatment was made on the XRD curves by the use of WinFit program [28] to substract the peaks corresponding to the mineral. This program was also used to calculate the FWHM of the (002) reflexions. Soluble iron compounds and minerals in Jari and Rio Negro samples were extracted by the citrate-bicarbonate-dithionite (CBD) procedure [29] prior to XRD analysis.

To evaluate the amount of defects of the kaolinites ("high" or "low" amounts) from the XRD curves, four methods were used; these methods are not equally sensitive to all structural defects of kaolinite crystals [7,30]; also, they may suffer the influence of the percentage of some associated minerals present in the sample, as pointed before [9,10,12].

3.1. Hinckley's Method [1,2,6,30]

It is extensively used due to its simplicity, using a scanning of only a short interval of $(20°-23°)2\theta$. More than 150 data from literature [4,8,10,11,30, this work] show that the majority of the values of this index varies between 0.2 and 1.6 ; "low-defect kaolinites" present values of Hinckley Index (HI) greater than 1.0; "high-defect kaolinites" present HI smaller than 0.5. The HI values of kaolinites has been correlated with some of their industrial uses [1,2,30]. However, the use of the HI has some disadvantages, described in the literature [6,30].

3.2. Liètard's Method ("R2 Index") [4, 30]

The Liétard's R2 Index is complementary to HI, since it is only sensitive to small aleatory defects, specially those due to octahedral vacancies [4]. It is based in (1 3 1) and (1 $\bar{3}$ 1) reflexions in the region $37°$ to $40°(2\theta)$. More than 150 data from literature [4,8,10,11,23, this work] show that the majority of the values of this index varies between 0.2 and 1.2, the lower values been found for "high-defect kaolinites". The method presents the same disadvantages as Hinckley's.

3.3. Width at half-height of (0 0 2) reflexion [9,11,13]

The determination of the width at half-height (FWHM) of the (0 0 2) reflexion, also named in the literature "Amigó Index" [13], gives informations about the extent of the stacking order along de $\underline{c}$ axis: larger the 'thickness'of the coherent domains along the axis, smaller must be the width at half-height. The main advantage of the method is in its easy execution; its treatment may be made, either by hand, as well by aid of programs for the deconvolution of XRD curves [28]. An additional advantage is that the method is not sensible to preferential orientation problems during sample preparation, as occurs in other methods. The disadvantages are in the limitation of the received information (they consider only order along c axis) and in the sensitibity to accessory minerals (like anatase: more intense reflexion near to 002 kaolinite reflexion) and to 7Å-halloysite tubes.

3.4. Plançon and Zacharie's Method ("Expert System") [7]

Plançon and Zacharie's "Expert System" is based on eleven parameters measured in the region of $19°$ to $23°(2\theta)$ of (02,11) band and $34°$ to $40°(2\theta)$ of (20,13) bands; it allows an evaluation of different types of structural defects in kaolinite. The authors developed, in a single model, the informations treated, in separate, in the former three methods and also a free calculation program which indicates if the sample under analysis must be represented by a single phase or by a mixture of two "types" of kaolinite crystals, each with a different amount of structural defects. For a binary mixture, the model allows the evaluation of the quantity of the component with smaller percentage of crystalline defects (%ldp). For a one component kaolinite, the model allows the evaluation of some parameters related to the different defect types, such as : percentage of layers with type C octahedral vacancies (Wc); the proportion of big translation defects between adjacent layers (p); the variation of small aleatory translations between adjacent layers (δ); it also evaluates the average number of layers (M) in the coherent domains along $\underline{c}$ axis.

4. RESULTS AND DISCUSSION

The values obtained with the four mentionned methods in the 30 samples of Brazilian kaolinitic clays are presented in Table 1.

4.1. Correlations between the indices

Plançon and Zacharie's "expert system" allowed to divide the clay samples into two groups: (a) one group containing only one type of kaolinite defect- structure; (b) another one containing samples with two types of kaolinite defect- structure. Only three samples were classified in the first group, and so no linear correlations were calculated. In the group containing samples with two kaolinite types, the linear correlation coefficient (r) between %ldp index (quantity of the component with smaller percentage of crystalline defects) and the other ones (indices HI, R2 and FWHM) was calculated and is also shown in Table 2.

Table 1
Experimental results for the indices by four methods. Lines in light gray indicates samples of sedimentary origin.

SAMPLE	HI	R2	FWHM (2θ)	phases	% ldp	M	Wc	δ	p
Ja19	0.21	0.68	0.224			54	0.01	0.04	0.35
Ja23	0.22	0.75	0.238	1		53	0.01	0.03	0.35
RM-15 / 3	1.63	1.12	0.180	2	65				
RM-15 / 6	1.31	0.85	0.236	2	53				
RM-15 / 9	1.59	1.16	0.201	2	69				
RM-35 / 1	0.88	0.60	0.253	2	33				
RM-35 / 8	1.63	1.21	0.179	2	74				
RM-35/12	1.55	1.24	0.171	2	73				
RM-35/14	1.63	1.19	0.178	2	67				
C-1 / 1	1.59	1.08	0.191	2	66				
C-1 / 3	1.60	1.10	0.181	2	71				
C-1 / 9	0.69	0.51	0.347	1		35	0.13	0.06	0.31
AC1	1.29	0.92	0.248	2	47				
AC2	1.52	1.03	0.249	2	64				
AC3	1.44	0.93	0.275	2	59				
Capim 1	1.17	0.99	0.200	2	40				
Capim 2	0.97	0.92	0.311	2	26				
Tanquinho	1.37	1.19	0.161	2	53				
Favela	1.27	1.13	0.197	2	49				
Giz	1.29	1.14	0.187	2	48				
Redonda	1.38	1.17	0.207	2	55				
Noruega	1.12	1.14	0.184	2	43				
Olho d'Água	1.03	0.91	0.267	2	36				
Cajazeiras	1.32	1.15	0.193	2	52				
Aldeia II	1.45	1.18	0.158	2	63				
Amarelo	0.93	0.81	0.363	2	27				
Mar de Espanha	0.84	0.85	0.305	2	29				
Horii	1.24	1.12	0.222	2	46				
MM	1.23	1.10	0.238	2	37				
Floresta	0.36	1.36	0.598	1		21	0.00	0.02	0.30

4.2. Discussion

Accepting the value **0.7** or higher as significant for the linear correlation coefficient (r), the following observations can be made :

- The (r) values for the correlations tested with all samples are all smaller than 0.7; those values present a gain when the data from Floresta kaolin is not considered, probably due to its special geological origin plus high quartz content.
- The residual group clays present values for (r) higher than 0.7 in all correlations tested, when the data from Floresta sample are not included.
- Considering all the sedimentary clays, there's no (r) values higher than 0.7; such (r) values are observed in some correlations when those clays are grouped, either by origin – Manaus+Rio Negro clays – or by type after Plançon and Zacharie's method – "two-type kaolinite" clays.
- All samples with two kaolinite types for %ldp vs HI, including the correlations with the data taken from literature, present values for (r) higher than 0.7.
- The value of (r) for the "two-type kaolinite" samples for %ldp vs R2 is smaller than for sedimentary Manaus + Rio Negro and residual groups, both showing (r) values higher than 0.7.

Table 2
Values of the linear correlation coefficient (r) between the indices. Lines in light gray indicates the values of (r) calculated between the "two-type kaolinites", classified according to the "expert system" of Plançon and Zacharie.

Samples Group	Number of Samples	HI vs R2	HI vs FWHM	R2 vs FWHM	%ldp vs HI	%ldp vs R2	%ldp vs FWHM
all samples	30	0.314	0.370	0.035			
all except Floresta	29	0.614	0.284	0.569			
	25	0.503	0.484	0.596	0.933	0.436	0.507
sedimentar	17	0.684	0.264	0.566			
	13	0.721	0.506	0.506	0.933	0.638	0.555
sedimentar Manaus + Rio Negro	13	0.922	0.708	0.781			
	11	0.878	0.451	0.622	0.914	0.909	0.530
residual except Floresta	12	0.857	0.746	0.903			
	12	0.857	0.746	0.903	0.887	0.786	0.802
this work + literature data [4,8,10,30]	over 150	0.673					
this work + literature data [5,7,10,30]	36	0.696	0.555	0.674	0.958	0.639	0.564

The Hinckley index has presented values higher than 1.0 for 22 Brazilian kaolinitic clays, residual or sedimentary. Only three presented HI smaller than 0.5, the two samples from Jari and the Floresta clay, formed from volcanic tuffs. Other five clays showed HI smaller than 1.0. Some clays may have some quartz as an accessory mineral; some residual kaolins from Southern Region contain some 7Å-halloysite tubes; these minerals decrease the value of HI index, and halloysite can influence the FWHM of (002) reflexion [9, 12]:

5. CONCLUSIONS

a) The Hinckley index has presented values higher than 1.0 for 22 Brazilian kaolinitic clays, residual or sedimentary.

b) Values of the linear correlation coefficient (r) higher than 0.7 were observed for the several indices, either for all the 30 samples, either for the sedimentary and the residual groups.

c) Values of the linear correlation coefficient (r) higher than 0.9 were found between some indices in the following groups of samples : HI vs R2 for Manaus + Rio Negro (r = 0.922); R2 vs FWHM for the 12 "two-type kaolinite" residual clays (r = 0.903); %ldp vs HI for the 25 "two-type kaolinites" and for almost all the groups of these samples (except for the residual clays); %ldp vs R2 for the Manaus + Rio Negro clays (r = 0.922).

d) These results suggest that, for an evaluation of the amount of defects of a kaolinitic clay, with a small amount of accessory minerals or 7Å-halloysite, residual or sedimentary, if two types of kaolinite are present, HI and %ldp indices are equivalent.

e) There is a good agreement with literature on the correlation between HI vs %ldp for the majority of the studied samples from different Regions of Brazil: the linear correlation coeficient (r) considering the samples of this work and some data from literature also resulted higher than 0.9.

f) There is a quite good agreement with literature on the correlation between HI vs R2 for the "two-type kaolinite" samples from different Regions of Brazil : the linear correlation coeficient (r) was very near from 0.7 (r=0.697).

g) The experimental results have shown that, when all 30 samples are considered as one group, in general, high values of the linear correlation coefficient are not found. However, high values for (r) were found in some instances when the samples were grouped according their origin, residual or sedimentary. That observation shows the importance of analyzing the linear correlation coefficient between indices from groups based on specific characteristics and / or properties of the kaolinitic clays.

REFERENCES

1. H.H. Murray; S.C. Lyons, Clays and Clay Minerals **4** (1956) 31.
2. D.N. Hinckley, Clays and Clay Minerals **13** (1963) 229.
3. M. Thiry, Bull. Minéral. **105** (1982) 521.
4. J.M. Cases; O. Lièrard; J. Yvon; J.F. Delon , Bull. Minéral. **105** (1982) 439.
5. G.W. Brindley; C.-C. Kao; J.L. Harrison; M. Lipsicas; R. Raythatha, Clays and Clay Minerals **34** (1986) 239.
6. A.Plançon ; R.F. Giese; R. Snyder, Clay Minerals **23** (1988) 249.
7. A.Plançon; C. Zacharie, Clay Minerals **25** (1990) 249.
8. R. Delgado; G. Delgado; A. Ruiz; V. Gallardo;E. Gamiz, Clay Minerals **29** (1994) 785.
9. E. Galán; P. Aparicio;I. González; A. La Iglesia, Geologica Carpathica – Series Clays **45**, (1994), Bratislava, 59.
10. E. Galán; P. Aparicio;I. González;A. Miras, Clay Minerals **33** (1998) 65.
11. I.González; P. Aparicio; E. Galán, in H. Kodama (ed.) – Proceedings of the 11th International Clay Conference. Ottawa, Canadá (1997) 367.
12. P. Aparício; E. Galán , Clays and Clay Minerals **47** (1999) 12.

13. J.M. Amigó; J. Bastida; M.J. Agramut; M. Sanz; J. Galván, Proceedings Euroclay Conference Sevilla'87 (1987) 74.
14. G.W. Brindley, Order-disorder in clay minerals structures, in Crystal Structures of Clay Minerals and their X-Ray Identification, G.W. Brindley, G. Brown eds, London (1980).
15. R.L. Frost, G.N. Paroz, S.J.Van der Gaast, in H. Kodama (ed) – Proceedings of the 11th International Clay Conference. Ottawa, Canada (1997) 393.
16. R.L. Frost, T.H.T. Tran, J. Kristof, in H. Kodama (ed) – Proceedings of the 11th International Clay Conference. Ottawa, Canada (1997) 397.
17. R.L. Frost, G.N. Paroz, in H. Kodama (ed) – Proceedings of the 11th International Clay Conference. Ottawa, Canada (1997) 403.
18. S.Guggenheim (chairman), Report of the AIPEA Nomenclature Committee for 2001, to be published in Clays Clay Min. (2002).
19. P. Souza Santos, I. Wilson, Proceedings 10th International Clay Conference, Adelaide, Australia, (1995) 116.
20. I.R. Wilson; H. Souza Santos; P. Souza Santos, Cerâmica **44** (1998) 118.
21. C.R. Montes; A.J. Melfi; A.Carvalho; A.C. Vieira Coelho; M.L.L. Formoso, Genesis, mineralogy and geochemistry of kaolin deposits of Jari river, Amapá State, Brazil, Clay Minerals (2001) (in press).
22. M.L. Costa; E.L. Moraes, Mineralium Deposita **33**(1998) 283.
23. C.R Montes-Lauar; E. Balan; E. Fritsch; A.J. Melfi; R. Boulet; Ph. Magat; A. Carvalho, CD-ROM Proceedings of the 11th International Clay Conference. Ottawa, Canadá (1997) Abst. 220.
24. F.B. Angeleri, S.R.F. Cardoso, P. Souza Santos, H. Souza Santos, Cerâmica **31** (1985) 5.
25. P. Souza Santos, H. Souza Santos, Acta Microscopica **7** (1998) 197.
26. S.P. Toledo, H. Souza Santos, P. Souza Santos, Acta Microscopica **9** (2000) 69.
27. M.L.L. Formoso, Cerâmica **12** (1966), 132.
28. S. Krumm, S. , Acta Universitatis Carolinae Geologica **38** (1994) 253. WinFit Program can be taken on www.geol.uni-erlange.de.
29. O. P. Mehra; M.L. Jackson, Clays and Clay Minerals **7** (1960) 317.
30. T. Delineau - Les Argiles Kaoliniques du Bassin des Charentes (France) : Analyses Typologique, Cristallo-Chimique, Spéciation du Fer et Applications, Institut National Polytechnique de Lorraine / École Nationale Supérièure de Géologie de Nancy, France (1994).

ACKNOWLEDGEMENTS

This paper is part of Projeto Temático FAPESP 1995/00544-0 and Projeto FAPESP 2001/04681-5. The authors are thankfull to Prof. Dr. Adolpho José Melfi, Prof. Dr. Célia Regina Montes (ESALQ-USP) and Prof. Dr. Adilson Carvalho (IG-USP) for the Amapá and Amazonas samples; to Prof. Dr. Heber Carlos Ferreira (DEMa-UFPb) for the Northeast kaolins.

Geology and evaluation of the Yarmouth Kaolin Deposit, Nova Scotia, Canada

I.R. Wilson[a], H.H. Murray[b], G. MacGillivray[c] and J.Keating[c]

[a]Consultant, Withielgoose Farmhouse, Withiel, Bodmin, Cornwall, PL30 5NW, United Kingdom (E-mail: ian.r.wilson@btinternet.com)

[b]Department of Geological Sciences, Indiana University, Bloomington, Indiana, USA. Zip Code: 42405

[c]Black Bull Resources Inc., Suite 303, Sun Tower, 100 West Pender Street, Vancouver, British Columbia, Canada. V6B 1RB.

A recent exploration programme involving geophysics (resistivity and gravity) and a detailed drilling campaign has identified a kaolin deposit in the Yarmouth area of Nova Scotia, Canada. At present Canada is a major paper producer and imports approximately 0.8Mt of high quality filler and coating kaolin, there being no Canadian source available. Resources of 5Mt have been identified and, based on the gravity survey, this tonnage can be significantly increased. The kaolin is derived from a granite and the yield is >30wt.% at – 325# refining. Work is currently being carried out on beneficiation by hydrocyclones and centrifugal separation. In examining the coarser fractions a stacky kaolinite is identified. Some flotation trials to separate out the kaolin stacks, followed by delamination to produce a high aspect ratio kaolin is being explored for supercalendered (SC) papers. Brightness improvement trials are being carried out utilizing reductive bleaching and magnetting and initial results are encouraging. Other markets are being studied for use in other types of paper, ceramics and specialities (paint, rubber and plastics). The quartz from the deposit is currently being mined and flotation trials on the mica residue from the kaolin beneficiation processes are being studied for their commercial properties. The Yarmouth kaolin deposit represents a good opportunity for Canada to have its first commercial production of high quality material for various markets.

1. INTRODUCTION

Canada is a major paper producer and imports approximately 0.8Mt per annum of high quality kaolin for paper filler and coating. In this paper a description will be given of the exploration and evaluation of the Yarmouth kaolin prospect of Nova Scotia. The deposit is situated in the south-western part of Nova Scotia 40km east of the port of Yarmouth and 45 km north of Shelburne (Figure 1).

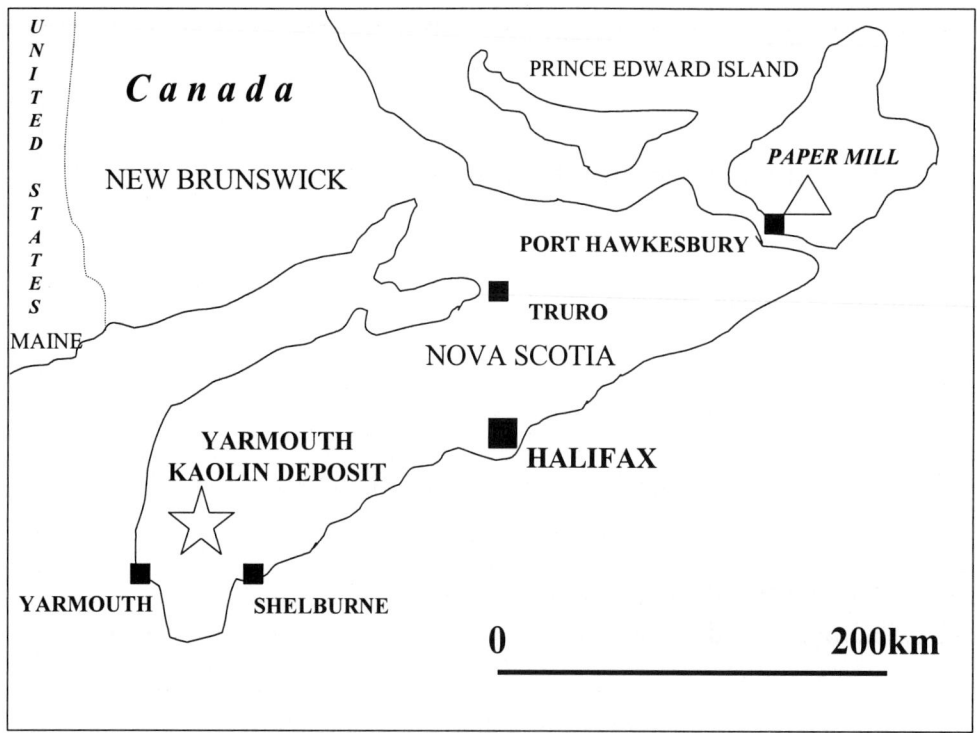

Figure 1. Sketch map of the location of the Yarmouth kaolin deposit, Nova Scotia

2. GEOLOGY AND EXPLORATION

The Yarmouth kaolin property is located within the Meguma Formation of the Canadian Appalachians. The area consists of regionally deformed Cambro-Ordovician turbidites that have been intruded by Late Devonian to Early Carboniferous plutons. The central portion of the property is underlain by a 100 to 200 metres wide, northeast-trending, quartz-kaolinite breccia zone, which dips at approximately 40 to 70 degrees to the southeast. The zone is emplaced along a major shear (Tobeatic Shear Zones) and has been traced north to the Clyde River and south to Frog Pond, a distance in excess of 7 kilometres (Figure 2). The core of the zone is occupied by a 25 to 75 metres wide zone of high purity brecciated quartz. The kaolinised zone forms the footwall to the quartz breccia, dipping to the southeast and varying in width form 15 to 30 metres. Kaolinitie bearing zones also overlay the quartz breccia, forming a hanging wall along the south-easterly portion of the zone. The kaolin zone varies in width from 5 to 60 metres and has a whitish colour. The zone of the quartz brecciation has been traced along the strike of the shear zone by resistivity and the kaolin zone shows a distinct gravity anomaly which, will be useful in identifying new areas for drilling. To date 30 holes have been drilled, some to depths of 150m and more, and kaolinised granite is still being encountered at these depths. Provisional resources of 5Mt of kaolin have so far been identified and, based on the gravity survey, this tonnage can be significantly increased. Samples of the kaolin core have been collected and tested by a number of laboratories in the world for their commercial clay properties.

In addition to the kaolin, quartz is already being mined and evaluated for a number of markets. Work is also continuing on the market potential of the mica.

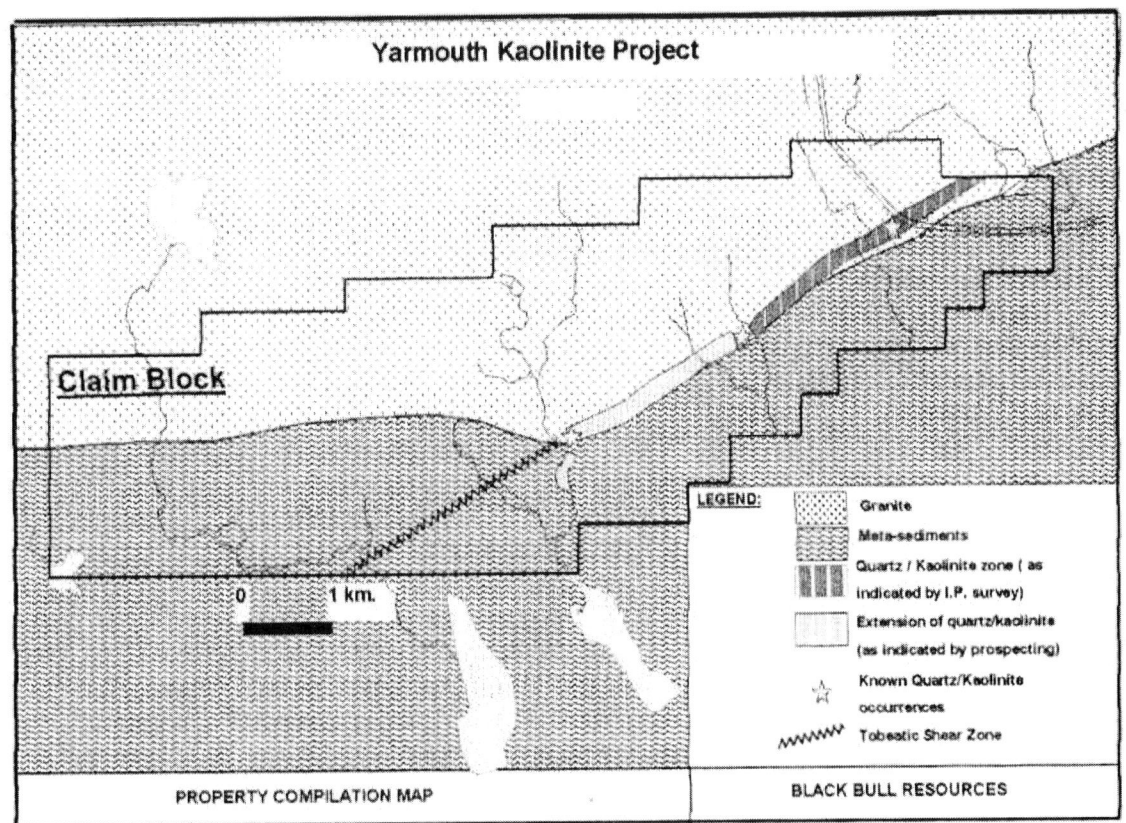

Figure 2. Geological map of the Yarmouth Kaolin Deposit, Nova Scotia

3. MORPHOLOGY OF THE KAOLIN

The coarse fraction <10microns shows the presence of kaolinite stacks (Figure 3):

Figure 3. Kaolinite Stack (width 12 microns)

4. TEST RESULTS OF THE KAOLIN

Some twenty samples have been tested so far to determine the quality of the Yarmouth kaolin. A typical result of some of the physical and chemical test results are in Table 1.

Table 1.
Test Results of the Yarmouth Kaolin

Yield of kaolin at –325# Refining	32%
Wt.% +10 microns	15%
Wt.% -2 microns	50%
ISO Brightness (unbleached)	82.0
ISO Brightness (bleached)	84.0
ISO Brightness (magnetted)	87.5
Flow (wt.% solids)	69.4
Viscosity Concentration (wt.% solids)	71.7
Chemistry (wt.%)	
SiO_2	48.2
Al_2O_3	36.4
Fe_2O_3	0.9
TiO_2	0.1
CaO	0.1
MgO	0.3
K_2O	2.1
Na_2O	0.1
L.O.I.	11.5
Mineralogy (wt.%)	
Kaolinite	84
Mica	11
Quartz	2
Feldspar	2

The kaolin exhibits a pseudohexagonal morphology typical of kaolinite and at the coarser fractions (>10 microns) there are a large number of stacky kaolinites (Figure 3). The kaolin shows a good response to bleaching and magnetting indicating that some of the iron is present as discrete particles, rather than within the lattice. Chemically, the kaolin is high in potash which is reflected in the high mica content of 11%. There are small amounts of quartz and feldspar present but these will be removed by further refining and processing.

The yield of kaolin from the matrix is high for a kaolinised granite at 32% which is a good feature. Work is continuing on evaluating a further 70 borehole and bulk trench samples.

5. PROCESSING TRIALS

A 60kg representative sample of kaolinised granite taken from the trial pit is currently being investigated to determine the commercial viability of the deposit. The first steps will be to make-down the matrix into a slurry and separate out the sand and coarse mica. The resultant mixture of dominantly kaolin, some mica and fine sand will be subjected to a series of refining trials to separate and collect particles of different size. Hydrocyclones will be used to separate the fine kaolin from the sand and mica while a centrifuge will be used to separate the finer kaolin from the coarser kaolin generated from the hydrocyclone stage. These separation processes will generate a series of samples that will be further evaluated physically and chemically for their suitability in paper, ceramics and other uses. The coarser underflows from the centrifuge and hydrocyclone stages will be subjected to a series of flotation trials to remove the coarser, very fine sand and some mica and leave dominantly kaolin. If the morphology of the kaolin is stacky a sand grinding state may be required to delaminate the kaolin into platier material. This form of delaminated kaolin, with a high aspect ratio, is suitable for use in super calendered paper production. A summary of the tests for the proposed flow-sheet is shown in Figure 4.

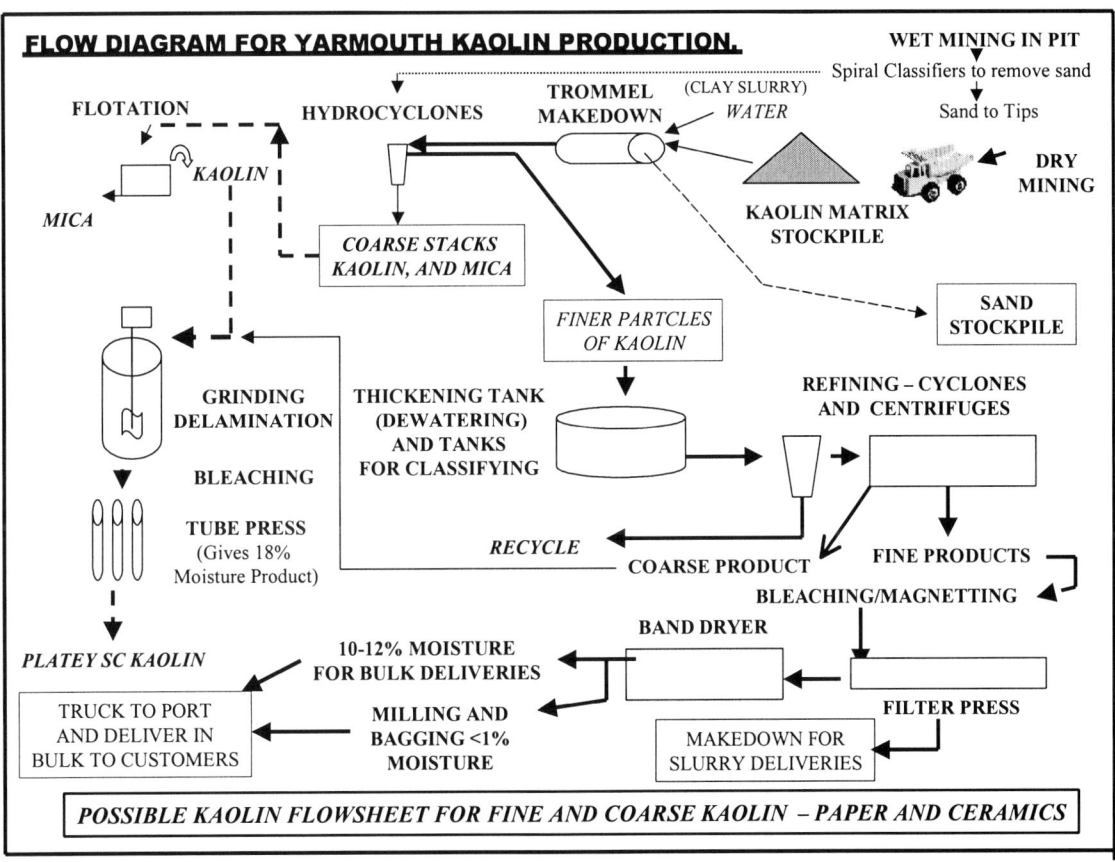

Figure 4. Potential Flow Sheet for the Yarmouth Kaolin

6. SUMMARY. The Yarmouth kaolin deposit represents a good opportunity for Canada to have its first significant commercial production of high quality material for various markets.

Demonstration of the existence of subcritical growth cracks in sintered kaolin

C. Xavier[a]; P.K. Kiyohara[b] and P. Souza Santos[a].

[a]Depto. de Engenharia Metalúrgica e de Materiais - Escola Politécnica da USP - Av. Prof. Lucioano Gualberto, Travessa 3, 380 CEP 05508-970 São Paulo, SP, Brasil Email: cexavier @usp.br
[b]Lab. Microscopia Eletrônica – Instituto de Física da USP.

The subcritical growth of cracks in test pieces of advanced ceramic materials has been studied by the observation of the change of crack velocity, $\dot{c}$, in function of the stress intensity factor, K_I, that leads to the equation of subcritical crack growth: $\dot{c} = A\, K_I^N$. The study of the material is conducted under tension, in a corrosive environment and, usually under normal atmospheric conditions, room temperature, and several values of the air humidity. The materials used are alpha-alumina, silicon carbide, silicon nitride and lime-soda glasses. The occurrence of subcritical growth crack in such types of ceramic bodies affects negatively the performance of the ceramic products at work; this then leads to fracture of the pieces and the effet is called static fatigue, dynamic fatigue, delayed fracture and stress corrosion craking. The purpose of this paper is to prove the existence of subcritical growth of cracks in pressed pieces of a Brazilian kaolinitic clay, fired and sintered at 1600°C; it is a ceramic body composed dominantly of mullite and a glassy phase. The subcritical growth was proved by the use of the curve probability of fracture *vs* strength using Weibull statistics. The strength was measured in a three and four loading fixture. The N parameter for the equation of subcritical crack growth was calculated on basis of the dynamic fatigue model.

1. INTRODUCTION

The use of ceramic materials in structural components has shown an increase in growth; however, their use is only made when their rupture during work does not lead to irreparable damages, since the failure is not always predictable. However, there are instances in which the required properties in the component parts of the project make necessary the use of ceramic materials and they cannot break while at work. In these cases, the ceramic pieces, being under the working conditions are not supposed to break during work and need to have an assurance of a minimum working life.

---

[1] * This paper is part of Projetos Temático FAPESP 1995/0844-0 and FAPESP 1998/09095-2.

Examples are the inspection windows of nuclear reactors, of controlled atmosphere laboratories; windows of space ships and space laboratories. Fragile materials break without previous warning, opposite to ductile materials to which it is possible to apply non-destructive methods to follow up the mechanical strength degradation. The assurance of a minimum working life for ceramic components may be obtained, if some intrinsic properties of the ceramic material are known, the environmental conditions in which the component is going to work, the applied stress during operation and using an analysis method known as *proof test* [1].

The ceramic pieces when used as components of an engineering structure are subject to stresses smaller than the critic stress (σ_c), otherwise they will break immediately. Working with service stress (σ_s) smaller than (σ_c) allows that these ceramic pieces can work for some time until suddenly they break; this happens because the defects start to grow under the stress and by the corrosive action of the environment and for at least one of the defects reach the critical size of the crack and the critical stress intensity factor (K_{Ic}) of the material. The slow crack growth, called sub-critical, occurs under the effect of a static stress producing "static fatigue" [2,3,7,8]; dynamic stress and "dynamic fatigue" [2,3,9,10]; cyclic stress and "cyclic fatigue" [7] or stress resulting from corrosive environment, producing "stress corrosion fatigue" [8].

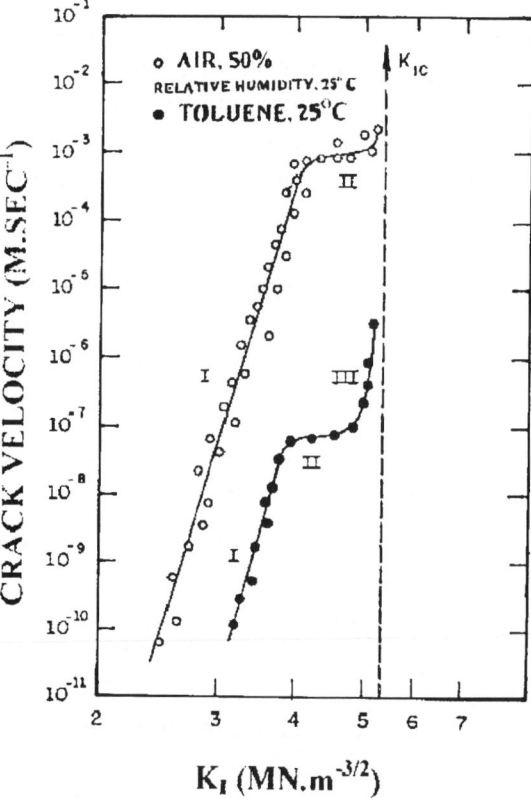

Fig.1 - Crack velocity in function of the stress intensity factor (K_I) for an alumina body [2,4]

Several researchers [4,6] have studied the subcritical growth of cracks by observation of the variation of crack velocity ($\dot{c}$) as function of the stress intensity factor (K_I) and getting graphic relationships, like Figure 1, that presents three distinct regions; they can be explained as resulting from the following mechanisms:

Region I - The crack velocity depends on the reaction rate of the vapour of the corrosive substance at the extremity of the crack. It is the longest and varies exponentially with the stress intensity factor, according to equation:

$$(\dot{c}) = A * K_I^N \tag{1}$$

where: $(\dot{c})$ = velocity of sub-critical crack growth; K_I = stress intensity factor; A and N are parameters of sub-critical crack growth, for a given material in a specific corrosive environment; A nd N are known also as fatigue parameters.

Region II - The crack velocity depends on the diffusion of the corrosive substance in the environment in direction of the crack tip; it is constant and exists only for a short time.

Region III - The crack velocity is independent from the environment, increases very rapidly and reaches the velocity of sound in solids; the values of K_I are close to the value of critical stress intensity factor (K_{Ic}).

1.1. Principle of the method

The experimental procedure to determine A and N, by equation (1), measuring $\dot{c}$ in funciton of K_I, is difficult and complicated, specially in opaque ceramic materials. Due to these facts, a possibility is based on fatigue tests (static, dinamic or cyclic) conducted on several specimens and in a series of equations [7,9]:

$$t_f = B S_I^{(N-2)} \sigma_a^{(-N)} \tag{2}$$

$$\sigma_f^{(N+1)} = B(N+1) S_I^{(N-1)} \dot{\sigma}_a \tag{3}$$

$$t_f = [2/N-2][Y^{(-N)} \pi^{(-N/2)} A^{(-1)} c_i^{(2-N)/2} \sigma_a^{(-N)} w] \{\int [t(t)](N)dt\}^{(-1)} \tag{4}$$

In these equations, t_f = time for the ceramic material to break; S_I = inert strength of the ceramic material, measured in an inert environment, that is, not producing slow crack growth (such as silicone oil, vacuum, nitrogen, argon); Y = shape-dependent factor of the test pieces used; σ_a = applied stress rate; w = time interval for application of the cyclic stress, in which:

$$B = 2/[AY^2(N-2)K_{Ic}^{(N-2)}] \tag{5}$$

Using a linear regression between the experimental results in one of equations (2), (3) or (4), the values of A and N are evaluated.

Having the values of A and N, it is possible to predict a working lifetime under the conditions in which these parameters were evaluated, using the equation (2). The calculated time will be the working expected lifetime of 50% of the ceramic pieces industrially produced under the same conditions as used for the test pieces to determine S_I, because the obtained S_I values present a normal statistical distribution and, therefore, the arithmetic mean of a series test results is considered. A higher percentage than 50% for the breaking time can be obtained by using, instead of the arithmetic mean, the smallest of the values of the bending tests for evaluation of S_I. In order to assure of 100%, that is, in all produced pieces a minimum working lifetime, without any failure during work, a test can be made called Proof Test.

The proof test is performed on the pieces that are going to be used as components to stresses greater than the stress at work (it must be known precisely) and in an inert environment (silicone oil, nitrogen, vacuum, argon). Some specimens will break, others won't, in the specimens that do not break, the biggest defect is calculate by equation:

$$K_{Ic} = Y \sigma_a c^{1/2} \qquad (6)$$

where: K_{Ic} = critical stress intensity factor; Y = parameter which depends of the whole "geometry" of the test, especially of the pieces; σ_a = applied stress and c = diameter of the bigger defect, to be evaluated. That procedure has been applied to materials for advanced ceramic products, like alpha-alumina [2].

1.2. Aim

The aim of this paper is to show that the proof test can be applied to high temperature sintered kaolin test pieces as used in traditional ceramics.

2. MATERIALS AND METHOD

2.1. Kaolin

Commercial HH-86 kaolin, produced by Mineração Horri, Jundiapeba, State of São Paulo, Brazil. Its chemical composition is shown in Table 1.

Table 1
Chemical composition of HH-86 Kaolin (dried 105° C-110° C)

Loss-on-Ignition	13.00%	Magnesia (as MgO)	0.08%
Silica (as SiO_2)	47.30%	Calcia (as CaO)	0.05%
Alumina (as Al_2O_3)	37.20%	Sodium oxide (as Na_2O)	0.16%
Iron (as Fe_2O_3 total)	0.70%	Potassium oxide (as K_2O)	1.35%
Titania (as TiO_2)	0.09%	Total	99.93%

2.2. Subcritical crack growth

Test pieces, size 60x5x4mm^3, were prepared by uniaxial pressing (50MPa). They were fired in a programmed electric furnace, air atmosphere, at 1600°C maximum temperature during 2 hours. Five groups were tested under flexion, one at 4 points and the other at 3 points. The test pieces used for 4 points were reused also for 3 point testing; they were called "stressed" test pieces. The 4 points, the "stressed" and one 3 points were tested in laboratory environment (25°C ± 1°C; 70% relative humidity in São Paulo) and velocity for the equipment (Vm) of 5.0mm/min. Other 3 points group was tested in the same laboratory environment using three different velocities for the equipment: Vm=5.0, 25.0 and 50.0mm/min. The applied Stress Rate σ_a is related to the equipment velocity Vm by the equation:

$$\sigma_a = (12Eh/L^2)Vm \qquad (7)$$

where: E = elasticity modulus of the sintered kaolin; L = distance between supports in the 3 points test and h = height of the test piece.

Other group of test pieces, also 3 points, was tested in an inert atmosphere (silicone oil); as pointed before; the inert atmosphere (IN AT) is used to avoid subcritical crack propagation.

3. RESULTS AND DISCUSSION

3.1. Kaolin

The chemical analysis from Table 1 shows that it is a low iron kaolin, with a composition close to that of theoretical kaolinite. Its X-ray diffraction (XRD) curve presents the reflections of the kaolin group; the scanning electron micrographs (Figure 2) show that the crystals are platy, with hexagonal profile; the presence of booklets is frequent. So, the sample is characterized as constituted by kaolinite and, from its XRD curve, of the well ordered type; the fair amount of K_2O suggests the presence of muscovite.

Fig. 2 - Scanning electron micrograph of kaolin HH 86 powder. Bar in figure = 10µm.

The 1600°C fired test piece was transversally fractured and examined by SEM; at low magnification (Figure 3): the glassy matrix shows the presence of elliptically shaped closed pores; at higher magnification (Figure 4), the elongated mullite crystals can be seen covering the upper internal surface of a closed pore.

The test pieces, after sintering at 1600°C, presented a linear contraction of 8.3% and an apparent open porosity of 2.8%. Closed pore volume was measured in SEM micrographs; the average closed pore volume was 3.0%.

3.2. Flexural Strength Tests

The results of the flexural tests are shown in the Table 2. The flexural test in 4, 3 and 3 points after stress are shown in Figure 5, with their straight lines drawn after Weibull statistical theory. It can be noted that the three lines are almost parallel; that observation indicates that: a) the tests did not produced significant changes in the defects distribution and that the Weibull's parameters are very close each other; b) the displacement of the straight lines in the 4 and 3 points shows values of the strength for the 4 points, that is due to the large volume of the test piece subjected to the maximum stress; c) there was a decrease of strength in the pre-stressed pieces in relation to non pre-stressed ones, shown by a shift to the left of the respective straight line; it was due to a sub-critical growth of the cracks, from the pre-stressing.

Table 2

Results of the flexural tests (MPa)

4 Points	3 Points (ST)	3 Points (ST)	3 Points (ST)	3 Points	3 Points	3 Points	Vel. 5	Vel. 25	Vel. 50	Vel.5 (INAT)
60.90	43.62	86.01	109.48	71.53	101.00	120.80	34.33	45.43	44.92	100.99
61.41	52.36	86.63	109.98	80.84	101.89	122.08	67.67	59.34	53.87	129.15
73.06	72.21	86.70	111.05	80.91	102.57	122.23	68.91	63.89	63.41	137.58
74.25	74.07	91.28	111.63	83.67	105.75	122.31	74.18	70.33	66.62	148.69
79.86	76.64	91.74	113.84	84.61	107.25	125.37	75.59	73.69	67.56	152.69
95.09	79.47	94.93	114.41	88.68	107.58	125.93	79.69	77.98	69.28	155.06
	79.47	95.36	115.42	88.99	107.60	127.33	76.93	78.69	71.90	162.33
	80.79	98.62	116.82	93.74	109.07	127.79	82.56	79.70	72.52	168.50
	81.57	99.36	117.01	95.13	109.49	132.96	86.64	81.13	76.90	168.93
	82.93	102.57	117.68	96.06	110.17	133.74	87.63	89.49	83.51	171.92
	84.12	103.56	118.58	97.34	116.92	134.63	102.23	95.86	88.22	180.15
	84.17	103.76	119.93	97.82	116.99	152.03	109.02	101.95	91.56	181.36
	84.17	105.41		98.49	119.64		113.60	102.23	91.56	183.47
	85.62	107.07		99.32	120.22		116.45	114.48	101.29	190.18
										194.88

Using the pairs of values (σ_f, σ_a), obtained on Figure 6, for a probability of failure of 50% and equation (5), the value of 18.9 for N was calculated. The next stage is the determination of the value of K_{Ic} for evaluation of the A parameter.

The results of the flexural strength tests with different velocities are shown in Figure 6, also plotted as a Weibull's diagram and showing the four straight lines, one for each velocity and one for the inert strength without subcritical growth of cracks; the lines are almost parallel and show a decrease of strength at increasing the velocity of the applied stress. These results are different from those previously obtained in similar test pieces of sintered alpha-alumina [2]. Thay may be attributed to differences in behavior under flexural stress of the two types of test
pieces: alumina, with just one crystalline component plus pores, while the sintered kaolin is a continuous glassy matrix with embedded rod-shaped crystals of mullite plus pores.

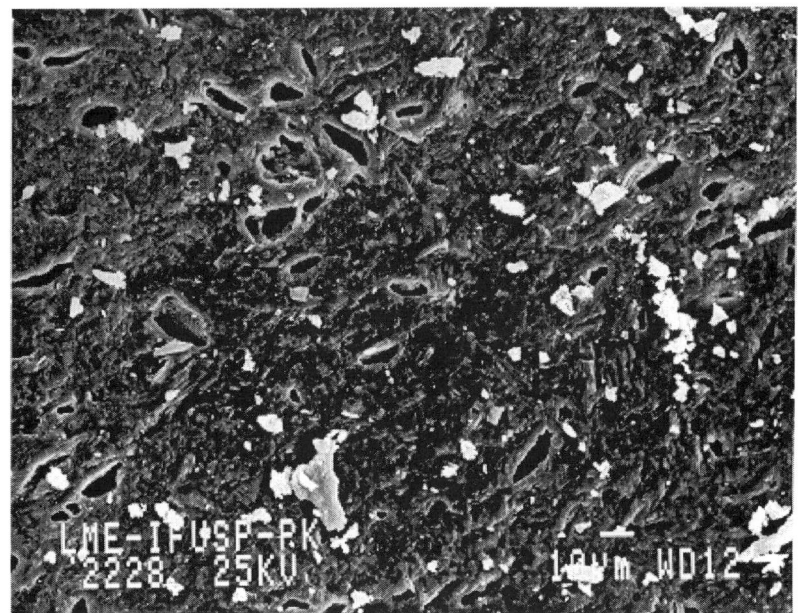

Fig. 3 - Scanning electron micrograph of kaolin HH 86, sintered at 1600°C. Bar in figure = 10µm.

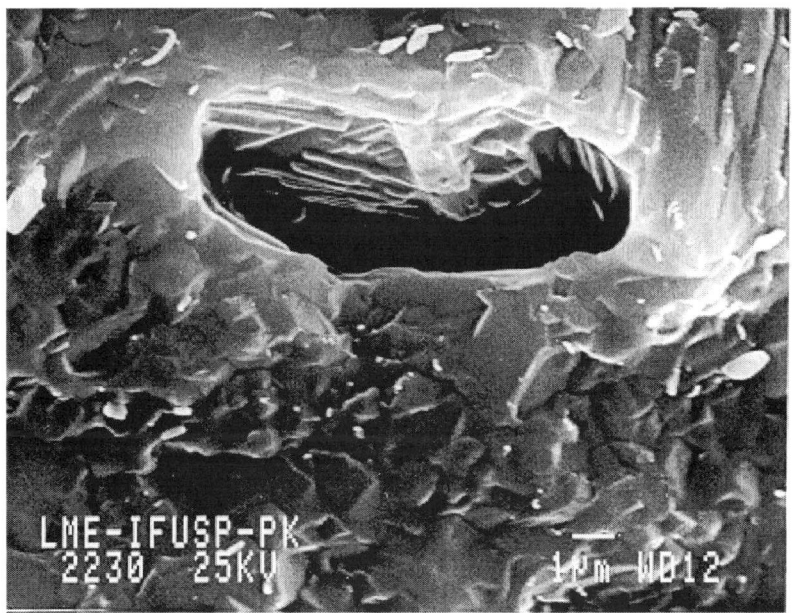

Fig. 4 - Scanning electron micrograph of kaolin HH 86, sintered at 1600°C. Bar in figure = 10µm.

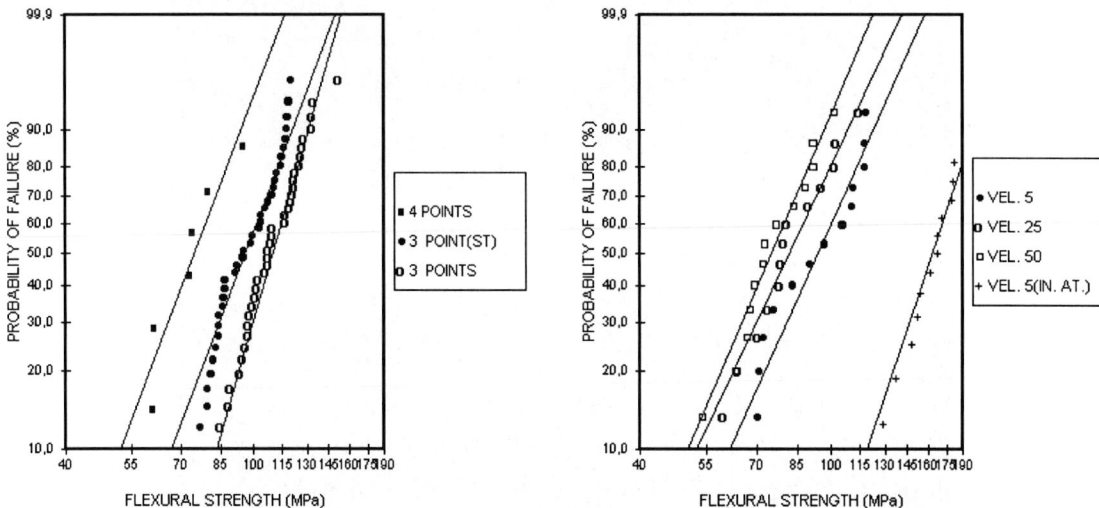

Fig. 5 – Results of the flexural tests of kaolin HH 86, sintered at 1600°C.

Fig. 6 – Results of hte flexural test of kaolin HH86, sintered at 1600°C with several velocities for the applied stress.

5. CONCLUSIONS

1) The evaluation of the sub-critical growth of cracks in test pieces made of a kaolin sintered at 1600°C has shown that the procedure, as used in advanced ceramics, may be used in high kaolin traditional ceramic bodies for products rich in mullite.

2) It is possible to evaluate the sub-critical parameters and thus to establish and prediction of the lifetime in service for sintered kaolin test pieces and similar ceramic products.

REFERENCES

1. A.G. Evans and S.M. Wiederhorn, Int. J. of Fracture, 10 (1974) 379.
2. C. Xavier and H.W. Hubner, Cerâmica, 28 (149) May (1982) 161.
3. J.E. Ritter Jr.and J.N. Humenik, J. Mater. Sci. 14 (1979) 626-632.
4. A.G. Evans, J. Mater. Sci. **7**, (1972) 1137.
5. J.L Henshall and D.J. Rowcliffe, J. Am. Ceram. Soc, 62 (1979) 36.
6. J.B. Wachtman Jr, J. Am. Ceram. Soc. 57 (1974) 509.
7. T. Kawakubo and K. Komeya, J. Am.Ceram. Soc., 70 (1987) 400.
8. S.M. Wiederhorn and L.H. Bolz, J. Am. Ceram. Soc. 53 (1970) 543.
9. S.M. Barinov, N.V. Ivanov, S.V. Orlov and V. Shevchenko, Ceram Int 24 (1998) 4321.
10. S. Raynoud, E. Champion, D. Vernache-Assolant and D. Tetard, J. Mater. Sci. Med. 9 (1998) 221.

Geology and physical properties of Palygorskite from Central China and Southeastern United States

H. Zhou and H.H. Murray

Dept. of Geological Sciences. Indiana University. Bloomington, IN 47405 USA

Relatively new and large palygorskite deposits of Miocene age have been discovered in Anhui and Jiangsu provinces in Central China. The Guanshan deposit is used as an example to compare the geologic occurrence and physical properties of the Chinese palygorskites with the palygorskites produced from the Miocene Florida-Georgia deposits in the United States. The thickness of the Chinese palygorskite beds range from 3 to 6 meters and the Florida-Georgia palygorskite beds range from 2 to 4 meters in thickness. Both the Chinese palygorskites and the Florida-Georgia palygorskites are Middle Miocene in age.

The Guanshan palygorskite was formed as the alteration product of basaltic ash deposits in a lacustrine environment. By contrast, the Florida-Georgia palygorskite was precipitated in restricted lagoonal and tidal environments. These different conditions of formation resulted in mineralogical and physical property differences. For example, the Guanshan palygorskite deposit contains a lower quartz and carbonate content than the Florida-Georgia deposits. Also, no phosphate minerals are present in the Guanshan deposit. As a result, the overall palygorskite content is higher in the Chinese deposits.

Chemically, the Guanshan deposit has a higher content of Mg0 and also the viscosity and absorption capacity are higher. In addition, because of the higher palygorskite content, the bulk density is lower. Both the Florida-Georgia and Chinese deposits are high quality and large. However, because they were formed in different depositional environments, there are differences in mineral content and physical properties. (240)

1. INTRODUCTION

Palygorskite and attapulgite are synonymous terms for the same hydrated magnesium aluminum silicate mineral. The preferred name is palygorskite as specified by the International Nomenclature Committee. However, the name attapulgite is so much used in trade circles that it continues to be used in literature. In this paper, the term palygorskite will be used.

Palygorskite has been produced in southeastern United States for nearly 100 years near the Georgia-Florida border (Figure 1).

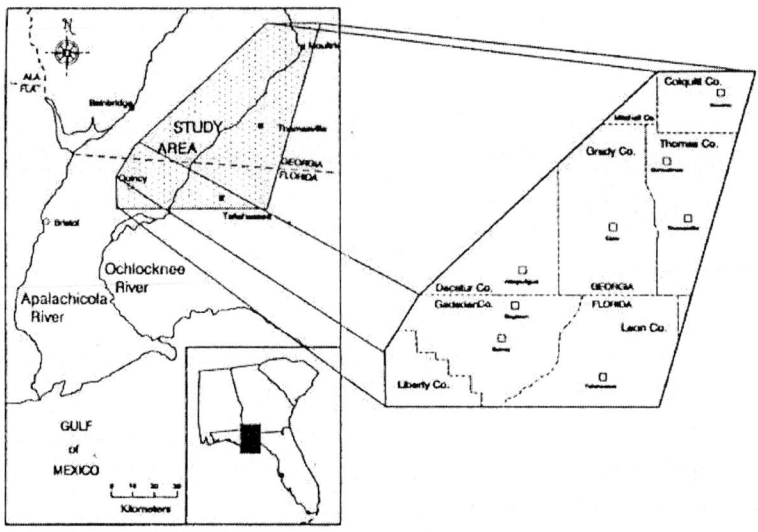

Figure 1 - Location Map of Georgia-Florida Palygorskite District

Within the last 20 years (Zhou, 1996) more than twenty high quality palygorskite deposits have been discovered in China near the provincial boundary of Anhui and Jiangsu (Figure 2). In this paper, the Guanshan deposit in Anhui province is used as the example in the comparison with the Georgia-Florida deposits.

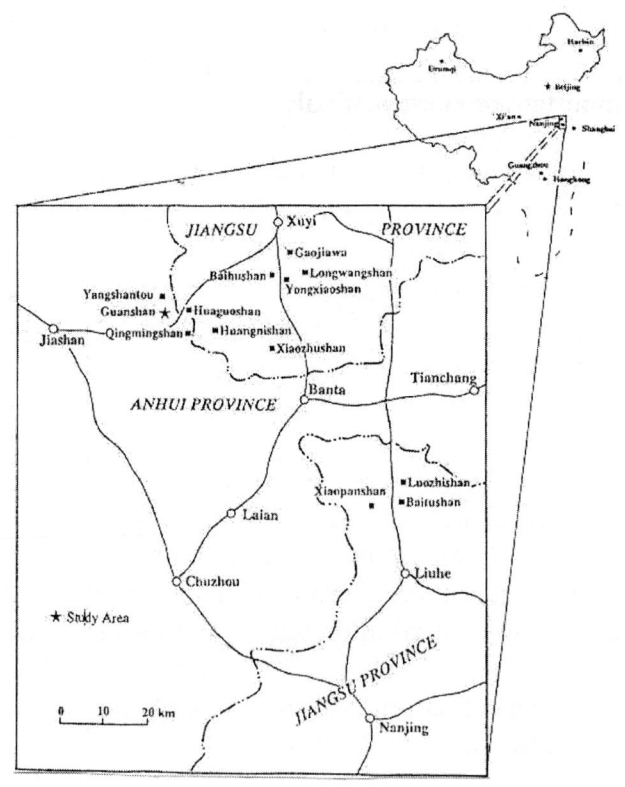

Figure 2 - Location Map of China Palygorskite Deposits

Palygorskite is an elongate particle (Figure 3), which along with its other physical and chemical properties, results in some special industrial applications.

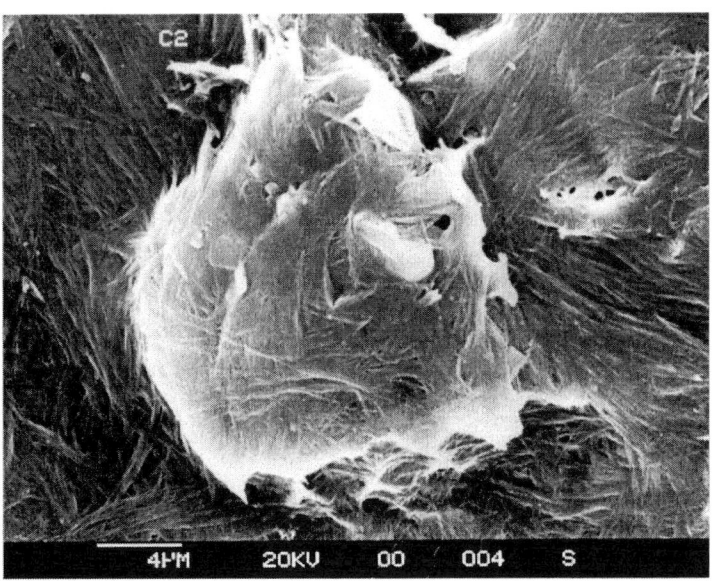

Figure 3 - SEM of a Palygorskite from Quincy, Florida

A brief outline of some of the properties of palygorskite that relate to its application is shown in Table 1.

Table 1 - Some important properties of palygorskite
- 2:1 Layer inverted
- Light tan or gray in color
- Substitutions in the octahedral layer
- Moderate layer charge
- Moderate base exchange capacity
- Elongate in shape
- High surface area
- High absorption capacity
- High viscosity

There is considerable substitution of magnesium and iron for aluminum in the octahedral layer which results in a moderate layer charge. This layer charge and the high surface area usually of the order of 160 to 200 m2/gm results in a moderate cation exchange capacity normally about 30 to 40 meq./100 gs. The high surface area, the charge on the lattice, and the inverted structure which leaves parallel channels through the lattice (Grim, 1962) gives palygorskite a high absorption capacity (Figure 4).

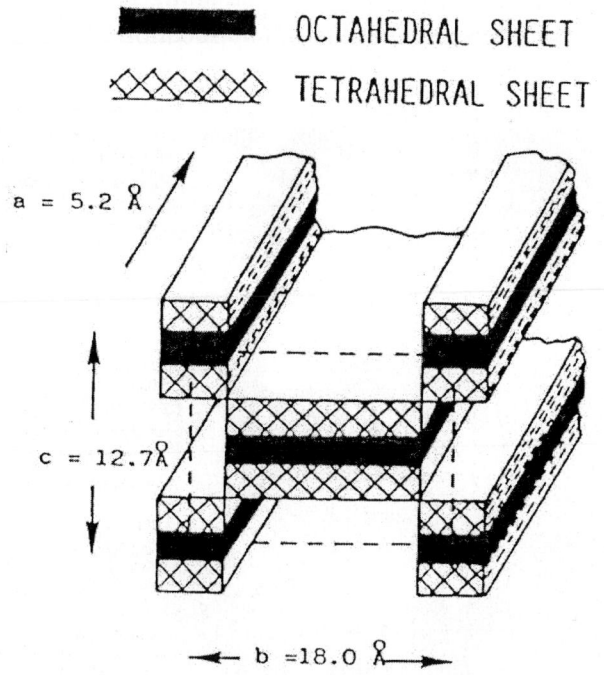

Figure 4 - Schematic Representation of the Structure of Palygorskite

These physical and chemical properties, along with the elongate habit of palygorskite makes it particularly useful in many industrial applications. Some important uses are listed in Table 4.

Table 4 - Some important uses of palygorskite		
Drilling fluids	Cat box absorbents	Paper
Paint	Suspension fertilizers	Pharmaceuticals
Agricultural carriers	Animal feed bondants	Anti-caking agent
Industrial floor absorbents	Catalyst supports	Reinforcing fillers
Tape joint compounds	Adhesives	Environmental absorbent

A description of these uses is given by Heivilin and Murray (1994) and Galán (1996). Because of the elongate shape, palygorskite maintains its viscosity and suspension properties in salt water brine and electrolytes.

2. GEOLOGY AND MINERALOGY

Both the palygorskite deposits in China and the United States are Middle Miocene in age. The Guanshan deposit is part of the upper member of the Huaguoshan Formation (Figure 5). The thickness of the palygorskite bed ranges from 3 to 6 meters in thickness (Zhou, 1996). The palygorskite was formed in a lacustrine environment as the alteration product of basaltic ash deposited in the lake. A thick basalt overlies the palygorskite deposit. It is estimated that there is well over 200 million tons of palygorskite reserves in Anhui and Jiangsu provinces.

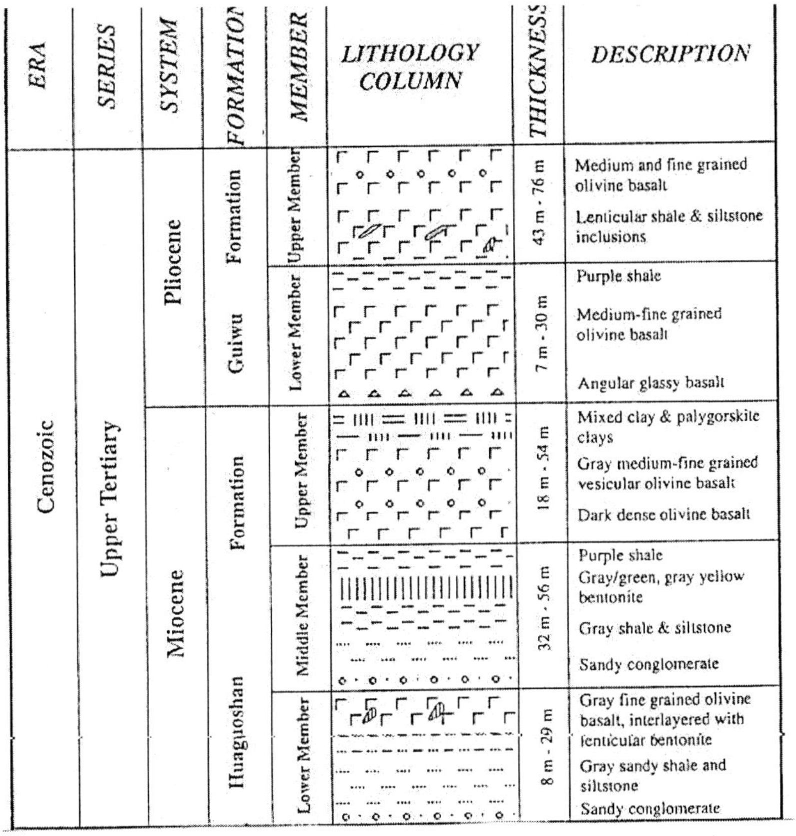

ERA	SERIES	SYSTEM	FORMATION	MEMBER	LITHOLOGY COLUMN	THICKNESS	DESCRIPTION
Cenozoic	Upper Tertiary	Pliocene	Guiwu Formation	Upper Member		43 m - 76 m	Medium and fine grained olivine basalt / Lenticular shale & siltstone inclusions
				Lower Member		7 m - 30 m	Purple shale / Medium-fine grained olivine basalt / Angular glassy basalt
		Miocene	Hunguoshan Formation	Upper Member		18 m - 54 m	Mixed clay & palygorskite clays / Gray medium-fine grained vesicular olivine basalt / Dark dense olivine basalt
				Middle Member		32 m - 56 m	Purple shale / Gray/green, gray yellow bentonite / Gray shale & siltstone / Sandy conglomerate
				Lower Member		8 m - 29 m	Gray fine grained olivine basalt, interlayered with lenticular bentonite / Gray sandy shale and siltstone / Sandy conglomerate

Figure 5 - Stratigraphic Column Showing the Age of the Guanshan Palygorskite in te Upper Member of the Hungushan Formation

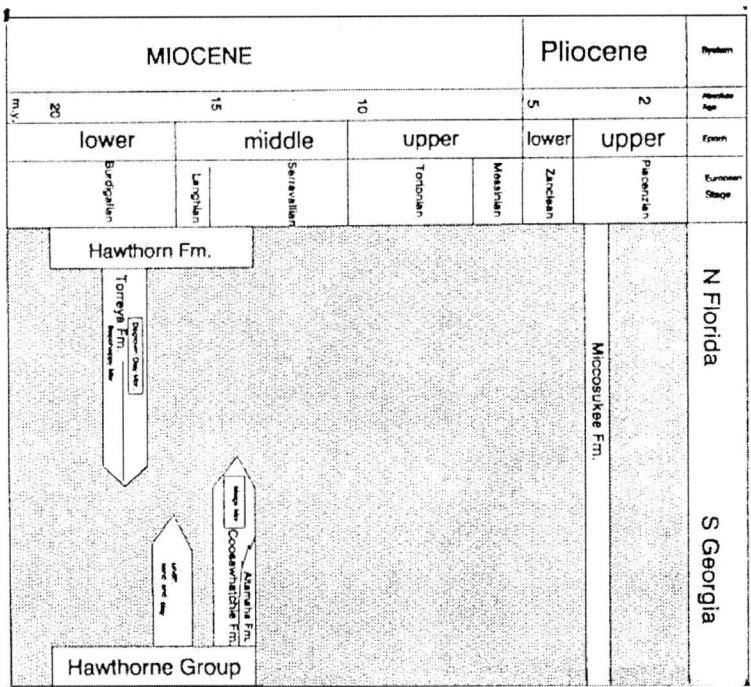

Figure 6 - Stratigraphic Column Showing the Age of the Georgia-Florida Palygorskite in the Dogtown Member

The palygorskite in Southeastern U.S. is part of the Hawthorne Formation of Middle Miocene age (Figure 6). This palygorskite was deposited in a tidal-lagoonal environment in a highly saline marine setting (Patterson, 1974). The palygorskite deposits range in thickness from 2 to 4 meters (Merkl, 1989)

The mineralogy of the Guanshan and the U.S. palygorskite deposits are very similar. The Guanshan deposit has a lower content of quartz and dolomite generally less than 5% in total. Most of the palygorskite contains no dolomite and has less than 2% quartz. The palygorskite deposits in South Georgia and North Florida has about 10% quartz and dolomite. The mineralogy was determined by x-ray diffraction analysis.

3. COMPARATIVE PHYSICAL AND CHEMICAL PROPERTIES

Several laboratory tests were performed on the palygorskites from the Guanshan deposit and a deposit near Quincy, Florida. These were pH, percent plus 200 mesh, percent plus 325 mesh, bulk density, water absorption, oil absorption, dispersion viscosity, API yield, thixotropic index, and chemical analysis. The physical property results are shown in Table 1 and the comparative chemical analysis are shown in Table 2.

Table 1 - Comparative physical properties of the Guanshan and Quincy, Florida Palygorskites

Test	Guanshan palygorskite	Quincy, Florida palygorskites
pH	7.4	8.2
Grit (plus 200 mesh) percentage	1.2	8.0
Grit (plus 325 mesh) percentage	1.8	9.2
Bulk density (g/cm3)	0.46	0.58
Water absorption (ml/100g)	84	85
Oil absorption (ml/100g)	49	42
Dispersion viscosity (cps)1	3494	3384
API yield (bbls/ton)2	128	125
Thixotropic index	7.6	7.1

(1) Florida Co. (1993)
(2) Amer. Petrol. Inst. (1985)

Table 2 - Comparison of major oxide percentage of the Guanshan and Quincy, FL palygorskites

Elements	SiO2	MgO	Al2O3	Fe2O3	CaO	P2O5	Na2O	K2O	LO1	Total
Guanshan	56.09	11.16	8.7	4.5	0.39	0.09	0.04	0.9	17.4	99.27
Quincy, Fl	59.06	9.79	9.12	3.06	1.97	1.10	0.06	0.79	15.1	100.05

The results in Table 1 and Table 2 show that the Guanshan and Quincy, FL deposits are very similar. The physical property results are the average of 16 tests for each property and 15 chemical analyses of the Guanshan palygorskites and 2 chemical analyses of the Quincy, Florida palygorskites. The pH of the Florida palygorskite is higher probably because of the higher calcium content. As noted in the previous section, the grit percentage of the Florida palygorskite is higher. The remaining properties are very similar. The thixotropic index is the ratio of the viscosity measured at 10 rpm to the viscosity measured at 100 rpm.

Because of the higher grit percentage, the crude palygorskite from Florida must be pulverized to a fine particle size in order to reduce the grit content. The Guanshan palygorskite does not need to be pulverized as fine because of its high purity because the palygorskite will disperse to fine particle very easily in water.

In processing to make a high gel grade of palygorskite MgO is added by most producers in the U.S. Because the Guanshan palygorskite has a higher MgO content, no MgO addition is needed.

4. SUMMARY

The geologic conditions in which the Guanshan and Florida palygorskites formed is very different. The deposits in China formed in a fluvial-lacustrine system whereas the Georgia-Florida deposits formed in a starved marine lagoonal system. The reserves of palygorskite in China are very large and much larger production capacity will be forthcoming in the immediate future. The physical properties of both the U.S. and China palygorskites are very good and high quality products are produced from both areas.

REFERENCES

1. API (American Petroleum Institute), API specification for oil-well drilling fluid material. API specifications 13A (1983) 17
2. Floridan Company, Dispersion viscosity test procedure. Method FLQ-VLT 4100 (1993) 2
3. E. Galan, Properties and applications of palygorskite-sepiolite clays. Clay Miner. 31 (1996) 443-445
4. R.E. Grim, Applied Clay Mineralogy. McGraw-Hill, N.Y. (1962) 596
5. F.G. Heivilin and H.H. Murray, Hormites: Palygorskite (Attapulgite) and Sepiolite; In: D.D. Carr (ed) Industrial Minerals and Rocks. 6th edn. SME, Littleton, CO(1994) 249-254
6. R.S. Merkl. A Sedimentological, Mineralogical and Geochemical Study of the Fuller's Earth Deposits of the Miocene Hawthorne Group of South Georgia-North Florida. Ph.D. Thesis, Indiana University (1989) 182
7. S.H. Patterson, Fuller's Earth and Other Industrial Mineral Resources of the Meigs-Attapulgus. Quincy District, Georgia and Florida. U.S. Geol. Survey Prof. Paper 828 (1974) 45
8. H. Zhou, Mineralogical and Industrial Evaluation of a Palygorskite Deposit from Guashan, Anhui Province, P.R. China, Ph.D. Thesis, Indiana University (1996)197

V
Mineral Structure and Investigational Methods

Crystal Chemistry

Distribution and characterization of forms of Fe and Al in particle-size fractions of an Entic Haplustoll by selective dissolution techniques, X-ray powder diffraction and Mössbauer spectroscopy

S.G. Acebal[a], M.E. Aguirre[b], R.M. Santamaría[c], A. Mijovilovich[d], S. Petrick[e] and C. Saragovi[f]

[a] Department of Chemistry and Chemical Engineering. Universidad Nacional del Sur Av. Alem 1253. 8000 - Bahía Blanca, Argentina.

[b] Department of Agronomy. Universidad Nacional del Sur. Altos del Palihue. 8000 - Bahía Blanca, Argentina.

[c] CIC. Comisión de Investigaciones Científicas de la Provincia de Buenos Aires.

[d] EMBL c/o DESY, Notkestrasse 85 Geb 25 A, 22607 Hamburg, Germany.

[e] Facultad de Ciencias. Universidad Nacional de Ingeniería, Av. Tupac Amarú 210 Lima, Perú.

[f] Department of Physics. Comisión Nacional de Energía Atómica. Avda. Libertador 8250. 1429 - Buenos Aires, Argentina

Aggregate stability depends on soil properties such as organic matter content, type and quantity of clay and poorly crystalline Fe and Al oxide and oxy-hydroxides. Then, the quantification and characterization of forms of Fe and Al are of major importance for understanding soil structure. Particle-size fractions (5-2 μm, 2-1 μm, and <1 μm) from the Ap horizon of an Entic Haplustoll from Argentina, treated with the selective-dissolution techniques ammonium oxalate (OX), dithionite-citrate-bicarbonate (DCB), NaOH, and Na-pyrophosphate (PY), are studied by X-ray powder diffraction (XRD) and ^{57}Fe Mössbauer spectroscopy (MS). Quartz, feldspar, smectite, illite and interstratified illite-smectite are predominant whereas Fe and Al oxides and oxy-hydroxides are present in low concentration. Al-substituted hematite and poorly crystallized hematite and goethite are present in similar proportions.

1. INTRODUCTION

Previous studies of Mollisols from the province of Buenos Aires, Argentine, have indicated that Fe and Al amorphous colloid compounds play an important role in soil-aggregation processes [1]. In these soils, the silt fraction contains particles which, can be considered as pseudo-silts. They were formed from the association of Fe and Al amorphous colloid compounds with clay minerals.

Structural stability index and distribution curves of aggregates have shown a natural process of structural organization, though soil-aggregates have very low stability. The relative contribution and significance of organic matter, type and quantity of clay and Fe and Al oxide and oxy-hydroxides are difficult to determine and seem to vary among different size fractions.

For a better understanding of soil-aggregation processes, three particle-size fractions (5-2 µm, 2-1 µm, and <1 µm) were studied. Selective-dissolution techniques using chemical extractant agents such as Na-pyrophosphate (PY), acid ammonium oxalate (OX), dithionite-citrate-bicarbonate (DCB) and NaOH were applied. XRD and MS were used on the untreated and chemically treated soil samples to identify and characterize non-clay and clay minerals, Fe and Al oxides and oxy-hydroxides.

MS allows a specific analysis of the Fe-bearing phases, in particular oxides and oxy-hydroxides, since it is very sensitive to the microscopical Fe environments. MS uses a nuclear effect and the electronic and magnetic environment on each Fe site can be determined from the measured hyperfine parameters: isomer shift (IS), quadrupole splitting (QS) and hyperfine magnetic field (H). The oxidation state of Fe is easily obtained, different sites distinguished and their relative population determined using the corresponding subspectral areas. The effect does not depend on crystallinity, and allows to distinguish between crystalline and amorphous states [2].

2. MATERIALS AND METHODS

Surface soil sample (0-12 cm), representing the Ap horizon of an Entic Haplustoll from Bordenave, province of Buenos Aires, Argentina, was used. Soil sample was air-dried, ground and sieved to obtain the < 2-mm size fraction. Then, the soil sample was chemically specify treated ($CO_3^=$ and organic matter removed and soil dispersed) and the different fractions (5-2 µm, 2-1 µm, and <1 µm) separated by sedimentation techniques. Some physico-chemical properties of the soil are listed in Table 1.

Table 1
Some physical and chemical properties of the A_p horizon studied.

Parameter	Values
pH[a]	8.0
pH[b]	8.3
Organic Carbon (g.kg^{-1})[c]	11.4
CEC (cmol$_c$kg^{-1})[d]	23.36
Sand (g.kg^{-1})[e]	433.1
Coarse Silt (20 - 50 µm) (g.kg^{-1})[e]	241.5
Fine Silt (2 - 20 µm) (g.kg^{-1})[e]	174.8
Clay (<2 µm) (g.kg^{-1})[e]	148.0

[a] Measured in water suspension
[b] Measured in 1 mol L^{-1} KCl 1:2.5 soil/solution ratio
[c] Walkey (1946) [3]
[d] Bower et al. (1952) [4]
[e] Robinson (1922) Pipette method. [5]

2.1. Chemical Analysis

The three particle-size fractions obtained were treated with chemical extractant reagents as described below, producing five subsamples for each fraction: untreated (UT), PY treated, OX treated, DCB treated, and NaOH treated.

To determine associated organic matter Fe, Al and Si contents, a 16-h treatment with PY (0.1M, pH10) was carried out [6]. To extract poorly crystallized Fe and Al oxides, a single 2-h treatment in darkness with OX was performed [7, 8]. For complete dissolution of all crystalline Fe and Al oxides and oxy-hydroxides, the DCB procedure of Mehra and Jackson [9] was performed. NaOH was used for dissolution of amorphous Al and Si oxides [10].

Total Fe, Al and Si were determined after complete dissolution of soil fractions in a Pt crucible using the alkaline fusion method [11]. All treatments were performed in duplicate and Fe, Al and Si contents in extracts were analyzed by atomic absorption spectroscopy (AAS).

2.2. X-ray diffraction

Patterns from (randomly oriented) bulk specimens were collected using a Philips X-Pert PW3710 diffractometer. This unit is equipped with a Cu tube, with a generator setting of 45 kV and 35 mA, and a graphite monochromator in the diffracted beam. Patterns were collected from 2 to 70 °2θ using a step size of 0.02 °2θ and a count time of 2s per step. From these patterns the non-clay minerals such as quartz, feldspar, hematite, goethite, etc. were identified.

For clay mineral analysis oriented specimens were made after Mg-exchange and K-exchange of a part of the untreated and treated size fractions. After measuring both types of specimens at room conditions the Mg-exchanged specimens were treated with glycerol and measured again. The K-exchanged specimens were again measured after a heat treatment at 823 K. Patterns were collected using a Philips PW1011/00 diffractometer. This unit is also equipped with a Cu tube, with a generator setting of 36 kV and 18 mA, using a Ni filter in the diffracted beam. Patterns were collected from 2 to 30 °2θ using a step size of 0.01 °2θ and a count time of 30s per step.

2.3. Mössbauer Spectroscopy

Mössbauer spectra were collected with the source at room temperature (RT) and the absorbers at RT and 15 K. Samples were carefully ground in an agate mortar, weighted and mixed with nonrefined sugar to avoid texture effects [12] before assembling the "thin absorbers" [13]. All the spectra were fitted with combinations of six-lines spectra (sextets S) and two-lines spectra (doublets D) using the NORMOS program [14]. Crystalline sites and distribution of very similar sites were taken into account accordingly to the experimental spectra. The quality of each computer fit was checked by a χ^2 test. The relative populations of the Fe-bearing compounds were computed from the corresponding fitted areas at 15 K in each sample.

3. RESULTS AND DISCUSSION

3.1. Chemical Analysis

Chemical data of the different soil fractions are summarized in Table 2 a) and Table 2 b).

Fe and Al oxides, comprising oxides and oxy-hydroxides, are present at low concentration. Total Fe, Al and Si contents decrease as soil-particle size decreases. DCB-insoluble Fe and Al

Table 2 a)
Dissolution analysis of the soil fractions used in this study

Soil Fraction	Fe_{ox} (%)	Al_{ox} (%)	Si_{ox} (%)	Fe_{DCB} (%)	Al_{DCB} (%)	Si_{DCB} (%)
5 – 2 µm	0.900	1.730	0.240	3.350	1.500	1.410
2 – 1 µm	0.901	2.550	0.250	3.250	1.680	1.540
< 1 µm	0.992	2.790	0.180	2.920	1.407	1.750

OX: Oxalate-extractable Fe, Al and Si expressed as % Fe_2O_3, Al_2O_3 and SiO_2
DCB: Dithionite-citrate-carbonate-extractable Fe, Al and Si expressed as % Fe_2O_3, Al_2O_3 and SiO_2

are either a structural component of the silicates or an interlayer constituent of the hydroxy-interlayered 2:1 type minerals present.

Crystalline and amorphous Fe and Al oxides and/or oxy-hydroxides content are slightly higher in larger particle-size fractions. In the 5–2 µm fraction these compounds may be acting as cementing agents coating the surface of small clay particles, while in the < 1 µm fraction they could be single particles. Previous disaggregation studies of this soil [1] showed that most of the clay particles (2-1 µm) are aggregated into secondary particles larger than 2 µm in size. The removal of Fe oxides with DCB in larger particle sized fractions caused an increment (~18%) in the 5-2, 2-1 and <1 µm fractions. These results are in agreement with our present observation.

Acid–oxalate-extractable Fe (Fe_{OX}) values indicate that all samples contain small amounts of poorly-ordered, oxalate-soluble Fe-oxides. Acid-oxalate-extractable Al (Al_{OX}) is larger than DCB-extractable Al (Al_{DCB}). These values may possibly be attributed to the acid-oxalate ability to dissolve certain short-range-order minerals as noticed by Acebal et al., [15] for Mollisols.

The molar ratio Al / (Al + Fe) ranges from 0.47 (in the 5-2 µm fraction) to 0.49 (in the <1 µm fraction) mol mol^{-1}, according to the DCB values (Table 2 a). These results indicate a medium to high Al-substitution in the different soil-fractions [16].

Pyrophosphate-extractable Fe (Fe_{PY}) and pyrophosphate-extractable Al (Al_{PY}) values increase as particle size decreases. Al_{PY} corresponds to Al in humus complexes. Fe_{PY} can be used to estimate Fe in such organic complexes, but in some soil samples Fe_{PY} could also arise from Fe in ferrihydrite and goethite dispersed by the dissolving reagent. However, the Si levels extracted by PY are very low indicating that PY in these soils dissolves very little amorphous material.

Table 2 b)
Dissolution analysis of the soil fractions used in this study

Soil Fraction	Fe_{Py} (%)	Al_{Py} (%)	Si_{Py} (%)	Al_{NaOH} (%)	Si_{NaOH} (%)	Fe_t (%)	Al_t (%)	Si_t (%)
5 – 2 µm	0.097	0.152	0.280	4.502	5.400	12.800	17.280	59.400
2 – 1 µm	0.136	0.246	0.468	5.180	5.800	8.930	14.640	51.110
< 1 µm	0.312	0.994	1.044	9.730	7.830	7.010	13.000	41.300

PY: Na-Pyrophosphate-extractable Fe, Al and Si expressed as % Fe_2O_3, Al_2O_3 and SiO_2
NaOH: NaOH-extractable Al and Si expressed as % Al_2O_3 and SiO_2
t: Total Fe, Al and Si expressed as % Fe_2O_3, Al_2O_3 and SiO_2

Alkali-soluble Si and alkali-soluble Al values increase as particle size decrease. The amounts of Si and Al removed from the different soil fractions indicate the great influence of the dissolving agent over small clay particles, rather than the presence of low crystallinity compounds. Fe-oxides and oxyhydroxides are highly insoluble in alkali.

3.2. X-ray diffraction

The X-ray diffraction patterns of the 5-2-μm size fraction are shown in Figure 1. These patterns show the presence of poorly crystallized materials such as expansible clays from smectite group and irregular and no quantitative amounts of interstratified illite-smectite (10.00 – 15.06 Å). Also, illite (10.00 Å) is identified. Non-clay minerals like quartz (4.27, 3.34 Å) and Na-feldespars (6.40, 4.05, 3.24, 3.21 Å) are present.

The identification of iron oxides was difficult owing to their low concentration and the overlap of reflections from other phases present in the soil samples (XRD patterns are not shown). In all samples, hematite and goethite are hardly identified, goethite in minor amounts. The intensities of both hematite (0.269 and 0.368 nm) and goethite (0.421 and 0.269 nm) peaks seem to decrease with chemical treatments. The 0.297 nm reflection slightly decreased in intensity after the chemical treatments, suggesting the presence of maghemite and/or magnetite. The most intense 0.253-0.251 nm peaks of these Fe compounds could be overlapped with those corresponding to feldspars. The presence of lepidocrocite is not consistent with the assumed environmental conditions of the soil [1].

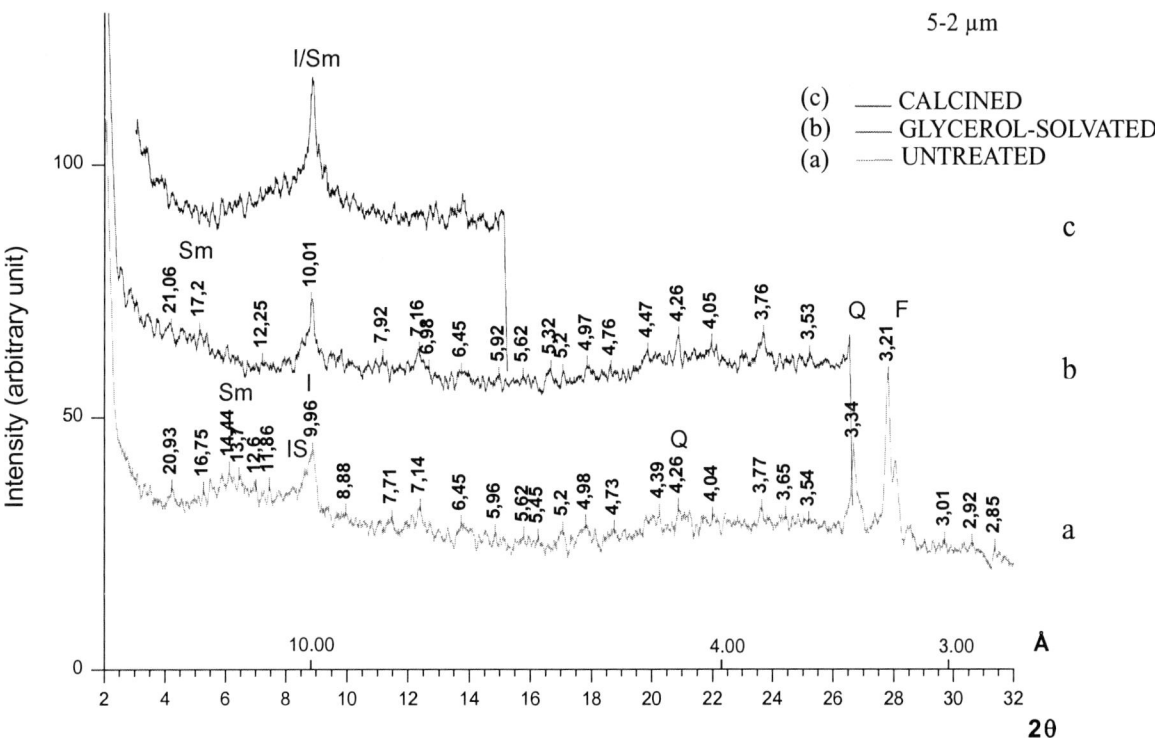

Figure 1. X-ray diffractograms of the 5-2-μm size soil fractions [oriented specimen (a), glycerol-solvated (b) and calcined (c)]. I, illite; Q, quartz; F, Na-rich feldspar; Sm, smectite; IS, interstratified illite-smectite.

3.3. Mössbauer Spectroscopy

Mössbauer spectra were fitted using a narrow sextet S1 and a broad distribution of sextets, S_D, in the magnetic region and two doublets, D1 and D2, in the paramagnetic one. The values of the corresponding parameters and their behavior when lowering the temperature allowed the identification of the Fe oxides [17]. A characteristic fitted spectrum is displayed in Fig 2.

In the magnetic region, S1 is assigned to hematite, which remains weakly ferromagnetic (WF) in all the used temperature range; no Morin transition occurred, as it could have been the case with a pure "bulk" hematite. In general, the absence of this transition is due to very small hematite sizes (< 20 nm) and/or Al substitution [18, 19]. In the present soil fractions, Al substituted hematite is very likely to be present.

The distribution of hyperfine fields S_D is consistent with the presence of highly Al substituted and small sized hematite, goethite and magnetite and/or maghemite. The identification of magnetite and/or maghemite is very difficult in soils. Since no powder was attracted by a hand magnet, their presence seems unlikely unless they were coating non-magnetic particles. The just mentioned hematite and mainly the goethite are not well crystallized and will be referred as poorly crystallized Fe oxides.

In general, ferrihydrite is closely related to non-well crystallized goethite but in this soil there are no reports about its presence.

In the paramagnetic region, D1 doublet parameters indicate the presence of Fe^{3+} while those from D2 doublet indicate the presence of Fe^{2+}. The iron ions are located both in the illite and in the interstratified illite/smectite detected by XRD [20-22]. Contribution to D1 from very small sized (<2 nm) and/or very highly substituted Fe^{3+} oxide particles cannot be totally discarded, especially in the <1-µm fraction, where a slight increase of the magnetic/paramagnetic signals ratio for decreasing temperature was observed. Fe associated to organic matter could also be contributing to these signals.

The relative composition of the above-mentioned Fe phases is discussed from the corresponding areas at 15 K. The relative amount (%) of each Fe-bearing phase are shown in Table 3.

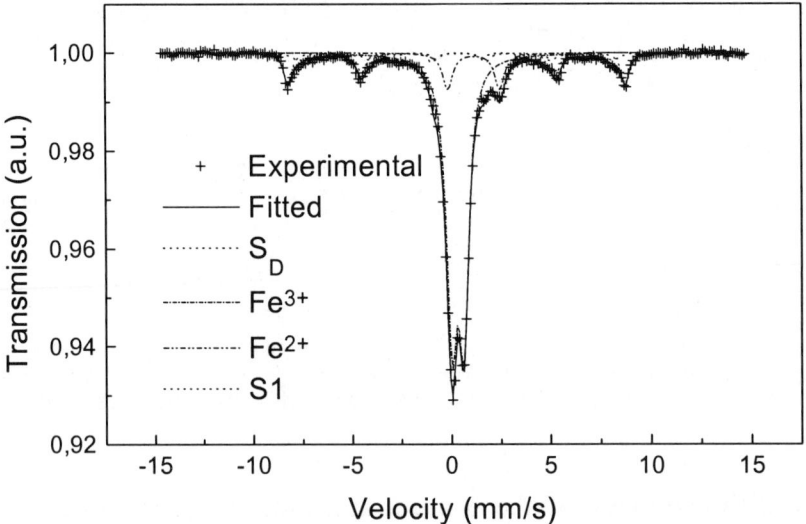

Figure 2. Mössbauer spectrum of the UT 5-2-µm sample at RT.

Table 3
Percentil amount of Fe-bearing phases from MS at 15 K

Size Fraction	S1	S_D	D1 Fe^{3+}	D2 Fe^{2+}
5-2 µm	13.0	7.0	71.0	9.0
2-1 µm	6.0	7.0	81.0	6.0
< 1 µm	3.0	10.0	84.0	3.0

S1 correspond to WF hematite; S_D corresponds to poorly crystallized Fe-oxides.

The relative areas of the magnetic signals indicate the proportion of Fe-oxides compared to the Fe-illite and Fe-interstratified illite/smectite in each fraction. The Fe-oxides are more significant in the 5-2 µm fraction than in the smaller ones (~20% of the total area in this case, ~13% in the intermediate size and ~12% in the small size). The proportion of Fe-bearing clay minerals is predominant in all fractions and rather increases as fraction size decreases. The WF hematite (S1) is well crystallized but less than a "bulk" one and the signals from Fe-oxides disappeared completely with DCB treatments. Poorly crystallized oxides (S_D) amounts, that is goethite and highly substituted hematite amounts are lower to, comparable to and higher to the WF hematite amounts in the 5-2 µm, 2-1 µm and <1 µm fractions respectively. In the smaller fraction, OX treatment produced the total disappearance of the corresponding signal.

The amount of paramagnetic Fe^{3+} from the illite and the interstratified illite/smectite is higher than that of Fe^{2+} (almost 8-9 times). A small amount of these Fe^{3+} may belong to very small goethites and Fe associated to organic matter. A small reduction of the D1 areas is observed in the PY treated samples, which is coincident with Fe_{PY} values. Furthermore, chemical treatments, in particular DCB technique, extracted part of this Fe^{3+} (D1) associated to the clay minerals and/or hydroxy-interlayer material. This last extraction is reflected by the differences found between the Fe_{DCB}/Fe_t values (~26-42%) and the proportions of Fe-oxides from Table 3 (20-13%). The result also suggests that DCB has extracted more Fe from the hydroxy-interlayers in the smallest fraction. There is little Fe^{2+} (D2), which was not affected by chemical treatments These finding are in agreement with previous results for argentine Mollisols [15].

REFERENCES

1. M.E. Aguirre, Thesis Mg. Universidad Nacional del Sur, Bahía Blanca, Argentina (1987).
2. e.g. N.N. Greenwood and T.C. Gibb, Mössbauer Spectroscopy, Chapman & Hall, London, 1971 or P. Gütlich, R. Link, A. Trautwein, Mössbauer Spectroscopy and Transition Metal Chemistry, Inorganic Chemistry Concepts 3, Springer-Verlag, Berlin Heidelberg, 1978.
3. A. Walkley, Soil Sci., 63 (1946) 251.
4. C.A. Bower, R.F. Reitemeier and M. Fireman, Soil Sci., 73 (1952) 251.
5. G.W. Robinson, J. of Agricultural Sci., 12 (1922) 306.
6. P.J. Loveland and P. Digby, J. Soil Sci., 35 (1984) 243.
7. U. Schwertmann, Z. Pflanzenernähr. Düng. Bodenkd.,. 105 (1964) 194.
8. U. Schwertmann, Can. J. Soil Sc., 53 (1973) 244.
9. O.P. Mehra and M.L. Jackson, Clays and Clay Miner., 7 (1960) 317.
10. J. Hashimoto and M.L Jackson, Clays and Clay Miner., 7 (1960) 102.

11. M.L. Jackson, Análisis Químico de Suelos. Ediciones Omega, Barcelona, (1964).
12. T. Ericsson and R. Wäppling, J. Phys., Colloque C6 (1976) 719.
13. G.J. Long, T.E. Cranshaw and G. Longworth, Mössbauer Effect Data Ref. J 6(2) (1983).
14. R.A. Brand, NORMOS program, Internat. Rep. Angewandte Phys., Universität Duisberg (1989).
15. S.G. Acebal, A. Mijovilovich, E.H. Rueda, M.E. Aguirre and C. Saragovi, Clays and Clay Miner., 48 (2000) 322.
16. R.M. Cornell and U. Schwertmann, The iron oxides: Structure, Properties, Reactions, occurrence and Uses. Verlag Chemie, Weinheim, (1996).
17. S. Petrick, C. Saragovi private communication, to be published elsewhere.
18. E. Murad and U. Schwertmann, Clays and Clay Miner., 34 (1986) 1.
19. R.E. Vandenberghe, E. De Grave, L.H. Bowen and C. Landuydt, Hyp. Int., 53 (1990)175.
20. I. Rozenson and L. Heller-Kallai, .Clays and Clay Miner., 24 (1976a) 271.
21. E. Murad and U. Wagner, Clay Miner., 29 (1994) 1.
22. J.M.D. Coey, At. Energy Rev., 18 (1980) 73.

ACKNOWLEDGEMENTS

Authors are grateful to A. Petragalli and A. Carnero (CNEA) and to Dr. D. Poiré (CIG-UNPL) for technical support. S. P. and C. S. are indebted to CLAF (Centro Latinoamericano de Física) for financial support.

Thermal transformation of synthetic bayerite and nordstrandite as studied by electron-optical methods

M.L.P. Antunes and H. Souza Santos

Electron Microscopy Laboratory, Physics Institute, University of SãoPaulo
C.P. 66318 – Cep 05389-970 – São Paulo – S.P. - Brazil

Bayerite and nordstrandite were prepared by the reaction of amalgamated aluminum foils with distilled water (bayerite) and ethylene glycol/water solution (nordstrandite) at room temperature. The Al-hydroxides were heated up to 1300^0C to change in the various alumina types. The Al-hydroxides and alumina powders were characterized by XRD, SAED, EDS/TEM, TEM and SEM. The morphological changes during growth of the bayerite somatoids and many-sides stars of nordstrandite are presented in details. Both hydroxides have the thermal sequence of transformation into eta (300^0C), theta (900^0C) and alpha-aluminas (1300^0C).

1. INTRODUCTION

The crystallization of aluminum trihydroxides $Al(OH)_3$ can occur in three different forms: gibbsite (monoclinic structure a=8.684Å, b= 5.078Å, c=9.736Å, β= 94.54^0 [1]), bayerite (monoclinic structure a=5.062Å, b= 8.671Å, c=4.713 Å, β= 90.27^0 [2]) and nordstrandite (triclinic structure a=5.082Å, b=5.127 Å, c=4.980 Å, α=93.67^0 β= 118.92^0 γ=70.27^0 [3]). These three crystalline aluminum hydroxides have crystals varying from micrometric to millimetric size; dehydroxilation of the hydroxides occurs by heating them between 300^0C and 600^0C and several transition aluminas are formed. Irreversibly crystallization into α-alumina occurs during the sintering of the fine grained transition aluminas between 1100^0C and 1500^0C.

Stumpf [4] showed that, between the dehydroxilation of the hydroxide and α-alumina crystallization, a number of intermediate crystalline structures of aluminas are formed, which are reproducible, each one having a different crystalline structure and being stable at a given temperature range.

Those aluminas are, in general way, called "transition aluminas"; all are always designated by Greek letter to identify them: alpha, gamma, delta and others [4].
By progressive elevation of the temperature, each transition alumina can suffer transformation into another transition alumina with a different crystalline structure, indicated by another Greek letter and so on; however, all of them, above the temperature of 1100^0C [5] recrystallize into α-alumina, which has a very stable crystalline structure, and remains unchanged up to its fusion point.

The aluminum oxide, in its several forms, has a large use in the Chemical Process Industry, as a catalyst, a catalyst support or as adsorbent, due to the variety of surface properties, or in the Advanced and Traditional Ceramics Industries, due to the remarkable thermal, mechanical and chemical properties of the α-alumina structure.

Extensive literature exist on the thermal dehydroxilation of the crystalline hydroxides in the transition aluminas until the alpha-alumina, [4–11], specially by ALCOA group [5]. However, the majority of these studies were concentrated on Bayer gibbsite, due to its industrial importance; few studies exist on bayerite and less in nordstrandite.

The aim of the present paper is to present the results of a comparative study of the thermal transformation of synthetic microcrystals of bayerite and nordstrandite [9].

2. MATERIALS AND METHODS

2.1. Preparation of the Al(OH)$_3$

The aluminum hydroxide bayerite is obtained by reacting slightly amalgamated aluminum foils, at room temperature, with distilled water and with ethylene glycol/water solution to produce nordstrandite.

Aluminum foils were amalgamated in a 1% mercury-II chloride solution for three minutes. The amalgamated aluminum foils were washed in distilled water and immersed in different medium (distilled water to produce bayerite [12] and aqueous ethylene glycol to produce nordstrandite).

2.2. Heat Treatment

The hydroxides produced were dried at 70^0C and the powder fired in a programmed electric furnace EDG, **under air**, for 3 hours, at the maximum temperature from 300^0C up to 1300^0C.

2.3. Methods of Characterization

The hydroxides powders were characterized by X-ray powder diffraction (XRD), selected area diffraction (SAED) and X-ray microanalysis using EDS/MET to identify the hydroxide produced; scanning electron microscopy (SEM) and transmission electron microscopy to visualize the morphology of the microcrystals. For structure and particle shape characterization of the alumina phases formed by the thermal decomposition of the bayerite and nordstrandite crystals; SEM,TEM , EDS, and SAED were used to identify the different aluminas transition.

X-Ray Diffraction - The equipment used was a Philips Diffractometer, X′Pert Model MPD (PW 3050/10), using copper K-alpha radiation operating at 40kV and 40mA. The scanning was made in the range of 2θ (1^0) and 2θ (90^0). The X-ray data were compared with ICDD Files and reference [5] for aluminum hidroxides and aluminas.

Scanning Electron Microscopy - The equipment used was a scanning electron microscope JEOL model JSM 840A, operating at 25kV. The sample powder was placed upon SEM stubs covered with gold.

Transmission Electron Microscopy - Each powder was dry and prepared on a carbon covered grids for transmission electron microscopy (SEM), selected area diffraction (SAED) and X-ray microanalysis using EDS/MET. The preparation were examined in a Philips CM200 transmission electron microscope operating at 200kV.

3. RESULTS AND DISCUSSION

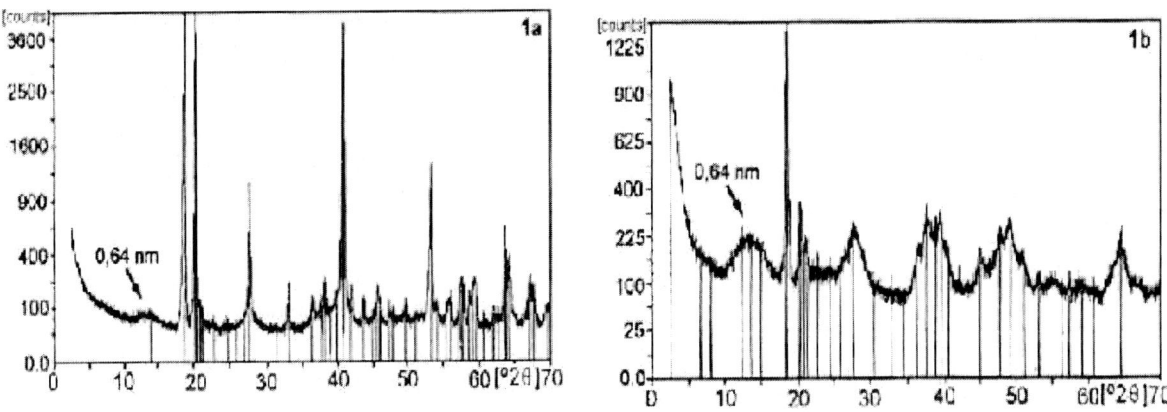

Figure 1 - XRD curve of crystals of Al(OH)$_3$: (a) Bayerite (b) Nordstrandite

3.1. X-Ray Diffraction

The XRD patterns in Figure 1a shows bayerite peaks formed by reaction of aluminum foils and distilled water. Figure 1b presents the XRD pattern of the hydroxide resulting from the aging of the product of reaction of aluminum foil and ethylene glycol/water solution: the nordstrandite peaks are present. In the beginning of the aging process for nordstrandite crystallization, the XRD pattern presents a broad peak corresponding to d=0,64nm identified as pseudoboehmite. The term pseudoboehmite refers to the poorly crystallized Al^{3+} compound with interplanar spacing increasing in the (020) direction up to a value of 0,68nm in comparison with the value 0,612nm for d$_{020}$ for well crystallized boehmite [13]. The presence of the broad peak suggests that, in a first stage of aging, occurs the conversion of amorphous Al(OH)$_3$ into pseudoboehmite, followed, in a second stage, by the pseudoboehmite into bayerite or nordstrandite recrystallization.

3.2. Results at room temperature

Figure 2a shows a typical SAED of bayerite obtained from particles produced by distilled water. The SAED results are compatible with the XRD's one, but pseudoboehmite could not be identified by this methods. Figure 2b presents the morphology of bayerite particles by TEM, and Figure 3 by SEM. Bayerite occurs as crystals of uniform shapes know in the literature as somatoids [14]. The EDS confirms only the presence of Al and O, collected at the particles, no Hg peak was found.

Figure 4a shows a typical SAED of nordstrandite. The morphology of nordstrandite is observed in figure 4b (TEM) and 5 (SEM): it crystallizes as particles many points star-shaped. The EDS (Figure 6) of the particles confirm only the presence of Al and O.

3.3. Phases at 500^{0}C

Bayerite: After firing at 500^{0}C, SAED characterizes the formation of the transition cubic eta-alumina (cubic structure a=b=c=7.9Å, $\alpha=\beta=\gamma=90^0$) from monoclinic bayerite dehydroxilation. Pore formation by dehydroxilation may be observed in the particles.

Nordtrandite: Firing at 500^{0}C changes triclinic nordstrandite into transition eta-alumina, with pore formation in the platy stars similar to what occurs in bayerite.

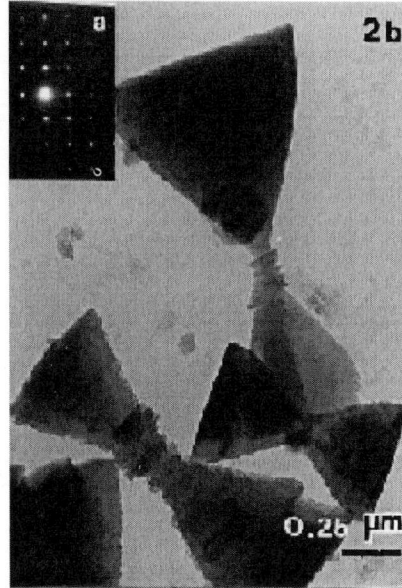

Figure 2 a – Typical SAED of bayerite.
Figure 2 b – The morphology of bayerite particle by TEM.

Figure 3 – The morphology of bayerite particle by SEM.

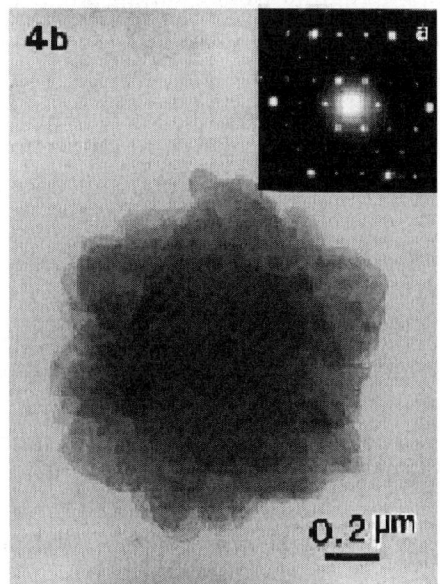
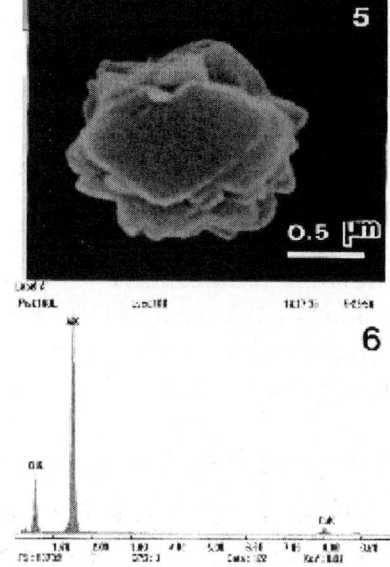

Figure 4 a – Typical SAED of nordstrandite.
Figure 4 b – The morphology of nordstrandite by TEM

Figure 5 – The morphology of nordstrandite by SEM.
Figure 6 – The EDS of the particle of nordstrandite confirm the presence of Al and O

3.4. Phases at 700°C

Bayerite: Eta-alumina is still present, as characterized by SAED (Figure 7a). The somatoids shape of the pseudomorphs has not changed, but inside the pores have ordered themselves into rows (Figure 7b).

Figure 7 a – Eta-alumina characterized by SAED.
Figure 7 b – Bayerite somatoids at 700°C (Eta-alumina).

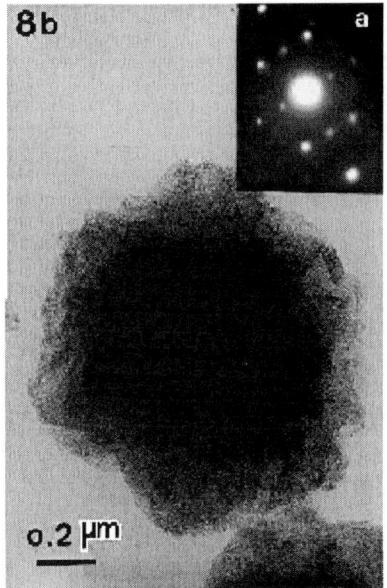

Figure 8 a – Eta-alumina characterized by SAED.
Figure 8 b – Nordstrandite at 700°C (Eta-alumina).

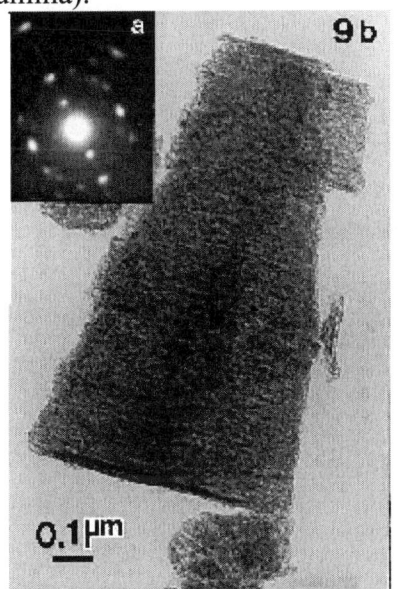

Figure 9 a – Theta-alumina characterized by SAED.
Figure 9 b – Bayerite (Theta-alumina) at 900°C.

Figure 10 a – Mixture of eta and theta-aluminas characterized by SAED.
Figure 10 b – Nordstrandite at 900°C (mixture of eta and theta aluminas).

Nordstrandite: Also, eta-alumina is characterized by SAED (Figure 8a) at that temperature. Internally, a number of small pores may be observed in the pseudomorphs by the dehydroxilation (Figure 8b). The plate star-shaped pseudomorph has maintained its morphology. Some gamma-Al_2O_3 is detected, but they are from fibrillar pseudoboehmite, which exists as an intermediary during the aging for the nordstrandite recrystallization.

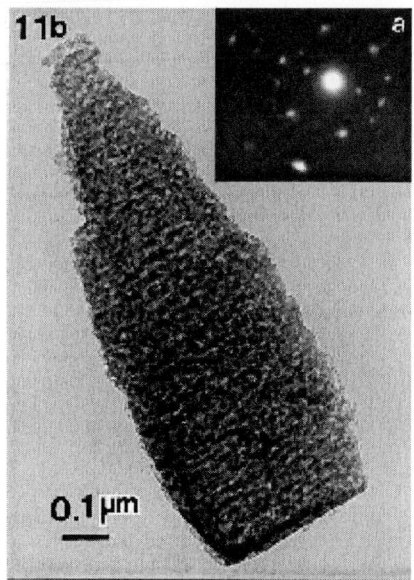

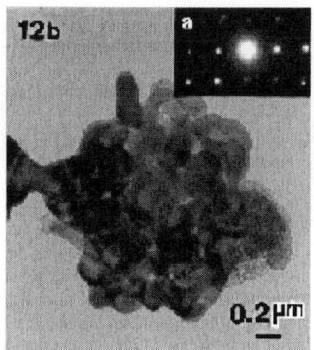

Figure 12 a – Alpha-alumina characterized by SAED.
Figure 12 b – Alpha-alumina from nordstrandite precursor at 1300°C.

Figure 11 a – Theta-alumina characterized by SAED.
Figure 11 b – Bayerite (Theta-alumina) at 1100°C.

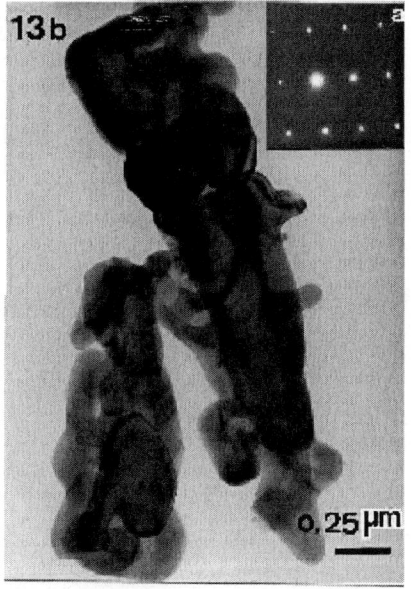

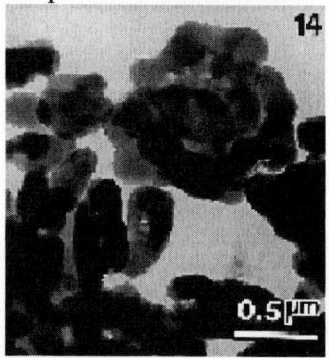

Figure 13 a – SAED pattern of alpha-alumina.
Figure 13 b – Alpha-alumina from bayerite precursor at 1300°C.

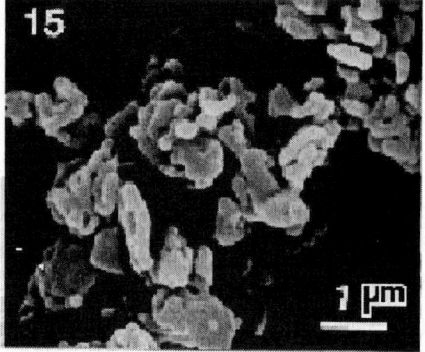

Figure 14 – Alpha-alumina morphology (precursor nordstrandite) by TEM.

Figure 15 – Alpha-alumina morphology (precursor nordstrandite) by SEM

3.5. Phases at 900°C

Bayerite: By SAED (Figure 9a), the particles in the somatoid pseudomorphs are characterized as the transition monoclinic theta-alumina (a=11.24 Å b= 5.721 Å c= 11.74Å, $\alpha=\beta= 90^0$ $\gamma= 103.33^0$) formed by eta to theta-alumina recrystallization. One can notice a better orientation of parallel rows of pores, what gives a striated appearance to the somatoid pseudomorph (Figure 9b).

Nordstrandite: By SAED, mixtures of eta- and theta-aluminas are characterized is the star-shaped pseudomorphs (Figure 10a). The coexistence of eta- and theta-phases in nordstrandite fired at 900^0C indicates that the transformation is happening at that temperature, while in bayerite all eta-alumina has changed to theta at a temperature, lower than 900^0C as it should be expected from reference [5]; what is another difference between them. Figure 10b is a micrograph of the 900^0C pseudomorph, showing an increased organization in the micropores development.

3.7. Phases at 1100^0C
Bayerite: All indexation of the SAED pattern after firing at 1100^0C only characterizes the transition theta-alumina (Figure 11a); that is in disagreement with data from reference [5], where firing bayerite at 1030^0C changes eta completely into alpha-alumina. Figure 11b shows the pseudomorphs in which coalescence between the rows of lamellar is beginning.

3.8. Phases at 1200^0C
Nordstrandite: By firing at 1200^0C, there is alpha- Al_2O_3 formation, identified by SAED (Figure 12a); however, the theta-Al_2O_3 pseudomorphs, during the transition, have sintered and coalesced, thus losing the original star-shape of the nordtrandite precursor (Figure 12b).

3.9. Phases at 1300^0C
Bayerite: The SAED pattern (Figure 13a) characterizes only cubic alpha-alumina (hexagoanal structure a=b=4.759Å, c= 12.992Å, $\alpha=\beta= 90^0 \gamma=120^0$); curiously, in spite of some coalescence between pseudomorphs, the original hour-glass shape is mostly maintained (Figure 13 b).
Nordstrandite: All indexation of the SAED pattern after firing at 1300^0C only characterizes the alpha-Al_2O_3. By TEM (Figure 14) and by SEM (figure 15), it can be observed that the star pseudomorphs and the small crystals have coalesced and sintered in agglomerates of some particles of alpha- Al_2O_3.

4. CONCLUSIONS

(a) The alumina phases characterized by XRD and SAED from firing bayerite up to 1300^0C are in agreement with reference [5], the alumina phases from firing nordstrandite are the same as from bayerite.

(b) By using SAED, differences were found from the alumina transition temperatures from bayerite and nordstrandite.

(c) The following sequences are for bayerite from reference [5] and for bayerite somatoids and nordstrandite stars from data from the present paper.

(d) Bayerite somatoids and nordstrandite platy stars in spite being both crystalline aluminum trihydroxides, present completely different particle shape and texture of their alumina pseudomorphs after firing from 300^0C to 1300^0C; the difference are probably due, to the great difference in original crystal shapes.

	300	500	700	900	1100	1300 °C
bayerite		η			θ	α
bayerite (Wefers)		η		θ		α
nordstrandite		η			θ	α

REFERENCES

1. Saafeld, H.; Wedde, M., Z. Kristallogr. 139 (1974) 129.
2. Routhbauer, R.; Zigan, F.; O'Daniel, H., Z. Kristallogr. 125 (1967) 317.
3. Bosmans, J.H., Acta Crystallogr. B26 (1970) 649.
4. Stumpf, H.C.; Russel, A S.; Newsome, J.W. and Tucker, C.M., Ind. Eng. Chem. 42 (1950) 1398.
5. Wefers, K. and Misra, C., Oxides and Hydroxides of Aluminium, Alcoa Techinical Paper N° 19, Alcoa Pittsburgh, 1987.
6. Brindley, G.W., Progress in Ceramic Science, Oxford:Pergamon Press. Vol. 3, 1963.
7. Day, M.K.B. and Hill, V.J., J. Phys. Chem. 57 (1953) 946.
8. Sato, T., J. Appl. Chem. 12 (1962) 533.
9. Aldcroft, D. and Bye, G.C., Science Ceramics 3 (1967) 76.
10. Tanev, P.T. and Vlaev, L.T., Catalysis Letters 19 (1993) 351.
11. Souza Santos, H. Kiyohara, P.K. and Souza Santos, P., Materials Research Bulletin 31 (1996) 799.
12. Schmäch, H., Z. Naturforschg 1 (1946) 323.
13. Krivoruchko, O.P.; Buyanov, R.A. and Playasova, L.M., Russian Journal Of Inorganic Chemistry 23 (1978) 988.
14. H. Souza Santos, A. Vallejo-Freire, P. Souza Santos, Kolloid Z. 133 (1953) 101; 135 (1954) 56.

Proton binding at clay surfaces in aqueous media

Marcelo J. Avena, Marcelo Mariscal and Carlos P. De Pauli

INFIQC. Departamento de Fisicoquímica, Facultad de Ciencias Químicas, Universidad Nacional de Córdoba, Ciudad Universitaria, 5000 Córdoba, ARGENTINA.

Proton binding at a phyllosilcate clay surface takes place at oxygen-containing groups. As in the case of oxygen-containing molecules, the groups behave as monoprotic entities and undergo only one protonation-deprotonation step in a pH range 3-11. The protonation behavior depends on the intrinsic affinity of surface groups and long range electrostatics. The intrinsic affinities of siloxane groups and gibssite surface groups located at the basal surfaces (001 planes) of 2:1 or 1:1 clays are very low. They might be protonated only under very acidic condition. On the contrary, the intrinsic affinity of surface groups at the broken edges is sufficient to undergo protonation-deprotonation reactions under weakly acid, neutral or weakly basic conditions. The negative charges originated by isomorphic substitutions within the clay structure produce an electric field that affect the electrical potential at both, the basal and edge surfaces. When these surfaces have no net electrical charge, the electrical potential is negative due to the presence of the negative charges at the edges. This potential increases the affinity of basal surface for cations and also the affinity of surface groups at the edges for protons. Evidences for this can be obtained from proton adsorption curves performed at different supporting electrolyte concentration.

1. INTRODUCTION

Among all the processes that take place at a clay surface, ion binding and ion exchange in aqueous media have been intensively investigated. Proton binding, for instance, is particularly important because it can directly affect the charging of the surface and indirectly affect the attachment of other substances, the stability of clay dispersions as well as the environmental behaviour and the technological applications of clays.

Protonation of a clay surface can be understood in terms of conventional acid-base and electrostatic theories based on simple concepts such as surface charge, surface potential and surface functional groups[1]. This article deals mainly with these concepts and intends to discuss the points that have significant impact in the protonation-deprotonation behaviour of a phyllosilicate clay surface. The key roles of intrinsic affinity and the particular way by which electrostatics acts in solids with structural charges are highlighted.

2. THE PROTONATION REACTION

A protonation reaction at a surface occurs when a proton from the bulk solution reacts with the surface. The reactive species in clays, oxides and other oxygen-containing minerals are oxygen atoms that form part of surface groups. It has been customary so far to assume that protonating oxygen atoms can bind two protons in a normal pH range. However, both theoretical and experimental evidences indicate that two protonation steps seldom occur at oxygen atoms in aqueous media[2,3]. Table 1 schematises the two protonation steps that could take place at the protonating atom of several molecules.

The logarithm of the protonation constant is given, and some data for protonating nitrogen are also included. The difference between two consecutive constants is, in the case of oxygen, between 11 and 18 orders of magnitude. This difference is so large that basically only one reaction is "active" in a normal pH range. Consider the case of acetic acid, for instance: the protonation of acetate ions to form acetic acid molecules takes place at pH around 5 (log $K_{H,1} \cong 5$). According to the value of $\log K_{H,2}$ ($\cong -6$), the second step would take place at pH around –6, which is impossible. The oxygen group in acetate can thus be considered as monoprotic[2]. This situation seems to be normal for oxygen-containing groups belonging to molecules in true solutions or to the surface of solids. According to Borkovec et al.[2] the monofunctionality of oxygen containing groups also hold for protonation reactions on the other elements in the second row of the periodic table. This is supported at least for the case of molecules containing a protonating nitrogen, as shown by the data in Table 1.

The binding of a proton ion to a monoprotic surface group can be represented by:

$$A^x + H^+ = AH^{x+1} \tag{1}$$

Where A^x denotes a functional surface group carrying a charge x (fractional or integer, negative or positive) and AH^{x+1} is the protonated group. The mass action law of this reaction can be written as:

Table 1
Proton affinity of oxygen-containing molecules

	$LogK_{H,1}$			$LogK_{H,2}$		$\Delta LogK_H$
OH^-	14 →	H_2O		0 →	H_3O^+	14
RO^-	16 →	ROH		-2 →	ROH_2^+	18
ArO^-	9-11 →	$ArOH$		-7 →	$ArOH_2^+$	16-18
$CH_3\text{-}COO^-$	5 →	$CH_3\text{-}COOH$		-6 →	$CH_3\text{-}COOH_2^+$	11
NH_2^-	34 →	NH_3		9 →	NH_4^+	25
$ArNH^-$	25 →	$ArNH_2$		3-5 →	$ArNH_3^+$	20-22

R, aliphatic group; Ar, aromatic group. Taken from Borkovec et al.[2] and Allinger et al.[3]

$$K_H^{int} = \frac{\Gamma_{AH}}{\Gamma_A a_{H,0}} \quad (2)$$

where Γ_{AH} and Γ_A are the amount per unit area of the protonated group (AH^{x+1} in this case) and the unprotonated one (A^x), respectively, K_H^{int} is the intrinsic protonation constant, and $a_{H,0}$ is an expression for the proton activity in the solution at the location of the adsorption site. $a_{H,0}$ is defined as

$$a_{H,0} = a_H e^{-\frac{F\psi_0}{RT}} \quad (3)$$

where ψ_0 is the smeared out surface potential and represents the difference in the electrical potential between the surface and the bulk solution, and a_H represents the activity of protons in the bulk.

Combination of eq. 2 and 3 gives

$$K_H^{int} e^{-F\psi_0/RT} = \frac{\Gamma_{AH}}{\Gamma_A a_H} \quad (4)$$

In dilute solutions K_H^{int} is independent on the electrical potential and surface charge and the left-hand side of eq. 4 depends on the magnitude of the surface potential, and it is often termed effective or apparent constant, K_H^{eff},

$$K_H^{eff} = K_H^{int} e^{-F\psi_0/RT} \quad (5)$$

In a logarithmic form:

$$Log K_H^{eff} = Log K_H^{int} - \frac{F\psi_0}{2.303 RT} \quad (6)$$

Thus, the protonation can be seen as resulting from two different contributions: a chemical or intrinsic contribution and a long-range electrostatic contribution. The chemical contribution to $Log K_H^{eff}$ is given by $Log K_H^{int}$, which is a measure of the intrinsic (chemical, non electrostatic) affinity of the group for protons. The other contribution to $Log K_H^{eff}$ is given by the second term of equation 6 and results from long-range electrostatics. Both, intrinsic and electrostatic contributions will be treated separately.

3. THE INTRINSIC AFFINITY

The intrinsic affinity arises from the ability of orbitals and electrons in the site to overlap the 1s orbital of the attaching proton, and from short-range electrostatics. Hiemstra et al.[4] have suggested a rather general approach to treat protonation reactions at the surface of oxides, that can be extended to other minerals and clays. The treatment is called the MUSIC model. By using this model it is possible to estimate the value of $Log K_H^{int}$ for different surface groups. Several values are listed in Table 2. $Log K_H^{int}$ depends on the number and type of metal ions that are surrounding the reactive oxygen atom and on the structure of the surface. The model also predicts that the difference between the $Log K_H^{int}$ values of the two consecutive protonation steps is 13.8, showing that oxygen-containing groups at a solid surface are also monoprotic.

Table 2. $LogK_H^{int}$ values for the protonation of different surface groups. The location of the groups in phyllosilicate layers is indicated

Surface group	location	$LogK_H^{int}$
$Al\text{-}OH^{1/2-}$	Edge surface	10.0
$Al_2\text{-}O^{1-}$	Edge surface	12.3
$Al_2\text{-}OH$	Basal surface	-1.5
$Al_3\text{-}O^{1/2-}$		2.2
$Fe\text{-}OH^{1/2-}$		10.7
$Fe_2\text{-}O^{1-}$		13.7
$Fe_2\text{-}OH$		-0.1
$Fe_3\text{-}O^{1/2-}$		4.3
$Ti\text{-}OH^{1/3-}$		6.3
$Ti_2\text{-}O^{2/3-}$		5.3
$SiAl\text{-}O^x$		-16.9 - 12.3
$Si\text{-}O^{1-}$	Edge surface	11.9
$Si\text{-}OH$	Edge surface	-1.9
$Si_2\text{-}O$	Basal surface	-16.9

Modified from Hiemstra et al[4].

Since the MUSIC model was developed for ionic solids it was modified to explain the acid base behavior of surface groups in non-ionic oxides[5]. The new approach predicts $LogK_H^{int}$ values that are around 1 or 2 units different from those predicted by the original version of the MUSIC model and the difference between $LogK_H^{int}$ of two consecutive steps becomes now 11.9 instead of 13.8. In spite of this, the values presented in Table 2 can still be considered as reasonable first estimates of the constants.

Data in Table 2 indicate that siloxane groups ($Si_2\text{-}O$) and gibbsite surface groups ($Al_2\text{-}OH$) are very stable species. They do not protonate or deprotonate in a normal pH range. These groups are the only ones that populate the basal surfaces of 2:1 and 1:1 phyllosilicates layers. Therefore, the unreactivity of basal surfaces in a normal pH range emerges directly from the application of the model and does not need to be assumed in advance. Edge Si-OH groups could become deprotonated at high pH and $Al\text{-}OH^{-1/2}$ groups could become protonated at intermediate and low pH. The intrinsic protonation constant of the transitional site $AlSi\text{-}O^x$ at edges cannot be estimated with the MUSIC model but it is possible that one protonation reaction takes place at these groups in a normal pH range. Thus, the explanation of the edge reactivity also emerges directly from the application of the model.

The very low affinity for protons of $Si_2\text{-}O$ and $Al_2\text{-}OH$ groups does not mean that they will never be protonated. Perhaps very acidic conditions such as those given by hydrochloric acid or sulphuric acid media can induce some protonation.

4. THE ELECTROSTATIC INTERACTION

Due to their different reactivity, basal surfaces and edges have different charging behavior and different surface potential. The charge at the basal surface is zero in a normal pH range because of the neutrality and unreactivity of siloxane and gibbsite surface groups; the charge at the edges can be either positive, zero or negative depending on the pH because groups at this location can undergo protonation or deprotonation. Underneath the basal plane, and depending on the clay structure, there may be negative structural charges resulting from isomorphic substitutions. These structural charges do not reside at the surface plane but produce an electrical field that emanates in all the directions. This potential, together with the electrical potential generated by the attachment-detachment of protons and other ions at the surface will significantly influence the proton binding.

The effects of long range electrostatics are illustrated in Figure 1. A 2:1 phyllosilicate layer is represented and a proton ion is allowed to bind a surface group at the edge. The arrows pointing radially from the proton represent lines of force that schematise the electric field generated by the proton charge. As soon as the site is protonated, the charge changes and the potential is modified, leading to a change in the effective constant for the protonation of the next site, such as equation 6 indicates. This effect is called electrostatic interaction between sites, since proton affinity of a given site changes when another site is being protonated in its neighbourhood[2]. Indeed, equation 6 indicates that there is a 10-fold decrease in the value of the effective protonation constant for each 59 mV increase in ψ_0.

The electrostatic interaction takes place not only between sites at the surface but also between structural charge and surface sites. Consider the negative structural charge represented in the drawing of Figure 1. The electrical field emanating from this charge affects the potential and reactivity of the basal planes as well as that of the edge surface. The magnitude of the effect will depend on the distance between the structural charge site and either the basal surface or the edge surface. For example, in a dioctahedral smectite having 25% of Al^{3+} substituted by Mg^{2+} in the octahedral sheet (as it could be the case of montmorillonite), there will be substitution sites regularly distributed at a distance of 5 to 6 Angstroms from the edge surface, which is about the same distance between the structural charge and the basal surface.

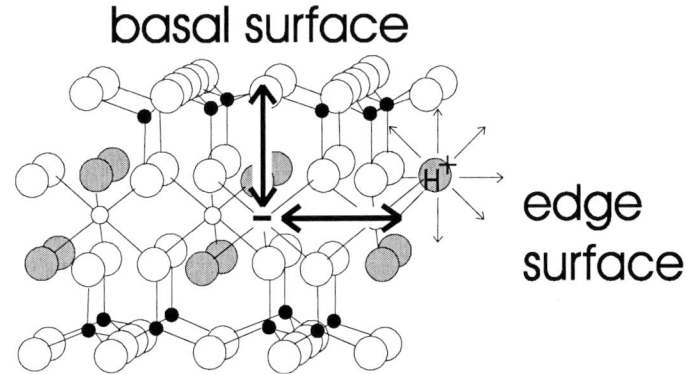

Figure 1. Scheme of a 2:1 dioctahedral phylosilicate layer indicating protonation at the edge surface and comparing distances between a structural charge in an average position and both, basal and edge surfaces.

These distances are represented by the double-headed arrows in the drawing of Figure 1. The effects of structural charges on the reactivity of the basal planes are well known and are responsible for the important cation exchange capacity of clays. Although the surface charge is zero at these planes, there is a net negative potential caused by the charges within the clay layer[6]. Monte Carlo and molecular dynamic calculations show that the surface binds cations in a stronger way when structural charges are present[7,8]. The effects of structural charges on the reactivity of edges have been seldom considered[9]. They can change the affinity for protons of edge groups because they can modify the potential at the edge surface. Indeed, if there is a pH where the edges carry no net charge, there will be a net negative potential caused by the presence of structural charges. This property, as it will be shown in the following section, differences clearly the protonation behavior of clays with that of simple metal oxides, which usually have a zero net surface potential when the net surface charge is zero.

5. EXPERIMENTAL EVIDENCE

The main experimental evidence comes from proton adsorption (σ_H) vs pH curves performed at different supporting electrolyte concentrations. Figure 2 shows the curves for a Na$^+$-montmorillonite obtained from the Province of Neuquén, Argentina[10], a Na$^+$-illite[11] and a kaolinitic soil[12] and compares them with typical curves that can be obtained with simple oxides. The main difference between the behavior of the three considered clays and that of the oxide is the effect of changing the electrolyte concentration on the pH value where the surface charge becomes zero ($pH_{\sigma_H=0}$). This value decreases by increasing the electrolyte concentration in the case of clays, but does not change in the case of the oxide. In order to explain these important differences, and for sake of simplicity, a surface only populated with $A^{-1/2}$ groups that can become protonated to form $AH^{+1/2}$ groups will be considered, and equation 4 will be used. Taking logarithm of this equation and rearranging it:

$$pH = Log K_H^{int} - Log \frac{\Gamma_{AH}}{\Gamma_A} - \frac{F\psi_0}{2.303RT} \qquad (7)$$

what in the case of $\sigma_H=0$ ($\Gamma_{AH} = \Gamma_A$) becomes:

$$pH_{\sigma_H=0} = Log K_H^{int} - \frac{F\psi_{0,\sigma_H=0}}{2.303RT} \qquad (8)$$

where $\psi_{0,\sigma_H=0}$ is the value of the surface potential when $\sigma_H=0$.

Oxide surfaces have zero net surface charge and zero net surface potential at $pH_{\sigma_H=0}$[13,14], and thus the second term of the right hand side of the equation vanishes. A change in the electrolyte concentration does not change the potential under these conditions and $pH_{\sigma_H=0}$ is independent on the electrolyte concentration. This independence is evidenced by a crossing point of the proton adsorption curves at different electrolyte concentration, such as shown in Figure 2d.

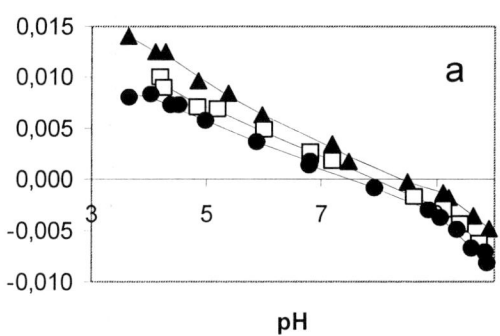
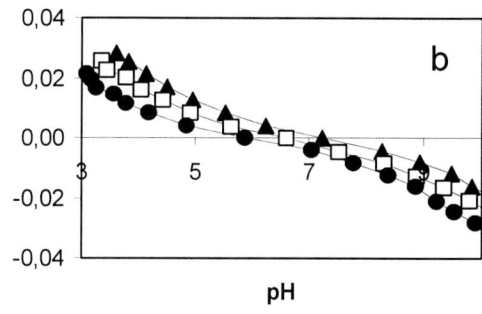
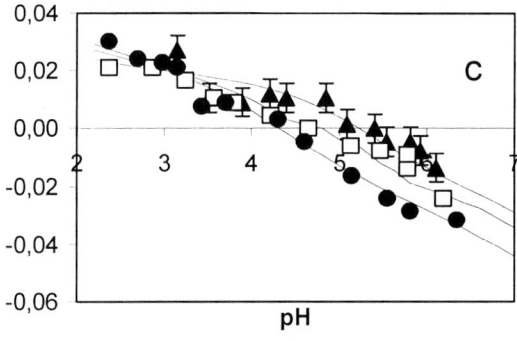
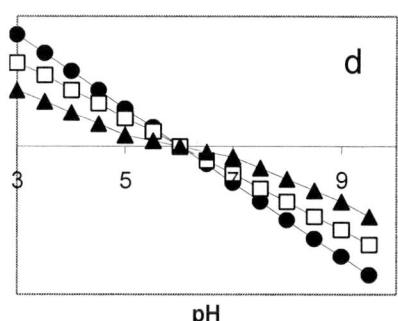

Figure 2. σ_H(C/m^2) vs pH curves of:
a) Na$^+$-montmorillonite in NaCl solutions, circles, 0.12M; squares, 0.01M; triangles, 0.002M; b) illite in NaCl solutions, circles, 0.1M, squares, 0.01M, triangles, 0.002M; c) kaolinitic soil in LiCl solutions, circles, 0.01M; squares, 0.005M; triangles, 0.001M; d) oxide in an indifferent electrolyte, circles, 0.1M, squares, 0.01M, triangles, 0.001M

The case of phyllosilicate clays having negative charges within the layer is somewhat different. Although they have a zero net edge charge, the net edge potential is negative at $pH_{\sigma_H=0}$. This potential becomes less negative by increasing the electrolyte concentration and thus $pH_{\sigma_H=0}$ decreases by increasing electrolyte concentration.

As a short summary of this section, equations 7 and 8 indicate that in a simple oxide $pH_{\sigma_H=0}$ equals $LogK_H^{int}$, which is independent on the electrolyte concentration, whereas in a clay with structural charges $pH_{\sigma_H=0}$ equals $LogK_H^{eff}$ at that pH, which depends on the electrolyte concentration.

6. SUMMARY AND GENERAL CONCLUSIONS

The proton adsorption-desorption properties of a phyllosilicate surface depend on the intrinsic affinity of surface groups and long-range electrostatics. The intrinsic affinity can be estimated with the MUSIC model or similar models and it turns out that siloxane groups and gibssite surface groups located at the basal planes of 2:1 or 1:1 clays are unreactive in aqueous media under normal pH conditions. They might be protonated only under very acidic condition. On the contrary, the intrinsic affinity of surface groups at the broken edges is enough to undergo protonation-deprotonation reactions under weakly acid, neutral or weakly basic pH conditions.

The negative charges due to isomorphic substitution affect not only the behavior of the basal surfaces, increasing their affinity for cations. They also affect the behavior of surface groups at the edges, which increase their effective proton affinity due to the negative edge surface potential caused by the presence of the mentioned negative charges.

As in the case of protons, the presence of structural charges within the clay layer increases the affinity of edges for any other cation and decreases the affinity for any anion present in the bulk solution.

7. ACKNOWLEDGEMENTS

CONICET, Agencia Córdoba Ciencia and SECyT-UNC are thanked for financial support. J. Chorover is specially thanked for providing the experimental data of the kaolinitic soil.

REFERENCES

[1] M.J. Avena, *Acid-base behavior of clay surfaces in aqueous media*, Encyclopedia of Surface and Colloid Science; A. Hubbard, Ed., in press.

[2] M Borkovec; B. Jönsson; G.J.M. Koper. *Ionization Processes and Proton Binding in Polyprotic Systems: Small Molecules, Proteins, Interfaces, and Polyélectrolytes*. Surface and Colloid Science; E. Matijevic, Ed. To be published.

[3] N.L. Allinger; M.P. Cava; D.C De Jongh; C.R. Johnson; N.A. Lebel; C.L. Stevens. Química Orgánica. 2nd edition, Reverté, Barcelona, 1983.

[4] T. Hiemstra; J.C.M. De Wit; W.H. Van Riemsdijk. J. Colloid Interface Sci. 1989, 133 (1), 91-104.

[5] T. Hiemstra ; P. Venema; W.H. VanRiemsdijk. J. Colloid Interface Sci. 1996, 184 (2), 680-692.

[6] W.F. Bleam. Clays Clay Miner. 1990, 38 (5), 527-536.

[7] J. Greathouse ; G. Sposito. J. Phys. Chem. B 1998, 102, 2406-2414.

[8] R-F. C. Chang ; N.T. Skipper ; G. Sposito. Langmuir 1998, 14 (5), 1201-1207.

[9] A.M.L. Kraepiel ; K. Keller ; F.M.M. Morel. Environ. Sci. Technol. 1998, 32, 2829-2838.

[10] N. Peinemann; E.A. Ferreyro; S.G. De Bussetti. Rev. Asoc. Geolog. Argent. 1972, 27, 399-405.

[11] W.H. Hendershot ; L.M. Lavkulich. Soil. Sci. Soc. Am. J. 1983, 47, 1252-1260.

[12] J. Chorover ; G. Sposito. Geochim. Cosmochim. Acta. 1995, 59 (5), 875-884.

[13] J.C. Westall. *Adsorption Mechanisms in Aquatic Surface Chemistry*. Aquatic Surface Chemistry; W. Stumm, Ed.; John Wiley: New York, 1987; 3-32.

[14] J. A. Davis and D.B. Kent. *Surface Complexation Models in Aqueous Chemistry*. Reviews in Mineralogy. Mineral Water Interface Geochemistry; M.F. Hochella, Jr., A. F White, Eds.; Mineralogical Soc. Am.: Washington, 1990; Vol 23, 176-260.

Hydrothermal transformation of kaolinite into 2:1 expandable minerals

M. Bentabol[a]; M.D. Ruiz Cruz[a]; F.J. Huertas[b] and J. Linares[b]

[a]Departamento de Química Inorgánica, Cristalografía y Mineralogía. Facultad de Ciencias. Campus de Teatinos. 28971 Málaga (Spain)

[b]Departamento de Ciencias de la Tierra. Estación Experimental del Zaidín. C.S.I.C. Profesor Albareda. 18008 Granada (Spain)

1. INTRODUCTION

In the recent years a considerable attention has been devoted to the stability of the kaolin-group minerals under weathering, hydrothermal and diagenetic conditions. Some of the mineralogical transformations observed in the natural systems, such as the smectite-, mixed-layer- and illite-formation, can be interpreted today on the basis of experimental studies. Thus, kaolinite to K-mica transformation has been observed in laboratory experiments, either directly using KOH solutions (Huang, 1993; Bauer et al., 1998) or through the initial formation of serpentine-like phases, which reacted with KCl to form dioctahedral mica (Kotov et al., 1980; Frank-Kamenetski et al., 1983, 1990). In a similar way, the formation of Na-bearing phases from kaolinite has also been studied (Chatterjee, 1973). Starting from albite, kaolinite and quartz, Chatterjee (1973) observed the formation of Na-smectite, followed by rectorite and paragonite, at increasing run duration and temperature.

On the contrary, less attention has been devoted to other kaolinite transformations observed in natural systems, such as the transformation either into tri- or di,trioctahedral phases (such as chlorite or mixed-layer minerals) or into dioctahedral chlorites and tosudite.

The evolution of kaolinite toward the several cited phases occurs at relatively low temperature, being controlled by temperature, pressure, surface area, and solution composition, including pH and reaction affinity (deviation from the equilibrium) (Bauer et al., 1998). Nevertheless, in experimental studies small differences in run times are also decisive. The present study was designed to investigate the transformations of kaolinite at 200 °C in aqueous solutions, within a wide range of pH values, controlled by different concentrations of NaOH, KOH, and SiO_2.

2. METHODOLOGY

Poorly crystalline kaolinite from Georgia (standard KGa-2) was used in the experiments, after an intense grinding. The modified kaolinite was characterized by X-ray diffraction (XRD), X-ray fluorescence (XRF), scanning (SEM) and transmission (TEM) electron microscopy, and granulometric analyses. Grinding produced highly disordered kaolinite consisting of rounded particles with mean size <0.1 µm. (Gonzalez Jesús et al., 2000).

The experiments were performed in 50 cm^3 Teflon-lined reactors (Parr 4744), which were oven-heated at a constant temperature of 200 °C (within ±3 °C). Five different experiments were carried out by addition of 1.7 g kaolinite to 25 mL of aqueous solutions containing KOH, NaOH, and MgCl$_2$, and variable amounts of silica gel, in the following proportions:

(1)   1kaolinite + 2SiO$_2$ + 1NaOH + 0.5KOH + 0.6MgCl$_2$
(2)   1kaolinite + 1SiO$_2$ + 1NaOH + 0.5KOH + 0.6MgCl$_2$
(3)   1kaolinite + 1SiO$_2$ + 1NaOH + 1KOH + 0.6MgCl$_2$
(4)   1kaolinite + 1NaOH + 0.5KOH + 0.6MgCl$_2$
(5)   1kaolinite + 2NaOH + 1KOH + 0.6MgCl$_2$

So, the solid/solution ratio was maintained constant whereas the initial pH changed within a wide interval. The experiments were run for 15 days. Additional tests of experiment (1) were performed varying the reaction time between 1 and 180 days.

After quenching the reactors, solid and solution were immediately separated by centrifuging. The solution pH was measured in the supernatant, which was stored for chemical analysis. Al and Si analyses were carried out using UV-Visible spectrophotometry, by means of the molybdate blue method and red S-alizarine complex, respectively. Na, K, and Mg were analyzed by atomic absorption spectrometry. Saturation indexes in the hydrothermal solutions were calculated using EQ3/6 geochemical code (Wolery, 1992).

The solid products of the reactions were characterized by XRD (Siemens D-5000 diffractometer; with CuKα radiation, at 40 kV and 40 mA, step size = 0.02 and counting time = 1 s), SEM (Jeol JSM-840 microscope), transmission and analytical electron microscopy (TEM/AEM) (Philips CM-20 microscope, fitted with a ultrathin window, solid-state Si(Li) detector for energy dispersive X-ray analysis), and thermal analysis (DTA-TG) (Rigaku-Thermoflex TG-8110, provided with a TASS 1000 station).

3. RESULTS AND DISCUSSION

3.1. Study of the solid products

XRD patterns obtained for unoriented samples after 15 days of reaction are shown in Fig. 1, together with the untreated kaolinite. The solid products of the reactions (1) and (2) are very similar, indicating the recrystallization of kaolinite, as revealed by the decrease in intensity of the amorphous band (~3.5 Å, in the kaolinite pattern), as well as the formation of a new phase (band at 12 - 14 Å). The presence of small peaks at 3.34 Å and 3.21 Å may suggest the formation of small amounts of quartz and feldspar. The XRD pattern of the solid products of reaction (3) reveals the presence of a new phase (band at ~12 Å), similar to that observed in the patterns of reactions (1) and (2). A set of new peaks, the most intense at 9.30 Å, 4.99 Å, 3.18 Å, and 2.93 Å, could be ascribed to albite and zeolites. The solid product of reaction (4) shows a XRD pattern where the kaolinite reflections are accompanied by a band at ~ 10Å, which suggest the presence of illite or, more probably, illite/smectite mixed-layers. Finally, the XRD pattern of the solid product of reaction (5) is clearly different, because only zeolite reflections are present.

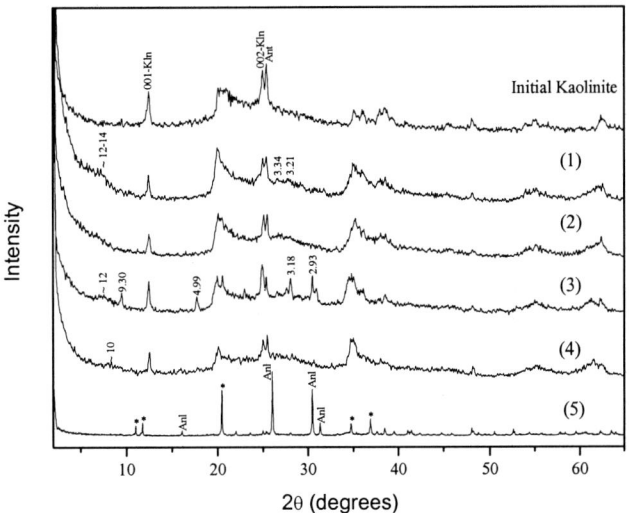

Figure 1. Unoriented X-ray diffraction patterns of the initial kaolinite and the solid products of the reactions (1)-(5) after 15 days. Kln: Kaolinite; Ant: Anatase; Anl: Analcime; *Mg-rich zeolite.

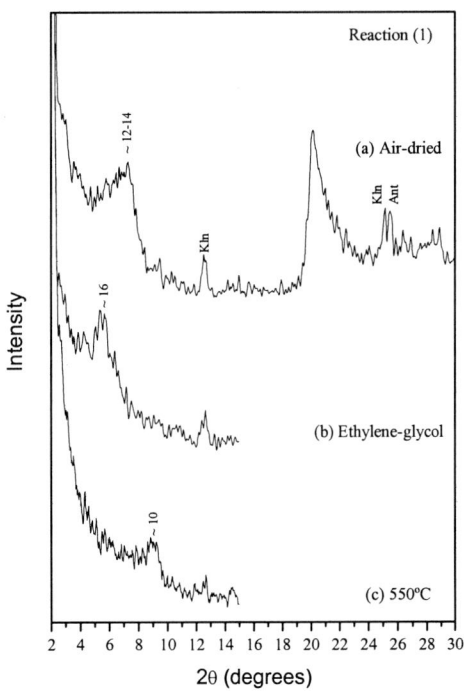

Figure 2. XRD patterns obtained from oriented mounts of the solid products of reaction (1) after 30 days run time.

The XRD patterns obtained for oriented mounts of the products of the reactions (1), (2) and (3) are very similar. The 12-14 Å band observed in the pattern obtained from the air-dried sample, expanded at ~ 14-16 Å after ethylene-glycol treatment, and contracted at ~10-12 Å after heating. This behaviour suggests the presence of disordered smectite/vermiculite or smectite/chlorite mixed-layers (Reynolds, 1985).

TEM observation of the products from reactions (1) to (4) reveals the presence of two characteristic morphologies: Spherical particles of kaolinite, with mean size in the order of 0.5 µm, and plate particles with bad-defined morphology, corresponding to the 14 Å and the 10 Å phases (Fig. 3).

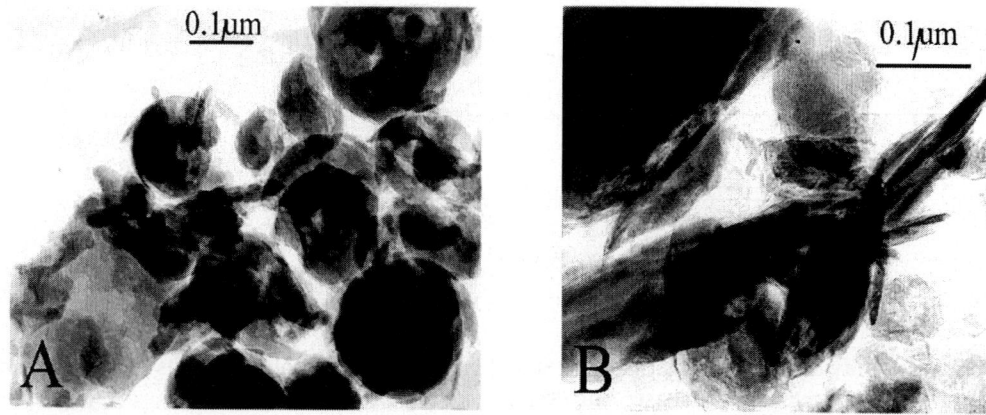

Figure 3. Characteristic morphologies of the solid products of the reactions (1) to (4). A: Spherical kaolinite. B: Platy particles of the expandable phase.

Spherical kaolinite is similar to that described by Tomura et al. (1983; 1985), Fiore et al. (1995) and González Jesús et al. (2000). High-resolution images of the plate particles reveal important stacking disorder and a mean basal spacing of ~ 10 Å, suggesting the presence of smectite or vermiculite layers in the products of reaction (1), (2) and (4). Some of the structural formulae calculated for 11 O, from the available AEM data, for this phase are shown in Table 1.

These analyses reveal that Mg has been, in part, incorporated into the octahedral sites in the expandable phase structure. Moreover, K has been preferentially incorporated, relative to Na, in these structures. On the other hand, the low K content in some of the analyses suggests the presence of vermiculitic or chloritic layers.

Table 1. Structural formulae for the expandable phase

	(a)	(b)	(c)	(d)	(e)	(f)
Si	3.89	3.85	3.82	3.58	3.56	3.54
^{IV}Al	0.11	0.15	0.18	0.42	0.44	0.46
^{VI}Al	1.49	1.73	1.49	1.31	1.88	1.69
Mg	0.60	0.33	0.62	0.84	0.20	0.62
Na	0.00	0.00	0.00	0.00	0.07	0.00
K	0.38	0.14	0.31	0.33	0.26	0.09
O	11	11	11	11	11	11

In the case of products from reaction (5), SEM and TEM observations show the presence of two main morphologies (prismatic and hexagonal), corresponding to two different zeolites (Fig. 4). Prismatic crystals show XRD and SAED reflections, and AEM data characteristic of analcime, whereas the hexagonal morphology corresponds to a Mg-rich zeolite, which has not been identified at the moment.

Figure 4. SEM images of the solid products of reaction (5). Prismatic crystals are analcime (Anl) and hexagonal crystals a Mg-rich zeolite.

3. 2. Characteristics of the solutions

The pH values measured in the initial solutions and after 15 days run time are shown in Table 2. The high initial pH dropped during the hydrothermal treatment, except for reaction (5). Differences in the final pH values and cation availability explain the different types of solid products formed. So, the high pH measured in reaction (5) explains the complete dissolution of kaolinite and the formation of zeolites exclusively. In spite of the high pH of reaction (4), this did not originate zeolitic phases probably due to the scarcity of available SiO_2. These data also indicate that the formation of expandable phases is restricted to the pH interval 6.24-8.

Table 2. pH values for the initial and final solutions

Reaction	pH (initial)	pH (15 days)
(1)	10.40	7.01
(2)	11.03	6.24
(3)	12.82	7.97
(4)	12.75	9.64
(5)	13.96	13.71

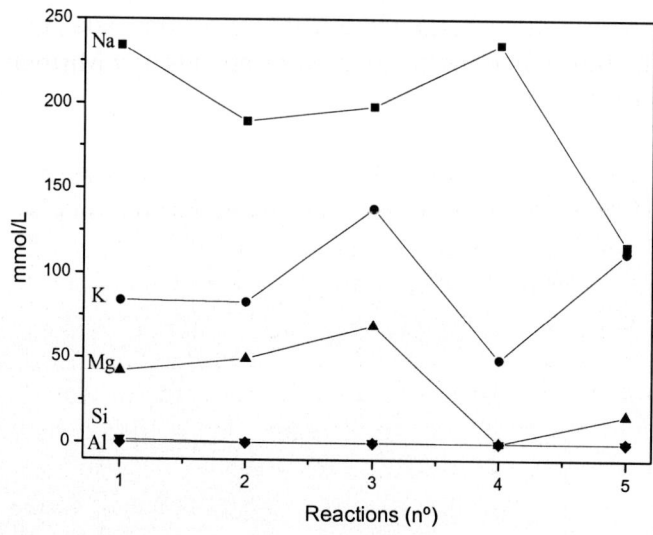

Figure 5. Variations of the Na, K, Mg, Si and Al contents in the several studied reactions.

Table 3. Composition of the solutions (mmol/L)

Reaction	Si	Al	Na	K	Mg
(1)	1.70	0.028	234	83.8	42.1
(2)	0.284	< dl	190	82.9	49.4
(3)	0.278	0.021	198	138	69.1
(4)	0.026	0.024	235	49.1	<dl
(5)	0.107	0.064	117	113	16.5

dl: detection limit

The compositions of the solutions are shown in Table 3, and Fig. 5. These data reveal, in spite of the different initial composition, some significant features:

i) Si and Al contents in solution are low, indicating that kaolinite dissolution is accompanied by precipitation of new phases.
ii) In spite of the different initial concentration in K, Mg and K follow a similar trend, which may suggest that the incorporation rate into the new phases (other than kaolinite) for both cations is similar.
iii) K is preferentially consumed in the reactions, relative to Na, except for reaction (5).
iv) The minimum K content corresponds to reaction (4), which is in agreement with the formation of a mica-like phase.
v) Mg is preferentially consumed in reactions (4) and (5) in the formation of either the mica-like structure (reaction 4) or the Mg-rich zeolite (reaction 5).

The results of EQ3/6 calculations generally agree with the mineral phases present in the solid and detected by XRD. The solution of experiment (1) was supersaturated with respect to clinochlore, kaolinite, trioctahedral smectites, muscovite, illite, talc, kaolinite, dioctahedral smectites, and K-feldspar, and slightly undersaturated with respect to quartz. For reaction (3), solution was supersaturated with respect to clinochlore, talc, trioctahedral smectites, and sepiolite. For reactions (2) and (4), estimated values of Al = 10 nmol L^{-1} and Mg = 1 ppm were used for calculations, but the outputs were not very consistent with the result of the analysis of the solid. The high pH for solution (5), and the confidence on the measure, renders the code output few reliable.

Chemical analyses of the solutions of reaction (1) as well as pH values measured after increasing time intervals (not shown) indicate, on the other hand, that the cation content and the pH stabilised after 15 days, which suggests that the reaction is approaching to equilibrium. Nevertheless, longer reaction times originate a better-defined 14 Å phase.

This behaviour suggests that much longer reaction times and especially higher temperatures would be necessary to form ordered mixed-layer structures. So, although the 14-Å phase is chemically similar to tosudite, the response of this phase against the several treatments does not correspond to an ordered 1:1 mixed-layer. Smectitic phases are frequently a step previous to the formation of corrensite-like mixed-layer minerals (Tomura *et al.*, 1983), whereas the ordered chlorite/smectite structure (either di- or trioctahedral) appears to form at higher temperatures (Eberl, 1978; Robinson *et al.*, 1999).

Formation of an illitic phase from reaction (4) indicates, however, that temperature was high enough for kaolinite illitization. These results are consistent with most of the experimental data, which indicates that the formation of illite-like layers from smectite is easier than the formation of 14 Å layers (Eberl, 1978; Eberl *et al.*, 1993).

4. CONCLUSIONS

The results of the experiments conducted indicate that the hydrothermal reaction of kaolinite at 200 ºC, in aqueous solutions containing NaOH, KOH, and MgCl$_2$, originate three main types of products, depending on the pH of the solution:

1) In the pH range of 6.24-8, dissolution and recrystallization of kaolinite is accompanied by the formation of a phyllosilicate phase, whose behaviour against the several treatment indicates a composition near smectite/vermiculite or chlorite/smectite randomly ordered mixed-layers, with high smectite proportion.
2) At higher pH (in the order of 9.5), recrystallization of kaolinite is accompanied by the formation of ~ 10 Å phases consisting probably of illite and illite/smectite mixed-layers.
3) At the highest pH values (~13.5), recrystallization of kaolinite does not occur and only zeolite phases (analcime and an unknown Mg-rich zeolite) were formed.

At the conditions of the experiments, 1:1 ordered mixed-layers such as dioctahedral smectite/vermiculite or chlorite/smectite were not formed, even though reaction time lasted 180 days. These results suggest that higher temperatures (in the order of 350 ºC) may be necessary for a further evolution of the smectitic phase.

ACKNOWLEDGMENTS

The authors are grateful to Tsutsumi Sadao and to C. Alfonso Pinto, whose suggestions have improved the manuscript. The authors are also grateful to M.M. Abad for help in obtaining the TEM/AEM data. This study received financial support from project BTE-2000-1150 (Ministerio de Educación y Cultura de España).

REFERENCES

Bauer, A.; Velde, B. and Berger, G. (1998) Kaolinite transformation in high molar KOH solutions. *App. Geoch.,* 13, 619-629.

Chatterjee, N.D. (1973) Low-temperature compatibility relations of the assamblage quartz-paragonite and the thermodynamic status of the phase rectorite. *Contrib. Mineral. Petrol.,* 42, 259-271.

Eberl, D.D. (1978) The reaction of montmorillonite to mixed-layer clay: the effect of interlayer alkali and alkaline earth cations. *Geochim. Cosmoch. Acta.,* 42, 1-7.

Eberl, D.D. Velde, B. and McCormick, T. (1993) Synthesis of illite-smectite from smectite at earth surface temperatures and high pH. *Clay Miner.,* 28, 49-60.

Fiore, S.; Huertas, F.J.; Huertas, F. and Linares, J. (1995) Morphology of kaolinite crystals synthesized under hydrothermal conditions. *Clays Clay Miner.,* 43, 353-360.

Frank-Kamenetskii, V.A.; Goilo, E.A.; Kotov, N.V. and Rieder, M. (1990) Structural transformations of kaolins into (Ni, Al) serpentine-like phases and subsequently into trioctahedral micas under hydrothermal conditions. *Clay Miner.,* 25, 121-125.

Frank-Kamenetskii, V.A.; Kotov, N.V. and Goilo, E.A. (1983) Structural transformations of layer silicates at elevated p, T parameters. *Nedra Pub.,* 151 pp.

González Jesús, J.; Huertas, F.J.; Linares J. and Ruiz Cruz, M.D. (2000) Textural and structural transformations of kaolinites in aqueous solutions at 200ºC. *App. Clay Sci.,* 17, 245-263.

Herman, E.; Roberson, R.C.; Reynolds, Jr. and Jenkin, D.M. (1999) Hydrothermal synthesis of corrensite: a study of the transformation of saponite to corrensite. *Clays Clay Miner.,* 47, 212-218.

Huang, W.J. (1993) The formation of illitic clays from kaolinite in KOH solution from 225ºC to 350ºC. *Clays Clay Mineral.,* 6, 645-654.

Kotov, N.V.; Soboleva, S.V.; Goilo, E.A.; Zvyagin, B.B. and Frank-Kamenetskii, V.A. (1980) Structural inheritance in the course of formation of mica after kaolins under hydrothermal conditions. *Izv. Akad. Nauk SSSR Ser. Geol.,* 12, 68-80.

Reynolds, R.C. (1985) NEWMOD. A computer program for the calculation of the basal difracction intensities of mixed-layered clay minerals. *R.C. Reynolds ed.*

Tomura, S.; Shibasaki, Y.; Mizuta, H. and Kitamura, M. (1983) Spherical kaolinite: synthesis and mineralogical properties. *Clays Clay Miner.,* 31, 413-421.

Tomura, S.; Shibasaki, Y.; Mizuta, H. and Kitamura, M. (1985) Growth conditions and genesis of spherical and platy kaolinite. *Clays Clay Miner.,* 33, 200-206.

Wolery, T.J. (1992) EQ3/6, a software package for geochemical modeling of aqueous systems. *Lawrence LivermoreNational Laboratory,* URCL-MA-1100662.

Synthesis and Characterization of Novel Inorganic/Organic Hybrid Materials Prepared from Layered Double Hydroxides.

G. A. Caravaggio, A. Moser, C. Detellier

University of Ottawa, Center for Research & Innovation in Catalysis, Ottawa, ON, K1N 6N5, Canada

Layered double hydroxides of Zn and Al of varying ratios have been synthesized and reacted with phenylphosphonic acid. The materials were characterized by inductively couple plasma (ICP), elemental analysis (EA) and powder XRD. The results show that the phenylphosphonic acid has been grafted inside the layers of the layered double hydroxides.

1. INTRODUCTION

Layered double hydroxides (LDHs) are hydrotalcite-type materials composed of positively charged di and trivalent metal hydroxide sheets with charge balancing anions and water molecules dispersed between the layers. The chemical composition of the compounds is generally expressed as $M(II)_{1-x}M(III)_x(OH)_2 A^-_x B^{2-}_{x/2}$ where $M(II)$ can be Ni^{2+}, Zn^{2+}, Mg^{2+}, Mn^{2+}, etc...; $M(III)$: Al^{3+}, V^{3+}, Co^{3+}, Cr^{3+}, etc. ; A^-: NO_3^-, Cl^-, etc... and B^{2-}: SO_4^{2-}, CO_3^{2-}, etc. In the past few years, LDHs have attracted considerable attention due to their many possible applications such as adsorbents,[1] hosts for intercalated species,[2] catalysts,[3,4,5] catalysts precursors,[6,7] anionic exchangers[8] and as electrode material.[9,10,11]. Furthermore, grafting of inorganic[12,13] and organic anions[14] inside LDHs have been reported in the literature. The grafting of organic phosphonate moieties inside the layers could provide materials with microporosity characteristics. In the following study, zinc aluminium layered double hydroxides of varying M(II)/M(III) ratios have been treated with phenylphosphonic acid and the materials have been characterized by ICP, elemental analysis, and powder X-ray diffraction (XRD).

2. EXPERIMENTAL

The preparation of the layered double hydroxides was performed under a nitrogen atmosphere. Distilled water was deionised to 18.3 MW/cm in resistivity through a Barnstead B-pure deioniser and then was boiled (refluxed) under nitrogen for at least a day before being used for the LDHs synthesis. The pH values for the LDH synthesis were measured with a VWR scientific model 2000 pH meter fitted with a VWR low maintenance Triode pH electrode with Ag/AgCl internal reference system, sealed reference and built in thermistor for automatic temperature compensation. The pH meter was always calibrated with a pH 7 and pH 10 buffer prior to any measurements. Reagents used were from commercial sources.

2.1 Synthesis of the LDHs and the reacted LDH/Phenylphosphonic acid (PPA)

2.1.1 Zn$_2$Al, Zn$_3$Al, Zn$_4$Al LDH

All LDHs were prepared using the method reported by Wang et al.[15] A mixture consisting of Zn(NO$_3$)$_2$·6H$_2$O (29.7 g, 0.10 mol) and of Al(NO$_3$)$_3$·9H$_2$O (18.89 g, 0.05 mol) in 300 mL of freshly boiled deionised water was added together with a 2 M NaOH solution (~200 mL), to a flask containing 150 mL of deionised water. Rapid stirring was kept throughout the addition and the rate of addition was such that the pH was kept at 10.0 ± 0.5. After the addition was completed, the mixture was heated at ~75-80°C for 16 hrs under N$_2$. It was then filtered, washed with water and ethanol then dried at 70°C for 24hrs. Zn$_3$Al LDH and Zn$_4$Al LDH were prepared in a similar fashion but the zinc and aluminum nitrates were weighed to obtain the appropriate molecular ratios.

2.1.2 Phenylphosphonic acid/LDH reaction

The LDH (1.00g) was dispersed in a 50 mL phenylphosphonic acid (0.1 M, Aldrich, 98%) solution, mixed and heated at 75°C for 48 hrs. The pH of this solution was initially 2 and increased to approximately 5 to 8 depending on the LDH being intercalated. The mixture was filtered and the product was transferred to another phenylphosphonic acid solution (50 mL, 0.1 M, Aldrich, 98%), mixed and heated at 75°C for another 48 hrs. The initial pH of this solution was also 2 and did not change after 48 hrs. The final product was filtered and dried for 18 hrs at 70°C.

2.2 Characterization

XRD patterns were obtained using a Philips PW3710 diffractometer with Cu-Kα radiation (λ = 1.54059 Å). Elemental analysis of the metal and phosphorus atoms were carried out on a Jarrell Ash Atom Scan inductively coupled plasma emission spectrometer (ICPES). C, H, and N elemental analysis were done on a Perkin Elmer Series II CHNS/O 2400 analyser.

3. RESULTS AND DISCUSSION

3.1 Powder XRD analysis

Figure 1 shows the XRD patterns of the Zn$_2$Al, Zn$_3$Al and Zn$_4$Al LDHs. The d-spacing and the a parameter of each compound are found in Table 1. The value of the d-spacing of the Zn$_2$Al LDH was obtained by calculating the average of the first three reflections (003, 006 and 009 reflections) found in the compounds diffractogram (Figure 1a). The d-spacings of the Zn$_3$Al, Zn$_4$Al LDHs were obtained from the average of the first 2 reflections (003 and 006 reflections) of the material's diffractograms (Figure 1b and 1c respectively).[15] The d-spacings of these LDHs are similar to those found in the literature for ZnAl LDHs containing nitrates.[16,17] Furthermore, Table 1 shows that the a parameter decreases as the value of x of the LDHs increases. The a parameter is related to the mean radius of the metals found in the layers of the LDH, and as the substitution of a metal with a smaller radius inside the layer increases, the a parameter decreases.[18] Therefore, since the radius of Al^{3+} is smaller than the one of Zn^{2+}, the a parameter decreases with increasing content of Al^{3+} (or an increase of x).

Table 1: d-spacing and "a" parameter of ZnAl LDHs

ZnAl LDH ratio	d-spacing (D)	a parameter (Å)
Zn_2Al	8.77	3.06
Zn_3Al	8.83	3.07
Zn_4Al	8.89	3.08

In the diffractogram of each LDH, extra reflections were noted and were labeled in the Figure 1 as ZnO. These reflections were attributed to a zinc oxide contamination that was identified from the JCPDS database. This contamination was however considered insignificant because the deviation between the calculated and experimental data of each LDH was small (Table 2). Finally, other important peaks that are indexed on the spectrum correspond to the ones found in the literature for ZnAl layered double hydroxides.[19]

The XRD patterns of the Zn_2Al, Zn_3Al and Zn_4Al LDHs that have been treated with phenyl phosphonic acid (PPA) are found in Figure 2. The successful reactions can be noted by the disappearance of the original LDH patterns which are replaced with the patterns of new layered compounds that have 3 orders of (00l) reflections present at 2θ values of ~ 6.0°, 12.3° and 18.5°. The d-spacing calculated from the average of these 3 reflections corresponds to 14.5 Å. This value is very similar to the interlayer spacing of zirconium phenyl phosphonate (ca 14.7 Å)

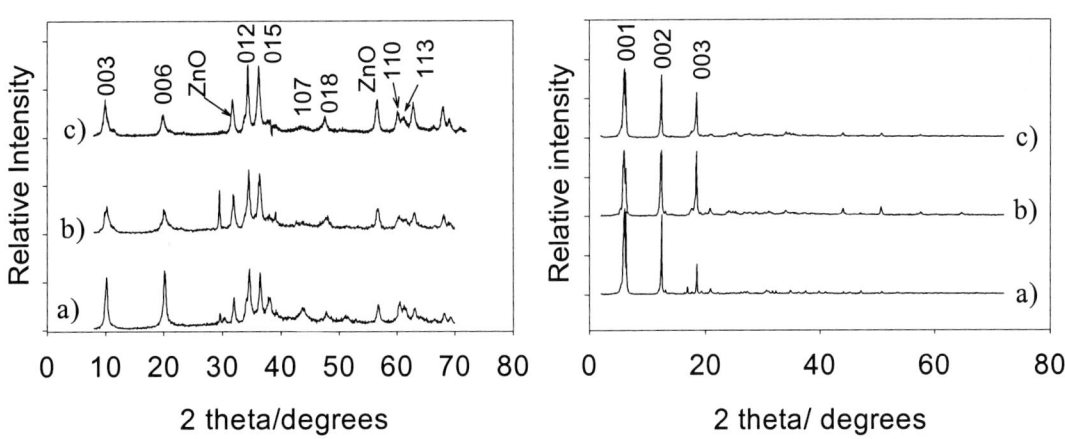

Figure 1: XRD of a) Zn_2Al LDH b) Zn_3Al LDH and c) Zn_4Al LDH

Figure 2: LDHs treated with phenylphosphonic acid; a) Zn_2Al/PPA, b) Zn_3Al/PPA, c) Zn_4Al/PPA

reported by

Alberti et al.[20] and Poojary et al..[21] In these materials, the three oxygens of the phosphonate group of the phenylphosphonic acid molecules are directly bound to the zirconium atoms. In addition, the benzene rings are pointing inside the layers and form a bilayer arrangement that is perpendicular to the zirconium layers. The observed d-spacings of the new materials are also comparable to the interlayer distance (ca 14.7 Å) reported by Carlino *et al.* for a product obtained from the thermal reaction of phenylphosphonic acid with a MgAl LDH.[22] In the latter article the following structure was proposed for the product: one oxygen group of the phenylphosphonic acid is grafted to every second metal atom in the LDHs brucite-like sheets and the remaining metal atoms are bound to hydroxyl groups. Moreover, the phenyl groups are semi perpendicular to the metal layers. The structure proposed for the compounds in this study is a ZnAl metal phosphonate where the phenyl groups are perpendicular to the metal layers and one or two of the oxygens atoms of the phosphonate groups of the acid are bound to the metals.

3.2 Chemical analysis

Table 2: ICP and elemental analysis of the ZnAl LDHs

M(II)/M(III) ratio	Elements	Calculated weight %	Exp. Weight %	Molecular formula
Zn_2Al	Zn	40.7	39.0	$Zn_{0.67}Al_{0.33}(OH)_2(NO_3)_{0.33}$ · $0.01H_2O$
	Al	8.27	7.91	
	C	0	0.17	
	H	1.90	1.90	
	N	4.29	4.39	
Zn_3Al	Zn	47.7	42.1	$Zn_{0.76}Al_{0.24}(OH)_2(NO_3)_{0.24}$ · $0.02H_2O$
	Al	6.21	5.56	
	C	0	0.21	
	H	1.90	1.87	
	N	3.23	3.90	
Zn_4Al	Zn	52.4	52.4	$Zn_{0.81}Al_{0.19}(OH)_2(NO_3)_{0.1}$ · $0.01H_2O$
	Al	5.07	5.21	
	C	0	0.15	
	H	1.84	1.84	
	N	2.56	2.18	

Table 2 shows the results obtained from ICP and elemental analysis of the synthesized LDHs. The molecular formula of each compound was determined by calculating the value of x in the LDH formula below from the zinc and aluminum molar ratio determined by the ICP analysis.

$$Zn_{1-x}Al_x(OH)_2(NO_3^-)_{2x} \cdot y\ H_2O \qquad (1)$$

These results coupled with the powder XRD pattern (Figure 1) confirm that LDHs were obtained. The deviation between the calculated and the experimental values is most likely due to the zinc oxide contamination that was identified in the powder XRD pattern.

3.2.1 Molecular formula determination of the reacted LDHs

A program was written in visual basic 6.0 to determine the molecular formulas of the reacted compounds. This program varied the coefficient "a", "z" and "y" in equation 2 below, and the value of x was calculated from the Zn:Al ratio determined by the ICP analysis.

$$Zn_{1-x}Al_x(OH)_a(C_6H_5PO_3H_z)^{2-z}_{((2-a)+x)/(2-z)} \cdot yH_2O \qquad (2)$$

The coefficient "a" was varied from 0 to 2, and for each step, "z" was increased from 0 to 1.99 and "y" from 0 to 2. By using these values all different valid possibilities of chemical formulas were considered. For each variation of the coefficients, the weight percent of the elements were

Table 3: ICP and elemental analysis of the phosphonate grafted ZnAl materials

Sample name/ M(II)/M(III) ratio	Method of analysis		Cal. Weight %	Exp. Weight %	Dev. %	Molecular formula
ZnAl09IN Zn$_2$Al + PPA	ICP[1]	Zn	14.9	14.9	0.1	Zn$_{0.63}$Al$_{0.37}$(OH)$_{0.05}$
	ICP	Al	3.65	3.65	0.08	(C$_6$H$_5$PO$_3$H$_{0.23}$)$_{1.32}$
	EA[2]	C	34.4	35.0	1.7	·0.99H$_2$O
	EA	H	3.24	3.23	0.2	
	ICP	P	14.8	13.9	6.4	
	TGA[3]	H$_2$O	6.51	6.50	0.08	
ZnAl15IN Zn$_3$Al + PPA	ICP	Zn	18.4	18.4	0.2	Zn$_{0.72}$Al$_{0.28}$(OH)$_{0.02}$
	ICP	Al	3.00	3.00	0.13	(C$_6$H$_5$PO$_3$H$_{0.01}$)$_{1.16}$
	EA	C	32.8	35.3	7.1	·1.04H$_2$O
	EA	H	3.14	2.89	8.7	
	ICP	P	14.1	13.9	1.5	
	TGA	H$_2$O	7.36	7.60	3.14	
ZnAl14IN Zn$_4$Al + PPA	ICP	Zn	20.7	20.8	0.5	Zn$_{0.78}$Al$_{0.22}$(OH)$_{0.03}$
	ICP	Al	2.40	2.40	0.2	(C$_6$H$_5$PO$_3$H$_{0.01}$)$_{1.10}$
	EA	C	32.2	34.8	7.4	·0.95H$_2$O
	EA	H	3.04	2.71	12.4	
	ICP	P	13.8	14.2	2.5	
	TGA	H$_2$O	6.95	7.00	0.8	

Notes: 1 ICP: Inductively coupled plasma
2 EA : Elemental analysis
3 TGA: Thermogravimetric analysis

calculated and compared to the values obtained from the experimental chemical analysis. The molecular formulas having coefficients resulting in the lowest overall deviation between the experimental and the calculated values were kept. The formulas with the optimum results are shown in Table 3. Molecular formulas are indicated, where the hydroxyl group coefficient are lower than 0.05 and the coefficient for the protons are lower than 0.23. This indicates that the compounds are almost devoid of hydroxyl groups and that the great majority of interlayer acids are completely deprotonated. The absence of OH groups also indicates that the phenylphosphonic acid molecules are grafted directly to the metal layers of the material. This outcome has been described previously in the studies of the intercalation of phosphonate molecules in bayerite[23] and in LDHs.[24,25]

4. CONCLUSION

Three ZnAl layered double hydroxides of varying ratios have been synthesized and treated with phenylphosphonic acid. Phenylphosphonic acid moieties were grafted inside the layers of the ZnAl layered double hydroxide. The powder XRD pattern of each material shows a d-spacing of ~14.5 Å, which is similar to the d-spacing of zirconium phosphonates. The results from the chemical analysis are in agreement with a molecular formula in which very few OH groups are present and most of the PPA molecules are deprotonated. Further studies are in progress to describe more precisely the structure of the LDH/PPA materials.

5. ACKNOWLEDGMENTS

The Natural Science and Engineering Research Council of Canada and the Ontario Graduate Scholarships in Science and Technology are gratefully acknowledged for a research grant (C.D.) and a scholarship (G.C.) respectively.

REFERENCES

1. Cavani, F.; Trifiro, F.; Vaccari, A. Hydrotalcite-Type Anionic Clays: Preparation, Properties and Applications. *Catal. Today* 1991, *11,* 173-301.
2. Raki, L.; Rancourt, D. G.; Detellier, C. Preparation, Characterization, and Mossbauer Spectroscopy of Organic Anion Intercalated Pyroaurite-Like Layered Double Hydroxides. *Chem. Mater.* 1995, *7,* 221-224.
3. Kruissink E.C.; Van Reijen L.L. Coprecipitated Nickel-Alumina Catalyst for Methanation at High Temperature. *J. Chem. Soc. Faraday Trans.* 1981, *77,* 649-663.
4. Mckenzie, A. L.; Fishel, C. T.; Davis, R. J. Investigation of the Surface Structure and Basic Properties of Calcined Hydrotalcites. *J. Catal.* 1992, *138,* 547-561.
5. Nakatsuka, T.; Kwasaki, H.; Yamashita, S.; Kohjiya, S. The Polymerization of B-

Propiolactone by Calcined Synthetic Hydrotalcite. *Bull. Chem. Soc. Jpn.* 1979, *52(8)*, 2449-2450.
6. Gusi, S.; Pizzoli, F.; Trifiro, F.; Vaccari, A.; Del Piero, G. Preparation of Milticomponent Catalysis for the Hydrogenation of Carbon Monoxide Via Hydrotalcite-Like Precursors. *Prep. Cata* 1987, 753-765.
7. Busetto, C.; Del Piero, G.; Manara, G.; Trifirò, F.; Vaccari, A. Catalysis for Low-Temperature Methanol Synthesis. Preparation of Cu-Zn-Al Mixed Oxides Via Hydrotalcite-Like Precursors. *J. Catal.* 1984, *85,* 260-266.
8. Meyn, M.; Beneke, K.; Lagaly, G. Anion-Exchange Reactions of Layered Double Hydroxides. *Inorg. Chem.* 1990, *29,* 5201-5207.
9. Caravaggio, G. A.; Detellier, C.; Wronski, Z. S. Synthesis, Stability and Electrochemical Properties of NiAl and NiV Layered Double Hydroxide. *J. Mater. Chem.* 2001, *11,* 912-921.
10. Kamath, V. P.; Dixit, M.; Indira, L.; Hukla, A. K.; Kumar, V. G.; Nichandraiah, N. Stabilized α-Ni(OH)$_2$ As Electrode Material for Alkaline Secondary Cells. *J Electrochem Soc* 1994, *141,* 2956-2959.
11. Sugimoto, A.; Ishida, S.; Hanawa, K. Preparation and Characterization of Ni/Al-Layered Double Hydroxides. *J. Electrochem. Soc.* 1999, *146,* 1251-1255.
12. El Malki, K.; De Roy, A.; Besse, J. P. New Cu-Cr Layered Double Hydroxide Compound: Discussion of Pillaring With Intercalated Tetrahedral Anions. *Eur. J Solid. State. Inorg. Chem* 1989, *26,* 339-351.
13. Depege, C.; El Metoui, F. Z.; Forano, C.; De Roy, A.; Dupuis, J.; Besse, J. P. Polymerization of Silicates in Layered Double Hydroxides. *Chem. Mater.* 1996, *8,* 952-960.
14. Morioka, H.; Tagaya, H.; Karasu, M.; Kadokawa, J.; Chiba, K. Preparation of New Useful Materials by Surface Modification of Inorganic Layered Compounds. *J. Solid State Chem.* 1995, *117,* 337-342.
15. Wang, J. D.; Tian, Y.; Wang, R. C.; Clearfield, A. Pillaring of Layered Double Hydroxides With Polyoxometalates in Aqueous Solution Without Use of Preswelling Agents. *Chem. Mater.* 1992, *4,* 1276-1282.
16. Cervilla, A.; Liopis, E.; Ribera, A.; Corma, A.; Fornés, V.; Rey, F. Intercalation of the Oxo-Transfer Molybdenum(VI) Complex [MoO$_2$]{O$_2$CC(S)Ph$_2$}$_2$]$^{2-}$ into a Zinc(I)-Aluminium(III) Layered Double Hydroxide Host. Catalysis of the Air Oxidation Odf Thiols. *J. Chem. Soc. ,Dalton Trans.* 1994, 2953-2957.
17. Kwon, T.; Pinnavaia, T. J. Pillaring of a Layered Double Hydroxide by Polyoxometalates With Keggin-Ion Structures. *Chem. Mater.* 1989, *1,* 381-383.
18. Brindley, G. W.; Kikkawa, S. A Crystal-Chemical Study of Mg,Al and Ni,Al Hydroxy-Perchlorates and Hydroxy-Carbonates. *Amer. Mineral.* 1979, *64,* 836-843.
19. Badreddine, M.; Legrouri, A.; Barroug, A.; De Roy, A.; Bessé, J. P. Influence of PH on Phosphate Intercalation in Zinc-Aluminium Layered Double Hydroxide. *Collect. Czech. Chem. Commun.* 1998, *63,* 741-748.
20. Alberti, G.; Costantino, U. Crystalline Zr(R-PO3)2 and Zr(R-OPO3)2 Compounds (R=Organic Radical). A New Class of Materiala Having Layered Structure of Zirconium Phosphate Type. *J. Inorg. Nucl. Chem.* 1978, *40,* 1113-1117.
21. Poojary, D. M.; Hu, H. L.; Campbell III, F. L.; Clearfield, A. Determination of Crystal

Structures From Limited Powder Data Sets: Crystal Structure of Zirconium Phenylphosphonate. *Acta. Cryst.* 2000, *b49*, 996-1001.

22. Carlino, S.; Hudson, M. J.; Husain, S. W.; Knowles, J. A. The Reaction of Molten Phenylphosphonic Acid With a Layered Doube Hydroxide and Its Calcined Oxide. *Solid State Ionics* 1996, *84*, 117-129.

23. Raki, L.; Detellier, C. Lamellar Organominerals: Intercalation of Phenylphosphonate into Layers of Bayerite. *Chem. Commun.* 1996, 2475-2476.

24. Nijs, H.; Clearfield, A.; Vansant, E. F. The Intercalation of Phenylphosphonic Acid in Layered Double Hydroxides. *Microp. Mesop. Mater.* 1998, *23*, 97-108.

25. Don Wang, J.; Serrette, G.; Tian, Y.; Clearfield, A. Synthetic and Catalytic Studies of Inorganically Pillared and Organically Pillared Layered Double Hydroxides. *Appl. Clay Sci.* 1995, *10*, 103-115.

EPR characterisation of iron in various clay minerals : Montmorillonites and Layered Double Hydroxides

F. Dijoux[a], B Deroide[a] and D. Tichit[b]

[a] Laboratoire de Physico-chimie de la Matière Condensée, UMR CNRS 5617, Université de Montpellier II, place Eugène Bataillon, 34095 Montpellier cedex 5, France.

[b] Laboratoire des Matériaux Catalytiques et Catalyse en Chimie Organique, UMR CNRS 5618, Ecole Nationale Supérieure de Chimie de Montpellier, 8 rue de l'Ecole Normale 34296 Montpellier Cedex 5, France.

The local environment of Fe^{3+} in natural clays has been analysed by using electronic paramagnetic resonance (EPR) and Mössbauer spectroscopies, in addition to classical methods of investigation. For various raw montmorillonites, the X-band-EPR ($\approx$9.5GHz) allows the identification of structural Fe^{3+} and other phases (iron oxides and hydroxides), as well as other defect species, (E' centres in quartz, 3d-elements...). We also used high frequency EPR (95GHz and 285GHz) and Mössbauer spectroscopies to obtain accurate values of the fine structure parameter D, which is sensitive to the local ordering. Distribution laws P(D), correlated with the distribution of distortions of iron sites, were deduced from this analysis. The electric field gradient model, appropriate for amorphous or glassy solids was not suitable for simulations of P(D) in clays. The experimental laws P(D), obtained from the distribution of Mössbauer quadrupole splittings, reproduced correctly the EPR spectra. These distributions are narrower than in glasses, due to the lamellar organisation of clays and suggest the presence of only one family of distortions from octahedral symmetry .In order to avoid the paramagnetism of secondary phases or species (quartz, Mn^{2+}, etc), we have studied similar synthetic lamellar compounds. Layered double hydroxides, which are anionic clays, were selected for this comparison. X-band spectra of compounds with various Fe^{3+} molar contents, show a wide absorption around g=2 and no significant absorption near g=4. Iron is in a more symmetrical situation than in natural clays and strong spin-spin interactions were observed. Multi-frequency EPR and Mössbauer spectroscopies at room and low temperatures, were used to describe the actual nature of couplings between d-elements.

1. INTRODUCTION

Transition metals have been widely used as paramagnetic probes for studying the local structure of various materials, especially poorly crystalline or amorphous materials. In the case of natural clays, Fe atoms are often present either as oxides or as substitutes for Al atoms in the clay hexagonal framework [1]. Electron Paramagnetic Resonance (EPR) [2] and Mössbauer spectroscopies [3] have been already used to understand the Fe^{3+} local environment. Concerning EPR study of kaolinites, there have been many papers which concluded that Fe^{3+} occupies different sites [4]. Indeed, for most of the kaolinites, the EPR signal with a spectroscopic factor g near 4, can be separated in two sub signals ascribed to different sites. Nevertheless, some kaolinites are more similar to smectites [5], at least concerning their EPR characteristics.

Even if EPR spectra of smectites are well known, the local structure of Fe is difficult to discuss based on results obtained by classical X-band spectroscopy [6]. The environment of Fe is very different from that in kaolinites: the latter have a TO structure (one tetrahedral layer + one octahedral layer), but smectites have a TOT structure. This structural difference could explain the complex form of the X-band EPR spectra, and the difficulty in analysing these spectra. More precise information could be obtained by performing EPR measurements at various frequencies. Actually, EPR experiments carried out at higher frequencies, enable a better analysis of the Fe^{3+} local structure.

We have selected several montmorillonites from different countries, with different Fe contents. In order to get more general information on the Fe environment in clays, we have prepared Layered Double Hydroxides (LDH). We have studied LDH which are hydrotalcite-like materials. These anionic clays have the general composition $[M^{2+}_{1-x} M^{3+}_x (OH)_2]^{x+}[A^{n-}]_{x/n}$ m H_2O where M^{2+} and M^{3+} are metal cations and A^{n-} denotes anions. Their structure consists of brucite-like layers, $Mg(OH)_2$, with a partial M^{2+}/M^{3+} substitution resulting in a net positive charge which is balanced by interlayer anions associated with variable amounts of water molecules. These compounds can be prepared with a controlled and variable Fe^{3+} content and can be used as references samples with Ni as M^{2+}, and Al and Fe as M^{3+}; these LDH are well known, especially for their catalytic properties[7].

For all the samples, the results in X-Band (around 9.8 GHz) were very complex and could not be explained using a classical perturbation method. By recording EPR spectra at different frequencies: X-Band ($\approx$9.8 GHz), W-Band ($\approx$95 GHz) and "3W"-Band ($\approx$285 GHz), other information could be obtained and the multi-frequency study appears to be a way to more correctly analyse iron sites in these materials. Mössbauer spectroscopy of ^{57}Fe provides some local information, principally concerning the electric field distribution.

2. SAMPLES AND METHODS

Samples :

We only discuss results obtained for 6 natural samples with different structural properties, which were previously well characterised. These samples are labelled as:
a) ARG sample: The raw ARG sample is a clay coming from Argelliers near Montpellier, in the north of Hérault (France). The X-band EPR spectra and chemical analysis of this clay and its <2µm fraction have been recently reported [6];
b) ALM1 sample: from Almeria "Los Troncos", Spain;
c) ALM2 sample: from Almeria, Spain, same origin as ALM1 but with different compound ratio (presence of Mn);
d) WYO sample is the well known Wyoming montmorillonite, from Upton, Wyoming, USA;
e) OTO montmorillonite from Otog, USA (ref WARD's 49 E2602);
f) CHE montmorillonite from Cheto, USA (ref WARD's 49 E2603);
g) LDH1 is a layered double hydroxide synthesised with $[Fe^{3+}] = 0.4\%$;
h) LDH2, the same as g) but with $[Fe^{3+}] = 1.5\%$;
i) LDH3 the same as g) but with $[Fe^{3+}] = 6.2\%$;

LDH samples were prepared by direct mixing of aqueous solutions of pure (3 N) nitrates $Fe(NO_3)_3$, $Al(NO_3)_3$ and $Ni(NO_3)_2$, and NaOH in adequate amounts ($M^{3+}/M^{2+} = 3$) at pH 10. After refluxing for 15 hours samples were dried at 80°C.

The samples were synthesised with the same nickel molar ratio ($\approx$40%) and three different Fe molar ratios: 6.2%, 1.5% and 0.4%. These three different Fe ratios cover the domain currently observed in many natural clays. LDH consist of layers built up with octahedral units. In these compounds no secondary phases are observed and Fe can substitute for Ni in octahedral sites,

Table 1 Results of local analysis of raw samples from SEM-EDS measurements (%mol).

Sample	Si	Al	Fe	Ni	Mg	Ca	Na	K
ARG	22.8	9.1	1.8	0	0.5	0.4	0.9	0.2
ALM1	26.4	5.3	0.7	0	0.8	0.5	1.2	1.2
ALM2	23.3	7.9	1.3	0	2.4	0.4	1.6	0.7
WYO	24.9	7.9	1	0	1.1	0.7	2.1	0.2
OTO	30.2	8.4	4.5	0	2.3	0.9	1.7	2.5
CHE	31.6	8.6	1.2	0	3.1	0.7	1.5	0.6
LDH 1	0	9.1	0.4	43.6	0	0	0	0
LDH 2	0	7.9	1.5	38.3	0	0	0	0
LDH 3	0	6.0	6.2	39.4	0	0	0	0

with variable ratios depending on the synthesis. For these reasons LDH could be used as reference samples for EPR experiments and comparisons with natural clays. Results of chemical analysis obtained by scanning electron microscopy using energy dispersive spectroscopy (SEM-EDS) are given in Table 1.

Sample preparation:

All the natural samples were studied at the raw state and after chemical treatment. These treatments were performed in order to remove organic compounds and non structural d elements (especially iron oxides and Mn^{2+} phases) often observed by EPR in natural clays. After removing organic compounds by preliminary washing with H_2O_2 (33%) at room temperature, the classical DCB treatment [8] was performed to remove iron oxides from clays. After drying samples were kept at 60°C in a drying oven to limit their hydration.

Experimental:

X-band EPR measurements were performed on a Bruker Elexys ER049 spectrometer and on a Bruker ER100. The experimental parameters were the following: 100KHz modulation frequency, 0.4mT modulation amplitude and a time constant of 20ms. High magnetic field measurements were performed at the High Magnetic Fields Laboratory of Grenoble (France) in a superconductoring magnet, permitting measurements at variable frequencies (100GHz≤ν≤300GHz) with 100KHz modulation frequency, 0.4mT modulation amplitude and a time constant of 20ms.

Powdered samples, for Mössbauer measurements were mixed in Apiezon grease in order to get an effective thickness of about 1mg/cm² of ^{57}Fe. The Mössbauer spectra were obtained in transmission mode at room temperature using a constant acceleration spectrometer with a 10mCi source of ^{57}Co in a Rh matrix. Data were transferred to a computer using EG&G Ortec multichannel scaling hardware and software. Metallic iron was used to calibrate the velocity scale and as a reference for the isomer shifts.

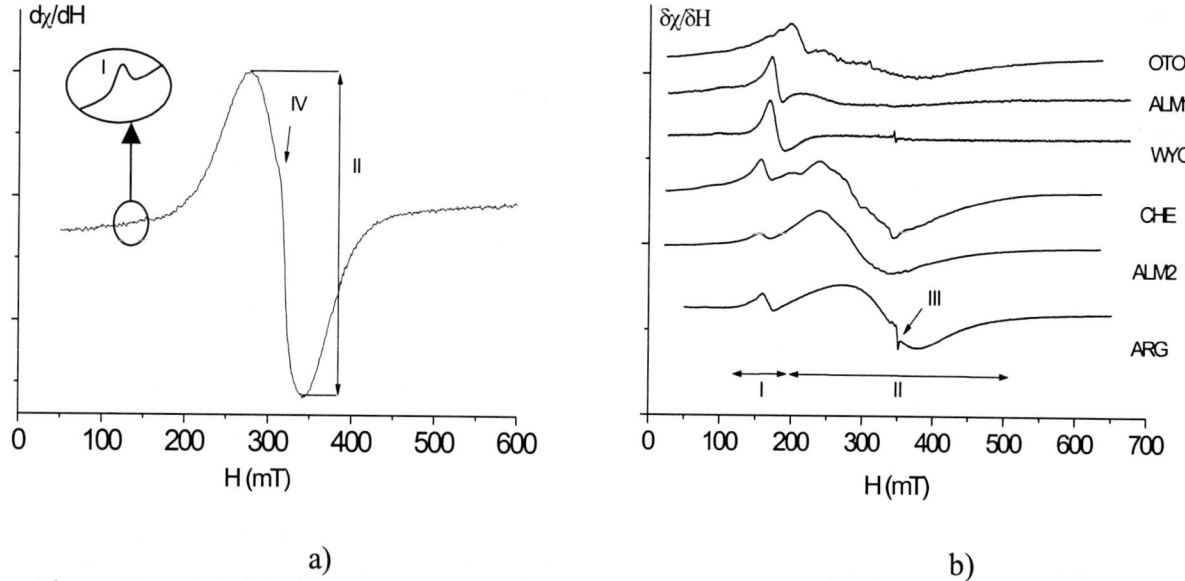

Figure 1.   a. X-band EPR spectra at room temperature of raw montmorillonite samples.
b. X-band EPR spectra at room temperature of a LDH3 sample.

3. RESULTS

Figure 1 shows the room temperature X-Band EPR spectra of the six raw montmorillonite. All montmorillonite samples give rise to three signals with variable relative amplitudes: the well resolved signal near g=4, arises from Fe in a low symmetry state (signal I) and is present in each sample. A large signal near g=2, that arises from the interaction between iron atoms (signal II), is present in all samples with variable intensity (high for ALM2 and ARG but smaller for WYO or ALM1). In the domain near g=2 we observed more complex signals with various origins such as 3d elements (Mn) in ALM2 or defects in quartz which were clearly observed in ARG and WYO (thin signal at g=2). The EPR characteristics of these secondary phases (signals III) have been described in detail in a previous study [9].

Concerning LDH samples (Fig.1b) 3 signals are also observed. Signals I and II have the same nature as in raw clays. An other signal (not well resolved) termed signal IV, is also observed near g=2. This signal is ascribed to Ni species present in this compound [10].
In LDH samples, signal I is much weaker which indicates that Fe occupy mainly sites with high symmetry. This signal for LDH3 sample is indicated in figure1b, but is not observed in LDH1 and LDH2 samples. Signal III is not observed due to the synthetic nature of the LDH (no impurities). Low temperature (4 K) X-band EPR measurements did not show significant change in EPR spectra shapes.

Measurements were systematically performed on all samples at 95 GHz and at 285 GHz. Because of the lower sensitivity of the spectrometer at these frequencies, the spectra were recorded at low temperature in order to increase the signal quality.
High frequency (285 GHz) EPR spectra are shown on figure 2 for samples WYO and LDH3. W-band results (ν=95 GHz) will be discussed in the following chapter.

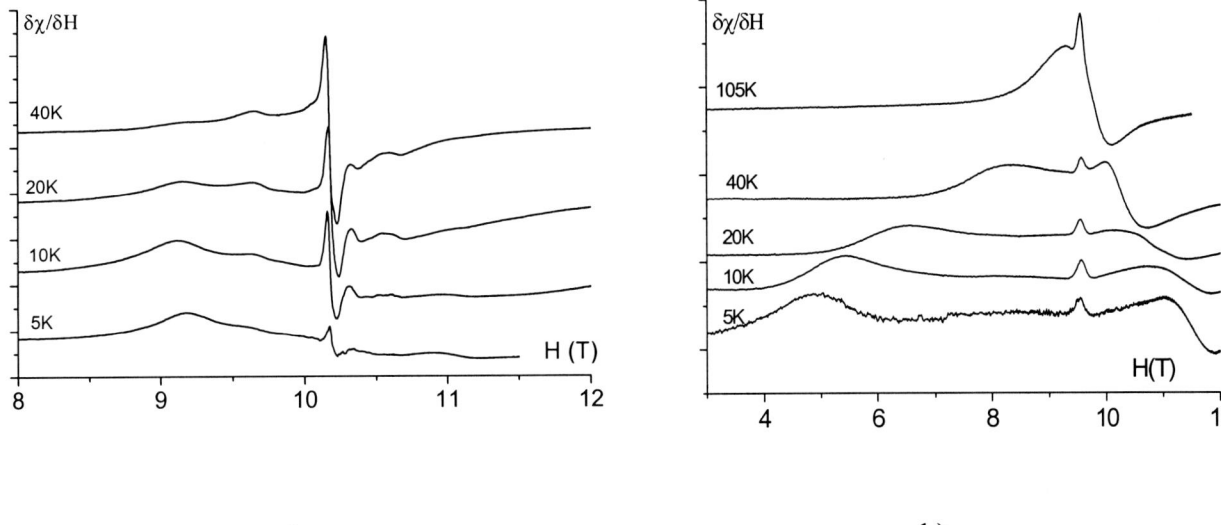

a) b)

Figure 2 "3W"-band EPR spectra at low temperature of a) WYO sample after chemical treatment and b) LDH 3 samples

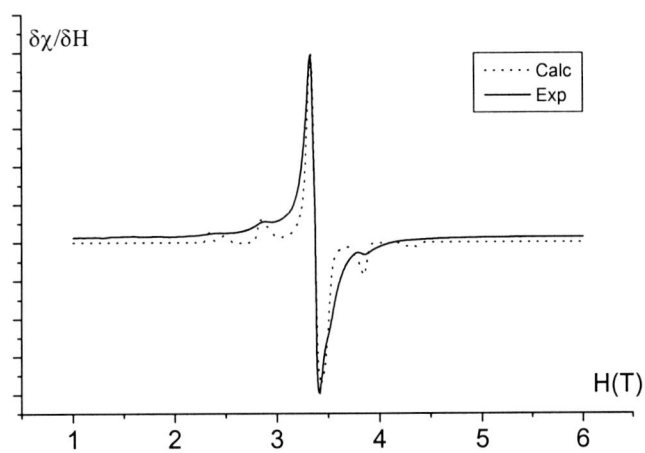

Figure 3 W-band EPR WYO spectra measured at 10K and calculated (perturbation theory) with one single D value : D=0.27cm^{-1} and : g_x=2.01 g_y=2.025 g_z=2.04

4. DISCUSSION

The classical Hamiltonian used to describe the EPR spectra of high spin Fe^{3+} (d^5, S=5/2) includes the electron Zeeman effect and the fine structure effect:

$$H = g \cdot \beta \cdot \vec{B}_0 \cdot S + D\left[S_z^2 - \frac{1}{3}(S(S+1))\right] + E(S_x^2 - S_y^2) \qquad (1)$$

Higher order terms (not shown in equation 1) are generally assumed to be small compared to the quadratic terms.

Parameters D and E measure the deviation of the ion crystal field from ideal tetrahedral or octahedral symmetries for which both D=0 and E=0. The ratio λ=E/D is such that λ=0 for axial symmetry, and rhombic distortion is measured by λ with 0<λ<1/3.

Concerning Fe^{3+}, values of D have often the same magnitude as the spectrometer energy in the X-Band (near 0.3cm^{-1}), whereas for Mn^{2+}, D is around 0.02cm^{-1}. In this context, a perturbation method cannot be used for simulations of X-band spectra. The D value is about 10% of hν at 95 GHz and about 3.5% of hν at 285 GHz. In both cases, the classical simulation software using perturbation theory approximations can be successfully used.

For natural clays, five extrema are clearly observed in spectra recorded at 285 GHz (Fig.2a). These signals are due to the 5/2 spin of Fe^{3+}. At 95 GHz (Fig.3), these signals are also observed, but are not so clearly separated as at higher frequency. The energy levels corresponding to these absorptions can be calculated according to the following expressions [11]:

$$E|\pm 5/2\rangle = \pm 5/2\ g\beta H + 10D/3$$
$$E|\pm 3/2\rangle = \pm 3/2\ g\beta H - 2D/3 \quad (2)$$
$$E|\pm 1/2\rangle = \pm 1/2\ g\beta H - 8D/3$$

The central line of the spectra corresponds to the transition $E_{(+1/2)} \leftrightarrow E_{(-1/2)}$ and does not depend on D whereas all the other transitions depend on D. We obtain the result shown in Figure 3 by calculating the spectrum with a single value of D=0.27cm^{-1}.

The central line is well explained using an axial g tensor ($g_{//}$=2.01 and $g_{\perp}$=2.04). The field positions of the other extrema are also well simulated with D=0.27cm^{-1}. Nevertheless, this hypothesis cannot successfully explain the full shape of the spectra. It was verified that two sites (like in kaolinites [3]), can no longer produce a good simulation. The only way to succeed is to take into account a distribution of D values. Some previous studies have considered this approach. A distribution law was established by Czjzek [12,13], for amorphous solids with random ionic coordination yielding an analytic expression of the distribution of the electric field gradient (EFG). This distribution law has been used for fluoride glasses, considering the analogy between the EFG and the fine structure interaction in EPR tensor component expressions [14].

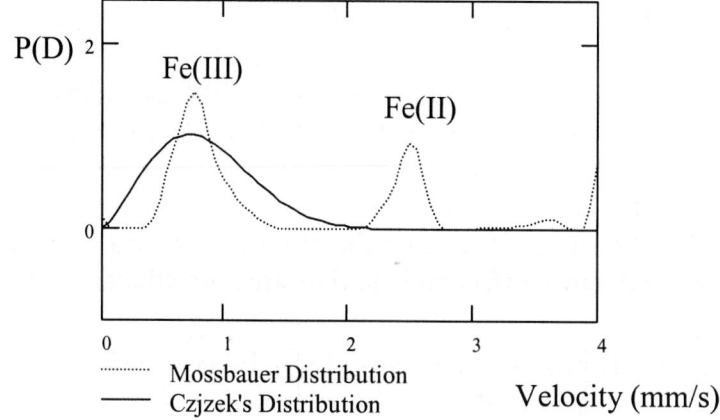

Figure 4 WYO sample: Distribution law of Mössbauer quadrupolar splitting and Czjzek's distribution law

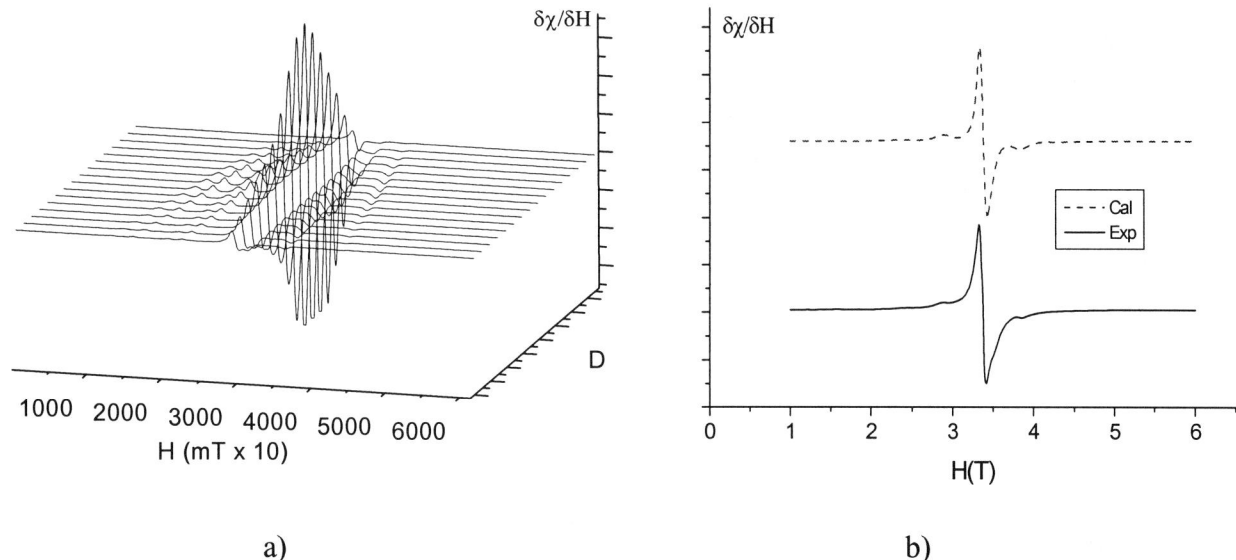

Figure 5: a) : Distribution of EPR spectra with Mössbauer distribution as lineshape
b): EPR Simulation (95GHz) with a distribution of D values with
$D_m = 0.24 cm^{-1}$, $D_{min} = 0.15 cm^{-1}$ and $E/D = 0.25$. The g components are: $g_x=2.01$ $g_y=2.025$ $g_z=2.04$ and line width : $\sigma = 0.6$ T

Figure 4 compares the Czjzek distribution and the experimental Mössbauer quadrupole splitting distribution in the case of the WYO sample. The Mössbauer distribution shows 2 signals: the first one corresponds to Fe^{3+}, the second which corresponds to Fe^{2+}, is not used for EPR distributions, because Fe(II) cannot be detected by EPR spectroscopy. As montmorillonites are more organised than glasses, Figure 4 shows that this distribution law is too large for the WYO sample (the same conclusion is valid for the other clays). According to previous papers [15] it is known that EPR and Mössbauer studies yield a similar distribution of crystal field parameters and of quadrupole splittings. We used a distribution law P(D) based on Mössbauer results. The experimental law in Figure 4 was used to obtain the distribution law P(D). For this purpose this law is centred at the mean value D_m, which is the D value used for simulation in Figure 3 (corresponding to 50% of the distribution). This distribution was used with a EPR simulation performed using $0 < D < 0.5 cm^{-1}$ values. This distribution applied to the EPR spectra is the final distribution curve used to simulate our spectra (Fig.5a). For this simulation $\lambda = 0.25$ was determinated using the Frequency Independent Resonance Diagram, (see for example [16]). The law P(D) is obtained from the Mössbauer results and the minimum value D_{min} is fitted after simulation of the spectra. The two parameters D_m and D_{min} are characteristic of the sample. As an example, Figure 5a shows the different elemental spectra used in the case of WYO sample; the final spectrum, shown in Figure 5b, is similar to the experimental one. Using the same method, D_m and D_{min} can be obtained for all the natural clays samples. These results indicate the possibility to classify montmorillonite clays as a function of their D distribution, which reflect the state of distortion around Fe sites. At 285 GHz and T<20K (Fig.2a), the spin population of the levels are not equal and the Boltzmann law must be taken into consideration for the simulation of the spectra. However, the thermal difference of population allows the sign of the D parameter to be determinated. The presence of signals, which are more intense for transitions at low magnetic field indicates D<0. The graphical estimation of D, obtained by measuring the distance between two signals (see equation 2) agrees with the values used for simulation at 95 GHz. LDH spectra recorded at 285GHz (figure 2b) are explained by the simultaneous presence of paramagnetic iron and nickel species. When the temperature is reduced below 110K, the anisotropy of the g of the Ni signals increases as T→0, whereas the signal near g=2 is not modified. This phenomenon, already observed in $Li_{1-z}Ni_{1+z}O_2$ compounds [17] confirms the presence of Ni^{2+}-O-Ni^{3+} in LDH materials. The wide g=2 signal observed at X-band, is due to Ni^{2+} species, and is too wide to be observed at higher frequencies. In contrast, the g=2 signal confirms the high symmetry of the Fe sites, in these materials. Mössbauer spectrometry also indicated the presence of Fe^{2+} in LDH, as 10% of total Fe. These observations confirm an exchange between Ni and Fe leading to local Ni^{2+}-O-Ni^{3+} species.

5. CONCLUSIONS

We have shown by analysis of the high frequency spectra of montmorillonite clays that only one type of Fe sites are observed, characterised by the mean value of parameter D. Moreover, the D distribution around this value is correlated to the local ordering around Fe sites and a possible correlation with the Fe^{3+} ratio has been shown. These clays could be classified by their distortion level. This should be confirmed by further Mössbauer and EPR studies. The difference between EPR results of kaolinite and bentonite is obvious, and multi-frequency EPR indicates the local ordering around Fe sites, in montmorillonites. A first comparison between anionic and cationic clays has been made and we have shown that high symmetry Fe sites are present in HDL samples. The presence of Ni species modifies the EPR spectra and the high frequency spectra confirmed the presence of Ni^{3+}-O-Ni^{2+} units.

REFERENCES

1. B.R Angel and W.E.J Vincent, Clays Clay Min 26 (1978) 4 263.
2. J.P. Jones, B.R. Angel and P.L. Hall, Clay Min 10 (1974) 257.
3. R.E. Meads and P.J. Malden, Clay Min 10 (1975) 313.
4. J.M. Gaite, P. Ermakov and J.P. Muller, Phys. Chem. Mineral. 20 (1993) 242.
5. D. Bonnin, S. Muller and G. Calas, Bull. Minéral. 105 (1982) 467.
6. Y. Bensimon, B. Deroide and J.V. Zanchetta, J. Phys. Chem. Solids. 60 (1999) 813.
7. N.T. Dung, D. Tichit, B.H. Chiche and B. Coq, Appl. Catal. A: General. 169 (1998) 179.
8. O.P. Mehra and M.L. Jackson, Clays and Clay Minerals 7 (1960) 317.
9. Y. Bensimon, B. Deroide, F. Dijoux and M. Martineau, J. Phys. Chem. Solids. 61 (2000) 1623.
10. J.W. Orton, P. Auzins, J.H.E. Griffiths and E. Wertz, Proc. Phys. Soc. 78 (1961) 554.
11. J. Vedrine, Advanced Chemicals Methods for Soil and Clay Minerals Research (1980) 331.
12. G. Czjzek. Phys. Rev. B 25 (1982) 4908
13. G. Czjzek, J. Fink, F. Gotz, H. Schmidt, J.M.D. Coey, J.P. Rebouillat and A. Lienard; Phys. Rev. B 23(1981) 2513.
14. C. Legein, J.Y. Buzaré, J. Silly and C. Jacoboni. J. Phys. Condens. Matter. 8 (1996) 4339.
15. C. Legein, J.Y. Buzaré, B. Boulard, C. Jacoboni, J. Phys. Condens. Matter. 7 (1995) 7 4829.
16. F.E. Mabbs and D. Collison, EPR of d Transition Metal Compounds. Elsevier (1992).
17. A.L. Barra, G. Chouteau, A. Stepanov, A. Rougier and C. Delmas, Eur. Phys. J. B. 7 (1999) 551.

Synchrotron x-ray study of hydration dynamics in the synthetic swelling clay Na-fluorohectorite

E. DiMasi,[a,*] J. O. Fossum,[b] G. J. da Silva[c,©]

[a]Physics Department, Brookhaven National Laboratory, Upton NY 11973-5000, USA [1]
[b]Department of Physics, Norwegian University of Science and Technology, N-7491 Trondheim, Norway
[c]Instituto de Física, Universidade de Brasília, 70910, Brasília-DF, Brazil [2]

We present time-resolved synchrotron x-ray diffraction from the swelling clay Na fluorohectorite during controlled hydration and dehydration. A comparison of bulk and surface scattering reveals that the time dependence of basal Bragg peak positions and intensities has two components, which come into play at distinct temperatures. Intercalation of water into the crystal structure commences at a temperature of about 60°C, on a time scale on the order of 1 hour. By contrast, percolation of water into the porous medium has a characteristic time constant of 3-4 h. This is the rate limiting process for hydration of the interior of the clay for T>40°C. We suggest that the temperature-dependent percolation step may account for some of the hysteresis reported in earlier diffraction studies of the hydration of similar systems.

1. INTRODUCTION

The ability of smectite clays to absorb water and swell is long known, and some of the earliest x-ray diffraction studies focussed on the changes in basal spacing that accompanied this absorption [1]. It was soon concluded that water was able to enter between the silicate layers in these clays, intercalating into the crystal structure. In the hydration of a bulk clay sample, this process accompanies the adsorption onto the exterior of the particles. Due to their small size, the available external surface area of the clay particles may be comparable to the internal area. For this reason, water adsorption studies of bulk clay samples must always attempt to take both processes into account. This is often complicated by sensitivity to inital conditions or hydration history [2]. And of course, internal and external water adsorption may not have the same dependence on well controlled variables such as the size or charge of the exchangeable cation in the interlayer space. This fact can make systematic trends in cation substitution difficult to observe.

In this paper we present time-resolved synchrotron x-ray diffraction studies of a synthetic swelling clay, Na-fluorohectorite, conducted during controlled hydration and dehydration. Two aspects of this experiment distinguish it from previous work. First, the x-ray scans can

---

[*]Author to whom correspondence should be addressed. Email: dimasi@bnl.gov
[©]Present address: Dept. of Physics, Norwegian University of Science and Technology

be taken quickly (several per minute), revealing the time dependence of Bragg peak positions and intensities as water enters the clay. Second, we can conduct the scattering measurement either in transmission through the bulk sample, or in reflection from the surface. By this means we can detect intercalation either as it occurs at the surface, a direct result of the prevailing environment, or as it occurs in the bulk. We will show that in the latter case the percolation of water or vapor through the porous medium affects the apparent time scale and temperature dependence for the clay hydration.

2. EXPERIMENTAL DETAILS

Fluorohectorite clay was purchased from Corning [3] and ion-exchanged through dialysis to produce Na-fluorohectorite having the nominal chemical formula per half unit cell $Na_{0.6}$-$(Mg_{2.4}Li_{0.6})Si_4O_{10}F_2$. The wet clay powder was compressed in a furnace to make monolithic solid samples with the flat clay particles aligned along their stacking directions to within 5-10 degrees. Rectangular pieces were cut from the furnace-pressed disk for x-ray diffraction along the stacking direction (c axis), either in reflection from a flat surface 5-10 mm long in the beam direction, or in transmission through 2-3 mm of material. For the transmission measurements, the surface signal was blocked using a thin cover of Pb tape.

The samples were mounted on a copper heating/cooling block with a thermocouple imbedded near the sample. Temperature control in the range 0-130°C was achieved with a combination of a thermoelectric cooling element, a commercial heating element connected to a Lakeshore 321 temperature controller, and external cooling water circulated through the bottom of the block. The sample block was contained within a vacuum-tight beryllium metal can which served as the x-ray window. A peristaltic pump was used to circulate air through the chamber and subsequently either filter it through a dessicant column, or bubble through saturated K_2SO_4 salt water, in a closed loop. Under these conditions the relative humidity as measured by a sensor in the room temperature section of the system stabilized at 0.5% and 97% respectively.

X-ray scattering measurements were carried out at beamline X22A at the National Synchrotron Light Source at Brookhaven National Laboratory (Upton NY, USA), with an x-ray wavelength of 1.24 Å. The beam was focussed to a spot size of 0.5×0.5 mm^2. The resolution in momentum transfer q_z of 0.005-0.010 Å^{-1} was established by 0.5 or 1.0 mm slits before the NaI single channel detector, 600 mm from the sample. This arrangement was optimal to get the maximum intensity with minimal instrumental broadening.

3. RESULTS

Diffraction patterns along the (001) direction show that three stable hydration states are found in Na-fluorohectorite, as has been reported previously [4]. From the dehydrated state with basal spacing c = 9.7 Å (Fig. 1a), the lattice spacing swells to 12.4 Å when the

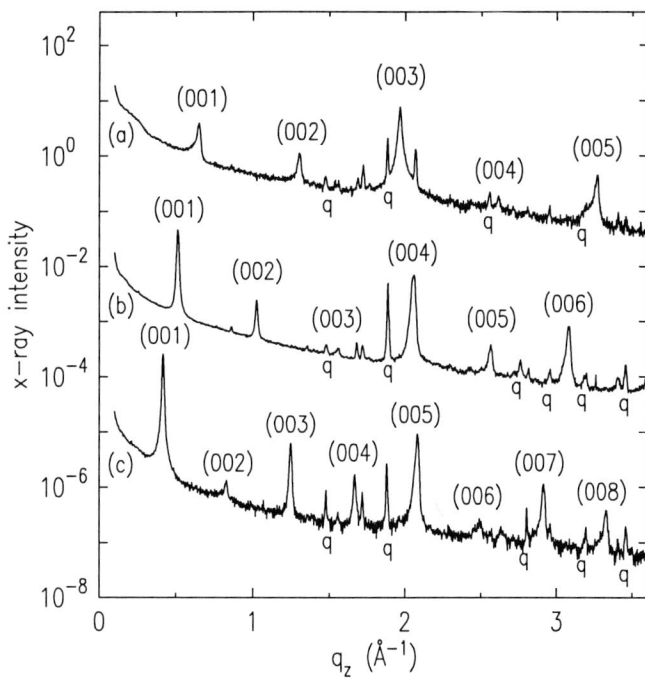

Figure 1: X-ray diffraction patterns taken in transmission through the bulk Na fluorohectorite samples in three different hydration states. Intensities are normalized on the same scale, but shifted for clarity. Basal peaks are indexed. Sharp peaks labeled "q" are from quartz impurities. (a): dry air, 130°C. (b): room temperature and humidity. (c): humidified air, 10°C.

temperature is reduced to the range 30-60°C, depending on how much water vapor is available (Fig. 1b). At lower temperatures under humid conditions, additional intercalated water increases the basal spacing to 15.0 Å (Fig. 1c). And as with other swelling clays [5], the temperatures at which the various hydration states may be found is variable, dependent on the sample environment and history.

These hydration states are generally referred to in terms of the number of intercalated water layers which are supposed to incorporate into the structure — nominally zero, one or two layers in the case of Na-fluorohectorite. This identification would be adequate in the case of static, stable hydration states. However, it is recognized that structural changes occur during hydration [6]. These may include local rearrangements in water molecule positions [7-9], as well as the alternating intercalation patterns originally described by Hendricks and Teller [10,11].

Evidence for both, in some cases fleeting evidence, is obtained by in-situ diffraction during hydration. Figure 2 shows scans taken through the (001) peak positions at low q_z, while a sample in humid atmosphere is allowed to cool from a dry state at 130°C to below room temperature. In most of these scans, intensity is found simultaneously at two different (001) peak positions. This shows the extent to which regions of clay, probably separate domains or particles, coexist in different hydration states. During the process of hydration, intensity on "dryer" peaks at higher q_z gives way to those corresponding to "wetter" states. We can also see that the peak at $q_z = 0.51$ Å^{-1} shifts in position during

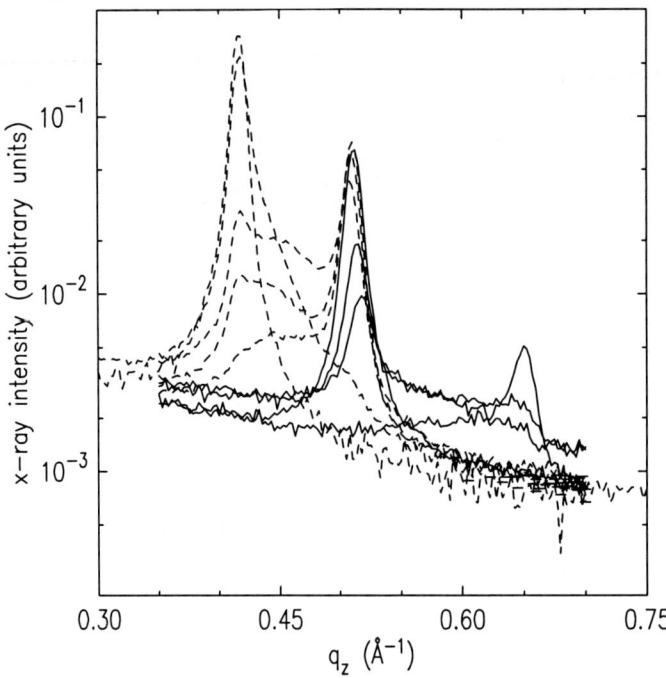

Figure 2: X-ray diffraction scans taken in transmission during hydration. Solid lines: initial, high temperature traces, with intensity at zero-layer and one-layer (001) peak positions. Dashed lines: low temperature scans indicating hydration from one- to two-water-layer states.

hydration, showing that the lattice constant associated with the one-water-layer hydration state is changing to accommodate some structural reorganization. And only for the one- and two-water-layer states do we find the intensity between the Bragg peaks (at $q_z = 0.45$ Å^{-1}) that is indicative of irregular alternate layer intercalation. In short, the complete crystallographic picture promises to be quite involved, and to contain a wealth of information regarding motions of water molecules during the hydration process — analysis is underway.

The present paper will focus on the time scales for hydration, which we obtain from the time-dependent peak intensities measured during hydration. For this purpose, we take the total integrated intensity of each Bragg peak to be proportional to the volume of clay in the corresponding hydration state. This simplification ignores changes in the structure factor which may accompany hydration: for example, we assume that although the peak at $q_z = 0.51$ Å^{-1} shifts in position, its magnitude relative to the corresponding sample volume is unchanged. Our analysis also requires that hydrated regions be structurally independent of each other. This is clearly not the case for the transition between the one- and two-water-layer states, which show evidence for Hendricks--Teller type alternating intercalation. But it is a plausible description of the hydration from the zero- state to one-water-layer state, where no peak at an intermediate position is observed.

Hence, we are concerned principally with the behavior of zero-water-layer (001) and (003) peaks, corresponding to the dry regions in the sample. We also investigate the

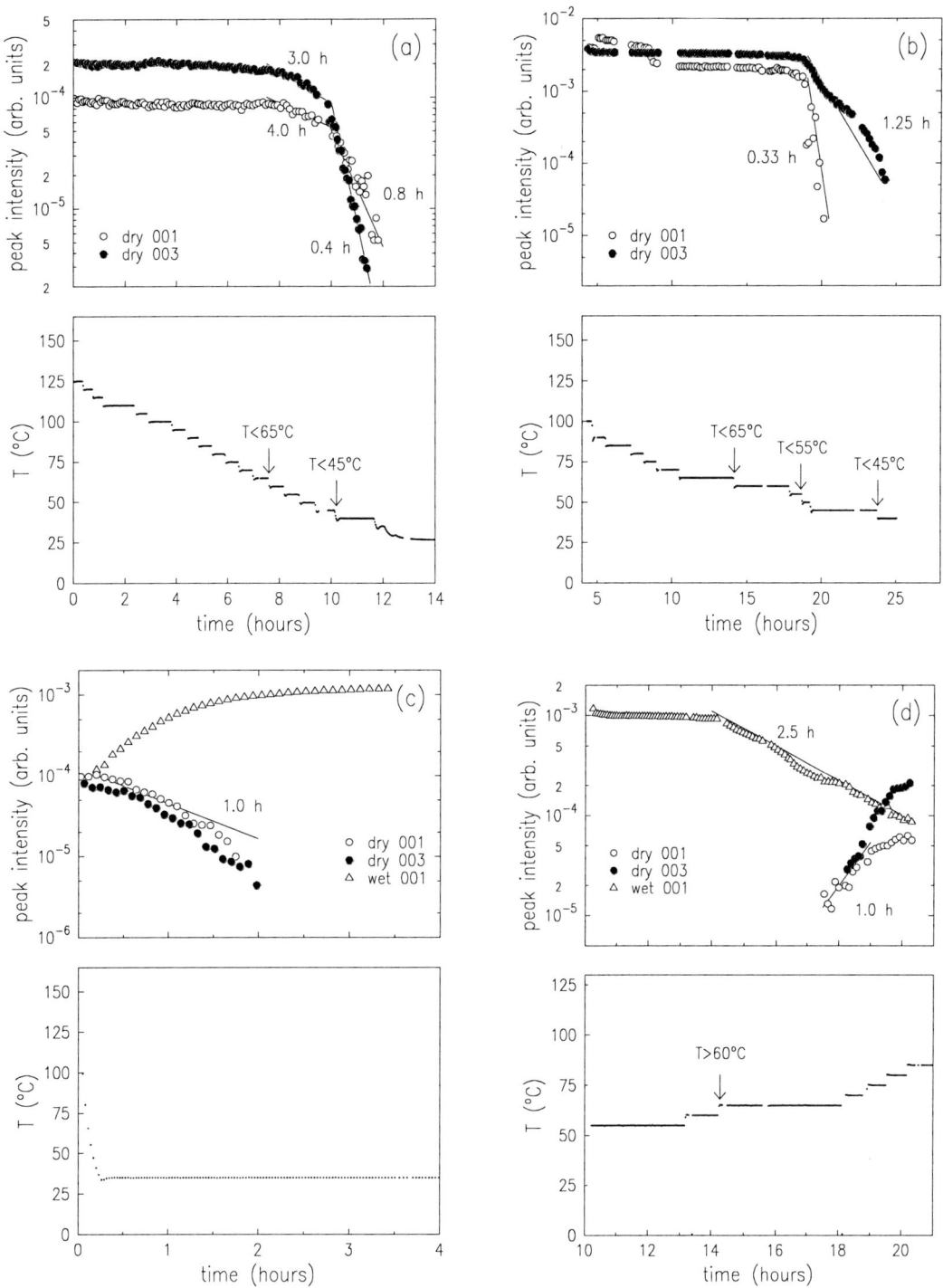

Figure 3: Time dependence of Bragg peak integrated intensities (top panels) as temperature is controlled (lower panels). Solid lines are guides for the eye, labelled with exponential time constants in hours. Labels "dry" and "wet" refer to zero- and one-water-layer Bragg peaks. (a) Slow cool, bulk. (b) Slow cool, surface. (c) Fast cool, bulk. (d) Slow heat, bulk.

one-water-layer (001) peak in the case of dehydration towards the zero-water-layer state. Integrated peak intensities were determined from Lorentzian fits on a linear background.

Figure 3 shows examples of integrated peak intensities monitored as a function of time (upper panels), while hydration was controlled by stepping the temperature (lower panels). Our aim is first to identify the temperatures at which hydration proceeds under these conditions, and then to determine the time constant for hydration in that interval. Figure 3(a) shows data taken in transmission through the bulk sample during a slow cool. For temperatures above 65°C, the Bragg peak intensities at the zero-water-layer (001) and (003) positions change very slowly, requiring >30 hours to fall to 1/e of the initial value. When T is reduced below 65°C, however, intensity coming from these dry regions in the sample begins to decrease. As shown by guides in the figure, the time dependence can be roughly described by an exponential decay on the order of 3-4 hours. Surprisingly, further decrease of temperature has no obvious effect until T<45°C, at which point hydration proceeds much more rapidly, with time scales <1 hour.

This observation of two significant temperatures, each with their own time constant, prompts us to consider how two different mechanisms may contribute to the hydration. An obvious consideration is the transport of water or vapor through the pores between the particles that make up the bulk clay sample. Such percolation is necessary to bring water into the sample, independent of the microscopic intercalation process that allows water molecules to enter the interlayer spaces in the clay crystallites. To test this idea, we conducted a similar experiment in reflection from the surface of the sample. Here, we directly observe the microscopic intercalation that influences the lattice spacing, without the limitation of transport through the bulk. Figure 3(b) shows that fast hydration begins at 55°C, with no need to wait until the temperature has been lowered to 45°C. This strongly suggests that the slower time scale belongs to bulk water transport through the porous medium, a process that limits the rate of bulk hydration at temperatures above 45°C.

To ensure that fast hydration was really the result of the drop below 45°C, and not simply the passage of sufficient time, we quenched the dry sample rapidly to 35°C in humid air, while measuring scattered intensity from the bulk (Figure 3(c)). It is clear that fast hydration commences right away at the lower temperature, even in the interior of the sample. The time dependence of the one-water-layer (001) peak shows that in about two hours, hydration of the sample is essentially complete.

Finally, to assess the temperature hysteresis, we subjected the hydrated sample to a slow heating profile, while taking data in transmission through the bulk. As Figure 3(d) shows, dehydration begins at about 60-65°C. This is evident from the decrease in intensity on the one-water-layer (001) peak, though it is several hours before any peaks corresponding to dry clay become visible.

4. DISCUSSION

We characterized the time dependence of the peak intensities in terms of exponential time constants. This assessment has no precise quantitative basis at present, since some data are nonlinear, and since Bragg peaks nominally belonging to the same clay volumes — the zero-water-layer (001) and (003) peaks — do not have identical time dependences in most regions. We expect that the discrepancies will be resolved by crystallographic modelling, in which all peaks will be described by a single structure at each time point. By such analysis, the time dependence of structural parameters can be tracked and more profitably interpreted.

Nevertheless, our qualitative observations are quite clear. Water is found to intercalate on time scales of about an hour for temperatures below 55-60°C. But for temperatures above

45°C, the hydration is rate limited by percolation into the bulk, and proceeds three to four times more slowly. At the present, we have no physical model to explain this lower temperature limit. We are currently conducting similar measurements from Ni-fluorohectorite for comparison.

On heating, hydration also commences at about 60°. Therefore, we observe a temperature hysteresis of at most about 5°C. By contrast, an x-ray diffraction study of Na vermiculite has revealed a much larger hysteresis of about 26°C in temperature cycling: the zero-water-layer peak appeared at 58°C upon heating, and persisted to 32°C on subsequent cooling (at a water vapor pressure of about 20 torr) [5]. However, the notable difference between the two studies is that for the Na vermiculite experiment, the system was allowed to stabilize for 6--12 hours at each temperature. Our experience with Na-fluorohectorite calls into question the notion of "stability" of a hydration state at a given temperature. As Figure 3(a), shows, the peak intensities are always changing; the temperature interval simply determines how fast this happens. Observation of temperature hysteresis is such systems probably depends strongly upon the time-temperature profile. It would be interesting to test this generalization with a greater variety of comparable studies.

Much more information is available in the literature regarding hysteresis in water adsorption isotherms. Indeed, the combination of water adsorption isotherm and x-ray diffraction measurements has laid a good foundation for the understanding of long-studied clays such as montmorillonites [2,12-14]. Even when the isotherms are hysteretic or have shapes that cannot be described by simple models, it has been possible to correlate steps or kinks in the isotherm with stepwise changes in the c-axis lattice constant. In the interpretation of these data, previous workers have also described two contributions to hysteresis [15]: that intrinsic to the process of physically pushing the smectite layers apart on the microscopic scale; and that resulting from other physical factors on the scale of the particle morphology. It seems possible that the latter, extrinsic hysteresis could be related to the percolation effect we describe here. It is difficult to follow up on this idea since hysteresis in temperature might not be directly comparable to hysteresis in adsorption isotherms. We suggest that time-resolved bulk and surface diffraction conducted during isotherm measurements may shed some light on the processes involved.

ACKNOWLEDGEMENTS

The National Synchrotron Light Source is supported under USDOE Contract No. DE-AC02-98CH10886. J.O.F. acknowledges support from the Norwegian Research Council Project for Synchrotron Related Support, Project # 138368-431.

REFERENCES

1. S. B. Hendricks, R. A. Nelson, and L. T. Alexander, J. Am. Chem. Soc. 2 (1940) 1457, and references therein.
2. R. W. Mooney, A. G. Keenan, and L. A. Wood, J. Am. Chem. Soc. 74 (1952) 1367.
3. Corning Incorporated, Corning NY 14831.
4. P. D. Kaviratna, T. J. Pinnavaia, and P. A. Schroeder, J. Phys. Chem. Solids 57 (1996) 1897.
5. N. Wada, D. R. Hines, and S. P. Ahrenkiel, Phys. Rev. B 41 (1990) 12895.
6. R. Calvet and R. Prost, Clays Clay Miner. 19(1971) 175.

7. G. Sposito, R. Prost and J.-P. Gaultier, Clays Clay Miner. 31 (1983) 9.
8. J. J. Tuck, P. L. Hall, M. H. B. Hayes, D. K. Ross, and J. B. Hayter, J. Chem. Soc. Faraday Trans. 81 (1985) 833.
9. N. T. Skipper, M. V. Smalley, G. D. Williams, A. K. Soper, and C. H. Thompson, J. Phys. Chem. 99 (1995) 14201.
10. S. Hendricks and E. Teller, J. Chem. Phys. 10 (1942) 147.
11. P. G. Slade, P. A. Stone, and E. W. Radoslovich, Clays Clay Miner. 33 (1985) 51.
12. R. W. Mooney, A. G. Keenan, and L. A. Wood, J. Am. Chem. Soc. 74 (1952) 1371.
13. K. Norrish, Discus. Faraday Society 18 (1954) 120.
14. A. C. Zettlemoyer, G. J. Young, and J. J. Chessick, J. Phys. Chem. 59 (1955) 962.
15. D. A. Laird, C. Shang, and M. L. Thompson, J. Colloid Interface Sci., 171 (1995) 240, and references therein.

Experimental weathering of biotite, muscovite and vermiculite: a Mössbauer spectroscopy study.

Embaie A. Ferrow

Lund University, Institute of Geology, Department of Mineralogy & Petrology, Sölvegatan 13, SE-22362 Lund, Sweden

Biotite, muscovite and vermiculite were treated in HCl and H_2SO_4 solutions at pH ranging between 7 and 1. The solid residue was studied using Mössbauer spectroscopy at room temperature. The spectra were fitted assuming a quadrupole splitting distribution, QSD. Different fitting models with different combinations of Gaussian components for the Fe sites were tested and the models with χ^2 near 1.0 ± 0.1237 and with the least number of free parameters were adopted as working models. The models were then used to fit the experimental products of each group. If a model fails to describe a spectrum of a sample in the group then this is taken to signal a significant crystal chemical change induced by weathering.

All three minerals are oxidized during experimental weathering and the correlation between oxidation ratio and pH is modeled using a Weibull function. The rate of oxidation in biotite is higher than that in muscovite and vermiculite. The average center shift, CS, and quadrupole splitting, QS, of the Fe sites in a group remain more or less constant during the experiment. However, the position of the peak QS of $^{[6]}Fe^{2+}$ in biotite is shifted to low energies with decreasing pH, a measure of the decreasing covalence character of the Fe/Mg-O bonds. In muscovite the major feature observed is the enhancement of the bimodal distribution of the QSD of $^{[6]}Fe^{2+}$ with decreasing pH.

1. INTRODUCTION

Recently, a series of studies on the products of experimentally weathered biotite, muscovite and vermiculite have been reported [1-3]. The studies focused on identifying the dissolution kinetics and mechanisms of weathering in biotite by monitoring the compositional change of the fluid and by high-resolution TEM imaging of the structure of the solid residue at equilibrium. In this study, the studies mentioned above are complemented by presenting a complete report on the Mössbauer analysis of natural biotite, muscovite, vermiculite and their weathering products. Different fitting models are discussed and the validity of the models for the different experimental products is shown.

2. MATERIALS AND EXPERIMENTAL TECHNIQUE

The biotite, muscovite and vermiculite samples used in this study are the same samples that were used for kinetic and TEM studies [1-3]. The biotite and the muscovite come from Moen and Tuftane in Norway. The vermiculite comes from an unidentified locality in Chile.

The dissolution experiments were performed in open systems in which the dissolved reaction products were continuously removed from the reactor loaded with the mineral of interest. The minerals were suspended in an acidified eluent and were kept in contact with an external eluent through a dialysis membrane. The experimental conditions used for the three mineral series are listed in table 1.

Sample	pH	Solvent	Time (hrs)
Biotite			
D0Bio	7	H_2O	12
D12Bio	4	HCl	1100
D31Bio	3	HCl	5217
D10Bio	3	HCl	2800
D9Bio	2	HCl	2800
D24Bio	3	H_2SO_4	2020
D23Bio	1	H_2SO_4	618
Muscovite			
D0Mus	7	H_2O	12
D6Mus	4	HCl	2800
D2Mus	3	HCl	1450
D1Mus	2	HCl	1450
D5Mus	1	HCl	2800
Vermiculite			
D0Ver	7	H_2O	12
D43Ver	5	HCl	4571
D35Ver	4	HCl	5217
D34Ver	3	HCl	5217
D42Ver	2	HCl	4571

Table 1. General experimental conditions for the weathering experiment in the biotite, muscovite and vermiculite series.

Mössbauer spectra were recorded at room temperature on a Mössbauer spectrometer with a 57Co/Rh source. A simple spectrometer interfaced to a personal computer is used for collecting the data [4]. Velocities were calibrated using a 25 μm thick Fe-foil. Because sheet silicates have preferred orientation, the tablets were set at an angle of 54.7^0 to obtain as symmetric absorption doublets as possible. The spectra were computer-fitted using the Voigt-based quadrupole splitting distributions, QSD, in program Recoil, a commercially available Mössbauer spectral analysis software package.

3. SPECTRAL ANALYSIS

A pair of broad but asymmetric absorption lines characterize the Mössbauer spectrum of biotite. For the same number of free parameters, Rancourt [5] demonstrated that in a biotite spectrum fitting is best obtained by assuming a distribution of quadrupole splitting. This produces better results than fitting assuming a Lorentzian doublet of the absorption lines. As in biotite, both muscovite and vermiculite spectra are composed of broad and asymmetric lines. Consequently, in this study all three minerals are fitted using QSD.

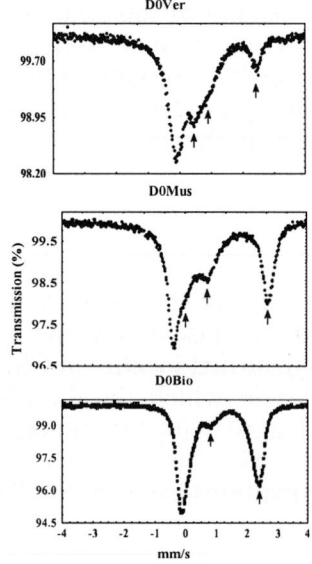

Figure 1. Mössbauer spectra of D0Bio, D0Mus and D0Ver

In QSD spectral analysis the shape and the parameters obtained depend on the Lorentzian width, Γ, the absorber texture and absorber thickness [6]. Γ is a well-defined physical parameter with a constant value provided the spectrum obtained is free from experimental artifacts such as absorber texture and thickness effects. In this study Γ is fixed at 0.224 mm/s for all sites, a value reported for the source in our laboratory by the suppliers. Texture effects are minimized by orienting the absorber at 54.7^o to the direction of the γ-rays and thickness effects are eliminated by using 5-10 mg/cm^2 of absorber. However, some textural effects, especially in muscovite, cannot be avoided fully and these are taken into account during fitting.

The fitting adopted here does not attempt to constrain the solution in parameter space for a given spectrum, but it attempts to use a consistent fitting model for all the experimental runs in a given series based on a model constructed with the fewest free parameters possible and with the reduced chi square, χ^2, near 1.0.

As the arrows in figure 1 show, in D0Bio only two absorption lines, one at 0.8 mm/s and the other at 2.4 mm/s, could be visually identified at the high-energy side of the spectrum. These are assigned to $^{[6]}Fe^{3+}$ and $^{[6]}Fe^{2+}$, respectively. In D0Mus two distinct absorption lines, a broad one at 0.70 mm/s and a sharp one at 2.73 mm/s, assigned to the high energy components of $^{[6]}Fe^{3+}$ and $^{[6]}Fe^{2+}$ and a shoulder at lower velocity, 0.13 mm/s, are identified. The shoulder could be related to $^{[4]}Fe^{3+}$ but it could also be related to contribution of the low-energy component of either $^{[6]}Fe^{3+}$ or $^{[6]}Fe^{2+}$. In D0Ver, three absorption lines are visible, two well-resolved at 0.46 mm/s and 2.44 mm/s and a broad shoulder at 0.77 mm/s. The first is assigned to $^{[4]}Fe^{3+}$, the second to $^{[6]}Fe^{2+}$, and the third to $^{[6]}Fe^{3+}$. Thus, there are two generalized sites in biotite, two or possibly three in muscovite and three in vermiculite. Each generalized site has its own continuous QSD with one or more Gaussian components. Different QSD fitting models based on the general sites defined above but having a different number of components have been tried to explain the spectra of D0Bio, D0Mus, and D0Ver.

In each series, only the model with χ^2 close to 1.0 ± 0.1237 and with the least number of free parameters is accepted as a starting model for the whole series. In fitting the spectrum of D0Bio, a model with one component for $^{[6]}Fe^{3+}$ and three sub-spectral components for $^{[6]}Fe^{2+}$ was taken. In fitting the spectrum of D0Mus, a model with one component each for $^{[4]}Fe^{3+}$ and $^{[6]}Fe^{3+}$ and two components for $^{[6]}Fe^{2+}$ was adopted. Finally, in D0Ver a model with one component each for $^{[4]}Fe^{3+}$ and $^{[6]}Fe^{2+}$ and two components for $^{[6]}Fe^{3+}$ gave the best fit.

Table 2. Mössbauer parameters of D0Bio and its oxidation products

Sample/pH	Fe-site	CS (mm/s)	QS (mm/s)	W (mm/s)	SP (%)	χ^2
D0Bio/7	Fe^{3+}	0.414	0.833	0.43	19.00	1.129
	$Fe^{2+}(1)$	1.072	2.680	0.11	81.00	
	$Fe^{2+}(2)$		2.423	0.19		
	$Fe^{2+}(3)$		2.025	0.35		
D12Bio/4	Fe^{3+}	0.449	0.799	0.38	20.91	1.048
	$Fe^{2+}(1)$	1.066	2.643	0.13	79.09	
	$Fe^{2+}(2)$		2.356	0.23		
	$Fe^{2+}(3)$		1.940	0.42		
D31Bio/3	Fe^{3+}	0.428	0.828	0.35	30.17	0.901
	$Fe^{2+}(1)$	1.062	2.586	0.17	69.83	
	$Fe^{2+}(2)$		2.240	0.28		
	$Fe^{2+}(3)$		1.801	0.58		
D10Bio/3	Fe^{3+}	0.419	0.848	0.36	30.09	0.964
	$Fe^{2+}(1)$	1.065	2.585	0.18	69.91	
	$Fe^{2+}(2)$		2.221	0.32		
	$Fe^{2+}(3)$		1.660	0.73		
D9Bio/2	Fe^{3+}	0.369	0.879	0.36	77.20	1.928
	$Fe^{2+}(1)$	1.130	2.376	0.33	22.80	
D24Bio/3	Fe^{3+}	0.437	0.832	0.37	26.61	1.058
	$Fe^{2+}(1)$	1.067	2.590	0.19	73.39	
	$Fe^{2+}(2)$		2.226	0.28		
	$Fe^{2+}(3)$		1.691	0.60		
D23Bio/1	$Fe3+$	0.358	0.859	0.37	54.80	1.224

	Fe^{2+}(1)	1.137	2.570	0.14	45.20
	Fe^{2+}(2)		2.192	0.26	
	Fe^{2+}(3)		1.690	0.79	

4. RESULTS

The Mössbauer parameter of D0Bio and its weathering products are listed in Table 2. The χ^2 shows that the model adopted describes the spectra of D0Bio and its weathering products quite well for the whole series except for D9bio, the sample where magnetite has been observed in the spectrum [3]. The major systematic variation registered in the weathering experiment of biotite is, however, the change of the oxidation ratio, $^{[6]}Fe^{3+}/(^{[6]}Fe^{2+}+ {}^{[6]}Fe^{3+})$ within the series. With decreasing pH the oxidation ratio increases. The Mössbauer parameters of D0Mus and its weathering products are listed in Table 3. The Table shows that, as in biotite, the oxidation ratio increases with decreasing pH. The model adopted for the series works well for D0Mus and its weathering products down to pH 2. For pH of 1 in D5Mus, however, the positions of the two Gaussian components in $^{[6]}Fe^{2+}$, approached each other closely such that a simpler model with a single component for the $^{[6]}Fe^{2+}$ site was sufficient for fitting its spectrum. Moreover, the shape of the distribution pattern changes with pH, as reflected in the enhanced bi-modality of the QSDs with decreasing pH (Fig. 2).

Table 3. Mössbauer parameters of D0Mus and its oxidation products

Sample/pH	Fe-site	CS (mm/s)	QS (mm/s)	W (mm/s)	SP (%)	χ^2
D0Mus/7	$^{[4]}Fe^{3+}$	0.231	0.300	1.85	26.70	1.049
	$^{[6]}Fe^{3+}$	0.310	0.828	0.49	26.50	
	Fe^{2+}(1)	1.167	3.123	0.18	46.80	
	Fe^{2+}(2)		2.821	0.65		
D6Mus/4	$^{[4]}Fe^{3+}$	0.231	0.300	1.99	26.65	1.065
	$^{[6]}Fe^{3+}$	0.352	0.751	0.41	26.50	
	Fe^{2+}(1)	1.168	3.121	0.20	46.80	
	Fe^{2+}(2)		2.200	0.42		
D2Mus/3	$^{[4]}Fe^{3+}$	0.231	0.300	1.89	26.58	1.135
	$^{[6]}Fe^{3+}$	0.343	0.779	0.42	26.94	
	Fe^{2+}(1)	1.165	3.120	0.19	46.47	
	Fe^{2+}(2)		2.232	0.47		
D1Mus/2	$^{[4]}Fe^{3+}$	0.231	0.300	2.03	26.79	1.054
	$^{[6]}Fe^{3+}$	0.315	0.784	0.46	30.90	
	Fe^{2+}(1)	1.166	3.128	0.19	42.30	
	Fe^{2+}(2)		2.211	0.15		
D5Mus/1	$^{[4]}Fe^{3+}$	0.231	0.300	2.81	26.58	1.215
	$^{[6]}Fe^{3+}$	0.375	0.310	1.52	39.91	
	Fe^{2+}(1)	1.191	3.063	0.65	33.52	

The Mössbauer parameters of the vermiculite weathering series are listed in Table 4. As in the biotite and muscovite series, the oxidation ratio in vermiculite increases with decreasing pH, although the degree of oxidation is very low. In fitting D0Ver, 16 parameters out of a total of 35 were refined. Fitting in batch, using the parameters of D0Ver as input and keeping all parameters fixed except for the Gaussian line width and the absorption areas gave acceptable χ^2 only for D43Ver. Further refinement of the center shift and keeping all other parameters fixed for all three sites reduced the χ^2 for the rest to acceptable values (Table 4).

5. DISCUSSION

We consider that in any of the three mineral groups studied, oxidation proceeds by the same mechanism(s) so long as the fitting model that was adopted holds. Once the model fails, then an important crystal chemical change has taken place that differs in its oxidation mechanism. With these assumptions in mind, the change in the concentration of ferrous and ferric iron, [Fe^{2+}] and [Fe^{3+}], has been plotted against the equilibrium pH in figure 3.

Previous studies have shown that during weathering, oxidation by deprotonation is a common mechanism [7, 3]. Assuming that this is the dominant mechanism in the three minerals studied and using the chemical composition of biotite as an example, the redox reaction at equilibrium can be written as:

$$(Ca_{0.02}Na_{0.04}K_{1.80})(Al0_{0.22}Fe^{3+}_{0.53}Fe^{2+}_{2.49}Mn_{0.01}Mg_{2.70})Si_{5.72}Al_{2.28})O_{20.20}(OH)_{3.8}F_{0.03} + Fe^{2+}_{-n}H_{-n}Fe^{3+}_{n} <\text{-}> (Ca_{0.02}Na_{0.04}K_{1.80})(Al0_{0.22}Fe^{3+}_{0.53+n}Fe^{2+}_{2.49-n}Mn_{0.01}Mg_{2.70})Si_{5.72}Al_{2.28})O_{20.20+n}(OH)_{3.8-n}F_{0.03} \quad (1)$$

Table 4. Mössbauer parameters of D0Ver and its oxidation products

Sample/pH	Fe-site	CS (mm/s)	QS (mm/s)	w (mm/s)	SP (%)	χ^2
D0Ver/7	$^{[4]}Fe^{3+}(1)$	0.197	0.538	0.31	26.40	1.012
	$^{[6]}Fe^{3+}(1)$	0.357	1.103	0.37	58.50	
	$^{[6]}Fe^{3+}(2)$		0.821	1.34		
	$Fe^{2+}(1)$	1.147	2.552	0.23	15.10	
D43Ver/5	$^{[4]}Fe^{3+}(1)$	0.197	0.538	0.31	26.34	1.096
	$^{[6]}Fe^{3+}(1)$	0.357	1.103	0.37	61.18	
	$^{[6]}Fe^{3+}(2)$		0.821	1.53		
	$Fe^{2+}(1)$	1.147	2.552	0.23	12.48	
D35Ver/4	$^{[4]}Fe^{3+}(1)$	0.197	0.538	0.34	26.67	1.104
	$^{[6]}Fe^{3+}(1)$	0.357	1.103	0.30	67.18	
	$^{[6]}Fe^{3+}(2)$		0.821	1.30		
	$Fe^{2+}(1)$	1.147	2.552	0.23	6.15	
D34Ver/3	$^{[4]}Fe^{3+}(1)$	0.197	0.538	0.31	26.20	0.983
	$^{[6]}Fe^{3+}(1)$	0.357	1.103	0.49	69.57	
	$^{[6]}Fe^{3+}(2)$		0.821	1.76		
	$Fe^{2+}(1)$	1.147	2.552	0.28	4.23	
D42Ver/2	$^{[4]}Fe^{3+}(1)$	0.197	0.538	0.32	26.87	1.081
	$^{[6]}Fe^{3+}(1)$	0.357	1.103	0.52	69.48	
	$^{[6]}Fe^{3+}(2)$		0.821	1.84		
	$Fe^{2+}(1)$	1.147	2.552	0.40	3.65	

Although in D0Bio the maximum magnitude of the exchange component, n, is 2.49, less than 50% of the available ferrous iron was oxidized during the experiment (Fig. 3).

There is a certain amount of octahedral ferrous iron, $[Fe^{2+}]^*$, in biotite that is not available for oxidation by reaction 1. Inversely, there is a certain amount of octahedral ferric iron, $[Fe^{3+}]^*$, that is not available for reduction as well. The observations are also valid for muscovite and vermiculite. If the initial concentration of $^{[6]}Fe^{2+}$ is given by $[Fe^{2+}]^0$, then the amount of $^{[6]}Fe^{2+}$ left after oxidation at a given pH, $[Fe^{2+}]$, can be modeled using the Weibull function b
$$[Fe^{2+}] = [Fe^{2+}]^0 - ([Fe^{2+}]^0-[Fe^{2+}]^*)\exp(-f(pH)^\beta) \quad (2)$$
Similarly if the initial amount of $^{[6]}Fe^{3+}$ before reduction is given by $[Fe^{3+}]^0$, then the amount of $^{[6]}Fe^{3+}$ left after reduction at a given pH, $[Fe^{3+}]$, can be written by the reciprocal Weibull function as: $[Fe^{3+}] = [Fe^{3+}]^0 - ([Fe^{3+}]^0-[Fe^{3+}]^*)(1 - \exp(-f(pH)^\beta))$ (3)
where the scale parameter f corresponds to the oxidation rate constant and the shape parameter β is a measure of the extent by which the reaction rate is accelerated or decelerated. The factor β is a measure of the order of reaction. If $\beta = 1$, as in first order reactions, the Weibull function is identical to an exponential function, if $\beta = 2$ then the Weibull function is similar to a Rayleigh function, if $\beta > 2$ the Weibull function approaches a normal function and for $\beta > 4$ the Weibull function and the normal function are indistinguishable from each other [8].

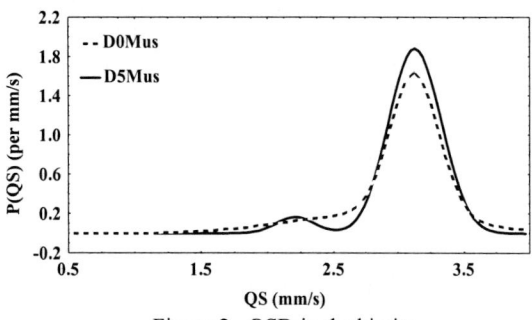

Figure 2. QSD in the biotite weathering series for D0Bio and D23Bio.

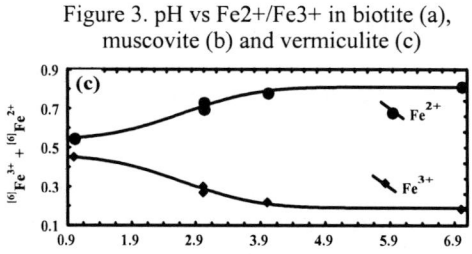

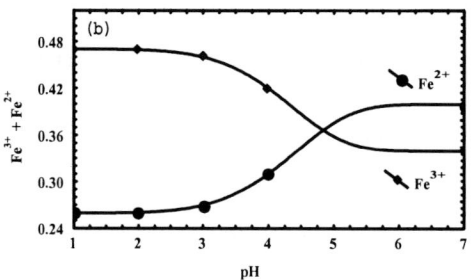

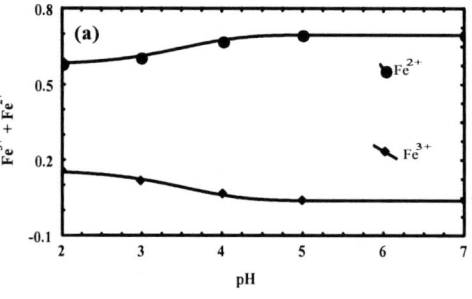

Figure 3. pH vs Fe2+/Fe3+ in biotite (a), muscovite (b) and vermiculite (c)

not as a function of time, t, as usually is the case for Weibull functions, but as a function of pH. Because at equilibrium the rate of oxidation is proportional to both t and pH and b to a power of pH, we argue that the Weibull function can be expected to be applied to this study as well. The initial values of $[Fe^{2+}]^0$ and $[Fe^{2+}]^*$ were estimated from the concentrations of $^{[6]}[Fe^{2+}]$ at pH 7 and 1, and the initial values of $[Fe^{3+}]^0$ and $[Fe^{3+}]^*$ from the concentrations of $^{[6]}[Fe^{3+}]$ at pH 1 and 7, respectively (Fig. 3).
Thus, there is a system involving two equations, 2 and 3, with three unknowns $[Fe^{2+}]$ or $[Fe^{3+}]$, β and f to solve.
The results of such simultaneous fitting are shown in Fig. 3. The fitting data show that the oxidation rate constant, f, for biotite is 35 times higher than that of vermiculite and 263 times higher than that of muscovite.
From wet chemical analysis, Kalinowski and Schweda [1] estimated that the apparent dissolution rate constant for biotite is 188 times higher than that of muscovite, showing the presence of other weathering mechanisms besides oxidation of structural iron. At the same time β, which is a measure of the extent by which the oxidation rate is accelerated, is higher in muscovite and vermiculite but lower in biotite. In general, chemical weathering is significantly easier in

biotite than it is in vermiculite or muscovite. The high rate of oxidation in biotite indicates that additional reaction mechanism(s) may be operating, mechanism(s) that are not possible in muscovite and vermiculite. One such mechanism is oxidation by $Fe^{2+}_{-n}K_{-n}Fe^{3+}_{n}$ where the excess charge introduced through the oxidation of Fe^{2+} is balanced by release of K ions to the solution. The preferred release of K in biotite over the release of K in muscovite reported by Kalinowski and Schweda [1] confirm the presence of more than two reaction mechanisms in biotite. The difference in chemistry and structure between biotite on the one hand and muscovite and vermiculite on the other supports the contention that Fe is easily oxidized in biotite by $Fe^{2+}_{-n}K_{-n}Fe^{3+}_{n}$. The interlayer K in muscovite is tightly bonded to the layers [9], limiting its leaching by $Fe^{2+}_{-n}K_{-n}Fe^{3+}_{n}$. In vermiculite the amount of K in the interlayer is low and consequently the contribution of $Fe^{2+}_{-n}K_{-n}Fe^{3+}_{n}$ to the rate of oxidation is marginal.

6. CONCLUSIONS

The study shows that it is possible to monitor the chemical changes occuring during experimental weathering of biotite, muscovite and vermiculite using Mössbauer spectroscopy. By choosing a statistically valid QSD fitting model for a mineral, Mössbauer parameters could be correlated to the crystal chemical changes experienced by the mineral during weathering. The correlation between weathering pH and oxidation ratio was modeled using Weibull function for all three samples with R^2 close to 1. The fitting parameters obtained indicate that biotite possesses high redox-rate constants, f, and low shape parameters, β, compared to muscovite and vermiculite. The shape of QSDs for $^{[6]}Fe^{2+}$ in biotite and muscovite change with pH but the changes are different. In biotite it is the position of the peak QS that shifts to low energies with decreasing pH while in muscovite it is the bi-modality of the QSD that is enhanced by decreasing pH. The first behavior is attributed to decrease in the covalence character of the Fe/Mg-O bond in biotite with increasing degree of oxidation and the latter is attributed to the absence of chemical and structural perturbations in muscovite.

Acknowledgments: The work has been supported by a grant from the Swedish Research Council. I am thankful to Steven Guggenheim for his review of the manuscript.

REFERENCES

[1] Kalinowski, B.E., Schweda, P. (1996): Kinetics of muscovite, phlogopite, and biotite dissolution and alteration at pH 1-4, room temperature. *Geochem. Cosmochem. Acta*, 60, 3, 367-385.
[2] Schweda, P., Kalinowski, B.E., Ferrow, E.A. (1997): Rates and non-stoichiometry of vermiculite dissolution at low temperature. In: Kalinowski, B.E. Dissolution kinetics and alteration products of micas and epidote in acidic solutions at room temperature. *Meddelanden från Stockholms Universitet, Institution för Geologi och Geokemi*, 294.
[3] Ferrow, E.A., Kalinowski, B.E., Veblen, D.R., Schweda, P. (1999): Alteration products of experimentally weathered biotite studied by high-resolution TEM and Mössbauer spectroscopy. *Eur. J. Mineral.*, 11, 999-1010.
[4] Mashlan, M., Zak, D., Cholmeckij, A., Evdokimov, V., Misevic, O., Fedorov, A., Lopatik, A., Snasel, V. (1994): The PC-AT based Mössbauer spectrometer. *Acta Universitaris Palackinae Olomoucensis. Facultas Rerum Naturalium, Physica XXXIII*, 116.
[5] Rancourt, D.G., (1994): Inadequacy of Lorentzian-line doublets in fitting spectra arising from quadrupole splitting distributions. *Phys. Chem. Mineral.*, 21:244-249.
[6] Rancourt, D.G, Ping, J.Y, Berman, R.G. (1994): Mössbauer spectroscopy of minerals. III Octahedral-site Fe2+ quadrupole splitting distributions in the phlogopite-annite series. *Phys. Chem. Mineral.* 21:258-267.

[7] Amonette, J.E. (1988): The role of structural iron oxidation in the weathering of tri-octahedral micas by acqueous solutions. *Ph.D. thesis*, Iowa State University, Ames, Iowa.
[8] McCormick N.J. (1981): Reliability and Risk Analysis: Methods and Nuclear Power Applications. Academic Press.
[9] Bassett, W.A. (1960): Role of hydroxyl orientation in mica alteration. *Bull. GSA,* 71, 449-45

Syntheses of smectite-analogue/coumarin composites

K. Fujii[a] and S. Hayashi[b]

[a]Advanced Materials Lab., National Institute for Materials Science, 1-1 Namiki, Tsukuba, Ibaraki 305-0044, Japan

[b]Institute for Materials & Chemical Process, National Institute of Advanced Industrial Science and Technology (AIST), Central 5, 1-1-1 Higashi, Tsukuba, Ibaraki 305-8565, Japan

Smectite-analogue/coumarin composites were synthesized from suspensions consisting of silylated coumarin dye (derCoum), silica sol, $Mg(OH)_2$ and LiF. X-ray diffraction (XRD) and transmission electron microscopy (TEM) revealed that the inorganic portion is a layered structure analogous to that of magnesium phyllosilicate, probably a smectite-like phase. XRD also showed that the spacing between the inorganic layers is 1.45~1.6 nm, suggesting that the coumarin chromophore lies nearly parallel to the Si-O-Si surface plane of the smectite-like layers. Elemental analysis, infrared (IR) spectra and high-resolution solid-state ^{13}C nuclear magnetic resonance (NMR) indicated the presence of organic components in the composites. High-resolution solid-state ^{29}Si NMR was applied to examine the local structures of the smectite-like layers. The signal intensities at about -60 and -85 ppm increase as the [derCoum]/[Si] molar ratio increases. The smectite-like layers in this study have $CSi(OZ)_3$ (Z = Si, H and/or Mg) species in a tetrahedral sheet as well as silanol groups adjacent to the $CSi(OZ)3$, whereas smectites have neither $CSi(OZ)_3$ species nor the surface silanol groups. Chemical forms and local environments of the coumarin chromophore were investigated by means of ultraviolet-visible (UV-VIS) and fluorescence spectroscopies. Both a neutral form and a cationic form of the coumarin chromophore are present in the smectite-analogue/coumarin composites. Furthermore, the wavelength of the maximum fluorescence emission in the cationic form is red-shifted, indicating stabilization effects caused by the smectite-like layers.

1. Introduction

Considerable attention has been paid to inorganic-organic hybrid compounds.1-15 Various compounds and synthesis methods have been reported. Fukushima and Tani successfully synthesized inorganic-organic hybrid layered compounds using organofunctional trialkoxysilanes and metal chloride hydrates.1,2

Coumarin dyes are attractive materials for use as organic dyes in a tunable solid-state laser because the fluorescence bandwidth is large. The incorporation of laser dye molecules into inorganic hosts has been studied because the inorganic hosts have relatively high laser-damage thresholds. Dunn et al. reported the incorporation of coumarin in xerogel hosts.3,4 Suratwala et al. reported the synthesis of silylated coumarin dyes which were chemically altered to provide alkoxysilane functionality and fluorescence properties of the dyes incorporated into SiO2 xerogels and into various solvents.5,6 Endo et al. reported an intercalation of coumarin dye in swelling clays.7

In the present work, we have synthesized a new magnesium smectite-analogue/coumarin composite material by using silylated coumarin, silica sol and inorganic salts, and we have studied their structures and fluorescence properties. Because the chemical environment affects the fluorescence property, this study can provide information about the environment of the coumarin chromophore and the interaction between the coumarin chromophore and the smectite-like portion.

2. EXPERIMENTAL

7-(3-Triethoxysilylpropyl)-O-(4-methyl-coumarin)urethane (Scheme 1), abbreviated to derCoum, was synthesized from 7-hydroxy-4-methylcoumarin (Wako Pure Chemical Industries, Ltd.) and isocyanatopropyltriethoxysilane (Shin-Etsu Chemical Co., Ltd) as described by Suratwala et al.[5] The structure of the synthesized derCoum was confirmed by ^{1}H and ^{13}C NMR and IR. Mg(OH)$_2$ gel was prepared from MgCl$_2 \cdot$6H$_2$O (Wako Pure Chemical Industries, Ltd.) and NH$_4$OH.[15] Both dried and fresh Mg(OH)$_2$ gels were used for the synthesis of smectite-analogue/coumarin composites. Silylated coumarin, Mg(OH)$_2$ gel, silica sol (Ludox HS-30, Du Pont) and LiF (Wako Pure Chemical Industries, Ltd.) were mixed in water to obtain starting suspensions. Because derCoum had ethoxysilyl groups, both the hydrolysis reaction and the condensation reaction with silanol groups of the silica sol surface proceeded under the reaction conditions. Molar ratios of [Si] : [Mg(OH)$_2$] : [LiF] were 1.47 : 1 : 0.2 and 1.52 : 1 : 0.2 and the [derCoum]/[Si] molar ratio was 0.02 to 0.1. The chemical formula of the silica sol was assumed to be SiO$_2$ in calculating molar ratios. The starting suspensions were kept at 70 to 150°C for 3 to 8 days. The synthesis conditions are summarized in Table1. Two types of reaction vessels were employed; one was a Morey-type pressure vessel (a closed system) without stirring and the other was a round-bottom flask (an open system) with

Table 1

Synthesis conditions

Sample	Reaction vessel	[derCoum]/[Si]^1	[Si]/[Mg]^1	Mg(OH)$_2$ gel	Temperature [°C]	Time [days]
M-1	Morey	0.057	1.52	dried	150	3
M-2	Morey	0.057	1.52	dried	150	8
M-3	Morey	0.057	1.52	dried	80	3
M-4	Morey	0.057	1.52	fresh	80	3
M-5	Morey	0.057	1.52	dried	70	3
F-1	flask	0.02	1.52	fresh	reflux condition	4
F-2	flask	0.02	1.47	dried	80	3
F-3	flask	0.06	1.47	dried	80	3
F-4	flask	0.1	1.47	dried	80	3

[1] Chemical formula of silica sol was assumed to be SiO_2 in calculating molar ratios.

stirring. The obtained mixtures after the heat treatments were washed in tetrahydrofuran to remove free coumarin dye molecules and then washed in water. Precipitates were filtered and dried under reduced pressure.

XRD measurements were performed at room temperature at ambient humidity using a Rigaku Rint 2000S diffractometer with CuKα radiation. TEM images and electron diffraction (ED) were obtained with a JEOL electron microscope at an accelerating voltage of 100 kV. High-resolution solid-state ^{29}Si and ^{13}C NMR spectra were measured using a Bruker ASX400 spectrometer at room temperature. The ordinary cross-polarization (CP) pulse sequence as well as a single-pulse sequence with ^1H decoupling (HD) were used together with magic angle spinning (MAS) of the samples. CHN analysis was performed in a Perkin Elmer CHNS/O analyzer SeriesII 2400. Elemental analysis was performed using an inductively coupled plasma (ICP) atomic absorption. IR spectra were recorded with a Bio-Rad FTS-65 FT-IR spectrometer. UV-VIS spectroscopy was measured by the reflection method using a Shimadzu UV-VIS-NIR scanning spectrophotometer UV-3100PC. Fluorescence emission spectra and excitation spectra were recorded with a Shimadzu spectrofluorophotometer RF-5300 for powder specimens.

3. RESULTS AND DISCUSSION

3.1. Synthesis conditions for smectite-analogue/coumarin composite

Figs. 1(a)~(d) show XRD patterns of samples M-1 and M-3 to M-5, which are obtained at 70 to 150°C for 3 days from the starting suspensions with the molar ratio of [Si] : [Mg(OH)$_2$] : [LiF] of 1.52 : 1 : 0.2 and the [derCoum]/[Si] molar ratio of 0.057 in the Morey-type pressure vessel. The XRD pattern of M-1 is similar to that of a magnesium phyllosilicate such as hectorite or stevensite and suggests that the phyllosilicate belongs to the smectite group. An asymmetrical peak at about $d = 0.26$ nm ($2\theta = 35°$) is indexed as 13 and 20,[16] a reflection peak at about $2\theta = 20°$ ($d = 0.45$ nm) as 02 and 11 and a reflection peak at about $2\theta = 60°$ ($d = 0.15$ nm) as 06 and 33.[16] The XRD pattern of M-2 is similar to that of M-1. On the other hand, all peaks in Fig. 1(c) are indexed to brucite, indicating that the starting reagents did not react at 70°C. In Fig. 1(b) there are both peaks that can be indexed to smectite and brucite, indicating that a smectite-like structure is formed at 80°C. To accelerate the reaction of Mg(OH)$_2$, fresh Mg(OH)$_2$ gel has been used instead of its dried counterpart. The sample (M-4) thus obtained has an XRD pattern similar to that of magnesium smectite, as shown in Fig. 1(d). TEM shows images of layered materials for M-4. Furthermore, the ED pattern of the layered material agrees with the d values of the XRD pattern. Results of the elemental analysis indicate that the sum of Si and Mg makes up about 40 wt% of contents of M-4. The above results

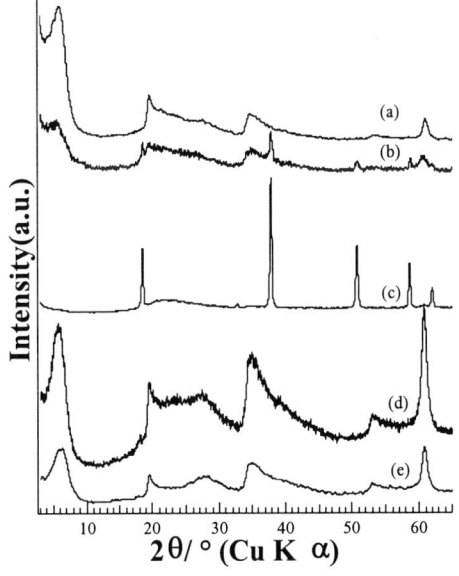

Figure 1. XRD patterns of samples M-1 (a), M-3 (b), M-5 (c), M-4 (d) and F-2 (e).

indicate that a structure similar to magnesium smectite is formed.

Reaction conditions to form the magnesium smectite-like structure have further been examined using dried Mg(OH)$_2$ gel. The reactivity of Mg(OH)$_2$ increases when the starting suspension is allowed to react in a round-bottom flask with stirring. Fig. 1(e) shows the XRD pattern of sample F-2. A magnesium smectite-like structure is formed at 80°C as it does for samples F-1 and F-3. In contrast, there are reflection peaks indexed to unreacted Mg(OH)$_2$ in the XRD pattern of F-4. CHN analyses give carbon contents of 1 to 3.5 wt%.

Figs. 2(a) and (b) show the IR spectra of the obtained samples M-2 and M-4. The IR spectrum of derCoum (Fig. 2(c)) is assigned as follows. The strong bands at 2980-2880 cm^{-1} are due to the C-H symmetrical and asymmetrical stretching vibrations of CH$_2$ and CH$_3$ groups, and a strong absorption at about 1735 cm^{-1} to a carbonyl group. C=C stretching and N-H bending vibrations are observed at 1600-1500 cm^{-1}, associated with both the coumarin chromophore and the urethane-junction portion, -O-CONH-. In Fig. 2(b) there are broad bands at near 3680 cm^{-1} (OH groups of the magnesium smectite-like portion), and at about 1000 cm^{-1} (Si-O-Si). Absorption bands related to H$_2$O are observed at about 3400 cm^{-1} and 1630 cm^{-1}. Because of strong absorptions from the inorganic portion and H$_2$O, many bands owing to the organic portion are often obscured, or only occur as shoulders. The C-H stretching vibrations at 2980-2880 cm^{-1} in Fig. 2(b) suggest the presence of alkyl chains in M-4. Expanded IR spectra are shown in Fig. 3. Absorption peaks at 1700-1350 cm^{-1} in Fig. 3(b) suggest that M-4 contains C=C and/or N-H groups. These absorptions are difficult to confirm in M-2 (Figs. 2(a) and 3(a)), suggesting that the coumarin dye has decomposed. The IR pattern associated with the organic portion for F-1 to F-3 is similar to that of M-4 (Figs. 2(b) and 3(b)).

3.2. Structure of the magnesium smectite-analogue/coumarin composite

The low-angle ($2\theta < 10°$) reflections in Fig. 1 are assigned to the basal spacings of the magnesium smectite-analogue/coumarin composites. The corresponding d values of 1.4 to 1.6 nm suggest that the coumarin chromophore is located between the layers of the magnesium smectite-like portion,

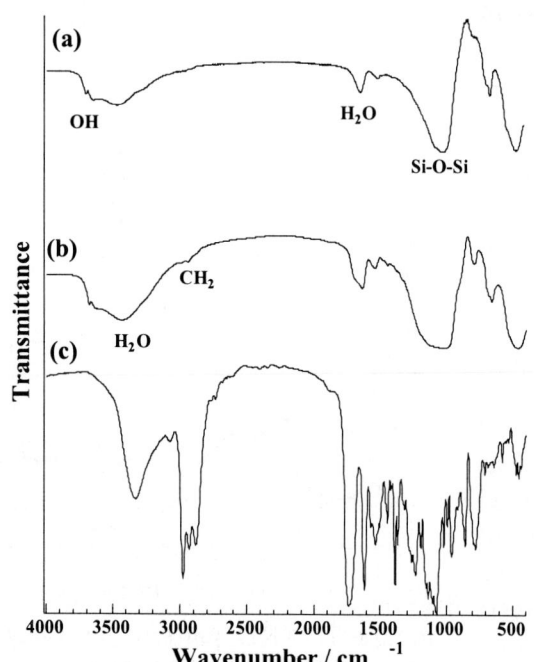

Figure 2. IR spectra of samples M-2 (a) and M-4 (b) and derCoum (c).

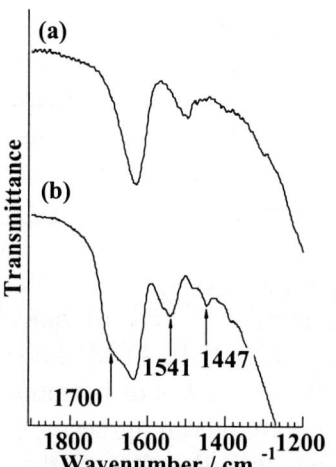

Figure 3. IR spectra of samples M-2 (a) and M-4 (b) in the range of 1900 -1200 cm^{-1}.

lying nearly parallel to the Si-O-Si surface. Endo et al.[7] reported that saponite-coumarin composites, obtained by intercalation, had basal spacings of 1.30 to 1.85 nm. For the composite with the *d* value of 1.30 nm the clearance space (interlayer separation) is about 0.34 nm since the thickness of the clay layer is about 0.96 nm. This value is almost equal to the thickness of the coumarin molecule. The coumarin molecules are lying parallel to the aluminosilicate surfaces of the clay. Carrado et al.[15] prepared porphyrin-containing layered silicates with *d* values of 1.50 ~ 1.67 nm. They concluded that the organic macrocycles were intercalated parallel to the clay layers.

IR spectra indicate the existence of organic species (Figs. 2 and 3) and absorption and fluorescence emission spectra described below show the existence of coumarin chromophore. Fig. 4 shows a high-resolution solid-state ^{13}C NMR spectrum of F-3. The signals at 11, 24 and 44 ppm are attributed to carbons in the SiCH$_2$CH$_2$CH$_2$ group, the signal at 160 ppm to the carbonyl group carbon, the signal at 15 ppm to $\underline{C}$H$_3$ and the signal at 63 ppm to $\underline{C}$H$_2$ bonded to oxygen. For reference, ethyl alcohol has signals at 17.6 ($\underline{C}$H$_3$) and 57.0 ppm ($\underline{C}$H$_2$–O). For ethylcarbamate,[17] signals at 157.8, 60.9 and 14.5 ppm are assigned to $\underline{C}$=O, $\underline{C}$H$_2$ and $\underline{C}$H$_3$, respectively. It is difficult to confirm the ^{13}C NMR signals caused by the coumarin chromophore, although absorption and fluorescence emission spectra demonstrate its existence. We suggest that most of the derCoum have decomposed during the synthesis reactions and that the major organic components represent the degradation products. However, a small amount of the coumarin chromophore is present in the composites.

Fig. 5 shows high-resolution solid-state ^{29}Si NMR spectra for F-1 and F-3. The signal at -95 ppm is assigned to (MgO)$\underline{Si}$(OSi)$_3$ in the magnesium smectite-like portion.[18] The signals at about -86 and -79 ppm are in the region of the signals of (SiO)$_{4-n}\underline{Si}$(OX)$_n$. The signal at about -86 ppm is assigned to (SiO)$_2$(MgO)$\underline{Si}$(OH) or (SiO)$_2$(MgO)$\underline{Si}$(OCH$_2$CH$_3$). We have measured HD/MAS NMR spectra as well as the CP/MAS NMR spectra. In principle, the Si signal in the neighborhood of H is enhanced in the CP/MAS NMR spectra when compared with the HD/MAS NMR spectra. The signal intensity at about -86 ppm is not as enhanced in

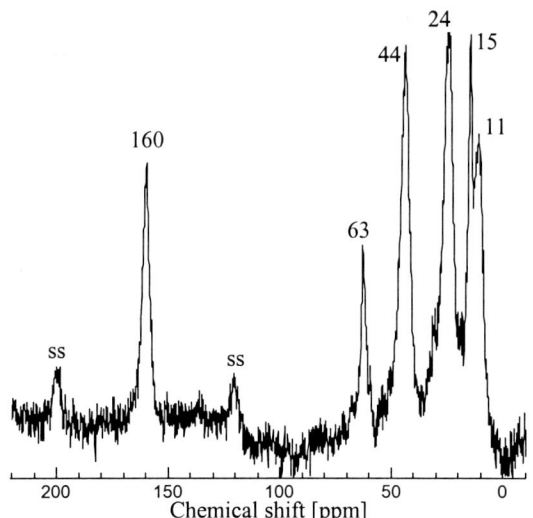

Figure 4. ^{13}C CP/MAS NMR spectrum of sample F-3. Spinning sidebands are marked with "ss".

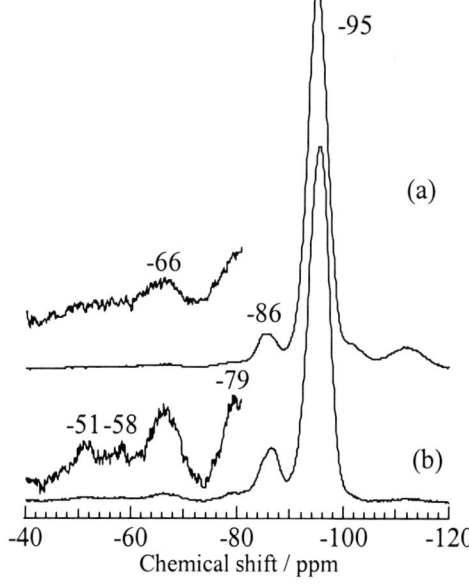

Figure 5. ^{29}Si CP/MAS NMR spectra of samples F-1 (a) and F-3 (b). Enlarged spectra (10 times) are inserted.

the CP/MAS spectra. Thus, the signal at about -86 ppm is attributable to $(SiO)_2(MgO)\underline{Si}(OCH_2CH_3)$ or a mixture of $(SiO)_2(MgO)\underline{Si}(OH)$ and $(SiO)_2(MgO)\underline{Si}(OCH_2CH_3)$. Three signals are observed near -66, -58 and -51 ppm in the region of the signals of $(SiO)_{3-m}C\underline{Si}(OY)_m$ (Y is H, Mg or -CH$_2$CH$_3$ in this study). The signals related to the $(SiO)_{3-m}C\underline{Si}(OY)_m$ species, as well as the signal near -86 ppm increase as the [derCoum]/[Si] molar ratio in the starting suspension increases. The latter increase can be interpreted as follows: For the magnesium phyllosilicates as in the smectites, the individual SiO$_4$ tetrahedron in the two-dimensional tetrahedral sheets are linked with three neighboring tetrahedra by sharing three corners, that is, the basal oxygen atoms, to form a hexagonal mesh pattern. The fourth tetrahedral corner is shared with an adjacent octahedral sheet. In contrast, when a CSi(OH)$_3$ species undergo the condensation reaction with silanol groups of the silica sol and the CSiO$_3$ is located in place of the SiO$_4$ tetrahedron in the tetrahedral sheet, the hexagonal mesh pattern is not formed completely. One of the Si adjacent to the CSi(OZ)$_3$ species (Z is Si, H and/or Mg) cannot share one of the three corners and a $(SiO)_2(MgO)Si(OX_b)$ species is formed. Because the composites are synthesized under coexistence of ethoxy group, X_b is considered to be H or -CH$_2$CH$_3$. Thus, the signal at –86 ppm is attributed to $(SiO)_2(MgO)\underline{Si}(OX_b)$.

3.3. Absorption and fluorescence spectroscopies

UV-VIS spectra have been measured by the reflection method to confirm the presence of the coumarin chromophore and chemical forms of it in the magnesium smectite-analogue/coumarin composites. The peak near 325 nm in the spectra for the magnesium smectite-analogue/coumarin composites (not shown) agrees with these of a neutral form of the coumarin chromophore. The spectra of 7-hydroxy-4-mthylcoumarin and derCoum were reported by Dienes et al.[19] and Suratwala et al.[5] The solution spectra showed peaks at about 322 nm for the neutral forms of coumarin and derCoum, at 350 nm for the cationic forms, and at 365 nm for the anionic forms. The presence of a neutral form of the coumarin chromophore is confirmed in the magnesium smectite-analogue/coumarin composites. An absorption band at near 275 nm is observed for the composites obtained in the round-bottom flasks. Its assignment has not been established, although the absorption may be attributed to the magnesium smectite-like portions of the composites.

Fig. 6 shows the fluorescence emission spectra with excitation wavelengths of 320, 337 and 365 nm for F-1. The fluorescence emission maximum is observed at near 385 nm when the coumarin chromophore is pumped at 320 and 337 nm, and this maximum agrees with the neutral form of coumarin chromophore. The fluorescence emission bandwidth is larger than that of 7-hydroxy-4-methylcoumarin powder. The large bandwidth may be ascribed to interactions between the coumarin chromophore and the magnesium smectite-like phyllosilicate portion. When the coumarin chromophore is pumped with the excitation wavelength of 365 nm, a weak fluorescence emission near 440 nm is observed.

Fig. 7 shows the fluorescence emission spectra for F-3. The peak near 385 nm and a shoulder near 435 nm are observed when the coumarin chromophore is pumped at 320 and 337 nm. On the other hand, only the fluorescence emission near 440 nm is observed in the fluorescence emission spectra pumped with the excitation wavelength of 365 nm. Dienes et al.[19] and Suratwala et al.[5] reported a peak at 385 nm for the neutral form of 7-hydroxy-4-methylcoumarin in neutral and slightly acidic solutions. A peak at 415-430 nm was observed for the cationic form in strongly acidic solutions, and one at 455 nm for the anionic form in basic solutions.

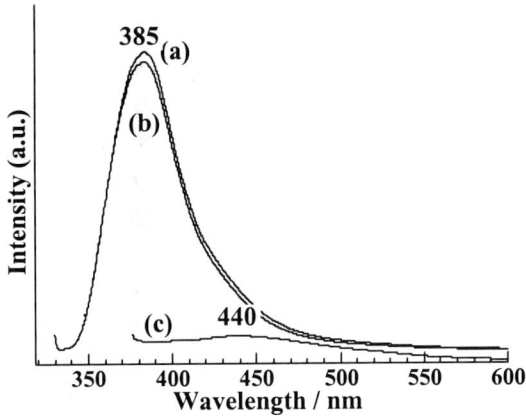

Figure 6. Fluorescence spectra of sample F-1 pumped at (a) 320, (b) 337 and (c) 365 nm.

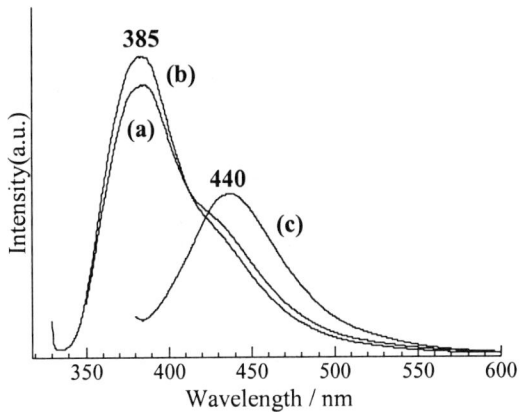

Figure 7. Fluorescence spectra of sample F-3 pumped at (a) 320, (b) 337 and (c) 365 nm.

The fluorescence emission peak near 385 nm in the present work is ascribed to the neutral form of the coumarin chromophore. The peak near 440 nm is attributed to the cationic form of the coumarin chromophore and the emission maximum wavelength is slightly red-shifted. The cationic form of 7-hydroxy-4-mthylcoumarin was observed only in concentrated H_2SO_4.[5] It is therefore interesting to note that the cationic form is present in the composites because magnesium phyllosilicates are not strong acids.[5] It is known that cationic species can intercalate more readily into smectites than non-ionic species. The coumarin chromophore is believed to be incorporated between the magnesium smectite-like layers in the cationic form as well as in the neutral form.

The polar environment may cause the red shift of the fluorescence emission maximum in the cationic form. The lowest energy transition of coumarin chromophore is (π, π^*). It was reported[5,20,21] that the (π, π^*) excited state is more polar and more polarizable than the ground state so that the transition energy decreases in polar solvents. The red shift is considered to be caused by the interactions of the coumarin chromophore with the surface oxygen atoms and the silanol groups of the magnesium smectite-like portion of the composites. The composites in this study have silanol groups adjacent to $CSi(OZ)_3$, whereas smectites have no surface silanol groups. The red shift of the fluorescence emission maximum was also reported for derCoum incorporated in xerogel[5] and for coumarin dye intercalated in a swelling clay.[7]

In conclusion, the fluorescence spectroscopy has revealed that the neutral form and the cationic forms of the coumarin chromophore occur in the magnesium smectite-analogue/coumarin composites and that there are interactions between the coumarin chromophore and the magnesium smectite-like portion of the structure.

Acknowledgments

The authors are grateful to Dr. T. Fujita and Mr. S. Takenouchi, NIMS for elemental analyses and to Mr. Y. Kitami, NIMS for the TEM observation.

REFERENCES

1. Y. Fukushima and M. Tani, Bull. Chem. Soc. Jpn., 69 (1996) 3667.

2. Y. Fukushima and M. Tani, J. Chem. Soc., Chem. Commun., (1995) 241.
3. B. Dunn and J. Zink, J. Mater. Chem., 1 (1991) 903.
4. J. M. Mackiernan, S. A. Tamanaka, E. Knobbe, J. Pouxviel, S. Parvaneh, B. Dunn and J. Zink, J. Inorg. Organomet. Polym., 1 (1991) 87.
5. T. Suratwala, Z. Gardlund, K. Davidson, D. R. Uhlmann, J. Watson and N. Peyghambarian, Chem. Mat., 10 (1998) 190.
6. T. Suratwala, Z. Gardlund, K. Davidson, D. R. Uhlmann, J. Watson, S. Bonilla and N. Peyghambarian, Chem. Mat., 10 (1998) 199.
7. T. Endo, N. Nakada, T. Sato and M. Shimada, J. Phys. Chem. Solids, 50 (1989) 133.
8. T. Fujita, N. Iyi, T. Kosugi, A. Ando, T. Deguchi and T. Sota, Clays Clay Miner., 45 (1997) 77.
9. K. Takagi, H. Usami, H. Fukaya and Y. Sawaki, J. Chem. Soc. Chem. Commun., (1989) 1174.
10. M. G. da Fonseca, C. R. Silva and C. Airolodi, Langmuir, 15 (1999) 5048.
11. A. N. Parikh, M. A. Schivley, E. Koo, K. Seshadri, D. Aurentz, K. Mueller and D. L. Allara, J. Am. Chem. Soc., 119 (1997) 3135.
12. A. Shimojima, Y. Sugawara and K. Kuroda, Bull. Chem. Soc. Jpn., 70 (1997) 2847.
13. Y. Sugahara, T. Inoue and K. Kuroda, J. Mater. Chem., 7 (1997) 53.
14. E. Ruiz-Hitsky and J. M. Rojo, Nature, 287 (1980) 28.
15. K. A. Carrado, P. Thiyagarajan, R. E. Winans and R. E. Botto, Inorg. Chem., 30 (1991) 794.
16. G. W. Brindley and G. Brown (eds.), Crystal Structures of Clay Minerals and Their X-ray Identification, Mineralogical Society, London, 1980.
17. R. M. Silverstein, G. C. Bassler and T. C. Morrill, Spectrometric Identification of Organic Compounds, John Wiley & Sons, Inc., 1991.
18. S. Hayashi, unpublished data measured for hectorite.
19. A. Dienes, C. V. Shank and R. L. Kohn, IEEE J. Quantum Electron, QE-9 (1973) 833.
20. E. Wehry (ed.), Effects of Molecular Structure on Fluorescence and Phosphorescence, Marcel Dekker, New York, 1990.
21. C. Nichols and P. Leermakers, Adv.Photochem., 8 (1971) 315.

Thermogravimetric analysis-mass spectrometry (TGA-MS) of hydrotalcites containing CO_3^{2-}, NO_3^-, Cl^-, SO_4^{2-} or ClO_4^-

J. Theo Kloprogge[a], János Kristóf[b] and Ray L. Frost[a]

[a] Centre for Instrumental and Developmental Chemistry, Queensland University of Technology, 2 George Street, GPO Box 2434, Brisbane, Qld 4001, Australia.

[b] Department of Analytical Chemistry, University of Veszprém, H8201 Veszprém, P.O. Box 158, Hungary.

Mg/Al-hydrotalcites containing NO_3^-, Cl^-, SO_4^{2-} or ClO_4^- were synthesised under N_2 to prevent incorporation of CO_3^{2-}. The presence of the anions in the hydrotalcite structure was confirmed by infrared and Raman spectroscopy. The CO_3^- and the NO_3^- and CO_3^{2-}, while the Cl-hydrotalcite also contained some CO_3^{2-}. It is known that during thermal treatment of hydrotalcites dehydroxylation and decarbonisation strongly overlap. Mass spectrometry following TGA enables one to identify both reactions. For CO_3-hydrotalcite CO_2 is released simultaneously with water (dehydroxylation) around 335°C followed by NO around 365 and 500°C. The stability of the NO_3-hydrotalcite is different showing a major loss of CO_2 and H_2O (dehydroxylation) around 410°C with losses of NO around 345 and 450°C. The Cl-hydrotalcite shows a similar behaviour for the H_2O loss (dehydroxylation), but Cl is lost over a range from 400 to 900°C and CO_2 comes off in steps around 360 and 500°C. Completely different is the thermal behaviour of SO_4- and ClO_4-hydrotalcites. SO_4-hydrotalcite shows a gradual weight-loss due to dehydroxylation with two minor water peaks around 260 and 375°C, while the sulphate remains in the structure. The sulphate is not lost until heated to 900°C. The ClO_4-hydrotalcite shows a complex thermal behaviour with 2 steps of water loss around 375 and 440°C, where the second step is accompanied by the loss of O_2. A possible explanation is a redox reaction between perchlorate and the cations giving metal-chlorides and O_2.

1. INTRODUCTION

Layered double hydroxides (LDHs) are also known as hydrotalcites or anionic clays, due to their layered structure with a charge opposite to that of smectites. The structure of hydrotalcite can be visualised as positively charged OH-layers comparable to those in brucite ($Mg(OH)_2$) in which a part of the Mg_{2+} is substituted by a trivalent metal such as Al_{3+} separated by charge compensating, mostly hydrated, anions between the hydroxide sheets. In layered double hydroxides a large range of compositions is possible based on a general formula of $[M^{2+}_{1-x}M^{3+}_x(OH)_2][A^{n-}]_{x/n} \cdot yH_2O$, where M_{2+} and M_{3+} are the di- and trivalent metals in the octahedral sites within the OH-sheets with x normally between 0.17 and 0.33.

The number of anions or anionic complexes in layered double hydroxides is essentially unlimited provided that the anion does not form a complex with the cations in the hydroxide sheets during the formation (Vaccari 1998). Therefore, the variety possible in both the cationic and anionic compositions of the hydrotalcite offers the possibility to prepare tailor-made materials for specific applications, such as basic catalysts, as a precursor for the preparation of mixed metal oxidic catalysts, absorbents, as for other specific powder properties such as filler, UV-radiation stabiliser, chloride scavenger and thermal stabiliser (Titulaer 1993).

In some recent publications (Kloprogge et al., 2000, 2002) we have described the incorporation of CO_3^{2-}, NO_3^-, SO_4^{2-} and ClO_4^- in hydrotalcites and the effects that the confinement of these anions in the interlayer space of hydrotalcites has on their vibrational spectra (infrared and Raman) in comparison to the anions in solution. It was shown that in comparison to free CO_3^{2-} a shift towards lower wavenumbers was observed. A band around 3000-3200 cm-1 has been attributed to the bridging mode H_2O-CO_3^{2-}. The IR spectrum of the CO_3-hydrotalcite clearly shows the split $\acute{\iota}_3$ around 1365 and 1400 cm-1 together with weak $\acute{\iota}_2$ and $\acute{\iota}_4$ modes around 870 and 667 cm-1. The $\acute{\iota}_1$ is activated and observed as a weak band around 1012 cm-1. The Raman spectrum shows a strong $\acute{\iota}_1$ at 1053 cm-1 plus weak $\acute{\iota}_3$ and $\acute{\iota}_4$ modes around 1403 and 695 cm-1. The symmetry of the carbonate anions is lowered from D_{3h} to C_{2v} resulting in activation of the IR inactive $\acute{\iota}_1$ mode around 1050-1060 cm-1. In addition, the $\acute{\iota}_3$ shows a splitting of 30-60 cm-1. Although the NO_3-hydrotalcite has incorporated some CO_3^{2-} the IR shows a strong $\acute{\iota}_3$ at 1360 cm-1 with a weak band at 827 cm-1, $\acute{\iota}_4$ is observed at 667 cm-1, although it is largely obscured by the hydrotalcite lattice modes. The Raman spectrum shows a strong $\acute{\iota}_1$ at 1044 cm-1 with a weaker $\acute{\iota}_4$ at 712 cm-1. The $\acute{\iota}_3$ at 1355 cm-1 is obscured by a broad band due to the presence of CO_3^{2-}. The symmetry of NO_3^- did not change when incorporated in the hydrotalcite. The IR spectrum of the SO_4-hydrotalcite shows a strong $\acute{\iota}_3$ at 1126, $\acute{\iota}_4$ at 614 and a weak $\acute{\iota}_1$ at 981 cm-1. The Raman spectrum is characterised by a strong $\acute{\iota}_1$ at 982 cm-1 plus medium $\acute{\iota}_2$ and $\acute{\iota}_4$ at 453 and 611 cm-1, $\acute{\iota}_3$ can not be identified as a separate band, although a broad band can be seen around 1134 cm-1. The site symmetry of SO_4^{2-} is lowered from T_d to C_{2v}. The distortion of ClO_4^- in the interlayer of hydrotalcite is reflected in the IR spectrum with both $\acute{\iota}_3$ and $\acute{\iota}_4$ split around 1096 + 1145 cm-1 and 626 + 635 cm-1, respectively. A weak $\acute{\iota}_1$ is observed at 935 cm-1. The Raman spectrum shows a strong $\acute{\iota}_1$ around 936 cm-1 plus $\acute{\iota}_2$ and $\acute{\iota}_4$ at 461 and 626 cm-1, respectively. $\acute{\iota}_3$ cannot be clearly recognised, but a broad band is visible around 1110 cm-1. These data indicative a symmetry lowering from T_d to C_s.

In this paper we describe the thermal behaviour of these hydrotalcites containing CO_3^{2-}, NO_3^-, SO_4^{2-} and ClO_4^-. In addition to the weight losses as function of temperature as measured with conventional thermogravimetric analysis and differential thermal analysis, the evolved gasses were analysed by mass spectrometry. This way it is possible to get a better understanding of the mechanisms involved in the decomposition of hydrotalcites containing various anions.

2. EXPERIMENTAL METHODS

2.1 Synthesis of the Mg/Al-hydrotalcites containing CO_3^{2-}, NO_3^-, Cl-, SO_4^{2-} and ClO_4^-

The hydrotalcite with theoretical composition of $Mg_6Al_2(OH)_{16}CO_3 \cdot nH_2O$ was synthesised according to the method described before by Kloprogge and Frost (1999). This method comprises the slow simultaneous addition of a mixed aluminium nitrate (0.25M)-magnesium nitrate (0.75M) and a mixed NaOH (2.00M)-Na_2CO_3 (0.125M) solution under vigorous stirring buffering the pH at approximately 10. The product was washed to eliminate excess salt and dried at 60°C.

The incorporation of carbonate during the synthesis of the hydrotalcites containing the other anions was as much as possible prevented by boiling the deionised water before use, by rinsing the NaOH pellets before use and by executing the synthesis under a nitrogen atmosphere. As sources for the sulphate and perchlorate anions in solution the corresponding magnesium and aluminium salts were used instead of the magnesium and aluminium nitrates as discussed above.

The crystalline nature of the resulting materials was checked by X-ray powder diffraction (XRD). The XRD analyses were carried out on a Philips wide angle PW 1050/25 vertical goniometer equipped with a graphite diffracted beam monochromator. The radiation applied was Co$K\alpha$ (1.7902 Å) from a long fine focus Co tube operating at 35 kV and 40 mA. The samples were measured at 50 % relative humidity in stepscan mode with steps of 0.02° 2θ and a counting time of 2s.

2.2. Analytical techniques

Thermoanalytical investigations were carried out in a Netzsch (Selb, Germany) TG 209 type thermobalance in a flowing argon atmosphere of 99.995% purity (Messer Griesheim, Hungary) at a heating rate of 10°C/min. To simultaneously follow the evolution of the gaseous decomposition products over the temperature range investigated, the thermobalance was connected to a Balzers MSC 200 Thermo-Cube type mass spectrometer (Balzers AG, Lichtenstein). The transfer line to introduce gaseous decomposition products into the mass spectrometer was a deactivated fused silica capillary (Infochroma AG, Zug, Switzerland; 0.23 mm o.d.) temperature controlled to 150°C to avoid possible condensation of the evolved gases. In this way the thermogravimetric (TG), derivative thermogravimetric (DTG) and mass spectrometric ion intensity curves of the selected ionic species could be recorded simultaneously.

The finely powdered samples were combined with oven dried spectroscopic grade KBr (containing approximately 1 wt% sample) and pressed into a disc under vacuum. The spectra were recorded in triplicate by accumulating 512 scans at 4 cm^{-1} resolution in the spectral range between 400 cm^{-1} and 4000 cm^{-1} using the Perkin-Elmer 1600 series Fourier transform infrared spectrometer equipped with a LITA detector and on a Nicolet Magna 750 Fourier transform infrared spectrometer equipped with a DTGS detector. The Fourier transform Raman spectroscopic (FT-Raman) analyses were performed on a Perkin Elmer System 2000 Fourier transform spectrometer equipped with a Raman accessory comprising a Spectron Laser Systems SL301 Nd:YAG laser operating a wavelength of 1064 nm. 1000 scans were obtained at a spectral resolution of 4 cm^{-1} in order to obtain an acceptable signal/noise ratio.

3. RESULTS AND DISCUSSION

3.1. Characterisation of the solid materials

All the samples synthesised for this study were identified as Mg/Al-hydrotalcites. Fig.1 shows the XRD patterns of the synthesised hydrotalcites. The hydrotalcites synthesised with carbonate, nitrate or sulphate as the interlayer anion clearly show one crystalline phase. The value of 7.8 Å is characteristic of the 003 reflection of carbonate containing hydrotalcite (Fig. 1) Replacement of carbonated by nitrate as the interlayer anion results in a slight expansion from 7.8

Å to around 8.1 Å, which is slightly smaller compared to the value of 8.36 Å reported by Marino and Masculo (1982). Incorporation of sulphate gives a 003 reflection of 8.9 Å that agrees well with data reported by e.g. Miyata and Okada (1977), Bish (1980) and Sato et al. (1986). The sample with the perchlorate as the interlayer anion however clearly shows a second phase with a 003 reflection similar to that of a carbonate containing hydrotalcite in addition to a stronger reflection at 9.3 Å, which agrees with data reported by Miyata and Kumura (1973) and Brindley and Kikkawa (1980). A similar observation of two separate phases containing perchlorate and carbonate was made by Brindley and Kikkawa (1979).

Infrared spectroscopy (Fig. 2) was used at this stage to identify the interlayer anions in the hydrotalcites synthesised. This confirmed clearly the presence of carbonate in the perchlorate system. So, even though greatest care was taken to exclude carbonate from the system, carbonate is sometimes found in the interlayer space. A similar observation was made in the infrared spectra of the nitrate containing hydrotalcite, which probably also explains the slightly smaller 003 value and the broadening of the reflection compared to that of the carbonate containing hydrotalcite. Opposite to these observations, the use of nitrate salts in the case of the carbonate containing hydrotalcite resulted in the incorporation of a minor amount of nitrate as well, as was earlier observed by Kloprogge and coworkers. (Hickey et al. 2000, Kloprogge

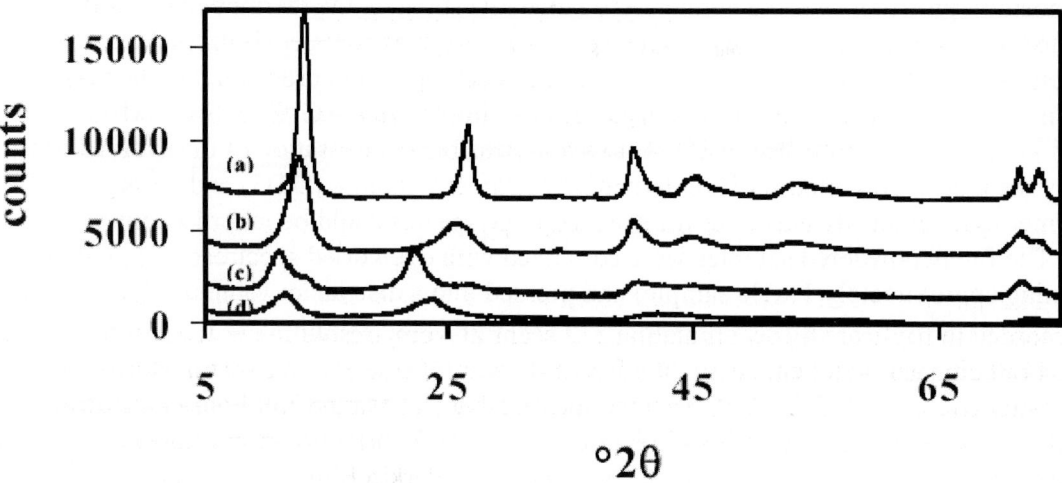

Fig. 1 XRD patterns of the hydrotalcites containing (a) CO_3^{2-}, (b) NO_3^-, (c) ClO_4^{2-} and (d) SO_{42-}.

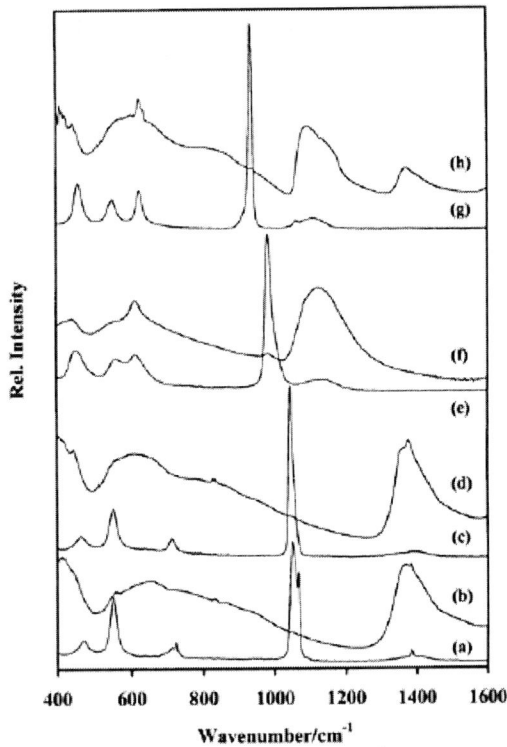

Fig. 2 Spectra of hydrotalcite containing CO_3^{2-} (a) Raman, (b) IR; NO_3^- (c) Raman, (d) IR; SO_4^{2-} (e) Raman, (f) IR and ClO_4^- (g) Raman and (h) IR.

3.2. Thermal behaviour of the anions in hydrotalcite

All the TGA patterns of the hydrotalcites, containing various anions, are characterized by a weight loss between 10 and 20 wt% due to loss of interlayer water with a maximum around 75-115°C. For CO_3-hydrotalcite (Fig. 3) The dehydration takes place in two minor steps followed by a larger step around 197°C. These steps can be interpreted as being due to the loss of adsorbed water followed by loosely bound interlayer water and finally water coordinated to the interlayer carbonate.

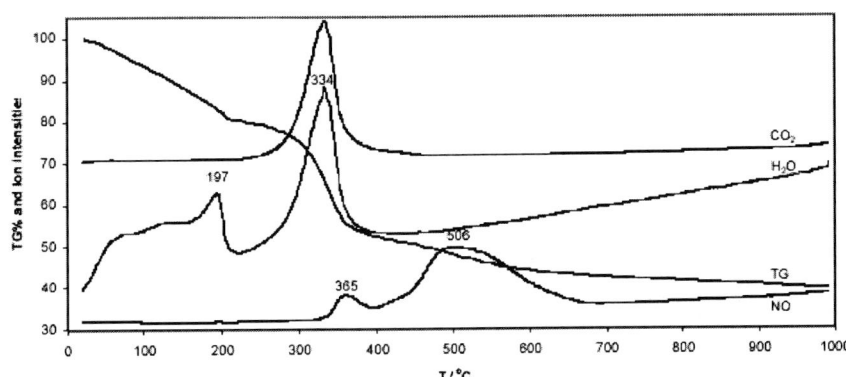

Fig. 3. TGA-MS patterns of CO_3-hydrotalcite.

The three steps of dehydration are also observed in the corresponding differential thermal analysis pattern. The interlayer carbonate is released as CO_2 simultaneously with water around 335°C followed by NO around 365 and 500°C. This water loss around 335°C is due to the dehydroxylation of the Mg/Al-hydroxide sheets.

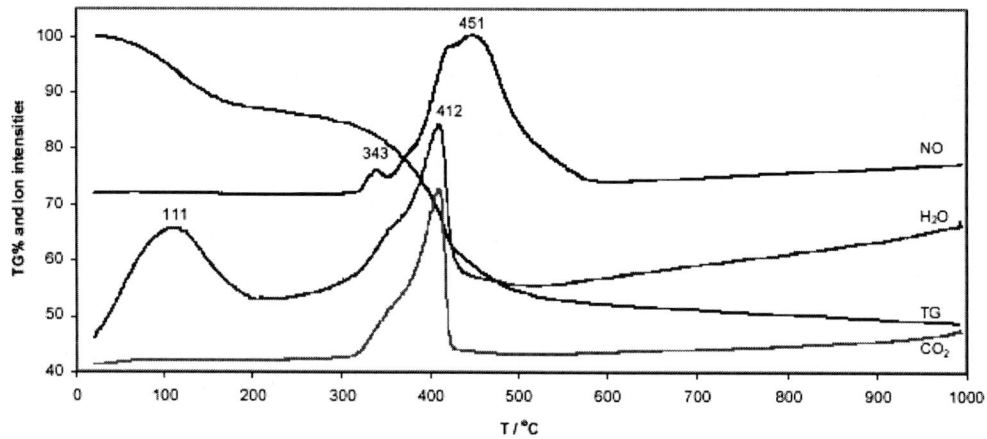

Fig. 4. TGA-MS pattern of NO_3-hydrotalcite.

The stability of the NO_3-hydrotalcite (Fig. 4) is higher than that of the CO_3-hydrotalcite (Fig. 3). Since some carbonate was taken up in the hydrotalcite structure during the synthesis as shown by the infrared spectrum a major loss of CO_2 is observed around 412°C. The dehydration in this case takes place as a single event over a large temperature range up to 200°C, while dehydroxylation and the associated loss of H_2O around 412°C. The interlayer nitrated decomposes to NO with a minor loss around 345°C followed by a major loss around 450°C. It seems that the presence of nitrate in the interlayer stabilizes the interlayer carbonate resulting in an increase in temperature of about 80°C before this carbonate is lost. So far, there is no clear explanation for the minor NO loss around 345°C. The presence of $NaNO_3$ as a cause can be excluded since $NaNO_3$ melts at a higher temperature and decomposition does not take place until 650°C

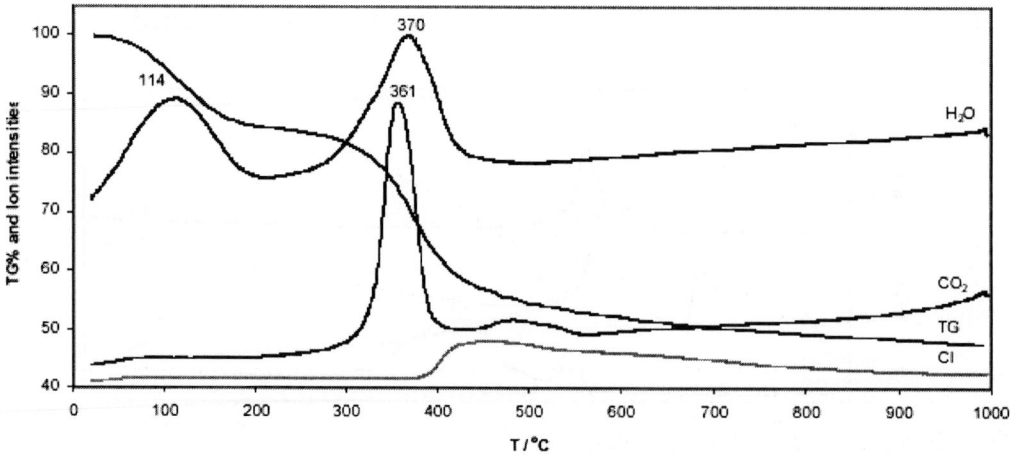

Fig. 5. TGA-MS patterns of Cl-hydrotalcite.

The Cl-hydrotalcite (Fig. 5) shows a similar behaviour for the H2O loss (dehydration) as the NO3-hydrotalcite, but Cl is lost over a larger and higher temperature range from 400 to 900°C. As with the nitrate the interlayer chloride anions seems to stabilize the interlayer carbonate as shown by the fact that after the major CO2 loss around 360°C a second step is observed around 500°C coinciding with the major release of Cl2. No HCl was detected.

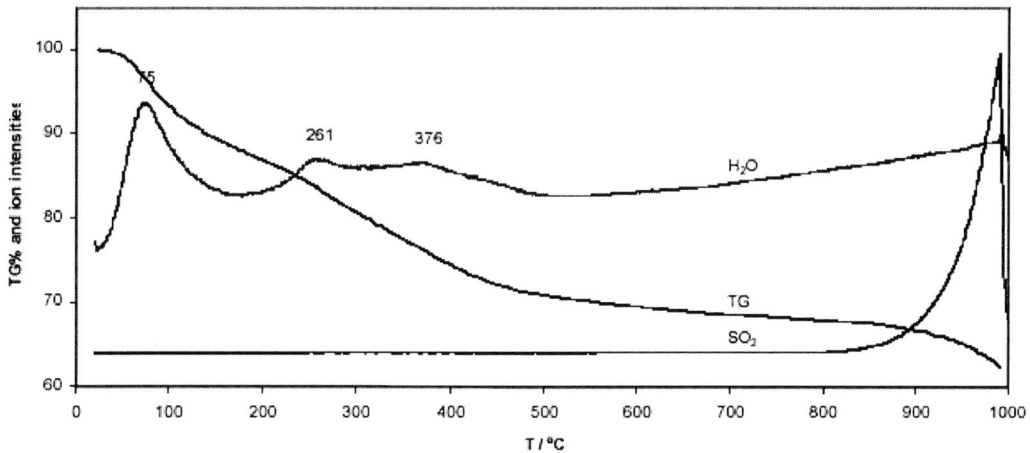

Fig. 6. TGA-MS pattern of SO4-hydrotalcite.

Completely different is the thermal behaviour of SO4- (Fig. 6) and ClO4-hydrotalcites (Fig. 7). SO4-hydrotalcite shows a gradual weight-loss due to dehydroxylation with two minor water peaks around 260 and 375°C, while the sulphate remains in the structure. The sulphate is not lost as SO2 until a temperature of 850°C is reached. This may be interpreted as being due to the formation of separate phases of magnesium and aluminium sulphate upon dehydroxylation. The ClO4-hydrotalcite shows a complex thermal behaviour with 2 steps of water loss around 375 and 440°C, where the second step is accompanied by the loss of O2.

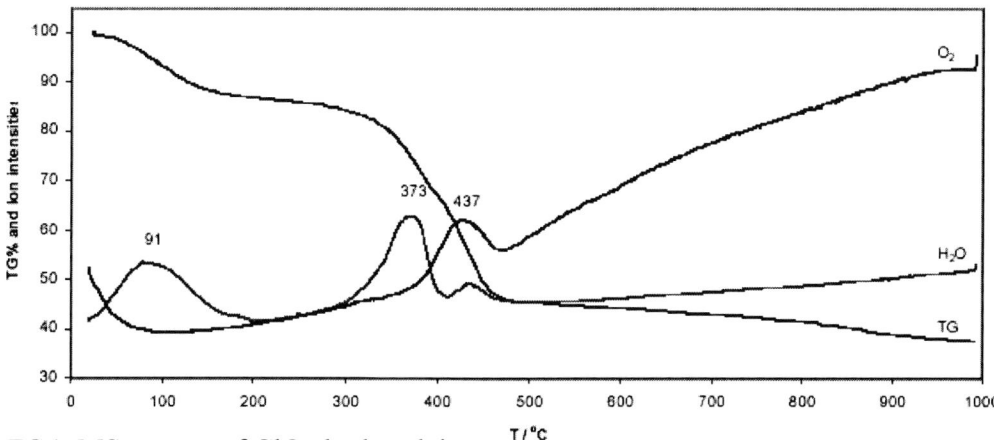

Fig. 7. TGA-MS pattern of ClO4-hydrotalcite.

REFERENCES

Bish, D.L. (1980) Anion-exchange in takovite: applications to other hydroxide minerals. Bulletin de Mineralogie, 103, 170-175.

Brindley, G.W. and Kikkawa, S. (1979) A crystal-chemical study of Mg,Al and Ni,Al hydroxy-perchlorates and hydroxy-carbonates. American Mineralogist, 64, 836.

Brindley, G.W. and Kikkawa, S. (1980) Thermal behavior of hydrotalcite and of anion exchanged forms of hydrotalcite. Clays and Clay Minerals, 28, 87-91.

Hickey, L., Kloprogge, J.T. and Frost, R.L. (2000) The effects of various hydrothermal treatments on Mg/Al - hydrotalcites. Journal of Materials Science, 35, 4347-4355.

Kloprogge, J.T. and Frost, R.L. (1999) Fourier Transform Infrared and Raman spectroscopic study of the local structure of Mg, Ni and Co - hydrotalcites. Journal of Solid State Chemistry, 146, 506-515.

Kloprogge, J.T., Wharton, D. and Frost, R.L. (2000) Raman and infrared spectroscopy of interlayer CO_3^{2-}, NO_3^-, SO_4^{2-} and ClO_4^- in hydrotalcites. ICORS 2000, p. 96, Beijing, China.

Kloprogge, J.T., Wharton, D., Hickey, L. and Frost, R.L. (2002) Infrared and Raman study of interlayer anions CO_3^{2-}, NO_3^-, SO_4^{2-} and ClO_4^- in Mg/Al-hydrotalcite. American Mineralogist, 87, 623-629.

Marino, O. and Mascolo, G. (1982) Thermal stability of MgAl double hydroxides modified by anionic exchange. Thermochimica Acta, 55, 377-383.

Miyata, S. and Kumura, T. (1973) Synthesis of new hydrotalcite-like compounds and their physico-chemical properties. Chemistry Letters, 843-848.

Miyata, S. and Okada, A. (1977) Synthesis of hydrotalcite-like compounds and their physico-chemical properties - the system Mg^{2+} - Al^{3+} - SO_4^{2-} and Mg^{2+} - Al^{3+} - CrO_4^{2-}. Clays and Clay Minerals, 25, 14-18.

Sato, T., Wakabayashi, T. and Shimada, M. (1986) Adsorption of various anions by magnesium aluminium oxide ($Mg_{0.7}Al_{0.3}O_{1.15}$). Industrial & Engineering Chemistry Product Research and Development, 25, 89-92.

Titulaer, M.K. (1993) Porous Structure and Particle Size of Silica and Hydrotalcite Catalyst Precursors, Geologica Ultraiectina 99, p. 268. Utrecht University, Utrecht, The Netherlands.

Vaccari, A. (1998) Preparation and Catalytic properties of cationic and anionic clays. Catalysis Today, 41, 53-71.

Spectroscopic analysis and X-ray diffraction of zinnwaldite

J. Theo Kloprogge[1], Sjerry van der Gaast[2], Peter M. Fredericks[1] and Ray L. Frost[1]

[1] Centre for Instrumental and Developmental Chemistry, Queensland University of Technology, 2 George Street, GPO Box 2434, Brisbane, Qld 4001, Australia.

[2] Netherlands Institute for Sea Research (NIOZ), P.O. Box 59, 1790 AB Den Burg, The Netherlands

This paper describes an X-ray diffraction and spectroscopic study, including infrared, near-infrared and Raman spectroscopy of some selected zinnwaldites. In general, zinnwaldite forms a member of the trioctahedral true micas with characteristically Li in the octahedral positions and low iron contents. Although the infrared spectrum of zinnwaldite has been described before, near infrared and Raman spectroscopy have not been used so far to study this mineral. X-ray diffraction showed that all the samples reported in this study have the 1M structure. The Raman spectra are characterised by a strong band at 700-705 cm^{-1} plus a broad band associated with the SiO modes around 1100 cm^{-1}. Less intense bands are observed around 560, 475, 403 and 305 cm^{-1}. The corresponding IR spectra show strong overlapping SiO modes around 1020 cm^{-1} plus less intense bands around 790, 745, 530, 470-475 and 440 cm^{-1}. Two overlapping OH-stretching modes can be observed around 3550-3650 cm^{-1}, in agreement with a broad band in the IR around 3450 cm^{-1} and a complex band around 3630 cm^{-1}. The near-IR spectra basically reflect combination and overtone bands associated with protons in the zinnwaldite structure. A very broad band observed around 5230 cm^{-1} is characteristic for adsorbed water while bands around 4530, 4435 and 4260 cm^{-1} can be ascribed to metal-hydroxyl groups.

1. INTRODUCTION

Zinnwaldite is a phyllosilicate of the group of the trioctahedral true micas. Its properties are similar to that of biotite, while it resembles lepidolite in its chemical aspects with high Si/Al ratio, high Li content and replacement of OH by F. In its structure it usually exhibits an ordering of the octahedral cations with Al in one of the M2 sites and Fe^{2+}, Li in the M1 and the other M2 site, causing a lowering of the symmetry of the space group from *C2/m* to *C2* (Rieder et al., 1996). The *1M* polytype seems to be characteristic although other polytypes are possible (Èerný and Burt, 1987). Zinnwaldites are commonly found in pegmatites rich in rare elements, especially Li, where they are associated with other lithium bearing minerals such as lepidolite and spodumene. The mineral got its name from the locality Zinnwald, Erzgebirge, Saksen (nowadays known as Cínovec in the Czech Republic).

This paper describes the X-ray diffraction and spectroscopic study, including infrared, near-infrared and Raman spectroscopy of some selected zinnwaldites. Although the infrared spectrum of zinnwaldite has been described before, the near-infrared and Raman spectroscopic techniques have not been used so far to study this mineral.

2. ANALYTICAL TECHNIQUES

The XRD analyses were carried out on a Philips wide angle PW 1050/25 vertical goniometer equipped with a graphite diffracted beam monochromator. The radiation applied was Cu$K\alpha$. The samples were measured at 50 % RH in stepscan mode with steps of 0.02° 2θ.

The finely powdered samples were combined with oven dried spectroscopic grade KBr (containing approximately 1 wt% sample) and pressed into a disc under vacuum. The spectra were recorded in triplicate by accumulating 512 scans at 4 cm^{-1} resolution in the spectral range between 400 cm^{-1} and 4000 cm^{-1} using either the Perkin-Elmer 1600 series Fourier transform infrared spectrometer equipped with a LITA detector or a Nicolet Magna 750 Fourier transform infrared spectrometer equipped with a DTGS detector.

The micromount and single crystal samples were placed on a polished metal surface on the stage of an Olympus BHSM microscope, which is equipped with 10x, 20x, and 50x objectives. The microscope is part of a Renishaw 1000 Raman microscope system, which also includes a monochromator, a filter system and a CCD detector (1024 pixels). The Raman spectra were excited by a Spectra-Physics model 127 He-Ne laser producing highly polarised light at 633 nm and collected at a resolution of 2 cm^{-1} and a precision of $\pm$ 1 cm^{-1} in the range between 200 and 4000 cm^{-1}. Repeated acquisition on the crystals using the highest magnification (50x) were accumulated to improve the signal to noise ratio in the spectra. Spectra were calibrated using the 520.5 cm^{-1} line of a silicon wafer.

Spectral manipulation such as baseline correction were performed using the Spectracalc software package GRAMS (Galactic Industries Corporation, NH, USA). Band component analysis was undertaken using the Jandel 'Peakfit' software package that enabled the type of fitting function to be selected and allows specific parameters to be fixed or varied accordingly. Band fitting was done using a Lorentzian-Gaussian cross-product function with the minimum number of component bands used for the fitting process. The Gaussian-Lorentzian ratio was maintained at values greater than 0.7 and fitting was undertaken until reproducible results were obtained with squared correlations of r^2 greater than 0.995.

Samples used in this study were taken from the Van der Marel Collection (located at the Netherlands Institute for Sea Research (NIOZ) under curatorship of SvdG):

871 Zinnwald, Saksen, $(K_{0.95}Na_{0.05})Li_{0.934}(Fe^{2+})_{0.595}Al_{1.139}(Si_{3.394}Al_{0.635})O_{10}(OH_{0.617}F_{1.383})$

1106 Saksen, $(K_{0.92}Na_{0.07})Li_{1.04}(Fe^{2+})_{0.60}(Fe^{3+})_{0.06}Mn_{0.04}Al_{1.05}(Si_{3.26}Al_{0.74})O_{10}(OH_{0.41}F_{1.66})$

2012 Hoegur, Algeria, no analysis

2133 Tordal, Norway, no analysis, contains 1 % Sc (van der Marel and Beutelspacher, 1976)

3. RESULTS AND DISCUSSION

The XRD patterns show that three of the four zinnwaldites used in this study belong to the 1M polytype. Sample 2133 belongs to the 2M1 polytype. Sample 2012 contains an unidentified impurity with sharp peaks at 8.01, 4.616, 4.297 and 2.873 Å and two broad peaks at 4.008 and 3.028 Å (Fig. 1)

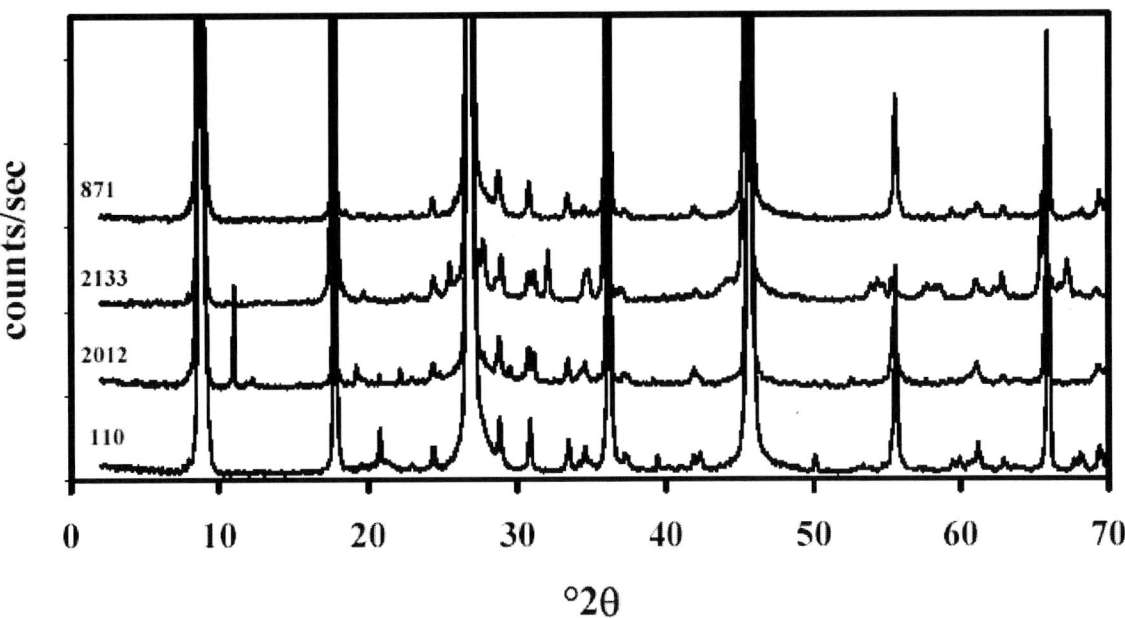

Fig. 1. XRD patterns of the zinnwaldite samples 871, 2133, 2012 and 1106.

The Raman spectra in the low wavenumber region are characterised by a strong OH-libration mode at 700-705 cm^{-1} plus a broad band associated with the SiO stretching vibratons around 1100 cm^{-1} (Fig. 2). Minor bands are observed around 560, 475, 403 and 305 cm^{-1}. The corresponding IR spectra show strong overlapping SiO modes around 1020 cm-1 plus less intense bands around 790, 745, 530, 470-475 and 440 cm^{-1}. Similar bands were reported by Gadsden (1975) for the infrared spectrum of zinnwaldite at 1100, 1040, 1010, 780, 740, 702, 530 and 435 cm^{-1}. The band associated with the AlO$_4$ unit in the tetrahedral sheet, which is coupled to the vibrations of SiO$_4$, is shifted to 790 cm^{-1} due to the high amount of Fe in the structure of zinnwaldite, similar to the shift observed for ferrous phlogopites (Vedder, 1964; Farmer, 1974). The other bands below 600 cm^{-1} are probably associated with the SiO bending modes strongly coupled with vibrations of the octahedral cations, while OH-librational modes can be observed between 600 and 950 cm^{-1} (Farmer, 1974).

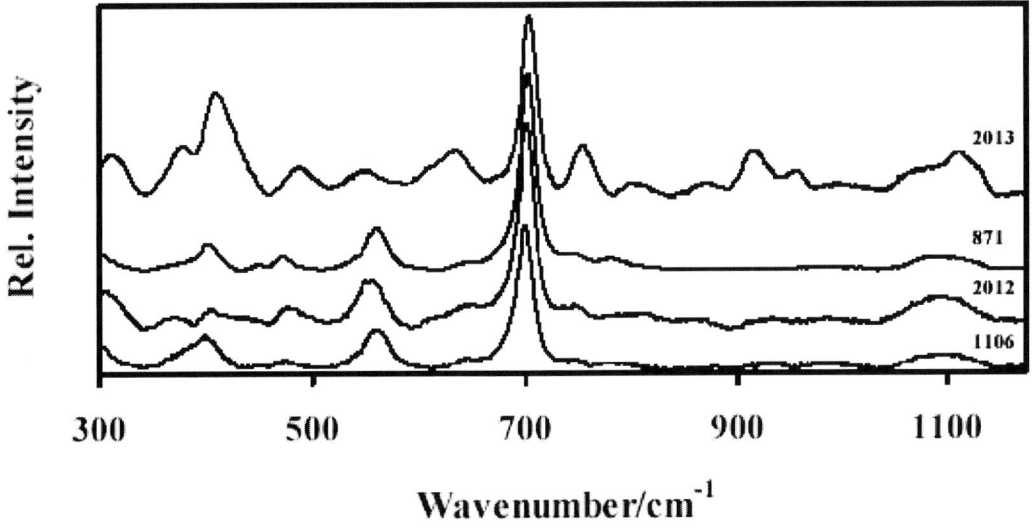

Fig. 2. Raman spectra in the region between 300 and 1175 cm^{-1} of the zinnwaldite samples 2013, 871, 2012 and 1106.

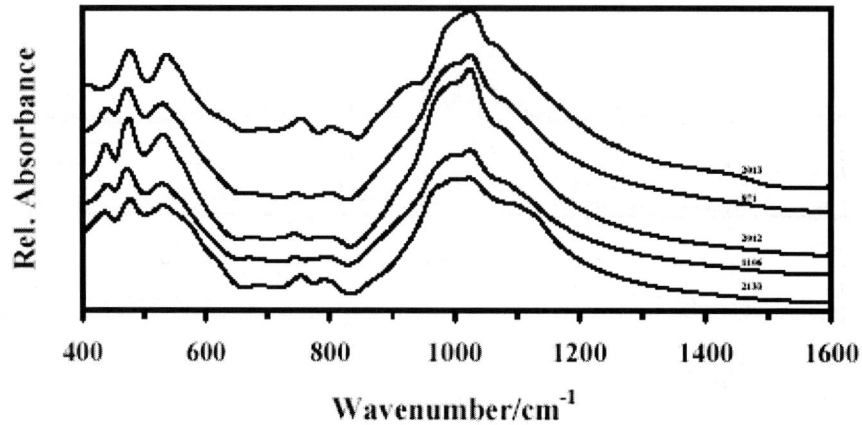

Fig. 3. FT-IR spectra in the range between 400 and 1600 cm^{-1} of the zinnwaldite samples 2013, 871, 2012 and 1106.

The effect of Li on the OH-stretching frequencies is not well known (Farmer, 1974). For zinnwaldite two overlapping OH-stretching modes can be observed in the Raman spectra (Fig. 4) around 3550-3650 cm^{-1}, in agreement with a broad band in the IR around 3450 cm$_{-1}$ and a complex band around 3630 cm^{-1} (Fig. 5). Gadsen (1975) reports also two bands around 3440 and 3571 cm^{-1}, while for another Li-mica, ephesite a characteristic band was observed around 3608 cm^{-1} and for Li-containing smectites a value around 3640 cm$_{-1}$ has been found (Calvet and Prost, 1971). The differences between the zinnwaldite spectra can be explained by the substitution of the OH-groups by F.

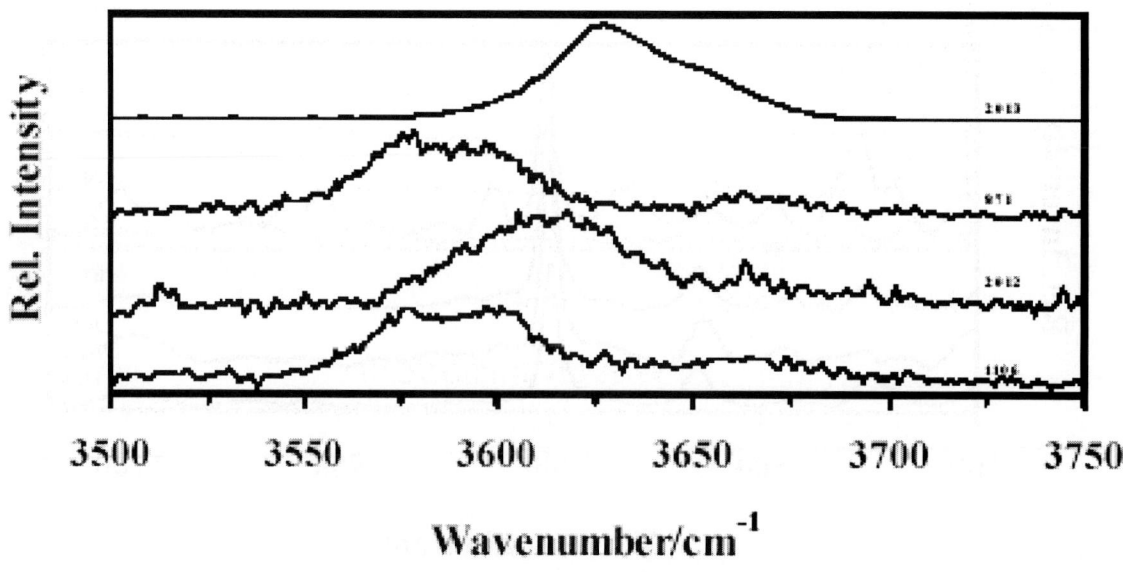

Fig. 4 and 5. Raman (top) and FT-IR (bottom) spectra in the range between 3500 and 3750 cm^{-1} of the zinnwaldite samples 2013, 871, 2012 and 1106.

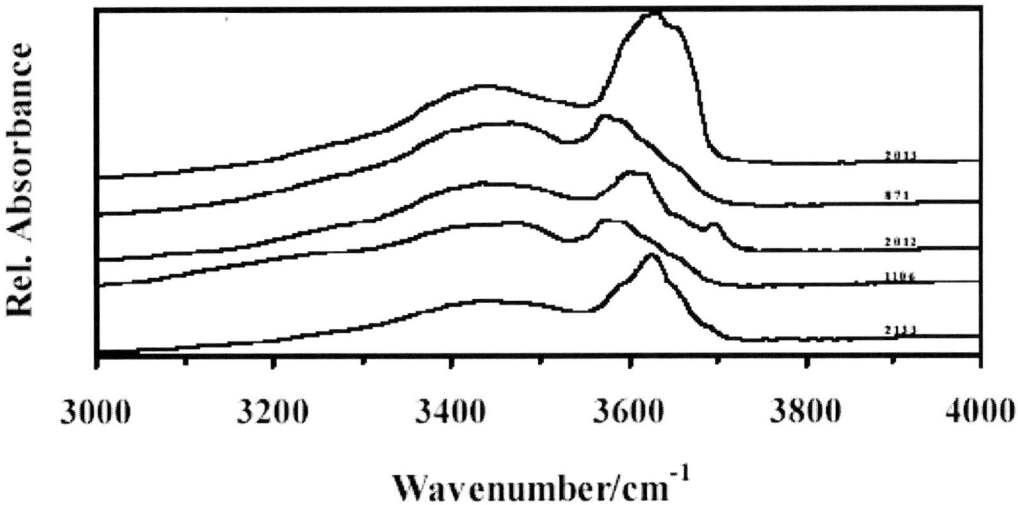

The near-IR spectra basically reflect combination and overtone bands associated with protons in the zinnwaldite structure. Characteristic are a very broad band around 5230 cm^{-1} characteristic for adsorbed water plus bands around 4530, 4435 and 4260 cm^{-1} that can be ascribed to metal-hydroxyl groups (Kloprogge et al., 2000).

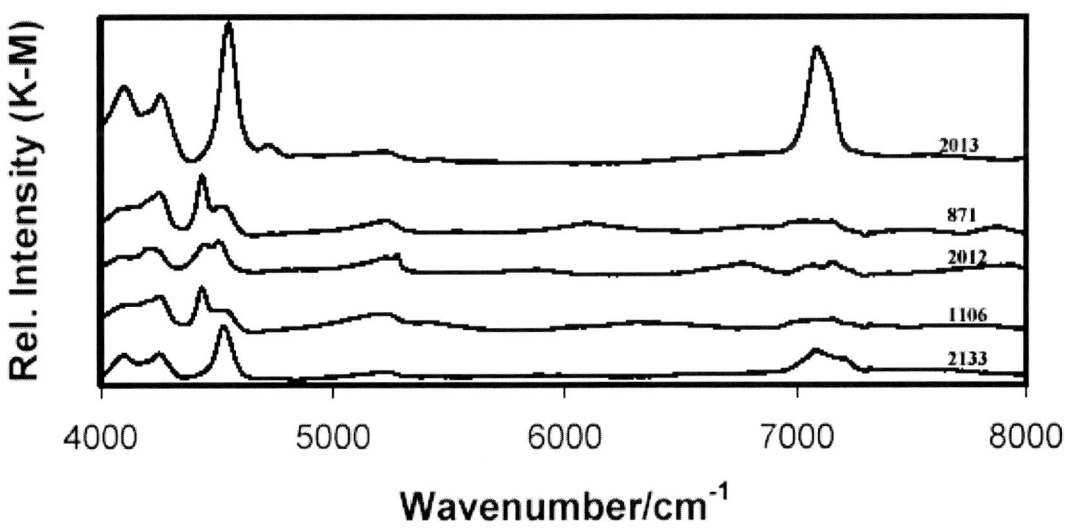

Fig. 6. Near-IR spectra in the range from 4000 to 8000 cm^{-1} of the zinnwaldite samples 2013, 871, 2012, 1106 and 2133.

REFERENCES

Calvet, R. and Prost, R. (1971) Cation migration into empty octahedral sites and surface properties of clays. Clays & Clay Minerals, 19, 195-186.

Èerný, P. and Burt, M.D.(1987) Paragenesis, crystallochemical characteristics, and geochemical evolution of micas in granite pegmatites. In: Micas. Reviews in Mineralogy vol.13, S. W. Bailey (ed.), 257-298.

Gadsen, J.A. (1975) Infrared spectra of minerals and related inorganic compounds. Butterworth. London, pp 277.

Farmer, V.C. (1974) The Layer silicates. In: The infrared spectra of minerals. V.C. Farmer (ed.) Mineralogical Society monograph 4, 331-363.

Kloprogge, Ruan and Frost (2000) Near-infrared spectroscopic study of synthetic and natural pyrophyllite. Neues Jahrbuch für Mineralogie, Monatshefte, 337-347.

Marel, van der H.W. and Beutelspacher, H. (1976) Atlas of infrared spectroscopy of clay minerals and their admixtures. Elsevier, Amsterdam, 397 pp.

Rieder, M., Hybler, J., Smrcok, L. and Weiss, Z. (1996) Refinement of the crystal structure of zinnwaldite 2M_1. European Journal of Mineralogy, 8, 1241-1248.

Vedder, W. (1964) Correlations between infrared spectrum and chemical composition of mica. American Mineralogist, 49, 736-768.

Clay mineralogy and tensile strength of different soils from the southwestern Buenos Aires province, Argentina

M del P. Moralejo [a], O. E. Soulages [a], S. G. Acebal [a], M. E. Aguirre [b] and R. M. Santamaría [c]

[a] Department of Chemistry and Chemical Engineering, Universidad Nacional del Sur, Avda. Alem 1253, (8000) Bahía Blanca, Argentina

[b] Department of Agronomy, Universidad Nacional del Sur, Altos del Palihue, (8000) Bahía Blanca, Argentina

[c] CIC. Comisión de Investigaciones Científicas de la Provincia de Buenos Aires.

Soil strength is a sensitive indicator of physical status of soils and influences agricultural uses in different ways.

Changes in tensile strength (Yc) with soil-particle size, clay mineralogy, organic carbon content and cation exchange capacity (CEC) were assessed for eight different soils from the southwestern Buenos Aires province, Argentina. A positive linear correlation was obtained when clay content and Yc, and clay plus fine silt and Yc were compared. Tensile strength increased as illite (I) content increased. The influence of dominant clay mineral on tensile strength also depended on clay mineral crystallinity. No relation was found between smectite (S) and randomly interestratified mineral (I/S) content and Yc. Other soil properties such as organic carbon and CEC were informed. Tensile strength and organic carbon (OC) contents of soils were related in a complex way. The potential regression between soil CEC and tensile strength was significant ($R^2 = 0.81$).

1. INTRODUCTION

Many of the agriculturally important soils in the world exhibit physical problems due to densification either in the upper layer or in the subsoil. Soils under dry land farming are also suffering from compaction of land during sowing, weed control and other agricultural activities by using heavy equipment on wet soils. Due to the suppressing effects of high strength on plant root and shoot growth, the strength of compact soils has become an increasing concern.

Soil strength is a sensitive indicator of physical status of soils and influences agricultural use of soils in different ways, viz. power requirement of tillage operations, and on root penetration and seedling emergence [1][2]. Soil strength is influenced by several factors associated with the soil clay fraction such as exchangeable cations [3][4], clay content [5][6], clay type [7], the amount of dispersible clay [4][8] and organic matter status of soils [9].

The results from many experiments showing an increased strength with increasing clay content for a given soil type have led to the conclusion that clay particles are involved in binding or cementing soil particles [5][6][10]. Clay mineralogy may also affect the nature of individual contact points and the strength of binding between soil particles [11]. The surface characteristics of clay particles are related to their morphology. High swelling clays like smectites always cause higher strength compared to other minerals [7]. In spite of considerable research on this topic, the specific effect on soil strength of clay with its variations in morphology, size and other surface properties is not clearly understood.

Soil OC enhances soil strength in a complex way [12]. Soil OC can act as an aggregating or segregating material or have unknown influence on aggregate stability, depending on its composition in soil and the relative contributions of other aggregating-stabilizing substances [13].

In this paper we report the influence on soil strength of soil-particle size, clay mineralogy, organic carbon content and cation exchange capacity of eight different soils from the southwestern Buenos Aires province, Argentina.

2. MATERIALS AND METHODS

2.1. Soils

Eight surface layers (0-10 cm) of soils collected from different locations in South West Buenos Aires province were used in this study. Six Mollisols (LC, Ma, Mu, Si, Bo, BB), one Entisol (AA) and one Aridisol (As) were air-dried, ground, and passed through a 2-mm stainless steel sieve.

Some physicochemical properties are listed in Table 1.

2.2. Separation of clay fraction

The clay fraction (< 2 µm) from eight soil samples was separated by mechanical dispersion (1:5 soil/solution ratio), sedimentation and decantation.

2.3. X – ray diffraction

Clay mineralogy was determined by XRD analysis. The untreated different fractions and residues from Mg saturation followed by glycerol solvatation, and K saturation followed by heating to 823 K were examined from 2° (2θ) to 30° (2θ) using a 0.01° step and a counting time of 30 s per 2θ incremented, using CuKα (1.5406 Å) radiation (36 kV, 18 mA) in an X-Pert Philips PW 3710 diffractometer.

The percentage of clay minerals was determined by means of the intensity of the diffraction peak. The evaluation is better characterised as being semiquantitative rather than quantitative [18].

2.4. Tensile strength determination

Separate, triplicate soil samples were placed in 56 mm diameter and 35 mm length aluminum cylinders and equilibrated at –100 kPa (ψ_m) moisture tension, using a pressure membrane plate, during 10 days. The water content was approximated the critical water

Table 1
Physicochemical properties of soils

Soil	Soil taxonomy	Particle size analysis [a]				OC[b]	pH		CEC[e]	ψ_m [f]
		Sand g.kg⁻¹	Coarse Silt g.kg⁻¹	Fine Silt g.kg⁻¹	Clay g.kg⁻¹	g.kg⁻¹	H₂O[c]	KCl[d]	cmol$_c$.kg⁻¹	(-100kPa)
Lc	Typic Argiudoll	185.2	203.0	309.2	302.6	33.5	5.9	5.2	19.8	275.3
Ma	Typic Argiudoll	331.0	196.0	237.0	236.0	26.9	6.3	5.3	18.0	246.5
Mu	Petrocalcic Paleustoll	504.4	144.0	133.2	218.4	29.0	7.5	6.3	17.1	195.4
Si	Pachic Argiudoll	420.4	237.3	202.2	140.0	47.8	6.4	5.2	15.0	252.7
Bo	Entic Haplustoll	464.3	241.5	174.8	148.0	11.4	8.0	8.3	13.2	199.3
Bb	Entic Haplustoll	739.3	43.7	75.0	142.0	9.0	6.8	5.9	11.2	99.6
Aa	Typic Ustipsamment	655.5	128.2	90.6	125.5	7.1	7.7	7.4	9.6	133.3
As	Typic Camborthid	836.7	42.9	46.9	73.5	4.0	6.5	5.7	8.0	75.6

[a] Robinson (1922): Pipette method [14].
[b] Organic carbon (OC). Walkley (1946): OC method.[15]
[c] Measured in water suspension. 1:2.5 soil/solution ratio
[d] Measured in 1mol L⁻¹ KCl 1:2.5 soil/solution ratio
[e] Cation exchange capacity (CEC). Bower et al. (1952): CEC method. [16]
[f] Matric Potencial (ψ). Richards (1949). Pressure membrane plates method. [17]

content for maximum compressibility [19]. The wetted soils were compacted at a pressure of 300 kPa in an oedometer during 3 minutes and oven dried at 40°C for one day. The individual soil disks, placed on their sides, were crushed between two parallel plates [20] and Yc was calculated using the formula:

$$Yc = (2F/\pi dL)$$

where F is the force applied at the point of failure, d and L are the diameter and height of the soil disks respectively.

3. RESULTS AND DISCUSION

3.1. Soil-particle size

Soil strength has been shown to be influenced by soil texture [21]. A feature of the soils was the wide range of the textures. Clay contents from 73.5 to 302.6 g.kg^{-1}, fine-silt (2-20 μm) from 75.0 to 309.2 g.kg^{-1}, coarse-silt (20-50 μm) from 42.9 to 241.5 g.kg^{-1} and sand from 185.2 to 836.7 g.kg^{-1}.

A positive relationship between Yc and clay content (Figure 1a) was obtained ($R^2 = 0.77$). The relative amount of clay fraction regulates the frequency of bonds within the fraction in comparison to bonds with other size fractions. Tensile strength increased with increasing clay plus fine-silt content ($R^2 = 0.87$) (Figure 1b). Both clay and fine-silt contribute to increase Yc because of their similar mineralogy. Linear correlation between Yc and sand content also was significant ($R^2 = 0.82$) (Figure 1c). Tensile strength decreased as total sand content increased. These results support the hypothesis that a greater number of contact points per unit volume leads to greater adhesion between particles and hence greater tensile strength [11].

3.2. Clay mineralogy

Different clay types contribute to different strengths development [11]. Clay mineralogy of studied soils is summarized in Table 2.

Tensile strength increased with increasing illite contents of clay fraction referred to soil clay contents ($R^2 = 0.70$) (Figure 2). When smectite contents and Yc were compared, no correlation was found. A similar result was obtained between I/S content and Yc. This suggests the influence of a dominant clay mineral and its crystallinity on tensile strength. Generally, high soil strength was observed whenever a 2:1 type of clay was present.

3.3. Organic carbon content

A direct relationship between Yc and OC content of eight soils could not be verified (Figure 3). A relationship was found when the studied soils were classified into low, intermediate and high clay plus fine-silt levels.

For Ma and Lc soils with clay plus fine-silt higher than 400 g.kg^{-1}, OC may need to be much higher to have any effect on Yc [22]. For As, Aa and Bb soils with clay plus fine-silt < 200 g.kg^{-1}, sand becomes more important influencing Yc.

Organic carbon content appears to have a great influence on Yc for Bo, Mu and Si soils with intermediate clay plus fine-silt levels (approximately 340 g.kg^{-1}). An increase in organic carbon content leads to decrease Yc. This suggests that organic material shows a higher degree of elasticity under compression forces than do mineral particles.

When clay OC contents and Yc were compared, no relationship was found.

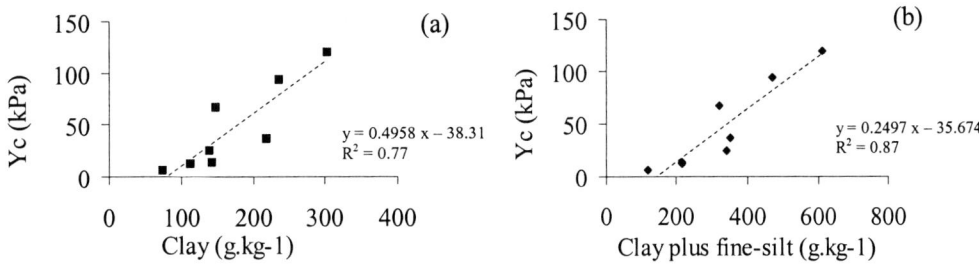

Figure 1. The effect on soil-tensile strength (Yc) of (a) clay contents, (b) clay plus fine-silt contents and (c) sand contents.

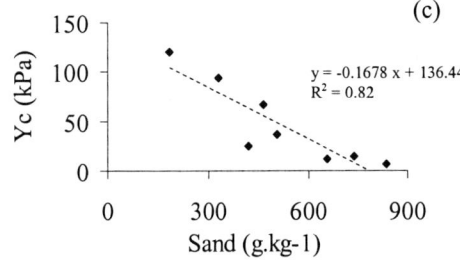

Figure 2. Relationship between tensile strength and illite content of soil-clay fraction.

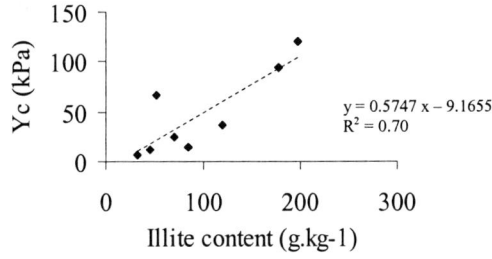

Figure 3. Tensile strength versus soil organic carbon content

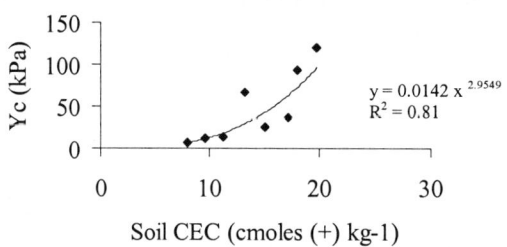

Figure 4. Relationship between tensile strength and soil cation exchange capacity.

Table 2
Measured variables and mean tensile strength of the soils studied.

Sample	Smectite		Interstratified		Illite		Impurities	Tensile Strength
	a %	c	a %	c	a %	c		kPa
Lc	10	P	25	P	65	G	ab. Q.F.	120.00
Ma	20	P	5	P	75	G	sc. Q.F.	93.60
Mu*	25	P	20	P	55	G	sc. Q.F.	36.50
Si	20	P	30	P	50	G	ab. Q.F.	24.60
Bo	35	P	30	P	35	G	sc. Q.F.	67.00
Bb	30	P	10	P	60	G	sc. Q.F.	13.97
Aa	50	P	10	P	40	M	sc. Q.F.	12.33
As	45	P	10	P	45	M	sc. Q.F.	6.27

a: abundance, c: crystallinity, G: Good, M: Moderate, P: Poor, Q.: Quartz, F.: Feldspars, ab.: abundant, sc.: scarce. Mu*: + traces of kaolinite.

3.4. Cation exchange capacity

The total charge on soil is related to the clay type and organic carbon content. High charge may increase the strength of bonding between particles. Therefore, we wanted to test the effect of soil-CEC on soil strength. A highly positive significant potential regression was found between Yc and soils CEC ($R^2 = 0,81$) (Figure 4). This suggests that soils containing greater amounts of illitic minerals of good crystallinity give rise to greater tensile strength.

4. CONCLUSIONS

In the present study, changes in tensile strength of different Argentinean soils were investigated due to soil properties. Although the wide range of soil properties, the relationships developed have shown that Yc is principally related to soil-particle size, clay contents, clay plus fine-silt contents and clay mineralogy. Yc value increased with increasing illite contents and is also dependent on clay mineral crystallinity.

Soil-particle size and clay mineralogy appear to be the most important parameters influencing soil tensile strength for the eight Argentinean soils.

REFERENCES

1. A.R. Dexter, Catena Suppl., 11 (1988[a]) 35.
2. A.R. Dexter, Soil Tillage Res., 20 (1991) 87.
3. A.R. Dexter and K.Y. Chan, J. Soil Sci., 44 (1991) 219.
4. A.R. Barzegar, R.S. Murria, G.J. Churchman and P. Rengasamy, Aust. J. Soil Res., 32 (1994[a]) 185.
5. W.D. Kemper, R.C. Rosenau and A.R. Dexter, Soil Sci. Soc. Am. J., 51 (1987) 860.
6. G.J. Ley, C.E. Mullins and R.Lal, Soil Tillage Res., 13 (1989) 365.
7. G. Spoor, G., P.B.Leeds-Harrison and R.J. Godwin, J. Soil Sci., 33 (1982) 427.
8. R.T. Shanmuganathan and J.P.Oades, Geoderma, 29 (1983) 257.
9. A.R.Barzegar, J.M. Oades, P. Rengasamy and L.Giles, Soil Tillage Res., 32 (1994[b]) 329.
10. K.J. Coughlan, R.J. Loch and W.E. Fox, Aust. J. Soil Res., 16 (1978) 283.
11. A.R. Barzegar, J.M. Oades, P. Rengasamy and R.S. Murray, Geoderma, 65 (1995) 93.
12. B.D. Soane, Soil Tillage Res., 16 (1990) 179.
13. S. Goldberg, B.S. Kapor and J.D. Rhoades, Soil Sci., 150 (1990) 588.
14. G.W. Robinson, J. Agr. Sci., 12 (1922) 306.
15. A. Walkley, Soil Sci., 63 (1946) 251.
16. C.A. Bower, R.F. Reitemeier and M. Fireman, Soil Sci, 73 (1952) 251.
17. L.A. Richards, Agron. J., 41 (1949) 489.
18. M.L. Jackson (eds), Soil Chemical Analysis, Advanced Course, 2nd Edition, 11th printing, Published by the author, Madison,Wis., 1979.
19. M.E. Aguirre, M.A. Commegna, R.M. Santamaria y L. Castro, XXVII Congresso Brasileiro de Ciencia do Solo, Brasilia, Brasil, 1999.
20. A.D. Dexter and B. Kroesbergen, J. Agric. Eng. Res., 31 (1985) 139.
21. C.J. Gerard, Soil Sci. Soc. Am. Proc., 29 (1965) 641.
22. C.W. Smith, M.A. Johnston and S. Lorentz, Geoderma, 78 (1997) 93.

Table 1
Physicochemical properties of soils

Soil	Soil taxonomy	Particle size analysis [a]				OC^b	pH		CEC^e	ψ_m^f
		Sand g.kg^{-1}	Coarse Silt g.kg^{-1}	Fine Silt g.kg^{-1}	Clay g.kg^{-1}	g.kg^{-1}	H_2O^c	KCl^d	cmol$_c$.kg^{-1}	(-100kPa)
Lc	Typic Argiudoll	185.2	203.0	309.2	302.6	33.5	5.9	5.2	19.8	275.3
Ma	Typic Argiudoll	331.0	196.0	237.0	236.0	26.9	6.3	5.3	18.0	246.5
Mu	Petrocalcic Paleustoll	504.4	144.0	133.2	218.4	29.0	7.5	6.3	17.1	195.4
Si	Pachic Argiudoll	420.4	237.3	202.2	140.0	47.8	6.4	5.2	15.0	252.7
Bo	Entic Haplustoll	464.3	241.5	174.8	148.0	11.4	8.0	8.3	13.2	199.3
Bb	Entic Haplustoll	739.3	43.7	75.0	142.0	9.0	6.8	5.9	11.2	99.6
Aa	Typic Ustipsamment	655.5	128.2	90.6	125.5	7.1	7.7	7.4	9.6	133.3
As	Typic Camborthid	836.7	42.9	46.9	73.5	4.0	6.5	5.7	8.0	75.6

[a] Robinson (1922): Pipette method [14].
[b] Organic carbon (OC). Walkley (1946): OC method.[15]
[c] Measured in water suspension. 1:2.5 soil/solution ratio
[d] Measured in 1mol L^{-1} KCl 1:2.5 soil/solution ratio
[e] Cation exchange capacity (CEC). Bower et al. (1952): CEC method. [16]
[f] Matric Potencial (ψ). Richards (1949). Pressure membrane plates method. [17]

Molecular and particulate organisation in dye-clay films prepared by the Langmuir-Blodgett method

Robin H. A. Ras[a], Bart van Duffel[a], Mark Van der Auweraer[b], Frans C. De Schryver[b] and Robert A. Schoonheydt[a]

[a]Centrum voor Oppervlaktechemie en Katalyse, Katholieke Universiteit Leuven, Kasteelpark Arenberg 23, B-3001 Heverlee (Leuven), Belgium

[b]Afdeling Fotochemie en Spectroscopie, departement Scheikunde, Katholieke Universiteit Leuven, Celestijnenlaan 200F, B-3001 Heverlee (Leuven), Belgium

Using the Langmuir-Blodgett method it is possible to construct ultrathin dye-clay films which possess a double organisation. The clay particles are organised in a closely packed monolayer as can be seen by atomic force microscopy. Fluorescence spectroscopy and fluorescence microscopy reveal that two different organic dye molecules are homogeneously distributed in the film. Energy transfer between two dye molecules is improved by the presence of clay particles in suspension as well as in clay films.

1. INTRODUCTION

Nanostructured organic/inorganic films are a promising new class of materials with interesting characteristics for technological applications such as light harvesting systems, sensors and non-linear optical devices. The combination of organics and inorganics in hybrid films, is considered to result in improved properties caused by a synergic effect[1].

An elegant manner for the preparation of nanostructured films is the Langmuir-Blodgett (LB) technique. LB films have been extensively studied because the controlled molecular architecture exhibits properties which cannot be observed in isotropic materials[2,3]. Recently, clay particles have been succesfully introduced in LB films[4,5].

When clay particles come into contact with a layer of amphiphilic cations at the water surface, adsorption of organic cations onto clay can occur via an ion exchange mechanism and a hybrid film of clay and organic molecules is formed. If the amphiphiles are dye molecules, functional units of cooperating molecules can be constructed that exhibit specific properties like excitation energy transfer.

Excitation energy transfer between two different dyes provides a sensitive tool for structure analysis. Because of the distance dependence of the efficiency of energy transfer at the

---

Robin H. A. Ras and Bart Van Duffel acknowledge a Ph.D. grant from *Instituut voor de aanmoediging van Innovatie door Wetenschap en Technologie in Vlaanderen*' (IWT). The authors acknowledge financial support from IUAP program 4-11 "Supramolecular Chemistry and Catalysis" and from FWO-Vlaanderen (Krediet aan Navorsers, n° 1.5.064.00).

nanometer scale, it is frequently applied to investigate biological and supramolecular systems[6]. To probe the structure of multilayered LB films, Kuhn et al. studied energy transfer between the layers[7]. Besides a tool for analysis, energy transfer is an important process in many applications including photosynthesis. For energy transfer to happen, dye molecules need to be organised close to each other. Avnir et al. reported enhanced energy transfer of dyes adsorbed onto clay particles in suspension[8].

In this work excitation energy transfer from an excited donor dye to an acceptor dye is investigated in solution, suspension and in LB films with and without clay. It is found that the presence of clay particles favours energy transfer. Hybrid films have been constructed with a double organisation: organisation of the clay particles and organisation of the dye molecules.

2. EXPERIMENTAL SECTION

2.1. Materials

The cationic dyes 3,3'-Diethyloxacarbocyanine iodide (OXA) (Aldrich) and RhodamineB (RhB) (Acros) were used in solution and suspension. 3,3'-Dioctadecyloxacarbocyanine perchlorate (OXA18) (Molecular Probes) and Octadecyl RhodamineB chloride (RhB18) (Molecular Probes) were used in films. The clay, natural saponite, was obtained from the Source Clay Minerals Repository of the Clay Minerals Society. The saponite was Na^+ saturated by repeated exchange with 1 M NaCl solutions followed by dialysis with water until Cl^- free (AgCl test). The particle size fraction between 0.5 and 2.0 µm was separated by centrifugation after which the clays were freeze-dried and stored in a dark room. The cation exchange capacity (CEC) was determined by the ^{22}Na method[9] to be 738 µeq/g. The water used as subphase for film preparation was MilliQ water (MilliPore) with a resistivity above 18 MΩ·cm. In all other experiments deionised water was used. As substrate for the films, microscopy glass slides were used, cleaned for 30 minutes with H_2SO_4/H_2O_2 (70/30 v/v%) at 90°C and thoroughly rinsed with deionised water. The hydrophilic glass slides were stored in deionised water.

2.2. Methods

Preparation of Dye-Clay Suspensions and Dye Solutions. The clay was suspended in deionised water (0.02 wt%) and stirred for at least 24 h. Stock solutions of 10^{-6} M OXA and RhB were prepared as well as a mixture of both dyes with the ratio 100:1 OXA:RhB (10^{-6} M OXA and 10^{-8} M RhB). Equal volumes of the stock solutions and suspensions were mixed and stirred for at least 1 h to obtain clay suspensions (0.01 wt%) with a dye concentration of 5×10^{-7} M. The coverage of the clay with dye was 0.7 % the CEC. Aqueous solutions with the same dye concentration were prepared for comparison.

Preparation of Films. Films were prepared by the Langmuir-Blodgett (LB) method (Figure 1). LB films and surface pressure versus molecular area (π-A) isotherms were made using a Nima LB trough model 611. As subphase MilliQ water and 0.005 wt% clay suspensions were

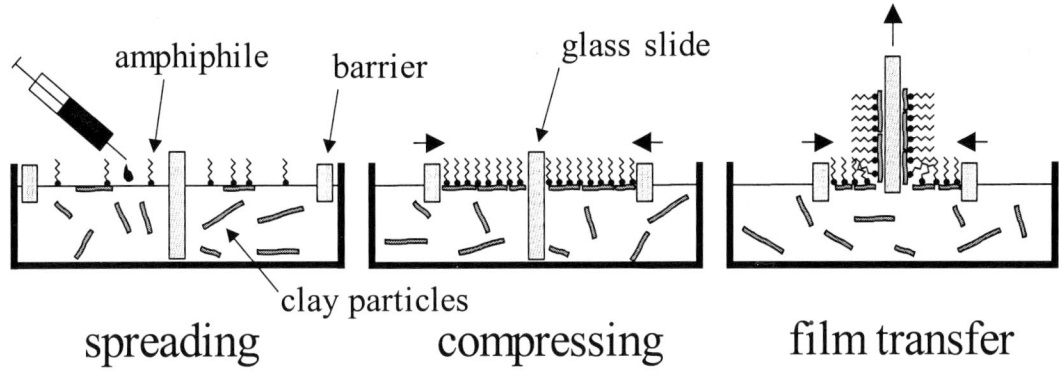

Figure 1. Schematic presentation of the preparation of dye-clay films by the Langmuir-Blodgett method.

used. 40 μl of a 10^{-3} M chloroform solution of the dyes (pure OXA18, pure RhB18 and the ratio 100:1 OXA18:RhB18) was spread on the subphase at a total area of 500 cm². The Langmuir film was compressed 15 minutes after spreading at a compression rate of 30 cm²/min. The surface pressure was measured with a Wilhelmy balance using a filter paper as plate. The floating film was transferred by vertical dipping in upstroke onto a hydrophilic glass slide at a constant surface pressure of 10.0 mN/m. Transfer ratios were determined to be 1.0 ± 0.1.

2.3. Measurements

Absorption spectra were recorded from 375 nm to 650 nm on a Perkin Elmer Lambda 12 spectrometer. Water was used as reference for solutions and suspensions, a glass slide as reference for films. Emission spectra were recorded on a SPEX Fluorolog spectrofluorimeter from 480 nm to 750 nm for solutions and suspensions and from 490 nm to 750 nm for films. The solutions and suspensions were excited at 470 nm whereas the films were excited at 480 nm. Solutions and suspensions were measured in a 10 mm quartz cell at right angle configuration. The films were measured in front face configuration. Atomic force microscopy (AFM) images were obtained on a Topometrix Discoverer system TMX 2010 with a Si_3N_4 tip in contact mode (spring constant = 0.032 mN/m). Fluorescence microscopy images were made on a Nikon Optiphot-2 with a Spot RT Slider camera from Diagnostic Instruments.

3. RESULTS & DISCUSSION

3.1. Characterisation of the monolayer at the air-water interface

3.1.1. Π-A isotherm

The surface pressure vs area (Π-A) isotherms of OXA18 and RhB18 on water and on the clay suspension are shown in figure 2. For OXA18 on water the surface pressure starts to rise at 150 Å² per molecule. The increase of the surface pressure is a two-step process, possibly indicating a phase transition. The second step starts at 15 mN/m and a molecular area of 95 Å². In the presence of clay particles the molecular area for lift-off of the surface pressure is 125 Å², significantly below the 150 Å² in the absence of clay.

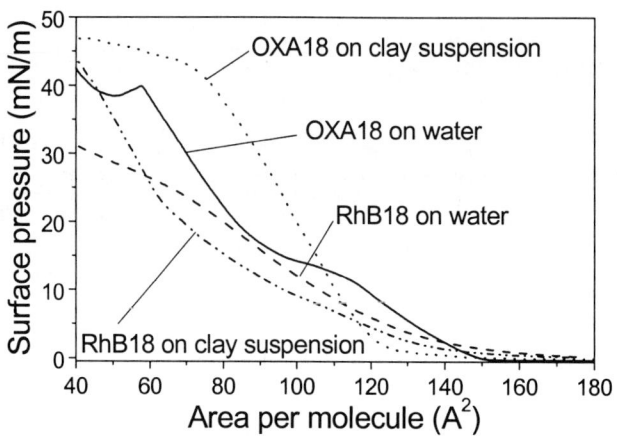

Figure 2. Surface pressure vs molecular area (Π-A) isotherms of OXA18 and RhB18 films on water and on 0.005 wt% saponite suspension.

Also the surface pressure continuously increases and there is no indication of a two-step process. Both films collapse at a surface pressure of 40 mN/m. For RhB18 the isotherms are different. The isotherms start to rise gradually at about 170 Å^2. The isotherm of RhB18 measured on clay suspension is shifted to lower molecular area than that on water. The RhB18 films collapse at about 20 mN/m. The isotherms of the films with ratio 100:1 OXA18:RhB18 are similar to those of pure OXA18.

3.2. Characterisation of the Langmuir-Blodgett film

3.2.1. Atomic force microscopy (AFM)

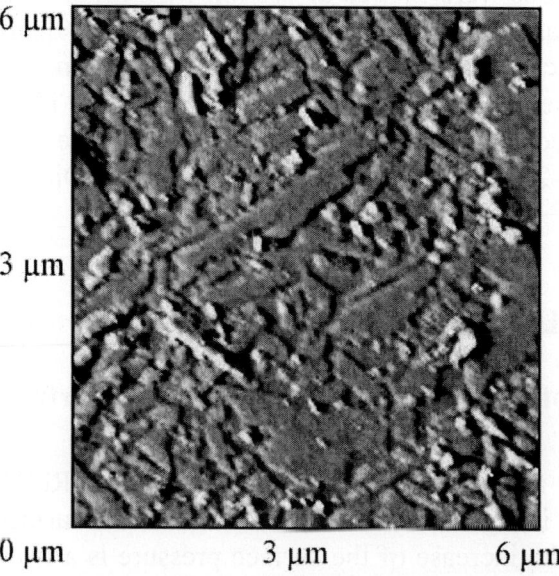

Figure 3. AFM image (cantilever deflection signal) of a saponite film with ratio 100:1 OXA18:RhB18 deposited at 10 mN/m.

The AFM image (figure 3) of the LB films prepared on the clay suspension clearly shows the presence of clay particles with a predominant rectangular shape. Sizes vary from about 100 nm to several µm. As a measure of surface roughness the standard deviation of the topography is used and is calculated to be between 0.8 and 1.5 nm. This is very low and indicates that the particles are lying with their flat surface parallel to the substrate. The clay particles are organised in the film in such a way that the particles touch each other at the edges to cover almost all the surface without considerable overlap.

3.2.2. UV-VIS absorption and fluorescence spectroscopy

Solution (5×10^{-7}M OXA, 5×10^{-7}M RhB and 100:1 ratio of these dyes)

Absorption and emission spectra of the dyes in solution are shown in figure 4. At this low concentration the dyes form only monomers in aqueous solution. Absorption maxima and emission maxima of OXA and RhB in solution, suspension and in films are represented in table 1. When the mixture 100:1 OXA:RhB is excited at 470 nm, no RhB emission is observed. The molecules OXA and RhB are too far separated to exhibit energy transfer in solution.

0.01 wt% saponite suspension (5×10^{-7}M OXA, 5×10^{-7}M RhB and 100:1 ratio of these dyes)

Absorption and emission spectra of the clay suspensions containing OXA, RhB and the 100:1 ratio are shown in figure 5. The spectroscopic properties of the dyes OXA and RhB in clay suspension change compared to solution where only monomers are present. An absorption band appears at 460 nm which can be dedicated to dimers of OXA[10]. The shoulder at 486 nm indicates the presence of monomers of OXA. The absorption band of RhB is broadened due to heterogeneous adsorption sites on the clay and is red shifted to 570 nm. Similar results are obtained by Chaudhuri et al. for the Rhodamine3B / Hectorite system[11].

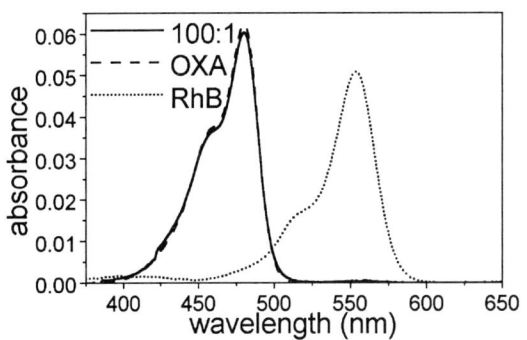 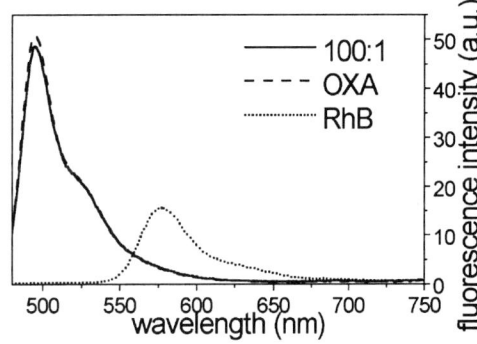

Figure 4. Absorption (left) and emission spectra (right; excitation at 470 nm) of solutions containing pure OXA, pure RhB and the 100:1 ratio of these dyes.

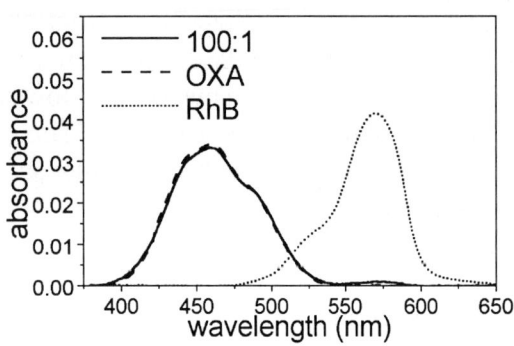 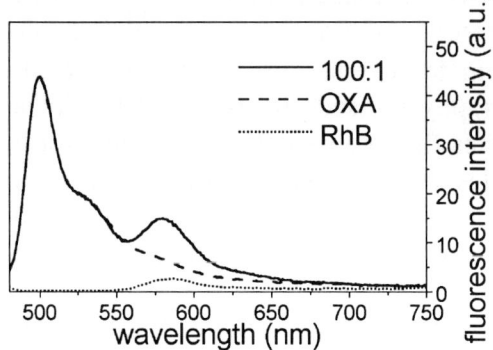

Figure 5. Absorption (left) and emission spectra (right; excitation at 470 nm) of 0.01 wt% saponite suspensions containing pure OXA, pure RhB and the 100:1 ratio of these dyes.

The emission spectra of both dyes are red shifted compared to solution (500 nm for OXA and 586 nm for RhB). When the two dyes (100:1 ratio) are simultaneously adsorbed onto the clay in suspension the emission of RhB is increased. This is evidence for energy transfer from the donor OXA to the acceptor RhB. Avnir et al. have reported similar enhanced energy transfer between cationic organic dyes adsorbed onto clay particles[8].

Table 1
Wavelengths of absorption (abs.) and emission (em.) maxima (nm) of OXA and RhB in solution, in suspension and of OXA18 and RhB18 in films with and without clay.

	solution		clay suspension		film without clay		clay film	
	abs.	em.	abs.	em.	abs.	em.	abs.	em.
OXA	480	495	460	500	499	508	496	510
RhB	554	577	570	586	575	586	580	592

LB films prepared on water and on clay suspension as subphase

Absorption and emission spectra of LB films containing OXA18, RhB18 and the 100:1 ratio are shown in figures 6 and 7 for films prepared on water and on a clay suspension respectively. RhB18 in the mixed LB films shows no noticeable absorbance because its concentration is too small to be detectable. However the presence of RhB18 in the mixed film can be seen by fluorescence spectroscopy (not shown, excitation at 550 nm) because this technique is very sensitive. When excited at 480 nm the film prepared on water containing the two dyes has a sensitised fluorescence of RhB18. This is evidence for energy transfer from OXA18 to RhB18. The distance between the two molecules in the monolayer is small enough to induce energy transfer. The clay film containing the two dyes even exhibits a stronger sensitised fluorescence of RhB18 (figure 7). Energy transfer in the clay film is more efficient than in the film without clay. This indicates that the average distance between the molecules in the clay film is even smaller hence the distribution of the dyes is more homogeneous.

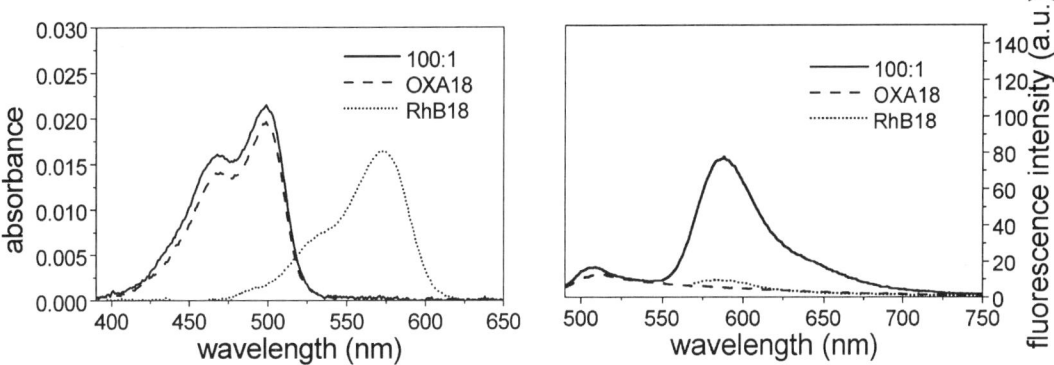

Figure 6. Absorption (left) and emission spectra (right; excitation of OXA18 and 100:1 at 480 nm and of RhB18 at 550 nm) of LB films *prepared on water* containing pure OXA18, pure RhB18 and the 100:1 ratio of these dyes.

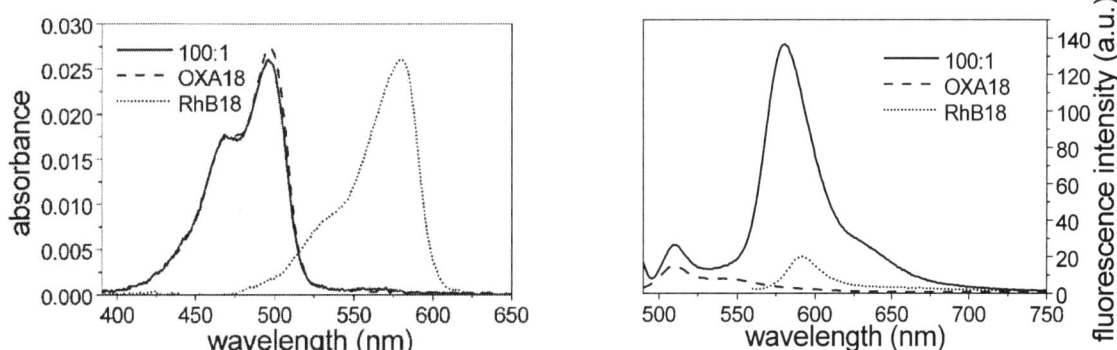

Figure 7. Absorption (left) and emission spectra (right; excitation of OXA18 and 100:1 at 480 nm and of RhB18 at 550 nm) of LB films *prepared on clay suspension* containing pure OXA18, pure RhB18 and the 100:1 ratio of these dyes.

3.2.3. Fluorescence microscopy

Fluorescence microscopy is a commonly used technique for imaging dye domains in monolayer films[12]. The 100:1 ratio film prepared on water (figure 8, left) contains domains with pure RhB18 that are embedded in a matrix of mainly OXA18. The clay film on the other hand (figure 8, right) has a much more homogeneous distribution of both dyes and no distinct dye domains are visible. Clay particles have a noticeable effect on the distribution of the dyes in the monolayer on a micron scale.

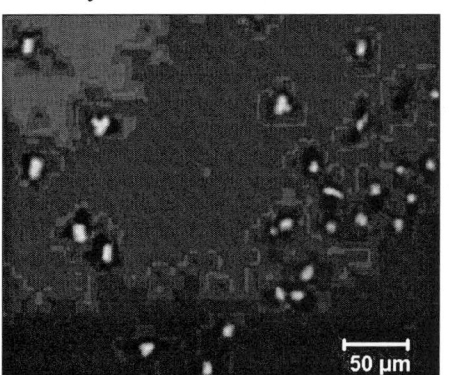

Figure 8. Fluorescence microscopy images of the 100:1 OXA18:RhB18 films prepared on water (left) and on clay suspension (right) (excitation 510 - 560 nm; emission > 590 nm).

4. CONCLUSIONS

Hybrid films consisting of clay and organic dye molecules have been constructed by the Langmuir-Blodgett method and a double organisation is created: organisation of the clay particles and organisation of organic molecules. AFM reveals that the clay sheets are arranged in a compact monolayer. Although RhB18 in the mixed film is present in a low concentration (100:1 ratio OXA18:RhB18) a large sensitised RhB18 fluorescence is visible in the films, indicating efficient energy transfer from OXA18 to RhB18. The clay films exhibit an even more pronounced sensitised fluorescence compared to films without clay. This is possible only if the two dyes are located close to each other. Hence RhB18 is homogeneously distributed in the film. Fluorescence microscopy pictures show the presence of domains in the film without clay, while none are visible in the clay film. This means that RhB18 is homogeneously distributed on a micrometer scale.

The effect of clay on energy transfer is also clear in suspension. Solutions show no sensitised fluorescence while suspensions do. Similar results with suspensions are presented by Avnir et al. who state that energy transfer is improved due to a concentration effect of dyes on clay sheets.

The Langmuir-Blodgett method can be applied for the preparation of hybrid nanostructured materials. In this work we investigated energy transfer, but by selecting the organic and inorganic components carefully, materials with other attractive properties might be constructed.

REFERENCES

1. P. Gomez-Romero, Adv. Mater., 13 (2001) 163-174.
2. H. Fuchs, H. Ohst and W. Prass, Adv. Mater., 3 (1991) 10-18.
3. G. Roberts, Langmuir-Blodgett films, Plenum, New York, 1990.
4. K. Tamura, H. Setsuda, M. Taniguchi and A. Yamagishi, Langmuir, 15 (1999) 6915-6920.
5. Y. Umemura, A. Yamagishi, R. Schoonheydt, A. Persoons and F. De Schryver, Langmuir, 17 (2001) 449-455.
6. J. Lakowicz, Principles of Fluorescence Spectroscopy, Plenum, New York, 1983.
7. H. Kuhn, D. Möbius and H. Bücher in Physical Methods of Chemistry, edited by A. Weissberger and B. Rossiter, vol.1, part 3B, Wiley, New York, 1972.
8. D. Avnir, Z. Grauer, S. Yariv, D. Huppert and D. Rojanski, New J. Chem., 10 (1986) 153-157.
9. P. Peigneur, A. Maes and A. Cremers, Clays Clay Miner., 23 (1975) 71-75.
10. H. Bücher, K.H. Drexhage, M. Fleck, H. Kuhn, D. Möbius, F.P. Schäfer, J. Sondermann, W. Sperling, P. Tillmann and J. Wiegand, Molecular Crystals, 2 (1967) 199-230.
11. F. Chaudhuri, F. López Arbeloa and I. López Arbeloa, Langmuir, 16 (2000) 1285-1291.
12. H. Möhwald, Annu. Rev. Phys. Chem., 41 (1990) 441-476.

Si-Al-Mg and Si-Al-Zn Montmorillonite : synthesis and characterization by EXAFS and quantitative ^{27}Al MAS-NMR

M.Reinholdt[a,b], J. Miehé-Brendlé[a], L. Delmotte[a], M.-H. Tuilier[b] and R. Le Dred[a]

[a] Laboratoire de Matériaux Minéraux, (UMR 7016), Ecole Nationale Supérieure de Chimie de Mulhouse, Université de Haute Alsace, 3 rue A. Werner F-68093 Mulhouse Cedex France

[b] Laboratoire de Physique et de Spectroscopie Electronique, (UMR 7014), Université de Haute Alsace, 4 rue des Frères Lumière F-68093 Mulhouse Cedex France

Two sets of montmorillonites with various contents of divalent cations (Mg or Zn) substituting Al in their octahedral sheets were prepared in acidic fluoride medium under hydrothermal conditions. All the samples were characterized by XRD, TGA-DTA, ^{29}Si, ^{19}F and ^{27}Al MAS NMR. Pure products were also studied by EXAFS at Al, Mg, Si and Zn K-edges and quantitative ^{27}Al MAS-NMR. The first oxygen shell coordination radii were deduced from EXAFS data. Moreover, the analysis of the more distant shells around Zn, Al and Si show that the distribution of Zn and Al in the octahedral sheet deviates from a statistical one and that Si and octahedral Al atoms seem to be in an environment close to the ideal montmorillonite lattice. The quantitative ^{27}Al MAS-NMR allows to determine the absolute Al content of the synthetic montmorillonites.

1. INTRODUCTION

Among the natural clay minerals, smectites and especially Montmorillonites are one of the most important groups in soils and sediments. Their layered structure consists of an octahedral sheet with various metallic elements (Al, Mg, Fe...) enclosed between two sheets of tetrahedra containing Si and sometimes Al. Isomorphous substitutions in octahedral or tetrahedral sheets lead to negative charges balanced by cations present in the interlayer spacing. Depending on the deposits, smectites have various structural compositions, various particle sizes and contain impurities such as Si-polymorphs. The best way to overcome this drawback is to synthesize such materials.

The aim of this work is to synthesize montmorillonites containing Al and only one additional divalent metallic element (Mg or Zn) in their octahedral sheets. Indeed, only few studies were performed in this field as recently mentioned by Kloprogge[1]. The fluorine route is chosen to perform the syntheses, this latter enables the formation of well crystallized clay minerals[2]. Two sets of syntheses were performed in the $MgO-Al_2O_3-SiO_2$ and $ZnO-Al_2O_3-SiO_2$ systems, in order that the highest amount is incorporated in the octahedral sheet. Beyond the usual characterization techniques (XRD, TGA – DTA, chemical analysis), solid state

Nuclear Magnetic Resonance (NMR) and Extended X-ray Absorption Fine Structure (EXAFS) are used in order to investigate the distribution of the metallic atoms in the structure. A peculiar attention is paid to the determination of the Al_{VI} to Al_{IV} ratio and quantitative aluminum-27 (^{27}Al) NMR experiments are carried out in this purpose.

2. EXPERIMENTAL

2.1. Syntheses

Two series of syntheses were performed in the $MgO-Al_2O_3-SiO_2$ and $ZnO-Al_2O_3-SiO_2$ systems in acidic and fluoride medium under hydrothermal conditions[3]. The compositions of hydrogel were based on a theoretical formula of montmorillonite : $Na_{2x}[(Al_{2(1-x)} M_{2x}) Si_4 O_{10} (OH,F)_2]$, n H_2O, where M represents the divalent element (Mg or Zn), x ranges from 0.0 to 1.0 and n equals 96.

The hydrogels were prepared starting from Mg or Zn acetates, Si and Al oxides, NaF, HF and H_2O. Mixtures were aged during 2 h at room temperature. At this stage, pH are in the range 4.5 – 6.0. The hydrogels were then hydrothermally treated in a PTFE lined stainless steel autoclave at 220 °C for 72 hours under autogenous water pressure. After crystallization, the products were separated by filtration. The pH of the surnatant equals 3.5 - 4.5. The dried samples were placed under controlled humidity ($P/P_0 = 80\%$).

2.2 Characterization

The X-ray powder diffraction patterns were recorded on a diffractometer using Cu K_α radiation with fixed divergence slits (PW1100). Diffractograms were performed on powders and under controlled humidity ($P/P_0 = 0{,}80$ - RH)[3,4].

Elemental analysis of Na-saturated clays was then performed by the Service Central d'Analyse of the Centre National de la Recherche Scientifique (Vernaison, France) by using Atomic Absorption Spectroscopy for Na, Mg, Zn, Al and Si element. The hydration of samples was determined by thermogravimetic analysis (TGA)[3,4].

The X-ray absorption experiments were realized at the Laboratoire pour l'Utilisation du Rayonnement Electromagnétique (LURE, Orsay, France). Measurements were performed on the D21 beam line of the DCI storage ring, at the Zn K-edge in transmission mode at 20 K. The soft X-ray absorption spectra were collected, at the Mg and Si K-edges in fluorescence mode and at the Al K-edge in total electron yield mode, on the SA32 beam line of the Super-Aco storage ring. Measurements were performed at room temperature in a chamber at a pressure of 10^{-9} bar. The Al and Si K-edges spectra were analyzed through the conventional procedure[5]. After background subtraction and conversion to k-space, the EXAFS data were k^3-weighted and Fourier transformed. Then the Fourier filtered contributions (FFC) of the nearest neighbors (NN) peak were simulated in the classical way in order to determine the structural information[4].

Aluminum-27 (^{27}Al) Nuclear Magnetic Resonance (NMR) experiments using Magic Angle Spinning technique (MAS) were performed on a Bruker MSL-300 spectrometer (magnetic field of 7 T) at 78.2 MHz. A standard 4 mm Cross Polarization and Magic Angle Spinning (CP-MAS) Bruker probe was used and the spinning speed was 10 kHz. The frequency of the

RF field (ν_{rf} = 14.3 kHz) corresponds to a $\pi/2$ pulse width of 17.5 µs for an aqueous solution (1.0 M Al(NO$_3$)$_3$). The excitation behavior of the radio-frequency pulse ν_{rf} is of major importance when a quantitative magnetic resonance study is required. It determines the changes in the spin population and consequently governs the intensity of the resulting signal[6,7].

3. RESULTS AND DISCUSSION

Figure 1 summarizes the results obtained after XRD analysis and indicates the phases as a function of the hydrogel compositions. It appears that a mixture of trioctahedral compound and amorphous solid was obtained for high contents of divalent element in the hydrogel. For intermediate composition of the hydrogel, mixtures of di- and trioctahedral compounds and amorphous solid were observed. Finally, pure montmorillonite was synthesized for a slight amount of Zn or Mg content. Only four samples will be considered later and were labeled Mg01 (x = 0.10), Mg02 (x = 0.20), Zn01 (x = 0.10) and Zn02 (x = 0.20).

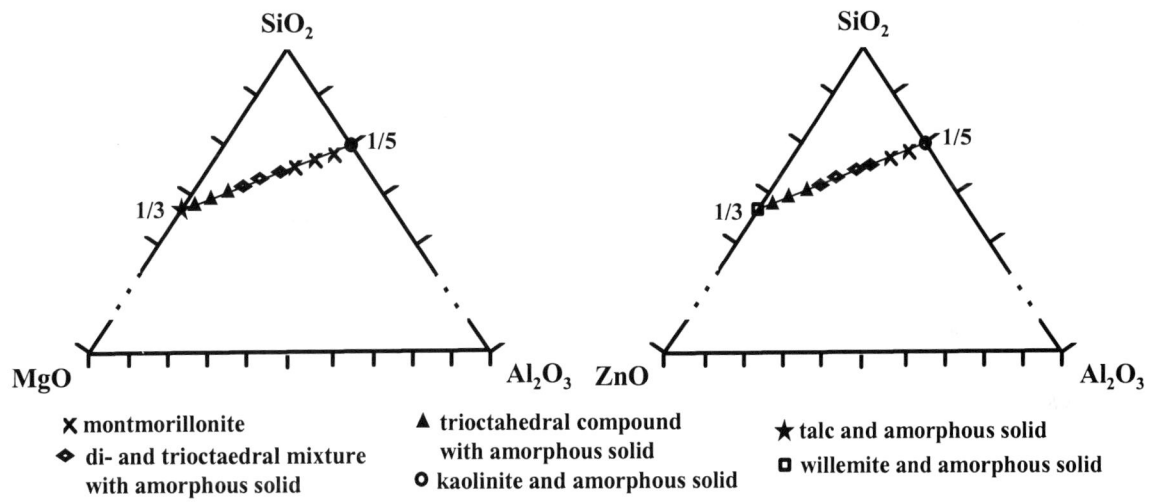

Figure 1 : The phases obtained according to the molar composition of the hydrogel (the space between two ticks represent 10%).

XRD patterns of each selected sample are typical of clays. For example, the diffractogram of Mg01 sample is shown on Figure 2. The (hkl) bands observed correspond to those of a smectite[8]. The d_{001} values are equal to 16.5 Å, 16.1 Å, 17.4 Å and 16.4 Å for samples Mg01, Mg02, Zn01 and Zn02 respectively. These values are slightly higher than those indexed in the literature (d_{001} = 15 - 16 Å), [9,10] but they are in good agreement with those determined, in the same conditions, for two natural montmorillonites from Camp Berteaux, Morocco (d_{001} = 15.7 Å) and Ivancice, Czech Republic (d_{001} = 16.5 Å). Each sample exhibits a (060) peak at about 1.49 Å corresponding to a dioctahedral compound.

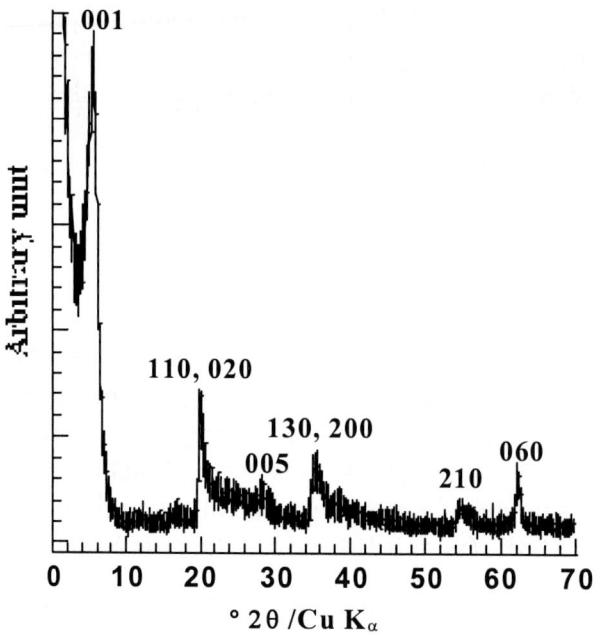

Figure 2 : Diffractogram of MgO$_2$ sample.

The k-weighted $\chi(k)$ Al K-edge EXAFS data are shown in Figure 3 together with the Fourier filtered contribution (FFC) of the nearest neighbors (NN) peak. The Al-O bond length derived from the simulations is $R_{Al-O} = 1.95 \pm 0.03$ Å. This value is identical to the distance from metal to six-fold coordinated oxygen in the ideal montmorillonite lattice[11]. It is about 0.15 Å shorter than the Zn-O ($R_{Zn-O} = 2.08 \pm 0.02$ Å) and Mg-O ($R_{Mg-O} = 2.11 \pm 0.02$ Å) distances derived from the previous Zn and Mg K-edge EXAFS studies of these samples[3]. The Al coordination polyhedra are much less distorted than Zn or Mg ones. The Zn next-nearest neighbors (NNN) have also been previously investigated and this study allowed determination of averages lengths of the Zn-Al ($R_{Zn-Al} = 2.98 \pm 0.04$ Å), Zn-Zn ($R_{Zn-Zn} = 3.11 \pm 0.04$ Å) and Zn-Si (R_{Zn-Si} stretching from 3.17 ± 0.04 Å to 3.25 ± 0.04 Å) bonds. The results of the Zn and Mg K-edges EXAFS study have been described elsewhere[3] The comparison of these values to the ideal metal-metal distances in the montmorillonite lattice ($R_{M-M} = 2.99$ Å $\pm$ 0.04 Å), has shown that insertion of Zn in the octahedral sheet induces strong distortions of the octahedra. Moreover, the surprisingly high values of the N_{Zn}/N_{Al} ratio indicate a clustering of Zn in the octahedral sheet[3]. The clustering of Mg was also demonstrated by ^{19}F MAS-NMR spectroscopy[3] and already observed in natural montmorillonite[12].

The Si K-edge EXAFS of Mg02 and Zn02 samples are presented together with the FFC of the NN environment in Figure 4. The results of the simulation confirm the tetrahedral coordination of Si fielding a Si-O distance at 1.60 ± 0.03 Å. The contribution of the Si NNN environment is very similar on the FT (Figure 4) for Mg02 and Zn02 samples, indicating that the average environment of Si is not disturbed by the nature of cations of the octahedral sheet.

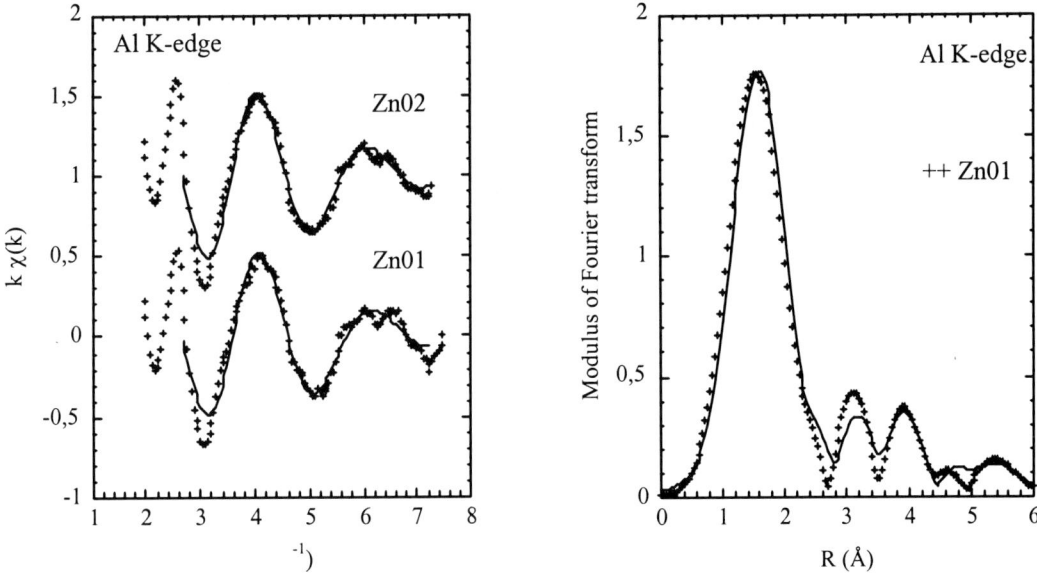

Figure 3 : Al K-edge EXAFS data together with the FFC of the NN environment and the FT of Zn01 and Zn02 samples.

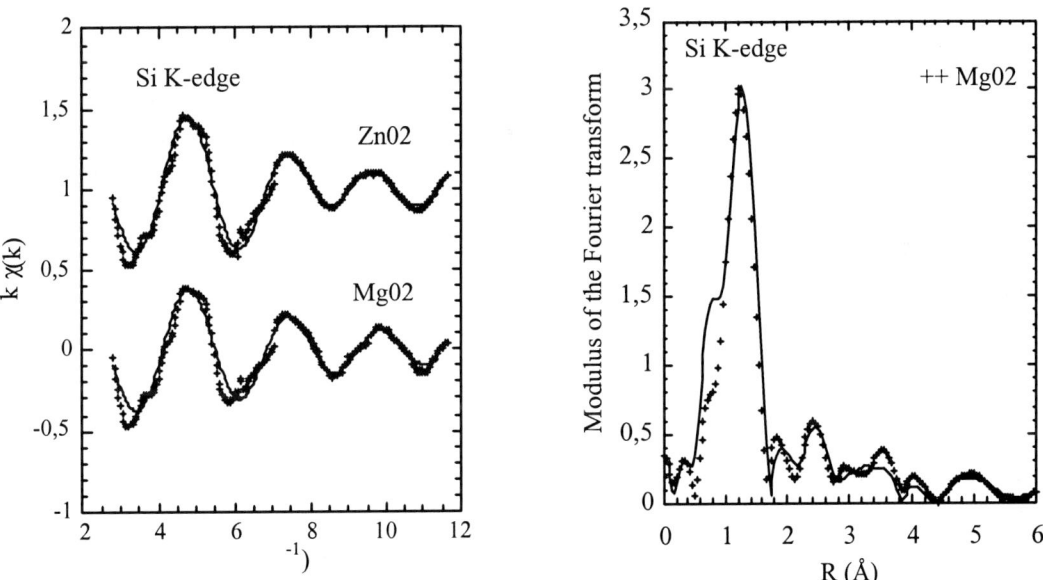

Figure 4 : Si K-edge EXAFS data together with the FFC of the NN environment and the FT of Mg02 and Zn02 samples.

^{27}Al MAS-NMR spectra (Figure 5) of the montmorillonites and beidellites, used as references, exhibit a main peak at about 0 -5 ppm corresponding to an octahedral Al (Al$_{VI}$) and a secondary peak at about 63 – 70 ppm corresponding to a tetrahedral Al (Al$_{IV}$)[13,14]. The areas of the octahedral and tetrahedral peaks were determined by using the WIN-NMR software (Bruker) and normalized to the unit mass of samples. This treatment, applied on the beidellites had allowed to calculate a standard line (Figure 6). Then the Al content of montmorillonites samples could be determined : the results are given and compared to those of chemical analyses in table 1. The values obtained by NMR and chemical analyses are in good agreement.

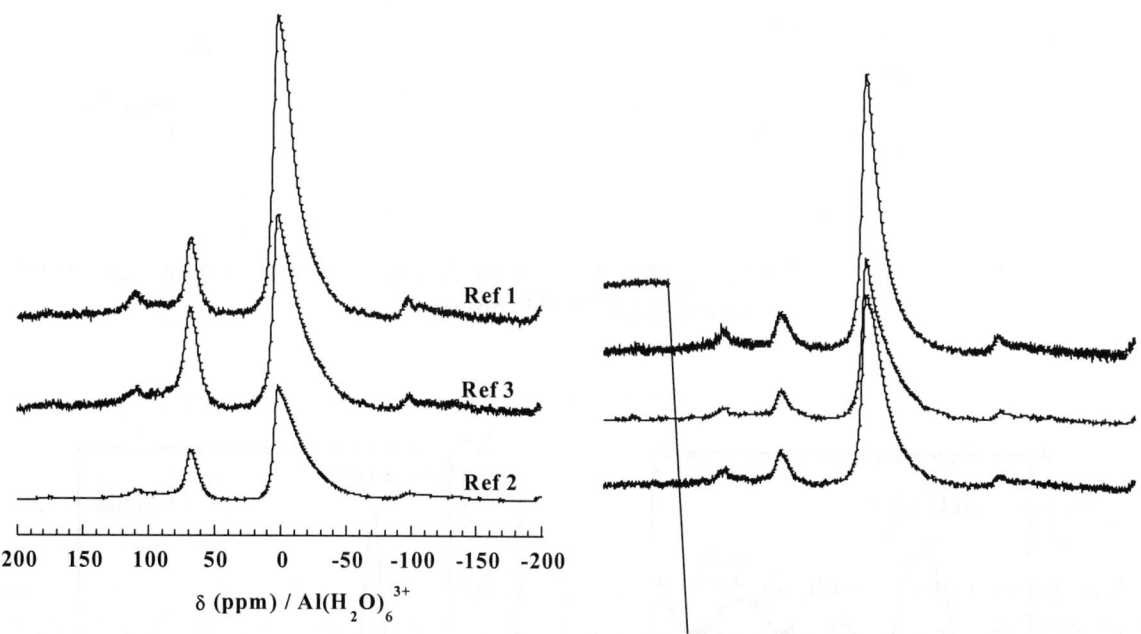

Figure 5 : Spectra of the beidellites used as references and of the montmorillonite samples.

Sample	x	Al content (weight %)		Half a unit cell formula (chemical analysis and TGA)
		Chemical analysis	NMR	
Mg01	0.10	13.89	13.37	$Na_{0.28}[Al_{1.81}Mg_{0.18}][Si_{3.93}Al_{0.07}]O_{10}(OH_{1.95}F_{0.05})$, 3.3 H_2O
Mg02	0.20	12.07	11.16	$Na_{0.34}[Al_{1.64}Mg_{0.37}][Si_{4.00}]O_{10}(OH_{1.95}F_{0.05})$, 4.0 H_2O
Zn01	0.10	13.36	12.77	$Na_{0.17}[Al_{1.83}Zn_{0.17}][Si_{4.00}]O_{10}(OH_{1.94}F_{0.06})$, 2.8 H_2O
Zn02	0.20	11.51	11.51	$Na_{0.3}[Al_{1.64}Zn_{0.39}][Si_{4.00}]O_{10}(OH_{1.95}F_{0.05})$, 4.0 H_2O

Table 1 : Al content and half a unit cell formula of samples.

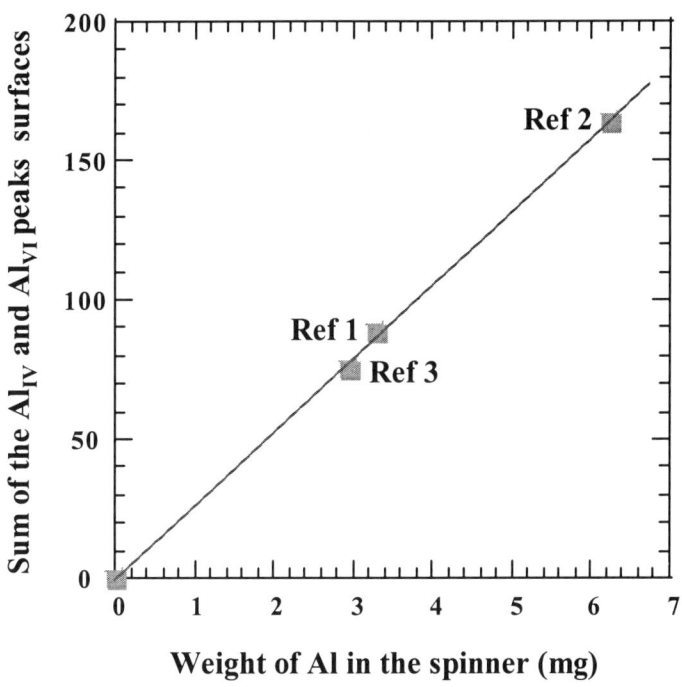

Figure 6 : Standard line calculated from the beidellites used as references.

4. CONCLUSION

Montmorillonites containing Al and exclusively Zn or Mg as divalent elements in the octahedral sheet were synthesized in acidic and fluoride medium. The substitution of Al was obtained for a narrow range of Mg or Zn contents. The EXAFS study of the pure products reveals strong distortions of the octahedra occupied by divalent cations (Mg and Zn), whereas the octahedral Al and tetrahedral Si environments are found to be nearly regular, in agreement with the description of the montmorillonite lattice used as a reference in this work[11]. A clustering of the octahedral elements was also revealed. These phenomena induce a mismatch between the octahedral and the tetrahedral sheets, which could give rise to local undulations of the layer. Al for Si substitutions in the tetrahedral sheet, are revealed by ^{27}Al MAS-NMR. Furthermore, this technique used in quantitative conditions yields the absolute Al contents of the synthetic montmorillonite clays. However, the accurate determination of the Al_{VI}/Al_{IV} ratio requires additional measurements[4]. Nevertheless, it appears from a first investigation that only a very small part of the total Al is located in the tetrahedral sheet. These results demonstrate that, beyond elemental analysis and usual characterizations, local probes like EXAFS and solid state NMR have to be carried out to establish accurately the unit cell formula of such complex materials.

REFERENCES

[1] Kloprogge J. T., Komarneni S., Amonette J. E., *Clay Clay Miner.* **1999**, *47(5)*, 529
[2] Joly J.-F., Huve L., Le Dred R., Saehr D., Baron J., French patent n° 2673930, **1992**
[3] Reinholdt M., Miehé-Brendlé J., Delmotte L., Tuilier M.-H., Le Dred R., Cortès R., Flank A.-M., *Eur. J. Inor. Chem.* **2001**, *11*, 2831
[4] Reinholdt M., Thesis, Université de Haute Alsace, Mulhouse, France, **2001**
[5] Michalowiz A., *J. Phys.* **1997**, *III - C2*, 235
[6] Fenzke D., Freude D., Fröhlich T., Haase J., *Chem. Phys. Let.* **1984**, *111*, 171
[7] Man P. P., Thesis, Université Pierre et Marie Curie, Paris, France, **1986**
[8] Maegdefrau E., Hofmann U., *Z. Krist.* **1937**, *98*, 299
[9] Watanabe T., Sato T., *Clay Sci.* **1988**, *7*, 129
[10] Yamada H., Nakazawa H., Hashizume H., Shimomura S., Watanabe T., *Clay Clay Miner.* **1994**, *42(1)*, 77
[11] Tsipursky S. I., Drits V. A., *Clay Miner.* **1984**, *19*, 177
[12] Muller F., Besson G., Manceau A., Drits V.-A., *Phys. Chem. Miner.* **1997**, *24*, 159
[13] Sanz J., Serratosa M., *J. Am. Chem. Soc.*, **1984**, *106*, 4790
[14] Drachman S. R., Roch G. E., Smith M. E., *Solid State Nucl. Magn.* **1997**, *9*, 257

Electron optical study of star shaped gibbsite microcrystals

H. Souza Santos[1] P. Souza Santos[2] and P.K. Kiyohara[1]

[1] Laboratório de Microscopia Eletrônica - Instituto de Física
Universidade de São Paulo, São Paulo, SP, Brazil

[2] Departamento de Engenharia Metalurgica e de Materiais - Escola Politécnica
Universidade de São Paulo, São Paulo, SP, Brazil

A new laboratory procedure is described for the preparation of microcrystals of gibbsite by the reaction of aluminum with water. Aluminum powder is used, activated by iodine dispersed in water, in place of the usual mercury II-chloride. The relative proportions of the reagents are: water - 500g; Al - 27.0g; iodine - 5.0g. Conditions - temperature of reaction: 50° to 60°C; minimum time for complete reaction: 21 days. The gibbsite crystals were characterized by X-ray diffraction (XRD), transmission electron microscopy (TEM), selected area electron diffraction (SAED) and elemental microanalysis (MA). The crystals are very thin hexagonal plates with diameters in the 0.2 - 0.6µm range. That morphology corresponds to the theoretical tabular pseudohexagonal habit of gibbsite crystals; however, it is different, in shape and size, from the synthetic Bayer gibbsite crystals.
If a smaller amount of iodine is used, such as 1.0g, a large number of platy hexagonal stars of gibbsite are formed, well characterized in their crystalline structure by SAED. These gibbsite hexagonal stars have in the TEM micrographs, an aspect completely different from the typical polygonal stars of nordstrandite.

1. INTRODUCTION

The crystal habit of synthetic gibbsite depends on the characteristics of the crystallization medium. Gibbsite is a frequent component of many high alumina clays and bauxites; its crystal habit is usually pseudo-hexagonal tabular [1].
The particle shape of synthetic gibbsite microcrystals usually is euhedral hexagonal plates. They may be prepared from ageing aluminum hydroxide gel from the reaction between aluminum chloride and sodium hydroxide solutions [2, 3]. The hexagonal platy morphology differs from the particle shape of the larger Bayer gibbsite grains (50µm to 100µm diameter) [4].
The aluminum hydroxide which crystallizes in water, at room temperature, in which no chemical is dissolved is the trihydroxide - bayerite [5]; that is the basis of the Schmah's method [6]; first non-crystalline Al (OH)$_3$ is prepared by the reaction, at room temperature, between water and amalgamated aluminum foil; second the non-crystalline Al(OH)$_3$ ages at room temperature and crystallizes into bayerite hour-glass shaped somatoids [7]. The mercury used to amalgam the aluminum is considered as an "activator" by Gitzen [8].

Recently, Souza Santos *et al* [9], have studied the reaction between aluminum powder and water, between 25ºC and 95ºC, using iodine as "activator" and maintaining constant weight proportions Al : I : H_2O = 27.0g : 5.0g : 500g in Na^+ free medium. So, by analogy with mercury as activator, bayerite somatoids should be produced from that reaction. However, euhedral thin hexagonal platy gibbsite microcrystals (0.2µm - 0.6µm in diameter) were formed in the narrow 50ºC - 60ºC temperature interval. These crystals disperse easily in distilled water forming white sols positively charged; they present different thichknesses and hexagonal profiles. The trihydroxide nordstrandite was characterized in small quantities, at higher temperature interval; that observation is important, because nordstrandite microcrystals may appear as many sided platy stars [7]. Continuing the study of the reaction, it was observed that the decrease of the amount of iodine induced changes of the crystal shape passing from a regular hexagon to six-point plates.

2. AIM

The purpose of this paper is to describe the preparation of gibbsite microcrystals having the unusual shape of platy six-point stars.

3. EXPERIMENTAL

3.1 Materials and Principle of the Method

Ion-free aluminum hydroxide gels are usually prepared by reaction of water with an aluminum alcoxide or with amalgamated aluminum. Gitzen [8] considers mercury and iodine as "activators" of the reaction of aluminum metal with water. In order to avoid the use of mercury for the last reaction, iodine was tried to produce an ion-free hydroxide because iodine attacks metals in presence of moisture. So, iodine and aluminum, in several conditions of aluminum to iodine weights, temperatures and times of reaction, were tried to obtain the reaction of the aluminum with water. These conditions were found, but no bayerite is formed, as it could be expected by analogy with amalgamated aluminum; different hydroxides and their mixtures could be prepared according to the temperature of reaction. In the narrow temperature range of 50ºC to 60ºC, pure microcrystals of gibbsite could be prepared. Surprisingly the proportion of iodine to aluminum directs the shape, hexagons or stars, of the gibbsite crystals formed. The precursor shapes carry a great responsibility in relation to the resultant products, mainly when the crystals are pseudomorphs, as occur with the aluminums. Compared with pure aluminum turnings, sheets, foils and other powders, the best pure aluminum for the reaction is the ALCOA 123 uncoated Al-powder (100% passing ASTM nº325 sieve; Al = 99.7%; Si = 0.15%; Fe = 0.17%); it is produced in Brazil. The iodine is added to the distilled water from an ethanol solution.

3.2 Method

In a two liters round bottom, three necked Pyrex flask, are placed in the following sequence: 500mL of water; 50mL of an iodine solution (5.0g to 1.0g crystals dissolved in 95% ethanol) and 27.0g of Al powder. The system is stirred for one hour for homogeneization and heated to 60ºC ± 1ºC by thermostat controlled heating mantle; a water-cooled condenser is fitted to the flask. It takes a minimum of three weeks for the aluminum to react completely. The condenser is retired in the last four hours to assure that if any remaining iodine is eliminated.

The white suspension presents strong dityndallism by stirring, an indication of anisodiametric shape of the particles. Then it is centrifuged in polyethylene tubes in a Servall centrifuge at 3500rpm for 30 minutes. The white precipitate is washed four times its original volume with 95% ethanol by centrifugation. The white precipitate of aluminum hydroxide gibbsite is dried at 60°C - 70°C and a white powder is then ready for characterization; the yield in $Al(OH)_3$ is 85% - 87%, based in the equation: $Al + 3H_2O \Rightarrow Al(OH)_3 + 1,5 H_2$.

3.3 Characterization

The X-ray diffraction (XRD) was conducted in a Philips equipment, copper K-alpha radiation, model X'Pert, operating at 40kV and 40mA, between 1°(2θ) and 90° (2θ): the dried powder was dispersed in distilled water and prepared in the conventional manner for TEM. A Philips CM-200 analytical microscope, operating at 200kV, was used, with facilities for EDS. The SAED studies were made in the same equipment.

4. RESULTS AND DISCUSSION

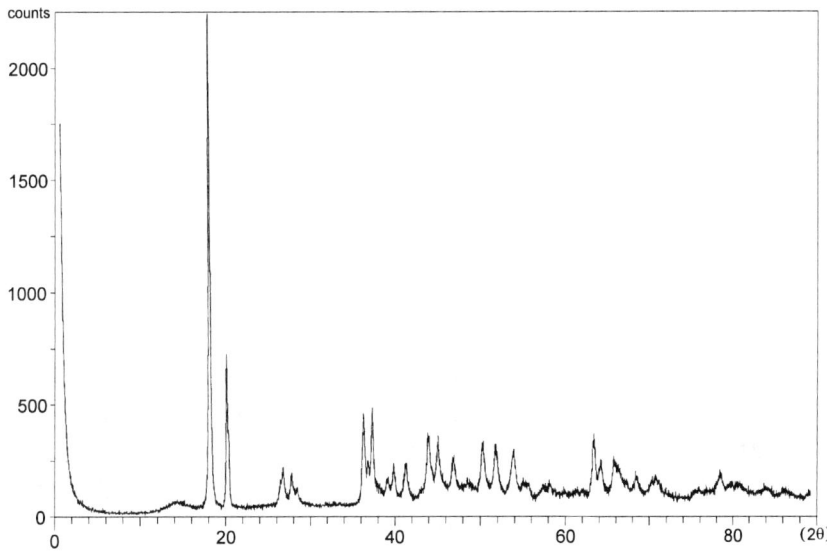

Figure 1. XRD curve for gibbsite prepared from Al (27,0g) + I (5,0g) + water (500g).

4.1 X-Ray Diffraction

Figure 1 shows the XRD curve of the dried aluminum hydroxide powder prepared at 60°C with 5.0g of iodine; the same curve is obtained with smaller iodine weights until 1.0g. All lines correspond to gibbsite lines listed in ICDD file n° 7 - 324, also 12 - 0460; 29 - 0041; 33 - 0018 [10]; the more intense line is the 4.85Å gibbsite reflexion. No other Al-hidroxide is detected by XRD at that temperature and with both iodine weights, 5.0g and 1.0g.

4.2 Electron Microscopy

Figure 2 and 3 are micrographs of the microcrystalline gibbsite crystals from 5.0g iodine preparation; they have regular hexagonal shapes, various thicknesses, the majority being very thin; their diameter are in the 0.2μm - 0.6μm interval.

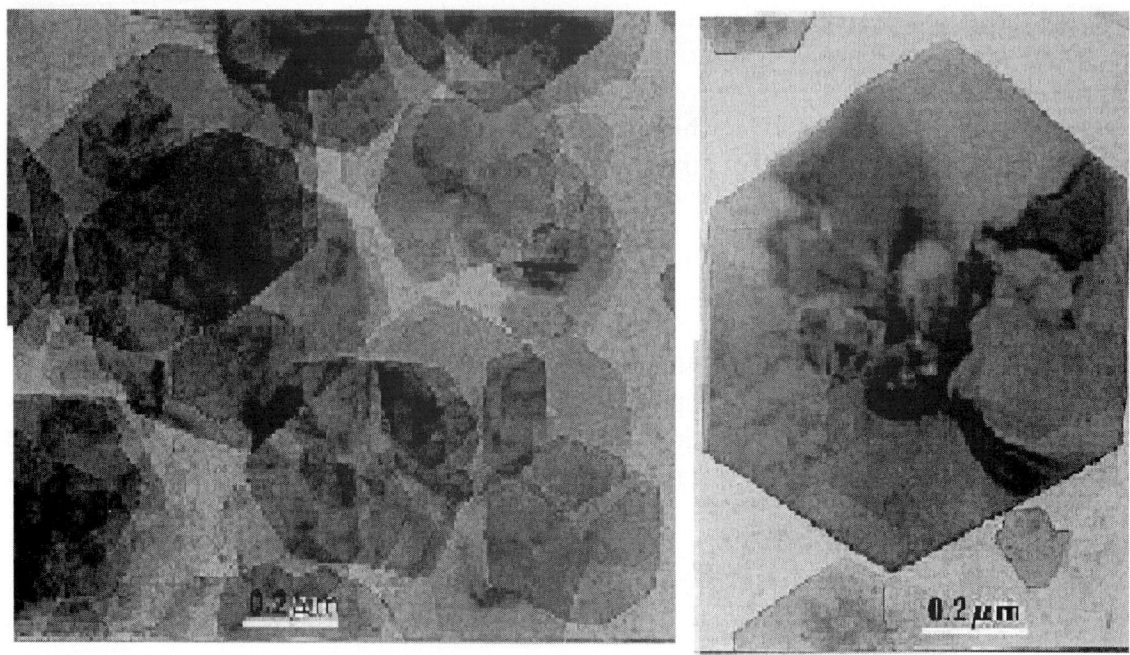

Figure 2 and 3 - microcrystalline gibbsite crystals (5.0 iodine).

The selected area electron diffraction (SAED) patterns of the stars confirm their gibbsite structure (Figure 4). EDS analysis shows absence of iodine in the crystals of gibbsite of both shapes.

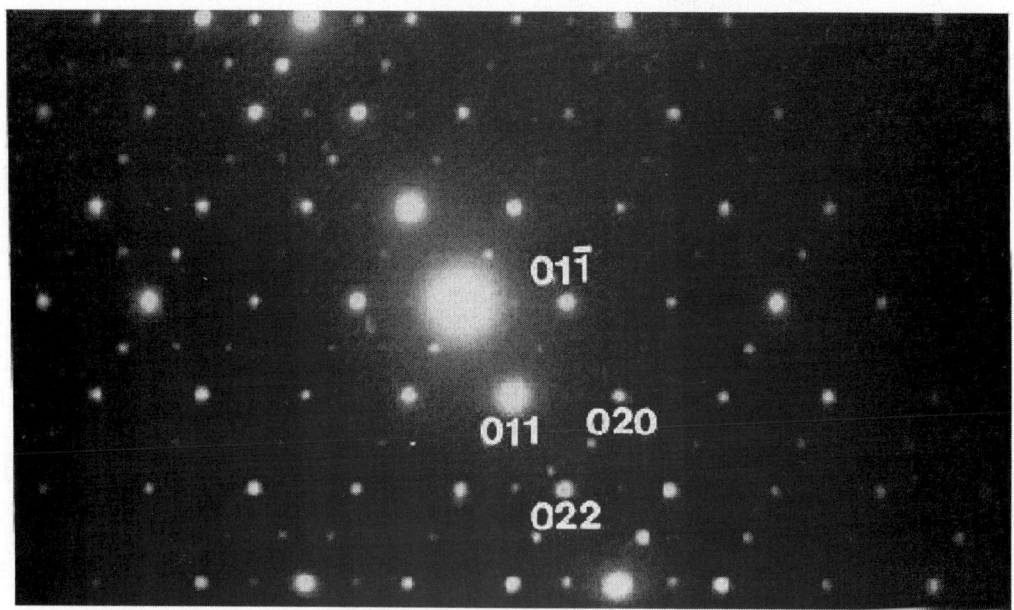

Figure 4 - SAED characterizing the gibbsite structure.

Figures 5 and 6 - star-shaped gibbsite crystal from Al (27.0g) + I (1,0g) + water (500g) at 60°C

By transmission electron microscopy very few thin hexagonal plates were observed (Figures 5 and 6); however, the majority of the crystals has the platy shape of six-point stars, with regular and irregular profiles. Due to the coexistence of some hexagonal plates with a majority of stars, it is tempting to suggest that the stars grew from the hexagons, but no explanation can be offered at the moment in relation to the mechanism of the iodine action, either directing the crystallization to gibbsite plates instead to bayerite somatoids or why the lower amount of iodine leads to the formation of the six-point star gibbsite plates. An explanation could be that the star formation is the result of the twinning of their three parallel faces, as may ocurr in natural gibbsite; also the growing of X-shaped twins may be observed (Figure 5).

4.3 Effect of the lower iodine weight

Several previous experiments were made varying the iodine to aluminum weight ratio, at 60ºC ± 1ºC; it was decreased from 25g : 27g Al to 5.0g iodine : 27.0g Al, the last one being used in this paper. Now, smaller ratios, at the same temperature, were tested, until 1.0g iodine : 27.0g Al : the reaction occurred at smaller rate than at the ratio 5.0g : 27.0g and was complete after a minimum of three weeks. Examined by XRD, the white powder presented only the gibbsite lines.

4.4 Effect of the temperatures between 20ºC and 50ºC

If the temperature is in that range, the reaction rate is slower and it takes a minimum of one month for complete reaction of the aluminum; that happens for iodine: Al proportions from 25.0g : 27.0g to 5.0g : 27.0g. By XRD, mixtures of gibbsite (basal 4.85Å) and bayerite (basal 4.69 Å) are formed, the amount of the former decreasing with decreasing temperatures. By TEM, it is observed that gibbsite crystals maintain the thin hexagonal platy shape; curiously, bayerite crystals do not appear as hour-glass or semi hour-glass shaped somatoids, but as irregular round shaped particles. At 20ºC and 30ºC, a small nordstrandite peak (4.79Å) is observed.

5. CONCLUSIONS

A new procedure is described to prepare microcrystalline gibbsite powder constituted by star shaped crystals by the reaction of aluminum powder with water at 60ºC in presence of iodine; the weight proportions aluminum : iodine : water are very specific in order to induce the formation of either hexagonal plates or six-point platy stars, always with gibbsite structure.

This paper is part of the "Projeto Temático FAPESP 1995 / 0544-0"

REFERENCES

1. W.D. Keller, Bauxittization of syenite and diabase illustrated in scanning electron micrographs. Econ. Geol. 74 (1979) 116.
2. M. C. Gastuche and A. Herbillon, Bull. Soc. Chim. France **5** (1962) 1404.
3. II. Hsu in Minerals in Soil Environments, 3rd Ed., edited by J.B. Dixon and S.B. Werd, Soil Science Society of America, Madison, 1989, p. 331.
4. C. Misra in Industrial Alumina Chemicals, American Chemical Society - Monograph 184, Washington, D.C.,1986, p. 46.

5. H. Ginsberg, W. Huttig and H. Stiehl. Z. anorg. alg. Chem. **318** (1962) 238.
6. H. Schmah. Z. Naturforsch. **1** (1942) 323.
7. M. L. P. Antunes and H. Souza Santos. Acta Microsc. **5B** (1996) 74.
8. W.H.Gitzen. Alumina as a Ceramic Material. American Ceramic Society, Columbus, 1970.
9. H. Souza Santos, P.K. Kiyohara and P. Souza Santos. J. Mat. Sci. Letters 19 (2000) 1525.
10. K. Wefers and C. Misra in Oxides and Hydroxides of Aluminum, Alcoa Technical Paper N°19, revised, Aluminum Company of America, Pennsylvania, 1987 p. 54.

A new method for the preparation of microcrystals of boehmite

P. Souza Santos[a] H. Souza Santos[b] and P.K. Kiyohara[b]

[a] Departamento de Engenharia Metalurgica e de Materiais - Escola Politécnica
Universidade de São Paulo, São Paulo, SP, Brazil

[b] Laboratório de Microscopia Eletrônica - Instituto de Física
Universidade de São Paulo, São Paulo, SP, Brazil

The purpose of the paper is to describe a new method for the laboratory preparation of boehmite by the reaction of aluminum powder with water at 90°C in presence of iodine as an activator, replacing mercury. By transmission electron microscopy (TEM), the boehmite microcrystals appear as elongated plates with many jagged or dendritic sides, or long flat fibrils, also with dendritic sides. The morphology of the final product can be confused with cylindrical fibrillar pseudoboehmite only at low magnifications. These latter morphologies are different both from boehmite and from pseudoboehmite microcrystals prepared by other methods.

1. INTRODUCTION

Aluminum monohydroxide boehmite - AlOOH - always occurs as crystals of micrometric size. Tabular-shaped naturally occurring boehmite monocrystals, with eight sides and "lentil" morphology, were described from bauxite from the Vihniovy mountains, Russia [1]. The tabular habit of natural boehmite crystals is due to the cleavage parallel to (020) planes. Well-ordered crystalline boehmite has a value for the basal reflection **d**(020) = 6.12 Å (0.612 nm). Pseudoboehmite is the "poorly crystallized" Al^{3+} compound, with the chemical composition of $Al_2O_3 \cdot xH_2O$ (with 2.0 > x > 1.0) and having the basal reflection of **d**(020) of values to 6.7 Å [2], whereas well-ordered boehmite has a value of 6.12 Å and x = 1.0. The structures and properties of the crystalline, non-crystalline and polymeric aluminum hydroxides were reviewed recently by Hsu [3].

Boehmite is usually prepared by the reaction of either amalgamated aluminum foils or turnings with water at 100°C or with water vapor at higher temperatures. Pure aluminum powder reacts with water vapor at 250°C to 375°C to produce boehmite. Aqueous suspension of -Al $(OH)_3$- crystalline aluminum trihydroxides, at acid or alkaline conditions, at 150°C to 250°C, also produces well-ordered boehmite.

Pseudoboehmite is formed by the addition polymerization of hydroxyls groups from non-crystalline -Al$(OH)_3$- aluminum hydroxide particles, in aqueous acidic media at 90°C - 100°C. The hydroxide may be formed from aluminum salts at several conditions. Formation also involves homogeneous precipitations, the reaction of amalgamated aluminum with water and the hydrolysis of aluminum organic salts or compounds. If the polymerization is predominantly linear, the pseudoboehmite microcrystals are fibrils. The length and aspect ratios of the fibrils increase with temperature and polymerization time, as shown by X-ray diffraction (XRD) [4, 5] and by TEM [6]. These fibrils are called "fibrillar pseudoboehmite" [7].

Boehmite and pseudoboehmite are the only direct precursors for the production of gamma-alumina for alumina catalysts, catalyst carriers, adsorbents, and as active binding agents for other catalysts like zeolites. Boehmite has been used in advanced ceramics for aluminum oxinitride -AlON- synthesis. Owing to the variety of methods and conditions for the synthetic production of well-ordered boehmite, pseudoboehmite and their aqueous dispersions, a variety of particle shapes has been observed in the microcrystal products by TEM [8]: (a) Boehmite - laths and plates of regular hexagonal profiles; spindle-shaped plates; rhombohedral-shaped crystals; needles; plates of irregular profile and rugose surfaces; thin triangular plates with dendritic sides. (b) Pseudoboehmite - cylindrical fibrils of various lengths; wrinkled or crumpled sheets.

2. AIM

The purpose of the paper is to describe a procedure for the preparation of powders by the reaction of aluminum powder with water to produce boehmite microcrystals with unique morphologies.

3. PRINCIPLE OF THE METHOD

Amalgamated aluminum foils react with water at 20ºC - 30ºC to produce a dispersion of non-crystalline or amorphous aluminum hydroxide. In the absence of electrolytes, at 20ºC - 30ºC, the aluminum hydroxide crystallizes into bayerite. Bayerite microcrystals are named somatoids and are hour-glass and semi hour-glass shaped. If the reaction of the amalgamated foils is maintained at 90ºC - 100ºC, wrinkled sheets of pseudoboehmite are formed [6]. Gitzen [9] considers mercury and iodine as "activators" of the reaction of aluminum with water. Apparently iodine reacts with moist metals [10]. Recently, it was shown that iodine is an excellent activator for that reaction, if a fine-grained pure aluminum powder is used: in the narrow 50ºC - 60ºC temperature range, no bayerite is formed and, only the aluminum trihydroxide gibbsite crystallizes [11].

We found that increasing the temperature of the reaction to 90ºC - 100ºC, and maintaining the same conditions for gibbsite production [11], only boehmite microcrystals crystallizes, but with a new morphology.

4. MATERIALS AND METHOD

4.1 Reagents

The aluminum for the reaction is the Alcoa 123 uncoated Al-powder (100% passing ASTM nº 325 sieve; Al = 99.7%; Si = 0.15%; Fe = 0.17%) produced in Brazil. The iodine is added to the distilled water in a 95% ethanol solution.

4.2 Procedure

In a two-liters round bottom, three necked Pyrex flask, the following chemicals are placed in sequence: 500 ml of distilled water; 50 ml of an iodine solution: (10.0 g to 1.0 g iodine crystals dissolved in 95% ethanol) and 27.0 g of Al powder. The system is stirred for one hour until homogenized and heated to 90°C ± 1°C by a thermostat-controlled heating mantle; a water-cooled condenser is fitted to the flask. It takes two to four weeks for the aluminum to react completely, depending upon the amount of iodine used. The condenser is retired in the last four hours to assure that all remaining iodine is eliminated. The white suspension shows strong dityndallism by stirring, an indication of the anisodiametric shape of the particles. The product is centrifuged in polyethylene tubes in a Servall centrifuge at 3500rpm for 30 minutes. The white precipitate is washed four times its original volume with 95% ethanol by centrifugation. The white precipitate of aluminum monohydroxide (boehmite) is dried at 60°C - 70°C and the white powder is then ready for characterization. The yield, calculated as Al OOH, is 82% - 85%, based in the equation: $Al + 2H_2O \Rightarrow Al\,OOH + 1.5H_2$.

4.3 Characterization

X-ray diffraction was conducted in a Philips diffractometer, with copper **K**-alpha radiation, using the model X'Pert program, operating at 40kV and 40mA, between 1°(2θ) and 90°(2θ). The dried powder was dispersed in distilled water and prepared in the conventional manner for TEM. A Philips CM-200 analytical microscope, operating at 200kV, was used, with an energy dispersive system (EDS). Selected area electron diffraction (SAED) was also used.

5. RESULTS AND DISCUSSION

5.1 X-Ray Diffraction

Figure 1 shows the XRD curve of the dried aluminum hydroxide powder prepared at 90°C using 10.0g of iodine. All lines correspond to boehmite lines listed in ICDD files n° 21 - 1307 and 17 - 0940 and in reference [12]. The basal boehmite reflection is **d** = 6.12 Å. No other Al-hydroxide was detected by XRD with that or another iodine weight.

Figure 2 is the XRD curve of a fibrillar pseudoboehmite sample after thermal aging at 90°C ± 1°C for 24 days, with **d**(020) = 6.2 Å.

Figure 3 is the XRD curve of a well-ordered commercial boehmite (D-50) which has **d**(020) = 6.12 Å, both taken as references for comparison. The sharpness of the lines of Figure 1 is more similar to the sharpness of the fibrillar pseudoboehmite of Figure 2 and is less sharp than the lines of the well-ordered hydrothermal boehmite of Figure 3. However, all products from different iodine weights showed the presence of the 6.12 Å basal reflection of well-ordered boehmite.

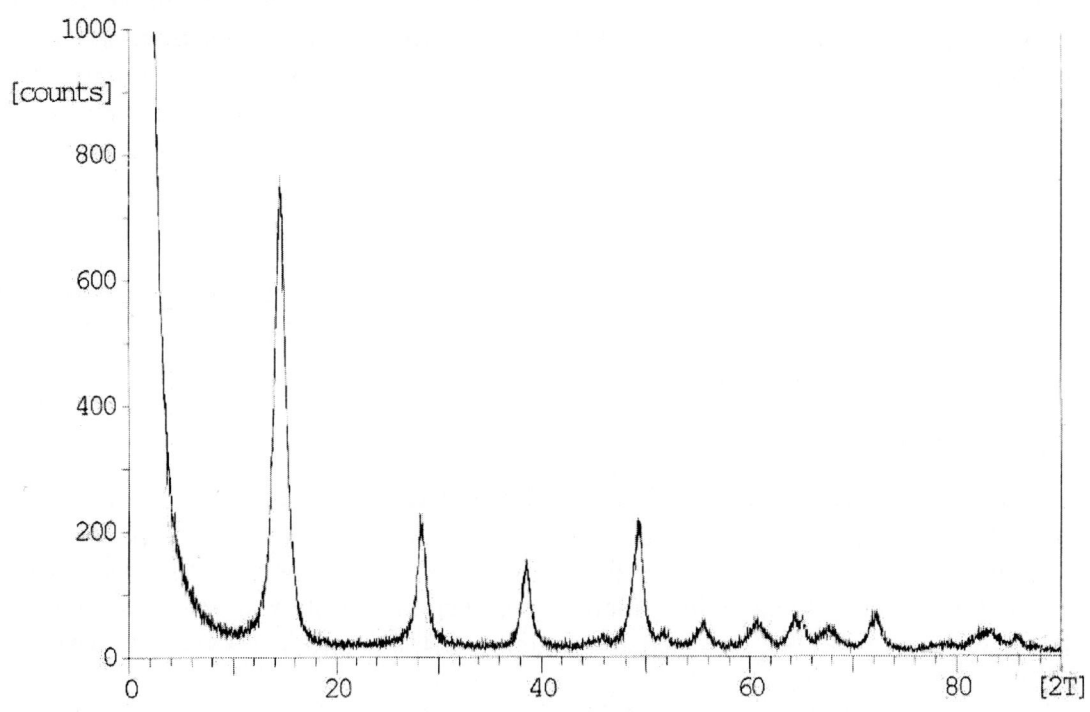

Figure 1. XRD curve (°2θ, CuK$_\alpha$) of the dried aluminum hydroxide powder prepared at 90°C using 10.0g of iodine, with **d**(020) = 6.12 Å.

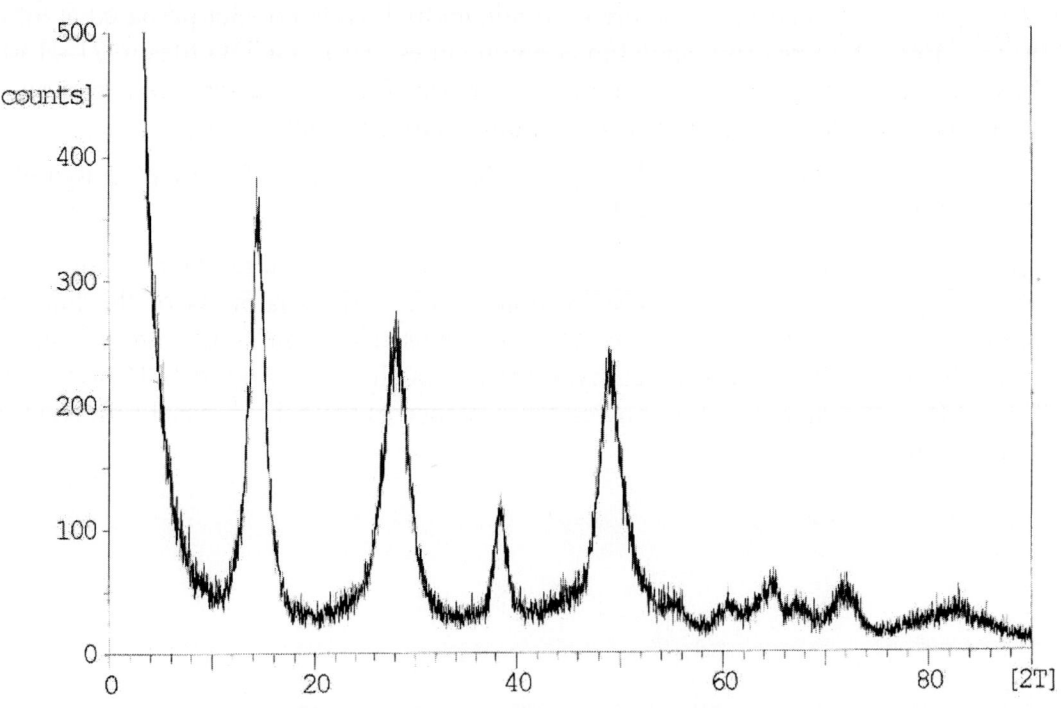

Figure 2. XRD curve (°2θ, CuK$_\alpha$) of a fibrillar pseudoboehmite after thermal aging at 90°C for 24 days, with **d**(020) = 6.2 Å.

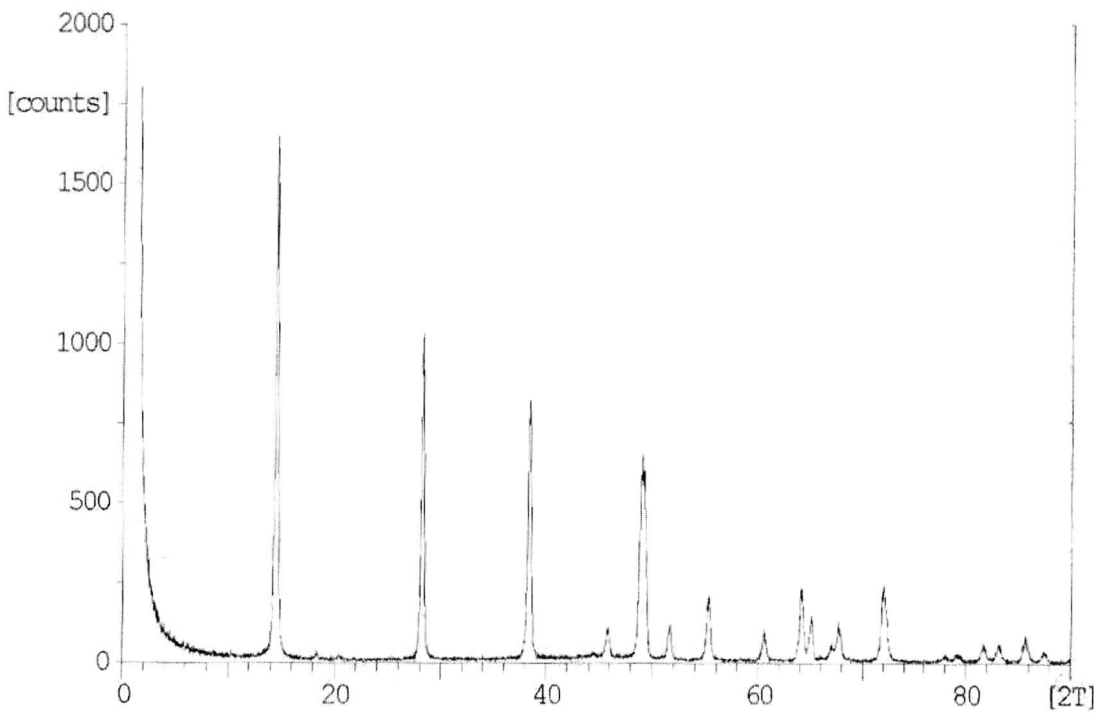

Figure 3. XRD curve of a well-ordered commercial boehmite with **d**(020) = 6.12 Å.

5.2 Electron Microscopy

Figure 4 shows a TEM image of a sample with flat fibrils with dendritic sides, prepared using 10.0g of iodine. Figure 5 is from pseudoboehmite with cylindrical fibrils.

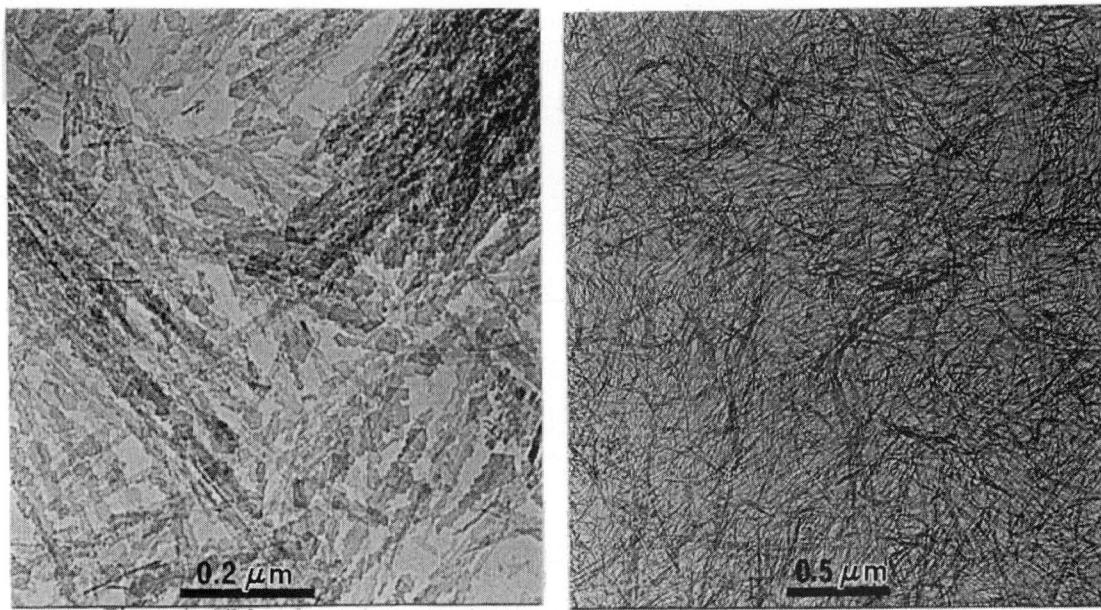

Figure 4. TEM of flat fibrils showing dendritic sides, prepared using 10.0g of iodine.
Figure 5. Pseudoboehmite with cylindrical fibrils.

The flat fibrils with dendritic or jagged sides from Figure 4 are clearly different from the smooth cylindrical microfibrills of Figure 5 and from the plates with dendritic or jagged side of hydrothermal [8] and well-ordered boehmite. The dendritic flat fibrils, with **d**(020) = 6.12 Å, shows the new morphology of boehmite microcrystals. The observed value of **d**(020) was the criterion to characterize boehmite.

5.3 Effect of Iodine to Aluminum ratio

Maintaining the reaction temperature at 90°C ± 1°C, a constant Al (27.0g) and water (500ml) weights, the iodine weight was varied from 27.0g to 1.0g. Boehmite as dendritic flat fibrils was obtained in the 27.0g to 5.0g interval, without any significant change in crystal shape, in the electron micrographs. All products showed **d**(020) = 6.12 Å. Decreasing the weight of iodine from 5.0g to 1.0g, significant changes were observed in the particle shape of the crystals, particularly for the sample with 1.0 gram of iodine: the initial dendritic and flat fibrillar morphology of Figure 4 changed to an elongated plate shape also with jagged outlines (Figure 6); the **d**(020) continues to equal 6.12 Å. Note that both types of platy crystals, the hydrothermal, well-ordered boehmite [8] and from 1.0g iodine sample (Fig. 6), have dendritic or jagged outlines and the 6.12 Å basal (020) reflections of boehmite.

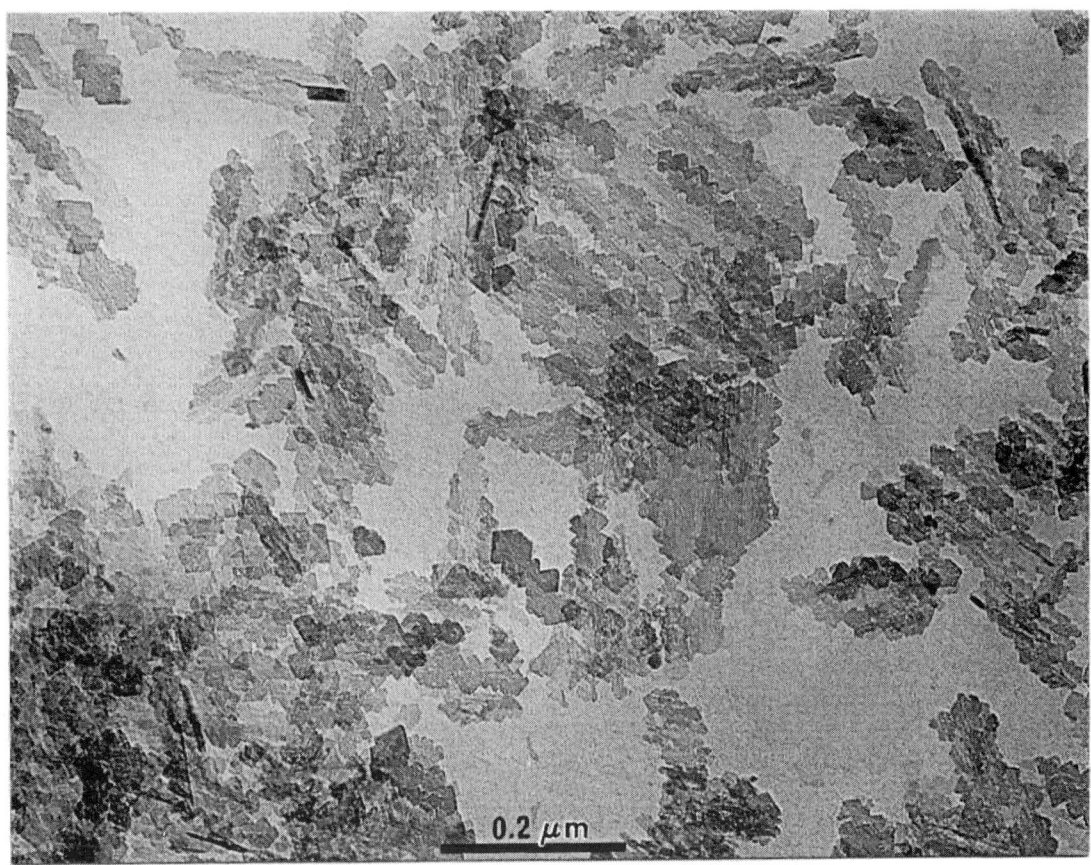

Figure 6. Dendritic flat fibrillar morphology has changed to an elongated plate shape also with jagged outlines in boehmite crystals.

Thus, the iodine used as an activator in place of mercury for the reaction of aluminum powder with water at 90°C, does not produce fibrillar pseudoboehmite. Instead boehmite microcrystals with acicular and flat fibrillar shapes, and dendritic profiles formed with a 6.12 Å basal reflection.

Therefore, iodine seems to have the effect of improving the internal order during the crystallization of the AlOOH at 90°C, thus preventing the formation and growth of "poorly crystallized" pseudoboehmite cylindrical fibrils. At this time, we do not understand why, varying the iodine weight in proportion to the aluminum powder weight, there is such a change from flat fibrils to plates, both with dendritic sides.

6. CONCLUSIONS AND SUMMARY

A method is described for the preparation of powders formed by microcrystals of the aluminum monohydroxide boehmite with **d**(020) = 6.12 Å to form new particles shapes (dendritic flat fibrils or plates). The morphology depend on the iodine weight in relation to the aluminum powder weight used for reaction with water at 90°C.

7. ACKNOWLEDGEMENTS

This paper is part of the "Projeto Temático FAPESP 1995 / 0544-0"

REFERENCES

1. E.M. Bohnstedf-Kupleskaya and N.L. Vlodavetz, Compte Rendus de l'Academie des Sciences de URSS, Leningrad, 49 (1945) 587.
2. O.P. Krivoruchko, R.A. Buyanov, M.A. Fedotov and L.M. Plyasova, Russian J. Inorg.Chem. 23 (1978) 988.
3. P.H. Hsu, in Minerals in Soil Environments, 3rd Ed., J.B.Dixon and S.B. Werd, Soil Science Society of America, Madison, 1989.
4. H. Ginsberg and K. Wefers, Aluminum and Magnesium, Ferdinand Enke Verlag, Stuttgart, 1971.
5. R. Tettenhorst and D.A. Hofmann, Clays Clay Min. 28 (1980) 373.
6. H. Souza Santos and P. Souza Santos, Materials Letters 13 (1992) 175.
7. J. Bugosh, R.L. Brown, J.R. McWhorter, G.W. Sears and R.J. Sippel, Ind.Eng.Chem. R.D. 1 (1962) 157.
8. P. Souza Santos, P.K. Kiyohara and H. Souza Santos, Bol. Tecn. Petrobrás 41 (1998) 45.
9. W.H. Gitzen, Alumina as a Ceramic Material, American Ceramic Society, Columbus, 1970.
10. Anonimous, The Merck Index of Chemicals and Drugs, 6th Ed., Merck Co., Rahway, 1952.
11. H. Souza Santos, P.K. Kiyohara and P. Souza Santos, J. Mat. Sci. Letters 19 (2000) 1525.
12. D. Wefers and C. Misra, Oxides and Hydroxides of Aluminum, Aluminum Company of America, Pennsylvania, 1987.

Studies of Synthetic Kaolinites Containing Copper and Zinc

A.R. Tong[a], B. J. Kennedy[a], and B. Singh[b]

[a]School of Chemistry, The University of Sydney, Sydney NSW 2006, Australia

[b]Department of Agricultural Chemistry and Soil Science, The University of Sydney, Sydney NSW 2006, Australia

The synthesis of copper and zinc metal substituted kaolinites was undertaken using a procedure designed to produce high quality samples in a relatively short time. Alumino-metal-silicate gels were hydrothermally treated with 0.1M KOH at 250 °C and 40 bars for varying time intervals. For the range of copper substituted materials, a crystalline copper bearing kaolinite was formed, along with varying amounts of poorly crystalline kaolinite, boehmite and copper. In contrast, it was found that zinc cations inhibited the formation of kaolinite. In such samples a rare zinc mineral, fraipontite, was formed along with boehmite. The samples were analysed by FTIR, x-ray diffraction and XAFS. A compositional analysis was also made using AAS.

1. Introduction

Natural kaolinite is usually formed by the *in-situ* weathering of feldspar-rich rocks or hydrothermal processes[1]. Since kaolinite is often the end product of weathering processes and transportation of transition metals by groundwater leaching or biological uptake occurs easily in the environment, deposits of transition metal substituted kaolinites are rare[2-5]. Anthropogenic uses of the biophysical environment have redistributed metal and clay particles, for instance in landfill and waste depositories there exists a combination of clay and waste material not found in natural environments. The structural transformations and chemical interaction between kaolinite and metal pollutants needs to be investigated in order to provide better management of waste materials. Cation substitution in kaolinite is best studied by doping synthetic kaolinites, and previous studies have shown that chromium[6], copper[7], iron[8,9] and gadolinium[10] can be incorporated into the clay framework. However, since there are many synthetic routes proposed for kaolinite[7,11-13], the maximum level of doping and the structural perturbations caused by such doping have not yet been comprehensively investigated. The purpose of this paper is to report preliminary research into the effects of copper and zinc cations on the formation and structure of kaolinite synthesised under hydrothermal conditions. Characterisation of the compounds was completed using chemical analysis, FTIR, powder x-ray diffraction, and x-ray adsorption spectroscopy.

2. Experimental

The preparation of doped kaolinites used a modification of the method in Huertas et al.[12] and de Kimpe[14]. Copper aluminosilicate and zinc aluminosilicate gels were precipitated from aqueous solutions containing copper acetate, zinc acetate, tetraethyl orthosilicate (TEO-Si), and aluminium tri-isopropoxide (Al-TIP). Heurtas et al.[12] found the optimum initial weight ratio (WR) of Al-TIP to total reagents was WR = Al-TIP/(Al-TIP + TEO-Si) = 0.402. In the present synthesis we used WR ~ 0.4, where WR = M acetate +Al-TIP/(M acetate +Al-TIP +TEO-Si). The weight ratio of M acetate to Al-TIP was either 1:9 or 1:19. The reagents were added to water (800 mL) in five equal fractions at hourly intervals to achieve a final suspension density of approximately 7.5 gL^{-1}. The suspension was then stirred for 24 hours, before heating the mixture to dryness at 150 °C. The powder obtained was then oven dried at 110 °C for 24 hours before being placed in a desiccator prior to hydrothermal treatment. Approximately ~ 2.5 g of gel and 10 mL of 0.1M KOH were hydrothermally treated in a 45 mL pressure vessel at 250 °C and ~ 40 bars for varying time intervals. When cool, the samples were extracted using a mixture of HCl and acetone, approximately pH 3. The samples were centrifuged and repeatedly washed to remove soluble impurities. A final mixture of product and acetone was heated to dryness at 75 °C and the powder then placed in a vacuum desiccator for 48 hours. The notation used for the samples in this work is AlNMT, where Al is aluminium, N is ratio of M acetate to Al-TIP, M is copper or zinc, and T is time of hydrothermal treatment in days. For example Al10Cu10 describes a sample with an Cu:Al ratio of 1:9 (i.e. a 10% copper substitution) hydrothermally treated for 10 days.

3. Compositional Analysis

The metal content of the samples were determined using a Varian FT220s instrument[15]. The samples (10 mg) were first digested[16] using a mixture of 5 mL HF, 0.5 mL HClO$_4$ and 0.1 mL H$_2$SO$_4$. The mixture was heated at 200 °C until dry and when cool 2 mL of water and a few drops of HClO$_4$ were added and the mixture heated to dryness at 100 °C. The samples were finally made to volume (25 mL) using 6M HCl. The weight percent of metal to aluminium cations is displayed in Table 1.

Table 1 Metal to aluminium weight percentage as determined by AAS.

Al10Cu10	Al10Cu2	Al5Cu10	Al5Cu2	Al10Zn10	Al10Zn2	Al5Zn10	Al5Zn2
10.54%	10.48%	5.48%	5.46%	12.32%	14.38%	6.45%	6.92%

4. Fourier-Transform Infra-red Spectroscopy

Infra-red spectra were collected using a Bio-Rad FTS - 40 spectrophotometer fitted with a drift attachment. The spectra were recorded over the range 4000-400 cm^{-1}, using potassium bromide as a background matrix (100 mg of sample : 300 mg of matrix), and at a resolution of 2 cm^{-1}. The samples and matrix were stored in a vacuum desiccator for 48 hours prior to analysis. The spectra for three of the Cu substituted samples (Figure 1) display stretches, albeit modest peak features due to the poor crystallinity of the samples, indicative of the Al-O octahedral and Si-O tetrahedral sheets of kaolinite [12,17-20]. However, there is considerable broadening of the Al$_2$-OH stretches at 3695, 3668, 3650, 3620 cm^{-1} that may be attributed to the incorporation of Cu atoms within the octahedral layer[7]. Further evidence for the presence of Cu in these layers is given by the presence of peaks at 3603, 850 and 535 cm^{-1}. By comparison with Ga and Fe containing kaolinites, the peak at 3603 cm^{-1} may be due to a Al-OH-Cu stretch[8,10,21], or as occurs in Cr substituted kaolinites a Cu-OH stretch[6]. Alternatively Filiaps et al.[22] has suggested such a peak is due to K$^+$ ions at the kaolinite surface. The peak at 850 cm^{-1} may be due to an Al-OH-Cu moiety[7]. Finally there is a slight shift in the position of the Si-O-Al peak, from 540 cm^{-1} to 535 cm^{-1}, which may be related to Cu ions which are heavy and larger than Al ions[10]. Near Infrared Spectroscopy (15000-4000cm^{-1}) and/or Raman spectroscopy may assist in this matter[21,23-26].

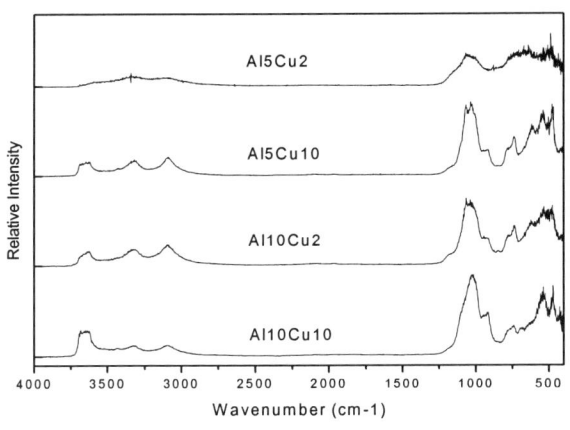
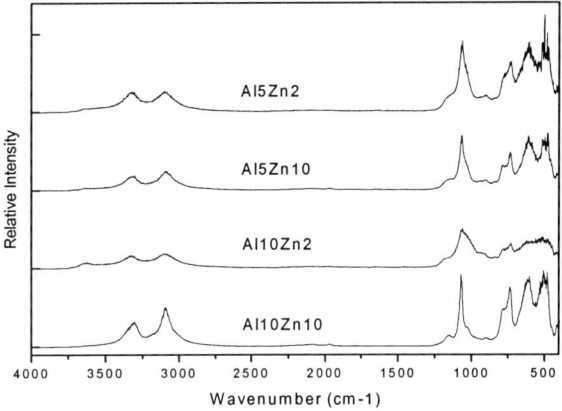

Figure 1 FTIR spectra of copper containing samples.   Figure 2 FTIR spectra of zinc containing samples.

The presence of two groups of peaks, around 3300 cm^{-1} and 3100 cm^{-1}, suggest boehmite (AlOOH) may also be present. The presence of this phase was confirmed by XRD in all samples except the Al10Cu10 (Section 5). The decrease in the intensity of these peaks between Al10Cu2 and Al10Cu10 suggests that boehmite may be an intermediate in the formation of kaolnite[27].

The spectra of Zn substituted kaolinite clearly indicate the presence of an additional as yet unidentified phase (Figure 2). The assignment of the following peaks or regions to bonds was proposed: 1060 cm^{-1} to Si-O-Si in a tetrahedral framework, 1030 cm^{-1} to Si-O in a tetrahedral framework, 901 cm^{-1} to Si-O in quartz, 780-730 cm^{-1} to Si-O in a tetrahedral framework, 650-605 cm^{-1} to Si-O-Al and Si-O-Zn where the silicon and zinc cation are in different layers of the material, 530-480 cm^{-1} to a mixture of Zn-OH, Si-O, Zn-O and Si-O-Al(Zn)[17,28].

5. Powder X-ray Diffraction

Diffraction patterns were obtained using a Shimadzu S-6000 Diffractometer (40kV, 30mA, divergence and anti-scatter slits 1 mm, receiver slit 0.3 mm) using Cu-Kα radiation (λ_1 = 1.5418 Å). Samples were ground into a fine powder and pressed into a flat plate holder for measurements. Data were collected in 0.02 ° steps over the range 10 ≤ 2θ ≤ 80 ° with 3⅓ seconds counting time per step. Figures 3 and 4 compare the diffraction patterns for three of the Cu substituted samples and the four zinc samples, respectively. Al5Cu10 was washed with acetone and then passed through a 200 μm copper sieve, in an attempt to remove any soluble impurities. Figure 3 shows this procedure reduced the intensity of the peak at 2θ = 43.3 ° which is due to metallic copper. Rietveld refinement[29,30] of the laboratory x-ray patterns using Rietica software[31], was hindered by the poor peak shapes inherent in such instruments. Consequently synchrotron x-ray diffraction patterns were collected at the Australian National Beamline Facility 20B, the Photon Factory, KEK, Tsukuba Japan. Such patterns afford better intrinsic peak shapes and permit more accurate measures of hkl reflection d spacings[32]. Figures 5 and 6 compare the copper and zinc patterns, and Figures 7 and 8 show the Rietveld fitted patterns for Al10Cu10 and Al10Zn10 samples. Table 2 summarises results from the x-ray powder diffraction analysis. By inspection of the phases it appears that the presence of zinc cations have inhibited the formation of kaolinite. Conversely, copper cations have been successfully incorporated into the kaolinite framework. The poorly ordered kaolinite phases (Hinckley Indices 0.71-0.84[5,33]) were refined in a C1 space group and the fraipontite in a CM space group[34,35]. There is much confusion over the exact atomic position and space group of fraipontite, so we assumed atomic coordinates similar to amesite another member of the berthierene phyllosilicates[35-37]. It is hoped that future work will determine the atomic coordinates of this rare but industrially important zinc mineral[28].

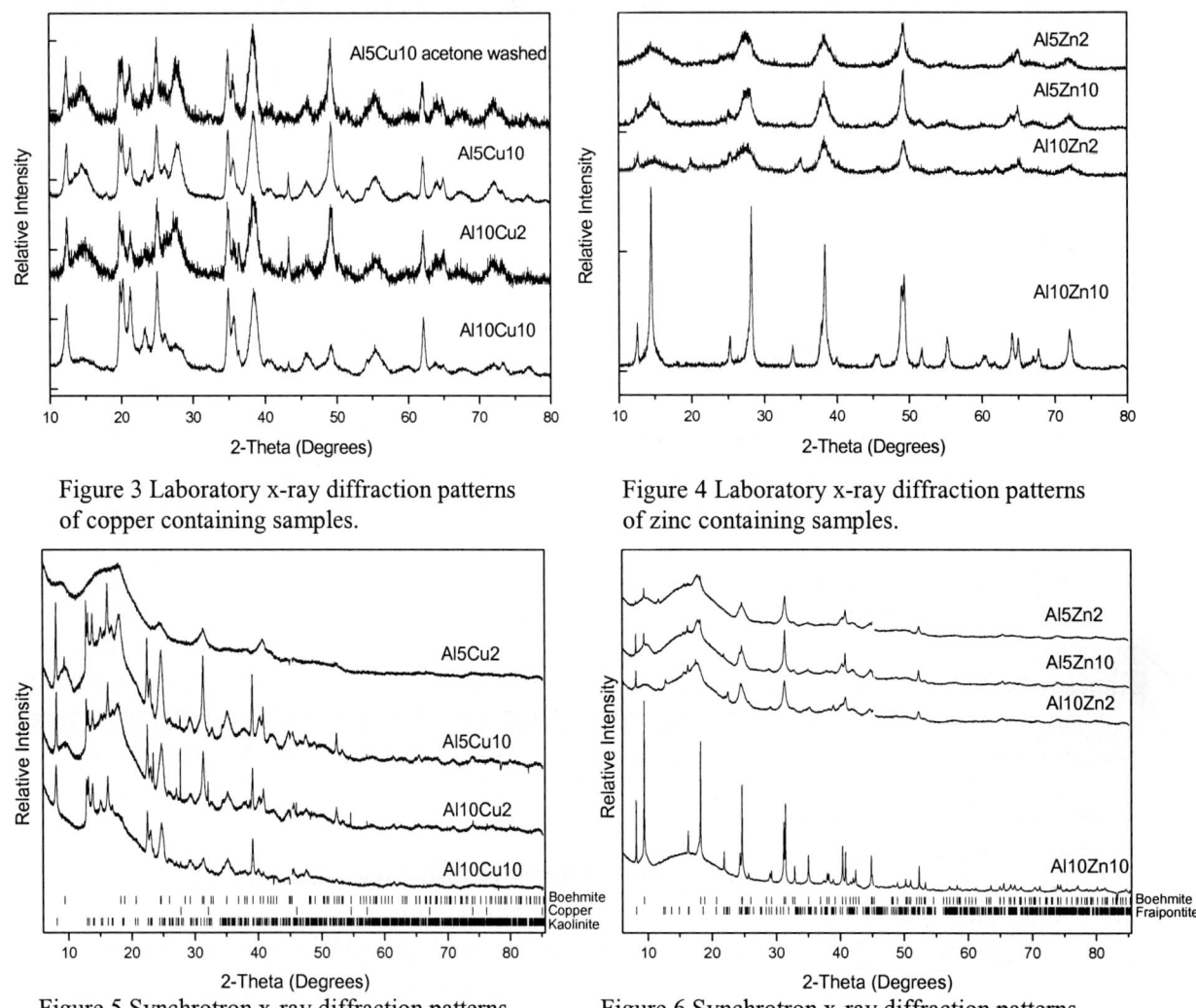

Figure 3 Laboratory x-ray diffraction patterns of copper containing samples.

Figure 4 Laboratory x-ray diffraction patterns of zinc containing samples.

Figure 5 Synchrotron x-ray diffraction patterns of copper containing samples.

Figure 6 Synchrotron x-ray diffraction patterns of zinc containing samples.

In addition to the normal least squares regression refinement, Le Bail refinements[31,38] were undertaken in order to calculate the lattice parameters. However, these did not significantly improve the precision of the analysis. Variation along the c-axis and the α angle between the copper samples reflects the difference in amounts of incorporated copper. There is also a general enlargement of the cell parameters of these copper doped kaolinites when compared to reference data for pure kaolinites (either the Keouk Iowa standard or Georgia KGa-1b standard (a ~ 5.139 Å, b ~ 8.932 Å, c ~ 7.371 Å, α ~ 91.6 °, β ~ 104.8 °, γ ~ 89.9 °[29,32,37,39]).

Two structural characteristics often determined in XRD of kaolinite are the position of the 060 and 1-31/1-3-1 reflections. The position of the 060 hkl reflection is an indication of whether the octahedral sheet contains di-octahedral (d(060) spacing in the 1.48-1.50 Å region) or tri-octahedral structures (d(060) spacing in the 1.53-1.55 Å region)[5,7,37]. Secondly, the relative positions of the 1-31 and 1-3-1 reflections is a measure of the distortion of the cell from a pseudo monoclinic cell, i.e. as $\gamma \rightarrow 90$ ° the 1-31 and 1-3-1 peaks overlap and the structure is described as having a monoclinic character. If the peaks are well resolved then the structure is clearly triclinic[7,40-42]. Considering these the copper doped kaolinites have dioctahedral species and the unit cells are monoclinic in character (d(060) spacing given in Table 2). However, caution must be taken in the simple assignment of these two characteristics, since there are a number of other hkl reflections near the 1-3-1/1-3-1 pair that have appreciable intensity.

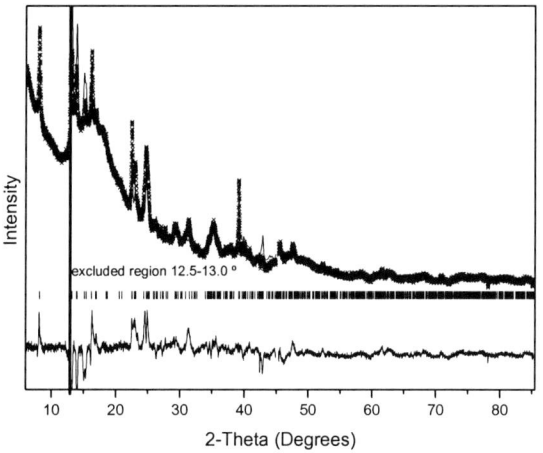

Figure 7 Observed (x), calculated (-), and difference data for Rietveld fitted synchrotron x-ray diffraction pattern of Al10Cu10.

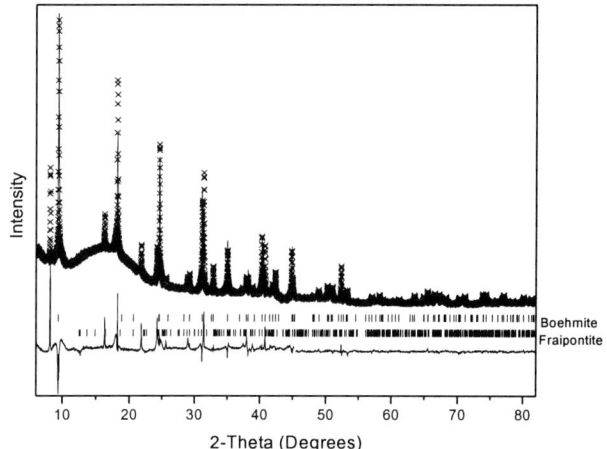

Figure 8 Observed (x), calculated (-), and difference data for Rietveld fitted synchrotron x-ray diffraction pattern of Al10Zn10.

Table 2 Summary of XRD analysis of samples.

Sample	Phase(s) (molar %)	a:b:c (Å), α:β:γ (°) major phase	Volume (Å^3)	Rp, Rwp, GOF	060 hkl	Hinckley Index
Al10Cu10	Kaolinite	5.18:8.96:7.34, 91.27:104.12:89.88	330.4±1.9	3.95, 4.55, 434.91	1.485	0.76
Al10Cu2	Kaolinite (98.91%) Boehmite Copper (1.09%)	5.19:8.99:7.35, 91.25:104.08:89.93	332.6±1.3	3.95, 4.37, 391.84	1.49175	0.84
Al5Cu10	Kaolinite (90.52%) Boehmite (9.39%) Copper (0.1%)	5.19:8.98:7.40, 90.72:104.18:89.90	334.4±1.5	4.21, 4.39, 340.98	1.4912	0.71
Al5Cu2	Boehmite?	-	-	-	-	-
Al10Zn10	Boehmite (89.89%) Fraipontite (10.11%)	2.87:12.23:3.69 5.34:9.22:7.24, β = 103.18°	129.5±0 346.9±0	3.06, 3.13, 198.6	1.53759	-

Al10Zn2, Al5Zn10, Al5Zn2 were composed of boehmite and fraipontite.
Reitveld Refinements were not completed for these samples.

Finally, comparison of Al10Cu2 and Al5Cu2 diffraction patterns suggest that the presence of copper cations may decrease the amount of intermediary phases in the formation of kaolinite, hence in effect promoting the reaction to completion. The formation process of kaolinite by the synthetic method used in this paper, has been studied and two stages have been identified[27,43].

6. X-ray Absorption Spectroscopy

Copper and zinc *K*-edge x-ray absorption fine structure spectroscopic measurements were recorded using a 10 element detector in fluorescence mode, at the Australian National

Beamline Facility 20B, the Photon Factory, KEK, Tsukuba Japan. The storage ring delivered a current of 250-400 mA at 2.5 GeV. The samples were pressed into aluminium sample holders and secured with Kapton tape windows. Six scans were made of each sample covering the pre-edge, edge, and post edge regions. The raw data was averaged and splined prior to X-ray absorption near edge structure (XANES) and extended x-ray absorption fine structure (EXAFS) studies. XANES is a powerful analysis that has been used to determine the oxidation state, coordination number (CN) and geometry of various metals and ligands in many chemical environments[10,44]. Some studies have also shown a link between XANES and metal-ligand covalency[44,45]. For copper, Cu^+ and Cu^{2+} are easily distinguished by the pre-edge feature at 8984 eV in the Cu^+ spectra. For Cu^+ compounds the nature of this pre-edge feature can be used to determine the CN of the complex[45]. Similarly, for Cu^{2+} compounds evidence for certain geometric complexes is provided by the occurrence and intensity of a pre-edge feature at 8979 eV. This feature is caused by a 1s → 3d electron transition, which is permitted in tetrahedral complexes but forbidden in octahedral complexes[46]. In practise though, the feature still exists in some octahedral copper structures, and so the relative intensities of the 8979 eV pre-edge peak, the 8986-8 eV shoulder feature, and the main absorption peaks must also be considered.

The XANES (Figure 9) clearly show that Cu^{2+} existed in the copper containing samples, since there was no Cu^+ 8984 eV feature. The strong 8979 eV pre-edge peak suggests the presence of tetrahedral copper in Al10Cu10 and Al10Cu2. However, this peak is shifted to a slightly higher energy in the Al5Cu2 sample and is comparatively very weak in the Al5Cu10 sample. This suggests the Cu geometry in these two samples is a less regular tetrahedron. The observed EXAFS spectrum (Figure 10) shows that there is a high degree of similarity between the copper samples with the exception of the Al10Cu10 sample. A single scattering analysis of the EXAFS was undertaken using the XFIT software[46,47] to confirm the geometry (Figure 11 and Table 3). For Al10Cu10 the proposed structure of this complex is CuO_3OH with three 1.94 Å Cu-O bonds and one 2.54 Å Cu-O bond. The next nearest atomic neighbours are at 3.03 Å, and these atoms are most likely aluminium. The XANES of the zinc containing samples (Figure 12) reveal that only octahedral arrangements (no pre-edge feature) of zinc exist. This is consistent with all of the four possible zinc sites used in the refinement of the x-ray diffraction patterns. EXAFS studies were not undertaken since reliable atomic positions and bond distances were not available for fraipontite, the EXAFS data were poor at k>9 Å (Figure13), and the presence of multiple zinc sites complicates both single and multiple scattering refinements.

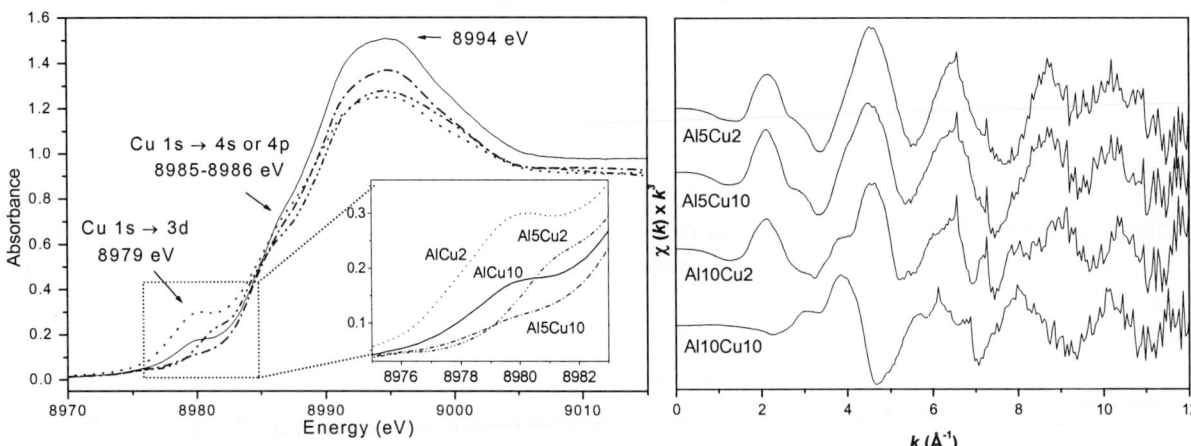

Figure 9 Cu K-edge XANES data observed for Al10Cu10 (—), Al10Cu2 (•••), Al5Cu10 (-•), Al5Cu2 (-••).

Figure 10 Observed Cu K-edge EXAFS.

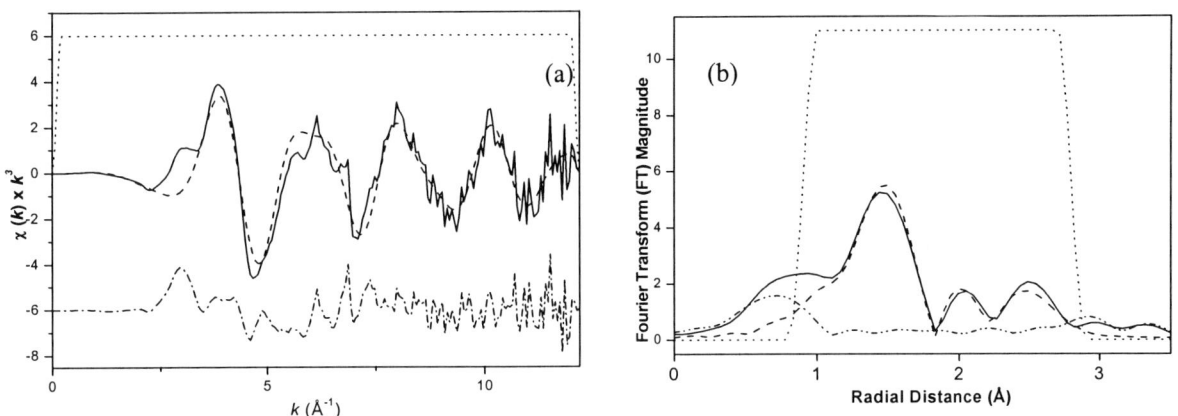

Figure 11 (a) Cu K-edge single scattering EXAFS data observed (–), simulated (---), and residual (-••) for Al10Cu10 (R_{xaf}= 13.9%). The region shown by (•••) is the k window used in the Fourier filter. (b) Subsequent single scattering Fourier transform observed (–), simulated (---), and residual (-•) of the EXAFS data from the EXAFS spectrum of Al10Cu10. The region shown by (•••) is the r window used in the Fourier filter.

Table 3 Summary of XFIT results for different models using Al10Cu10 data.

Shell Model	3O:1O:3Al	3O:1O:3:Si	2O:2O:3Al	3O:1O:3Al*
R_{xaf}(%)	23.8	25.7	21.2	13.9
Cu-O Å	1.94(2)	1.94(2)	1.95(2)	1.94(2)
Cu-O Å	2.54(2)	2.54(2)	2.12(2)	2.54(2)
Cu-Al Å	3.02(2)	2.99(2)	3.03(2)	3.03(2)
N_1	2.92	2.86	2.73	2.65
N_2	0.81	0.83	4.01	0.88
N_3	2.54	2.01	5.03	2.30
E_0 (eV)	-16.8	-16.4	-14.3	-16.1
S_0^2	0.90	0.9	0.90	0.90
S_1	0.005	0.005	0.005	0.005
S_2	0.006	0.005	0.054	0.007
S_3	0.011	0.010	0.019	0.009

* r window reduced.

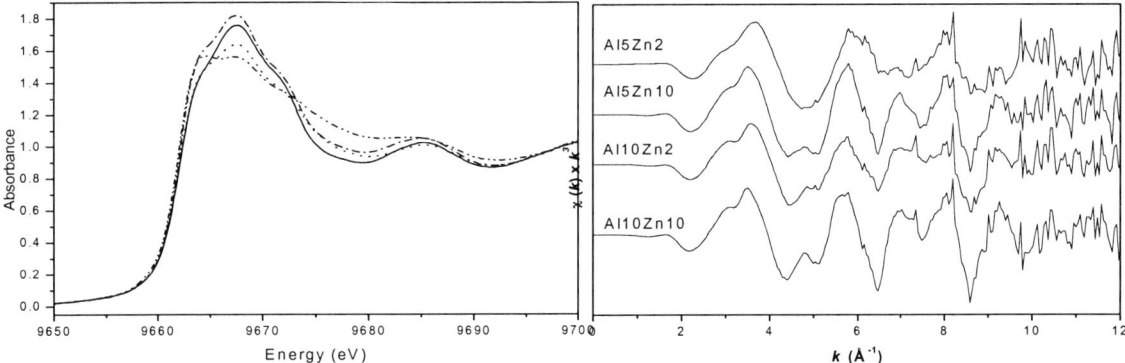

Figure 12 Zn K-edge XANES data observed for Al10Zn10 (---), Al10Zn2 (•••), Al5Zn10 (-•), Al5Zn2 (-••).

Figure 13 Observed Zn K-edge EXAFS.

7. Conclusions

The influence of copper and zinc cations on the formation and structure of hydrothermally synthesised kaolinite was studied. The diffraction data showed that copper cations promoted the formation of kaolinite whereas zinc cations poisoned the reaction, leading instead to the formation of fraipontite and boehmite. It was shown that copper was incorporated into the octahedral sheet, but existed as distorted tetrahedral copper (II) moeities. These exist without a high degree of perturbation to the ideal kaolinite structure and the precise lattice sites are yet to be determined.

8. Acknowledgement

The measurements performed at the Australian National Beamline Facility were supported by the Australian Synchrotron Research Program, which is funded by the Commonwealth of Australia under the Major National Research Facilities program.

References

(1) Wallis, D.S. In *Mineral Information Leaflet No.10*; Department of Mines and Energy, Queensland, 2000; p 2.
(2) Jackman, A.P., et al., *Journal of Hazardous Materials* **2001**, *82*, 27-41.
(3) Ledin, M., et al., *Soil Biology and Biochemistry* **1996**, *28*, 791-799.
(4) Lin, J.; Wang, K. *Toxicological and Environmental Chemistry* **1998**, *65*, 41-56.
(5) Bailey, S.W., Ed. *Hydrous Phyllosilicates (Exclusive of Micas)*, 1988.
(6) Mosser, C., et al., *Clay Minerals* **1993**, *28*, 353-364.
(7) Petit, S., et al., *Clays & Clay Minerals* **1995**, *43*, 482-494.
(8) Petit, S.; Decarreau, A. *Clay Minerals* **1990**, *25*, 181-196.
(9) Rengasamy, P. *Clays & Clay Minerals* **1976**, *24*, 265-266.
(10) Martin, F., et al., *Clay Minerals* **1998**, *33*, 231-241.
(11) Iglesia, A.L.; Oosterwyck-Gastuche, M.C.V. *Clays & Clay Minerals* **1978**, *26*, 397-408.
(12) Huertas, F.J., et al., *Applied Clay Science* **1993**, *7*, 345-356.
(13) Miyawaki, R., et al., *Clay Science* **1993**, *9*, 21-32.
(14) Kimpe, C.R.D. *Clays & Clay Minerals* **1976**, *24*, 200-207.
(15) Varian *Flame Atomic Absorption Spectroscopy Analytical Methods* Mulgrave, Victoria, Australia, 1989.
(16) Lim, C.H.; Jackson, M. In *Agronomy Monograph No. 9*; 2nd ed., 1982.
(17) Bougeard, D., et al., *Journal of Physical Chemistry B* **2000**, *104*, 9210-9217.
(18) Farmer, J.C. *The Infrared Spectra of Minerals*; Mineralogical Society: London, 1974.
(19) Russell, J.D. In *A Handbook of Determinative Clay Mineralogy*; Wilson, M. J., Ed.; Blackie: London, 1987; pp 133-173.
(20) Frost, R.L. *Clays & Clay Minerals* **1998**, *46*, 280-289.
(21) Petit, S., et al., *Clays & Clay Minerals* **1999**, *47*, 103-108.
(22) Fialips, C.I., et al., *Clays & Clay Minerals* **2000**, *48*, 173-184.
(23) Frost, R.L.; Kloprogge, J.T. *Spectrochimica Acta Part A-Molecular Spectroscopy* **2001**, *57*, 163-175.
(24) Frost, R.L.; Johansson, U. *Clays & Clay Minerals* **1998**, *46*, 486-477.
(25) Wada, K. *Clay Minerals* **1967**, *7*, 51-61.
(26) Shoval, S., et al., *Clay Minerals* **1999**, *34*, 551-563.
(27) Huertas, F.J., et al., *Chemical Geology* **1999**, *156*, 171-190.
(28) Kloprogge, J., et al., *Materials Research Bulletin* **2001**, *36*, 1091-1098.
(29) Bish, D.L.; Dreele, R.B.V. *Clays & Clay Minerals* **1989**, *37*, 289-296.
(30) Weidler, P.G., et al., *European Journal of Soil Science* **1998**, *49*, 95-105.
(31) Hunter, B., Rietica for Windows, 1.74, Sydney, Australia, 1997.
(32) Neder, R.B., et al., *Clays & Clay Minerals* **1999**, *47*, 487-494.
(33) Aparaicio, P.; Galan, E. *Clays & Clay Minerals* **1999**, *47*, 12-27.
(34) Fransolet, A.; Bourguignon, P. *Bulletin Société Française de Minéralogie et de Cristallographie* **1975**, *98*, 235-244.
(35) Mineralogical Database, http://www.webmineral.com, 2001.
(36) Russian Foundation of Basic Research, http://database.iem.ac.ru/mincryst/, 2001.
(37) Brindley, G.W.; Brown, G., Eds. *Crystal Structures of Clay Minerals and Their X-ray Indentification*; Mineralogical Society: London, 1980.
(38) Moore, D.M.; Jr., R.C.R. *X-ray Diffraction and the Identification and Analysis of Clay Minerals*; 2nd ed.; Oxford University Press: New York, 1997.
(39) Zvyagin, B.B. *Soviet Physics Crystallography* **1960**, *5*, 32-42.
(40) Fialips, C.I., et al., *Clay Minerals* **2000**, *35*, 559-572.
(41) Jesus, J.G., et al., *Applied Clay Science* **2000**, *17*, 245-263.
(42) Plançon, A.; Zacharie, C. *Clay Minerals* **1990**, *25*, 249-260.
(43) Fiore, S., et al., *Clays & Clay Minerals* **1995**, *43*, 353-360.
(44) Glaser, T., et al., *Accounts of Chemical Research* **2000**, *33*, 859-868.
(45) Kau, L., et al., *Journal of American Chemical Society* **1987**, *109*, 6433-6442.
(46) Weder, J.E., et al., *Inorganic Chemistry* **2001**, *40*, 1295-1302.
(47) Ellis, P., XFIT Software, 1.1 for WIN 32, Sydney, Australia, 1995.

Adsorption of quinoline on Na-sepiolite and Na-palygorskite

L. I. Vico and S. G. Acebal

Department of Chemistry and Chemical Engineering. Universidad Nacional del Sur. Avda. Alem 1253. 8000 - Bahía Blanca, Argentina.

Sorption of quinoline on homoionic Na-sepiolite and Na-palygorskite was determined. The adsorption of quinoline from aqueous solutions to Na-sepiolite and Na-palygorskite interface has been studied at 298 K as a function of both pH and ionic strength (μ) of the aqueous phase.

Adsorption of quinoline showed a plateau between 6-12 pH-values and exceeded the cation exchange capacity of Na-sepiolite and Na-palygorskite in aqueous suspension. Adsorption of quinoline by Na-palygorskite was lesser than Na-sepiolite adsorption. This observation would be related with the smaller size of the channels in Na-palygorskite than in Na-sepiolite.

A mechanism is suggested for the adsorption of quinoline by these clay minerals indicating both, an adsorption of ligand via surface hydroxyls and some ionic exchange processes associated with the existence of a small fraction of sites, which bind exchangeable cations, such as Na^+.

The adsorption reactions proposed are:

☐XNa + QH^+ ⇔ ☐XHQ + Na^+
☐XNa + Q + H_2O ⇔ ☐XHQ + Na^+ + OH^-
☐SOH + Q ⇔ ☐SOHQ
(☐ – indicate surface)

1. INTRODUCTION

Interest in the ability of clays to adsorb or release ions has increased greatly in recent years, and, while many aspects have been clarified, much remains to be learnt about relative affinities and the adsorption process. The study of sorption reactions of organic molecules onto clays, oxides and other solids produces information related to many useful applications [1-5]. Although clays in the soil environment often times contain adsorbed organic species, usually of biological origin, one must remember that clays are alteration products and that new surfaces are generated constantly. Hence any understanding of the reactivity of even simple organics on clay mineral surfaces, whether pristine or chemically modified, can aid in the prediction and perhaps prevention of contamination. Sepiolite and palygorskite are composed of trimorphic layers arranged in ribbons which are joined through oxygen ions [6,7]. The tetrahedral sheets are continuous but the apices of SiO_4 tetrahedra in adjacent ribbons point in opposite directions, giving rise to a corrugated surface with open channels running parallel to the ribbons. Palygorskite has a smaller trimorphic unit than sepiolite; it consequently has smaller channels widths. They both have the ability of selectivity sorbing polar and non-polar molecules on the external surfaces and in some cases (for small polar molecules) inside the channels [8].

Basic nitrogen-heterocyclic compounds are common to many products and, industrial and agricultural waste materials. In the absence of sorption processes, groundwater is particularly susceptible to contamination from these organic compounds [9-11]. Ionizable aromatic compounds are important environmental contaminants and ambiguities exist regarding their behaviour in soil and groundwater environments. In light of these considerations, the present study was undertaken to probe the exchange and molecular adsorption behaviour of quinoline over a wide range of pH and aqueous concentration using Na-sepiolite and Na-palygorskite.

2. MATERIALS AND METHODS

2.1. Clay minerals

A sepiolite from Vallecas (Madrid) and a palygorskite from Torrejón el Rubio (Cáceres), Spain, were used as adsorbents in this study. X R D patterns showed a highly pure sepiolite and a well-crystallised palygorskite containing quartz as impurity. Samples were prepared homoionic by saturating 35 g of natural sepiolite or palygorskite with 0.5 M NaCl. The fraction present as a suspension (<2 μm) was separated from residual coarser material. Several washings with bidistilled water removed excess salts until Cl⁻ detection by 0.5 M $AgNO_3$ became negative. Then, the air-dried Na-clay minerals were ground in a porcelain mortar.

Some surface and exchange properties of the selected homoionic samples are reported in Table 1. Specific surface area (S_{BET}) was determined by N_2 adsorption at the temperature of liquid N_2, after degassing samples overnight at 343 K. The classical BET equation was used for external surface area calculation. Total surface area was determined by ethylene glycol monoethyl ether (S_{EGME}) method. S_{BET} were smaller than S_{EGME} since the first method mentioned is used to determine only the external surface area. S_{BET} and S_{EGME} were determined after air-drying and grinding of the Na-clay minerals. Cation exchange capacity (CEC) was obtained by saturation with a buffered-pH 7 1 M NH_4–CH_3COO. Channel surfaces are also reported. The point of zero charge (PZC) was calculated as ½(pK_{a1} + pK_{a2}). K_{a1} and K_{a2} denote intrinsic constants.

Table 1
Some characteristics of the clay minerals studied.

Sample	Specific Surface area ($m^2 g^{-1}$)		Cation Exchange Capacity [b] ($cmol_c\ kg^{-1}$)	Channel Dimensions (Å)[c]	pK_{a1}	pK_{a2}	Point of Zero Charge (PZC)
	BET	EGME [a]					
Na-sepiolite	251.35	478.00	26.57	3.7 x 10.6	7.60	10.50	9.05
Na-palygorskite	147.60	354.32	17.01	3.7 x 6.4	6.20	11.50	8.85

[a] Carter et al. EGME method [12]
[b] Bower et al. [13]
[c] Singer (1989)[14]

2.2. Adsorption Measurements

Analytical grade quinoline (Q) was used in this study. The molecule acquires a proton in acid media producing a cation acid. The cation acid dissociation reaction:

$$C_9H_7NH^+ \Leftrightarrow C_9H_7N + H^+$$

has a pKa = 4.8. The dimensions of the molecule are: 9 x 7 x 3 Å^3 [4].

Experimental adsorption data were obtained as follows:
  a) Potentiometric titrations: A stock solution of 20 mM Q was prepared by dissolving the appropriate amount of Q in bidistilled water. The procedure involved a potentiometric titration of 1 g of Na-sepiolite or Na-palygorskite suspended in 100 mL of Q with 0.1M HClO$_4$ between 2-12 pH-range. The ionic strength (μ) was maintained at 1M, 0.1M and 0.01M with NaClO$_4$. The potentiometric measurements were made with a digital Ionalyzer Orion Research Model 701 A, fitted with a combined glass-calomel electrode and kept thermostatically at the required temperature (298 K). Each 1.5 pH-units and at equilibrium time (4h), 2 mL of suspension were extracted and Q was analysed at 280 nm in a UV-Visible Spectrometer CECIL 2021.
  b) Batch analysis: 0.20 g of each sample were placed into several stoppered conical flasks at 298 K and shaken for 4 hours (time required for equilibrium to be reached between adsorbed Q and Q in solution) with 20 mL of 20 mM Q aqueous solution. Three experiments at 2, 7 and 11 initial pH-units were carried out and μ was maintained constant at 0.1M and 1M NaClO$_4$. After shaking, suspensions were centrifuged and the supernatants were filtered through a 0.45-μm filter. Finally, Q concentration was determined in supernatant solutions using the same equipment as above. Amounts of Q adsorbed were calculated by the difference between initial and final suspension concentration. Blanks containing no Q were carried out in these series of experiments.

All titration experiments were corrected considering volume changes caused by sample extraction. Na and Mg levels were determined in supernatant solutions by atomic absorption spectrometry using a GBC 932 B computerized spectrometer.

3. RESULTS AND DISCUSSION

3.1. Influence of adsorbent

Q adsorption as a function of pH is shown in Figure 1 for Na-sepiolite and in Figure 2 for Na-palygorskite. As can be seen by comparing them adsorbed Q on Na-sepiolite was almost twice that on Na-palygorskite. This feature could be explained taking into account the almost double external specific surface area that exhibit sepiolite toward to palygorskite (Table 1), but the influence of channel relative dimensions should be considered in this case. Palygorskite channels are not big enough to host Q molecules (9x7x3 Å^3).

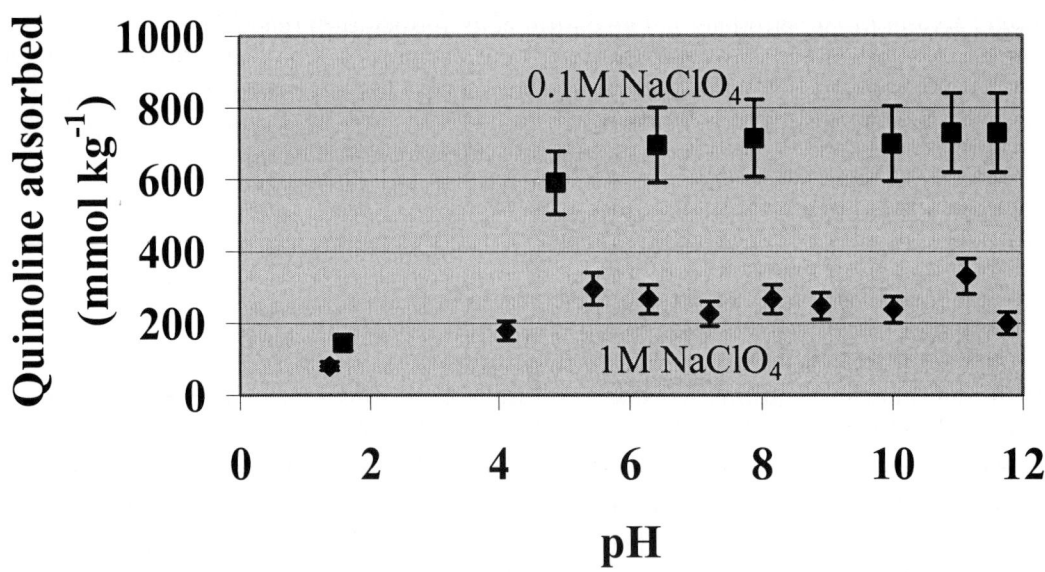

Figure 1. Adsorption of Q on Na-sepiolite as a function of pH.

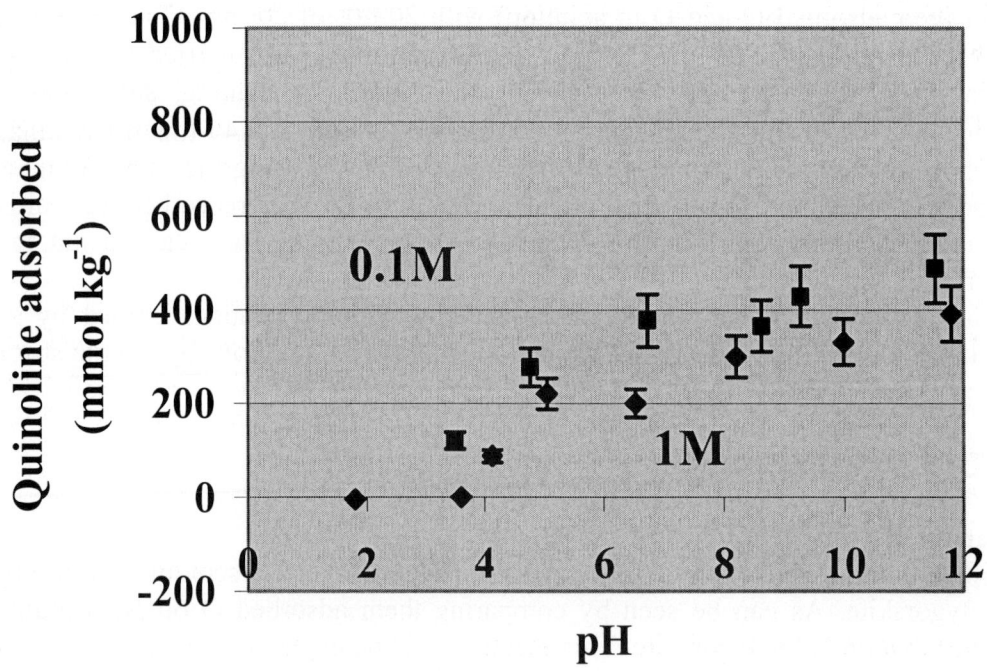

Figure 2. Adsorption of Q on Na-palygorskite as a function of pH.

3.2. Influence of pH

The effect of pH on the adsorption of Q ($\mu = 0.1M$) is shown in Figure 1 and 2. The shape of the curves is almost the same for both adsorbents. There is an increase in adsorption in 2-6-pH range. On Na-sepiolite the curve shows a plateau above pH 6. On Na-palygorskite the plateau and rise are less defined. The increase in adsorption in the range 2 to 6 occurs simultaneously with two important phenomena on clay mineral surface and in solution.

a) $\square SOH_2^+$ groups became in $\square SOH$ as pH-values increase because sepiolite-PZC and palygorskite-PZC values are higher than working-pHs (Table 1).
b) Concentration of neutral Q molecules in solution increases.

Above pH 6, the concentration of deprotonated Q is constant and a plateau in the curve is present. So, the preference of the neutral over the cationic form (HQ^+) adsorption could be explained from the shape of these curves.

3.3. Influence of ionic strength

Neither Na-sepiolite nor Na-palygorskite revealed influence of ionic strength on adsorption of Q when $NaClO_4$ was varied between 0.01M and 0.1M. In contrast, the enhancement of ionic strength to 1M affected Na-sepiolite and Na-palygorskite adsorption behaviour in two ways (Figure 1 and 2). On Na-palygorskite, this extremely high Na^+ concentration yielded a small reduction of adsorption over all pH-range studied. It was lesser in magnitude than CEC value (Table 1). As a consequence, this contribution to the overall sorption could be ascribed to a cationic exchange mechanism on permanent charge sites. In contrast, depletion of Q sorption on Na-sepiolite ($\mu = 1M\ NaClO_4$) exceeded widely that observed on Na-palygorskite and, it was greater than CEC value (Table 1). This behaviour was accompanied by Mg^{++} released ($\sim 1.2 \times 10^{-1}\ M.kg^{-1}$) and a concomitant Na^+ adsorption, suggesting a cationic exchange process on the silicate sheet. In order to find out the adsorption process that took place, Na^+ and Mg^{++} released were analysed in the Q-Na-sepiolite system in absent of 1M $NaClO_4$. The results showed a decrease in Mg^{++} amounts released ($\sim 1.2 \times 10^{-3}\ M.kg^{-1}$) concomitant with Q adsorption ($\sim 0.604\ M.kg^{-1}$). These data suggest that Q-adsorption process probably occurs in the broken bonds sites at the edge surfaces of the clay particles, closing the octahedral structural units protecting them of a subsequent dissolution.

3.4. Release of hydroxyl ions

Helmy et al.,[4, 5,] informed that Q adsorption on hydrous oxides is accompanied by an average increase of 0.3 pH-units. Ainsworth et al.,[11] found that concomitant with Q sorption the release of hydroxyl ions occurs. Their studies denoted one to one correspondences between Q-sorbed on Na-montmorillonite and H^+ consumed. Figure 3 shows acidimetric-alkalimetric titrations where:

- (Q-Na-sepiolite) represents titration of a suspension of Na-sepiolite (10 g.L^{-1}) in 20 mM Q ($\mu = 0.01M$),
- (Na-sepiolite) represents titration of a suspension of Na-sepiolite (10 g.L^{-1}, $\mu = 0.01M$),
- (Q) represents titration of 20 mM Q solution.

In order to determine possible release of OH^- ions concomitant to adsorption, the following equation is used:

$$\Delta C_a = H^+ \text{ consumed in Q sorption on Na-sepiolite} = \mathbf{a} - \mathbf{b} - \mathbf{c}$$

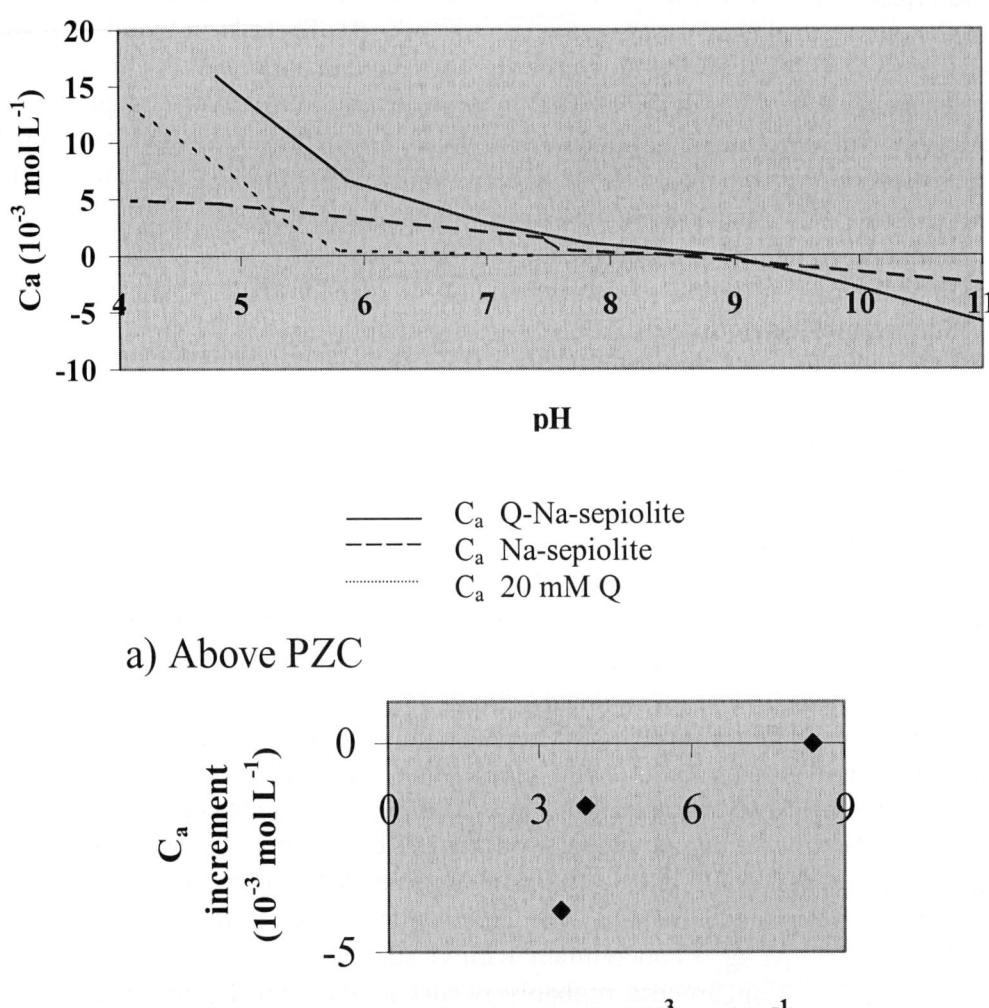

a) Above PZC

Figure 3. Alkalimetric-acidimetric titrations on Q-Na-sepiolite system.

b) Below PZC

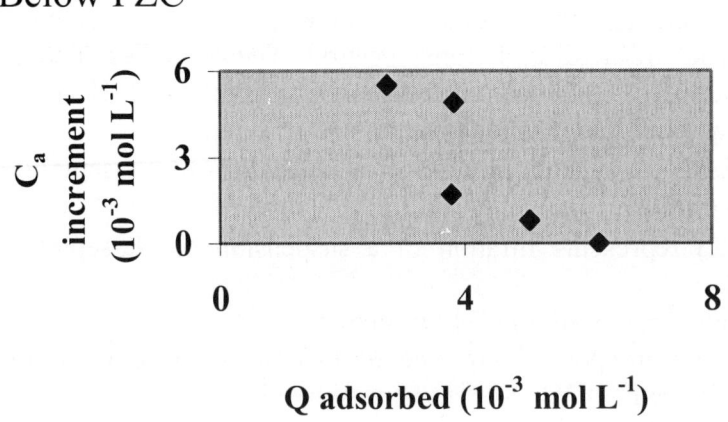

Figure 4 a) and b). Increment of acid added concomitant with quinoline sorption.

where C_a represents H^+ concentration in mol.L^{-1}; **a** indicates H^+ consumed in titration of Q-Na-sepiolite; **b** indicates H^+ consumed in titration of Na-sepiolite and, **c** denotes H^+ consumed in titration of Q solution.

Figures 4 a) and 4 b) display ΔC_a as a function of pH. Data collected indicates opposite behaviour below and above the PZC. Below PZC, the higher Q adsorption was the lower the consumption of H^+. Above PZC, the higher Q adsorption was the lower the consumption of OH^-.

CONCLUSIONS

The portion of adsorption assigned to ionic exchange may be represented by two reactions:

□XNa + QH$^+$ ⇔ □XHQ + Na$^+$

□XNa + Q + H$_2$O ⇔ □XHQ + Na$^+$ + OH$^-$

Other mechanism proposed may be related with Q adsorption via surface hydroxyls. The adsorption reaction proposed is:

□SOH + Q ⇔ □SOHQ
(□ – indicate surface)

REFERENCES

1. M.M. Mortland, Adv. Agron., 22 (1970) 75.
2. B.K.G. Theng, The chemistry of clay-organic reactions. Adam Higler, London, 1974.
3. J.A. Raussell-Colom and J.M. Serratosa, Reactions of clays with organic substances. pp.371-422. In A.C.D. Newman (ed.) Chemistry of clays and clay minerals. Mineralogical Sco. Monogr. 6. Longman Scientific and Technical, Essex, England, 1987.
4. A.K. Helmy, S.G de Bussetti, and E.A. Ferreiro, Clays Clay Miner., 31 (1983[a]) 29.
5. E.A. Ferreiro, S.G. de Bussetti, and A. K. Helmy, Z. Pflanzenernaehr. Bodenk., 146 (1983[b]) 369.
6. K. Brauner and A. Preisinger, Tschermaks Min. Petr. Mitt., 6 (1956) 120.
7. W.F. Bradley, Am. Mineral., 25 (1940) 405.
8. J.M. Serratosa, Proceedings of International Clay Conference, 1978, M.M. Mortland and V.C. Farmer, Eds. pp.99. Elsevier Amsterdam, 1979.
9. J.M.Zachara, C.C. Ainsworth, L.J. Felice, and C.T. Resch, Environ. Sci. Technol., 20 (1986) 620.
10. S. Chattopadhyay and S. Traina, Langmuir, 15 (1999) 1634.
11. C.C. Ainsworth, J.M. Zachara, and R.M. Schmidt, Clays Clay Miner., 35 (1987) 121.
12. D.L. Carter, M.C. Heilman, and C.L. González, Soil. Sci., 100 (1965), 356.
13. C.A. Bower, R. F. Reitemeier, and M. Fireman, Soil. Sci., 73 (1952) 251.

14. A. Singer, Palygorskite and sepiolite group minerals. In Minerals in Soil Environment. J.B. Dixon and S.B. Weed, Eds. pp 829-872. Soil Science Society of America. USA, 1989.

V
Mineral Structure and Investigational Methods

Methods

Controlled rate thermal analysis of formamide intercalated kaolinites

R. L. Frost*[a], J. Kristóf[b], Z. Ding[a] and Erzsébet Horváth[c]

[a]Centre for Instrumental and Developmental Chemistry, Queensland University of Technology, 2 George Street, GPO Box 2434, Brisbane, Queensland 4001, Australia.

[b]Department of Analytical Chemistry, University of Veszprém, H8201 Veszprém, PO Box 158, Hungary.

[c]Research Group for Analytical Chemistry, Hungarian Academy of Sciences, H 8201 Veszprém, PO Box 158, Hungary.

The use of controlled rate thermal analysis (CRTA) to separately identify the products of the adsorption by kaolinite from a pure formamide solution from those of an aqueous solution has been achieved by removal of the sample at the end of the controlled isothermal desorption step. The temperature of this isothermal desorption is dependent on the use of open or closed crucibles in the thermal analysis unit but is independent of the formamide/water ratio. X-ray diffraction shows that the formamide-intercalated kaolinite remains expanded with a d(001) spacing of 10.09 Å after formamide desorption. Further heating to 300 °C results in the de-intercalation of the formamide from kaolinite. DRIFT spectroscopy shows substantial and significant differences between the infrared spectra of the kaolinite from the different treatments. Raman spectroscopy supports the results of DRIFT spectroscopy. An intense band observed at 3629 cm^{-1} is attributed to the inner surface hydroxyls hydrogen bonded to the formamide. The kaolinites contain adsorbed water and show intensity in the 1705 cm^{-1} band, which is absent in the CRTA-treated formamide-intercalated kaolinites.

1. INTRODUCTION

Kaolinite and halloysite can be expanded with small organic compounds and with salts of short chain fatty acids [1-2]. This expansion occurs along the *c- axis*, is known as intercalation and is readily observed by X-ray diffraction (XRD). Kaolinite has a d(001) spacing of around 7.2 and expansion with formamide results in the basal spacing increasing from ~7.2Å to 10.09Å. When this intercalation occurs, changes are observed in the spectra of both the kaolinite and the inserting molecule. Upon intercalation additional infrared (IR) bands are observed for the formamide intercalated kaolinites at 3629 and 3606 cm^{-1} [3,4]. The 3629 cm^{-1} band is attributed to the hydroxyl-stretching frequency of the inner surface hydroxyl group hydrogen bonded to the carboxyl group of the formamide. The 3606 cm^{-1} band is ascribed to water in the interlayer. Upon intercalation changes are also observed in both the hydroxyl deformation modes and in the carboxyl bands. When kaolinite is reacted with formamide, two types of processes may occur: firstly the formamide is adsorbed on the outside of the kaolinite particles and secondly the formamide is inserted between the kaolinite sheets. This latter process is the intercalation. This paper reports a thermoanalytical process for separating these two types of adsorbed formamide.

Thermal analysis is normally carried out as a dynamic experiment with a constant and continuous heating rate. Such experimentation is often not able to determine phase changes, when these occur at close temperature intervals. New thermoanalytical techniques, which can separate thermal processes have been developed [5,6]. These techniques enable these closely overlapping phase phenomena to be separated. The method is known as constant rate thermal analysis (CRTA) and depends on the rate of mass loss, such that no heating occurs when the phase change occurs. Such thermoanalytical experiments are known as isothermal TGA or quasi-isothermal TGA. When undertaking these CRTA experiments, often simultaneous mass spectrometry is employed. One of the problems associated with the spectroscopic analysis of the intercalated kaolinites is uncertainty about the nature of the intercalate. The formamide-intercalated kaolinite may contain both adsorbed water as well as adsorbed formamide. Thus the infrared spectra of the intercalated kaolinites are probably determined by more than one phase i.e. the formamide-intercalated kaolinite and the formamide-adsorbed kaolinite. The technique of CRTA thermal analysis enables the separation of the adsorbed from the intercalated kaolinite. This is achieved through stopping the experiment between the desorption step and the de-intercalation step.

Thermoanalytical studies showed that the removal of intercalated formamide below 300°C is a complex process and that the liberation of formamide from the complex takes place in two overlapping stages [6]. Further research showed that a larger amount of formamide can be connected to the clay if intercalation is carried out in the presence of water. The complexity of the thermal decomposition patterns and the subtleties of the vibrational spectroscopic (FT-IR and Raman spectrometric) data require a detailed study of the thermal decomposition mechanism. The objective of this paper is to describe the differences between the adsorbed and formamide-intercalated kaolinite using a combination of thermal analysis, X-ray diffraction and infrared spectroscopic techniques.

2. EXPERIMENTAL

2.1 Materials

The kaolinite used in this study is from Királyhegy, Hungary. This mineral has been characterized both by X-ray diffraction and by Raman spectroscopy. The kaolinite is a low defect kaolinite with a Hinckley index of ~1.35. The kaolinite was intercalated by mixing 300 mg of the kaolinite with 5 cm^3 of a 100 % formamide or 50% formamide aqueous solution for 80 hours at room temperature. The excess solution was removed by centrifugation and the intercalated kaolinites kept in a desiccator before thermoanalytical, XRD and vibrational spectroscopic analyses.

2.2 Thermogravimetric analysis

Thermal decomposition of the intercalates was carried out in a Derivatograph PC-type thermoanalytical instrument (Hungarian Optical Works, Budapest, Hungary) in nitrogen atmosphere at a pre-set, constant decomposition rate of 0.20 mg/min. The samples were heated in an open platinum crucible at a rate of 1 °C/min up to 300°C. With the quasi-isothermal, quasi-isobaric heating program of the instrument the furnace temperature was regulated precisely to provide a uniform rate of decomposition in the main decomposition stage. Some of the samples were heated in a labyrinth-type crucible to provide a self-generated atmosphere and quasi-isobaric conditions during decomposition.

2.3 X-ray Diffraction

The X-ray diffraction (XRD) analyses were carried out on a Philips PW 3710 type diffractometer equipped with a PW 3020 vertical goniometer and curved graphite-diffracted beam monochromator. The radiation applied was $CuK\alpha_1$ from a long fine focus Cu tube, operating at 40 kV and 40 mA. The samples were measured in step scan mode with steps of 0.02° 2θ and a counting time of 1 s. Data collection and evaluation were performed with PC-APD 3.6 software.

2.4 Diffuse Reflectance Spectroscopy

Diffuse Reflectance Fourier Transform Infrared spectroscopic (commonly known as DRIFT) analyses were undertaken using a Bio-Rad 60A spectrometer. 512 scans were obtained at a resolution of 2 cm^{-1} with a mirror velocity of 0.3 cm/sec. Spectra were co-added to improve the signal to noise ratio. Approximately 3 weight % kaolinite or formamide-intercalated kaolinite was dispersed in 100 mg oven dried spectroscopic grade KBr with a refractive index of 1.559 and a particle size of 5-20 μm. Reflected radiation was collected at ~50% efficiency. Background KBr spectra were obtained and spectra ratio-ed to the background. The diffuse-reflectance accessory used was designed exclusively for Bio-Rad FTS spectrometers. It is of the so-called "praying monk" design, and is mounted on a kinematic baseplate. It includes two four-position sample slides and eight sample cups. The cup (3 mm deep, 6 mm in diameter) accommodates powdered samples mixed with KBr using an agate mortar and pestle in 1-3 % concentration. The collection efficiency of this adaptor is approximately 50 %. The reflectance spectra expressed as Kubelka-Mink unit versus wavenumber curves are very similar to absorbance spectra and can be evaluated accordingly. The advantage of using DRIFT measurements over the pellet technique is that in this case the likely interference of the mulling agent (intercalation of KBr in a liquid phase under pressure) can be avoided.

2.5 Raman spectroscopy

The formamide-intercalated kaolinite was placed on a polished stainless steel surface on the stage of an Olympus BHSM microscope, equipped with 5x, 20x and 50x objective lenses. The microscope is part of a Renishaw 1000 Raman microscope system, which also includes a monochromator, a filter system and a charge coupled device (CCD). Raman spectra were excited by a Spectra-Physics model 127 HeNe laser (633 nm), recorded at a resolution of 2 cm^{-1} in sections of 1000 cm^{-1} for 633 nm excitation. Repeated acquisitions using the highest magnification, were accumulated to improve the signal to noise ratio in the spectra. For the 298 K spectra, data were collected at 20 second intervals for 10 minutes at maximum magnification (X50). Spectra were calibrated using the 520.5 cm^{-1} line of a silicon wafer.

Spectral manipulation such as baseline adjustment, smoothing and normalization was performed using the Spectracalc software package GRAMS (Galactic Industries Corporation, NH, USA). Band component analysis was undertaken using the Jandel 'Peakfit' software package, which enabled the type of fitting, function to be selected and allows specific parameters to be fixed or varied accordingly. Band fitting was done using a Lorentz-Gauss cross-product function with the minimum number of component bands used for the fitting process. The Gauss-Lorentz ratio was maintained at values greater than 0.7 and fitting was undertaken until reproducible results were obtained with squared correlations of r^2 greater than 0.995. Graphics are presented using Microsoft Excel.

3. RESULTS AND DISCUSSION

3.1 Results of thermal analysis

The thermal decomposition curves (TG) of formamide-intercalated and formamide/water-intercalated kaolinite recorded under dynamic heating conditions are shown in Figure 1 a, b. For the formamide-intercalated kaolinite the differential analysis curves (DTA) show two endotherms at 65 and 150°C assigned to the loss of water and formamide respectively. A weak inflection is also observed at 560°C, attributed to kaolinite dehydroxylation. The thermogravimetric curve shows two weight losses at 150 and 130 °C. The differential thermogravimetric curves (DTG) are more informative and three inflections are observed at 65, 150 and 536 °C. These minima are attributed to the loss of water, formamide and kaolinite hydroxyls. Mass spectrometry confirms the loss of water and formamide at these temperatures. For the kaolinite intercalated with the water/formamide mixture, the DTA shows three endotherms at 65, 130 and 157 °C. The TGA curves show two weight losses at 50 and 150 °C. The DTGA curves are more informative and show three inflections at 50, 130 and 157 °C. The 130 °C inflection in the DTGA curves for the water/formamide-intercalated kaolinite is not observed for the formamide-intercalated kaolinite. This technique provides more useful information in the study of the structural changes of the formamide intercalated kaolinites. Importantly it is found that the use of the CRTA technique enables the weight loss steps to be more easily separated.

The thermoanalytical curves of the formamide intercalated kaolinite recorded under quasi-isothermal conditions are given in Figure 1 c, d. Under the controlled rate conditions two weight losses are observed (a) at 118 °C under quasi-isothermal conditions (b) at 153 °C. Thus under dynamic conditions only a single weight loss was observed but under quasi-isothermal conditions two weight losses are observed. According to TG-MS studies, the mass loss stage at 67°C belongs to the removal of 3.27 % by weight of water. Bonded formamide is released in two stages under quasi-isothermal conditions at 118°C (21.57% by weight) and under non-isothermal conditions at 153°C (4.24% by weight). Significant differences can be observed in the thermal behaviour of differently bonded formamide molecules. While the first type of bonded formamide is lost under isothermal conditions at 118°C, the second type of formamide molecules exhibits a drastically different behaviour on heating. The observation that the temperature remained spontaneously constant during the first formamide mass loss stage indicates an equilibrium reaction. The temperature of the experiment is controlled by the weight loss, such that when the weight loss exceeds a certain value (0.15mg/min), heating is stopped. An equilibrium reaction occurs as is indicated by the horizontal sections in the temperature-time plots. Stabilization of sample temperature in an equilibrium reaction also means the constant concentration (partial pressure) of gaseous decomposition products in the space among the particles in the open crucible used and the establishment of an equilibrium between the opposite diffusion processes of formamide vapour leaving the sample and nitrogen replacing formamide vapour. The liberation of strongly bonded formamide takes place under non-isothermal conditions in a similar fashion and at practically the same temperature (156 and 153°C) indicating the similar mechanism of formamide bonding to the clay structure. Since the best separation of the overlapping formamide decomposition stages can be obtained at 130°C, heating was stopped in a separate experiment at 130°C.

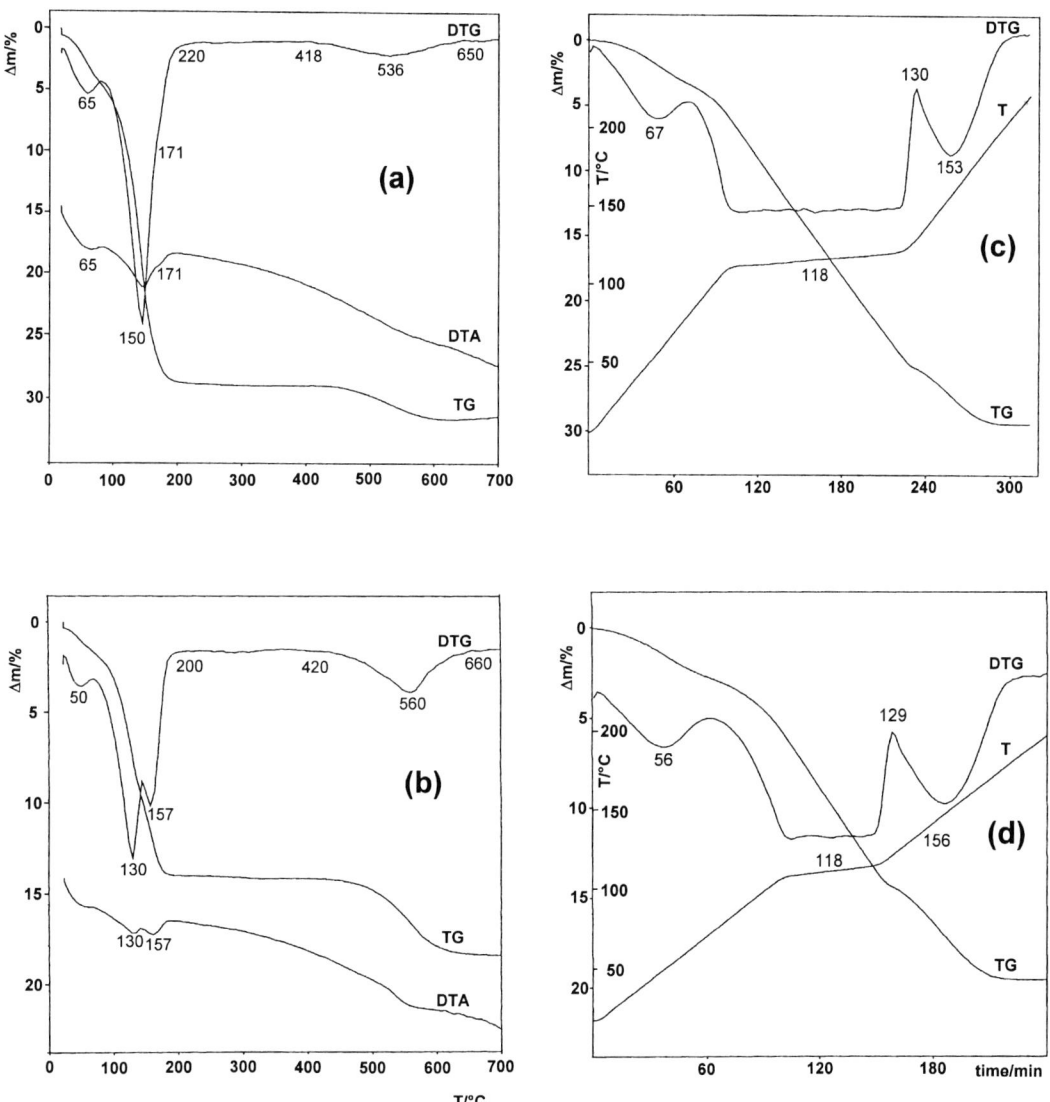

Figure 1 Thermal analysis of (a) formamide-intercalated kaolinite (b) formamide-water intercalated kaolinite (c) CRTA treated formamide-intercalated kaolinite (d) CRTA treated formamide-water intercalated kaolinite.

3.2 X-ray Diffraction

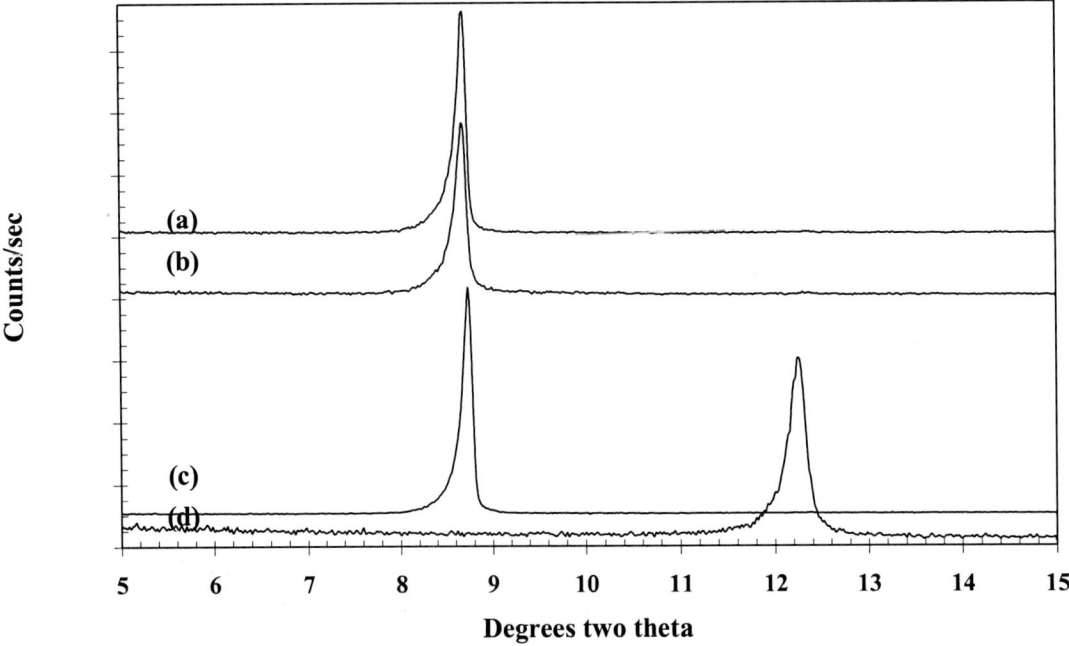

Figure 2 X-ray diffraction patterns of (a) formamide-intercalated kaolinite (b) water-formamide intercalated kaolinite (c) CRTA treated formamide intercalated kaolinite (d) CRTA treated intercalated kaolinite heated to 300°C.

Figure 2 displays the X-ray diffraction pattern of the formamide-intercalated kaolinite. Basal spacings were calculated from degrees two theta measurements using Bragg's law ($n\lambda=2d\sin\theta$). The XRD traces of the formamide and formamide/water intercalated kaolinites (a and b) show identical d(001) spacing of 10.2 Å. Trace (c) is that of the CRTA treated formamide intercalated kaolinite removed from the thermoanalytical instrument at 130°C. Importantly the XRD pattern of the CRTA treated formamide-intercalated kaolinite is identical to that of the untreated formamide-intercalated kaolinite. This proves that the CRTA treated formamide-intercalated kaolinite is still completely expanded at 130 °C. Upon heating the intercalated kaolinite to 300°C, the formamide is lost and the structure collapses to its original d spacing of ~7.2 Å (Figure 2d). The basal spacings are not the same with or without the formamide. Upon CRTA treatment to remove adsorbed formamide, the clay still remains expanded at 10.2Å. The adsorbed formamide is less strongly bound to the kaolinite hydroxyls sheets than the chemically bound formamide in the intercalation complex.

3.3 Drift spectroscopy

Figure 3 displays the DRIFT spectra of the CRTA treated and untreated formamide-intercalated kaolinite. The figure clearly illustrates the changes in the NH and OH stretching region of kaolinite-formamide complex spectra through intercalation with formamide and through the CRTA treatment. The differences in the DRIFT spectra between the formamide and CRTA treated formamide intercalated kaolinites are quite significant. The spectra of the CRTA treated samples show intensity in the 3300 to 3500 cm^{-1} region whereas the non-treated formamide intercalated kaolinite shows intensity in the 3100 to 3300 cm^{-1} region. This difference is attributed to the nature of the samples. The spectrum labelled b is that of a sample which contains both intercalated and adsorbed formamide. Spectrum a is that of the formamide-intercalated kaolinite only after CRTA treatment. The kaolinite shows five bands at 3693, 3681, 3667, 3651 and 3619 cm^{-1}. The first four bands are attributed to the kaolinite inner surface hydroxyls and the band at 3619 cm^{-1} to the inner hydroxyl. The relative intensities of these bands are 28.0, 9.3, 8.6, 31.9 and 21.1 % respectively.

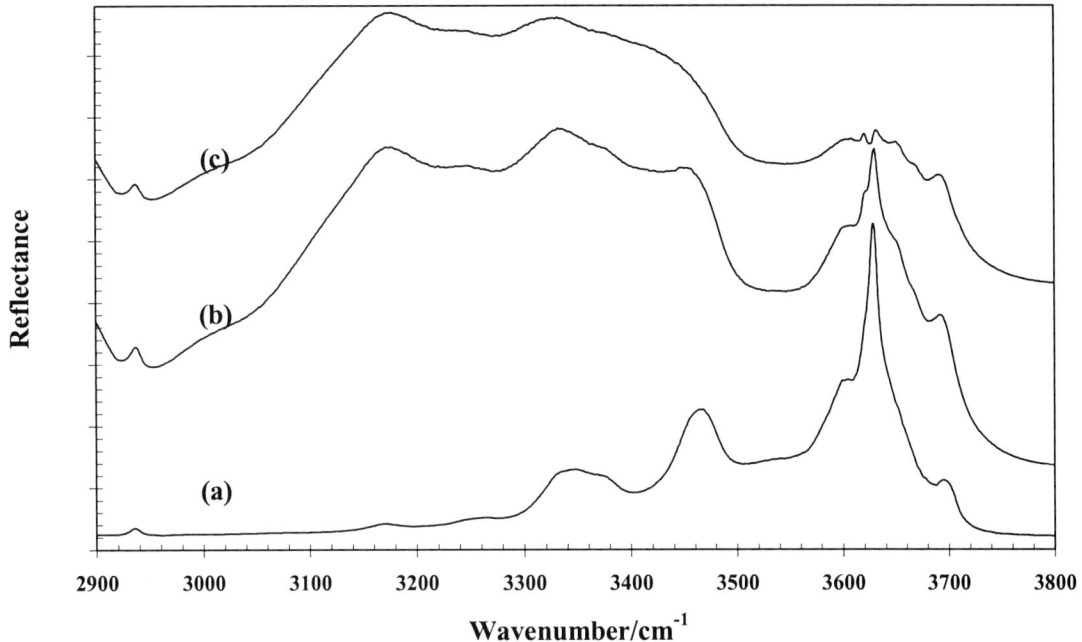

Figure 3 DRIFT spectra of the OH and NH stretching region of (a) CRTA treated formamide intercalated kaolinite (b) formamide intercalated kaolinite (c) The difference spectra between a and c.

Significant effects are observed in the intensities of the hydroxyl-stretching region of the formamide-intercalated kaolinites as a result of the CRTA treatment. Only two bands are resolved at 3695 and 3629 cm^{-1}. The relative intensities of these bands are 91.0 and 9.0% for the 100% CRTA treated formamide intercalated kaolinite and 94 and 6.0% for the CRTA treated 50%formamide intercalated kaolinite. The band at 3629 cm^{-1} is attributed to the inner surface hydroxyls hydrogen bonded to the formamide. The band at 3695 cm^{-1} is assigned to the non-bonded inner surface hydroxyls. In the NH stretching region bands are observed at around 3460, 3365, 3344, 3227 and 3148 cm^{-1}. The bands are broad and because of the complexity of the spectra in the NH and OH stretching regions, the bands are difficult to resolve with precision. What is significant, however, is that the bands in the 3344, 3227 and 3148 cm^{-1} region is attributable to adsorbed formamide and the bands at 3460 and 3365 cm^{-1} to the intercalated formamide.

4. CONCLUSIONS

Controlled rate thermal analysis has been used to separate adsorbed formamide from intercalated formamide in formamide-intercalated kaolinites. This separation is achieved by removal of the sample at the end of the controlled isothermal desorption of formamide. The differential analysis curves show two endotherms at 65 and 150°C ascribed to the loss of water and formamide respectively. A weak inflection is also observed at 560°C, which is attributed to the dehydroxylation of the kaolinite. Under quasi-isothermal conditions the liberation of loosely bonded formamide takes place in an equilibrium reaction at a constant temperature of 118°C, independently of the type of kaolinite used and the amount of water present in the intercalation solution. Mass spectrometry confirms the loss of formamide at this temperature. The liberation of the strongly bonded formamide takes place in a non-isothermal, equilibrium process between ~ 130 and 200°C. When the kaolinite is 100% expanded, the percentage amount of loosely bonded formamide is 1.87 times higher if intercalation is carried out in a formamide/water mixture, rather than in pure formamide. X-ray diffraction shows that after removal of the loosely bonded formamide, the completely expanded structure of the intercalated kaolinite still exists, i.e. no partial collapse of the structure results when some 70-80% of the reagent connected to the clay is removed. DRIFT spectroscopy shows that the spectra of the formamide intercalated kaolinite before and after the quasi-isothermal step are different. Significant differences exist between the DRIFT spectra of the inserting molecule before and after the removal of the adsorbed formamide. The spectra of the kaolinite before the quasi-isothermal desorption of formamide show significant remaining intensity in the bands attributed to the inner surface hydroxyls. The band observed at 3629 cm^{-1} is assigned to the inner surface hydroxyls hydrogen bonded to the formamide. Several significant differences are apparent between the spectra of the adsorbed and formamide intercalated kaolinites.

ACKNOWLEDGMENTS

Financial support from the Hungarian Scientific Research Fund under grant OTKA T25171 is acknowledged. The financial and infra-structural support of the Queensland University of Technology, Centre for Instrumental and Developmental Chemistry is gratefully acknowledged.

REFERENCES

1. R. L. Ledoux and J.L. White, J. Coll. Interf. Sc., 21 (1966) 127.
2. S. Olejnik, A. M. Posner, J. P. Quirk, Clay Miner., 8 (1970) 421.
3. R.L. Frost, W. Forsling, A. Holmgren, J.T. Kloprogge, J. Kristof, J. of Raman Spect., 29 (1998)1065.
4. R. L. Frost, J. Kristof, G.N. Paroz, T.H. Tran and J.T. Kloprogge, J. Coll. Interf. Sc., 208 (1998) 216.
5. F. Paulik and J. Paulik, Thermochim. Acta, 100 (1986) 23.
6. J. Kristof, E. Horvath, R. L. Frost and J. T. Kloprogge, J. of Thermal Analysis, 62 (2001) 279.

Cathodoluminescence of kaolin group minerals

M. Plötze[a] and J. Götze[b]

[a] Federal Institute of Technology Zurich, Division of Geotechnical Engineering, ClayLab, CH-8093 Zurich, Switzerland

[b] TU Bergakademie Freiberg, Institute of Mineralogy, D-09596 Freiberg, Germany

In the present study clay mineral samples of the serpentine-kaolin-group of worldwide occurrences were investigated by cathodoluminescence (CL) microscopy and spectroscopy complemented by electron paramagnetic resonance (EPR), X-ray diffraction, and other mineralogical studies. The study revealed for the first time that all dioctahedral minerals of the serpentine-kaolin group exhibit a deep-blue CL colour with emission bands at around $\lambda_{max} \sim 400$ nm (double peak with two maxima at 375 and 410 nm) with different irradiation-dependent intensity behaviour. The results of the CL and EPR measurements led to the conclusion that radiation induced centres (RID) are probably responsible for the blue CL of dioctahedral clay minerals of the serpentine-kaolin group. The blue CL and the typical luminescence kinetic enable to rapidly recognize these minerals with a CL microscope and to distinguish it from other minerals resulting in certain practical applications

1. INTRODUCTION

Clay minerals are the predominant constituents of sediments and sedimentary rocks and occur as alteration products in crystalline rocks as well. Because of the widespread occurrence and their physical and chemical properties, clay minerals serve as important indicators of specific facies conditions in different geological environments and are widely used in certain technical applications (e.g., ceramic industry, filling material, refractory materials, slurries, barrier material).

Despite numerous investigations of clay minerals there does not exist comprehensive knowledge about the luminescence behaviour. Up to know only few studies exist documenting the luminescence behaviour of kaolinite. One of the most conspicuous physical properties of kaolinite is its luminescence behaviour. Excited by an electron beam, kaolinite shows visible deep blue CL. A complex study by CL microscopy and spectroscopy complemented by EPR, X-ray diffraction, and other mineralogical studies should provide more information about the general luminescence behaviour of clay minerals.

2. EXPERIMENTAL

In the present study clay mineral samples of the serpentine-kaolin group of worldwide occurrences were investigated. The integrated study should provide information about the real structure and luminescence behaviour of the clay minerals and reveal factors influencing the cathodoluminescence of these minerals. Other clay minerals were investigated for comparison (CMS source clay montmorillonite SWy-1, STx-1, SAz-1; montmorillonite MX80, pyrophyllite/Belgium and North Carolina, illite Sárospatak/Hungary and Monte Caslano/Ticino).

Table 1: Samples

Mineral	Sample	Locality	Origin
Kaolinite	KGa-1	Washington county, USA	sedimentary (*CMS source clay*)
	KGa-2	Warren county, USA	sedimentary (*CMS source clay*)
	TOVO-3	Bahratal, Germany (Niederschöna layers)	diagenetic cement in cretaceous sandstone
	KREG-1	Regensburg, Germany	soil
	KAUG-1	Augustusburg, Germany	altered rhyolite
	KNAK-1	Rochlitz, Germany	hydrothermal (so called "Steinmark")
Dickite	KDICK-1	Barkly, R.S.A.	hydrothermal
	KDICKP-1	Huaraz, Perú	hydrothermal
	C1SQWL	Königstein, Germany (Niederschöna layers)	diagenetic cement in cretaceous sandstone
	ROTO204	Altenberg, Germany	hydrothermal
Nacrite	KNAK-2	Flöha, Germany	hydrothermal
Halloysite	KHAL-1	Entre-Sambre-et-Meuse, Belgium	karst formation
	KHAL-23	Djebel Debar, Algeria	sedimentary
Serpentine	SER-1	Snarum, Norge	

The crystal structure of the investigated clay samples was characterized by X-ray diffraction (BRUKER-AXS D8; CuKα, 4-70°2Θ, 0.02°, 3 s) using Rietveld refinement (AutoQuan/BGMN). The material was additionally studied by scanning electron microscopy (SEM JEOL 6400) to detect differences in crystal morphology and grain size of the samples. CL investigations were carried out on carbon coated polished thin sections of the samples using a "hot cathode" CL microscope (14 kV, 0.2 mA, current density of ~10 $\mu A/mm^2$). Luminescence images were captured during CL operations by means of an adapted digital video camera (Kappa 961-1138 CF 20 DXC with cooled CCD). CL spectra were obtained using an EG&G digital triple-grating spectrograph with liquid nitrogen cooled CCD detector in the range of 380 to 1000 nm (VIS-IR). To prevent distortion of the spectra by prolonged exposure to the electron beam, all spectra were taken on non-irradiated sample spots. Time-resolved (10 x 20 s) CL spectra were measured on selected samples to study the CL behaviour of the minerals during electron irradiation. In addition, spectral CL investigations were carried out using a JEOL JSM 6400 SEM equipped with an Oxford MonoCL system (20 kV, 0.6-1.6 nA). The CL spectra were detected over the range 200-800 nm (UV-VIS), with 1 nm steps and a dwell time of 1 s per step. For identification of point defects in the crystal structure, paramagnetic centres were investigated in powder samples by EPR (BRUKER ESP 300E; X-band, 295 and 70 K, $H_M = 5$ G, $p = 40$ mW) before and after gamma-irradiation (^{60}Co, 1.1 MGy). Additionally, the samples were investigated by trace-element analysis (ICP-MS Agilent 7500) and FT-IR spectroscopy (PERKIN-ELMER FT-IR 2000).

Figure 1a. CL-image of kaolinite KGa-1 (width 1.5 mm). The dark area is caused by the electron bombardment during the measurement.

Figure 1b. CL-image of dickite KDICK-1 (width 1.5 mm). The bright area is caused by the electron bombardment during the measurement.

3. RESULTS AND DISCUSSION

The study revealed for the first time that all dioctahedral minerals of the serpentine-kaolin group exhibit a deep-blue CL colour (Fig. 1). The visible deep blue cathodoluminescence enables to rapidly recognize these clay minerals with a CL microscope and to distinguish it from other minerals. The also investigated clay minerals (serpentine, talc, montmorillonite, and illite) show no visible CL.

In the CL spectra an emission band at $\lambda_{max} \sim 400$ nm can be established (Fig. 2). Spectral measurements with the SEM-CL revealed a badly resolved double peak consisting of two maxima at 375 nm and 410 nm. Furthermore a weak broad band emission between 700 nm and 800 nm was found. The observed emission bands at ~ 550 nm, e.g. in the time dependent spectra of powder samples (Fig. 3) are caused by emission of the used resin (araldite). No emission band in the UV was detected. Well-crystallized kaolinite shows the highest intensity, halloysite the lowest. The luminescence behaviour does not seem to depend on the genesis, crystallinity, morphology or grain size for the individual minerals.

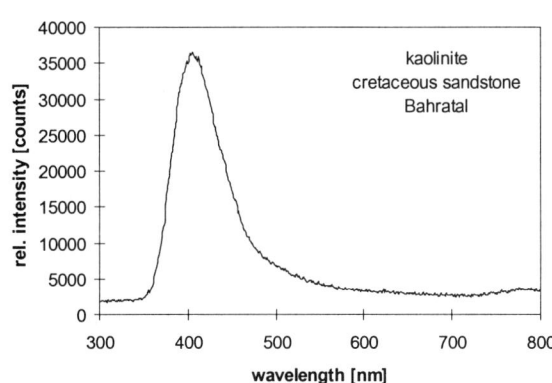

Figure 2. CL-spectra of kaolinite in cretaceous sandstone (TOVO-3, Elbe zone, Germany).

Different irradiation-dependent intensity behaviour can be determined in the time-resolved CL measurements (Fig. 1 and 3). Independently from origin (sedimentary or hydrothermal) and from crystalline order the

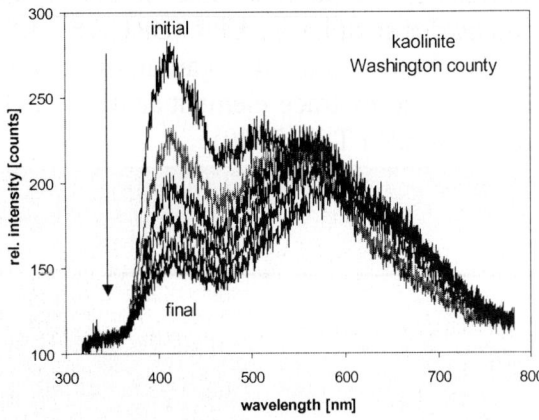
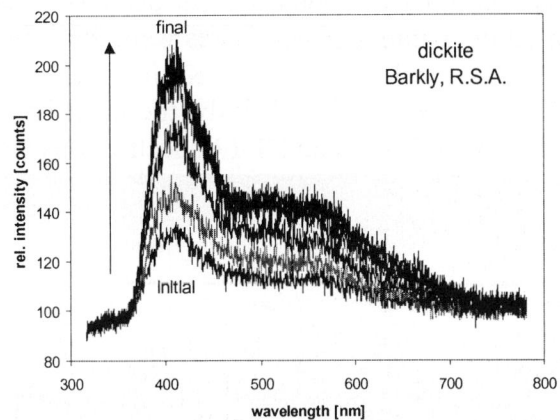

Figure 3a. CL-spectra of well crystallized kaolinite KGa-1.

Figure 3b. CL-spectra of dickite (KDICK-1).

CL of kaolinite shows a transient behaviour with decreasing emission intensity during electron irradiation. The CL of kaolinite shows short-lived behaviour, i.e. the intensity of CL emission rapidly decreases within 30-60 s during electron irradiation (Fig. 1a, 3a). In contrast, the intensity of the blue CL of dickite increases during CL measurement (Götze & Plötze 2000)

(Fig. 1b, 3b). In the samples of nacrite, halloysite and pyrophyllite low luminescence intensity with weak changes could be observed. Therefore a differentiation of kaolinite, dickite and nacrite according to the time dependent spectral characteristics of the CL appears possible also in thin sections.

By means of EPR spectroscopy some types of paramagnetic defects had been detected in non-irradiated and irradiated natural samples (Fig. 4). The signal intensity of centres varies for different kaolinite.

Fe^{3+} ions substituting for Al^{3+} at two octahedral positions cause signals in the EPR spectra at $g_{eff.}$ ~ 4.3 (Meads & Malden 1975). We found these spectra in all samples. The highest concentrations are in illite and montmorillonite and the lowest in dickite and nacrite samples. The kaolinite KGa-2 shows also high iron concentrations. This as well as the lower intensity of the RID is responsible for his low CL-intensity. One important group of paramagnetic centres are the radiation induced paramagnetic defect centres (RID) with spectra at $g_{eff.}$ ~ 2.0 (Clozel et al. 1994). There are centres assigned to electron holes trapped on apical oxygen ions (Si-O-centres) with different orientations. The other centre assigned to an electron hole located at the Al-O-Al group (Al substituting Si in the tetrahedron) shows an intense hyperfine structure with ^{27}Al nuclei (esp. in the spectra at 70 K). In illite and montmorillonite RID could not be detected. The highest concentration of RID was measured in well crystallized kaolinite. After gamma-irradiation the intensity of RID centres in kaolin group clay minerals strongly increases. The intensity of Al-O-Al-centre increases more strongly as those of the Si-O-centres. (Fig. 4b).

The results of the CL (wavelength of the luminescence emission (energy) as well as the time-resolved luminescence behaviour) and EPR measurements led to the conclusion that the

RID centres are probably responsible for the blue CL of the dioctahedral clay minerals of the serpentine-kaolin group. Substituting Fe^{3+} may act as quencher of the luminescence.

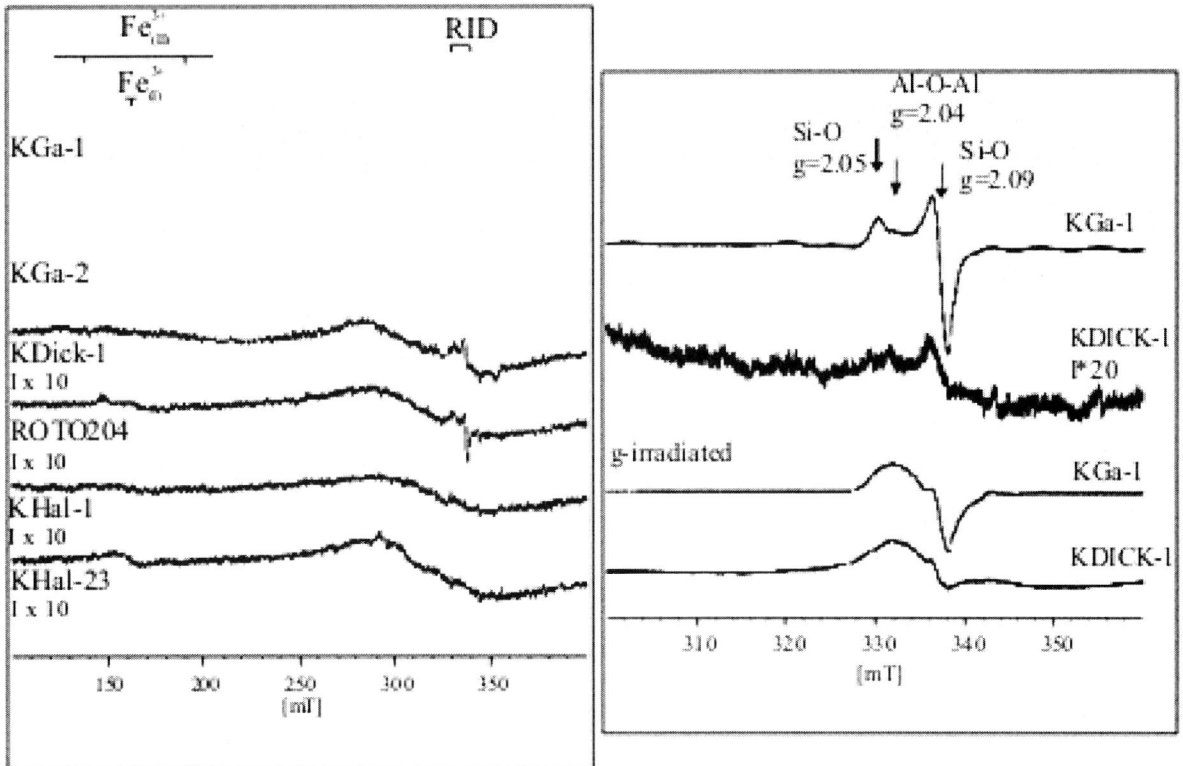

Figure 4a. EPR powder spectra (X-band, 295 K) of kaolin-group minerals.

Figure 4b. EPR powder spectra of kaolinite and dickite before and after gamma-irradiation.

The blue CL enables to rapidly recognize kaolinite with a CL microscope and to distinguish it from other minerals. These advantages resulted in certain practical applications.

CL microscopy is a useful tool in investigations of sandstones. The distinction of the different types of cements by means of their luminescence colour and the intensity behaviour is possible with CL microscopy of thin sections (kaolinite blue with decreasing intensity, dickite with increasing intensity, carbonate with orange colour and authigenic quartz with a very weak brownish to bluish CL) (Fig. 5). CL microscopy enables in combination with image analysis to quantify the amount of pore-filling cement (e.g. kaolinite in Götze & Magnus 1997).

Results are useful in several directions. The formation of different cement types (e.g., kaolinite, carbonate, quartz) reflects different facies conditions during sedimentation and diagenesis and thus, enables to reconstruct the sedimentary conditions (Magnus & Götze 1996). Kaolinite is occurring as typical clay mineral of the coast lithofacies with terrestrial provenance area. Increased carbonate content as well as the lack of kaolinite reflect formation conditions in the marine facies.

On the other hand, the amount of kaolinite significantly influences the physical properties of sandstones used as building materials. The detailed characterization of historical building materials is often necessary to reveal previous deposits of building materials and to select the material for reconstruction activities.

Other fields of application are the recognition of kaolinization effects due to weathering or hydrothermal alteration. With CL microscopy also smallest, with other methods not detectable amounts of kaolinite can be diagnosed.

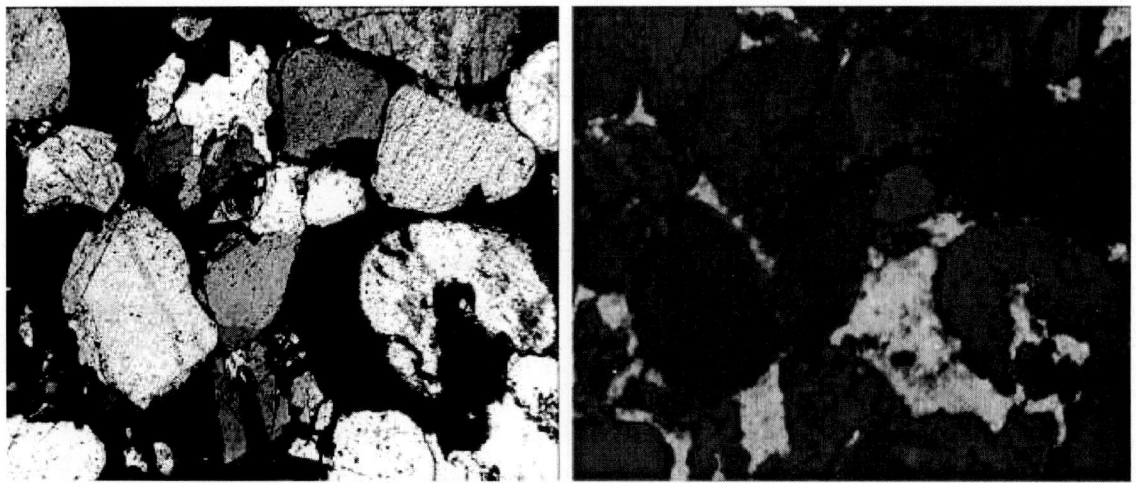

Figure 5. Polarizing microscopy (left) and CL image (right) of Cretaceous sandstone with kaolinitic cement (Elbe zone, Germany) width 1.5 mm.

Acknowledgements: We are grateful to R. Neuser (Bochum University) for taking the time-resolved CL emission spectra.

REFERENCES

Clozel, B., Allard, T., Muller, J.-P. (1994) Nature and stability of radiation induced defects in natural kaolinites: New results and a reappraisal of published works. Clay Clay Miner. **42**, 657-666.

Götze, J. & Magnus, M. (1997): Quantitative determination of mineral abundance in geological samples using combined cathodoluminescence microscopy and image analysis. Eur J Miner. **9**, 1207-1215.

Götze, J. & Plötze, M. (2000) Kathodolumineszenz von Kaolinit: Ursachen, Besonderheiten und praktische Anwendungen. Berichte DTTG **7**, 92-99.

Magnus, M. & Götze, J. (1996): Mineral composition of Lower Turonian sandstones of the Elbe zone (Germany): evidence from combined CL and image analysis. Zbl Geol. Paläont, part 1, n11/12, 1255-1262

Meads, R.E. & Malden, P.J. (1975) Electron spin resonance in natural kaolinites containing Fe^{3+} and other transition metal ions. Clay Miner. **15**, 1-13.

FT-IR photoacoustic spectroscopy of kaolinite and gibbsite surfaces

H. D. Ruan[a,b*], R. L. Frost[a], J. T. Kloprogge[a], L. Duong[a] and D. G. Schulze[b]

[a]Centre for Instrumental and Developmental Chemistry, Queensland University of Technology, GPO Box 2434 Brisbane 4001, Australia

[b]Agronomy Department, Purdue University, West Lafayette IN 47907-1150, U. S. A.

Fourier transform infrared photoacoustic spectroscopy (FTIR-PAS) has advantages in distinguishing hydroxyl surface spectra as a function of depth. FTIR-PAS spectra of kaolinite were recorded at mirror velocities of 0.05, 0.1, and 0.2 cm s^{-1}, and compared to gibbsite spectra recorded at mirror velocity of 0.2 cm s^{-1}. Four strong bands at 3694, 3670, 3654 and 3621 cm^{-1}, and a weak band at 3435 cm^{-1} in the hydroxyl stretching region were observed for kaolinite. The intensity of these bands increased as mirror velocity increased. Because an increase in mirror velocity means a shallower depth of the surface is measured, the increasing intensity of the hydroxyl stretching and bending vibrations is a consequence of more hydroxyl units on the outer surface than on the inner surface. The intensity of the kaolinite band at 1037 cm^{-1} was greater than that at 1010 cm^{-1} at 0.2 cm s^{-1} velocity but this order changed when mirror velocity was reduced to 0.05 cm s^{-1} in the hydroxyl deformation and water bending region. For the Si-O-Al modes of kaolinite in the low frequency region, one band at 473 cm^{-1} is observed at 0.2 cm s^{-1} velocity, while this band is split into two at 477 and 467 cm^{-1} at 0.05 cm s^{-1} velocity. Some vibrations of gibbsite resemble those of kaolinite.

1. INTRODUCTION

Since Pauling [1] and Gruner [2] determined the structures of kaolinite and related clay minerals in 1930 and 1932, research on the genesis and application of kaolinite, [Al$_2$Si$_2$O$_5$(OH)$_4$], and its weathering products such as gibbsite, [γ-Al(OH)$_3$], has attracted many researches in the world. The principal occurrences of kaolinite are as primary residual deposits formed by weathering or low-temperature hydrothermal alteration of feldspars, muscovite and other Al-rich silicates [3]. Laterization leads to the enrichment of oxides and oxyhydroxides of iron and aluminium plus kaolinite.

The fundamental unit of the kaolinite structure is an extended layer, which can be regarded as having two constituents [4, 5]. A sheet of composition (Si$_4$O$_{10}$)$^{4-}$ is formed by the linkage of SiO$_4$ tetrahedra in a hexagonal array, with the bases of the tetrahedra being approximately coplanar and all their vertices pointing in one direction. The apical oxygens, together with some additional (OH)$^-$ ions located over the centers of the hexagons, form the base of a gibbsite-like sheet of composition (OH$_6$)-Al$_4$-(OH)$_2$O$_4$. From plan and elevation views, there is a composite Al$_4$Si$_4$O$_{10}$(OH)$_8$ layer. Only two out of each set of three available

---

[*] Corresponding author. Tel: (765) 494-8786; Fax: (765) 496-2926; E-mail: hruan@purdue.edu

octahedral sites are occupied by Al ions. Successive 1:1 layers are superimposed so that, in general, the basal oxygens of the tetrahedral sheet are hydrogen-bonded to hydrogen ions of the neighbouring octahedral sheet, resulting in a single-layered triclinic unit cell.

Gibbsite (also known as hydrargillite) is a monoclinic mineral often found in the < 2 μm clay separate of various sediments, especially in areas with a hot rainy climate and a dry period (monsoon). It is also found concentrated in various bauxite deposits differing in purity and extent [6]. Gibbsite is often the predominant mineral, and kaolinite is the impurity second only to iron oxides in tropical bauxites, e.g. Weipa bauxite of Australia.

The fundamental unit of the gibbsite structure is a plane of Al ions sandwiched between two planes of hexagonally packed hydroxyl ions [3]. In the gibbsite structure, two out of three of the octahedrally coordinated sites between the planes of hydroxyls are occupied by cations. The octahedral sheets may be regarded as being built of octahedra linked laterally by sharing edges; the network so formed may be described by an orthogonal (pseudohexagonal) cell with parameters $a \sim 8.6$ Å, $b \sim 5$ Å ($\sim a/\sqrt{3}$). The planes of oxygen in gibbsite are in hexagonal close packing ABBAABBA…. In the ideal structure of gibbsite, oxygens at the bottom of the sheet lie directly above oxygens at the top of the sheet below. The structure of gibbsite is in fact somewhat distorted from the ideal, resulting in a monoclinic cell: there are two octahedral sheets in each cell [3].

FT-IR photoacoustic spectroscopy, in conjunction with scanning electron microscopy, has been used to determine the behavior of hydroxyl units near the crystal surfaces of minerals. This study is a part of a research project on the thermal activation of bauxite. The mineralogical and chemical properties of kaolinite and gibbsite are important to the understanding of dehydroxylation pathways of activated bauxite. We compare the FTIR-PAS spectra of kaolinite at various mirror velocities with spectra of gibbsite in order to provide information for the characterization of these minerals in activated bauxite.

2. EXPERIMENTAL

The kaolinite used in this study is a naturally occurring mineral obtained from Georgia, U. S. A. Sub-samples were preheated in an oven at 105 °C for about 15 hours in order to reduce the noise in the FT-IR spectrum resulted from adsorbed water. The gibbsite was synthesized from Al salts [7].

Fourier transform infrared photoacoustic spectroscopic (FTIR-PAS) spectra of kaolinite and gibbsite were obtained using a photoacoustic unit connected to a Digilab FTS-20/80 FT-IR spectrometer. Raw samples were used without preparation. Spectra at 8 cm^{-1} resolution were acquired by coaddition of 256 scans. Spectral manipulation such as baseline adjustment, smoothing and normalisation was performed using the Spectracalc software package (Galactic Industries Corporation, NH, USA). Band component analysis was undertaken with the Jandel "Peakfit" software package, which enables the type of fitting function to be selected and allows specific parameters to be fixed or varied. Band fitting was carried out using a Lorentz-Gauss cross-product function with the minimum number of component bands used for the fitting process. The Gauss-Lorentz ratio was maintained at values greater than 0.7, and fitting was undertaken until reproducible results were obtained with squared correlations of r^2 greater than 0.995.

Scanning electron microscopy was carried out using a JOEL-25 microscope. Powder samples were deposited onto tape on the specimen holder and coated with gold.

Fig. 1. Scanning electron micrographs of kaolinite (a, b), and gibbsite (c, d).

3. RESULTS AND DISCUSSION

The crystal size and shape of kaolinite and gibbsite obtained from scanning electron microscopy are illustrated in Fig. 1. Kaolinite crystals show pseudo hexagonal shapes (Fig. 1b) and a gibbsite crystal in Fig. 1-d resembles that of kaolinite. The particles of natural kaolinite show separate packages of layers (Fig. 1a) while the particles of synthetic gibbsite aggregate together (Fig. 1c).

The kaolinite samples were characterized between 400 and 4000 cm^{-1}. Kaolinite spectra were recorded at 0.2, 0.1 and 0.05 $cm\ s^{-1}$ mirror velocities (Fig. 2). Because an increase in mirror velocity means that a shallower depth of surface is measured, the hydroxyl stretching band intensities increase as mirror velocity increases, as shown in Fig. 2 and Fig. 3-a, indicating that more hydroxyl units resemble in the outer surfaces than in the inner surfaces. The effect of surface depth on the hydroxyl and Si-O-Al vibrations will be discussed below.

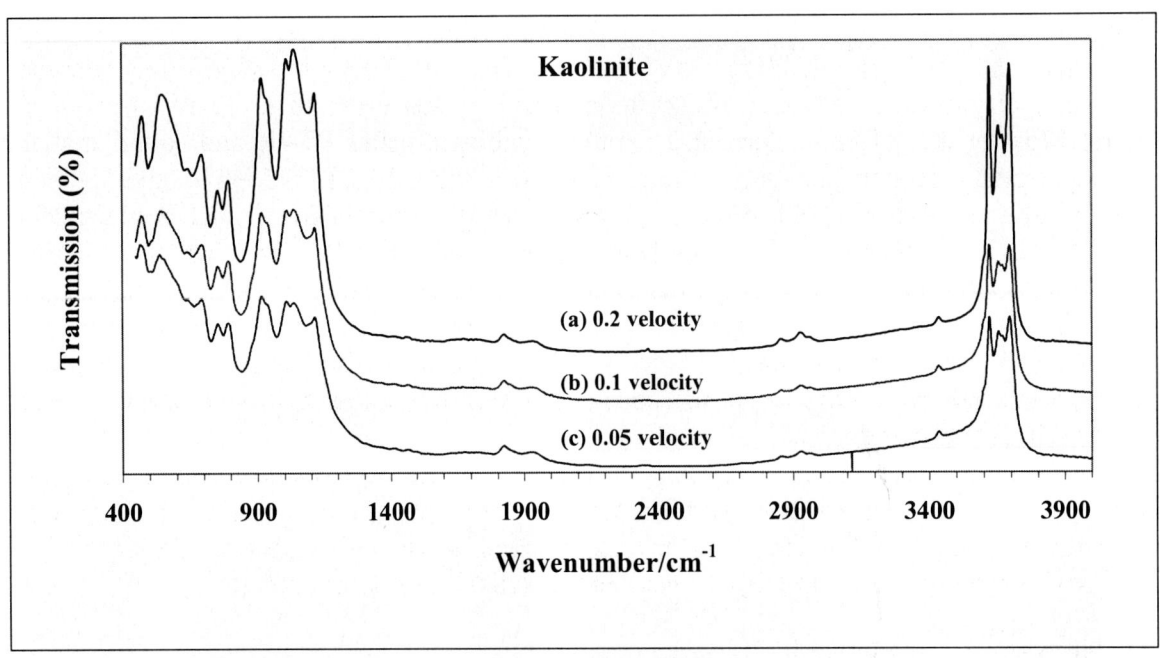

Fig. 2. FTIR-PAS spectra of kaolinite at various mirror velocities. Spectra are plotted with an offset on the y-axes.

3.1. Hydroxyl stretching

Four strong FTIR-PAS bands were observed for kaolinite at 3694, 3670, 3654, and 3621 cm^{-1} and a weak band at 3435 cm^{-1}. The band values reported here agree well with those previously reported from infrared absorption spectroscopy of natural kaolinite [4, 5]. Gadsden [8] reported the IR hydroxyl stretching bands of kaolinite at 3696, 3670, 3645, and 3630 cm^{-1}, whereas Farmer [9] reported 3697, 3669, 3652, and 3620 cm^{-1}. There are, however, no reported comparisons of the behaviour of hydroxyl units with depth below the surface. Figures 2 and 3a show that the intensities of the kaolinite bands decrease as mirror velocities decrease from 0.2 to 0.05 cm s^{-1}. The changes in band intensity and position may be indicative of the amount of hydroxyl units related to their bonding and location.

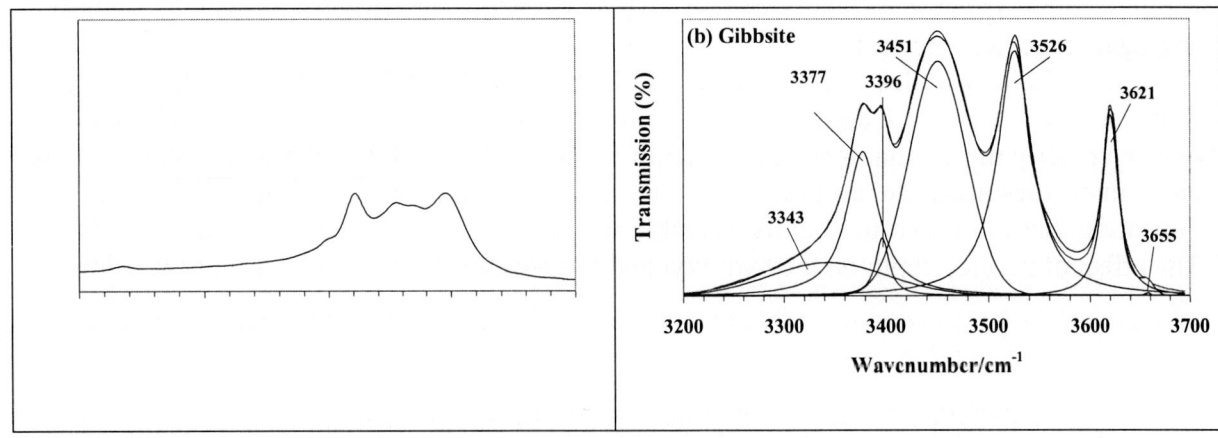

Fig. 3. FTIR-PAS spectra of (a) kaolinite, and (b) gibbsite in the hydroxyl stretching region. Kaolinite spectra of Figures 3, 4 and 5 are plotted with an offset on the y-axes.

For gibbsite five strong bands were observed at 3621, 3526, 3451, 3396 and 3377 cm^{-1}, with a less intense band at 3655 cm^{-1} (Fig. 3b). The broad band at 3343 cm^{-1} may reflect adsorbed water on the crystal surface. Our results are in good agreement with those reported in the literature. Frost et al. [7] characterized a synthetic gibbsite using FT-IR and found that the hydroxyl stretching frequencies were around 3670, 3620, 3524, 3452, 3395, 3375, and 3300 cm^{-1}. Their results are similar to those reported by Farmer [9], van der Marel and Beutelspacher [6], and Wang and Johnston [10]. It has been reported that the 3451 cm^{-1} band is polarized perpendicular to the 001 plane and that the bands at 3526 and 3621 cm^{-1} are polarized in this plane, which is parallel to the Al(OH)$_3$ sheets in the gibbsite structure [11]. It has been suggested by Frost et al. [7] that the 3452 cm^{-1} band is associated with hydrogen bonds between the layers, while the two higher frequency bands correspond to longer hydrogen bonds between hydroxyls lying in the same plane.

The gibbsite bands at 3655, 3621, and 3451 cm^{-1} are identical to the kaolinite bands at 3654, 3621, and 3435 cm^{-1} in the hydroxyl stretching region. Since kaolinite consists of a siloxane and a gibbsite-like sheet linked through an apical oxygen [4, 5], the similarity in FT-IR band positions (Fig. 3a to Fig. 3b) is based on the structure of both minerals.

3.2. Hydroxyl deformation and water bending

Spectra of kaolinite were record between 900 and 1200 cm^{-1} in the hydroxyl deformation and water bending region (Fig. 4a). Five bands at 1116, 1038, 1010, 938, and 915 cm^{-1} were observed for kaolinite in this region. The intensity of these bands decreases as mirror velocity decreases (Fig. 4a). The intensity of the band at 1038 cm^{-1} is greater than that at 1010 cm^{-1} at 0.2 cm s^{-1} mirror velocity but this intensity ratio is reversed when the mirror velocity is reduced to 0.05 cm s^{-1}. This may be interpreted as the band at 1038 cm^{-1} being an outer hydroxyl bending mode, whereas the band at 1010 cm^{-1} is an inner hydroxyl bending mode.

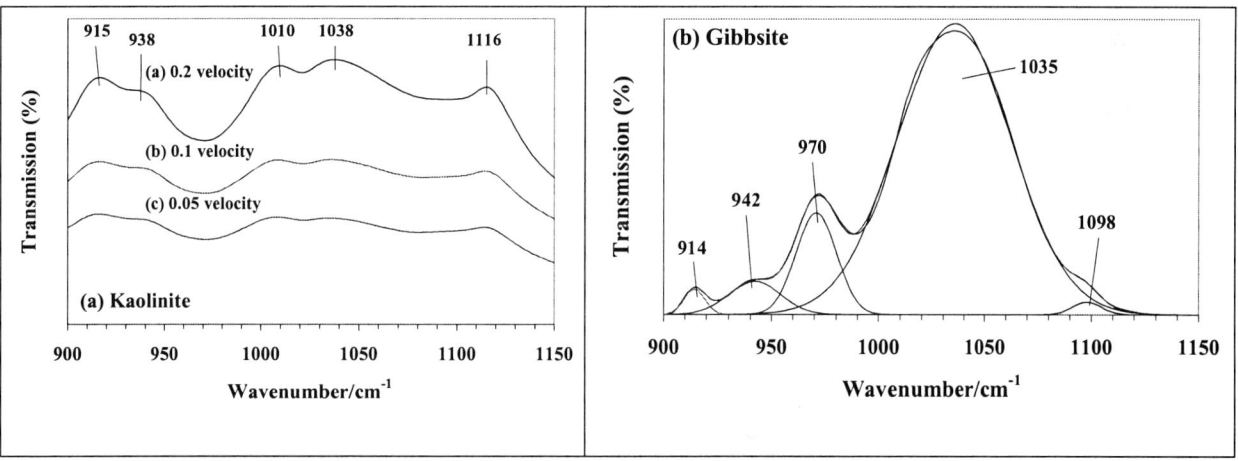

Fig. 4. FTIR-PAS spectra of (a) kaolinite, and (b) gibbsite in the hydroxyl deformation and water bending region.

In contrast to kaolinite, only one strong band at 1035 cm^{-1} and four weak bands at 1098, 970, 942, and 914 cm^{-1} were observed for gibbsite in the hydroxyl deformation and water bending region (Fig. 4b). These band values are different from those reported by Frost et al. [7] who found the IR gibbsite bands around 1060, 1024, and 969 cm^{-1}, Farmer [9] who reported bands at 1091, 1033, 1010, 938, and 915 cm^{-1}, van der Marel and Beutelspacher [6] who reported bands at 1102, 1020, 968, 940, and 915 cm^{-1} and Takamura and Koezuka [12] who observed four bands at 1060, 1020, 969, and 914 cm^{-1}. The difference in band position may be due to different samples used. The band at 1098 cm^{-1} is attributed to Al-O antisymmetric stretching. The three hydroxyl deformation modes at 1035, 970, and 914 cm^{-1} correspond to the three hydroxyl stretching vibrations at 3655, 3526, and 3396 cm^{-1}. The broad band at 1035 cm^{-1} may be due to the effect of interlayer adsorbed hydroxyl groups. The 914 cm^{-1} mode is assigned to the hydroxyl deformation vibration of Al(OH)Al groups free from hydrogen bonding. This band corresponds to the deformation mode of the inner hydroxyl units found in kaolinite [13] and thus the 914 cm^{-1} band is common to all the Al$_2$O$_3$-H$_2$O phases. The 914 cm^{-1} band also corresponds to the hydroxyl stretching frequency of 3621 cm^{-1} and both these two bands are reported to be found in the infrared and Raman spectra of kaolinite [4, 5] and Raman spectra of gibbsite [14]. The gibbsite bands at 1035, 942, and 914 cm^{-1}, are identical to the kaolinite bands at 1038, 938, and 915 cm^{-1} in the hydroxyl deformation and water bending region. This again supports the concept that the similar vibrations of kaolinite and gibbsite reflect the similarities of their crystal structures.

3.3. Low frequency region

Fig. 5 shows the spectra of kaolinite at different mirror velocities and of gibbsite at 0.2 cm s^{-1} velocity. Five intense bands at 792, 753, 694, 547, and 473 cm^{-1} were observed for kaolinite in this low frequency region. Similar observations have been reported previously [4, 5, 7, 8]. One of the important features in this region is that the kaolinite band at 473 cm^{-1} is split into two bands at 477 and 467 cm^{-1} at the 0.05 cm s^{-1} velocity (Fig. 5a). This may be because the bonds of the Si-O-Al skeleton are different on the outer and inner surfaces of a kaolinite crystal.

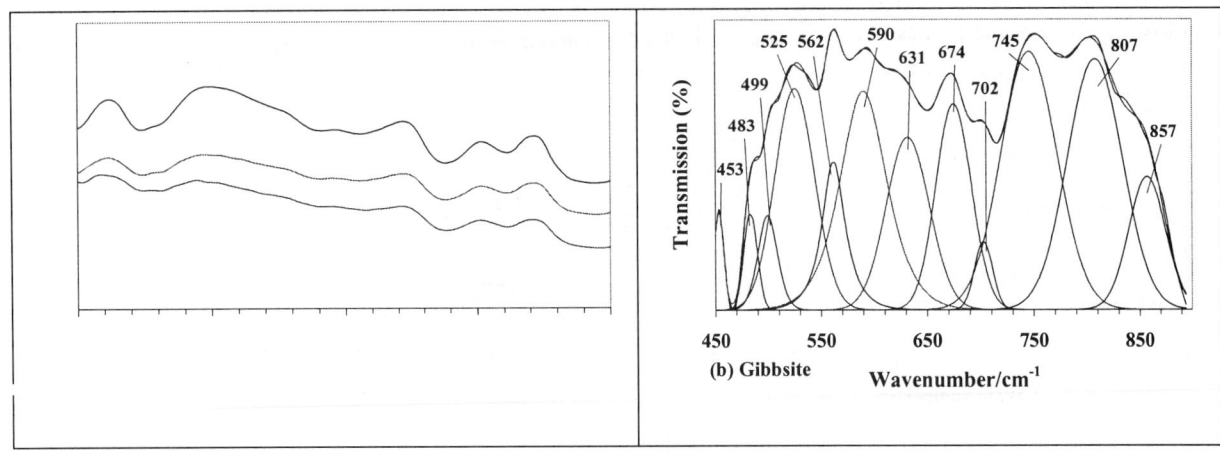

Fig. 5. FTIR-PAS spectra of (a) kaolinite, and (b) gibbsite in the hydroxyl deformation and water bending region.

The spectrum of gibbsite in this region is much more complicated, with most bands being weak and overlapping (Fig. 5b). Twelve bands are observed after manipulating the spectra using Peakfit software. The FTIR-PAS vibrations are found at 857, 807, 745, 702, 674, 631, 590, 562, 525, 499, 483, and 453 cm^{-1}. Weak and overlapped bands have been reported by some workers who characterized gibbsite by infrared and Raman spectroscopy [6, 9, 14, 15]. The bands at wavenumbers higher than 600 cm^{-1} may correspond to the Al-OH-Al translational vibrations, whereas those lower than 600 cm^{-1} may be assigned to the vibrations of the Al-O skeleton. Although the FTIR-PAS spectra of gibbsite are complex in the low frequency region, the gibbsite bands at 807, 745, 702, 525, and 483 cm^{-1} resemble the kaolinite bands at 792, 753, 694, 547, and 473 cm^{-1} in this region, indicating that these minerals have similar spectral characteristics.

4. REFERENCES

1. L. Pauling: The structure of micas and related materials. Proc. National Acad. Sci. U.S.A. 16 (1930) 123.
2. J. W. Gruner: The crystal structure of kaolinite. Z Kristallogr 83 (1932) 75.
3. Deer, W. A., Howie, R. A. Zussman, J. An Introduction to the Rock-Forming Minerals; John Wiley & Sons Inc: New York, 1992.
4. R. L. Frost, Clays Clay Miner., 43 (1995) 191.
5. R. L. Frost, Clay Miner., 32 (1997) 65.
6. H. W. van der Marel and H. Beutelspacher, Atlas of infrared spectroscopy of clay minerals and their admixtures. Elsevier, Amsterdam, 1976.
7. R. L. Frost, J. T. Kloprogge, S. C. Russell and J. L. Szetu, Applied Spectr., 53 (1999) 423.
8. J. A. Gadsden, Infrared spectra of minerals and related inorganic compounds, Butterworth, England, London, 1975.
9. V. C. Farmer (ed) The infrared spectra of minerals, Mineralogical Society, London, 1974.
10. S. Wang and C. Johnston, Am. Miner. 85 (2000) 739.
11. J. D. Russell, R. L. Parfitt, A. R. Fraser and V. C. Farmer, Nature 248 (1974) 220.
12. T. Takamura and J. Koezuka, Nature 207 (1965) 965.
13. R. L. Frost and U. Johansson, Clays Clay Miner.46 (1998) 466.
14. H. D. Ruan, R. L. Frost and J. T. Kloprogge, J. Raman Spectr., 32 (2001) 745.
15. H. D. Ruan, R. L. Frost, J. T. Kloprogge and L. Duong, Spectrochimica Acta Part A 58 (2002) 265.

FT-Raman spectroscopy and SEM of gibbsite, bayerite, boehmite and diaspore in relation to the characterization of bauxite

H. D. Ruan[a,b*], R. L. Frost[a], J. T. Kloprogge[a], D. G. Schulze[b] and L. Duong[a]

[a]Centre for Instrumental and Developmental Chemistry, Queensland University of Technology, GPO Box 2434 Brisbane 4001, Australia

[b]Agronomy Department, Purdue University, West Lafayette IN 47907-1150, U. S. A.

Gibbsite, bayerite, boehmite and diaspore were investigated using FT-Raman spectroscopy in conjunction with scanning electron microscopy (SEM) in order to provide information for the characterization of bauxite. The Raman spectrum of gibbsite shows four strong, sharp bands at 3617, 3522, 3433 and 3364 cm^{-1} and the spectrum of bayerite shows seven bands at 3664, 3652, 3552, 3542, 3450, 3438, and 3420 cm^{-1} in the hydroxyl stretching region. Four broad and weak bands are observed at 3371, 3220, 3085 and 2989 cm^{-1} for boehmite, and five broad bands are observed at 3445, 3363, 3226, 3119 and 2936 cm^{-1} for diaspore in the hydroxyl stretching region. The hydroxyl stretching bands are related to the surface structure of these minerals. The Raman spectra of bayerite, gibbsite and diaspore are complex while the Raman spectrum of boehmite only shows three major bands in the hydroxyl deformation and water bending, and low frequency regions. These bands are assigned to hydroxyl deformation and translational, and Al-O modes of the alumina phases. The present results in relation to the characterization of bauxite are interpreted.

1. INTRODUCTION

Gibbsite, boehmite and diaspore are the three principal hydrates of alumina and are found in bauxite, lateritic or terra rossa soils and various sediments [1] [2] [3] [4]. Bayerite is very rare in nature although it is an important alumina phase. It is found sometimes in veinlets in calcite rocks [5]. However, bayerite can be easily synthesized in the laboratory [3].

Gibbsite [γ-Al(OH)$_3$] and bayerite [α-Al(OH)$_3$] have the same structural sheetlike unit, which consists of a layer of aluminium ions sandwiched between two layers of hexagonally packed hydroxyl ions, in the plane defined by the *a* and *b* axes. The oxygens of one layer lie directly above the oxygens at the top of the layer below. The sheetlike units are stacked along the *c* axis [1] [6] [7]. Bayerite is a dimorph with similar layers of octahedra but these are stacked differently to yield a single-layer cell. The difference in structure between bayerite and gibbsite is that one can be generated from the other by a rotation about the *c* axis of one of the sheetlike units by 60° [7].

Boehmite (γ-AlOOH) is isostructural with lepidocrocite (γ-FeOOH). The structure of boehmite consists of double sheets of octahedra with Al ions at their centers, and the sheets

---

[*] Corresponding author. Tel: (765) 494-8786; Fax: (765) 496-2926; E-mail: hruan@purdue.edu

themselves are composed of chains of octahedra, the repeat distance of which defined the *a* parameter of the unit cell [4]. The stacking arrangement of the three oxygen layers is such that the double octahedral layer is in cubic close packing [8] [9].

The structure of diaspore (α-AlOOH) is based on layers of oxygen atoms in hexagonal close packing; the isostructure of diaspore is goethite (α-FeOOH) [10] [11] [12] [13]. Aluminium ions occupy octahedrally coordinated sites between layers in such a way as to form strips of octahedra, the direction of which defines the *c* parameter of the unit cell. The strips have the width of two octahedra and yield an orthorhombic cell in which is twice the distance between oxygen layers, and $b/2 \sim c\sqrt{3}$ [4]. Diaspore and boehmite are polymorphs of the same mineral and differ in structural relationships by the packing of the oxygens.

Fourier transform Raman spectroscopy is one of the advanced techniques and is complementary to infrared spectroscopy in the study of the mixture of aluminas. It has been extensively applied to characterize the phases of the Al_2O_3-H_2O system [1] [8] [12] [14]. In this paper we report the characterization of alumina phases of gibbsite, bayerite, diaspore and boehmite based on FT-Raman spectroscopy in conjunction with scanning electron microscopy (SEM) in order to provide information for the characterization of the mixture of alumina phases in bauxite.

2. EXPERIMENTAL

Gibbsite, bayerite and boehmite were synthesized from aluminium salts and the details of these minerals have been described elsewhere [1] [8] [10]. Diaspore was a naturally occurring mineral obtained from Nevada, USA.

FT-Raman spectra of gibbsite, bayerite, boehmite and diaspore were obtained using the Perkin-Elmer 2000 series Fourier Transform spectrometer fitted with a Raman accessory comprising of a Spectron Laser Systems SL301 Nd-YAG laser operating at a wavelength of 1064 nm. Raman shifts were observed in the spectral range 4000-200 cm^{-1}. Raman spectra were obtained by using a sample of the raw mineral directly in the incident beam at a resolution of 4 cm^{-1}. The spectra were recorded between 32 and 256 scans depending on signal to noise ratio. Manipulation of spectra including baseline adjustment, smoothing and normalization was performed with the use of the GRAMS software package (Galactic Industries Corporation). Band component analysis was undertaken with the Jandel "Peakfit" software package.

Scanning electron microscopy was carried out using a JOEL-25 microscope. Powder samples were deposited onto tape on the sample holder and coated with gold.

3. RESULTS AND DISCUSSION

Gibbsite crystals show a pseudo-hexagonal shape with a crystal size about 4-6 μm (Fig. 1a). Bayerite has a round crystal shape with a crystal size ~ 1 μm (Fig. 1b). The crystal shape of boehmite reveals sheet type with pseudo-hexagonal types with crystal size between 1-3 μm (Fig. 1c). Diaspore shows a sheet shape with crystal size around 25-30 μm (Fig. 1d).

The FT-Raman spectra of these four aluminas are well distinguished (Fig. 2) although some similarities are observed from the spectra between gibbsite and bayerite, and between boehmite and diaspore. These phenomena are shown by some bands in the hydroxyl stretching and the low frequency regions in Fig. 2 and will be discussed below.

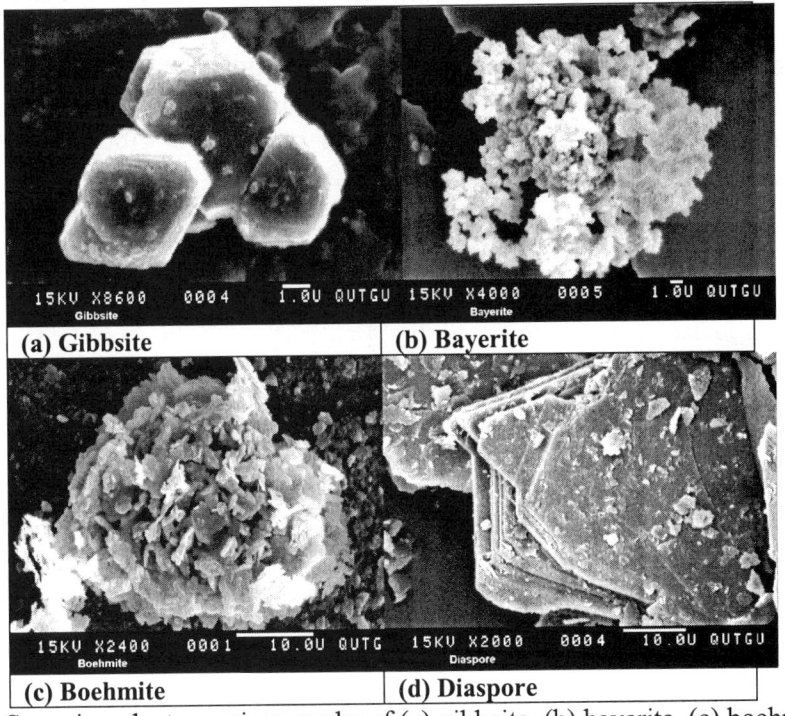

Fig. 1. Scanning electron micrographs of (a) gibbsite, (b) bayerite, (c) boehmite, and (c) diaspore.

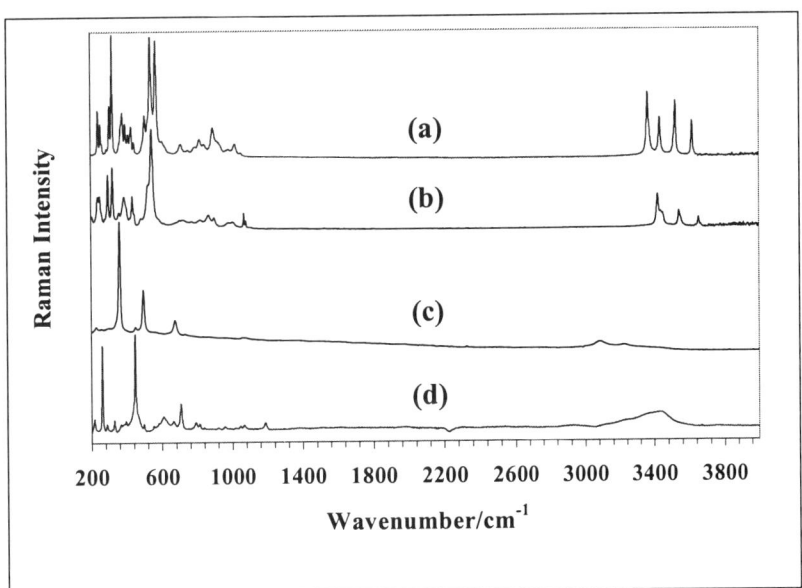

Fig. 2. A comparison of the Raman spectra of (a) gibbsite, (b) bayerite, (c) boehmite, and (d) diaspore.

3.1 Gibbsite and bayerite

Four strong and sharp bands are observed at 3617, 3522, 3433 and 3364 cm^{-1} for gibbsite at the hydroxyl stretching region and are assigned to ν(OH) (Fig. 3a). These results are in good agreement with those reported by other workers [15] [16] [17]. The band at 3364 cm^{-1} is associated with hydrogen bonds between the layers, whereas the two higher wavenumber bands correspond to longer hydrogen bonds between hydroxyls lying in the same plane [1]. It has been reported that the infrared absorption band at 3460 cm^{-1} is polarized perpendicular to the 001 plane and the bands at 3617 and 3522 cm^{-1} are polarized in this plane, which are parallel to the Al(OH)$_3$ layer in the gibbsite structure [1] [18]. The 3617 cm^{-1} band is also observed in the Raman spectra of kaolinite, which contains a gibbsite-like sheet and a siloxane sheet linked through an apical oxygen [19] [20].

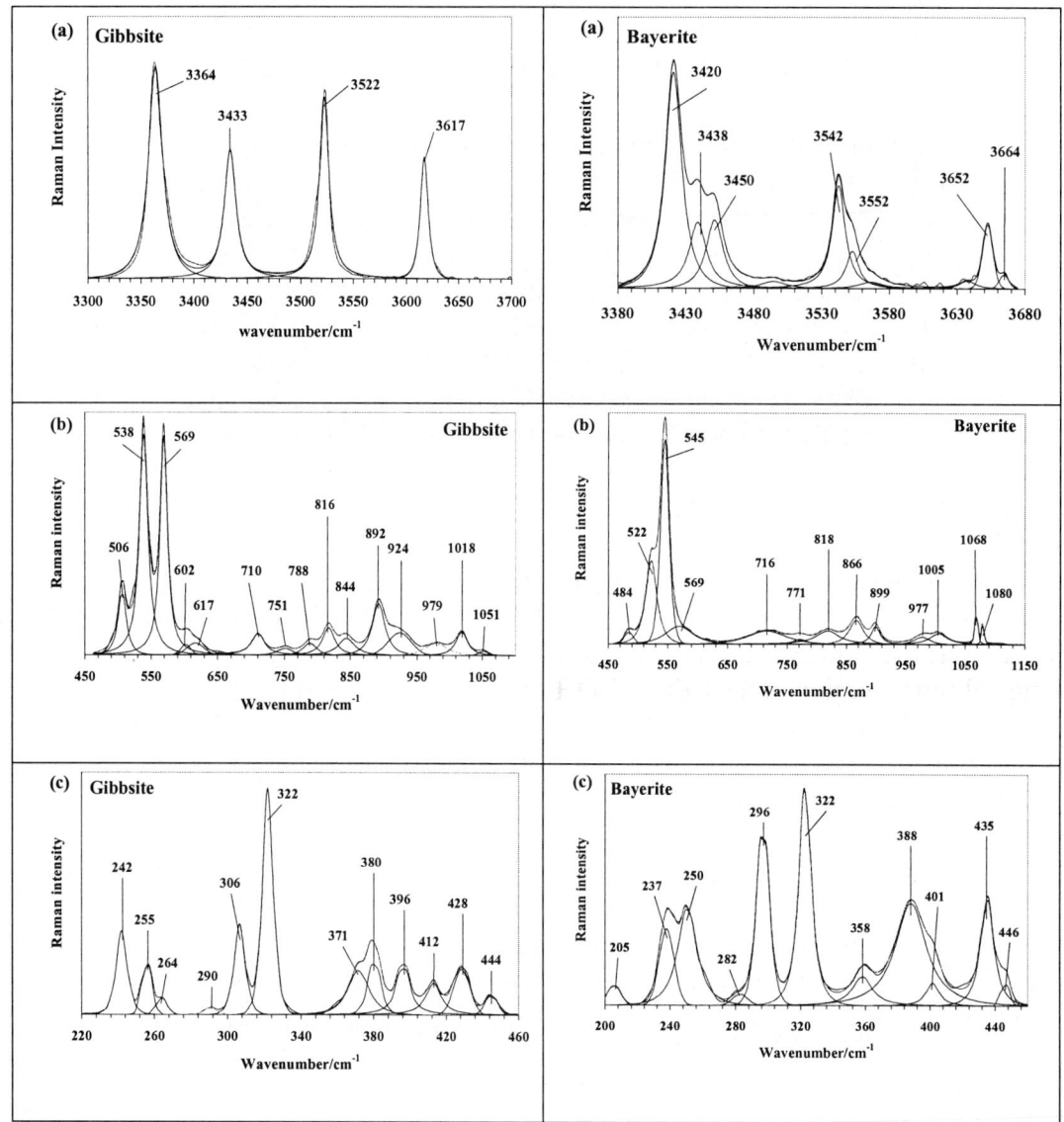

Fig. 3. Raman spectra of gibbsite, (a) hydroxyl stretching, (b) hydroxyl deformation and water bending and (c) low frequency regions.

Fig. 4. Raman spectra of bayerite, (a) hydroxyl stretching, (b) hydroxyl deformation and water bending and (c) low frequency regions.

Four hydroxyl deformation models of gibbsite observed at 924, 979, 1018 and 1051 cm^{-1} (Fig. 3b) correspond to 3617, 3522, 3433 and 3364 cm^{-1} bands in the hydroxyl stretching region. The band at around 924 cm^{-1} is ascribed to an Al(OH)Al group free from hydrogen bonding and is commonly assigned to the hydroxyl deformation vibration of aluminium oxyhydroxides [19] [20] [21] [22]. Bands in this position are observed in the Raman spectra of kaolinite [19] [20]. Kolesova and Ryskin [23] found that the 1020, 958 and 916 cm^{-1} bands disappeared in the infrared spectrum of Al(OD)$_3$ when compared with that of Al(OH)$_3$ and confirmed that these bands were assigned to the vibrations of hydroxyl groups. They assumed that the 1020 cm^{-1} band is evoked by bending δ(OH) vibrations of the hydroxyl groups associated under the influence of the hydrogen bond, while the 958 and 916 cm^{-1} bands are attributed to δ(OH) vibrations of weakly interacting hydroxyl groups. Our observation is identical to the Raman results of natural samples used by Huneke et al. [15] although their results showed poor resolution at 1051, 980 and 916 cm^{-1} bands. Rodgers [16] observed two to four Raman bands with variations of band positions depending on the occurrence of the natural gibbsite. This variation may be due to the amount of surface hydroxyl groups.

Gibbsite has intense bands at 1018, 892, 816, 710, 569, 538, 506, 428, 396, 380, 322, 306, 255 and 242 cm^{-1} in the hydroxyl deformation and water bending region (Fig. 3b), and in the low frequency region (Fig. 3c). Weak bands are also found at 1051, 979, 924, 844, 788, 751, 617, 602, 444, 412, 290 and 264 cm^{-1}. The bands with most intensity are at 569, 538 and 322 cm^{-1}. The 569 and 538 cm^{-1} bands are attributed to Al-O-Al deformation while the 322 cm^{-1} band with a shoulder at 306 cm^{-1} is ascribed to Al-O stretching vibrations.

The spectrum of bayerite is distinguished differently from that of gibbsite (Fig. 2). In the hydroxyl stretching region as shown in Fig. 4a, three major bands at 3652, 3542 and 3420 cm^{-1} are identical to those observed by Huneke et al. [15] at 3651, 3542 and 3421 cm^{-1}. Rodgers [16] found that Raman spectrum of bayerite from waste deposits had two shoulder bands at 3449 and 3436 cm^{-1} in addition to these three strong bands. Bayerite from the Raoul Island in New Zealand showed three strong bands at 3651, 3541 and 3420 cm^{-1} with three shoulder bands at about 3625, 3455 and 3441 cm^{-1} [24]. It is noted that bayerite consists of the same hexagonally packed hydroxyl group, as does gibbsite. Differences in the Raman spectra in the stretching vibration region reflect the Al-OH distance 1.74-2.06 Å for bayerite compared to 1.73-2.14 Å for gibbsite, and the OH-OH distance 2.92-3.24 Å for bayerite compared to 2.75-3.24 Å for gibbsite [17]. Three shoulder peaks at 3552, 3450 and 3438 cm^{-1} may be attributed to the surface hydroxyl groups in the bayerite structure.

The bayerite bands are observed at 1080, 1068, 1005, 977, 899, 866, 818, 771, 716, 569, 545, 522 and 484 cm^{-1} in the hydroxyl deformation and water bending region (Fig. 4b) and at 446, 435, 401, 388, 358, 322, 296, 282, 250, 237 and 205 cm^{-1} in the low frequency region (Fig. 4c). Bayerite bands at 1005, 977, 899, 444, 322 cm^{-1} may overlap with the gibbsite bands at 1018, 979, 892, 444 and 322 cm^{-1} compare Fig. 4b, c to Fig. 3b, c, indicating that there are some similarities, both for surface hydroxyl groups and packing of oxygens in these two minerals. However, other bands of bayerite are well resolved and are distinctly different from those of gibbsite.

3.2 Boehmite and diaspore

Weak and broad bands at 3371, 3220, 3085 and 2989 cm^{-1} are the distinguishing features for boehmite (Fig. 5a). Two broad bands at 3371 and 2989 cm^{-1} are probably due to surface hydroxyl groups. This type of hydroxyl groups is easily formed during the synthesis of boehmite [25]. The shift in these two bands to lower positions is attributed to the increase in surface hydroxyl groups. Recent work on FT-infrared spectroscopy further confirmed that a shoulder band at 3494 cm^{-1} and a broad band at 1640 cm^{-1} were assigned to the stretching and bending modes of adsorbed water respectively in addition to two distinct ν(OH) stretching modes at 3292 and 3097 cm^{-1} [26]. The bands at 3220 and 3085 cm^{-1} may be attributed to the hydroxyl stretching wavenumbers of the hydroxyl group within the structure.

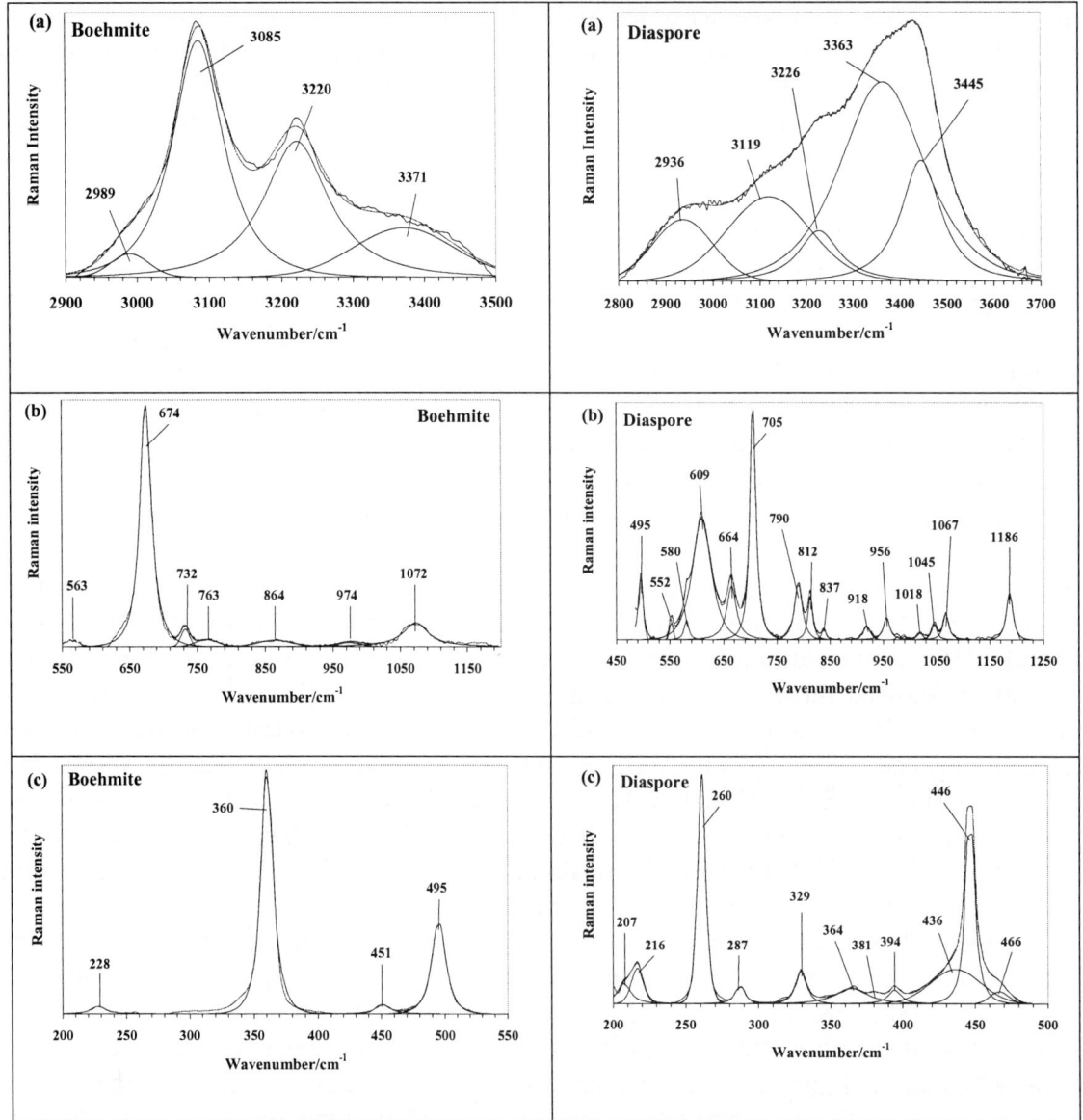

Fig. 5. Raman spectra of boehmite, (a) hydroxyl stretching, (b) hydroxyl deformation and water bending and (c) low frequency regions.

Fig. 6. Raman spectra of diaspore, (a) hydroxyl stretching, (b) hydroxyl deformation and water bending and (c) low frequency regions.

For the hydroxyl deformation and water bending modes and hydroxyl translation modes, three very strong bands are observed at 674, 495 and 360 cm^{-1} and are associated with very weak bands at 1072, 974, 864, 763, 732, 563, 451 and 228 cm^{-1} (Fig. 5b, c). The hydroxyl bending band at 1072 cm^{-1} is assigned to the hydroxyl deformation vibration and is similar to that at 1091 cm^{-1} of FT-Raman shift as reported in previous work [8]. The bands at 674, 495 and 360 cm^{-1} are assigned to the Al-O stretching vibrations.

For diaspore bands are observed at 3445, 3363, 3226, 3119 and 2936 cm^{-1} in the hydroxyl stretching region (Fig. 6a). Infrared investigation on diaspore has shown differences in band position. Farmer [17] reported two bands at 2994 and 2915 cm^{-1}, which are similar to 2985 and 2910 cm^{-1} obtained by van der Marel and Beutelspacher [2] but differed from 3365, 3284, 3095 and 2938 cm^{-1} observed by Frost et al. [10]. Some overlapped bands are well resolved using the peak fit technique (Fig. 6a).

The broad and weak bands at 1186, 1067, 1045 and 956 cm^{-1} are assigned to the Al(OH) bending modes of diaspore (Fig. 6b). In the hydroxyl deformation and water bending and low frequency regions, the bands at 1186, 812, 790, 664, 609, 495, 329 and 216 are moderate weak and bands at 1067, 1045, 956, 918, 837, 580, 552, 446, 436, 394, 381, 364, 287, and 207 cm^{-1} are much weaker (Fig. 6b, c). The bands at 705, 446 and 260 cm^{-1} are the most intense in the Raman spectrum and are assigned to the symmetric stretching modes, the 260 cm^{-1} band to Al-O-Al symmetric stretch and the 446 cm^{-1} band to the Al-O-Al bend. Although the spectrum of boehmite is distinguished from that of diaspore, some similarities between these two alumina phases are observed mainly for the vibrations of hydroxyl groups, as interpreted by comparing the bands at 3220 and 1072 cm^{-1} for boehmite (Fig. 5a, b) with the bands at 3226 and 1067 cm^{-1} for diaspore (Fig. 6a, b).

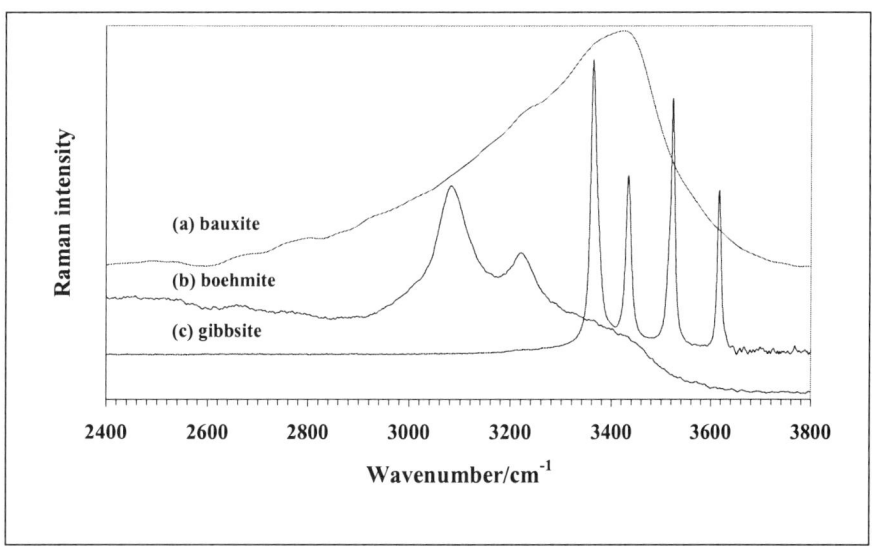

Fig. 7. Characterisation of Raman spectrum of (a) bauxite by comparing to the spectra of (b) boehmite and (c) gibbsite.

3.3 Characterization of bauxite

As mentioned earlier, the characterization of alumina phases are based mainly on the identification of hydroxyl groups. Although the hydroxyl groups are very sensitive to the influence of dehydroxylation temperature, the XRD provides no data on the loss of hydroxyl groups, and thus it is difficult to distinguish the formation of transitional phases in a mixture using XRD. Raman spectroscopy, along with near-, mid-, and far-infrared spectroscopy is therefore used to gain detailed data of phase transformations.

The Raman spectrum of the Australian bauxite is compared to those of boehmite and gibbsite shown in Fig. 7. In the hydroxyl stretching region, a broad multi-band was observed between 2950 and 3700 cm^{-1} for bauxite. This multi-band seems to consist of boehmite and gibbsite bands in addition to hydrohematite and kaolinite bands. The bands at ~ 3433 and ~ 3364 cm^{-1} correspond to the gibbsite bands at these wavenumbers. Two other gibbsite bands at ~ 3522 and ~ 3617 cm^{-1} are also identified (Fig. 7). The band at ~ 3220 cm^{-1} is indicative of the boehmite vibration although the ~ 3085 cm^{-1} is not clearly shown from the bauxite spectrum. Meanwhile, the similarity between the bauxite spectrum and the boehmite spectrum in the region between 2400 and 2900 cm^{-1} further indicates that those minor bands may correspond to the boehmite vibrations (Fig. 7). Our results have shown that there will be a possibility to use Raman spectroscopy for the characterization of alumina phases and their transformation during the thermal activation of bauxite.

REFERENCES

1. R.L. Frost, J.T. Kloprogge, S.C. Russell and J.L. Szetu, Appl. Spectr. 53 (1999) 423.
2. H.W. van der Marel and H. Beutelspacher, Atlas of Infrared Spectroscopy of Clay Minerals and their Admixtures, Elsevier: Amsterdam, The Netherlands, 1976.
3. J.T. Huneke, R.E. Cramer, R. Alvarez and S.A. El-Swaify, Soil Sci. Soc. Am. J. 44 (1980) 131.
4. W.A. Deer, R.A. Howie and J. Zussman, An Introduction to the Rock-Forming Minerals John Wiley & Sons Inc: New York, 1992.
5. S. Gross and L. Heller, Mineralog. Mag. 32 (1963) 723.
6. P.H. Hsu and T.F. Bates, *Mineralog. Mag.* 33 (1964) 749.
7. J.L. Bersillow, D.W. Brown, F. Fiessinger and J.D. Hem, J. *Res. U.S. Geol Survey* 6 (1978) 325
8. R.L. Frost, J.T. Kloprogge, S.C. Russell and J.L. Szetu, Appl. Spectr. 53 (1999) 572.
9. P.A. Buining, C. Pathmananoharan, J. B. H. Jansen and H. N. W. Lekkerkerker, J. Am. Cer. Soc. 74 (1992) 87.
10. R.L. Frost, J.T. Kloprogge, S.C. Russell and J.L. Szetu, J. L. Appl. Spectr. 53 (1999) 829.
11. H.D. Ruan, R.J. Gilkes, Clays Clay Miner. 43 (1995) 196.
12. D.G. Schulze, Clays Clay Miner. 32 (1984) 36.
13. M.H. Francombe and H. P. Rooksby, 1959; Clay Miner. Bull. 21 (1959) 1.
14. H.D. Ruan, R.L. Frost and J.T. Kloprogge, J. Raman Spectr. 32 (2001) (in press).
15. J.T. Huneke, R.E. Cramer, R. Alvarez and S.A. El-Swaify, Soil Sci. Soc. Am. J. 44 (1980) 131.
16. K.A. Rodgers, Clay Miner. 28 (1993) 85.
17. V.D. Farmer, Infrared Spectra of Minerals, Mineralogical Society: London, 1974; p149.
18. J.D. Russell, R.L. Parfitt, A.R. Fraser and V.C. Farmer, Nature 248 (1974) 220.
19. R.L. Frost, Clays Clay Miner. 43 (1995) 191.
20. R.L. Frost, Clay Miner. 32 (1997) 79.
21. R.L. Frost, Clays Clay Miner. 46 (1998) 280.
22. T. Takamura and J. Koezuka, Nature 207 (1965) 965.
23. V.A. Kolesova, I.A.I. Ryskin, Optics and Spectr. 7 (1959) 165.
24. K.A. Rodgers, M.R. Gregory and R.P. Cooney, Clay Miner. 24 (1989) 531.
25. Y.Y. Chen, J. Hyldtoft, C.J.H. Jacobsen and O.F. Nielsen, Spectr. Acta Part A-Molecular Spectr. 51 (1995) 2161.
26. G.K. Priya, P. Padmaja, K.G.K. Warrier, A.D. Damodaran and G. Aruldhas, J. Mater. Sci. Letters 16 (1997) 1584.

VI
Teaching Clay Mineralogy

Evaluation of colloidal properties using clay minerals: Experiments for undergraduate and introductory chemistry students

María Bárbara Lombardi and Miria Baschini

CIMAR – Departamento de Química, Facultad de Ingeniería, Universidad Nacional del Comahue, Buenos Aires 1400, 8300, Neuquén. TE: 0299 – 4485344
email: miria@uncoma.edu.ar

Colloids are an important subject in undergraduate and introductory chemistry courses, specially in Agronomy, Biology, Chemical Engineering, Geology and similar disciplines.

The aim of this work is to use clay minerals in laboratory experiences to introduce undergraduate and introductory university students in the colloidal thematic. The properties analysed are:
- Tyndall effect,
- adsorption process and
- hydrophilic/ hydrophobic behaviour.

The clay minerals used for the experiences were bentonites from the Argentine North Patagonia deposits. The use of regional minerals, due to the proximity of the deposits to the University location, increase the students enthusiasm.

1. INTRODUCTION

As it is known dispersed particles in a solution have molecular size, while dispersed clay particles in a clay – water system have larger size due to the association forming larger particles.

These systems, where dispersed particles settle down by gravity force alone and having a particle diameter from 10 to 2000 Å, classified as colloids (1).

Colloids is an important topic in General and Introductory chemistry courses. The colloidal properties easily allows experimental and theoretical discussions during the courses development.

The aim of this work is to use clay regional minerals in laboratory experiences to introduce undergraduates students in colloidal thematic. The properties analysed are (2):
- the Tyndall effect,
- adsorption process and
- hydrophilic / hydrophobic behaviour.

2. RESULTS AND DISCUSSION

In General and Introductory Chemistry courses, there is a lack in the analysis of colloidal properties. The proposed experiments are attractive for introductory university students since they allow them to relate Colloidal Chemistry with regional technological and industrial problems.

The laboratory experiences proposed utilize regional clay mineral. A 99 % Na – montmorillonite, obtained from North Patagonia bentonite deposits (3) with a cation exchange capacity (CEC) of 1.03 meq/g and a specific surface area between (S) 790 to 810 m^2/g.

2.1. The Tyndall effect

The known Tyndall effect (4) is the production of light scattering by colloidal suspensions.

The experiment was done with a colloidal clay dispersion, 0,01% w/v, and it was analysed by visible spectroscopy with a Shimadzu UV 240. Data values of % transmittance obtained in the range of wavelength between 350 to 700nm were presented as a curve (% transmittance (% T) versus wavelength (λ)) as it is shown in Figure 1.

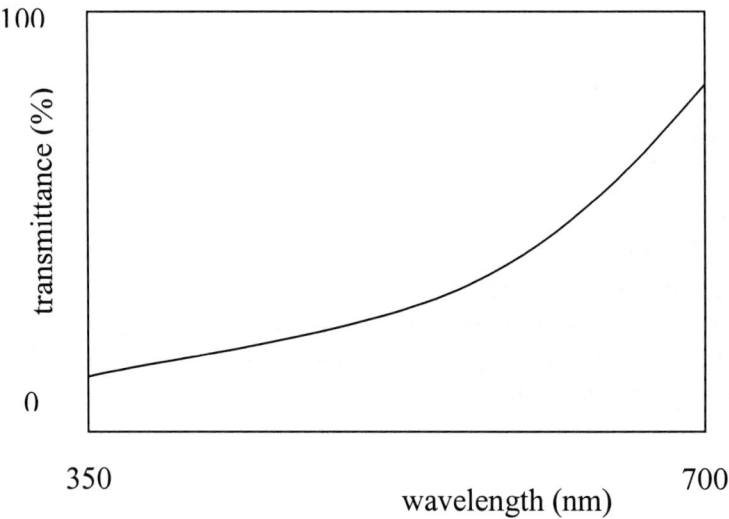

Figure 1. Visible spectrum of 0.01%m/v clay suspension

The % T increase with the λ increase, it can be explained by the light deviation ascribed to the colloidal particles interference (clay mineral). While λ increase (in the red region of the visible spectrum) the probability of light deviation decrease increasing the %T. On the other hand, while λ decrease (in the blue region) the %T also behaves like this (5).

In advanced chemistry courses (i.e. Analytical Chemistry), students can use the spectroscopic concepts to quantify the amount of clay mineral in colloidal systems. The experiment consists in making a calibration curve, it was done analysing by visible spectroscopy at a constant λ different concentrations of clay suspensions. The λ=420nm, used in the example, was selected because this is the wavelength were the maximum absorption was obtained.

The calibration curve is shown in Figure 2. A simplest version of this experience can be done, when a spectrophotometer is not available, by using a slide projector, a recipient containing the dilute clay colloidal dispersion and a mirror at the end of the pass of the light beam. When the slide projector is turned on, the light beam passing through the system deviates. It is easily observed blue light, short wavelenght, emerging laterally due to the colloidal particles collisions while the orange light emerges from the front of the recipient (5).

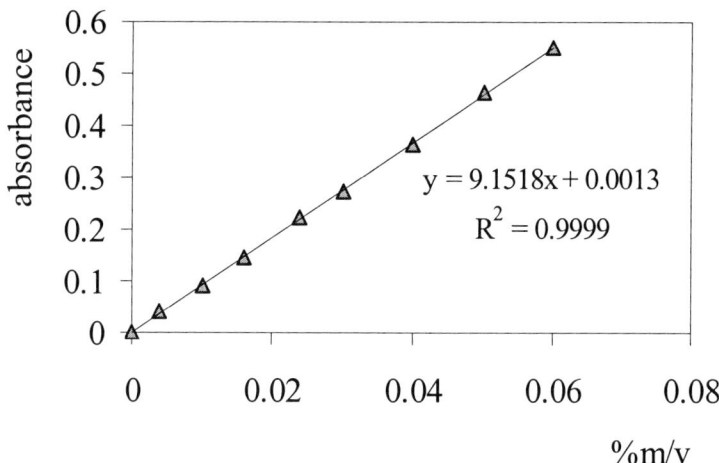

Figure 2. Quantification of clay suspension content at λ =420 nm.

2.2. Organic dye adsorption

Adsorption is a process where a substance accumulate between two continuos phases (4). The study of complexes formed by clay and organic substances is an important subject used in colloid science, soil science, petroleum technology, dyeing of textiles, among others (6,7).

Na bentonite delaminates in aqueous diluted suspensions, producing, by the addition of organic cations, a high degree of adsorption.

There are important chemical aspects related to adsorption process that generates students dicussion and research, as chemical bonding, intermolecular forces, associated energy, etc.

Adsorption process can be quantitative or qualitative evaluated. The experiment was done at constant temperature and preparing:
- 0.0010 N Malachite Green (A)
- 0.05 w/v % clay suspension (B)

Tubes were prepared adding different volumes of A (from 0 to 5 ml), constant volume of B (2 ml), and distilled water until 10 ml of final volume.

The test tubes were maintained at constant temperature for one hour. A variation from clear to coloured supernatant can visually observed indicating qualitatively the adsorption process.

The amount of dye adsorbed was estimated, by means of spectroscopic measurements of the dye supernatant contents, as follows:

Amount of dye adsorbed = Initial amount of dye added – amount of dye in the supernatant

The amount of malachite green adsorbed per gram of clay versus the initial concentration of dye added is shown in figure 3.

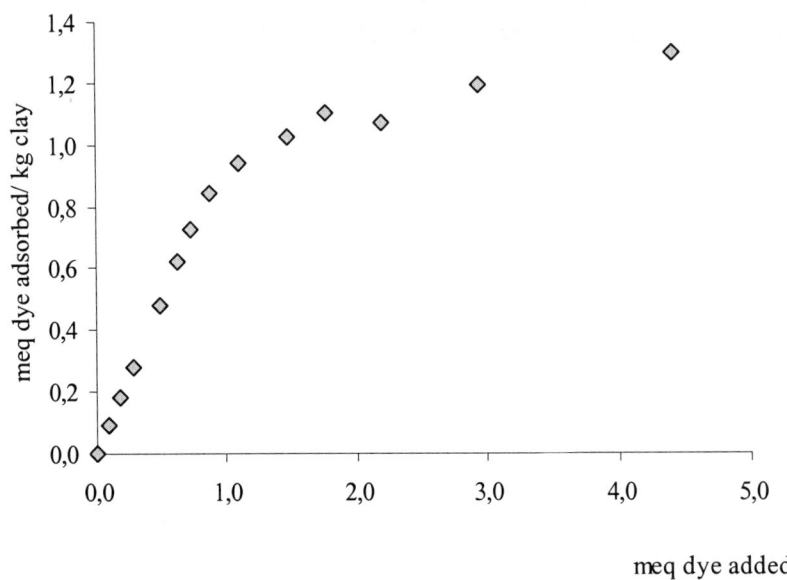

Figure 3. Adsorption isotherm of Malachite Green on bentonite.

The maximum adsorption capacity is around 1.10 meq/kg, slightly above the cation exchange capacity.

Adsorption experiences with organic compounds and clay minerals (i.e. bentonite) demonstrated that exchangeable inorganic cations could be easily replaced by organic ones.

2.3. Hydrophilic - hydrophobic behaviour

Dispersion of clay particles in water are not in a thermodynamic equilibrium state. They tend to minimise the Gibbs´ interfacial energy in a process called ageing (8).

Furthermore, at any given stage of ageing the interactions between particles and solvent are such that the Gibbs´ energy of the system is minimised by aggregation of the particles. This process is called flocculation or coagulation, properties used in the clarification process of juice, wines and oils.

The tendency of proteins, starches, and tannins is to aggregate in particles of colloidal dimensions, promoting hazes and turbidity in juice and wine (9).

The use of clays to clarified fruit juices, wines and oils is a common practice in the related industries. In this experience, wine (polar system) and oil (non polar) were selected to be clarified.

The experiment was done adding
- different amount of clay mineral (2 and 4 g) to 100 cm^3 of oil
- different volumes of 5%m/v clay suspensions (0 to 11 ml) to 100 cm^3 of wine.

In both cases destilled water was added until final volume. The clarification capacity of the clay used was analysing the supernatant solution (separation was done by centrifugation) by visible spectroscopy at $\lambda = 420$ nm. The systems clarification (transmittance) increase with the amount of mineral, as it is shown in Figure 4.

Clarification was followed in polar and non polar systems, in wine experiments the amount of clay (in a suspension) is lower than oil experiments (clay in dry state). The different behaviour can be explained by the highest affinity of clay for aqueous systems, called hidrophilicity. However, clays can adsorb appreciable amounts of nonpolar compounds (organics) under special conditions (11) and it is possible to greatly increase its surface hidrophobicity changing adsorbed inorganic cations for organic ones.

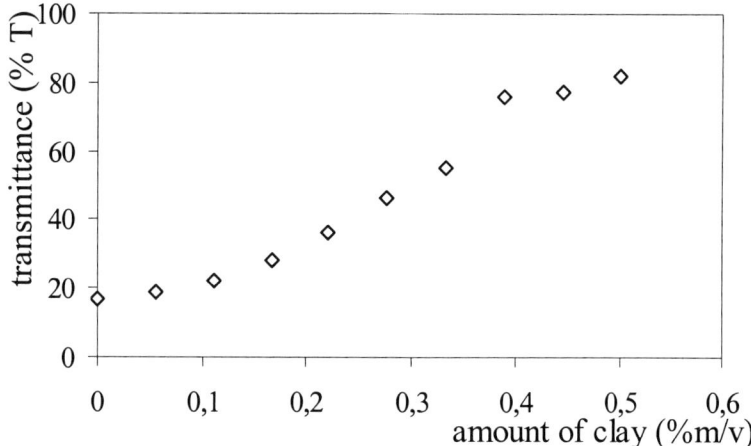

Figure 4. Clay Clarification capacity in a commercial and unclarified wine at a selected λ =420 nm.

3. CONCLUSIONS

Clay minerals might be used for laboratory experiments in order to introduce students to some colloidal properties.

The proposed experiences are attractive for introductory university students shich allow them to relate Colloidal Chemistry with technological and industrial problems. With those experiences, many chemistry concepts can be discussed, chemical bonds, bond polarity and electronegativity, intermolecular forces, properties of solutions, etc., and an integrant laboratory practice can be organized in several sessions.

The practice were proposed for General and Inorganic Chemistry course, but it can be adapted for advanced subjects like Analytical or Physical Chemistry.

REFERENCES

1. K.W. Whitten, R.E. Davis and M. Peck. Química General, quinta Edición, Mc Graw Hill, p529, 1998.
2. T.L. Brown, H.E. LeMay and B.E. Bursten. Chemistry: the Central Science, seventh Edition, Prentice Hall, p. 479, 1997.
3. B. Lombardi, M. Baschini y R.M.Torres Sánchez. Caracterización de bentonitas de la región NorpatAgonica Argentina. V Jornadas de Tratamiento de Minerales, p. 130-134, (1998).
4. S. Ross, I. Morrison. Colloidal Systems and Interfaces. John Wiley & Sons, p.20, 1988.
5. Z. Shakhashiri Bassam. Color of the sunset: the Tyndall Effect, in Chemical Demonstrations, a handbook for teachers of Chemistry, V 3, The University of Wisconsin Press, p 353, 1989.
6. G. Sposito. Surface chemistry of soils. Oxford University Press, p. 26-29, 1984.
7. J.A. Rausell-Colom and J.M. Serratosa. Reactions of Clays with Organic Substances, in Chemistyr of Clay and Clay Minerals, Mineralogical Society, A.C.D. Newman Editor, p. 371, 1987.
8. H. Van Olphen. Dispersion and Flocculation, in Chemistyr of Clay and Clay Minerals, Mineralogical Society, Monograph 6, A.C.D. Newman Editor, p. 371, 1987.
9. D.A. Heatherbell. Haze and Sediment Formation in clarified apple juice and apple wine, Alimenta 15: (1976) 151.
10. K. Wucherpfenning and P. Possmann. The use of a combination of Gelatin and Silica Sol for Clarification Flüss. Obst. 40, 81972) 488.
11. B.L. Shawhney. Sorption and Desorption of Organic Contaminants by Clays and Soils, in Organic Pollutants in the Environment, CMS Workshop 8, p. 53,1996.

Three simple experiments to visually demonstrate the impact of clay minerals on the behavior of organic pollutants in soils

G. Rytwo[a*]

[a] School of Environmental Sciences and Technology, Tel Hai Academic College, Upper Galilee 12210, Israel, rytwo@telhai.ac.il

We present some experiments designed to visually demonstrate the importance of clay minerals and its properties on the behavior of organic pollutants in soils. The first experiment shows the influence of the charge of the pollutant molecules, by simulating the behavior of cationic and anionic organic chemicals in soil. After a single addition of the simulated pollutants, aliquots of water are added, to simulate rain or irrigation. While the cations crystal violet (CV^+) and methylene blue (MB^+) remain in the upper layer of the soil column, the anions fast green (FG^-) and erythrosin-B (EB^-) flow with the water, and are leached out from the column. This emphasizes the environmental difference in movement of anions and cations in soils. A second experiment is designed to show how soils with different clay contents respond when exposed to consecutive aliquots of CV^+. As the amount of clay decreases the same pollutant leaches deeper into the column, demonstrating the influence of the amount of clay on the behavior of cationic pollutants from a constant source. A third experiment shows the velocity of flocculation as a function of the neutralization of the double layer. When CV^+ is adsorbed at 25% or 175% of the cation exchange capacity (CEC), the clay surface is charged, and the clay disperses. When the amount adsorbed is 100% of the CEC the charge of the clay is neutralized and fast flocculation occurs because there is no repulsion between clay particles.

1. INTRODUCTION

It is instructive to demonstrate to people lacking background in clay mineralogy, that clay minerals have a very significant influence on the behavior of organic pollutants. Most lecturers discover that it is difficult to convince even students with a broad environmental background of the importance of issues such as the diffuse double layer, cation exchange capacity, or clay-organic interactions in understanding the fate of pollutants in soils. Usually the question that arises is "Why do can such tiny particles that sometimes make up only a small percent of the soil, have such a crucial impact on their behavior?".

In my attempts to show my class how much can be done by a very small amount of clay, I have developed a series of very easy experiments. Their purpose is to demonstrate vividly that clay minerals and their very special structure can significantly affect the behavior of organic pollutants in soils. The most important point to be demonstrated is that a relatively small

---

[*]Giora Rytwo, phone +972-4-6868440, fax +972-4-6868499, rytwo@telhai.ac.il

difference in the amount of clay, or in the type of the pollutant, can change completely the fate of potential pollutants in a specific soil.

There are several definitions of the term "clay", and it is very important for students to understand the distinction of the different usages of the term. This is directly related to the unified teaching of Clay Science (Bergaya, 2000). On one hand, "clay" is used to define particles with an equivalent diameter smaller than 2 μm (Lahav, 1983). On the other hand, "clay" is used for clay minerals, which are essentially composed of crystalline particles of one or more members of a small group of minerals (Grim, 1953). Clay-sized particles contain clay minerals, but may also include calcite, metal oxides and organic matter.

In this paper, we will define clays following Manahan (2000), as "a group of microcrystalline secondary minerals consisting of hydrous aluminum silicates that have sheet like structures".

Clays have been important "since the first records of human existence were documented" (Murray, 1997), and interactions between clay minerals and organic compounds have been known for thousands of years. Theng (1974) mentions in this context the use of clay minerals to remove unwanted colors from olive oil or wine, and the use of mud to remove oil from wool. Such applications led to definitions such as "bleaching earth" or "fuller's earth". However, even though the effects were known, the environmental implications of such interactions are less obvious.

Clay minerals are usually negatively charged to some extent due to isomorphic substitutions in the structure (Sparks, 1995). These charges are compensated by inorganic cations that can be exchanged by other cations to an extent that is determined by their affinity to the binding sites of the clay minerals, and by their equilibrium activity in solution. Johnston (1996) develops the concept of "active sites" adopted from biological macromolecules, and defines several different types of sites that can appear in clay minerals. The six predominant active sites considered for clay-organic interactions are: (a) "broken edge" sites and exposed surface aluminol and silanol groups, (b) isomorphic substitutions, (c) exchangeable cations, (d) hydrophobic silanol surfaces, (e) hydration shell surrounding exchangeable cations, and (f) hydrophobic sites on organic molecules adsorbed on clays. All these sites contribute somehow to the interactions between organic pollutants and clay minerals in soils, giving unique results that lead to considerable implications even when the percentage of the clay minerals in the soil is low. In this paper, I present experiments that help illustrate such implications to people without a soil or colloidal chemistry background. The general purpose is to show (a) the behavior of a single-spill organic pollutant, (b) the behavior of a constant-source organic pollutant, and (c) changes in soil properties due to interactions between organic pollutants and clay minerals.

Table 1
Particle size distribution and organic matter content of the soils used (%)

Soil	Clay	Silt	Sand	Organic Matter
Rehovot Sand[1]	2.5	5.0	92.5	0.8
Nir Galim Sand[1]	11.2	3.0	85.8	0.6
Hulah Peat[2]	14.0	24.0	62.0	29.0

[1] Soil and data kindly provided by Mrs. D. Tropp and Ms. H. Zivner
[2] Soil and data kindly provided by the team of Kibbutz Yiron Fields Crop Farm.

2. MATERIALS AND METHODS

The particle size distributions of the soils used in the experiments are presented in Table 1. Wyoming montmorillonite SWy-1, was purchased from the Source Clays Repository of the Clay Minerals Society (Columbia, MO). The CEC of this clay was reported by van Olphen and Fripiat (1979), as 0.764 mol$_c$ kg^{-1}. The total specific surface area (SSA), 7.56x10^5 m^2 kg^{-1}, was measured as described in Carter et al. (1965). The initial exchangeable cations were the same as in Rytwo et al. 1995, i.e., 0.41 mol$_c$ Ca kg^{-1}, 0.14 mol$_c$ Mg kg^{-1}, and 0.26 mol$_c$ Na kg^{-1}.

Methylene blue (MB) was purchased from E. Merck, Darmstadt, Germany. Crystal violet (CV) and acriflavin (AF) were obtained as chloride salts from Fluka Chemie AG, Buchs, Switzerland. Erythrosin-B FD&C Red #3 (EB) and fast green FD&C Green #3 (FG) were purchased from Spectrum Inc. Gardena, Ca. All materials were used without as received.

3. SINGLE SPILL POLLUTANT - INFLUENCE OF CHARGE ON THE MOBILITY OF THE POLLUTANT

This very simple experiment aims to illustrate the different behavior of a cationic and an anionic pollutant spilled on the soil and later washed into the soil by rain or irrigation. For this purpose we fill four Pasteur micropipettes with red-brown sand from Rehovot, Israel (Table 1), after first placing a small piece of permeable paper or cotton at the bottom of the pipette, to enable the water to drain without letting the soil out. In general, the use of sandy soils allows the experiment to progress rapidly. Use of clayey soil should be avoided, since the low permeability will cause the experiment to last hours instead of a few minutes. Five mL of tap water are first dropped slowly in each micropipette, in order to wet the soil. The use of tap water is not essential, but it does more closely imitate the process of irrigation.

Separate solutions containing 5 mM of crystal violet (CV$^+$), methylene blue (MB$^+$), fast green (FG$^-$) and erythrosyn-B (EB$^-$) are prepared. Other organic cations or anions could be used, but

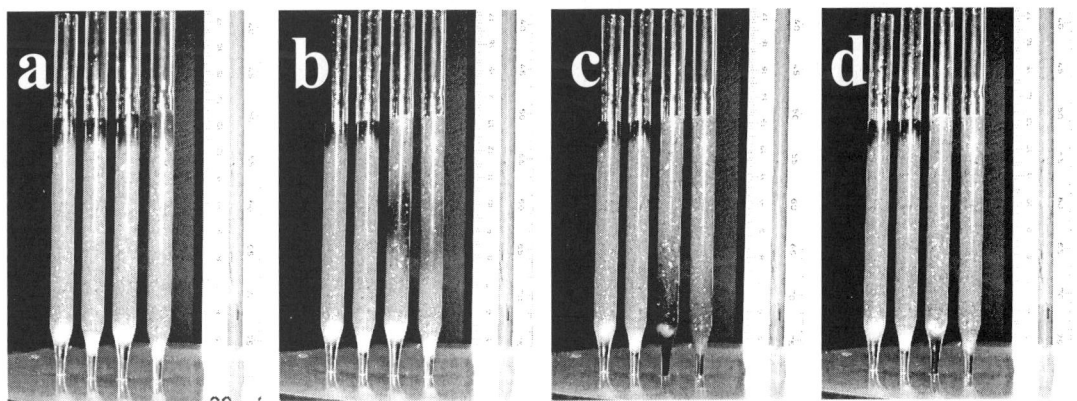

Figure 1. From left to right, behavior of CV$^+$, MB$^+$, FG$^-$ and EB$^-$ in a soil column, after washing with an equivalent to (a) 0, (b) 12.7 mm, (c) 25.4 mm, and (d) 38.1 mm of rain or irrigation.

the ones proposed here are colorful and can be seen by the naked eye even at µM concentrations. Their charge and color is not influenced in the pH range usually encountered in regular soils (6-8). Thus, their movement in the soil can be easily followed.

To begin the experiment, 100 µL of each "pollutant" solution is added at the top of each micropipette (Fig. 1a). Considering the area of each soil column, this amount is equivalent to 5.1 L m^{-2}, or a column of 5.1 mm high. This is a relatively low quantity of pollutant. Assuming an "average" molecular weight of 400 g, it amounts to approximately 10 g per m^2. An aliquot of 50 µl of tap water are added to each column, simulating a rainstorm of 2.54 mm, or an irrigation application of 25.4 m^3 per hectare. We wait a few minutes before adding more water, to let the column drain. Figure 1b shows the situation after adding 5 aliquots of water (12.7 mm of rain). The cationic CV^+ and MB^+ stay in the upper layer of the soil, while the anionic pollutants move almost half way down the column. The distance that the anions move increases after 25.4 mm of rain (Fig. 1c), and the anionic dyes start to leach out of the column. After 15 aliquots or 38.1 mm of rain (Fig. 1d), the anionic dyes are almost completely removed. This can be analogous to the pollutant reaching the underground water. After 50.8 mm of rain (results not shown), all the anionic pollutants were washed from the columns, while the cationic pollutants stayed immobilized due to their strong interactions with the clay minerals.

An experiment such as this raises the following question: is it really the clay that is responsible for this effect? Organic molecules may also be adsorbed by organic matter or metal oxides in the soil, but this interaction has more or less the same magnitude for both cations and anions. Figure 2 demonstrates that the leaching of FG^- in Hulah peat soil, a soil with a very high organic matter content, is reduced in comparison to Rehovot sand. As a

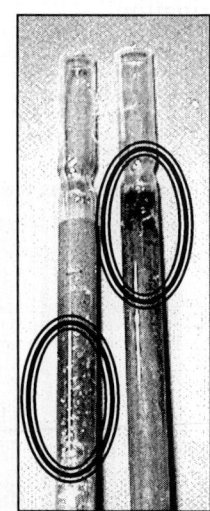

Figure 2. Leaching of a single spill of FG^- in Rehovot sand (left) and Hulah peat following an equivalent of 25.4 mm of rain

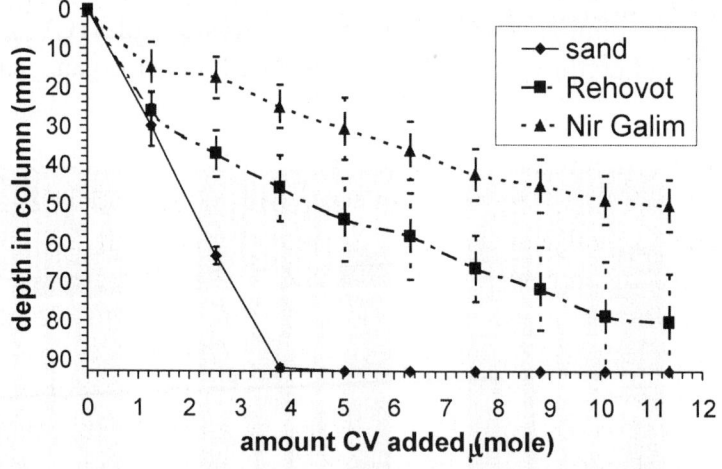

Figure 3. Depth of pollution from a constant source of CV^+ in quartz sand, Rehovot soil (2.5% clay) and Nir Galim soil (11.2% clay), as a function of the amount of CV^+ added.

matter of fact, the anions in the peat are confined to almost the same depth as the cationic CV^+ in the previous experiment. From the literature (Galan, 2000) it is known that metal oxides also strongly adsorb organic cations and anions. Thus, we may deduce that the mobility of the anionic pollutant on the one hand, and the immobilization of the cationic pollutants on the other, depends directly on the amount and type of clay or other adsorbers in the soil. This point will be emphasized in the next experiment.

This simple experiment demonstrates that clay minerals play a crucial role in the behavior of charged organic pollutants in soils: A positively charged organic chemical binds to the negative sites of the clay minerals, and remains in the upper layer, becoming a potential pollutant to crops and plants. A negatively charged organic compound is washed away and may eventually pollute the groundwater.

3. CONSTANT SOURCE POLLUTANT - INFLUENCE OF THE AMOUNT OF CLAY ON THE MOBILITY OF A CATIONIC POLLUTANT.

In order to demonstrate the influence of the amount of clay on the behavior of a cationic pollutant we simulated constant-source pollution by preparing three columns with different soils. The columns are prepared in micropipettes, as in the previous experiment. The soils used were Rehovot sand (2.5% clay), Nir Galim sand (11.2% clay), and pure quartz sand (no clay). Aliquots of 250 µL of a 5 mM CV^+ solution were added to each column, with five minutes intervals between each aliquot. Considering the area of the columns, each aliquot simulates a volume of pollutant equal to 12.7 L per m^2. Assuming a molecular weight of about 400 g, this is equivalent to approximately 25 g per m^2. Typical results for this experiment are presented in Fig. 3. It can be seen that for pure quartz sand, the pollutant leaches through the 90 mm distance of the column after less than 4 µmole are added (38 L pollutant per m^2). The same amount of pollutant reaches depths of 17 and 38 mm in the high clay (Nir Galim) or low clay (Rehovot) soils, respectively. In the Rehovot sand, the pollutant starts leaching out of the bottom of the column after adding an equivalent of 102 L pollutant per m^2 are added (10 µmole CV^+ added), while the same amount of CV^+ leaches in the Nir Galim soil to a depth of only 45 mm.

The differences between the three soils may be directly ascribed to the cation exchange capacity (CEC). The number of available sites for the cationic pollutant is a function of the amount of soil and its CEC. In general, most arid and semi-arid soils have relatively low amounts of organic matter. This can be also observed in the soils used in this experiment (Table 1). In Israel soils most of the clay is smectitic (A. Sandler, 2001 personal communication). According to that, the CEC of the soils used in the experiment depends almost entirely on the amount of clay. We can see that as the amount of clay increases the depth reached by the pollutant decreases. A direct implication of this effect is that the distance that a cationic pollutant travels from the source it is inversely dependent on the clay content of the soil. This relationship applies directly to agricultural use of the divalent contact herbicides diquat and paraquat, for example. In clayey soils they are considered to lose their herbicidal activity when they came in contact with soil or dust (Calderbank, 1968). The recommendation in Israel is to minimize their use in very sandy soils, due to the risk of damaging adjacent non-target plants.

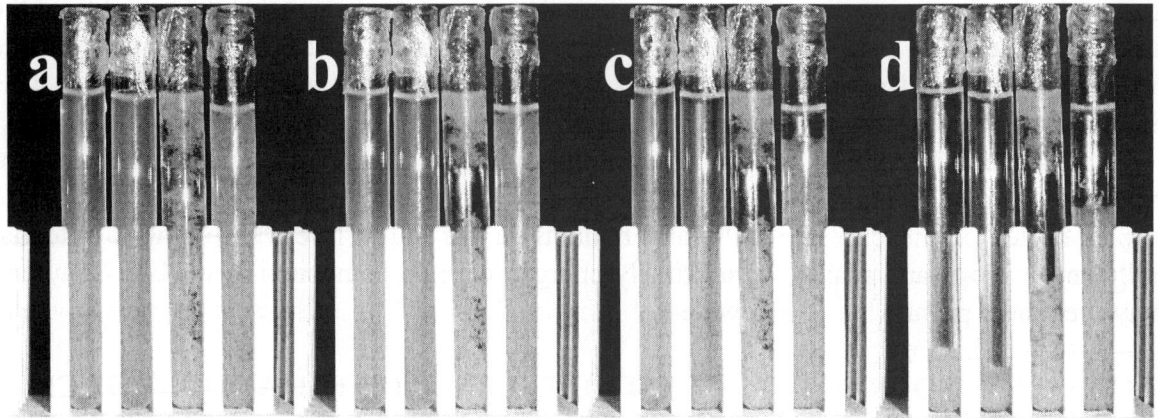

Figure 4. Influence of the amount of adsorbed AF^+ on the dispersion or flocculation of 0.02% montmorillonite suspension. Columns contain from left to right: no AF^+ added, 0.2, 0.8 and 1.4 mol AF^+ kg^{-1} clay. Figure (a) is immediately after shaking, (b) after 8 minutes, (c) after 28 minutes, and (d) after 114 minutes.

4. INFLUENCE OF CLAY-ORGANIC INTERACTIONS ON SOIL PROPERTIES

4.1. Dispersion and coagulation.

The third experiment shows how the amount of an adsorbed organic cation influences the dispersion and flocculation of a suspension. A 0.5% SWy-1 montmorillonite suspension is prepared. Four laboratory tubes are filled with suspensions containing 0, 0.2, 0.8 and 1.2 mmole AF^+ per g clay, corresponding to 0, 25, 100 and 150% of the CEC, respectively. From adsorption experiments (Rytwo et al., 1995) it is known that at these added amounts, all of the organic cation AF^+ is sorbed, and no traces of the yellow dye can be found in the uncolored supernatant. The tubes are shaken and left to settle. Figure 4 shows the results of such an experiment immediately after shaking (Fig. 4a) and 8, 28 and 114 minutes after shaking (Figs. 4b, c and d, respectively). It can be seen in Fig. 4a that when the amount of AF^+ adsorbed equals the CEC (the third tube from the left), clusters of aggregates are observed immediately after shaking. Coagulation and settling of this sample is noticed clearly after 8 min. (Fig. 4b). After 28 minutes (Fig.4c) we notice clustering and settling of the sample with 150% CEC (fourth tube from the left). In the first two tubes, where no dye or 25% of the CEC is added, some settling can be seen at the bottom only after 114 minutes.

Such an experiment illustrates the role of the diffuse double layer in coagulation, and the effect of the adsorbed ion on it. When no organic cation is added, the negative charges of the clay particles induce the existence of a diffuse positive layer in the suspension. These diffuse layers cause repulsion between the particles, leaving the suspension in a dispersed state even after one hour (results not shown). On the other hand, when the amount of organic cation adsorbed equals the CEC, the surface potential of the clay platelets is almost zero, thus no repulsion effect is observed, and the aggregation of the particles due to Van-der-Vaals attractive forces is almost immediate. A similar effect can also be obtained using inorganic cations, but only by adding considerably higher concentrations. But the more important effect caused by clay-organic interactions can be seen in the right tube of each figure: adsorption higher than the CEC causes the surface to be positively charged, leading to an anionic double layer, and a dispersion state is obtained. This effect is known as "charge reversal" (Margulies et al., 1988) and ordinarily cannot be observed with inorganic cations. Such interactions between organic cations and clay minerals can yield changes in the structure of the soil, leading to a good arrangement of the soil particles if the amount adsorbed is close to the CEC. But if the amounts adsorbed are higher than the CEC, the results can be similar to those observed in a sodic soil. The formation of a dispersed soil can result in plugging of pores, reducing the permeability and infiltration, and might even cause changes in the amount of water retained by the soil.

4.2. Influence on the soil water content.

In order to demonstrate this effect, SWy-1 clay was prepared with various amounts of adsorbed MB^+ between 0 and 1.2 $mmole_c$ MB^+ per g clay (0-150% of the CEC). At these added amounts, all the MB^+ sorbs (Rytwo et al., 1995). The clay complexes are air-dried for a week, and the residual water content is measured. The residual water content of the pure clay is about 10.2%, but that value decreases as the amount of organic cation sorbed increases, and when the adsorbed amount reaches the CEC, the residual water contents is only 3.7%. The explanation of the effect is that the residual water content depends highly on the hydration shell of the exchangeable cations (Lahav, 1983). Exchanging these inorganic cations by large organic molecules, with very few water molecules hydrating them, leads to the formation of hydrophobic organo-clay complexes, reducing the amount of hygroscopic water to one third of the original value.

This effect is also demonstrated in the Nir Galim soil, with a 11.2% of clay. By adding MB^+ up to a saturation level, the field capacity water content drops from 14.9% to 9.1%- a 40% decrease. Such a decrease can have crucial agricultural implications, even if the organic compound adsorbed is not toxic, by reducing the amount of water available to a crop.

5. CONCLUSIONS

Simple demonstrations can be very useful to illustrate the impact of clay minerals on the behavior and influence of organic pollutants in soils. The use of colorful anions and cations can show vividly the different fate of such molecules when added to the same soil; i.e, while cations remain in the upper root-containing layer, anions may easily pollute groundwater. The amount of clay in the soil has a large influence on the depth to which a cationic pollutant that is constantly spilled may migrate. Organic compounds adsorbed to clay minerals may have considerable effect on the structure and the water content of the soil. We may conclude by quoting Johnston (1996): "… the presence of even small amounts of these fine-grained materials on a mass basis will exert a large influence on the behavior of organic and inorganic solutes and subsurface environments."

ACKNOWLEDGMENT

This study was supported by a fellowship from the Sacta-Rashi Foundation. I am grateful to

Prof. Shlomo Nir and Mrs. Yael G. Mishael for their helpful remarks and comments.

REFERENCES

Bergaya, F. A. 2000. Clay science studies in the world, *Proceedings of the 1st Latin American Clay Conference* Funchal, Madeira, Vol I, pp 1-3.

Calderbank, A., 1968. The bipyridylium herbicides. Adv. Pest Control 8, 127-190.

Carter, D. L., Heilman M. D. Gonzalez, C. L, 1965. Ethylene glycol monoethyl ether for determining surface area of silicate minerals. Soil Sci. 100, 356-360.

Galan, E. 2000. The role of clay minerals in removing and immobilising heavy metals from contaminated soils, *Proceedings of the 1st Latin American Clay Conference* Funchal Madeira, Vol I, pp 351- 361.

Grim, R.E. 1953. *Clay Mineralogy*. McGraw-Hill, New York NY.

Johnston, C.T. 1996. Sorption of Organic Compounds on Clay Minerals: A Surface Functional Group Approach, in CMS Workshop Lectures, Vol. 8, Organic Pollutants in the Environment, Sahwney, B. ed. The Clay Mineral Society, Boulder CO pp. 1-44

Lahav, N. 1983, *"Fundamentals of Soil Science"* (in Hebrew). Academon Press, Rehovot Israel.

Manahan, S.E. 2000. *Environmental Chemistry* 7th Ed. CRC Press, Boca Raton FL.

Margulies, L., Rozen, H., Nir, S. (1988): Model for competitive adsorption of organic cations on clays. Clays Clay Miner. **36**: 270-276.

Murray H. H. 1997. Clays for our future *in* H. Kodama, A. R. Mermut and J. K. Torrance (Eds.) Proc. 11th Int. Clay Conf., Ottawa, Canada, pp 3- 11.

Sparks, D.L. 1995. *Environmental Soil Chemistry*. Academic Press Inc., London UK.

Theng, B.K.G. 1974. *Clay Organic Interactions*. Adam Hilger Ltd., London UK.

Rytwo, G.; Nir, S. and L. Margulies. 1995. Interactions of monovalent organic cations with montmorillonite: adsorption and model calculations. Soil. Sci. Soc. Am. J. **59**:554-564.

van Olphen, H., Fripiat, J. J. (eds.). 1979. Data handbook for clay materials and other non-metallic minerals, p. 19. Pergamon Press, Oxford.

Microbes associated with clay minerals- Formation of bio-halloysite -

Kazue Tazaki* and Ryuji Asada*

*Department of Earth Sciences, Faculty of Science, Kanazawa University, Kakuma, Kanazawa, Ishikawa 920-1192 Japan

Interaction between clays and microbes has investigated at microbial film of the cultured solution with natural sediments. Optical and electron microscopic observations of cultured microbes revealed that ultra thin films covered on the microbes. XRD of the thin films indicated 7Å d-spacing which is almost the same diffraction pattern as standard kaolin group clay minerals and natural sediment clay samples except its high background, suggesting existence of organic-inorganic complex. FT-IR analysis of the thin film exhibited characteristic adsorption of O-H ($3651 cm^{-1}$), C-H and C-N ($2925, 1454$ and $1420\ cm^{-1}$) bands, indicatives of organic origin. Atomic force microscope observation also disclosed orientated nano-meter scale clusters. TEM observation of the thin films revealed that spheroidal, hollow halloysite associated with coccus and bacillus types bacteria were formed. The electron diffraction of spheroidal hallow halloysite shows 4.43, 2.56, 2.49, 2.22, 1.48Å d-spacings. The present investigation support that thin film with spheroidal halloysite associated with bacteria is biogenic product. This biogenic halloysite, hereafter called bio-halloysite is susceptible evidence for common occurrences of bio-clay minerals.

1. INTRODUCTION

Interaction of clays with microorganisms and bacterial survival systems of soils have physicochemically discussed by Theng and Orchard (1995). We have reported the biogenic mineralization of kaolinite and halloysite on living cells of microorganism in natural and cultivated systems. The transcription or mineralization ranged from poorly crystalline granular to well orientated crystalline states (Tazaki, 1997; Ueshima et al., 2000; Asada and Tazaki, 2000; Tazaki, 2000). In this study, the natural biogenic occurrences of clay minerals in freshwater systems have been observed in nano to micron order and conducting laboratory cultivation experiments made clear the role of microbes in clay biomineralization systems. The spheroidal, hollow 7Å halloysite produced with bacterial activities in freshwater and natural sediments system under room temperature, is reported in this paper.

2. MATERIAL AND METHODS

The microbes were cultured for a few months to two years under pH 6.0 to 7.4 in order to make clear the clay mineralization process and to understand the role of microbes living in biofilm at room temperature.

A schematic figure of cultivation system shows a biofilm formation not only on a slide glass, container (glass beaker) wall and top surface of sediments, but also on the surface of solution (Figure 1).

The sediment samples collected from dam sediment, Pôrto Alegre, Brazil, were pre-treated as follows before establishment in the system.

1) The natural sediment sample rinsed with distilled water was kept in distilled water for cultivation. The suspension in the water almost sunk down to the bottom.
2) Standard kaolinite sample (AP162, Cornwall, UK) was dipped into pure ethanol and boiled water to control bacterial contamination. After control treatment, samples were set in the system.

In natural sediment system, biofilm grew on the glass surface after a few months aging, and the film grew from 0.01 to 0.1mm thick after a year aging. In controlled standard sample, however no biofilm grew.

The biofilms were observed under the polarizing optical, fluorescent and SEM-, TEM electron microscopies. Mineralogical and chemical investigations of biofilms were carried out by X-ray diffraction, X-ray fluorescence analysis and SEM-EDX. Infrared absorption spectroscopy was used for analysis of chemical bond of biofilm. Atomic force micrograph was used for observation of top surface structure of microbial films, under air condition.

3. RESULTS

Microbes in the system have coated with newly formed clays after few months aging at room temperature under pH 6-7. This clay has determined as halloysite and also biogenic product by the following investigations. Thus this 7Å-halloysite is properly called bio-halloysite.

3-1. Optical and fluorescence microscopic observation

The biofilms were observed under the polarizing optical and fluorescence microscope (Nikon EFD3) using acridine orange and 4'6-diamidino-2-phenylindole (DAPI) staining. Optical micrographs showed abundant microbes associated with clay particles (Figure 2). Epifluorescence micrographs of DAPI-stained samples indicated blue and red colors that suggest presence of DNA and chlorophyll in living microbes respectively. Various microbes such as filamentous algae, spherical algae, cyanobacteria and coccus / bacillus bacteria are in the colony associated with brown clay particles.

3-2. X-ray powder diffraction and X-ray fluorescence analysis

Mineralogical investigations of oriented clay-biofilm aggregates were performed by X-ray powder diffraction (XRD) with Rigaku Rinto 1200 X-ray diffractometer using CuKα radiation. Chemical analyses of the biofilms were carried out by energy dispersive X-ray fluorescence device JEOL JSX-3201. X-ray powder diffraction patterns of clays with the microbial mats indicated that clays are mainly composed of 7 Å clays with trace of quartz and crystobalite. The d-spacing of the newly formed clay is as same as that of kaolin group clay minerals with higher background suggesting presence of organic materials. ED-XRF analysis of culture solution (A) and cultured biogenic clay (B), in biofilms exhibited presence of various elements, such as C, O, Mg, Al, Si, P, S, K, Ca, Ti, Mn and Fe (Figure 3). Among these elements, Al, Si, S, and Fe are distinctly enriched in bio-halloysite bearing biofilms, suggesting iron and sulfur bacterial activities.

3-3. Scanning and transmission electron microscopic observation

Both biofilm and powder samples were investigated using a JEOL-JSM-5200LV scanning electron microscope, equipped with Philips-EDAX PV9800STV energy dispersive X-ray spectrometer and with a JEOL JEM 2000EX transmission electron microscope (TEM).

Scanning electron micrographs of biofilm after freeze-drying treatment showed filamentous bacteria as the main component (Figure 4A). The bacteria were 1-2 μm wide and 100 μm long (Figure 4B). Cocci and bacillus were also spreading out around the space between 1 and 2. The EDX spectra of living bacteria indicated main components of both Al and Si with high background, suggesting bio-clays formation (Figure 4B-1). EDX analysis of matrix around bacteria indicated also Al and Si with low background (Figure 4B-2). TEM observation revealed thin film of clays adherent to the bacillus type bacterial cell (Figures 5 and 6). The TEM images of clays showed formation process of spheroid with hollows about 1 μm in diameter that strongly suggest spheroidal halloysite. The spheroidal halloysite grew from elliptical to spheroidal shape (Figure 5). The flaky spheroid might proceed to high dense halloysite ball (Figure 6). The present observation revealed that the thin flaky halloysite rolled up to sphroidal type. The occurrences of spheroidal halloysite in weathered volcanic ash of Japan have reviewed by Sudo and Shimoda (1978). The present TEM observation made clear that bio-halloysite also grew spheroidal form on the surface of microbes. Growing process of spheroidal halloysite from flaky halloysite film to the spheroidal form have clearly shown in Figure 6. Electron diffraction pattern with 4.43, 2.56, 2.49, 2.22 and 1.48 Å d-spacings also supported mineralogical evidence of halloysite.

3-4. Atomic force micrograph

The surface of biofilm mounted on silicon wafer was observed using both contact and AC modes of atomic force micrograph (AFM: JEOL JSTM-4200D) under air condition. Atomic force micrograph using contact mode showed formation of bio-halloysite clusters on the surface of bacterial cell wall (Figure 7). Close-up image of the area enclosed by frame showed in net-stitch structure of organic materials (Figure 7, C), in the clusters with 50-500 nm wide and 2-20 nm thick (Figure 7, A and B). The clusters with 50-80 nm width and 6-30 nm thickness, oriented to the same direction stuck to biofilms (Figure 7, D and E).

3-5. FT-IR spectra

Mineralogical bonding condition of the biofilms was examined using Jasco FT-IR-610, MICRO-20 Fourier transform infrared absorption spectroscopy. FT-IR spectra of bio-halloysite and standard kaolinite AP162 (Cornwall, UK) are shown in Figure 8.

The O-H bands of bio-halloysite are recognized at 3697 and 3622 cm^{-1} with absorbed bands at 3405 and 1657 cm^{-1}. C-H band (2925 cm^{-1}), Si-O band (1034 cm^{-1}), Al-OH band (912 cm^{-1}) and Si-O band (537 and 471 cm^{-1}) were recognized in bio-halloysite (Figure 8, A). The 3651 cm^{-1} absorbed band is characteristic for bio-halloysite. The 2925, 1454 and 1420 cm^{-1} bands are also recognized for bio-halloysite. The bands at 3651, 2925, 1454 and 1420 cm^{-1} of bio-halloysite suggested the presence of mixture of OH_2 and OH_4 adsorption (Bish and Johnston,1993). The adsorption band at 1454 and 1420 cm^{-1} were due to abundant organic materials in bio-halloysite (Fialips et al., 2000).

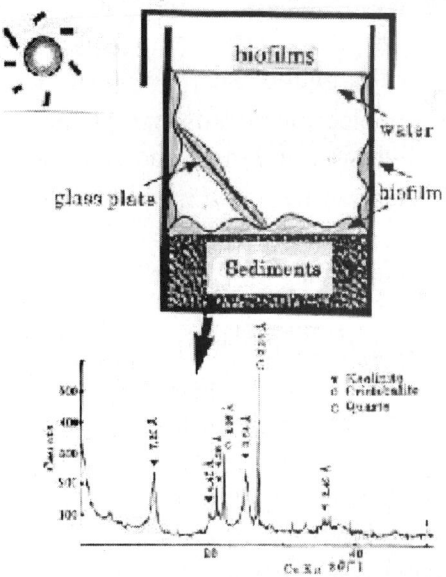
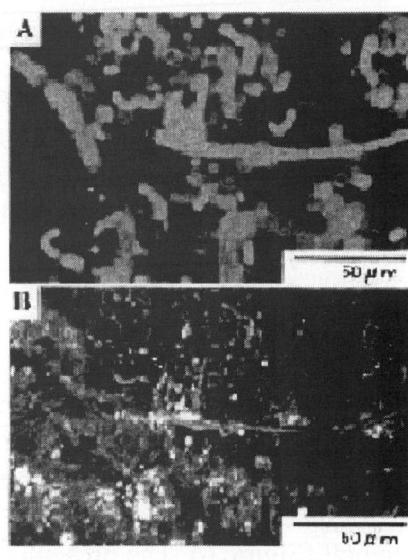

Figure 1: A schematic diagram of a natural cultivation system showing a biofilm formation on a glass slide, a wall of beaker and surface sediments

Figure 2: Optical micrographs of biofilm showing abundant microorganism associated with clay particles. A. microorganism excited with light from 510 to 560 nm, B;epifluorescence mirograph of DAPI-stained film

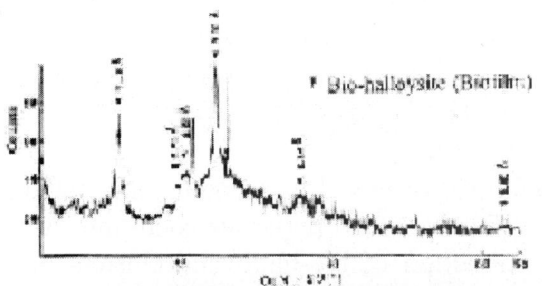
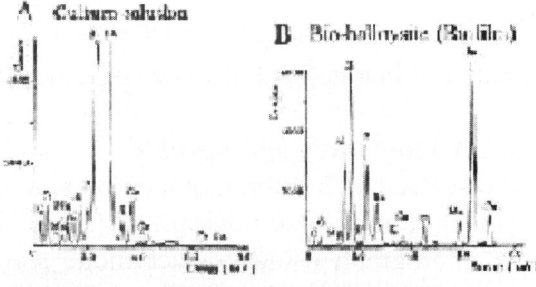

Figure 3: XRD and ED-XRF analyses of a cultured soution and biofilm

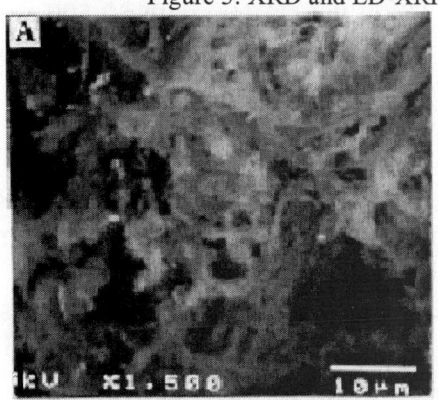

Figure 4. Scanning electron micrographs of biofilm after freeze-drying treatment showing filamentous bacteria as the main component (A). The EDAX spectrum of living filamentous bacteria indicated main components of Al and Si (B-1) whereas the background with spherical materials is composed of Al, Si and Fe (B-2)

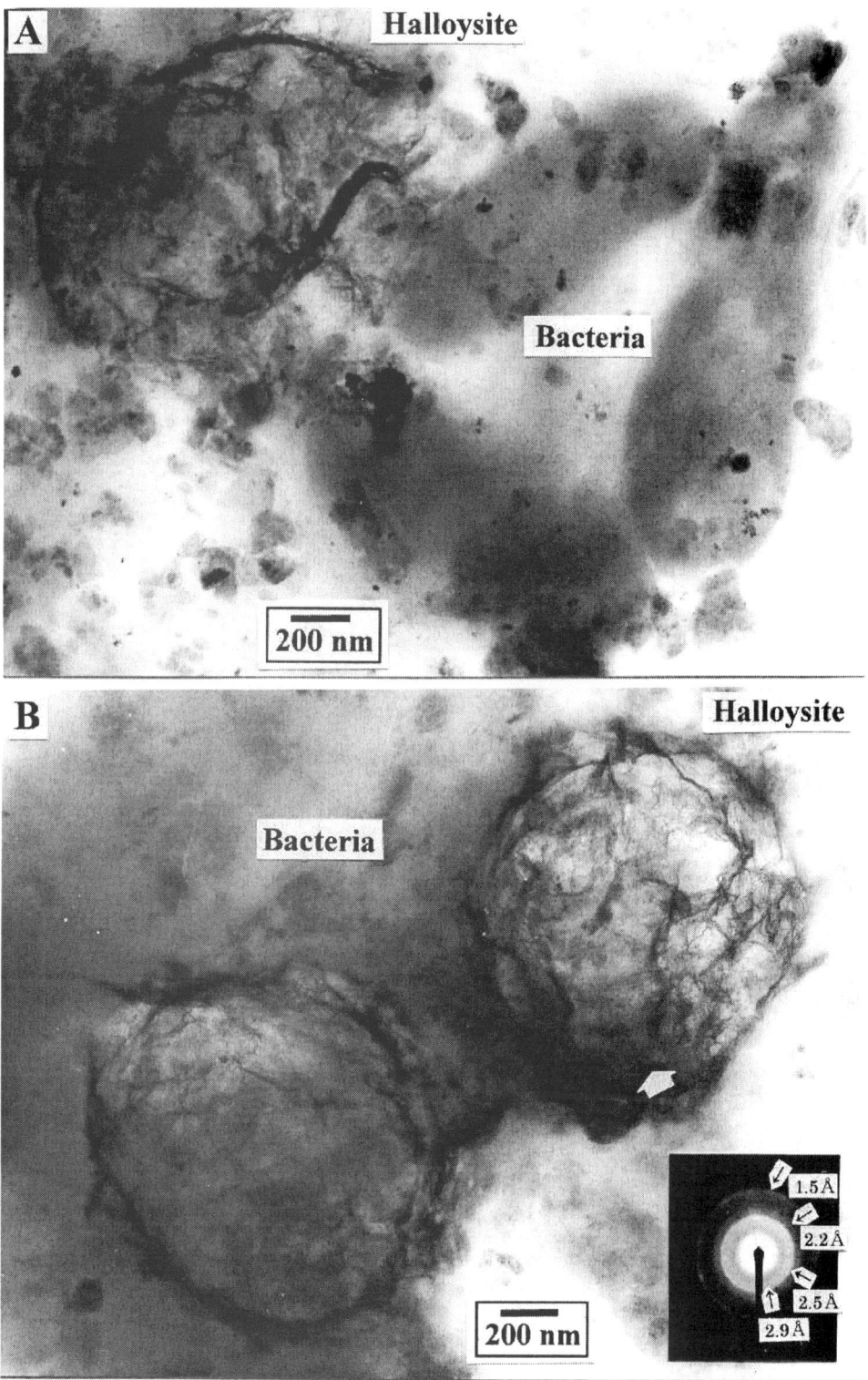

Figure 5: Transmission electron micrographs of bio-halloysite associated with bacillus type bacteria adhered to flaky materials (A). The hollow bio-halloysite developed to spheroidal halloysite ball with electron diffraction spots and rings (B)

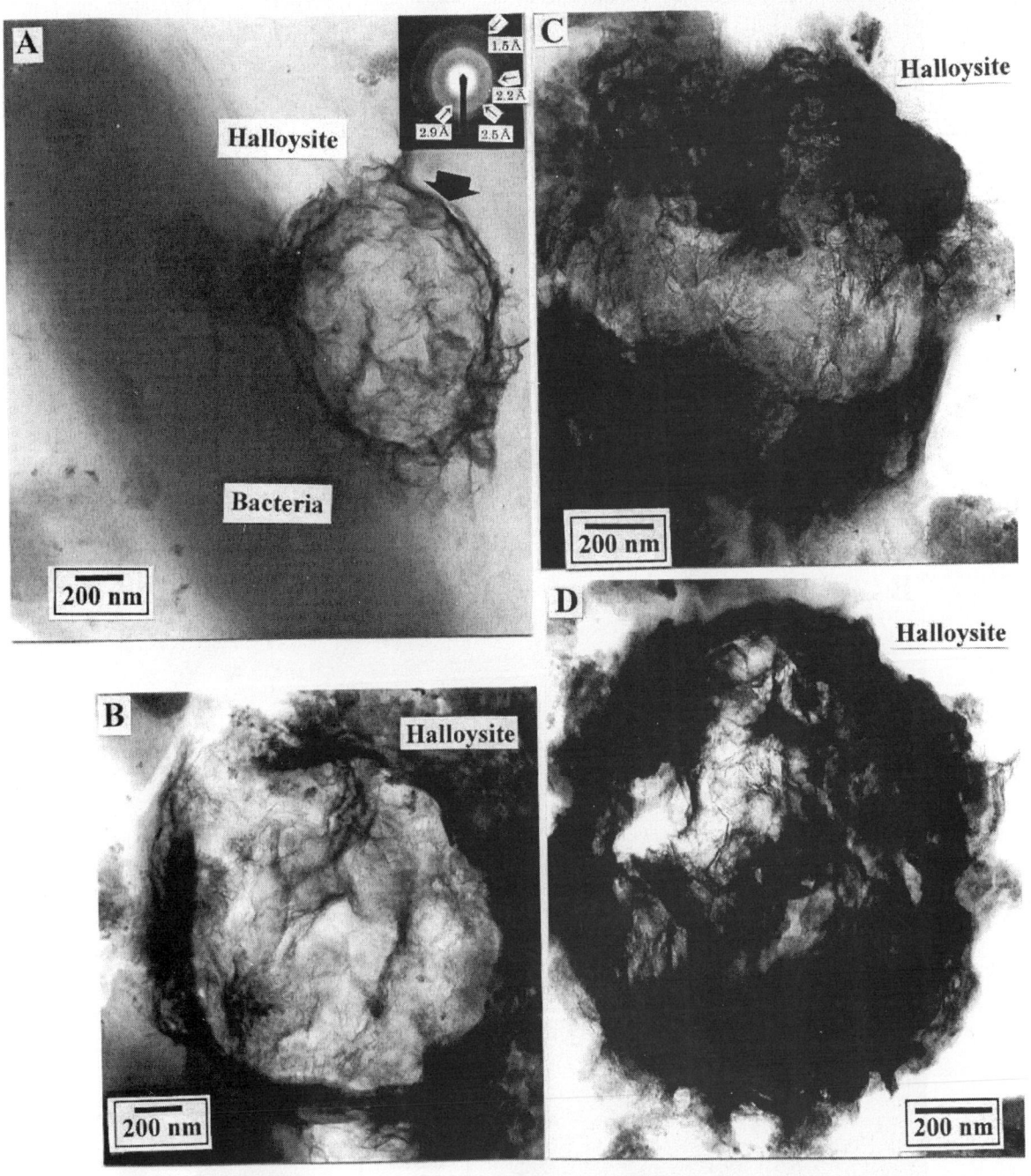

Figure 6: Transmission electron micrographs of bio-halloysite with bacteria. Spheroidal flaky halloysite grew on the cell wall of bacteria (A). The hollow bio-halloysite developed from rim part to center with high dense spheroidal halloysite ball (B, C and D). The diameter increase from 800 nm to about 1μm. The electron diffraction of bio-halloysite indicated d-spacings of 2.9, 2.5, 2.2 and 1.5 Å which reflect diffractions of 7 Å-halloysite.

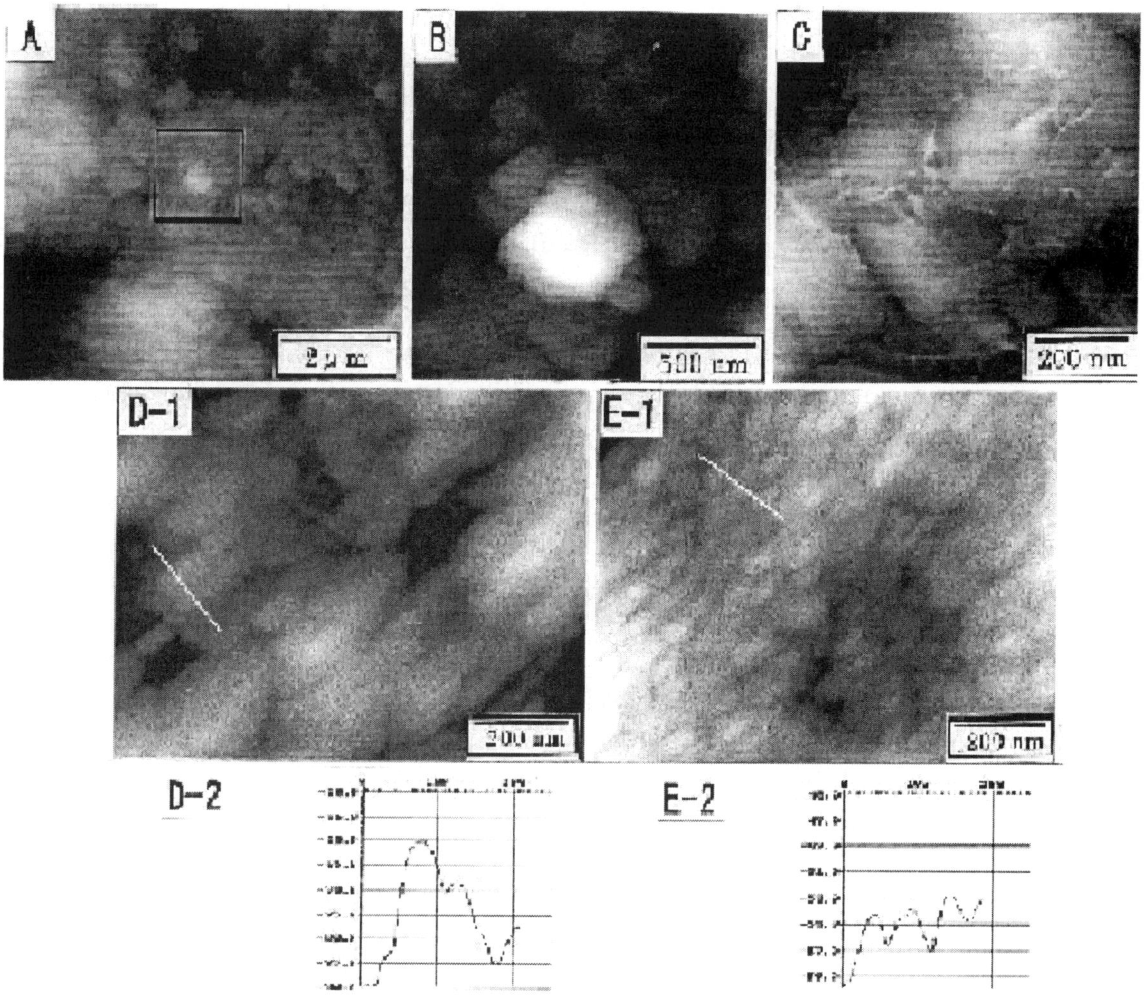

Figure 7: Atomic force micrographs using contact mode showed biofilm clusters (A). Close up images of the area enclosed by frame (B). Net-structure of organic materials on the clusters (C). Atomic force micrographs using AC mode showed oriented clusters on silicon wafer under air condition. Oriented clusters adhered side by side in biofilm (D, E).

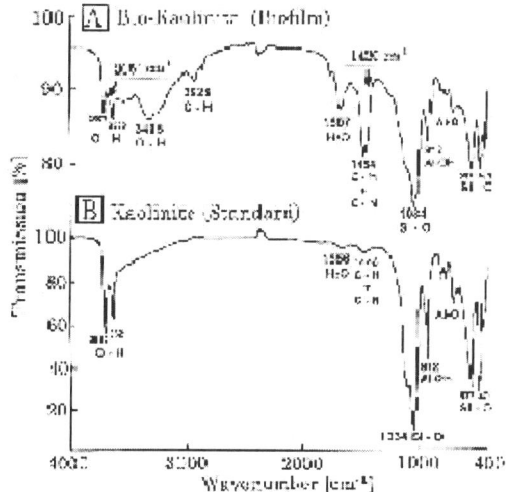

Figure 8: FT-IR spectra of biofilm (A) and standard kaolinite (AP162, Cornwall, UK) (B)

4. DISCUSSION

Bio-halloysite was successfully formed in cultivated biofilm after a few months under room temperature. The electron microscopic observations revealed that bio-halloysite grew as thin flaky film on the cell wall of bacillus type bacteria. The flaky bio-halloysite rolled up to form spheroidal form similar to that of volcanic ash soils.

Formation process of spheroidal halloysite started from flaky film on the bacterial cell wall. Then flaky halloysite rolled up to form spheroidal form. The EDX analysis of living bacteria indicated both Al and Si, the main components of halloysite with high background suggesting presence of organic materials. The electron diffraction pattern with 4.43, 2.56, 2.49, 2.22 and 1.48 Å d-spacings supported mineralogical evidence of halloysite. FT-IR data of bio-halloysite showed bands at 3697 and 3622 cm^{-1} suggesting characteristic O-H bond of structural water.

The atomic force microscopy measurement suggested that the clusters of bio-halloysite with 50-500 nm wide and 2-20 nm thick were oriented on the surface of cell wall with net-stitch structure of organic materials. The cell wall of the microbes was covered with organic adhesive that controlled the formation of oriented cluster associated with bio-halloysite particles. Clay mineral formation is linked together with microbes.

The present laboratory experiments indicate that halloysite formation in fresh water ascribed their initial crystallization to the biological origin.

ACKNOWLEDGMENTS

This study was partially supported by Grant-in-Aid for Science Research from the Ministry of Education, Science and Culture, Japan (Grant-in-Aid for Scientific Research B).

REFFERENCES

1. B.K.G. Theng and V.A. Orchard, Environmental impact of soil component interactions. (1995) 123.
2. K. Tazaki, Clays and Clay Minerals, 45 (1997) 203.
3. M. Ueshima, K. Mogi and K. Tazaki, Journal of the clays Science Society of Japan, 39 (2000) 171.
4. R. Asada and K. Tazaki, Journal of the clays Science Society of Japan, 40 (2000) 24.
5. K. Tazaki, Clays and Clay Minerals, 48 (2000) 511.
6. T. Sudo and S. Shimoda, (1978) Clays and clay minerals of Japan: Developments in Sedimentology 26, Elsevier.
7. D.L. Bish and C.T. Johnston, Clays and Clay Minerals, 41 (1993) 297.
8. C.F. Fialips, S. Petit, A. Decarreau and D. Beaufort, Clays and Clay Minerals, 48 (2000) 173.

3rd International Symposium on Activated Clays (ISAC)

Acid activation and bleaching capacity of bentonites from the Troodos ophiolite, Cyprus.

G.E. Christidis[a], S. Kosiari[a] and E. Petavratzi[b]

[a] Technical University of Crete, Department of Mineral Resources Engineering, 73133 Chania, Greece

[b] Camborne School of Mines, Redruth, Cornwall TR15 3SE, UK.

Acid activation with HCl and H_2SO_4 of a bentonite from the Troodos ophiolite complex, Cyprus, resulted in a 2.2-fold increase of the surface area of the raw materials. The activated materials have been rendered suitable for decolourization of crude maize oil and sunflower oil through removal of β-carotene. Activation increased microporosity and changed hysteresis of adsorption isotherms from H4-type to H3-type, suggesting aggregates of plate-like particles, which form slit-shaped pores. The coloring agents are not intercalated in interlayer sites, but are adsorbed in active centres present in crystal edges. Bleaching capacity is linearly related with BET specific surface area. STS diagrams combined with bleaching experiments are useful for evaluation of acid activated bentonites. The activated Cyprus bentonite cannot remove even small amounts of β-carotene selectively from sunflower oil, when other coloring agents are present.

1. INTRODUCTION

Acid activation increases specific surface area and modifies smectite structure [1-5]. Acid activated bentonites are used by the mineral oil and foodstuffs industry, in sulphur production, forest and water conservation, environmental protection, the beverages and sugar industry, the chemical industry, the paper industry and in cleaning and detergents [6].Their main application is the purification, decolorization, and stabilization of vegetable oils. They remove phospholipides, soaps, trace metals, organic compounds (carotenoids, such as β-carotene and their derivatives, xanthophylls, chlorophyll, pheophytin, tocopherols, and gossypol) and their degradation products, which impart undesirable colours to the edible oils [7-8]. Moreover, during bleaching they remove undesirable products of oxidation such as hydroperoxides, unsaturated fatty acids and glycerides.

Activation involves an initial replacement of the interlayer cations by H^+ followed by dissolution of the tetrahedral and octahedral sheets and subsequent release of the structural cations [4-5,9-10]. The depletion rate of octahedral cations follows the order $Mg^{2+}>Fe^{2+}>Fe^{3+}>Al^{3+}$ [11,12]. Hydrogen ions attack smectite structure releasing Mg and Al cations, via an interchange [14] or autotransformation [15] process.The release rates of tetrahedral and octahedral aluminium and iron are comparable in smectites and sepiolite [4,13]. Activated smectites are unstable.

The increase of specific surface area and average pore volume [2,3,16-17], depending on acid strength, time and temperature of treatment [2,3,8,17-19] are important physical changes occurring upon the activation of smectites. The specific surface area and surface acidity of activated smectites control the efficiency of decolorization. Bleaching is usually not associated with the maximum surface area [16,17,19] because it is not merely a physical adsorption process [7,16,20].

Upper Cretaceous bentonite beds up to 30 m thick occur above the Upper Pillow Lavas of the Troodos Ophiolite Complex of Cyprus. The bentonites have inferior swelling and rheological properties and cannot be used in traditional industrial applications such as drilling fluids and foundries [21]. The purpose of this contribution is a) to examine the mineralogical and physicochemical properties of the products of activation of the Cyprus bentonites with HCl and H_2SO_4 and b) to evaluate their capacity to decolorize crude maize and suflower oil.

2. MATERIALS AND METHODS

The samples were collected from bentonite beds at the area of Mitsero in the northern part of Troodos Ophiolite Complex. They contain 50-60% mixed layer illite-smectite (IS) with 80-85% expandable layers, minor clinoptilolite, quartz, opal-CT, carbonates (calcite and trace siderite) and trace amounts of illite, gypsum and kaolinite in places [21]. The raw material is described in Table 1. The materials were dried at 65°C overnight, ground in ball mill and passed through a 125μm sieve. Activation was carried out using HCl and H_2SO_4 of analytical grade, with concentrations between 1N and 8N for 1-6 h, at 70° ± 2°C and solid to liquid ratio 1:10. Activation with H_2SO_4 was carried out also at 85°C because, 70°C was not adequate for activation. After treatment the clay was washed until the pH reached 4, and the slurry was filtered. The material was dried at 60°C and was ground to pass through a 125μm sieve.

BET surface area measurements and full adsorption-desorption isotherms were obtained with a NOVA 2200 analyser (Quantachrome), at 77°K using N_2. BET surface area was determined in the P/P_o range 0-0.30. Mesopore size distribution was determined from the desorption branch. The structural changes of the activated bentonite were studied by XRD and FTIR. XRD was carried out on randomly orientated samples with a Siemens D500 diffractometer, (40KV, 30mA, CuKα radiation, graphite monochromator) at a scanning step of 0.02° and scanning time 1 s/step. FTIR spectra were obtained with a Perkin Elmer 1000 FTIR Spectrometer. 2 mg of each sample previously dried at 100°C were mixed with 200 mg of KBr and pressed on 13 mm disks [22].

The bleaching capacity of the end products was tested using crude maize and sunflower oil provided by ELAIS-UNILEVER S.A at a solid:liquid ratio 1:50. (0.5 g of clay in 25 ml of oil).

Table 1
Mineralogy, CEC specific surface area and dehydroxylation temperature of the Cyprus bentonite. M= major mineral phase, min= minor mineral phase, T= trace mineral phase

Mineralogical composition	CEC meq/100g	Spec. surface area $m^2.g^{-1}$	Dehydroxylation temperature (°C)
R0 Mixed layer I-S (20% illite) (M), quartz (M), opal CT (min), zeolite (min), calcite (min), illite (T) siderite (T) kaolinite (T), gypsum (T).	39-68	80	520-550

Atmospheric-type of bleaching [23] was carried out at 85° ± 2°C under stirring. Treatment time at maximum temperature was 20 minutes. The treated oils were filtered under vacuum and the bleaching capacity of the activated clays was evaluated by means of removal of β-carotene. Color changes in the treated oils were determined spectrophotometrically at 430nm (MILTON ROY SPECTRONIC 20D absorption spectrophotometer) by diluting 5ml of bleached oil in 15ml chloroform. The adsorbed amount of β-carotene was calculated by converting absorbance units to β-carotene concentration. The bleaching capacity of the activated clays was determined from the following equation:

Bleaching capacity=$(C_o - C / C_o) * 100$ (1)

where C_o=concentration (μg/ml) of β-carotene in the crude oil and
C=concentration of β-carotene in the bleached oil.

For terms of comparison the same bleaching experiment was performed using an industrial product available in the market (BENT EY-11 FF), provided by Bensan Company, Turkey.

3. RESULTS

3.1 Structural modification

Calcite was dissolved during activation. The effect of acid treatment is more significant in the prolonged treatments with strong acid (6h/5N) (Fig. 1). Activation has mainly affected the 001 reflection in prolonged treatments with strong acids while in milder treatments the effect is negligible. The remaining basal reflections and the prismatic reflections (110, 020 band and 060 reflection) were not affected. Intensively treated samples display an increase of the background between 20 and 30°2θ (a in Fig. 1) due to deposition of amorphous silica, produced by acid attack on the octahedral layer and the exposure of the tetrahedral layer.

The effects of acid activation on the FTIR spectra of bentonites (Fig. 2) are summarized as follows: The calcite band at 1426cm^{-1} disappeared after treatment. The intensity of the band at 3630cm^{-1} (AlAlOH coupled by AlMgOH stretching vibrations) decreases with increasing severity of treatment. With increasing intensity of treatment the Si-O band sifted to higher frequencies (1020cm^{-1} for untreated material, 1030cm^{-1} for 3N/1h and 1038cm^{-1} for 5N/6h treatment). The intensity of the Si-O band at 1088cm^{-1} increased due to formation of amorphous silica. The AlAlOH deformation band (916cm^{-1}) decreased while the band at 874cm^{-1} is almost eliminated with treatment. Although the position of this band suggests the presence of Fe-rich smectites with greater substitution of Fe for Al [22] it is most probable that it belongs to calcite, as suggested by the band at 1426cm^{-1}. The intensity of the band at

796cm^{-1} (disordered opal-CT) increases significantly with increasing intensity of treatment and so does the intensity of the band at 466 cm^{-1} (Si-O bending vibrations). On the other hand, the absorption band at 500 cm^{-1}, assigned to the same vibrations, decreases with increasing intensity of treatment.

3.2. Surface area and pore size distribution

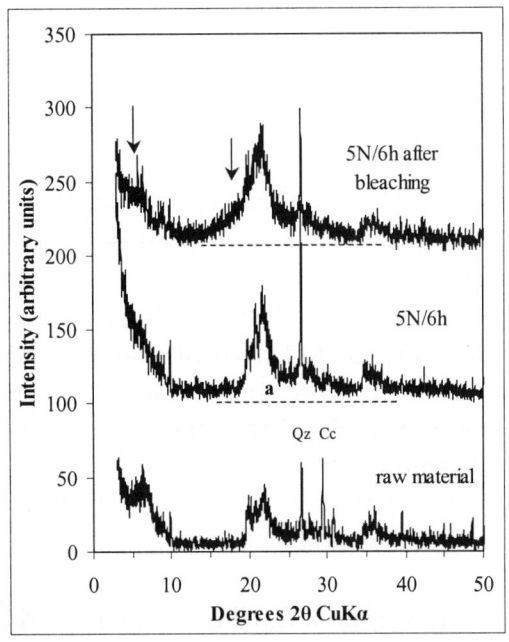

Figure 1. XRD traces of studied materials. a denotes background increase due to amorphous material. Arrows show changes after β-carotene adsorption and dashed lines denote background Qz=quartz, Cc=calcite.

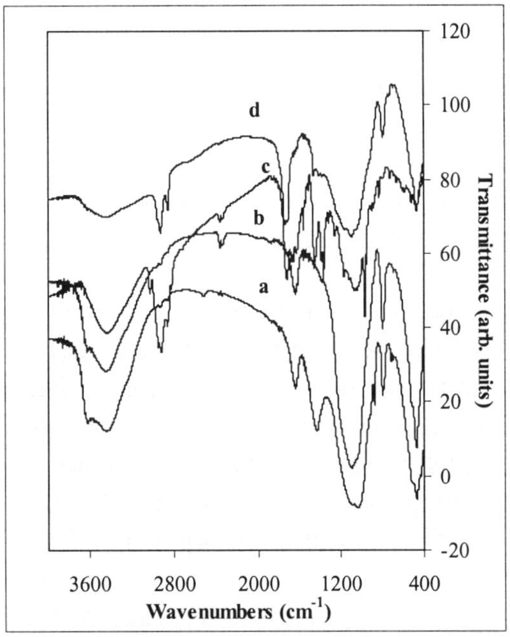

Figure 2. FTIR spectra of studied materials. a) original bentonite, b) bentonite activated with 5N HCl for 6h, c) β-carotene spectrum, d) 5N/6h after bleaching.

The evolution of specific surface area (SSA) is shown in the Surface area-Time-acid Strength (STS, [17]) diagram (Fig. 3). SSA displays a maximum 2.2 fold increase with increasing severity of treatment. SSA evolution paths depend both on the acid strength and time, but dependance on these factors is different for different bentonites (e.g. compare with

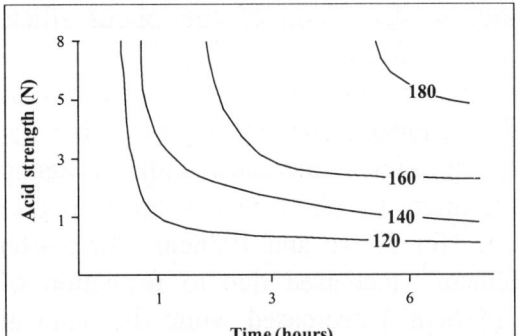

Figure 3. STS (Surface area-Time-acid Strength) diagram of the activated bentonite from Cyprus. The contours indicate evolution of specific surface area (m^2.g^{-1})

[18]). Thus STS diagrams are representative for each specific bentonite. Also, in comparison with other studies [8, 17-19] the surface area of the bentonite from Cyprus does not decrease after long treatments with strong acid, but seems to attain a rather constant value of 180 m^2/g. Thus, maximum surface area is attained with treatments characterized by various

combinations between acid strength and time. Complete adsorption-desorption isotherms of the original bentonite and acid treated materials are shown in Figure 4. They are type IV isotherms characterized by a hysteresis loop due to capillary condensation in mesopores [24]. The materials display H3-type hysteresis, which is associated with aggregates of plate-like particles yielding slit-shaped pores [24]. The untreated material displays hysteresis which is intermediate between H3- and H4-type [24]. Transition from H3-type to H2-type hysteresis in intense treatments [25] was not observed in this study.

Pore size distribution diagrams of the untreated bentonite and selected treated materials are shown in Figure 5. Acid activation reorganized meso and microporosity. Increasing treating time and acid strength enhanced the mesoporosity and microporosity of the bentonite. The most intensive treatments (5N/6h and 8N/6h), yielded almost identical pore size distribution curves, indicating that microstructural modifications affecting adsorption characteristics of the activated bentonite might occur to a certain extent.

3.3. Bleaching capacity.

The evolution of bleaching capacity of the Cyprus bentonite, activated with HCl, for maize oil and sunflower oil is shown in Figure 6. The activated materials display high bleaching capacity, comparable to that of the commercial bleaching earth (BENT EY-11 FF), reaching 80% for maize oil and 76% for sunflower oil. The shaded area in Figure 6, indicates the combinations of acid strength and residence time, which imparted bleaching capacity greater than 75% in the case of maize oil (Fig. 6a) and 70% in the case of sunflower oil (Fig. 6b) to

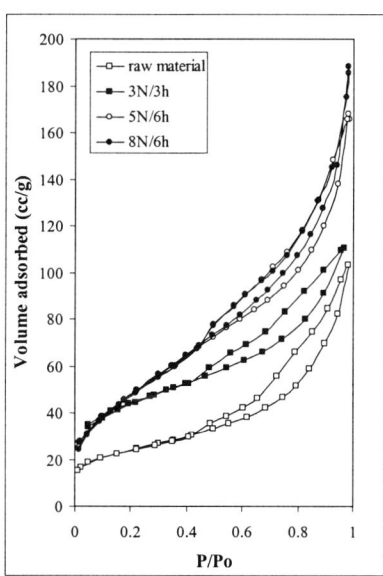
Figure 4. Adsorption isotherms of the acid activated bentonites.

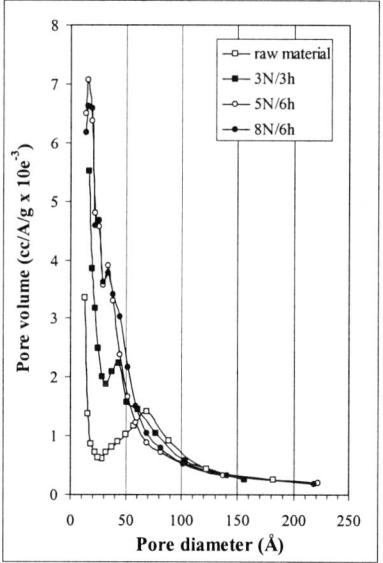
Figure 5. Pore size distribution of the acid activated bentonites.

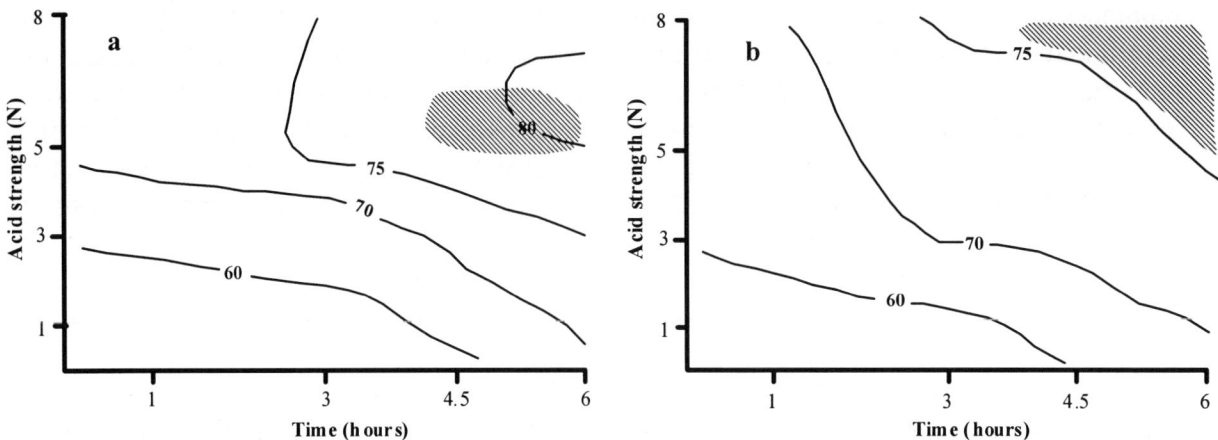

Figure 6. Bleaching capacity of the Cyprus bentonite as a function of time and acid strength a) maize oil b) for sunflower oil. The shaded area denotes the highest bleaching capacity.

each material. Thus the outcome of these diagrams is the determination of the combinations of acid strength and treating time, which produce optimum products for industrial applications.

In both oils and for constant activation time maximum bleaching capacity is attained with acids stronger than 5N (Fig. 6). Also, in both types of oil and for constant acid strength, maximum bleaching capacity is attained after 270 minutes of activation irrespective of acid strength. Longer treatment time does not improve decolorization efficiency. In contrast with other studies [16,17,19], optimum bleaching capacity is associated with maximum specific surface area (Fig. 3, 6), and mild treatments do not produce effective end products. This shows the different behaviour of the different bentonites during acid treatment.

Adsorption of β-carotene was monitored by XRD and FTIR. There are some indications of intercalation of β-carotene (Fig. 1), but further studies would be required to confirm this observation. The materials which adsorbed β-carotene display a small tail in the low-angle side of the 110, 020 band of smectite and a small hump at 4.7 °2θ (arrows in Fig. 1). The small tail in the low-angle side of the 110,020 band of smectite is located close to the XRD maxima of β-carotene (data not shown) and is thus considered an indication for adsorption. The FTIR spectra (Fig. 2) clearly demonstrate that β-carotene is the main compound adsorbed. There are also some additional absorption bands at about 1700 cm^{-1}, which are associated with carboxyl roots or aromatic rings. These bands probably belong to chlorophyll which contains both carboxyl roots and aromatic rings [7] and which is present in both crude oils.

Activation with H_2SO_4 was carried out at 85°C in order to obtain materials with comparable bleaching capacity for maize oil. At 70°C even the use of 12-15 N H_2SO_4 did not yield acceptable products. Although the activation temperature is higher for H_2SO_4, HCl is more expensive. However the price difference may counterbalance the use of more energy.

4. DISCUSSION

In contrast with other studies [16,17,19], bleaching capacity for both oils is correlated with surface area (Fig. 7) of acid activated low-grade bentonite from Cyprus This different behavior of the Cyprus bentonite is attributed to the structural features of the smectites. The bentonite contains I-S with 80-85% smectite layers, and has a low dehydroxylation temperature (Table 1), typical for non-ideal smectites [26]. Such temperatures are typical for montmorillonites with low OH- and high Fe-contents but are not common for Wyoming and Chambers type montmorillonites [26] present in Miloan bentonites [17]. The raw material was activated with acid strength-treatment time combinations that are comparable to those given to the Miloan bentonites [17]. The slightly lower bleaching capacity and the lower surface area obtained in this study is attributed to the lower solid:liquid ratio used (1:50 instead of 1:40 for the Miloan bentonites), the lower grade of the Cyprus bentonite and the different oil tested.

The slower activation of the Cyprus bentonite with H_2SO_4 than with HCl reflects the efficiency of acid to attack smectite particles and to dissolve mineral impurities. HCl is more efficient than H_2SO_4 for dissolving free Fe-oxides. Activation with HCl yielded yellow-brown supernatant solutions even at mild treatments, and activated products with lighter color than from H_2SO_4 treatment. Although part of the dissolved Fe is octahedral, the color of solution at mild treatments suggests that dissolution of free Fe-oxides was more important. With H_2SO_4, a slight change of the solution color occurred only after intense treatment, indicating release of octahedral Fe. Dissolution of Fe-oxides, usually covering smectite particles, facilitates formation of active sites.

The XRD traces suggest that intercalation of β-carotene at the interlayer space of smectites may only occur to a minor extent. Yet β-carotene is readily adsorbed, implying that the active sites (Brönsted acid centres) are located mainly at the edges of partially dissolved smectite quasicrystals, at disrupted octahedral and tetrahedral sheets. Disruption of smectite structure after activation is associated with curvature of the tetrahedral sheet before collapse and formation of amorphous silica [27]. Such a process explains the increase of meso and microporosity with increasing treatment. The lack of H2-type hysteresis [25] suggests that ink-bottle pores did not form [24].

Although crude sunflower oil contains considerably less β-carotene than maize oil the activated materials removed smaller amounts of the coloring agent. A kinetic study of

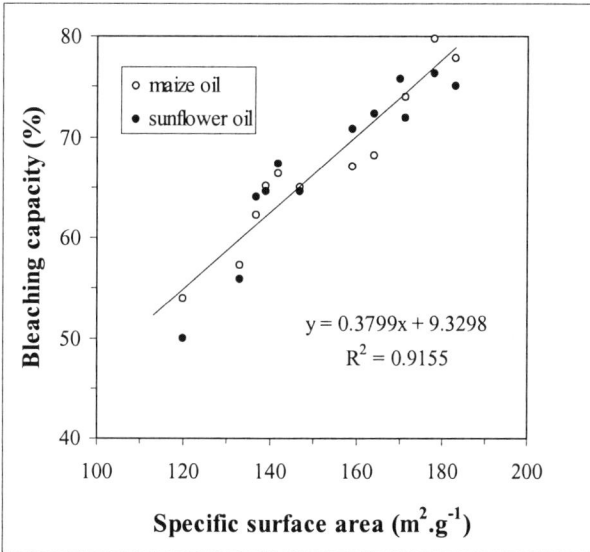

Figure 7. Projection of specific surface area vs bleaching capacity for the Cyprus bentonite.
adsorption of β-carotene yielded different activation energies for decolorization of the two oils [28]. This is attributed to the different composition of the oils. Absorption spectra for the sunflower oil (not shown) show greater absorption between 400 and 425 nm relative to 430 nm, indicating the influence of additional coloring agents, possibly chlorophyll. This suggests that a) there is competition for adsorption at the active sites among the compounds of the

crude oil and b) β-carotene is not selectively adsorbed. Therefore even small amounts of β-carotene might not be removed from sunflower oil in the presence of other coloring agents. However, it is not known if this is a general feature of activated bentonites.

REFERENCES

1. J.P. Rupert, W.T. Granquist and T.J. Pinnavaia, in A.C.D. Newman (ed) Chemistry of Clays and Clay Minerals, Mineralogical Society, London, (1987) 275.
2. E. Srasra, F. Bergaya, H. van Damme and N.K. Arguib, Appl. Clay Sci., 4 (1989) 411.
3. C.N. Rhodes and D.R. Brown, J. Chem. Faraday Trans., 88(15) (1992) 2269.
4. I. Tkac, P. Komadel and D. Mûller, Clay Miner., 29 (1994) 11.
5. H. Kaviratna, and T. Pinnavaia, Clays Clay Miner., 42 (1994) 717.
6. M. O' Driscoll, Ind.Miner., 250 (1988) 43.
7. M.K.H. Siddiqui, Bleaching earths, Pergamon Press, Oxford, 1968.
8. S.C. Kheok and E.E. Lim, J. Am. Oil Chem. Soc., 59 (1982) 129.
9. K. Brückman, J. Fijal, Z. Klapyta, T. Wiltowski, & W. Zabinski, Miner. Polon., 7 (1976) 5
10. I. Novak and B. Cicel, Clays Clay Miner., 26 (1978) 341.
11. R.W. Luce, R.W. Bartlett and G.A. Parks, Geochim. Cosmochim. Acta, 36 (1972) 35.
12. A. Corma, A. Misfud and E. Sanz, Clay Miner., 22 (1987) 225.
13. V. Luca and D.J. MacLachlan, Clays Clay Miner., 40 (1992) 1.
14. I. Barshad and A. E. Foscolos, Soil Sci., 110 (1970) 52.
15. M. Janek and P. Komadel, Geol. Carpathica-Ser. Clays, 44 (1993) 59.
16. D.A. Morgan, D.B. Shaw, T.C. Sidebottom, T.C. Soon and R.S. Taylor, J. Am. Oil Chem. Soc., 62 (1985) 292.
17. G. Christidis, P.W. Scott and A.C. Dunham, Appl. Clay Sci., 12 (1997) 329.
18. G.A. Kolta, I. Novak, Z. El-T. Samir and A. El-B. Kamilia, J. Appl. Chem. Biotechnol., 26 (1975) 355.
19. I. Zaki, M. Abdel-Khalik and G.M. Habashi, Colloids & Surfaces 17 (1986) 241.
20. L.E. Khoo, F. Morsingh and K.Y. Liew, J. Am. Oil Chem. Soc., 56 (1979) 672.
21. G. Christidis, Th. Marcopoulos and E. Petavratzi, Proc. 12ICC (Abstracts) (2001).
22. J.D. Russell, in: A handbook of determinative methods in clay mineralogy (M.J. Wilson ed) Blackie, Glasgow & London, (1987), 133.
23. L.L. Richardson, J. Am. Oil Chem. Soc., 55 (1978) 777.
24. K.S.W. Sing, D.H. Everett, R.A.W. Haul, L. Moscou, R.A. Pierotti, J. Rouquerol and T. Siemieniewska, Pure & Appl. Chem., 57 (1985) 603.
25. S. Mendioroz, J.A. Pajares, I. Benito, C. Pesquera, F. Gonzalez and C. Blanco, Langmuir, 3 (1987) 676.
26. L.G. Schultz, Clays Clay Min., 17 (1969) 115.
27. R. Fahn and K. Fenderl, Clay Miner., 18 (1983) 447.
28. G. Christidis and S. Kosiari (submitted to Clays Clay Min.).

Synthesis and Characterization of Copper-Loaded Titania Pillared Clays

Z. Ding[a*], R. L. Frost[a], J. T. Kloprogge[a], G. Q. Lu[b] and H. Y. Zhu[b]

[a]Centre for Instrumental and Developmental Chemistry, Queensland University of Technology, 2 George Street, GPO Box 2434, Brisbane, Qld 4001, Australia
[b]Department of Chemical Engineering, The University of Queensland, Brisbane, Qld 4072, Australia

Two series of copper-loaded titania pillared clays (Cu/Ti-PILCs) with various copper amounts were synthesized using an ion-exchange method. The starting materials were titania pillared clay prepared from normal sol-gel method (Ti-PILC) and from hydrothermal treatment method (Ti-PILC-H), respectively. Copper was loaded on these two samples (Ti-PILC and Ti-PILC-H) by suspending them in copper (II) nitrate solutions at pH of 6. It was found that for both supporting materials; the amounts of loaded copper were proportional to the starting concentrations of the copper (II) nitrate solutions. XRD results showed that for samples freshly prepared from the highest copper nitrate concentration, the loaded copper was mainly in gerhardtite phase, which transformed to copper (II) oxide phase upon calcination. In addition, Ti-PILC-H showed better-crystallized anatase phase but lower surface area than Ti-PILC and the copper-loading amount was more related to the surface area. More importantly, it was found that these loaded copper phases were stable against mild acid washing and even after being treated by concentrated nitric acid solution, there was still certain amount of copper remained in the samples.

1. INTRODUCTION

Smectite group of clays shows interesting features, such as swelling in water and possessing exchangeable cations on the silicate layers, which are the prerequisites for the successful synthesis of pillared clays (PILCs) [1]. The latter is generally produced by first intercalating the clay layers with metallic polyoxocations through ion-exchange. These metallic polyoxocations will then transform into metal oxide pillars upon calcination and thus expand the interlayer space permanently [2]. PILCs belong to two-dimensional porous materials, which show several attractive properties, such as dramatically enhanced surface area compared with the starting clay materials, tuneable porous structure by employing different types of intercalating species, strong surface acidity, good thermal stability and catalytic active metal oxide pillars. Ever since the first reports [3-5] on the synthesis of PILCs, great efforts have been put in the applications of these materials in catalytic reactions. The early studies were mainly focused on using PILCs in oil cracking reactions [1, 3, 4, 6] and fine chemical synthesis [7-9]. In more recent years, interest has been shown in applying PILCs as catalysts for treating environmental pollutants, both in gas phase, such as selective catalytic reduction of NO to nitrogen [10-13], and liquid phase, such as photocatalytic [14-16] or wet oxidation [17, 18] of organics to water and carbon dioxide. For the latter applications, it is well known that titania in anatase phase is the most efficient catalyst in photocatalytic reactions, while copper oxide and iron oxide are popular solid catalysts in wet oxidations. Since PILCs can provide high surface area with large pore size, such structure would be beneficial for organic reactants and intermediates to reach and leave the active sites on the surface. Therefore, PILCs can be used as potential support materials. More importantly, the catalytic active metal oxides, e.g. titania, can be prepared as pillars, e.g. Ti-PILC, where good dispersion of ultra-fine titania particles over clay surface can be obtained and can thus be used as catalysts directly. However, there are very limited reports on developing PILC based materials for this type of application.

The aim of this work is to synthesise Ti-PILCs and use them as supports for copper loading. Two series of Ti-PILCs, with different anatase crystallinity, were prepared and the effects of their physical properties on the copper loading were studied. The stability of these loaded copper species on Ti-PILCs was also investigated.

2. EXPERIMENTAL

2.1. Materials Synthesis

The supports, Ti-PILCs, were prepared from normal sol-gel method with titanium tetrachloride ($TiCl_4$) as Ti precursor. Firstly, $TiCl_4$ was dissolved in 0.5M of HCl solution with molar ratio of Ti/HCl of 8/5. The solution was continuously stirred and $TiCl_4$ was hydrolysed and peptised for 6h. The resulting titania sol was then mixed with 0.5wt% of Na-exchanged montmorillonite (supplied from Integrated Mineral Technology Ltd., Australia) suspension at Ti/Na-M ratio of 20mmol/g. After stirring the suspension for another 16h, the mixture was separated from water and washed by de-ionised water to get titania intercalated Na-M wet cake. Ti-PILC was obtained by drying the wet cake directly in oven at 80°C followed by calcination at 300°C for 12h. Or the wet cake was further hydro thermally treated at 200°C for 3h followed by washing, separation, drying at 80°C and finally calcining at 300°C for 12h. Thus obtained sample was labelled as hydro thermally modified titania pillared clay (Ti-PILC-H).

Copper was loaded onto these two supports by dispersing them in copper (II) nitrate solutions at pH of 6. The ratios of Cu/supports were varied from 1mmol/g, 2mmol/g to 4mmol/g. After stirring for 24h, the mixture was separated, washed and dried at 80°C. After calcination at 300°C for 3h, the final products were labelled as Cu/Ti-PILC-n and Cu/Ti-PILC-H-n, respectively. Where n (=1, 2, 3) represents different Cu/supports ratios (1, 2, 4mmol/g).

2.2. Materials Characterization

X-Ray Diffractometry (XRD) patterns of samples were recorded on ENRAF NONIUS DELFT DIFFRACTIS 583, using Co K_α radiation. To get better information on layered structure of PILCs, sample powders were dispersed in water and spread over the glass slides, then they were slowly dried at room temperature. The scans were conducted at speed of 1°/min, step size of 0.02° and between the range of 3° and 50°.

Nitrogen ads/desorption was performed on NOVA 1200 MICROANALYZER (Quanta Chrome) to measure the surface area. Samples were first degassed at 230°C for 3h and analysed under liquid nitrogen temperature. The total surface area was calculated by the BET equation, while the external surface area was determined by the t-plot method.

The concentrations of Cu in different samples were analysed by X-Ray Photoelectron Spectroscopy (XPS) using a PHI model 560 XPS/SAM/SIMS I multitechnique surface analysis system. The X-Ray source was provided by the monochromatic Mg K_α radiation, which was operated at 15kV. Powders were loaded onto sample holder by double face sticky tape. Since the thickness of single clay layer is around 9.8Å and XPS can detect surface information within 60Å depth, it is believed that for PILC samples, the concentrations observed from XPS are bulk concentrations.

Inductively Coupled Plasma – Atomic Emission Spectrometer (ICP-AES) (SPECTROFLAME) was also used to analyse the Cu concentration. Sample powders were first mixed with concentrated nitric acid under continuous stirring for 24h. Then the solution was filtered and diluted for ICP-AES analysis.

3. RESULTS AND DISCUSSION

3.1 Structure of the Ti-PILC and Ti-PILC-H

The XRD patterns for original Na-M, synthesized Ti-PILC and Ti-PILC-H samples are shown in Figure 1.

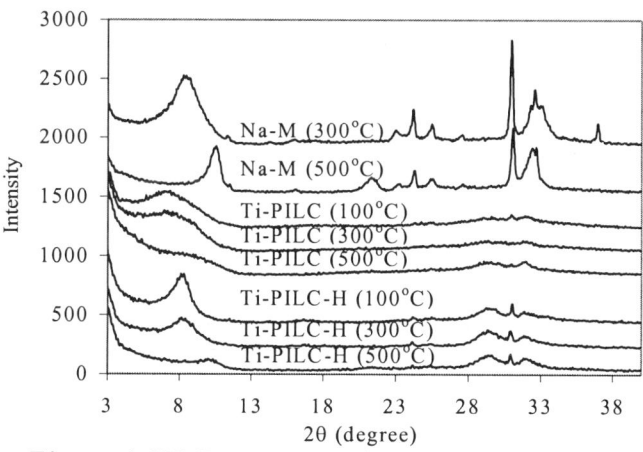

Figure 1 XRD patterns of Na-M and Ti-PILCs at different temperatures.

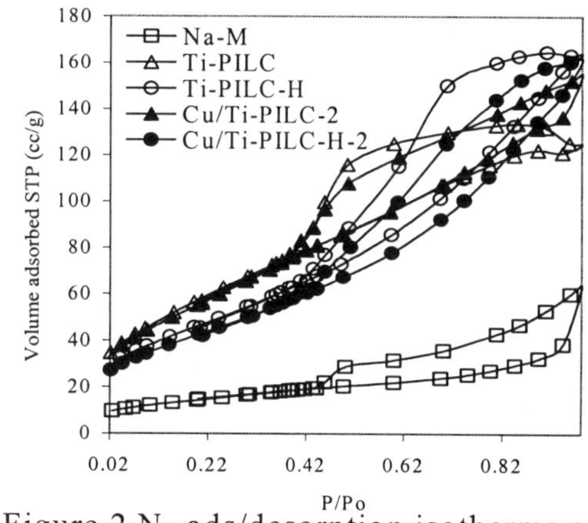

Figure 2 N_2 ads/desorption isotherms of Na-M, PILCs and Cu/PILCs.

Strictly speaking, samples prepared by heating at 100°C should be called titania intercalated clay. For simplicity, all the samples, obtained at 100°C, 300°C and 500°C, are labelled as Ti-PILC or Ti-PILC-H. From Figure 1, it is seen that for Na-M (300°C), there is a big peak centred at 8.4° with d-spacing of 12.2Å. Although the sample is calcined at 300°C, such d-spacing indicates that the adsorbed water between interlayer spaces does not disappear totally. After further calcined at 500°C, an obvious peak shift from 8.4° to 10.5° is observed, implying the complete dehydration and dehydroxylation. The d-spacing for Na-M (500°C) is 9.8Å, which agrees well with the literature report.

Compared with Na-M, Ti-PILC (100°C) shows a very broad peak around 7.1° (14.5Å). There is a little shift, to 7.1° (14.4Å), after calcined at 300°C. While for Ti-PILC (500°C), this peak is further broadened and shifted to 9.0° (11.4Å). The decrease in the intensity of d (001) peak along with the increase of calcination temperature suggests that during heating, the arrangements of clay layers tend to be more disordered. Such result might be caused by the irregular size of titania pillars or the distribution of titania pillars over the clay surface is not homogeneous, leading to more randomly orientated layer structure.

Similar phenomenon is also observed in Ti-PILC-H samples. The main difference from Ti-PILC samples is the d (001) peak position, which appears at much higher degree and is similar to that of Na-M, e.g. 8.1° (12.7Å), 8.1° (12.6Å) and 10.0° (10.3Å) for Ti-PILC-H heated at 100°C, 300°C and 500°C, respectively. In addition, the peak intensity, particularly for Ti-PILC-H (100°C) and Ti-PILC-H (300°C), is stronger than the correspondent Ti-PILC samples. Therefore, Ti-PILC-H (100°C and 300°C) samples have better layered structure. However, because the similarity of the d-spacings between starting Na-M and Ti-PILC-H samples, it is clear that there is no or very few titania pillars in between these short ranged overlayed silicate platelets. Such results are caused by hydrothermal treatment, which is helpful for getting better crystallized products. It is expected that during hydrothermal treatment, the titania fine particles are undergoing further agglomeration, thus generate bigger pillars and more interlayer spaces, which are filled by adsorbed water only. During drying, water starts to evaporate, leading to the collapse of the interlayer gaps and similar d-spacing as that of Na-M. Furthermore, the much less intensity of d (001) peak for Ti-PILC-H compared with Na-M also indicates that apart from such fully collapsed layered structure, more clay layers are in random orientation, similar to that of Ti-PILC samples and the dispersion of these bigger titania particles over silicate layers is good.

From Figure 1, it is also seen that for Ti-PILC-H samples, there is a broad but clear peak centred at 29.4°, which is ascribed to anatase phase. Such a peak is observed for Ti-PILC-H samples prepared at all temperatures, 100°C, 300°C and 500°C, with slightly increased intensity. However, for Ti-PILC samples, this peak is hardly detected. Even for Ti-PILC (500°C), the anatase peak is very weak. Such results confirm the above discussion, that hydrothermal treatment facilitates anatase crystallization, leading to bigger and better anatase crystallites.

Table 1 Surface areas of Na-M, PILCs and Cu/PILCs samples.

Sample	S_{total} (m^2/g)	$S_{external}$ (m^2/g)	Sample	S_{total} (m^2/g)	$S_{external}$ (m^2/g)
Na-M	51.5	26.7			
Ti-PILC	217	157	Ti-PILC-H	173	173
Cu/Ti-PILC-1	194	166	Cu/Ti-PILC-H-1	185	185
Cu/Ti-PILC-2	210	161	Cu/Ti-PILC-H-2	159	159
Cu/Ti-PILC-3	181	135	Cu/Ti-PILC-H-3	110	110

The nitrogen ads/desorption isotherms for the starting Na-M, Ti-PILC and Ti-PILC-H supports are shown in Figure 2, and the calculated total surface area (S_{total}) and external surface area ($S_{external}$) are listed in Table 1. From Figure 2, it is seen that compared with Na-M, a dramatic increase in the nitrogen adsorption amount is achieved after pillaring process. In addition, for Ti-PILC, such increase is more in the lower P/P_0 range, while for Ti-PILC-H, the enhancement happens more in the higher P/P_0 range. This difference suggests that Ti-PILC has larger microporous surface compared with Ti-PILC-H, which agrees well with the XRD analysis. Since Ti-PILC-H has better and bigger crystallized anatase, the pores generated between silicate layers and anatase particles are larger, which means surface area is mainly contributed from external / mesopore surface. In addition, the smaller total surface area of Ti-

PILC-H than that of Ti-PILC also proves that collapse of clay layers is more serious in Ti-PILC-H sample.

In summary, both Ti-PILC and Ti-PILC-H samples show much more disordered pillared structure compared normal PILC materials, e.g. Al-PILC, which has been studied extensively [2]. On the other hand, high surface area is obtained, which indicates the good dispersion of titania particles over the silicate layers. The main difference between Ti-PILC and Ti-PILC-H is in the extend of the titania crystallinity. Compared with Ti-PILC, Ti-PILC-H has better anatase crystallites, thus smaller surface area. The effect of such difference on copper loading will be discussed later.

3.2 Effects of Synthesis Conditions on Copper Loading

Table 1 also lists the surface area of different copper loaded PILC samples and their nitrogen ads/desorption isotherms are illustrated in Figure 2, using Cu/Ti-PILC-2 and Cu/Ti-PILC-H-2 as representatives. It can be seen that the shape of the isotherms does not change after copper loading and the calculated surface area shows a decreasing value with increasing concentration of the starting $Cu(NO_3)_2$ solution. Nevertheless, such a decrease is not critical and the final surface area is still above $150 m^2/g$ and $100 m^2/g$ for Cu/Ti-PILC series and Cu/Ti-PILC-H series, respectively.

The XRD patterns for these copper loaded samples, before and after calcination, are shown in Figure 3.

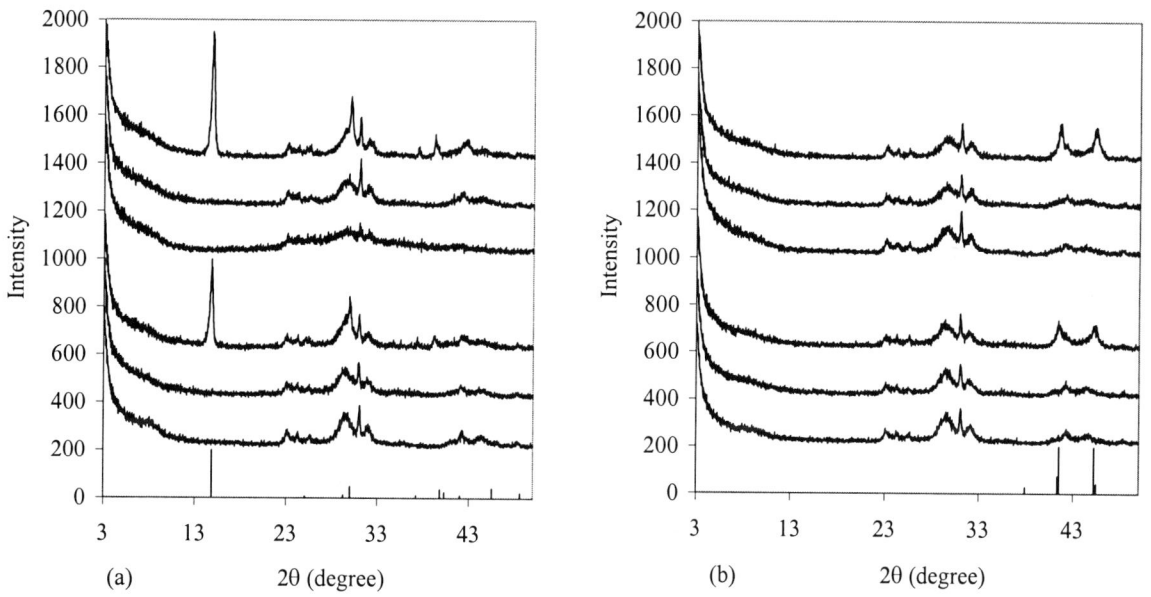

Figure 3 XRD patterns of Cu/Ti-PILCs and Cu/Ti-PILC-Hs. (a) 100°C (b) 300°C
From top: Cu/Ti-PILC-3; Cu/Ti-PILC-2; Cu/Ti-PILC-1; Cu/Ti-PILC-H-3; Cu/Ti-PILC-H-2; Cu/Ti-PILC-H-1; standard pattern for gerhardtite phase (a) or copper oxide phase (b).

From Figure 3, it is seen that copper loaded samples show similar XRD patterns as those of Ti-PILC and Ti-PILC-H supports, except Cu/Ti-PILC-3 and Cu/Ti-PILC-H-3. For these two samples, before calcination (Figure 3 (a)), they both have an intensive peak at 14.9° with a strong secondary diffraction peak at 30.0°, which is overlaid with anatase peak. These additional peaks match well with gerhardtite phase ($Cu_2NO_3(OH)_3$). After calcination at 300°C (Figure 3 (b)), the characteristic gerhardtite peaks disappear and are replaced by two broad peaks centred at 41.5° and 45.3°, which are ascribed to copper oxide (CuO) phase. Therefore, it is clear that during the copper loading process, under high Cu/support ratio, copper is firstly deposited on Ti-PILC or Ti-PILC-H supports as gerhardtite phase, which transforms into copper oxide phase upon calcination. The absence of the above peaks in samples prepared from the lower Cu/support ratios might be caused by not enough amount of copper or good dispersion of copper over supports. It is, thus, hard to determine the presence of copper and the state of the copper phase in these samples and techniques other than XRD were used to provide further information.

Table 2 shows the concentration of copper analysed by two methods, XPS and ICP-AES. The relation between the starting Cu/support ratios and the final deposited copper amounts is presented in Figure 4.

Table 2 Copper concentrations in the samples.

Sample	C_{Cu} (wt%) (XPS)	C_{Cu} (wt%) (ICP-AES)
Cu/Ti-PILC-1	4.87	1.10
Cu/Ti-PILC-2	10.84	3.23
Cu/Ti-PILC-3	26.29	10.52
Cu/Ti-PILC-H-1	4.86	1.20
Cu/Ti-PILC-H-2	9.80	3.60
Cu/Ti-PILC-H-3	20.91	10.95

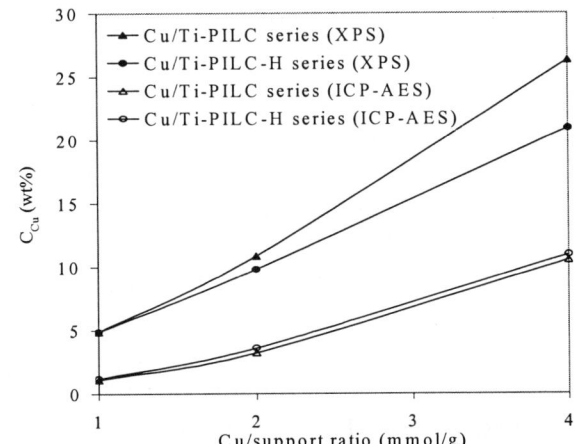

Figure 4 Concentrations of copper obtained from different methods.

It is clear that the copper loading amount is proportional to the starting $Cu(NO_3)_2$ concentration. The ICP-AES results show that there is little difference in the copper concentration between Cu/Ti-PILC and Cu/Ti-PILC-H series. Treating solid samples by concentrated nitric acid followed by ICP-AES analysis is a common method to measure the bulk copper concentration. It is believed that under such severe treatment, all the copper content will dissolve into the solution. However, what has been found in this study is that ICP-AES results always show much lower copper concentrations than XPS results. As mentioned before in experimental section, for clay samples, the XPS results reflect the bulk element concentrations. The disagreement between the two techniques in copper analysis suggests that the loaded copper is quite stable and even after washed by concentrated nitric acid, there is still certain amount of copper retained in the sample. The solid samples after treated by nitric acid were collected and washed by de-ionised water and dried. The XPS analysis of this sample shows 3.14wt% of copper concentration, which supports the above discussion.

Another experiment was also conducted to test the stability of these loaded copper species. In this case, samples with the highest copper amount, Cu/Ti-PILC-3 and Cu/Ti-PILC-H-3, were added to HCl solution at pH of 3. The mild acid condition was chosen because the catalytic reaction is normally conducted at pH of 4-5. The suspension was kept on stirring for two days followed by separation, washing and drying. Thus obtained samples were compared with the original samples by XRD analysis and the results are shown in Figure 5. It can be seen that after mild acid washing, the XRD patterns are very similar to those of before acid washing. The peak positions remain same. In addition, for copper oxide phase, there is no decrease in the peak intensity. On the contrary, a slight increase is observed for both samples after acid washing and the copper oxide peaks are also sharper. Therefore, the loaded copper species are stable under mild acid condition.

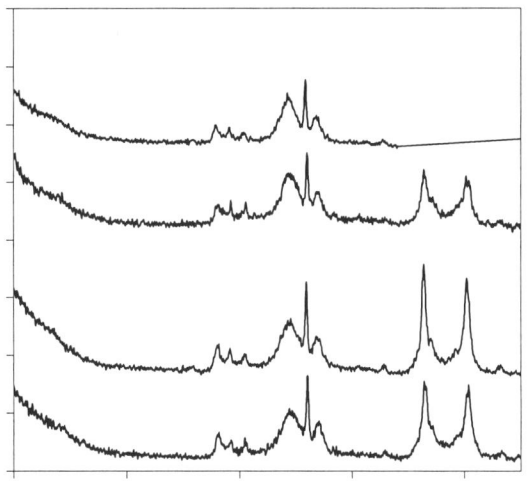 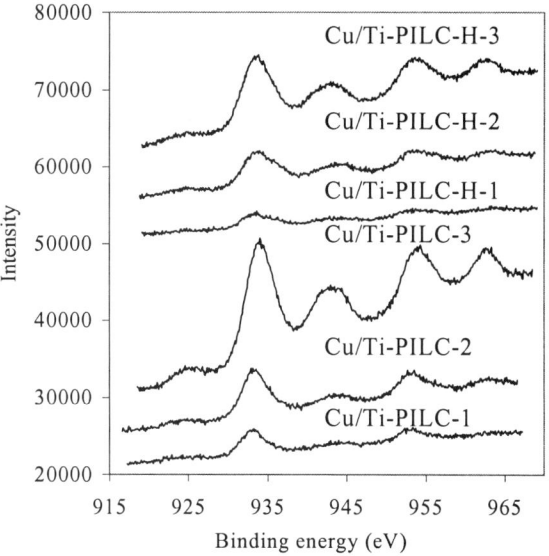

Figure 6 XPS diagram of copper loaded samples.

From XPS results in Table 2 and Figure 4, it is also seen that Ti-PILC tends to have more copper loading than Ti-PILC-H and the higher the starting Cu/support ratios, the larger the difference in the final copper concentration between Cu/Ti-PILC and Cu/Ti-PILC-H samples. Considering the higher anatase crystallinity but lower surface area of Ti-PILC-H than that of Ti-PILC, it can be concluded that large surface area is more important for copper loading.

As mentioned before, XRD results (Figure 3) show that only samples synthesized from the highest Cu/support ratio have peaks for copper phase. From Table 2, it is seen that the bulk copper concentrations in both series of the samples are quite high. Therefore, the reason for the absence of copper phase in XRD patterns should be the high dispersion of copper over PILC supports. It is known, from Figure 3, that for samples with the highest copper concentrations, copper exists as copper oxide phase. For samples with lower copper concentrations, the state of the deposited copper phase can be determined through XPS study, which is illustrated in Figure 6. It is seen that for all the samples, there are two main peaks centred at 933eV ($2p_{3/2}$) and 953eV ($2p_{1/2}$), coupled with two shake-up peaks centred at 943eV and 962eV. These are characteristic peaks for Cu in Cu (II) state. If Cu is in Cu (0) or Cu (I) state, there should be only two bands, $2p_{3/2}$ and $2p_{1/2}$. Since the samples were calcined at 300°C, it is expected that the copper is in CuO phase. Therefore, for samples prepared from lower Cu/support ratios, copper is deposited as well dispersed copper oxide phase.

4. CONCLUSIONS

In this study, two types of PILCs, Ti-PILC and Ti-PILC-H, have been synthesized and applied as supports for copper loading. Both materials show randomly pillared structure with high surface area, which is mainly contributed from external surface. Moreover, compared with Ti-PILC, Ti-PILC-H shows better crystallinity of anatase phase but lower surface area. It is found that the latter is more related to the copper loading amount, therefore, Ti-PILC has higher copper loading rate than Ti-PILC-H. For both supports, the final copper concentration is proportional to the starting Cu/support ratios and the loaded copper phases are well distributed over PILC supports. Only samples with the highest copper concentrations have copper phases been detected from XRD analysis. Based on XRD and XPS results, it is expected that the copper species tend to firstly deposit in gerhardtite phase, which is transformed into copper oxide phase upon calcination. More importantly, these supported copper oxides are found to be stable under mild acid condition. Therefore, combining their high surface areas and good dispersions of copper oxides, the synthesized materials would have great potential in the applications such as catalytic wet air / hydrogen peroxide oxidation process for wastewater treatment.

REFERENCES

1. T. J. Pinnavaia, Science, 220 (1983) 365.
2. J. T. Kloprogge, J. Porous Mater., 5 (1998) 5.
3. G. W. Brindley and R. E. Sempels, Clay Miner., 12 (1977) 229.
4. S. Yamanaka and G. W. Brindley, Clays Clay Miner., 26 (1978) 21.
5. D. E. Q. Vaughan, Catal. Today, 2 (1988) 187.
6. M. L. Occelli, in Proceedings of the International Clay Conference, L. G. Schultz, H. van Olphen and F. A. Mumpton (eds.), Denver: The Clay Minerals Society, 1985.
7. H. Auer and H. Hofmann, Appl. Catal. A, 97 (1993) 23.
8. A. Geatti, M. Lenarda, L. Storaro, R. Ganzerla and M. Perissinotto, J. Mol. Catal. A: Chem., 121 (1997) 111.
9. A. Gil, L. M. Gandia and M. A. Vicente, Catal. Rev. Sci. Eng., 42 (2000) 145.
10. D. T. B. Tennakoon, W. Jones and J. M. Thomas, J. Chem. Soc. Faraday Trans. 1, 82 (1986) 3081.
11. C. Philippopoulos, N. Gangas and N. Papayannakos, J. Mater. Sci. Lett., 15 (1996) 1940.
12. R. T. Yang, N. Tharappiwattananon and R. Q. Long, Appl. Catal. B, 19 (1998) 289.
13. R. Q. Long and R. T. Yang, Appl. Catal. B, 27 (2000) 87.
14. H. Yoneyama, S. Haga and S. Yamanaka, J. Phys. Chem., 93 (1989) 4833.
15. Z. Ding, H. Y. Zhu, G. Q. Lu and P. F. Greenfield, J. Colloid and Interface Sci., 209 (1999) 193.
16. C. Ooka, S. Akita, Y. Ohashi, T. Horiuchi, K. Suzuki, S. Komai, H. Yoshida and T. Hattori, J. Mater. Chem., 9 (1999) 2943.
17. J. Barrault, C. Bouchoule, K. Echachoui, N. Frini-Srasra, M. Trabelsi and F. Bergaya, Appl. Catal. B, 15 (1998) 269.
18. J. Barrault, M. Abdellaoui, C. Bouchoule, A. Majeste, J. M. Tatibouet, A. Louloudi, N. Papayannakos, and N. H. Gangas, Appl. Catal. B, 27 (2000) L225.

ACKNOWLEDGEMENTS

The authors are grateful for the financial support from ARC and Integrated Mineral Technology Ltd., and the infrastructure of the Centre for Instrumental and Developmental Chemistry in Queensland University of Technology.

Synthesis and characterization of SiO$_2$-Cr$_2$O$_3$ pillared montmorillonites

E.M. Farfán -Torres, A. G. Mercado, E. L. Sham[a] and M. Blesa[b]

[a] INIQUI-CONICET, Facultad de Ingeniería, Universidad Nacional de Salta, 4400-Salta, Argentina.

[b] CNEA, Unidad de Actividad Química, Avda. de los Constituyentes y Gral. Paz, Buenos Aires Argentina.

SiO$_2$-Cr$_2$O$_3$ pillared Argentinian montmorillonites were prepared by intercalation of oligosilsesquioxanes doped with Cr(III) ions. The basal spacing developed by intercalated materials was 17.0 Å. After calcination at 500°C a broad, low intensity 001 line at 15.30 Å was observed showing that long-range ordering along the c axis was loss. However a fraction of the material was still expanded as shown by the specific surface area of 200 m^2/g measured under this conditions.

1. INTRODUCTION

Although clays were useful for many applications, like catalysis, adsorption, etc., they had one main disadvantage: their lack of permanent porosity. In order to overcome this restriction many research efforts have been made looking for a way to permanently prop open the clay layers in a permanent way the clay layers. The intercalation of large inorganic cations between the layers of clays forms thermally stable microporous solids and overcomes this restriction (1-5). These new intercalate materials, called pillared clays (PILCs), are usually prepared by exchanging smectites with bulky metal polyhydroxy cations or positively charged colloidal particles.

Chromia pillared clays are materials of great catalyst interest because they exhibit high activity and selectivity in cyclohexane dehydrogenation, hydrocracking of n-decane (6-8), as well as toluene disproportionation and dehydrogenation of cumene (8). However these PILCs are unstable when calcined in air at elevated temperatures, because chromium is oxidised in air to produce more mobile species (6, 9-12). A promising alternative approach to thermal stable chromium containing pillared clays is doping with Cr(III) the pillar structure of Al$_{13}$ ions (13) or silica sols (14).

With the aim of preparing new microporous materials useful as catalysts and/or adsorbents, we have studied the synthesis of SiO$_2$-Cr$_2$O$_3$ pillared montmorillonites using oligosilsesquioxanes, as an alternative to SiO$_2$ sols, modified with Cr (III) cations.

2. EXPERIMENTAL

2.1 Preparation of the pillared clays

Two Argentinian sodium exchanged montmorillonites, Na-PESP and Na-PJ, from Minmarco Co. (particles < 2μm) with a cation exchange capacity of 94 and 130 meq/100g of clay were the starting materials. Aqueous 1.7-1.8 wt % pre-swelled clay suspensions were used for the intercalation reaction.

Two different intercalation methods were used. The first one, follows the technique reported by Han Y-S. et al. (14). In this case a silica sol solution was prepared by mixing silicon tetraethoxide (TEOS, $Si(OC_2H_5)_4$) with 2 M HCl and ethanol. Then a 0.25 M Cr^{3+} aqueous solution was dropwise added to the silica sol, and the pH adjusted to 2.1 with 0.2 M NaOH. This solution was kept at 85 °C during 1 h prior to mix with the clay suspension with a molar ratio of Si/Cr/CEC = 50/5/1. The obtained suspension was allow to react under stirring for 3 h at 85°C, and then the products were separated by centrifugation, washed several times with ethanol and water, and freeze dried. The dried samples were calcined in air at 100-800 °C.

For the second method 14.5 mL of 3-amino-propyl-triethoxysilane (APTEOS) were added to 4.0 mL of ethanol and 1.1 g of water and 3.3 mL of 2 M HCl were added dropwise and the resulting solution was stirred for 2 h at room temperature. After this a 1M aqueous $Cr(NO_3)_3$ solution was added and the final pH adjusted to 6.9 – 7.0 with 2 M HCl. Then the mix was reacted 2 h at 60°C prior to the addition to the clay suspension with a molar ratio of Si/Cr/CEC = 35/5/1. This suspension was stirred from 24 to 96 hours at 60 °C before products were separated and treated as in the first method.

Table 1 reports the details of preparation for the different samples.

Table 1.

SiO_2-Cr_2O_3 pillared clays, Si/Cr/CEC ratio, silica precursor, temperature and time of reaction.

Sample	Silica Precursor	Si/ Cr/ CEC ratio	Temperature of reaction (°C)	Time of reaction (h)
PESP-Si-Cr-01	TEOS	50/5/1	85	3
PESP-Si-Cr-03	APTEOS	35/5/1	60	96
PESP-Si-Cr-04	APTEOS	35/5/1	60	24
PESP-Si-Cr-05	APTEOS	35/5/1	60	48
PJ-Si-Cr-06	APTEOS	35/5/1	60	24

2.2 Characterization Techniques

X-ray diffraction (XRD) patterns were collected in a Rigaku diffractometer Dmax – II C using Cu K α radiation (λ= 0.1549 nm), using oriented films and powdered samples.

Fourier Transform Infrared (FTIR) spectra were recorded between 4000 and 400 cm-1 with a Bruker IFS 88 spectrometer. The samples were previously dried at 60°C overnight, crushed, mixed with KBr powder, and pressed into wafers. A typical wafer concentration was 10 mg of clay in 100 mg of KBr.

Simultaneous thermogravimetric (TG) and differential thermal analyses (DTA) were carried out with a Rigaku Thermoflex apparatus in air under dynamic conditions and a heating rate of 10 K min^{-1}.

Specific surface areas were determined at 77 K on samples degassed at 373 K for 2 hours with a Micromeritics Flow Sorb II 2300 single point apparatus.

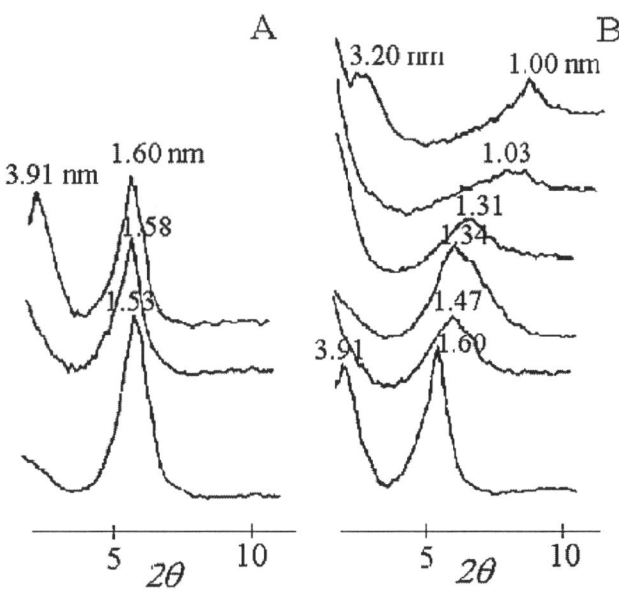

Fig.1. 001 XRD patterns of PESP-Si-Cr-01 clay:(A) Evolution with the number of washings: (a) without wash; (b) with 4 washes; (c) at constant conductivity; (B) Changes with the calcination temperature: (a) freeze dried; (b) 100°C; (c) 200°C; (d) 300°C; (e) 400°C and (f) 500°C.

3. RESULTS

The incorporation of Cr (III) species to the SiO_2 network before pillaring was confirmed by FTIR studies. For the two types of pillar precursors the bands characteristics of pure SiO_2 were shifted to lower wavenumbers as a result of consumption of free silanol groups. This was related with the coordination of these groups with chromium.

The intercalation of Na-PESP with Cr-modified sol solutions was accompanied by a shift of the 001 spacing from 12.44 to 1.53 nm (Fig. 1.A-a) for PESP-Si-Cr-01 before washing. After washing to constant conductivity, two intense diffraction lines at 3.91 nm and 1.60 nm were observed (Fig.1.A-c). After heat treatment the basal spacing decreases gradually and two broad lines were observed at 3.19 and 0.99 nm for oriented samples calcined at 500°C (Fig.1.B). The powdered random oriented samples show amorphous diffraction patterns indicating a poorly order stacking of layers.

In the case of samples prepared from APTEOS derived precursors the washing of the intercalated clay produces very few changes in the basal spacing, which shifted from 18.00 to 17.00 Å. The time of the intercalation reaction have some influence on the developed basal spacing. Higher values were obtained for samples prepared with times of reaction of 48 h of intercalation reaction (Fig.2.A).

After calcination the 001 line of PESP-Si-Cr-05 becomes broader and shifted to 16.00 Å at 300°C (Fig.2.B). For higher calcination temperatures the 001 line was very broad and disappears at 500°C showing a loss of long-range ordering along the c axis.

Simultaneously with the basal spacing changes the specific surface areas of the pillared clays calcined at 500°C increased from 50 m^2/g for the initial PESP-montmorillonite to 230

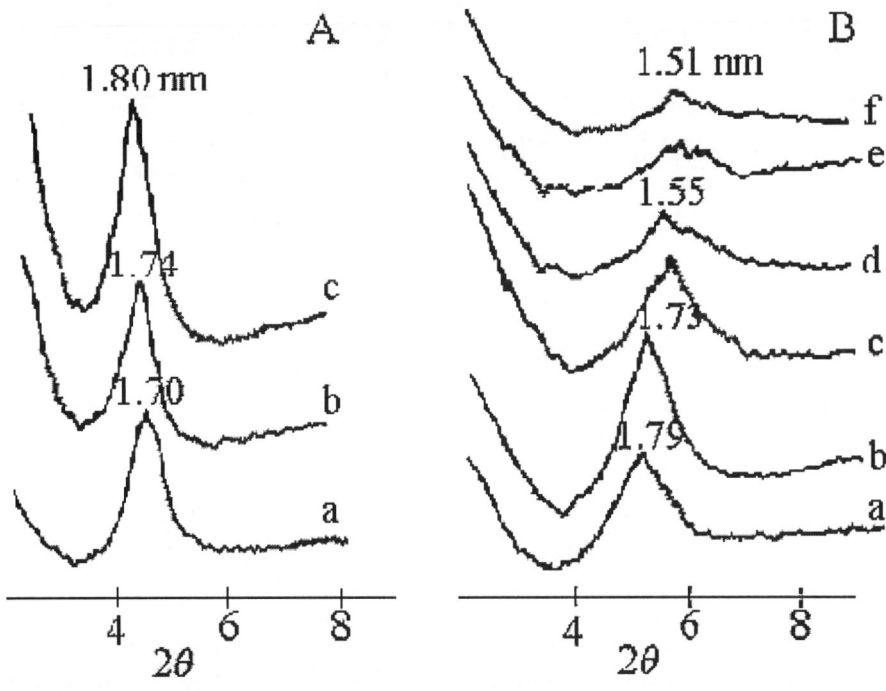

Fig.2. 001 XRD patterns of Na-PESP clay pillared with Cr(III)-doped-APTEOS:(A) Evolution with the time of reaction: after (a) 96 h; (b) 24 h; (c) 48 h; (B) Changes with the calcination temperature: (a) freeze dried; (b) 100°C; (c) 200°C; (d) 300°C; (e) 400°C and (f) 500°C.

and 200 m^2/g for PESP-Si-Cr-01 and PESP-Si-Cr-03 respectively. These results show that the two-intercalation methods lead to the formation of materials with very similar specific areas.

However the specific surface area changes of pillared clays derived from TEOS-sol-modified precursors were less marked than in the case of APTEOS derived intercalated clays as can be seen in Fig. 3.

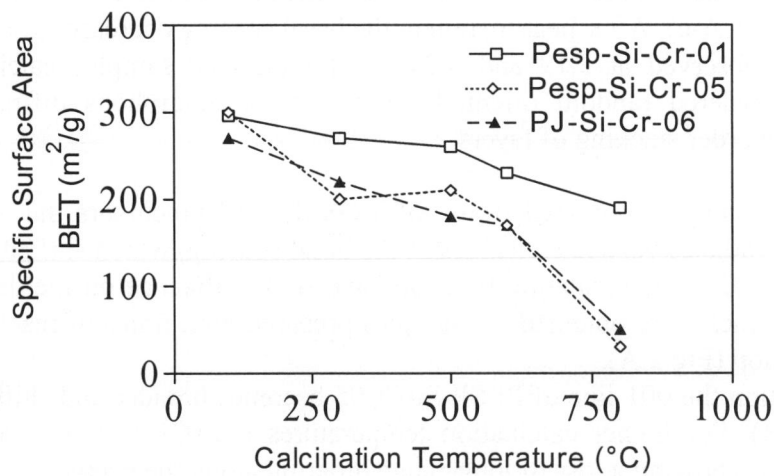

Fig.3. Thermal effect on the specific surface area of pillared SiO_2-Cr_2O_3 pillared clays.

TG and DTA curves for freeze dried samples of Na-PESP-montmorillonite, PESP-Si-Cr-01 and PESP-Si-Cr-03 materials are presented at Fig. 4. The large endothermic peak around 90°C observed for Na-PESP-monmorillonite was produced by dehydration of the clay (water intercalated between layers) (15). This peak moved to 100-120 °C for samples PESP-Si-Cr-01 and PESP-Si-Cr-03 respectively. At the same time an increase in the weight loss percentage was observed. These changes could be related with the dehydration of intercalated species (16). The second endothermic peak at 676 °C detected for the clay before pillaring was produced by the OH loss from the smectite layer. After pillaring this peak was shifted to 658°C for PESP-Si-Cr-01 and to 662 °C for PESP-Si-Cr-03 clays. According to Volzone (11), the position of this peak depends on the degree of order in the stacking of the layers but also as the intercalation process takes place in acidic medium, H^+ are simultaneously exchanged and could increase the rate of dehydroxylation reaction (16).

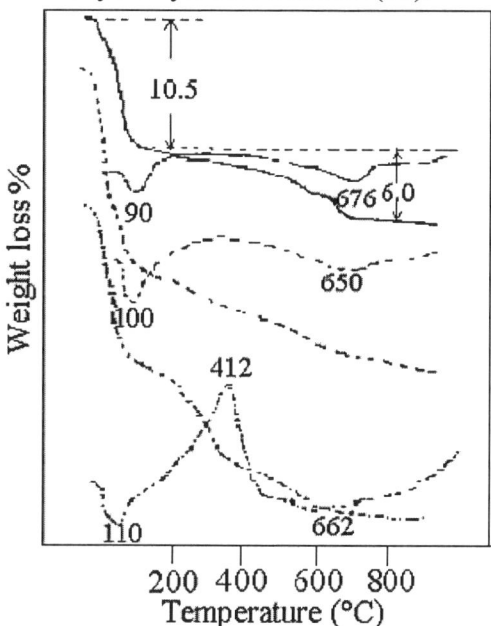

Fig. 4. TG and DTA curves for: (A) Na-PESP, (B) PESP-Si-Cr-01, (C) PESP-Si-Cr-03

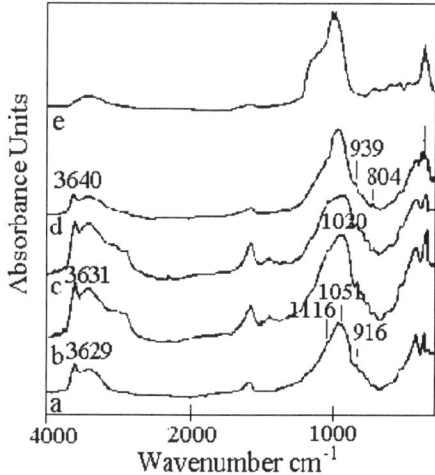

Fig.5. Infrared spectra of: (a) Na-PESP; and PESP-Si-Cr-01: (b) freeze-dried and calcined at (c) 100°C, (d) 500°C, (e) 800°C.

basal oxygens. The band at 916 cm^{-1} was related to bending vibrations of Al-(OH)-Al. The shoulders at 879, 842 and 798 cm^{-1} are attributed to bending vibrations of Al-(OH)-Fe(III), Fe(II)-(OH)-Fe(III), Al-(OH)-Mg and Mg-(OH)-Mg groups present in the octahedral layers. Bands in the range of 780 to 430 cm^{-1} were related with Si-O bending deformations.

For the samples calcined at 800 °C the bands at 3629 and 916 cm^{-1} were absent, and the band at 1051 cm^{-1} shift to 1041 cm^{-1} and becomes sharper. Shoulders at 879 and 842 cm^{-1} also disappear. These modifications were attributes to the dehydroxylatyon of the clay layers with distortions of the Si-O-Al bonds due to the structural changes that take place in the clay sheet.

The FTIR spectra of the PESP-Si-Cr-01 and PESP-Si-Cr-03 pillared clays are reported in Fig. 5 and 6. In the case of freeze dried PESP-Si-Cr-01 the shoulder at 1116 cm^{-1} was no longer resolved and the band at 1051 cm^{-1} shift to 1086 cm^{-1} becoming broader. These changes could be related with an overlapping of the Si-O-Si stretching of the clay layers with the Si-O vibration of intercalated silica species. The absorption at 798 cm^{-1} has not shifted but becomes a well-resolved and sharp band (Fig.8b). This could be indicative of the presence of free silica (17). After calcination at 500 °C the clay layers were dehydroxylated as confirmed by the absence of the band at 3629 cm^{-1}, the shifting to 1063 cm^{-1} of Si-O-Si stretching, the almost absent shoulder at 919 cm^{-1}. Together with these changes an increase of the intensity of the band at 800 cm^{-1} was also observed confirming the presence of free silica for this clay. At 800 °C the dehydroxylation process was completed.

FTIR spectrum of freeze-dried PESP-Si-Cr-03 pillared clay show that the absorption at 3631 cm^{-1} and the shoulder at 1116 cm^{-1} remain unmodified and the Si-O-Si stretching mode shift to 1051 cm^{-1} to 1020 cm^{-1}. Other bands related to OH bending vibrations of octahedral OH were not modified showing that the intercalation process takes place without alteration of the clay layer structure (Fig. 6b). After calcination at 500 °C the OH stretching of the clay lattice shifted to 3640 cm^{-1} and the shoulder at 916 cm^{-1} was shifted to 939 cm^{-1} being bad resolved. Only the absorption at 804 cm^{-1} belonging to Mg-(OH)-Mg bending was still observable indicating that the intercalated SiO_2-Cr_2O_3 species were still in interaction with the clay layers, and the clay was not yet dehydroxylated. Calcination at 800 °C produced the dehydroxylation of the clay as can bee seen in Fig. 6d.

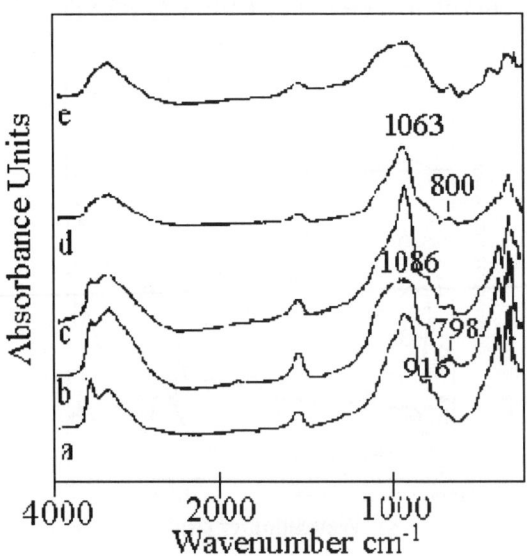

Fig.6. Infrared spectra of: (a) Na-PESP; and PESP-Si-Cr-03: (b) freeze-dried and calcined at (c) 100°C, (d) 500°C, (e) 800°C.

4. DISCUSSION

Pillaring increased the specific surface areas of the original montmorillonites from 50 to 300 m^2/g for the two intercalation methods used. This increase in the surface area is consistent with the expansion of the basal spacing observed by XRD (Fig.1B). However the thermal behavior of TEOS and APTEOS derived pillared montmorillonites is quite different.

Upon calcination in dry air the surface area of PESP-Si-Cr-01, TEOS derived pillared clay, decrease slowly, but at 800°C it is still a 63% of the initial value. In spite of this, XRD patterns of oriented films shows that a fraction of the clay was expanded at 32.0 Å and other important part was collapsed to the initial spacing value. Furthermore XRD patterns of powdered samples show that it was an amorphous solid. FTIR results confirm that clay layers were dehydroxylated after heating at 500°C and also confirm the presence of free SiO_2 species in the solids. This SiO_2 could come from free particles released from the clay interlayers during calcination due to weak interactions with the montmorillonite layers. The presence of this free SiO_2 could also explain the high surface area measured for the collapsed pillared structure. Further studies are necessary to confirm this hypothesis.

Calcination of APTEOS derived pillared clays produce a steady decrease of the surface area, which is consistent with basal space changes observed in XRD powder patterns. At 600°C the solids become amorphous and for higher calcination temperatures the pillared structures were completely collapsed (Fig.3). The combustion of residual organic groups does not affect strongly the interlayer spacing but produces modifications in the structure of the pillared clay as indicated by the shifting of the absorption band at 916 and 798 cm^{-1}. These modifications are consistent with the cross-linking mechanism proposed by Tennakoon et. Al (18). These authors suggested that terminal OH-groups on the pillar condensed with the lattice OH-groups of the clay. In the case of APTEOS derived clays FTIR measurements confirm that SiO_2-Cr_2O_3 pillars become directly linked via oxygen to the aluminium or magnesium atoms in the octahedral part of the montmorillonites. Further calcination to temperatures higher than 600°C results in the collapsing of the pillared structure probably due to the oxidation of chromium to more mobile species.

In conclusion, the use of oligosilsesquioxanes doped with Cr(III) ions yield solids which present good thermal stabilities without formation of free SiO_2-Cr_2O_3 species. These chromiasilica pillared clays could be good catalysts because gallery heights are higher than 5 Å at 400°C which provide a good accessibility to the chromium centers for alcanes adsorption.

Acknowledgements

We wish to thank Dr. Nilda Menegatti for her technical support in XRD measurements. The partial support of this work by the CONICET trough PICT 143 and Consejo de Investigación-UNSa through Project N°717 is gratefully acknowledged.

REFERENCES

1. Vaughan D. E. W., Catalysis Today, 2 (1988), 187.
2. Figueras F., Catal. Rev. Sci. Eng., 30 (1988), 457.
3. Plee D., Schutz, A., Poncelet G. And Fripiat J. J., Catalysis by Acids and Bases, Elsevier, Amsterdam (1985), 343.
4. Farfán-Torres E. M. and Grange P., J. Chym. Phys., 87 (1990), 1547.
5. Molinard A., PhD Thesis, University of Atwerp, 1994.
6. Tzou, M-S. And Pinnavaia, T. J., Catal. Today, 2(1988), 243.

7. Sychev M., De Beer V.H.J. and Van Santen R.A., Microporous Materials, 8(1997), 255.
8. Sychev M., De Beer V.H.J. and Van Santen R.A., J. Catal., 168(1997) 245.
9. Othsuka K., Chem. Mater., 9(1997), 2039.
10. Pálinkó I., Lázár K., Kiricsi I., 410-411(1997), 547.
11. Volzone, C. Clays &Clay Miner., 43(1995), 377.
12. Volzone C. And Cesio A.M., Mat.Chem.Phys., 481(1997), 216.
13. Carrado K.A., Suib S.L., Skoularakis N.D. and Coughlin R.W., Inorg.Chem, 25(1986), 4217.
14. Han Y-S., Yamanka S., Choy J-H., Appl.Catal.A. Gen., 174(1998), 83.
15. Grim, R.E. and Kulbicki B., Am. Mineralog., 46(1961), 1329.
16. Tichit D., Fajula F. And Figueras F., Clays &Clay Miner., 36(1988), 369.
17. Jones S.D., Pitchard T.N. and Lander D.F., Mocroporous Materials, 3(1995), 419.
18. Tenakoon D.T.B., Jones W., Thomas J. M., J. Chem. Soc., Faraday Trans. I., 82 (1986), 3081.

Castor, cottonseed, and soybean oil bleaching by activated bentonites

E.L. Foletto[a], C. Volzone[b], A.F. Morgado[a] and L.M. Porto[a]

[a] Departamento de Engenharia Química e Engenharia de Alimentos, Universidade Federal de Santa Catarina, C. P. 476 - 88040-900, Florianópolis-SC, Brasil

[b] Centro de Tecnología de Recursos Minerales y Cerámica, C.C. 49, Cno Centenario y 506 (1897) M. B. Gonnet, Prov. Buenos Aires, Argentina

Acid-treated clays are employed industrially as catalysts, catalyst supports and adsorbents. A relatively new market for activated clay is in animal feed where it absorbs mycotoxins. In terms of consumption, the most important use of the activated clays is in the purification, decolorization and stabilization of vegetable oils. They are used to remove phospholipids, soaps, trace metals, organic compounds (for example, chlorophyll) and their degradation products, which impart undesirable colors to the edible oils. The adsorptive capacity of these materials is significantly increased by treatment with strong inorganic acids, particularly sulfuric and hydrochloric acid. In this work, two bentonite clays with different mineralogical compositions from Mendoza, Argentine, were activated with H_2SO_4 solutions of 4 and 8 N at 90 °C for 3.5 hours. These treatment effects under clay structural properties were tested by thermogravimetry, infrared spectrometry and chemical analysis. The activated samples were tested in order to verify their bleaching capacity for castor, cottonseed, and soybean oils and were compared to a standard commercial bleaching clay. Bleaching efficiency was strongly dependent on acid concentration used for clay activation. The treated samples have shown a bleaching capacity compared to the standard. The mineralogical composition of natural clays have has influenced the properties of treated clays (or activated clays) and these properties increased its bleaching capacity.

1. INTRODUCTION

Acid treatment of clays is widely used in both research and industrial applications. Acid activated bentonites already have been used as solid acid catalysts and catalyst supports for a number of organic reactions of considerable industrial interest [1-4]. It has also been used in the foodstuffs industry, sulphur production, forest and water conservation, in chemical industry, environmental protection, the paper industry [5,6] as well as for bleaching of vegetable oils [7-9]. The bleaching of vegetable oils for edible purposes involves the removal of a variety of impurities wich include phosphatides, fatty acids, gums, trace metals, etc. followed by decolorization. The resultant product is a stable oil of acceptable color. The bleaching capacity of bentonites is greatly increased by activation treatment. While some of these clays are naturally bleaching, some have to be treated with mineral acids [10-12]. Activation proceeds with partial dissolution of smectite and is characterized by an initial replacement of the interlayer cations by H^+, followed by dissolution of the octahedral and tetrahedral sheets with subsequent release of structural cations. Despite numerous studies, no definite relationship exists between the performance of the acid-activated clay and the composition or properties of the original clay. Hence, each clay has to be specifically activated and tested for its performance [13]. The preparation of acid-activated bentonite requires an extreme degree of control of variables, in order to obtain maximum bleaching power.

The present study deals with the acid activation of Argentine bentonites and testing of their bleaching capacity for castor, cottonseed, and soybean oil, taking into consideration the bleaching capacity of standard commercial bleaching clay.

2. EXPERIMENTAL

2.1 Clay

Two natural bentonites (named as K and W) from two different deposits of the Mendoza province, Argentine, were used as the starting materials. The mineralogical compositions were determined by Foletto *et al.* [14]. Smectite was the main clay mineral in both samples (35 % in K and 47 % in W). The K bentonite contained quartz and feldspar as impurities, whereas in W bentonite also gypsum and kaolinite were found. Their chemical compositions are listed in Table 1.

2.2 Acid treatment

The bentonite sample (40 g) was treated with 400 mL of sulfuric acid (analytical grade) 4 N or 8 N at 90 °C during 3.5 h in an instantly stirred glass reaction vessel with reflux. After the acid treatment, the sample was filtered and washed with distilled water until free of SO_4^{2-}. The samples were dried at 60 °C for 12 h and ground to pass through a 0.074 mm sieve. Activated samples were designated as K4, K8, W4 and W8 where the numerical value indicates the concentration of acid solution used for the treatment.

2.3 Determination of bleaching capacity

The bleaching capacity of the run activated samples was tested using castor, cottonseed, and soybean oil (previously neutralized with soda). The bleaching process was carried out under a vacuum of 450 mmHg at the constant temperature of 100 °C with a contact time of 30 min. Stirring and heating was carried out by means of on electric heating band wrapped and a mechanic stirrer. A clay dosage of 1 % w/w of oil was used in all cases. During the heating of the oil and throughout the experiment, a stream of N_2 was maintained above oil surface. The hot oil and clay mixture was filtered under vacuum and the color of the bleached oil was measured spectrophotometrically (WFJ525-W UV-visible spectrophotometer). The bleaching capacity of the clays was determined from the following equation:

$$\text{Bleaching capacity (\%)} = \frac{A_o - A}{A_o} \times 100$$

where A_o and A are the absorbance of neutral oil and bleached oil, respectively, at the maximum absorbance of the neutral oil (at wavelength 415, 405 and 450 nm for castor, cottonseed, and soybean oil, respectively). A commercial product, Tonsil Actisil 220FF, was used for comparison of the bleaching experiments.

2.3 Sample characterization

The structural changes of the Argentine acid bentonites were examined by means of thermogravimetric (TGA) and infrared spectrometric (IR) analyses as well as chemical analysis (by XRF). TGA was performed on a Netzsch STA 409 instrument at a heating rate of 10 $^\circ$C.min^{-1} under a flow of atmospheric air of 35 mL min^{-1}, in the range 25-1000 $^\circ$C. IR spectra were recorded in the region 4000-350 cm^{-1} in a Perkin-Elmer 16 PC spectrophotometer, using the KBr pellet technique. Elemental compositions of the samples were determined by X-ray fluorescence technique with a Philips PW 2400 spectrometer.

3. RESULTS AND DISCUSSION

Table 1 shows the chemical analysis of the bentonites after different acid concentration treatments. The interlayer cations of the bentonites were removed, nevertheless the remained Ca^{2+}, Na^+, K^+ are due to the presence of impurities presence such as micas and feldspars, which are both very insoluble in acid medium. Part of the Mg^{2+}, $Fe^{2+/3+}$ and Al^{3+} was removed from octahedral sheet, except Ti^{4+} cations. The increase of Si^{4+} and the reduction of the octahedral cations originated by an increase of the acid concentration of the treatment resulted in an increase of the Si/(Al+ Mg +Fe) ratio.

Table 1
Results of XRF analysis, expressed in oxide form (%).

Sample	SiO$_2$	Al$_2$O$_3$	Fe$_2$O$_3$	MgO	CaO	Na$_2$O	K$_2$O	TiO$_2$	MnO	r
K	69.53	15.58	3.49	1.56	0.65	2.70	1.23	0.55	0.09	3.00
K4	77.45	12.55	2.03	1.02	0.35	0.85	1.10	0.55	0.01	4.35
K8	78.51	10.32	1.59	0.85	0.32	0.88	1.04	0.54	0.01	5.37
W	65.18	17.51	4.53	1.18	1.57	2.24	0.86	0.42	0.08	2.56
W4	73.08	14.70	3.43	0.64	0.18	0.57	0.71	0.42	0.01	3.50
W8	75.87	13.06	2.86	0.59	0.18	0.59	0.71	0.42	0.01	4.12

r: $Si^{4+}/(Al^{3+}+Mg^{2+}+Fe^{2+/3+})$ ratio.

The dehydroxylation weight loss (%) of the samples was obtained from TG curves (Figure 1) in the range 450-750 °C. These values were taken into account for evaluating the attack of the octahedral sheets after acid treatment, as shown in Table 2. The octahedral sheet destruction was calculated as a relative reduction percentage of the weight loss, with respect to original sample. The treated K samples showed 14 to 19 % of destruction of the octahedral sheet, while the treated W samples showed 12 to 15 % of destruction. This showed that even at high acid concentration (i.e., concentration of 8 N), their structures were not totally destroyed.

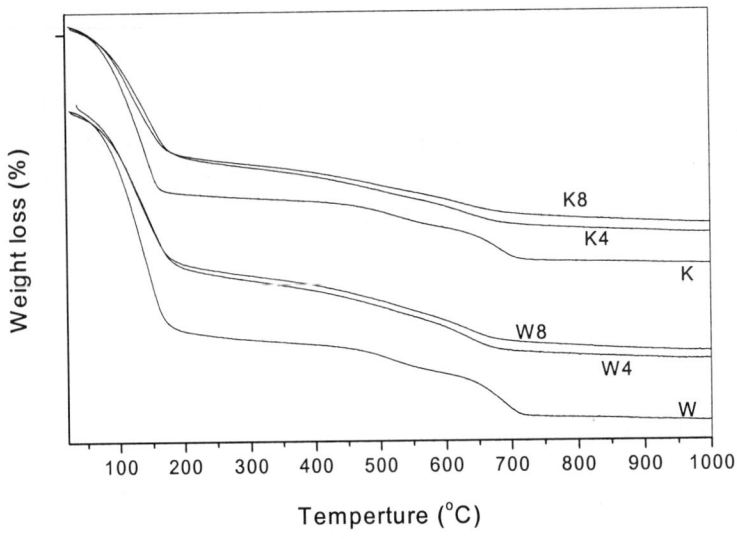

Figure 1. TGA curves of the natural and activated samples.

Table 2
Results obtained of curves TGA.

Sample	Weight loss range 450 - 750 °C	Octahedral sheet destruction (%)
K	2.69	0.00
K4	2.08	22.67
K8	1.80	33.08
W	3.28	0.00
W4	2.59	21.04
W8	2.44	25.61

Figure 2 shows the infrared spectra of the original bentonites and after acid treatments. In general, no important changes were observed, although the bands Si-O-Al (525 cm^{-1}) and Si-O-Si (465 cm^{-1}) were modified. The intensity ratio of both bands are shown in Table 3.

Table 3 presents the results of the characterizations of the natural and activated samples and bleaching of the soybean, cottonseed and castor oils. Figures 3, 4 and 5 show the correlation between the structural changes observed by different technical and bleaching capacity of the samples studied in this work.

The results of the analyses indicate that an increase of the acid concentration originates a bigger attack on the bentonite structure but, on the other hand, the percentage of attacks on the clay was bigger for the K that the W sample.

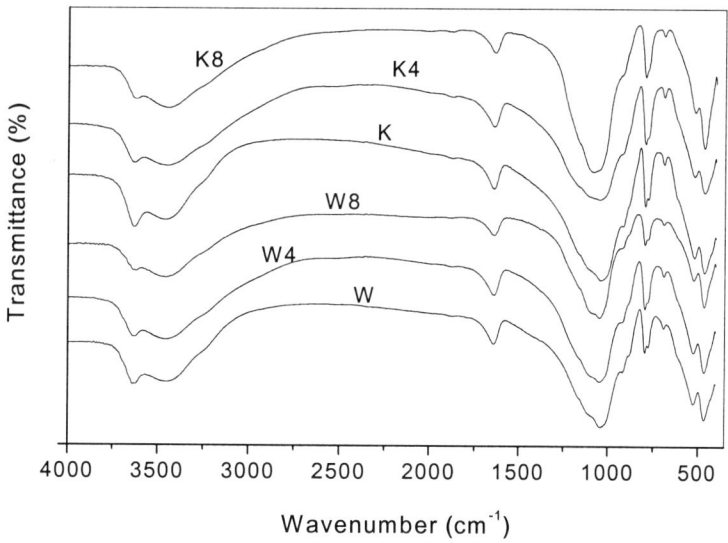

Figure 2. Infrared spectra of natural and activated samples.

Table 3
Structural changes and oil bleaching capacity of natural and activated samples.

Sample	IR [1]	AQ [2]	TG [3]	BC (%) Soybean [4]	BC (%) Cottonseed [5]	BC (%) Castor [6]
K	0.44	3.00	0.00	10.89	2.93	1.90
K4	0.26	4.35	22.67	88.55	13.73	30.16
K8	0.13	5.37	33.08	92.66	18.00	36.95
W	0.44	2.56	0.00	12.46	1.60	1.08
W4	0.34	3.50	21.04	86.25	10.93	29.34
W8	0.22	4.12	25.61	92.47	17.46	35.60

(1) Si-O-Al / Si-O-Si ratio (obtained from IR spectra, Figure 2)
(2) Si / (Al+Mg+Fe) ratio (obtained from chemical analysis, Table 1)
(3) Octahedral sheet destruction (%) (obtained from TGA curves, Table 2)
(4) Bleaching capacity (BC) in soybean oil
(5) Bleaching capacity (BC) in cottonseed oil
(6) Bleaching capacity (BC) in castor oil

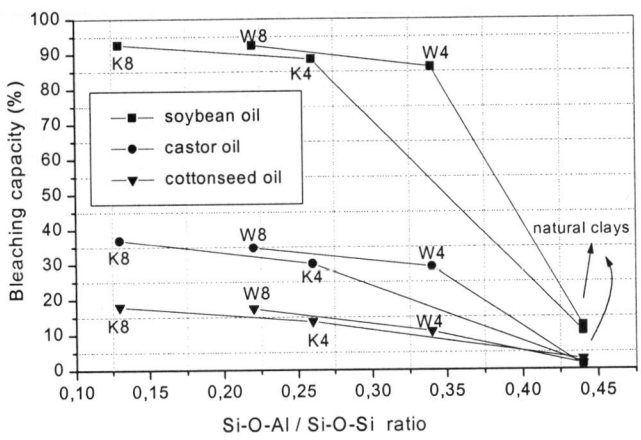

Figure 3. Bleaching capacity *vs.* Si-O-Al / Si-O-Si ratio.

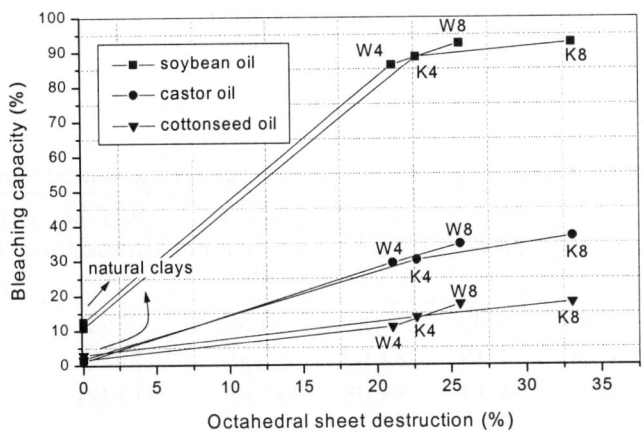

Figure 4. Bleaching capacity *vs.* octahedral sheet destruction.

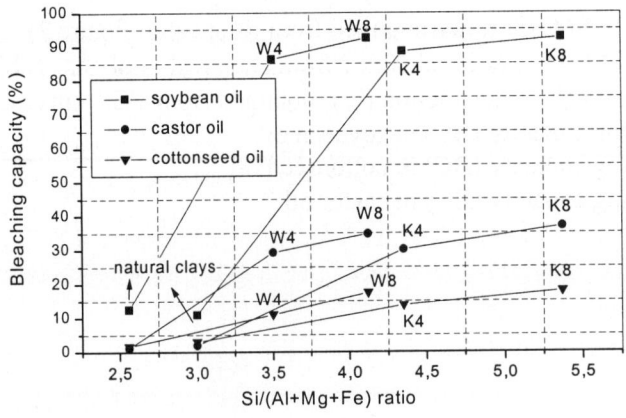

Figure 5. Bleaching capacity *vs.* Si / (Al+Mg+Fe) ratio.

The acid K bentonites showed a slightly superior bleaching capacity than the acid W bentonites and this behavior could be attributed to the higher acid attack of the K smectite structure than the W smectite. For comparison of the mineralogical compositions, the sample K has more MgO than sample W. It is seen that smectites with more MgO in their mineralogical compositions are more activated with mineral acids than those smectites with minor quantities [15]. An optimization in the acid treatment of the W bentonite probably would cause a higher attack in its structure and consequently would lead to a better adsorptive capacity, because this sample has higher clay content than K bentonite [14].

4. COMPARISON WITH COMMERCIAL BLEACHING CLAY

The best bleaching capacities of the activated bentonites obtained in this paper were compared with a commercial sample, as shown in Table 4. The retention of color impurities from soybean, cottonseed and castor oils by Tonsil (commercial sample) were similar to the acid bentonites prepared in this paper.

Table 4
Bleaching capacities of Tonsil (commercial) and prepared acid bentonites (K8 and W8).

Sample	BC (%) Soybean	BC (%) Cottonseed	BC (%) Castor
Tonsil Actisil 220FF	92.85	18.00	36.96
K8	92.66	18.00	36.95
W8	92.47	17.46	35.60

5. CONCLUSIONS

The sulphuric acid treatments on bentonites improved the bleaching of soybean, cottonseed and castor oils. The increase of the concentration of the acid used in the treatment improved the bleaching capacity. The acid treatment originated a structural change (observed by IR, TGA, and chemical analysis) in the bentonites and influenced directly the values of the oils bleaching. Bentonites treated with sulphuric acid at concentration of 8 N presented similar bleaching capacity for soybean, castor and cottonseed oils comparable with a commercial acid clay.

ACKNOWLEDGMENTS

The authors thank Susana Conconi and Dr. Luiz Fernando Dias Probst for determination of DTA/TGA and IR, respectively. We are grateful to Centro de Tecnologia em Cerâmica (Criciúma, Brazil) for determination of chemical analysis (XRF). The financial support of CAPES (scholarship provided to ELF) is also acknowledged.

REFERENCES

1. R. Mokaya, W. Jones, Pillared clays and pillared acid-activated clays: a study of physical, acidic, and catalytic properties, Journal of Catalysis, 153 (1995) 76-85.
2. C. Breen, F.D. Zahoor, J. Madejová, P. Komadel, Characterization and catalytic activity of acid-treated, size-fractionated smectites, J. Phys. Chem. B., v.101, n. 27 (1997) 5324-5331.
3. C. Volzone, O. Masini, N. Comelli, M. Grzona, E. Ponzi, M. Ponzi, M., Acción catalítica de arcillas en la obtención de canfeno, XI Jornadas Argentinas de Catálisis, San Luis, Argentina (1999) 319-321.
4. A. Valentini, E.L. Foletto, L.M. Porto, L.F.D. Probst, Isomerização do beta-pineno Catalisada por Argila Esmectítica Ativada, 23ª Reunião Anual da Sociedade Brasileira de Química, Poços de Caldas/MG, Brasil, vol.1, cód. CT-026 (resumo), 2000.
5. G.M. Clarke, Special clays, Ind. Miner., 216 (1985) 25-51.
6. M. O'Driscoll, Bentonite: overcapacity in need of markets, Ind. Miner., 250 (1988) 43-67.
7. S.C. Kheok, E.E. Lim, Mechanism of palm oil bleaching by montmorillonite clay activated at various acid concentrations, J. Am. Oil Chem. Soc., v. 59, n. 3 (1982) 129-131.
8. E. Srasra, F. Bergaya, H. van Damme, N.K. Ariguib, Surface properties of an activated bentonite: decolorisation of rape-seed oils, Appl. Clay Sci., 4, (1989) 411-421.
9. E.L. Foletto, C. Volzone, A.F. Morgado, L.M. Porto, Influêcia do tipo de ácido usado e da sua concentração na ativação de uma argila bentonítica, 14º Congresso Brasileiro de Engenharia e Ciência dos Materiais, São Pedro/SP, Brasil, p. 34 (resumo), 2000.
10. R.E. Grim, Applied clay mineralogy, McGraw Hill, New York, 1962.
11. A.J.C. Andersen, Refining of Oils and Fats for Edible Purposes, 2nd ed., Pergamon Press, London, 1962.
12. F.A., Norris, Bailey's Industrial Oil and Fat Products, 2nd ed., John Wiley & Sons, New York, 1982.
13. F.K. Hymore, Effects of some additives on performance of acid-activated clays in the bleaching of palm oil, Appl. Clay Sci., 10 (1996) 379-385.
14. E.L. Foletto, C. Volzone, A.F. Morgado, L.M. Porto, Aálise comparativa da ativação ácida de dois materiais argilosos com diferentes composições mineralógicas, VI Jornadas Argentinas de Tratamiento de Minerales, Salta, Argentina, p. 43-48, 2000.
15. C. Volzone, J. Ortiga, Retention gas O_2, CH_4 and CO_2 by acid smectites with and without thermal treatment, Journal Material Science, 35 (2000) 5291-5294.

OH-Al complexes on kaolinite surface and their effect on suspension properties

L. B. Garrido and C. Volzone

CETMIC (Centro de Tecnología de Recursos Minerales y Cerámica; CONICET-CIC-UNLP), C.C. 49 (1897) M. B. Gonnet, Provincia de Buenos Aires, ARGENTINA.

Different hydroxy aluminum species were adsorbed on a kaolin from partially hydrolyzed $AlCl_3$ solutions (0.03 M in Al) having NaOH/Al molar ratios between 0.5 and 2.5. Adsorptions were performed at 35.2 µmol Al/g kaolin at pH 4.5. The OH-Al kaolins powders were dispersed in water containing 0.3 g of Na hexametaphosphate per 100 g of kaolin to obtain a 50 wt% aqueous suspension. Stability and flow properties of suspensions of OH-Al samples with different exchangeable to non exchangeable Al ratios, which changed markedly with the NaOH/Al ratio of the solution, were compared. Low viscosity suspensions, which showed a slightly shear thickening behavior, were obtained after treating the kaolin with NaOH/Al ≥ 1.5 solutions. However, suspensions of OH-Al kaolin prepared from NaOH/Al<1.5 solutions exhibited a complex non Newtonian flow behavior and viscosity valued higher than that of the untreated sample.

1. INTRODUCTION

Rheological properties of kaolinite suspension are of interest for many industrial applications such as coating of paper, whiteware and sanitaryware production, etc. The processing steps like mixing, pumping or suspension consolidation are essential to control and improve the flow properties of concentrated suspensions.

Stability and rheology of concentrated suspensions depends largely on the chemistry of solid/solution interface. The surface chemistry of kaolinite can be significantly affected by formation of Al hydroxides or precipitation of Al complexes [1,2]. Jepson [3] indicated that OH-Al solution additions have been used for reducing the influence of smectites on the flow behavior of kaolin aqueous suspension.

However, hydrolysis of Al ions in aqueous solution is very complicated and variations in the composition of Al solution may directly affect the original surface and, in turn, the colloidal properties of soils and clay minerals.

Al ions form different Al hydroxylated species, precipitates and gels depending mainly on the OH/Al molar ratio, Al total concentration and aging time of the solution [4]. NMR data indicate that for 0.01 M $AlCl_3$ solutions having OH/Al= 0.5, 1 and 1.5, the amount of Al_{13} increases gradually from 10 to 35 and 60%, respectively. For OH/Al=2, Al_{13} is the prevailing species (90%) in the solution. The other ions recognized are mainly monomer and dimer species. Al_{13} polycations are isolated. For OH/Al = 2.5, the Al_{13} content decreases to 60% [4]. The decrease in Al_{13} ions was accompanied by an increase in aggregate polymers of $Al(OH)_3$-fragment structure. A small portion of this fraction appears to be submicron particulates of $Al(OH)_3$ [5].

In this work, kaolin powders containing different OH-Al species were prepared and dispersed in water to obtain 50 wt% suspensions using 0.3 wt% of Na hexametaphosphate HMP as dispersant. Rheological properties of the suspension at pH 5 were determined. The variation of dynamic mobility as a function of HMP additions was also measured.

2. MATERIALS AND METHODS

2.1. OH-Al solutions

The partially neutralized aluminum chloride solutions were prepared by following the method of Bottero et al. [4]. OH-Al solutions 0.03 M in Al were obtained as follows: 30 ml of 0.1M Al_3Cl solution were hydrolyzed with 70 ml of NaOH solutions of different concentrations to obtain NaOH/Al molar ratios varying in the range 0.5-2.5 (under N_2 atmosphere). The base addition rate was 1 ml min^{-1}. The hydrolysis time at room temperature was 1 day. Final pH of the solutions ranged from 3.7 to 4.8.

2.2. Preparation of the OH-Al kaolin

The kaolin used was from Chubut, Argentina was mainly composed of kaolinite with illite and quartz as impurities. Ethyleneglycol saturation showed that swelling minerals were absent. The particle size distribution indicated that 83 % of the material was finer than 2 μm and the average particle size d_{50} was 0.7 μm. Specific surface area BET was 20.6 m^2/g.

The OH-Al kaolin was prepared as follows: 47 ml of a 0.03M OH-Al solution was added dropwise to 20 wt% aqueous suspension of kaolin (40 g were dispersed in 200 g of water) while stirring. HCl was used to adjust the pH to 4.5, prior to the addition of the OH-Al solution.
The Al/clay ratio was 35.2 μmol Al/g dry clay, which was nearly equivalent to the addition of 0.1 g of Al per 100 g of clay. Stirring was maintained for 20 min. Then, the solid was separated by centrifugation and washed with 250 g of distilled water. Finally, the samples were dried at 60°C.
The samples were named K-R where K:kaolin, R: NaOH/Al molar ratio 0.5-2.5.

2.3. Preparation and rheological characterization of 50 wt% suspensions

The OH-Al kaolin powder was added to a solution containing 0.3 wt % Na hexametaphosphate HMP (29.4 μmol P/g dry clay) to prepare 50 wt% suspensions.

Flow curves of the 50 wt% suspensions were obtained using a Haake RV3 concentric cylinder viscometer and MVIP measure system at 25°C.

2.4. Aluminum and phosphate determinations

Aluminum and phosphate adsorbed were calculated by difference between the amount added and the amount measured in the supernatant liquid. Al and phosphate were determined by colorimetry [6,7] using a Hewellt Packard 8354 spectrophotometer.

2.5. Selective dissolution of OH-Al species adsorbed

The presence of different OH-Al species on the kaolin surface was confirmed by selective dissolution. A sequential procedure, which was similar to that of Robarge and Corey [8], was used. The method consisted of a sequence of washings made on 5 g of OH-Al kaolin (dried at 60°C) as follows: a) water at pH=5.5-6, b) 0.5 M KCl, c) 0.1 N HCl and following 1M KCl solution acidified at pH 3, and d) 1N HCl (20 min at room temperature) in order to remove free Al, exchangeable Al, Al polymers and non crystalline Al hydroxides and/or Al salts, respectively.

Preliminary washings on the original kaolin indicated no appreciable dissolution of kaolinite crystal.

The distribution of Al species in 0.5M KCl solutions was also determined using the kinetics of Al-Ferron reaction color development [9,10]. The adsorbance was read at 366 nm in a Hewellt Packard 8354 spectrophotometer as a function of time, for times longer than 30 s.

The influence of the distribution of Al species on further HMP adsorption was also examined: 5 g of OH-Al sample (dried at 60 °C) was dispersed in 5 ml of distilled water containing 0.3 wt% HMP. The suspension was stirred, after 24 h diluted and centrifuged. This wet sediment was used for the sequential extraction. Al and P were analyzed in each solution.

2.6. Particle size distribution

Particle size distribution PSD curves were obtained using a Sedigraph 5000D. Measurements were carried out on diluted (6 wt%) suspensions containing variable amounts of HMP for original and K-1.05 and K-2.50 samples.

2.7. Electroacoustic measurements

Electroacoustic measurements on 1 vol% suspensions for original and OH-Al kaolins were made by an ESA 8000 apparatus (Matec Applied Science). Dynamic mobility were determined from the ESA signal. The original, K-1.05 and K-2.50 samples were titrated using 0.00245 M HMP solution. Suspension pH decreased less than one unit due to HMP solution addition. HMP concentration varied between 6×10^{-6} M (at 0.05 ml added) and 1.7×10^{-4} M (at 1.5 ml added).

3. RESULTS AND DISCUSSION

3.1. Distribution of Al species adsorbed on kaolinite

Figure 1 shows the variation of dynamic mobility and conductivity as a function of pH for the original kaolin.
Negative dynamic mobility slightly increased from -1.6×10^{-8} at pHs 4-5 to -1.8×10^{-8} $m^2 V^{-1} s^{-1}$ at pH 10 due to negative charge increased on the edge surfaces. Kaolinite particles have a negative surface charge in the pH range studied.

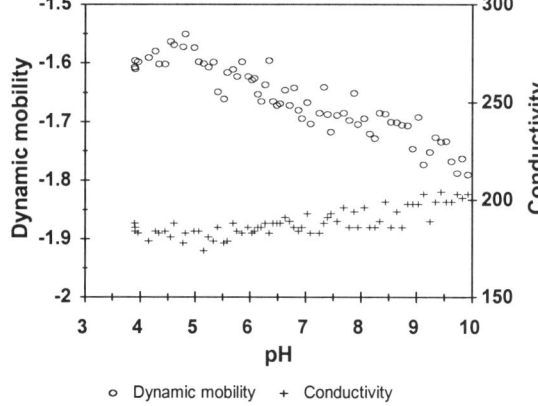

Figure 1. Dynamic mobility [10^{-8} $m^2 V^{-1} s^{-1}$] and conductivity vs. pH for original kaolin.

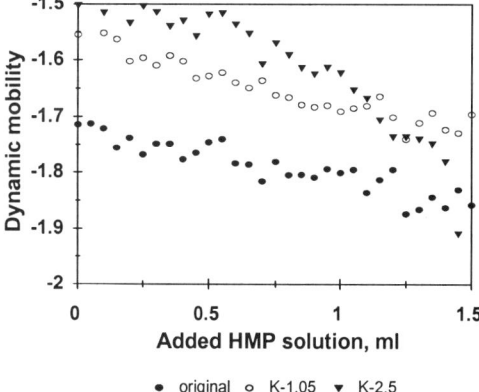

Figure 2. Dynamic mobility [10^{-8} $m^2 V^{-1} s^{-1}$] as a function of the volume of HMP solution added for original, K-1.05, K-2.5 samples.

Therefore, an electrostatic attraction between the negatively charged surface and the positively charged Al species in solution can occur. Moreover, at pH values below pH_{iep} of edges, adsorption of Al species occurred preferentially on the basal surfaces.

Table 1 shows the amount of Al adsorbed on the kaolin from partially neutralized $AlCl_3$ solutions having different NaOH/Al molar ratios by adding 35.2 μmol Al per gram of kaolin. More than 90% of the amount of Al added was adsorbed. Amount of Al adsorbed increased with increasing NaOH/Al ratio of the solutions. This may be attributed to the higher total positive charge of polymers with respect to monomers. Also, some Al polymers have a planar structures ($Al(OH)_3$-fragment structure) so that adsorption on kaolinite may be favored [10,11].

After OH-Al solution addition to kaolin suspension, the initial Al concentration was 0.006 M at pH 4.5. Then it was expected that formation of Al hydroxide solid phase in solution could be avoided because this concentration was below the precipitation threshold of Al hydroxide[12]. However, further hydrolysis or precipitation of Al species could take place on the kaolin surface. The relative content of the different Al species on the surface was determined from their solubility in: water, 0.5M KCl, 0.1N HCl –1M KCl at pH 3 and 1N HCl [8].

Table 1 shows exchangeable Al, Al polymer and also non crystalline Al hydroxides and/or Al salts contents of selected samples. Free Al content was 0.5% of the total amount of Al adsorbed indicating that solubility in water was very low.

Exchangeable Al content decreased from 66 to 5% for K-0.52 to K-2.50 samples, respectively.

Hsu [10] reported the presence of monomeric Al ions and some Al polymers in NaCl solutions after leaching OH-Al smectites. Then, the existence of Al polymers in 0.5 M KCl solution after washing OH-Al kaolin was tested using the kinetics of Al-ferron color development. Since, Al that reacted in 40 s was 98 % of the total Al amount, Al monomers predominated in the 0.5 M KCl solution. Consequently, Al exchangeable was monomeric.

Nevertheless, the non-exchangeable Al amounts for K-0.52 and K-1.05 were 34 and 59 % (Table 1), higher than Al polymer contents in the corresponding solutions (10 and 35%) according to Bottero[4], indicating some of Al monomers were hydrolyzed during adsorption process.

Table 1
Amounts of Al adsorbed on the kaolin. The amount of Al added was 35.2 μ mol/g.

Samples	Al ads. (μ mol/g)	Exch. Al (%)	Non-exchangeable Al		
			Al polymers (%)	Al soluble in 1N HCl (%)	Al insoluble* in 1N HCl (%)
K-0.52	31.9	66			
K-1.05	33.2	41	39	15	4.5
K-1.57	34.6	33			
K-2.10	34.9	15			
K-2.50	35.2	5	43	21	30.5

* % insoluble in 1N HCl = 100 - % recovered Al by the extraction procedure

For K-1.05 and K-2.50 samples, Table 1 shows that Al polymer contents as extracted by 0.1 N HCl and 1M KCl at pH 3 washings were 39 and 43 %, respectively. However, for K-2.50 sample, 21% of Al retained may be considered as non crystalline hydroxides and/or basic aluminum salts as was soluble in 1N HCl [11].For this sample, 30.5 % of total Al amount remained on the surface. This suggested that formation of Al(OH)$_3$ on the kaolinite surface, in addition to ion exchange, contributed to high Al retention from solutions containing dense and more aggregated polymers.

3.2. Colloidal stability of diluted suspensions

The initial pH of 6 wt% suspensions varied between 4.5 and 6 for K-0.52 to K-2.50 samples. Differences in the surface chemistry of the powders produced different initial suspension pH. Exchangeable Al content of the K-0.52 sample was higher than that of the K-2.50 (Table 1). Therefore, pH of the suspension decreased with increasing exchangeable Al content as more OH$^-$ was required to neutralize a high positive charge per Al atom adsorbed on the surface.

The negative dynamic mobility of treated samples (−1.5 to −1.55 x10^{-8} m^2 V^{-1} s^{-1}) were slightly lower than that of original kaolin due to the presence of positive surface sites (Figure 2). According to Arias [2] additions of Al above 0.43 wt% clearly change the surface electrical properties of kaolinite.

3.3. Effect of the HMP addition

Figures 3 a and b show the particle size distribution PSD curves for K-1.05 and K-2.5 samples with variable amounts of HMP added, respectively and for original kaolin with 0.3 wt% HMP.

Without dispersant, the PSD curves exhibited a sedimentation behavior of a flocculated suspension. Below pH$_{iep}$ of the edge surface, edge-face association and a small negative total charge produced a strongly flocculated suspension with gel structure, which prevented particle sedimentation.

At 0.03-0.12 wt% HMP, K-1.05 suspension showed a relatively large average particle size d$_{50}$ and low amount of particles finer than 0.5 µm due to low degree of dispersion of particles. Increasing HMP concentration to 0.3 wt%, d$_{50}$ decreased but the content of fines remained low, suggesting incomplete dispersion. For K-2.50, low d$_{50}$ and high content of fines were obtained at 0.06 wt% of HPM.

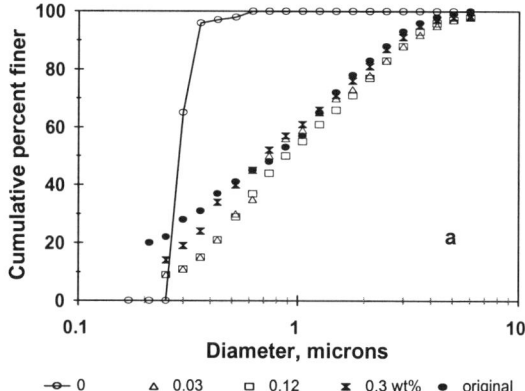
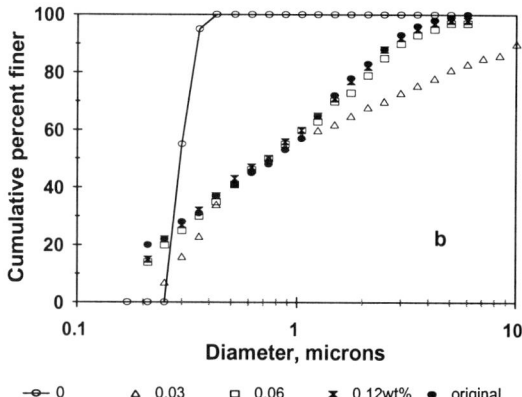

Figure 3. Cumulative percent finer than diameter vs. particle diameter curves for original and K-1.05 (a) and K-2.5 (b) samples with variable amounts of HMP added

Further additions of HMP had little effect on PSD curve indicating a well dispersed suspension. Contrarily, HMP concentrations higher than 0.3 wt% were required to stabilize the K-1.05 suspension

Figure 2 shows the dynamic mobility as a function of HMP additions for the original kaolin. After addition of 1ml, i.e. in the presence of 1.2×10^{-4} M of HMP, which corresponds to 0.3 % by weight of kaolin, the initially negative dynamic mobility reached a value of 1.8×10^{-8} m^2 V^{-1} s^{-1} because of increasing amounts of negatively charged HMP adsorbed.

Figure 2 also shows the dynamic mobility as a function of HMP additions for the K-1.05 and K-2.50 samples. Negative dynamic mobility gradually increased with the amount of HMP added. However, dynamic mobility of the K-2.50 sample remained nearly constant at low HMP additions while high amounts of HMP resulted in a rapidly increase in the dynamic mobility. Possibly, dynamic mobility behavior of the K-2.50 with increasing HMP additions (Figure 2) may be explained by a large amount of Al atoms adsorbed having a different HMP adsorption mechanism. Reactive sites involved in HMP adsorption were functional groups H_2O-Al-OH in the surface of clay, which were positively charged at low pH. However, most of the Al atoms adsorbed on the outer surface of the K-2.50 sample were in the form of Al-OH-Al, in which adsorption of HMP can not occur at low concentrations.

3.4. Rheological properties of 50 wt% aqueous suspensions with 0.3 wt% HMP.

Figure 4 shows flow curves of 50 wt% suspensions containing 0.3 wt% HMP at pH 5 for the original and OH-Al kaolins. Variation of apparent viscosity as a function of shear rate is shown in Figure 5.

The treatment of the kaolin with NaOH/Al < 1.5 solutions resulted in suspensions with high viscosity. Flow curves showed an important increase in the shear stress values than the one of the original suspension and a hysteresis loop. High apparent viscosity, shear thinning and a strong thixotropy are usually associated with flocculated suspensions. The liquid is immobilized in the flocs resulting in a high effective solid content and, therefore, an increasing viscosity of the suspension. An increase in viscosity indicated low degree of dispersion of the particles even in the presence of 0.3 wt% HMP.

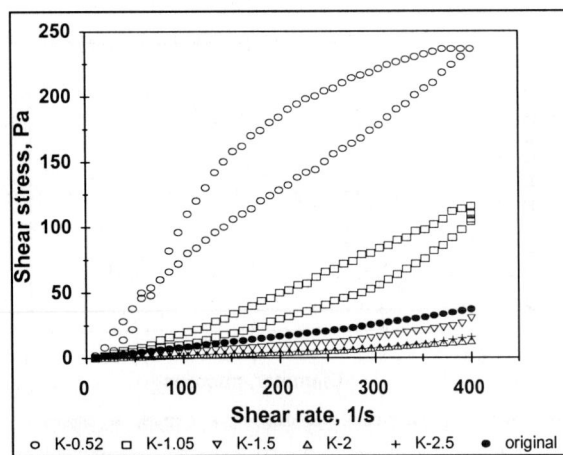

Figure 4. Flow curves of 50 wt% aqueous suspensions with 0.3 wt% HMP for original and treated kaolin.

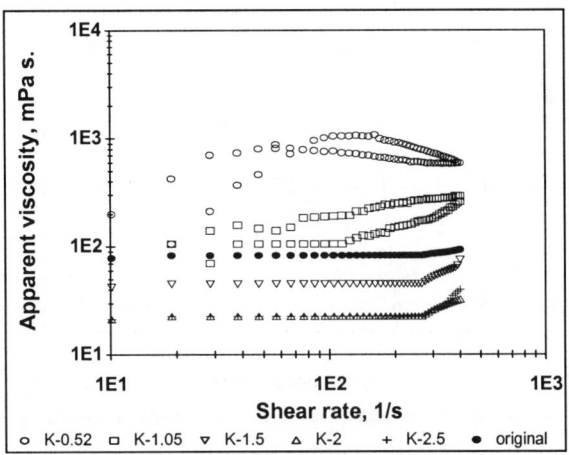

Figure 5. Apparent viscosity of 50 wt% aqueous suspensions with 0.3 wt% of HMP as a function of shear rate, for original and treated kaolins.

Table 2
Distribution of Al species adsorbed on the OH-Al kaolin and HMP amount adsorbed after addition of 0.3 wt% HMP. The Al/P molar ratio was determined in each solution.

Samples	P ads. (µmol/g)	Al in water (%)	Al in water Al/P	Exch. Al (%)	Exch. Al Al/P	Non-exchangeable Al Al polymers (%)	Al polymers Al/P	Al soluble in 1N HCl (%)	Al soluble in 1N HCl Al/P	Al insol.* in 1N HCl (%)	Al insol.* in 1N HCl Al/P
original	17.6										
K-1.05	20.0	9	0.3	41	6.0	33	2.5	11	2.4	6	n.d
K-2.50	21.8	7	0.3	5	2.6	42	3.1	12	2.1	34	n.d

* % insoluble in 1N HCl = 100 - % recovered Al by the extraction procedure

Suspensions of the kaolin prepared from NaOH/Al ≥1.5 solutions exhibited apparent viscosity values lower than that of the original suspension. These suspensions showed a nearly Newtonian behavior at low shear rates, although, a shear thickening behavior is observed beyond a critical shear rate. Low viscosity and shear thickening behavior are characteristics of well dispersed concentrated suspensions.

Different rheological behaviors may be explained by the relative content of exchangeable and non-exchangeable Al of the samples and by the amount of HMP adsorbed.

Table 2 shows free Al, exchangeable Al, Al polymers and also non crystalline Al hydroxides and/or Al salts contents for K-1.05 and K-2.5, the corresponding amounts of HMP adsorbed were reported as Al/P molar ratio.

Distributions of Al species on the surface with and without addition of HMP were similar. However, solubility of Al species increased in the presence of HMP. Equilibrium Al concentration in an aqueous 50 wt% suspension increased from 0.2 mM (calculated from the solubility of Al hydroxides [13]) to 3 mM due to formation of soluble Al-HMP complex.

For K-1.05 sample, the Al/P molar ratio 6 of exchangeable Al was higher than 2.6, the average Al/P ratio for Al polymers.

Table 2 shows that the total amount of HMP adsorbed (expressed as µmol P/g kaolin) for OH-Al samples was higher than that for the original due to the high affinity of HMP for the AlOH groups of the Al species adsorbed. The highest amount of HMP adsorbed for K-2.50 may be attributed to its large amount of Al atoms adsorbed, most of which were in the form of Al non exchangeable.

Sample K-1.05 contained high amount of exchangeable Al ions, which were monomers. Al monomers have the positive charge per Al atom higher than that of Al in the polymer. Although the amount of HMP adsorbed on this sample was greater than that of the original, negative surface charge was low because of the partial neutralization of exchangeable Al ions. Comparing dynamic mobility at high HMP additions, the K-1.05 sample showed the low negative value. Then, Al exchangeable had a negative effect on the rheological properties of the suspension.

For sample K-1.05, total amount of HMP adsorbed was lower than that for K-2.50. Thus, dynamic mobility at high HMP additions for K-1.05 sample was lower than that for K-2.50 sample and consequently, viscosity of the suspension increased.

Samples prepared from NaOH/Al>1.5 solutions contained more non-exchangeable Al species. With increasing NaOH/Al of the solution, hydrolysis is favored whereas the charge per Al atom in the polymers decreases. Moreover, Table 2 shows that the total amount of HMP adsorbed was high due to specific adsorption of HMP anion. Greater HMP adsorption increased dynamic mobility and probably electrostatic repulsion between kaolin particles, thus improving the degree of dispersion of the particles. Therefore, viscosity at low shear rates of 50 wt% suspensions decreased from 82 mPa s for the original to 23 mPa s for K-2.0 and K-2.50 samples.

4. CONCLUSIONS

Low amount (0.1 wt% of Al) of different OH-Al species were adsorbed on kaolinite at pH 4.5. Exchangeable Al (monomers), which have a positive charge per Al atom higher than the one in the polymer, produced more positively charged sites on the surface. However, low amount of HMP was adsorbed due to the low amount of Al atoms on the surface. When Al species retained were Al non exchangeable, most of them as polymers and hydroxides, the amount of reactive surface sites increased but their positive charge was probably not significant.

Concentrated 50 wt% suspensions were prepared by addition of OH-Al kaolin powders to water containing 0.3 wt% of HMP. Rheological properties of suspensions were also dependent on the relative contents of exchangeable and non-exchangeable Al on the surface. Suspensions of kaolin prepared from NaOH/Al <1.5 solutions showed complex non Newtonian flow behavior and high viscosity values. When kaolin was obtained from NaOH/Al $\geq$ 1.5 solutions, low viscosity suspensions with slight shear thickening flow behavior were produced. Consequently, these materials can be used to improve processing of concentrated suspensions with HMP as dispersant.

REFERENCES

1. J.M. Oades, Clays and Clay Miner., 32 (1984) 49.
2. M. Arias; T. Barral; F. Diaz-Fierros, Clays and Clay Miner.,43 (1995) 406.
3. W.B. Jepson, Phil. Trans.R. Soc. London, A 311 (1984) 411.
4. J.Y. Bottero; D. Tchoubar; J.M. Cases; J.J. Fripiat; F. Fiessinger. In: Interfacial phenomena in biotechnology and material processing. Y.A. Attia.; B.M. Moudgil and S. Chander (eds.) Elsevier. Amsterdam, 1988.
5. W.Z. Wang and P.H. Hsu, Clays Clay Miner.,42 (1994) 356.
6. P.H. Hsu, Soil Sci., 96 (1963) 230.
7. J. Murphy; J.P. Riley , Anal.Chim. Acta, 27 (1962) 31.
8. W.P. Robarge; R.B. Corey , Soil Sci.Soc. Am.J., 43 (1979) 481.
9. P.P. Tsai and P.H. Hsu , Soil. Sci. Soc. Amer. J., 48 (1984) 59.
10. P.H. Hsu, Clays and Clay Miner., 40 (1992) 301.
11. P.H. Hsu, In: Minerals in Soil Environments. J.B. Dixon and S.W. Weed (eds.), 2nd ed. Soil.Sc.Soc.Am , Madison, 1989.
12. N. Dezelic; H. Bilinski; R. Wolf , J. Inorg. Nucl. Chem.,33 (1971) 791.
13. R.M. Taylor, In: Chemistry of clays and clay minerals, A.C.D. Newman (ed.), Mineral Soc. Monogr. 6, John Wiley, New York , 1987.

Remotion of chromium from water by treated altered tuffaceous materials

A. Hülsken[a], D. Hall Gómez[a], J. Venaruzzo[a], J. Ortiga[b] and C. Volzone[b]

a. Asentamiento Universitario Zapala – Facultad de Ingeniería Universidad Nacional del mahue – UNC. 12 de Julio y Rahue, (8340) Zapala, Prov. Neuquén. Argentina, Tel. 54 2942-421847. e-mail: unc@zapala.edu.ar

b. Centro de Tecnología de Recursos Minerales y Cerámica- CETMIC-CONICET- C.C.49, Cno. Centenario y 506, (1897) M.B.Gonnet, Prov. Buenos Aires, Argentina.
e-mail: volzcris@netverk.com.ar

Two altered tuffaceous materials, containing different amount of illite-smectite clays, from Neuquén province, Argentina, were selected in order to evaluate the chromium retention from solutions. The capacity of chromium retention was measured on natural and after modified materials. The treatments were carried out in alkaline medium at boiling temperature, in order to modify the adsorption characteristics of the materials. Whole rock samples and smaller than two micron fractions were treated to determine the effect of the clay contents on retained chromium. The natural and treated samples were characterized by X-ray diffraction (XRD) and chemical analysis using ICP. The chromium in solution was measured using a spectrophotometer in the UV-visible range. The mineralogical composition, mainly the clay contents, of the tuffaceous materials and the conditions of alkaline treatment influenced the amount of chromium retention.

1. INTRODUCTION

Many industrial wastewaters contain heavy metals. Chromium can be present in different oxidation states. Cr^{3+} in trace amounts is necessary for the human health, however small proportions of Cr^{6+} are highly toxic. The remotion of Cr^{3+} from the wastewater prevents its possible oxidation. The elimination of the chromium from industrial wastewater is a topic of significant environmental interest. Previous work indicates that natural clays retain chromium from solution; and its efficiency mainly depends on the clay types and the polymerization states of the chromium species [1-3]. In the Neuquén province, Argentina, extensive deposits of altered tuffaceous material are found, which will be of economic significance once a use has been identified.

The aim of the paper is to evaluate the retention chromium (III) from water using altered-tuffaceous materials treated in alkaline media.

2. EXPERIMENTAL

The raw tuffaceous materials used in this paper were from two deposits of Neuquén province: the Toba Rosada (Pink Tuff) deposit, and a tuff belonging to the Barda Negra formation have named M1 and M6, respectively. The samples were milled passing a 200 mesh sieve (<74 μm). Furthermore, the < 2 μm size fraction has been separated after passing 200 mesh sieve.

The mineral components were modified by means of 3N KOH during 3 hours at boiling temperature. After treatment the solid was washed several times until pH 7. The solid/liquid ratio was 2 % wt./vol. The modified tuffaceous materials were named: M1K, M1fK, M6K, M6fK where f corresponds to the samples of the < 2μm size fraction.

The natural and modified samples were characterized by X-ray diffraction (XRD), using a Phillips 3020 equipment, at 40 kW and 20 mA and by chemical analysis (ICP method).

The chromium(III) retention by samples was carried out in a batch system, exposing the material with a 2000 ppm Cr sulfate solution during 24 hours. The ratio of solid/liquid was of 5.5% wt./vol. The chromium retained by the solid was calculated by difference between the chromium content in the original solution and the chromium content in the supernatant. The measurments in solutions were carried out using Hewlett Packard Model 8453 spectrophotometer, reading of the maximum of the band at 425 nm.

3. RESULTS

3.1. X- ray diffraction

Quartz, feldspar and clay mineral (smectite, illite-smectite) are present in the (M1) tuffaceous material, Figure 1. In < 2 μm size fraction of this material (M1f), Figure 2, an increment in the clay components and an important decrease of quartz and feldspar are observed.

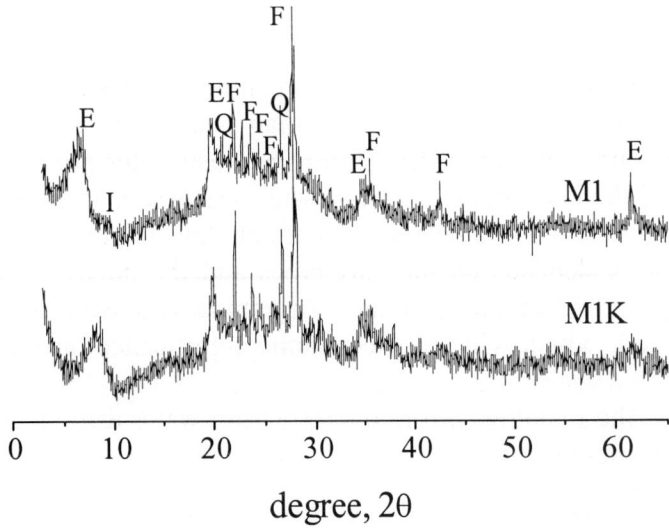

Figure 1. XRD patterns of M1 and M1K samples in air dried conditions.

The M6 sample is composed mainly by feldspar and cristobalite, with a small contents of illite-smectite clay phases, Figure 3. The M6f sample did not show a significant change in its mineralogical composition with regarding M6 sample, Figure 4.

The diffractograms of the Figure 1 show that the alkaline treatment had not destroyed the clay structure of the material. However, the d(001) spacing of the smectite was shifted similar to values to an illite [4] as well modifying the feldspar and quartz contents.

Figure 2 shows the changes of the M1 clay fraction (M1f) after alkaline treatment. The small proportion of quartz in the fine fraction was diminished by the lixiviation with the potassium hydroxide [5].

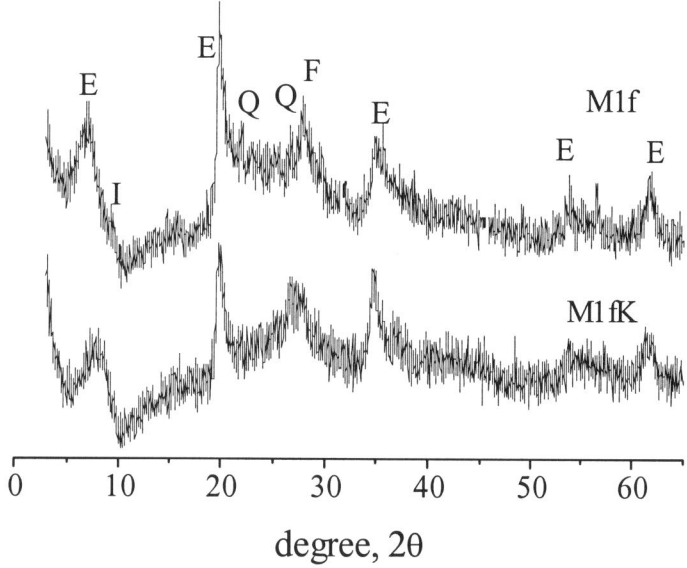

Figure 2. XRD patterns of M1f and M1fK samples in air dried conditions.

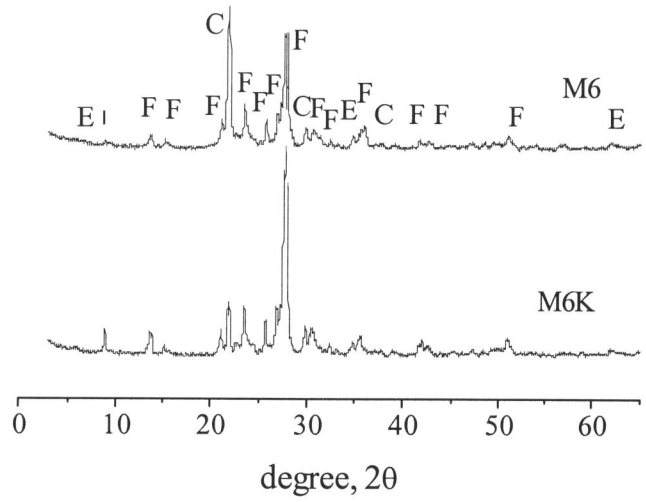

Figure 3. XRD patterns of M6 and M6K samples in air dried conditions.

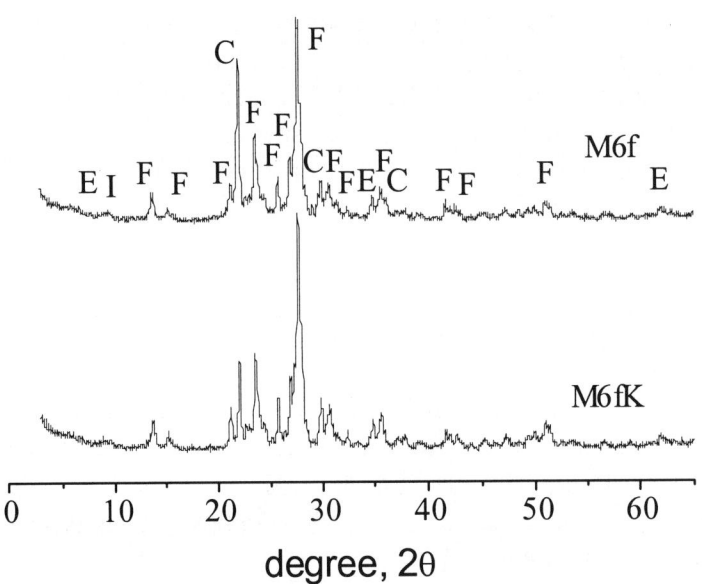

Figure 4. XRD patterns of M6f and M6fK samples in air dried conditions.

The M6 sample has been modified in its mineralogical composition after alkaline treatment (Figure 3). A decrease in the cristobalite content [5] and an increase of the relative intensities of the feldspar peaks were observed. In spite of the small clay content in the M6 sample, it can be appreciated that the d(001) spacing of the smectite was shifted to angles corresponding to that of an illite. Perhaps, a certain degree of illitization of smectite occurred in presence of potassium and the silica content in the original material, Figure 2 [4,5].

Figure 4 shows the changes in the fine fraction of M6 after alkaline treatment (M6fK), where a similar behaviour to M6 sample was observed. However, the reduction of the cristobalite was smaller. A smaller degree of illitization of the M6fK sample was also observed regarding the sample M6K, and it could be attributed to smaller silica content in the M6f sample.

3.2. Chemical analysis

Table 1 shows the chemical analysis of the original sample and fine fraction of M1 before and after the alkaline treatment. The increased concentration of the clay material in the M1f sample regarding to M1 is reflected by a lower $SiO_2/(Al_2O_3+Fe_2O_3+MgO+TiO_2+CaO+Na_2O+K_2O)$ ratio, due to an important decrease of the silica content (as quartz and also as feldspar) and in lower proportion of the Al_2O_3, CaO and K_2O, also components of the feldspar (Figs. 1,2). KOH treatment probably originated a reduction in the silica percentage due to the alkaline lixiviation.

An important increase can be appreciated in the amount of potassium after the treatment, Table 1, with simultaneous reduction of the sodium content. This behavior was accentuated in the fine fraction, and it would justify the progressive illitization process by transformation of the interstratified illite-smectite.

Table 1
Chemical analyses of M1 and M1f samples before and after alkaline treatment

SAMPLES	SiO_2 %	Al_2O_3 %	Fe_2O_3 %	MgO %	CaO %	Na_2O %	K_2O %	TiO_2 %	LOI %	SiO_2/MO*
M1	62.19	12.93	4.86	0.99	1.92	2.69	1.37	0.66	12.00	2.47
M1K	54.62	14.57	7.49	1.56	3.26	1.82	4.13	1.00	11.00	1.61
M1f	56.21	12.84	7.93	1.68	1.22	1.94	0.89	0.88	16.10	2.39
M1fK	43.78	14.13	11.64	2.57	1.96	0.51	6.96	1.39	16.40	1.47

*MO = ($Al_2O_3+Fe_2O_3+MgO+TiO_2+CaO+Na_2O+K_2O$)

The chemical analyses of M6 and M6f with and without treatment are shown in the Table 2. The main changes of chemical composition are due to the higher concentration of the clay material in the < 2μm fraction. They were expressed as a reduction in the silica content, due of the reduction of the cristobalite content (Figs. 3,4) in the fine material (M6f). This was also reflected in the silica/oxides ratio.

Table 2
Chemical analyses of M6 and M6f samples before and after alkaline treatment

SAMPLES	SiO_2 %	Al_2O_3 %	Fe_2O_3 %	MgO %	CaO %	Na_2O %	K_2O %	TiO_2 %	LOI %	SiO_2/MO*
M6	71.68	13.62	2.33	0.24	0.79	4.87	4.36	0.29	1.50	2.70
M6K	64.85	16.99	2.85	0.44	1.42	5.86	5.96	0.40	0.90	1.91
M6f	68.22	15.65	1.51	0.35	0.81	5.93	4.45	0.15	2.50	2.36
M6fK	65.59	17.17	1.64	0.45	1.07	6.41	5.41	0.18	1.60	2.03

*MO = ($Al_2O_3+Fe_2O_3+MgO+TiO_2+CaO+Na_2O+K_2O$)

Furthermore, the results from Table 2 indicate that the alkaline treatment resulted a significant reduction of the silica content in the sample M6K and a lesser reduction of this oxide in M6fK.

3.3. Adsorption of chromium

The retention of chromium by the M1 and M6 samples (original, clay fraction and after alkaline treatments) are shown in the Figure 5. The retention of chromium by M1 is five times greater than the M6, with a 30.6% and 6.0 % retention respectively. The difference could be attributed to the larger clay content in sample M1.

The < 2μm fractions of these samples increases the retention capacity to 40.1% and 18.0%, respectively.

The products obtained after the treatment with potassium hydroxide has even larger retention capacity than the original samples. The percentages of 30.6% and 6% for M1 and M6, respectively, increase to 76.3% and 20.1% for M1K and M6K. These values represent an increase of 149% for M1 and 135% for M6. The behavior of the alkaline treated fine materials was similar to that of the alkaline treated original samples.

This way the sample M1f increased its capacity of chromium retention in 102%, while the M6f an increase of only 29% was observed. Although the increase in percent of chromium retention was larger in the alkaline treated original samples; the maximum retention capacity was found for both samples of the treated fine fraction, Figure 5.

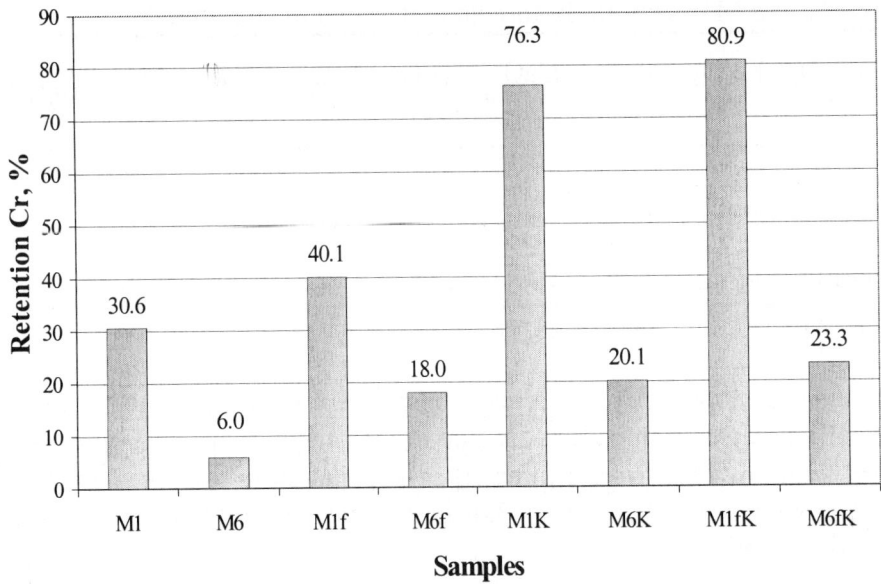

Figure 5. Retention chromium on M1, M1f and M6, M6f before and after alkaline treatment

4. CONCLUSIONS

The alkaline treatment on the altered tuffaceous minerals modified the capacity of the retention of chromium from solution. This capacity was highly influenced by the presence of clays. The modified tuff with a high clay contents showed the best retention of chromium. Due to the alkaline treatment it has been possible to increase the chromium retention from 30.6% to 80.9% in sample M1 and from 6% to 23.3% in sample M6. In the case of M1, a product with high potential for the chromium adsorption from solutions has been obtained.

The capacity of adsorption is fundamentally related to the mineralogy of the samples.

To evaluate the correlation between the textural and structural characteristics of these materials and the capacity of chromium adsorption from solutions, further research will be necessary.

REFERENCES

1. C. Volzone, In Catalizadores y Adsorbentes Iberoamericanos para la Remoción de Metales Pesados en Efluentes Industriales, Editores: G. Rodríguez Fuentes y P. Avila García. Ediciones CYTED, Madrid, España. 1ra Edición, 2000. ISBN: 84-931538-3-4, pp 83-90.
2. C. Volzone, Australian Journal of Soil Research, Vol 36 (1998) 423.
3. C. Volzone, Materials Chemistry and Physics, Vol 47 (1997) 13.
4. S. De la Fuente, Clays & Clay Minerals, Vol. 48 (2000) 339.
5. A. Bauer and B. Velde, Vol. 34 (1999) 259.
6. Olson, R.H., et al., Zeolites, Industrial Minerals and Rocks, 5th Edition, S.J. Lefond, ed., AIME, New York, NY, (1983) 333 – 366.

Clay-Iron oxide magnetic composites for the adsorption of contaminants in water

L. C. A. Oliveira[1], R. M. Lago[1], R. V. R. A. Rios[1], J. D. Fabris[1], C. Solar[2] and K. Sapag[2]

[1]Departamento de Química, UFMG, Belo Horizonte, MG, Brazil

[2]Lab.Cs. Superficie Medios Porosos, Univ. Nacional de San Luis, Argentina

In this work the adsorption features of clays with the magnetic properties of iron oxides have been combined in a composite to produce a magnetic adsorbent. These magnetic particles can be used as adsorbent for contaminants in water and subsequently removed from the medium by a simple magnetic device. The bentonite-iron oxide magnetic composites have been prepared with weight ratios of 2:1, 1.5:1 and 1:1 and characterized by powder XRD, magnetization measurements, chemical analyses, TPR, N_2 adsorption-desorption isotherms and Mössbauer spectroscopy. The results suggest that the main magnetic phase present is maghemite (γ-Fe_2O_3) with small amounts of magnetite (Fe_3O_4). A magnetization enhancement can be produced by treatment with H_2 at 400°C to reduce maghemite to magnetite. N_2 adsorption measurements showed that the presence of iron oxides did not affect significantly the surface area or the pore structure of the bentonite. The adsorption isotherms of metal ions Ni^{2+}, Cu^{2+}, Cd^{2+} and Zn^{2+} from aqueous solution onto the composites also showed that the presence of iron oxide did not affect the adsorption capacity of the bentonite.

1. INTRODUCTION

New and innovative methods are of great importance to devise technologies to deal with environmental problems. The application of magnetic particle technology to solve environmental problems is one of these methods that have received considerable attention in recent years. Magnetic particles can be used to adsorb contaminants from aqueous or gaseous effluents and after the adsorption separated from the medium by a simple magnetic process. Some examples of this technology is the use of magnetite particles to accelerate the coagulation of sewages [1], magnetite coated functionalized polymer such as resin to remove radionuclides from milk [2], poly(oxy-2,6-dimethyl-1,4-phenylene) for the adsorption of organic dyes [3] and polymer coated magnetic particles for oil spill remediation [4]. However, all these materials suffer from small surface area or small adsorption capacity, which limit their application. In our group we have been developing high surface area and high adsorption capacity magnetic composites based on active carbon/iron oxide or clay/iron oxide to remove contaminants from aqueous effluents.

Clays offer an attractive and inexpensive option for the removal of organic and inorganic contaminants [5]. The adsorption of several organic contaminants in water such as pesticides, phenols and chlorophenols has been reported recently in the literature [6-12]. Natural clays are cheap and available materials functioning as excellent cation exchange materials, which have been often used to adsorb metallic contaminants. The adsorption capacity of clays results from a net negative charge on their structure, which attracts and holds cations such as heavy metals. Their relatively high surface area (up to 800 m^2/g) also contributes to the adsorption capacity [13]. There are three basic classes of clays: micas (e.g. illite) and smectites (e.g. montmorillonite). The montmorillonites have the smallest crystals, the largest surface area and the highest cation exchange capacity [14]. For these reasons bentonite in different forms such as organo-tailored, acid and base treated, china clay, and wallastonite, has been extensively studied as sorbent for heavy metals [15-18]

2. EXPERIMENTAL

The composites were prepared from a suspension of bentonite (Vetec) in 400 mL solution of FeCl$_3$ (7.8 g, 28 mmol) and FeSO$_4$ (3.9 g, 14 mmol) at 70°C. A solution of NaOH (100 mL, 5 mol L^{-1}) was added dropwise to precipitate the iron oxides. The amount of bentonite was adjusted in order to obtain the following clay:iron oxide weight ratios 1:1, 1.5:1 and 2:1. The obtained materials were dried in a oven at 100°C for 3 h and were characterized. Powder XRD diffractograms were obtained with a Rigaku Geigerflex with Ni filtered Cu-Kα (λ=1.5418 Å). The magnetization measurements were carried out in a portable magnetometer with a fixed magnetic field of 0.3 tesla [19]. The samples were also chemically analyzed for Fe^{2+} and Fe^{+3}. TPR (Temperature Programmed Reduction) profiles were obtained in Chembet 3000 Quantachrome equipment. Mössbauer spectroscopy (with a ^{57}Co/Rh source at room temperature) calibrated with (α)-Fe. The adsorption isotherms have been obtained in a batch equilibrium experiment with 50 mg of the composites in 50 mL of 10, 20, 50 and 100 mg/L standard solutions of Cu^{2+}, Ni^{2+}, Cd^{2+} and Zn^{2+}. All the solutions were kept for 24 h at 28 ±2°C with a temperature-controlled bath and pH 5.0 adjusted with diluted HCl. The metal concentrations were determined by atomic absorption (Carls Zeiss Jena AAS).

3. RESULTS

The composites were prepared by the precipitation of iron oxides or hydroxides onto the clay surface. After the preparation a test with a 0.3 T magnet showed that the whole material was completely attracted to the magnet with no non-magnetic material observed. The chemical analyses and magnetization measurements obtained for the prepared composites are shown in Table 1.

As the magnetization measurements in Table 1 shows magnetic clay:iron oxide composites were formed. It can be observed that bulky sigma magnetization increases for higher Fe oxide wt% in the composite. However, this increase is not proportional to the Fe oxide content (Figure 1) indicating that the concentration of non-magnetic iron oxides increases for the composites with higher clay content. The pure Fe oxide prepared without the clay showed magnetization of 62 JT^{-1}kg^{-1}, which is similar to gama-Fe$_2$O$_3$ maghemite.

Table 1. Chemical analyses and magnetization measurements for the bentonite:iron oxide composites.

Composite	Bulky magnetization (J T^{-1}kg^{-1})	Fe oxide[a] wt %
Pure Fe oxide	62	ca. 100
Clay:Fe oxide 1:1	27	55
Clay:Fe oxide 1.5:1	18	40
Clay:Fe oxide 2:1	8	33

a- (as FeO + Fe$_2$O$_3$)

Chemical analyses showed a Fe^{+3}/Fe^{2+} ratio of approximately 7, which is much higher than the original ratio used (Fe^{+3}/Fe^{2+}=2), probably due to oxidation of Fe^{2+} during preparation. These results suggest the formation of mainly Fe^{+3} compounds in the composite.

The nitrogen adsorption isotherms (Figure 2) showed for the composite clay:Fe oxide 2:1 and pure bentonite clay BET surface areas of 55 (V$_{micropores}$ 0.009 cm^3.g^{-1}) and 7 m^2.g^{-1} (V$_{micropores}$ 0.003 cm^3 g^{-1}), respectively. It is interesting to note that no significant decrease in the surface area or in the porosity of the bentonite was caused by the presence of the Fe oxides.

The XRD analyses of the Fe oxide (Figure 3a) showed a cubic iron oxide phase (d= 2.50; 2.91; 2.07; 1.60; 3.28 Å) which can be related to the presence of maghemite or magnetite. For the composite these peaks appear broader suggesting smaller particle size for pure Fe oxide and for the composite. Besides the diffraction peaks for bentonite the composites show a weak diffraction peak at d=2.70 Å which might be related to the presence of small amounts of hematite.

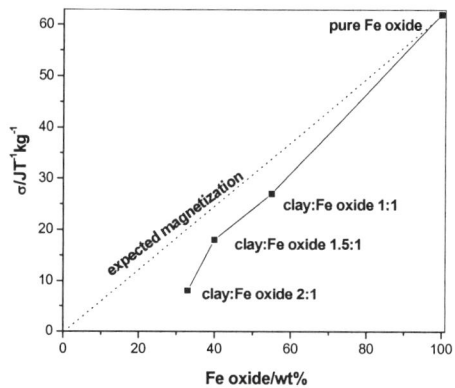

Figure 1. Bulky sigma magnetization for the diferent clay:Fe oxide composites

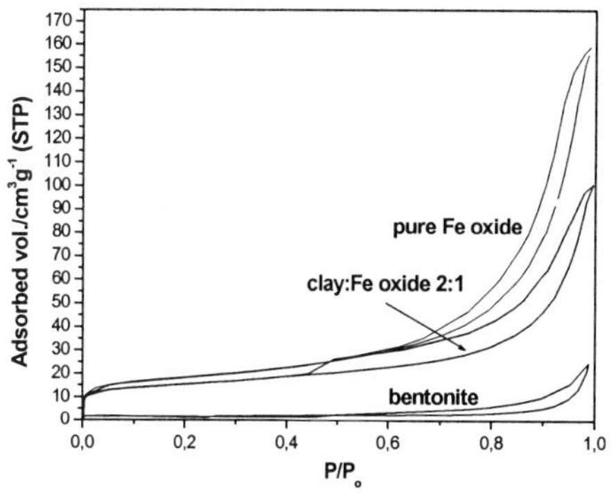

Figure 2. Nitrogen adsorption-desorption isotherms

Room temperature ^{57}Fe Mössbauer spectrum for the pure Fe oxide (Figure 4a) showed broad sextets, which suggests the presence mainly maghemite or hematite. It also seems to appear in high fields doublets, which might be related to the presence of small amounts of magnetite with its tetrahedral (A) and octahedral (B) sites. For the composite (Figure 4b) complex spectra were obtained probably representing superparamagnetic phases due to small particles of Fe oxide [20].

Temperature programmed reduction (TPR) profile of the composite clay:Fe oxide 1:2 (Figure 5a) showed a shoulder at approximately 460°C and three peaks centered at 550 and 650 and 740°C. TPR studies of iron oxides in literature [21,22], suggests that the reductions occurring at 460 and 550°C are probably due to the reduction of part of Fe^{+3} to produce Fe_3O_4 and the complete reduction to Fe^o, respectively.

The peaks at 650 and 740°C are likely related to iron cations present in the clay net as it was observed in the TPR for pure bentonite (Figure 5b). It can be observed that these peaks for the composite are more intense compared to the pure clay. This is due to a higher concentration of iron ions in the clay probably exchanged from the Fe^{2+} and Fe^{+3} solution used during the preparation of the composites.

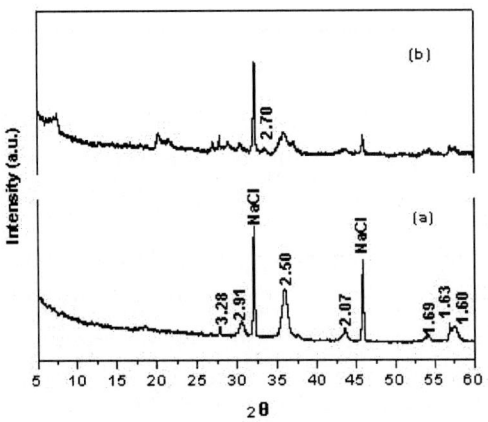

Figure 3. XRD for the clay: Fe oxides composites (b) and Fe oxide (a)

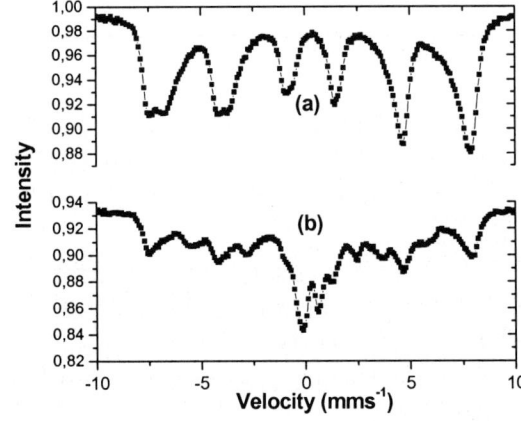

Figure 4. Room temperature Mössbauer spectra for the Fe oxide (a) and clay:Fe oxide composites (b).

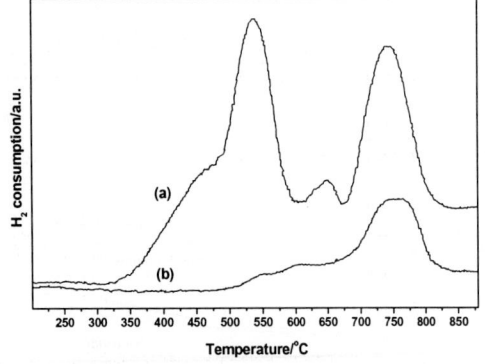

Figure 5. TPR analyses for the (a) clay: Fe oxide composites (b) pure bentonite.

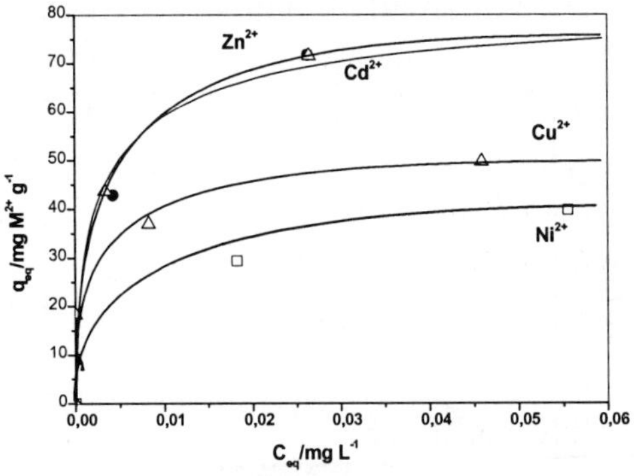

Figure 6. Adsorption isotherms at 28±1°C for Ni^{2+}, Cu^{2+}, Cd^{2+} and Zn^{2+} over clay: Fe oxide 2:1 composite.

The clay:Fe oxide 3:1 composite was exposed to pHs 11, 9, 7, 5, 3 and 1 for 72 h. After this period the composites were dried and its magnetization and Fe oxide content determined. It was observed no significant change in magnetization or Fe oxide content in the pH range 5-11. For pH 3 a small decrease in magnetization occurred probably due to the partial attack of the acid to the iron oxide whereas upon treatment at pH 1 the Fe oxide from the composite was completely dissolved. Moreover, at pH lower than 3 the bentonite is probably also attacked by acid [23].

The adsorption of Ni^{2+}, Cu^{2+}, Cd^{2+} and Zn^{2+} from aqueous solutions on the 3:1 clay: Fe oxide composite was studied. The adsorption isotherms are shown in Figure 6. It can be observed that the adsorption capacity increases in the order $Ni^{2+}<Cu^{2+}< Cd^{2+} <Zn^{2+}$. For metal concentrations higher than ca. 0.03 mg/L the adsorbent saturates showing adsorption capacities of approximately 40, 50, 74 and 75 mg/$g_{composite}$ of Ni^{2+}, Cu^{2+}, Cd^{2+} and Zn^{2+}, respectively. Adsorption tests of Zn^{2+} with the different composites showed the following adsorption order pure bentonite 72 mg/g, clay:Fe oxide 2:1 (75 mg/g) whereas the composite clay:Fe oxide 1:1 showed 77 mg/g.

4. DISCUSSION

Four iron oxides are commonly formed under the reaction conditions employed i.e. Fe_3O_4 (magnetite), γ-Fe_2O_3 (maghemite), α-Fe_2O_3 (hematite) and α-FeO(OH) (goethite) [15]. From these oxides, only the first two, magnetite and maghemite, are magnetic with magnetization of 100 and 60 J $T^{-1}kg^{-1}$, respectively. The magnetization measurements, Fe^{3+}/Fe^{2+} ratios, XRD and Mössbauer suggest that the main magnetic phase formed is maghemite, possibly with small amounts of magnetite and hematite. TPR controlled experiments showed that the Fe_2O_3 oxides in the materials could be selectively reduce to produce Fe_3O_4 enhancing the magnetization of the composites. Clay:Fe oxide showed much higher surface area compared to the pure bentonite clay. The composites showed high adsorption capacities for the Ni^{2+}, Cu^{2+}, Cd^{2+} and Zn^{2+} in aqueous solution and more importantly no reduction of the adsorption was produced by the formation of the composite. Interestingly, the Fe oxide seems to contribute to the adsorption of metal ions from the solution.

5. CONCLUSIONS

The adsorption features of natural clays and the magnetic properties of Fe oxides can be combined in a composite to produce magnetic adsorbents. The preparation is simple and does not require any special chemical or procedure. These magnetic adsorbents show good adsorption capacity for metal ions contaminants in water and can be used in a wide pH range.

REFERENCES

1. N.A. Booker, D. Keir, A. Priestley, C.D. Rithchie, D.L. Sudarmana, and M.A. Woods, Water Sci. Technol., 123 (1991) 1703.
2. K.S. Sing, Technology Profile, Ground Water Monitor, 21 (1994) 60.
3. I. Safarik, M. Safarikova, V. Buricova, Sorption of Water Soluble Organic Dyes on Magnetic Poly(oxy-2,6-dimethyl-1,4-phenylene), Collection, 60 (1995) 1448-1456.
4. J.D. Orbell, L. Godhino, S.W. Bigger, T.M. Nguyen and L.N. Ngeh, Oil Spill Remediation using Magnetic Particles, J. Chem. Educ. 74 (1997) 1446.
5. H.H. Murray, Tradial and New Applications for Kaolin, Smectite and alygorskite: a General Overview, Appl. Clay Sci., 17 (2000) 207-221.

6. T. G. Danis, T.A. Albanis, D.E. Petrakis and P.J. Pomonis, Removal of Chlorinated Phenols Aqueous Solutions by Adsorption on Alumina Pillared Clays and Mesoporous Alumina Aluminum Phosphates, Wat. Res., 32 (1998) 295-302.

7. C.C. Ainsworth, J.M. Zachara and R.L. Schimidt, Quinoline sorption on Na-montmorillonite: contribuition of protonated and neutral species, Clays and Clay Min., 35 (1987) 121-128.

8. J.M. Rodriguez, A.J. Lopez ans S. Bruque, Interaction of Phenamiphos with montmorillonite, Clays and Clay Min., 36 (1988) 284-288.

9. C. M. Sanchez and M.M.J. Sanchez, Factors Influencing interactions of organophosphorus pesticides with montmorillonite, Geoderma, 29 (1983) 107-118.

10. Torrents and S. Jayasundera, The sorption of non-ionic pesticides onto clays and influence of natural organic carbon, Chemosphere, 35 (1997) 1549-1565.

11. I.K. Konstantinou, T.A. Albanis, D.E. Petrakis and P.J. Pomonis, Removal of Herbicides from Aqueos Solutions by Adsorption on Al-Pillared Clays", Wat. Res., 34 (2000) 3123-3136.

12. H.T. Shu, D.Li, A.A. Scala and Y.H. Ma, Adsorption of Small Organic Pollutant from Aqueous Streams by Aluminosilicalite-based Microporous Materials, Sep. Purif. Technol., 11 (1997) 27-36.

13. F. Cadena, R. Rizvi and R.W. Peter, Feasibility Studies for the Removal of Heavy Metal from Solution using Tailored Bentonite, Harzadous and Industrial Wastes, Proceedins 22nd Midi Atlantic Industrial Waste Conference, Drexel University, 1990, p.77.

14. S.E. Bailey, T.J. Olin, R. M. Bricka and D.D. Adrian, A Review of Potentially Low Cost Sorbents for Heavy Metals, Wat. Res., 33 (1999) 2469.

15. E.G. Pradas, M. Sanchez, F.C. Cruz, M.S. Viciana, M.F. Perez, , J. Chem. Tech. Biotechnol., 59 (1994) 289.

16. C. Sharma, G.S. Gupta, G. Prasad and D.C. Rupainwar, Use of Wallostonite to Remove Ni(II) from Aqueous Solution, Water Air Soil Poll., 49 (1990) 69.

17. Gier and W.D. Johns, Heavy Metal Adsorption on Micas and Clay Minerals Studied by XPS, Appl. Clay Sci., 16 (2000) 289-299.

18. F. Barbier, G.Duc, M.P. Ramel, Adsorption of Pb and Cd ions from aqueous Sollution to the Montmorillonite/water Interface, Col. Surf. A: Phys. Eng., 166 (2000) 153-159.

19. J.M.D. Coey, O. Cugat, J. MaCauley and J.D. Fabris, Revista de Física Aplicada e Instrumental, 7 (1992) 25.

20. K.R.P.M Rao, F.E. Huggins, V. Mahajan, G.P. Huffman, V.U.S. Rao, R.J. O'Brien, Top. Catal., 2 (1995) 71.

21. G. Munteanu, L. Ilieva and D. Andreeva, Kinetic Parameter Obtained from TPR data for alfa-Fe_2O_3 and Au-alfa-Fe_2O_3, Thermochim. Acta, 291 (1997) 171.

22. L.C. A. Oliveira, J.D. Fabris. W.N. Mussel, R.V.R.A. Rios and R. M. Lago, The Effect of Mn Substitution on the Catalytic Properties of Ferrites, Stud. Surf. Sci. Catal., Espanha, 2000.

23. S. Mendioroz, J.A. Pajares, I. Benito, C. Pesquera, F. Gonzanles and C. Blanco, Texture Evolution of Montmorillonite under Progressive Acid Treatment: Change from H3 to H2 Type of Hysteresis, Langmuir, 3 (1987) 676-681.

Thermal decomposition of layered Co-Al hydrotalcite: An *in situ* study

J. Pérez-Ramírez*,[1], G. Mul, J.B. Taboada, F. Kapteijn and J.A. Moulijn

Section Industrial Catalysis, DelftChemTech, Delft University of Technology, Julianalaan 136, 2628 BL Delft (The Netherlands), *j.perezramirez@tnw.tudelft.nl.

High-Temperature X-Ray Diffraction (HT-XRD), Thermal Analysis (TGA-DTA), Mass Spectrometry (MS), and vibrational spectroscopies (*in situ* Fourier Transform Infrared (FT-IR), *in situ* Raman, and UV-Vis/DR) have been used to investigate the thermal decomposition of layered Co-Al hydrotalcite, $[Co_6Al_2(OH)_{16}]CO_3 \cdot 4H_2O$. The transition of the single groups (water, hydroxyls, and carbonates), structural phases, and metal oxidation state upon decomposition in air have been analyzed. Based on the analytical results, a simplified model for the decomposition process is presented. The presence of oxidizable Co^{2+} cations in the octahedral sheets of the anionic clay and the stability of the solid solution of Co-spinels formed, induce a low thermal stability of the hydrotalcite precursor in air and the absence of retrotopotactic reconstruction after decomposition at temperatures of only 175-200°C, where significant amount of hydroxyls and carbonates are still present in the sample.

1. INTRODUCTION

Hydrotalcite-like compounds (HTlcs) have been subject of many studies: synthesis, preparation, and structural studies, as well as application as sorbents, ion-exchangers, catalyst precursors, and catalyst supports have been reported [1,2]. These anionic clays, which consist of layered double hydroxides with a general formula $[M^{2+}_{1-x}M^{3+}_x(OH)_2]^{x+}[A^{m-}]_{x/m} \cdot nH_2O$, can be visualized as brucite-type octahedral sheets, in which M^{3+} cations partially substitute for M^{2+} cations. The positive charge resulting from this substitution is balanced by anions (often carbonates) and water molecules arranged in interlayers alternating with the octahedral sheets.

Heat treatment of these materials yields multicomponent mixed oxides with unusual properties such as a homogeneous mixture of cations, high thermal stability, and high surface area, which are attractive for catalytic applications [1,2]. These properties are a consequence of the nature of the hydrotalcite precursor and the decomposition mechanism [3].

Interpretation of the decomposition process has been often based merely on the information obtained from *ex situ* X-ray diffraction and thermal analysis [4-6], but an integration of suitable techniques is usually not applied. Furthermore, most of the studies on decomposed hydrotalcites are carried out at room temperature, where the state of the sample may have changed during cooling, exposure to atmosphere, and handling. Recently, *in situ* high-temperature XRD has been used to investigate phase transitions in Mg-based HTlcs,

---

[1] current address: Norsk Hydro Research Centre, Section Hydrocarbon Processes and Catalysis.
e-mail: javier.perez.ramirez@hydro.com.

which revealed interesting aspects related to the decomposition and reconstruction mechanism [3,7,8]. Other spectroscopic techniques, like infrared or Raman spectroscopy, have been rarely used to study the thermal decomposition mechanism [9-12], and experiments reported were usually performed in vacuum or *ex situ* [13-15].

In this paper, the physico-chemical changes upon thermal decomposition of Co-Al hydrotalcite have been investigated by a combination of different *in situ* techniques (HT-XRD, TGA-DTA, MS, FT-IR, Raman), and UV-Vis/DR. Based on these results, the key factors in the thermal decomposition and reconstruction mechanism are identified and a simplified model for these processes is proposed.

2. EXPERIMENTAL

Co-Al hydrotalcite (Co/Al = 3/1) was prepared by co-precipitation at constant pH and temperature at low supersaturation conditions, as described elsewhere [9]. The final slurry was aged at 65°C for 18 h under mild stirring, using a reflux to prevent water evaporation. Finally, the material was cooled to room temperature, filtered, washed with of warm (30°C) deionized water, and dried at 90°C for 12 h. The chemical composition of the as-synthesized material was determined by ICP-OES (Perkin-Elmer Plasma 40 (Si)). The thermal decomposition in air was investigated by High-Temperature XRD (STOE diffractometer, CuKα radiation), TGA-DTA (Mettler-Toledo TGA/SDTA 851) coupled to MS (Balzers Quadstar 421), *in situ* FT-IR (Nicolet Magna 550 spectrometer) using a Spectratech DRIFT accessory, *in situ* Visible Raman (Renishaw Raman Imaging Microscope, system 2000) using a 20 mW Ar^+ laser (λ = 514 nm), and UV-Vis/DR (Shimadzu UV-2010 PC spectrometer).

3. RESULTS AND DISCUSSION

3.1. HT-XRD

The analysis of the as-synthesized HTlc revealed that the ratio between cations in the sample (Co/Al = 2.9/1) is close to that in the starting solution (Co/Al = 3/1), hence the co-precipitation step was carried out effectively. The HT-XRD pattern of the as-synthesized material (Fig. 1) shows the hydrotalcite structure as the only crystalline component (JCPDS 22-700). Assuming a 3R stacking of the layers, the lattice parameters can be calculated as a = 0.308 nm and c = 2.286 nm. The shift of the basal spacing d(003) to lower d values upon heating from 25 to 150°C, indicates removal of physically adsorbed water and interlayer water of the hydrotalcite. At 150°C the sample retains a layered structure, although the disappearance of the (006) and (009) planes indicates disorder in the stacking of the layers [16]. In the diffraction pattern at this temperature, new broad reflections at 2θ values around 36° and 65° can be also identified, which suggests the presence of a mixture of the emerging mixed oxide phase and the dehydrated HTlc. This metastable intermediate recovers to the original layered material after exposure to the ambient for 12 h. At 200°C, the hydrotalcite structure collapses completely, leading to a poorly crystallized spinel-like cobalt mixed oxide, in agreement with the decomposition of monometallic Co(II)-Co(III) hydrotalcite [15]. No reconstruction of this sample is observed, even after 12 h in contact with an aqueous solution of 0.5 M Na_2CO_3 [17]. No amorphous phase could be observed in the TEM micrographs of the sample decomposed at 200°C for 16 h (not shown), which suggests the intimate dissolution of Al^{3+} ions in the main Co_3O_4 phase, forming a solid solution of cobalt spinels $(Co(Co,Al)_2O_4)$. Upon increasing the temperature, diffraction lines of the oxide phase sharpen as a consequence of enhanced crystallinity. The cell parameter a of the oxide phase is below the cell parameter of pure Co_3O_4 and $CoAl_2O_4$ in the temperature range investigated [9], which indicates the formation of non-stochiometry, in the form of anion vacancies and/or dissolution of Al^{3+} ions in the final phase.

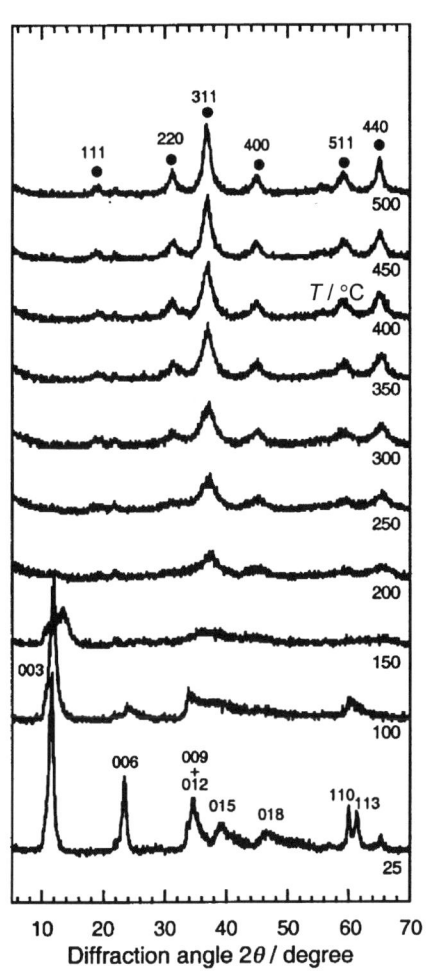

Figure 1. HT-XRD patterns during thermal decomposition of Co-Al hydrotalcite in air at different temperatures; (●) Co-spinel.

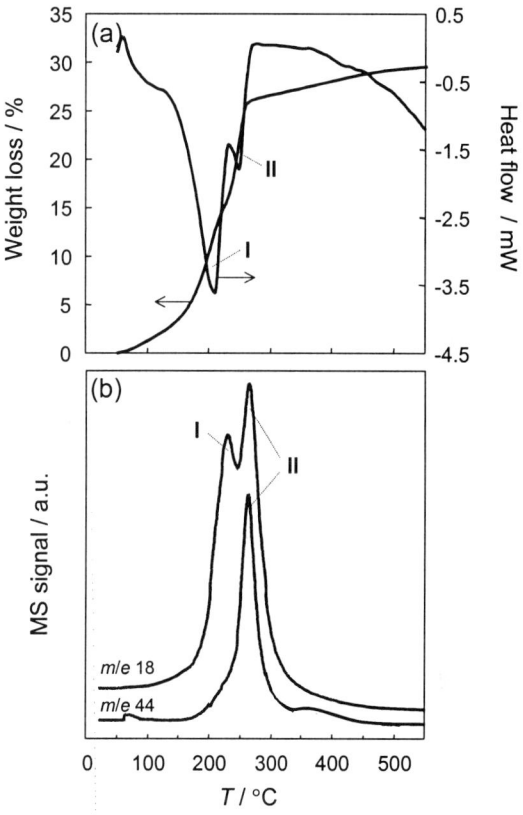

Figure 2. TGA-DTA and MS profiles during thermal decomposition of Co-Al hydrotalcite in air; heating rate = 10°C/min.

3.2. TGA-DTA-MS

The thermogravimetric profiles of Co-Al-HTlc shows that decomposition occurs in two main steps (I and II in Fig. 2a), as typically observed for layered materials [6,10]. Interpretation of the TGA-DTA profiles is based on the MS analysis of the evolved gases (Fig. 2b). The two overlapping m/e 18 signals are assigned to the removal of interlayer water (I) and dehydroxylation of the brucite-like layers (II), respectively. CO_2 was also produced in various temperature stages. A strong m/e 44 signal is observed around 265°C (transition II), suggesting decarbonation, which is coupled with the dehydroxylation stage. At around 375°C, an additional weak and broad m/e 44 signal is found, which is probably related to the residual weight loss observed above 350°C in the TGA profile.

3.3. *In situ* FT-IR

In situ infrared spectra during the decomposition of Co-Al-HTlc in air are shown in Fig. 3. Spectral assignments have been extensively discussed elsewhere [9,12]. In the spectrum at room temperature, the absorption band at 3545 cm^{-1} is attributed to the H-bonding stretching vibrations of the hydroxyl group and the stretching vibration of water molecules. The shoulder present at 3050 cm^{-1} has been assigned to hydrogen bonding between H$_2$O and the carbonate anion in the interlayer space. The H$_2$O bending vibration of free water molecules (physically adsorbed) is located at 1650 cm^{-1}, while the shoulder at 1750 cm^{-1} is assigned to a restricted bending mode of water in the interlayer. The main absorption bands of the carbonate anion in the interlayer are located at 1414 cm^{-1} (ν_3, asymmetric stretching) and at 833 cm^{-1} (ν_2, out-of-plane deformation), which are typical of carbonates in a symmetric environment (D$_{3h}$ symmetry). The presence of a shoulder at 1350 cm^{-1} (splitting of the ν_3 mode) indicates a low symmetry of the carbonates in the interlayer space [18]. In the low-wavenumber region information about the octahedral layers of the HT structure can be obtained. The bands centered at 625 and 471 cm^{-1} are assigned to Co-O and Al-O stretching in the brucite-like sheet, respectively, while the band at 955 cm^{-1} is assigned to a Al-OH deformation [11,19].

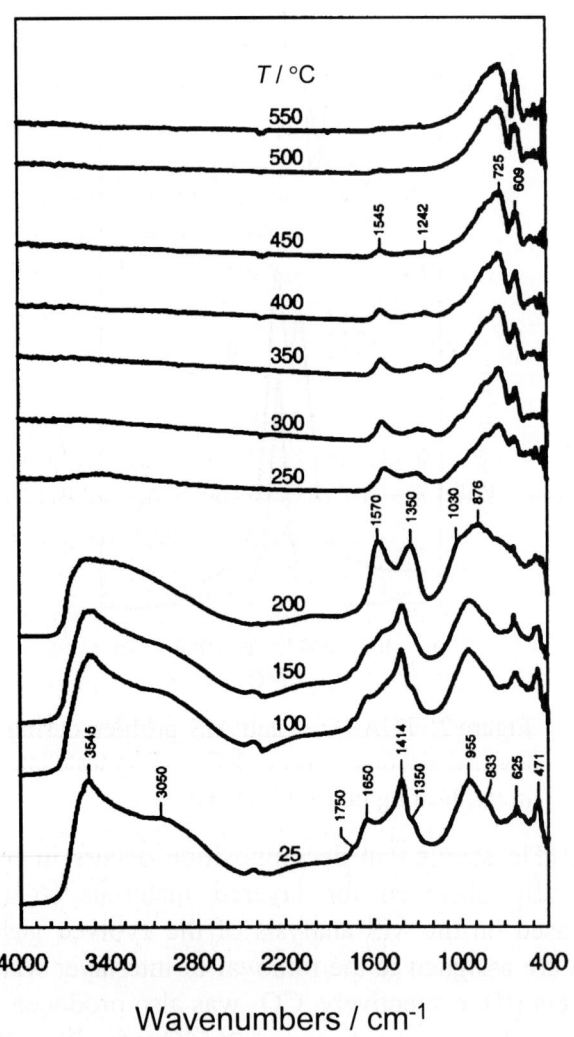

Figure 3. *In situ* FT-IR spectra for the thermal decomposition of Co-Al hydrotalcite in air at different temperatures.

Upon heating from 25 to 150°C, the band centered at 1650 cm^{-1} disappears. In this temperature range, there is also a decrease in intensity of the shoulders at 1750 and 3050 cm^{-1}, indicating partial removal of interlayer water. At 200°C interlayer water is completely removed and the carbonate band at 1414 cm^{-1} splits into two new bands at 1350 and 1570 cm^{-1}, indicating a rearrangement of carbonate in the interlamellar space and lowering of the carbonate symmetry (to C$_s$ or C$_{2v}$). This process also induces absorption at the position of the ν_1 mode of the carbonates (symmetrical stretching) at 1030 cm^{-1}, which is infrared inactive when the carbonate ion retains its full symmetry [20].

In the temperature range of 200-250°C, significant dehydroxylation and also decarbonation occurs, indicating coupling of both processes. This result confirms a crucial role of the intermediate dehydrated structure in the carbonate symmetry and intralayer

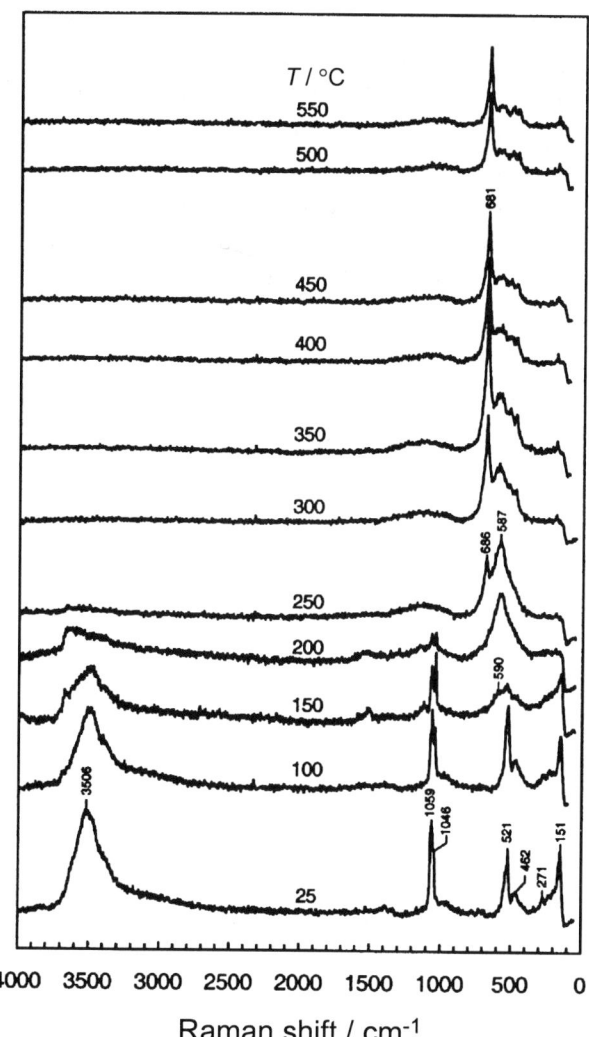

Figure 4. *In situ* Raman spectra for the thermal decomposition of Co-Al hydrotalcite in air at different temperatures.

bonding to approach the final decomposed material [3]. Complete decarbonation requires higher temperatures (500°C) than complete dehydroxylation (250°C), explaining the residual weight loss and CO_2 signal in the TGA-DTA-MS profiles above 350°C. The original band at 625 cm^{-1} is overlapped by the 700 cm^{-1} band of the emerging phase at 150°C, while the band at 471 cm^{-1} is hardly visible. The simultaneous presence of absorption bands of hydrotalcite and mixed oxide suggests the presence of the intermediate mixture of phases, as concluded from the HT-XRD pattern at 150°C. At 250°C a doublet centered at 725 cm^{-1} and 609 cm^{-1} appears, typical of a normal II-III spinel compound. These bands are shifted with respect to the data reported for pure spinel compounds, which indicates the presence of a solid solution of cobalt spinels [21].

3.4. *In situ* Raman

In situ visible Raman spectra during the decomposition of Co-Al-HTlc in air are shown in Fig. 4. Analogous to the infrared analysis, the absorption band around 3500 cm^{-1} in the spectrum at room temperature is attributed to the H-bonded stretching vibrations of the hydroxyl groups of the brucite-like layers, and stretching vibrations of water molecules. This spectrum contains a strong and sharp band at 1059 cm^{-1} (in-plane OH bending) with a second band at slightly lower Raman shift (1046 cm^{-1}, ν_1 mode of the carbonate ions). The low frequency region of the spectrum at room temperature contains bands at 521 and 462 cm^{-1}, which are assigned to hydroxyl groups associated with mainly Al, both also influenced by probably one Co in its coordination. The band at 462 cm^{-1} is only Raman active while the band at 521 cm^{-1} has an equivalent in the infrared spectrum at 625 cm^{-1}. The bands at 151 and 271 cm^{-1} have been also related to the hydrotalcite framework [9,12].

From 25 to 150°C, the bands at 1046 and 1059 cm^{-1} decrease in intensity, suggesting dehydroxylation and decarbonation, and the intensity ratio of the two peaks is reversed compared to the situation in the as-synthesized material. At 200°C, the characteristic bands of the hydrotalcite framework are overlapped by a new sharp band around 590 cm^{-1}. The bands at 151 and 271 cm^{-1} are no longer visible at this temperature. The bands at 1046 cm^{-1} and

1059 cm^{-1} further decrease in intensity and the shoulder at 3400 cm^{-1} disappears, indicating complete dehydration of the interlayer space. The above mentioned band at 590 cm^{-1} should be related to the emerging oxide phase and was assigned to a dispersed surface cobalt oxide phase [12]. Apparently, the layered hydrotalcite losses hydroxyls upon heating, leading to a first (preliminary) oxide phase, where the cobalt is dispersed upon emerging cobalt aluminate. At 250°C the dehydroxylation of the sample is complete. The intensity of the carbonate band decreases and a new band appears at 681 cm^{-1}, which is related to the formation of the spinel phase. Complete decarbonation of the material is identified around 400°C.

3.5. Combination of *in situ* techniques

Table 1 summarizes the transition temperatures for the different processes involved during thermal decomposition in air as derived from the different analytical techniques. The results of the various *in situ* techniques substantiate each other and in general are in good agreement.

Table 1.
Summary of the transition temperatures (in °C) during the decomposition of Co-Al hydrotalcite in air obtained by different techniques.

Process	HT-XRD	TGA-DTA	MS	FT-IR	Raman
Complete dehydration	150[a]	183	190	200	n.i.
Collapse of the HTlc	200[b]	n.i	n.i	250	150-200
Complete dehydroxylation	n.i	249[c]	249	300	250
Complete decarbonation	n.i		375	500	400

n.i. not identified by the technique.
[a] coexistance of dehydrated layered structure and emerging cobalt oxide.
[b] leading ultimately to a solid solution of Co-spinels (Co(Co,Al)$_2$O$_4$).
[c] dehydroxylation and decarbonation cannot be identified separately.

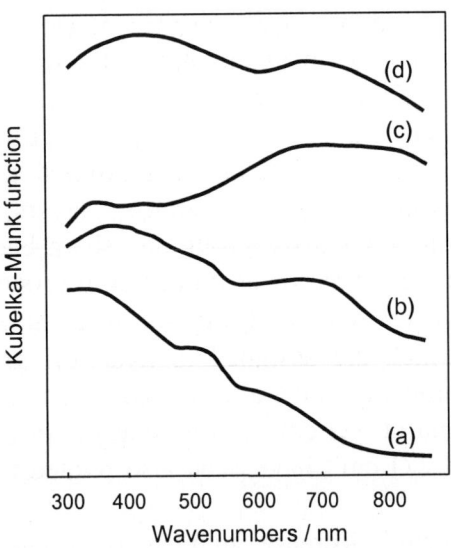

Figure 5. Diffuse reflectance UV-Vis spectra of (a) Co-Al hydrotalcite, (b) after treatment at 150°C for 1 h, (c) after treatment at 200°C for 1 h, and (d) Co$_3$O$_4$. Reference: MgO.

3.6. UV-Vis/DR

UV-Vis/DR spectra provide relevant information concerning to the oxidation of Co^{2+} to Co^{3+} upon thermal decomposition (Fig. 5). The spectrum of the sample decomposed at 150°C produced a broad absorption band at 600-800 nm, probably due to a Co^{3+} state, which is absent in the spectrum of the as-synthesized hydrotalcite. This suggests that the thermal treatment at 150°C partially oxidizes Co^{2+} species, which is in agreement with the relatively broad diffraction lines ascribed to the solid solution Co(Co,Al)$_2$O$_4$ in the HT-XRD pattern at this temperature. The UV-Vis/DR spectrum of the sample treated at 200°C shows an almost continuous absorption above 500 nm, very similar to that shown for Co$_3$O$_4$ (used as reference). The shoulder recorded around

500 nm in the as-synthesized material where the mixed oxide was absent, arises from Co^{2+} in an octahedral environment [CoO_6] in the brucite-like layers [22]. This shoulder is also present in the sample treated at 150°C, in agreement with the identification of the dehydrated hydrotalcite phase in the XRD pattern. The shoulder disappears after treatment of the sample at 200°C. These results are in excellent agreement with [15], where the oxidation of Co^{2+} was monitored in the range of 150-180°C by temperature-programmed oxidation.

3.7. Decomposition and reconstruction model

Co-based HTlcs decompose at lower temperatures than other hydrotalcites (*e.g.* Ni-Al, Mg-Al, or Mg-Fe systems), where no collapsing of the layered structure is observed below 523 K and 573 K [5-8]. The low thermal stability of Co-Al-HTlc has been related to the partial oxidation of Co^{2+} to Co^{3+}, which was even observed during the preparation of the layered material [5,23,24].

The decomposition and reconstruction processes can be schematically represented by the model in Fig. 6. The transition from (a) to (b) at 150°C, showing a decreasing $d(003)$ distance, is caused by the removal of interlayer water. This induces the diffusion of the trivalent cation from the octahedral layers to the interlayer space in tetrahedral coordination [3]. Simultaneously, there is a certain dehydroxylation of the octahedral layers, as observed in the infrared spectra, leading to the coexistence of both the dehydrated layered hydrotalcite and an emerging Co-spinel phase. This situation is recoverable since dehydroxylation is not severe, and few Co^{2+} ions have been oxidized to form metal-oxygen bonds between layers. Above 150°C the memory effect starts to fade and above 175-200°C is no longer observed. The transition to (c) at 200°C describes to the collapse of the layered structure by progressive intralayer dehydroxylation, thus increasing the amount of O-Co-O bridges between octahedral sheets. At higher decomposition temperatures (> 200°C) the reconstruction is even more difficult due to the high stability of the solid solution of spinels formed. The infrared spectrum at 400°C proves that, both dehydroxylation and decarbonation are almost complete, explaining the absence of a reconstruction process [25]. However, we additionally show that reconstruction is no longer possible after decomposition at 200°C, even though the sample still contains a very significant amount of hydroxyls and carbonates.

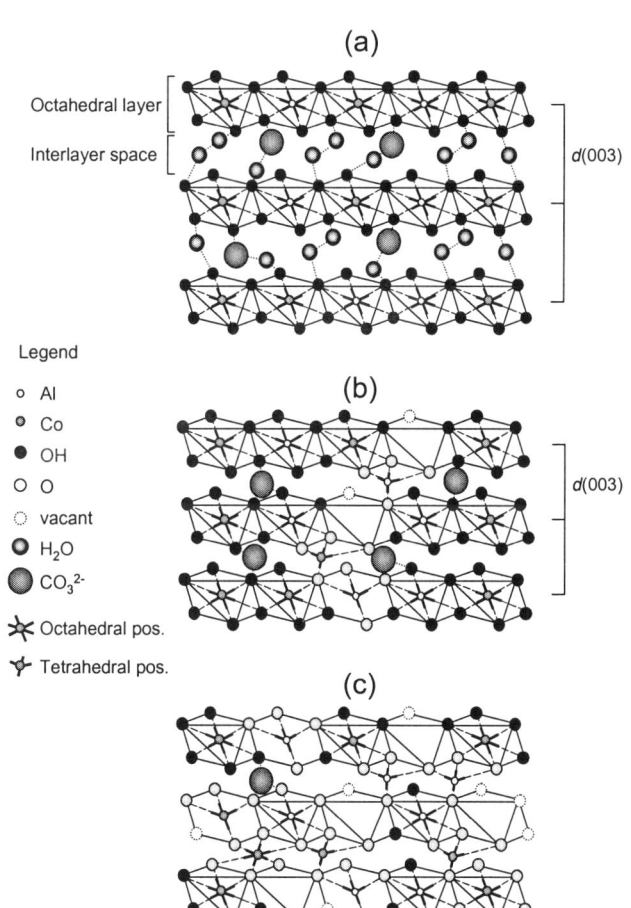

Figure 6. Schematic representation of the thermal decomposition of Co-Al hydrotalcite in air.

4. CONCLUSIONS

A combination of *in situ* techniques, including HT-XRD, TGA-DTA, MS, and FT-IR and Raman spectroscopies, provide a detailed description of the structural and physico-chemical transformations upon heating of the Co-Al hydrotalcite. Based on the information obtained from the various techniques a structural model describing the decomposition mechanism has been proposed. Partial oxidation of Co^{2+} to Co^{3+} during decomposition in air, assessed by UV-Vis/DR spectroscopy, leads to the formation of a highly stable and homogeneous solid solution of cobalt spinels, already identified at 150°C. The diffusion of trivalent cations (mainly Co^{3+}) to the interlayer space after removal of interlayer water favours the collapse of the layered structure by formation of strong O-Co-O bonds between layers, which makes the recrystallization of the layered structure very difficult, at decomposition temperatures even as low as 175-200°C.

REFERENCES

1. F. Cavani, F. Trifirò and A. Vaccari, Catal. Today, 11 (1991) 173.
2. A. Vaccari, Appl. Clay Sci., 14 (1999) 161, and references therein.
3. M. Belloto, B. Rebours, O. Clause, J. Lynch, D. Bazin and E. Elkaïm, J. Phys. Chem., 100 (1996) 8535.
4. W. T. Reichle, S. Y. Chang and D. S. Everhardt, J. Catal., 101 (1986) 352.
5. M. A. Ulibarri, J. M. Fernández, F. Labajos and V. Rives, Chem. Mater., 3 (1991) 626.
6. T. Sato, H. Fujita, T. Endo, M. Shimada and A. Tsunashima, React. Solids, 5 (1988) 219.
7. J. Rocha, M. del Arco, V. Rives and M. A. Ulibarri, J. Mater. Chem., 9 (1999) 2499.
8. F. Millange, R.I. Walton, and D. O'Hare, J. Mater. Chem., 10 (2000) 1713.
9. J. Pérez-Ramírez, G. Mul, F. Kapteijn and J.A. Moulijn, J. Mater. Chem., 9 (2001) 25.
10. J. Pérez-Ramírez, G. Mul, F. Kapteijn and J.A. Moulijn, Appl. Catal. A., 204 (2000) 265.
11. J. T. Kloprogge and R. L. Frost, Appl. Catal., A., 184 (1999) 61.
12. J. Pérez-Ramírez, G. Mul and J.A. Moulijn, Vib. Spectroscop., 27 (2001) 75.
13. F. Rey, V. Fornés and J. M. Rojo, J. Chem. Soc. Faraday Trans., 88 (1992) 2233.
14. B. Wu, S. Wang and Y. Zhuang, Spectrosc. Lett., 30 (1997) 1165.
15. B. Zapata, P. Bosch, G. Fetter, M.A. Valenzuela, J. Navarrete and V.H. Lara, Int. J. Inorg. Mat., 3 (2001) 23.
16. M. del Arco, P. Malet, R. Trujillano and V. Rives, Chem. Mater., 11 (1999) 624.
17. J. Pérez-Ramírez, G. Mul, F. Kapteijn and J.A. Moulijn, Mater. Res. Bull., 36 (2001) 1767.
18. D. L. Bish and G. W. Brindley, Amer. Min., 62 (1977) 458.
19. J. T. Kloprogge and R. L. Frost, J. Solid. State Chem., 146 (1999) 506.
20. F. M. Labajos, V. Rives and M. A. Ulibarri, J. Mater. Sci., 27 (1992) 1546.
21. J. Preudhomme and P. Tarte, Spectrochim. Acta, 27A (1971) 1817.
22. R.A. Schoonheydt, R. De Vos, J. Pelgrins and H. Leeman, Stud. Surf. Sci. Catal., 49A (1989) 559.
23. M. del Arco, R. Trujillano and V. Rives, J. Mater. Chem., 8 (1998) 761.
24. Z.P. Xu and H.C. Zeng, J. Mater. Chem., 8 (1998) 2499.
25. N.S. Puttaswamy and M. Rajamathi, J. Mater. Chem., 7 (1997) 1941.

Some structural properties of an Al-PILC influenced by grinding the starting montmorillonitic clay

M. Sergio[1,2,*], S. Cardozo[2], C. Froche[2], M. Bentancor[2], M. Musso[3,4], J. Medina[2], W. Diano[1]

[1] Laboratorio de Fisicoquímica de Superficies, Facultad de Química, Universidad de la República, C.C. 1157, Montevideo, Uruguay.

[2] Laboratorio de Fisicoquímica, [3] UNCIEP, Facultad de Ciencias, Universidad de la República, Iguá 4225. CP 11400. Montevideo, Uruguay.

[4] Departamento de Geotécnica, Facultad de Ingeniería, Universidad de la República, J.M. Reissig 565, CP 11300. Montevideo, Uruguay.

The influence of the type of grinding of the clay mineral, before Al_{13}-pillaring on the textural parameters of the solids was studied. A pink Uruguayan montmorillonite was ground using two different blenders: a commercial home coffee blender and a Lab cutting mill. For each of the milled samples a fraction less than 200 μm and the whole mineral were selected for pillaring. Yields were obtained of 90% relative to the starting raw mineral. Sample characterization was performed by nitrogen adsorption isotherms and XRD. The results indicate that an increase of the shear stress in the milling process produces structures more suitable for pillaring. In fact when the Lab cutting mill was used, specific micropore volumes of 0.179 mL/g and specific surface areas of 430 m^2/g were obtained for both samples when the solids were calcined at 550°C for two hours. On the other hand the milder milling conditions gave smaller textural parameters that were different for the two particle sizes selected.

1. INTRODUCTION

Interest in pillared clays developed in the eighties because they were considered as substitutes for zeolites as cracking catalysts. PILCs are prepared exchanging the interlayer cations of smectites by voluminous cations that are fixed to the clay layer by gentle calcination. A three-dimentional pore structure, with pore openings in the range of micropores is then produced. Pore diameters depend on the size of the cation used for exchange as well as on the pillar density per unit area of the interlamellar space achieved

---

[*] E-mail: msergio@bilbo.edu.uy

during the pillaring process. Both parameters can be tailored within certain range. The possibility of producing solids with favourable pore openings to meet process requirements (1) appeared especially attractive. The Al_{13} oxocation was the first inorganic species used as a pillaring agent (2, 3) and has since been one of those most studied. Thermal and hydrothermal stability of the structure hindered their use as cracking catalysts from the very beginning. Today a very wide field of applications is considered for pillared clays.

The classic preparation method has involved pillaring of the clay fraction, which was selected by Stokes gravity sedimentation after converting the mineral to the homoionic form normally, the sodium or calcium one. In this method the pillaring agent solution was added to the homoionic clay suspension, and as diluted solutions were used, very large volumes need to be handled. Following Vaugham's suggestion (4) concentrated clay suspensions have been used (5, 6, 7) for pillaring. Sodium montmorillonite slurries form gels at concentrations above 4% and become impossible to handle. Diano et al. (8) have reported pillaring of concentrated calcium montmorillonite slurries with 14% of solids; in this case the clay suspension was added to the pillaring agent solution. In any case, the process of obtaining the homoionic form, washing it to discard the excess of salt and selecting the clay fraction (<2μm) is troublesome, time consuming and expensive.

In this paper the results obtained when the clay mineral was pillared without any previous exchange are presented and the influence of the grinding process on the textural properties of the final product is analyzed.

2. EXPERIMENTAL

The montmorillonitic clay. A Uruguayan montmorillonite (Cerro Largo-Uruguay) was used. The mineral is pink and contains black spots in which manganese was recognized. Taylor (9) has attributed the rose color in those minerals to manganese. The composition of the Uruguayan clay mineral is shown in Table 1.

Quagliatta and Valentín (10) determined iron using Mössbauer spectroscopy; all of it was present as Fe^{+3}, with 98.2% as Fe (VI) and 1.8% as Fe (IV).

The cation exchange capacity (CEC) was 1.1 meq/g of dry mineral, determined by the acetate method, pH = 7. The interlayer cations were analyzed by atomic absorption spectroscopy and atomic emission spectroscopy. The results in meq/g were as follows: Ca^{+2} 0.83; Mg^{+2} 0.29; Na^+ 0.012; K^+ 0.002. Basal space for the mineral was 1.47 nm.

Mineral pretreatment. The mineral was oven dried at 105°C for 24 hours and ground. Samples A were prepared using a home coffee blender and samples B were obtained using a Lab cutting mill (Ketschmühle, tip WRb 90 Lb/4p). In both cases the whole ground mineral

Table 1
Uruguayan montmorillonite composition.

SiO_2	Al_2O_3	Fe_2O_3	MnO	MgO	CaO	Na_2O	K_2O	TiO_2	P_2O_5
55.55	16.17	1.86	0.304	4.50	2.38	0.09	0.20	0.095	0.02

and a fraction of less than 200 μm were selected. A 10% suspension of the clay mineral was prepared and left overnight under stirring. A commercial suspendent agent (2%), Kelzan KC

(Merck & Co, Inc.. Kelco Division), was added to the slurry, and stirred for 30 minutes before pillaring.

Pillaring solution. The pillaring agent was the Al_{13} oxocation. A 0.2 M NaOH (Carlo Erba, 98%) solution was added dropwise to a 0.2 M $AlCl_3$ (Merck, anhydrous, 97%) solution with vigorous stirring until the OH/Al molar ratio was 2.0. Aging for an hour at 50°C and under stirring gave a final pH of 3.95. According to Bottero (11), 95% of the aluminium is present as the Al_{13} cation under these conditions.

Pillaring process. In all cases the pillaring agent solution was added dropwise to the clay suspension while under vigorous stirring. The final Al : clay ratio was 5 mmol/ g and the pH of 3.5 was adjusted to 5.5 with ammonium hydroxide. The slurry was then stirred for 3 hours at 80°C and left overnight at room temperature. The solids were vacuum-filtered and washed with hot deionised water with a conductivity of < 1μS/cm. The washing process was repeated with the conductivity controlled after the solids had been vacuum filtered. The final conductivity was < 20μS/cm. The solids were oven dried at 60°C for 48 hours and kept in a desiccator with silica gel. Calcinations were performed in an horizontal furnace (Carbolite CTF-12/75 with a Eurotherm temperature controller); the temperature was raised by 1.0°C/min and then maintained at 550°C for 2 hours. Samples were labeled after the type of grinding as A or B and a subscript was used to designate the pillared fraction: g identifies the whole ground mineral and 200 the fraction under 200 μm. A_{200}, A_g, B_{200} and B_g represent PILC samples calcinated in layers of about 2 mm thickness as is usually done in the preparation of many catalysts. $B_{g\,(powder)}$ and $B_{g\,(core)}$ derived from pillared clays calcined in layers of about 2-3 cm thick. $B_{g(powder)}$ were obtained by scratching the external surface of the calcined pieces and the $B_{g\,(core)}$ were selected from the cores of these pieces.

Cation exchange capacity. The cation exchange capacity (CEC) was determined using the ammonium acetate pH = 7 centrifuge method. An ammonium acetate solution, pH = 7, was prepared using acetic acid (Merck, glacial) and ammonium hydroxide (Merck, 28%). Samples of 200 mg were exchanged with ammonium three times, washed three times with deionised water and a last wash was performed with anhydrous ethanol. The amount of retained ammonium was determined with a Kjeldtec System, 1002 Distilling Unit. The CEC is expressed as miliequivalents per gram of dry sample.

Adsorption isotherms. These were carried out at –196°C in an Autosorb-1 equipment (Quantachrome) with micropore option. BET (12) specific surface areas (s.s.a.) are presented. Specific micropore volumes were determined after Dubinin-Radushkevich ($V_{\mu pore\,DR}$) diagrams (13). Specific micropore volumes ($V_{\mu pore}$) and external specific surface area (S_{ext}) were derived from t-plots (14). Specific total pore volume (V_{total}) was determined from the nitrogen adsorption of the isotherm at $p/p_o = 0.95$; 0.808 g/mL was used as the specific weight for the liquid nitrogen. Mesopore size distribution analysis was performed after Barret, Joyner and Halenda's (15) method (BJH). Micropore size distribution was analyzed after Horvath and Kawazoe's (HK) proposal (16). Data derived with BET, Langmuir, t-plot

Table 2 .
Range of p/p_o used for analysis of nitrogen adsorption data.

	BET	Langmuir	t-plot	Dubinin-Radushkevich
p/p_o	$7\times10^{-3} - 0.1$	$2\times10^{-3} - 5\times10^{-2}$	$0.80 - 0.95$	$0.01 - 0.1$

and DR theories from nitrogen adsorption data give good linearity in the p/p_o ranges listed in Table 2.

3. RESULTS

Figure 1 shows the nitrogen adsorption isotherms for the pillared samples and Table 3 the textural parameters for the pillared clays.

The mesopore size distribution diagrams (BJH) show no peak for any of the samples. BET and Langmuir s.s.a are listed; the values derived with Langmuir equation are slightly higher (8%). Specific total pore volumes for B samples are twice those of A samples. Specific micropore volumes determined after DR and t-plots show no significant difference for each sample. They are approximately half of the specific total pore volumes calculated from the adsorption at $p/p_o = 0.95$. The effective pore size determined following the HK method is 0.62 and 0.72 nm for pillared samples A and B, respectively. Cation Exchange Capacity represents $\frac{1}{3}$ and $\frac{1}{4}$ of the value determined for the clay mineral for A and B pillared samples respectively.

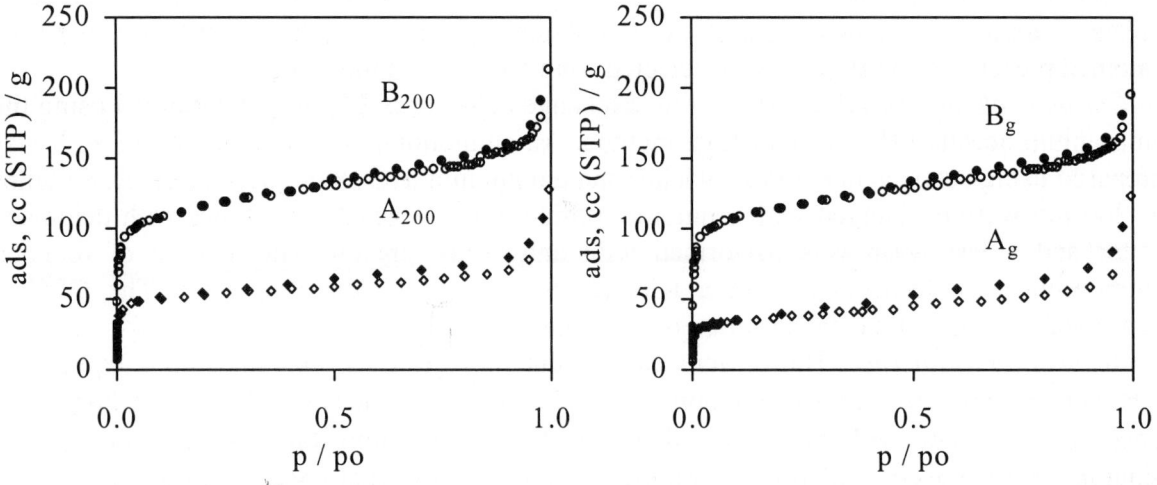

Figure 1. Nitrogen adsorption isotherms for pillared clays. Open marks: adsorption. Filled marks: desorption. References are in the text.

Figure 2 (a) shows the XR Diffraction d_{001} peaks of the clay mineral ground with the Lab cutting mill (1.47 nm) and that of the pillared sample B_g (1.81 nm). The d_{001} peak for B_{200} (1.85 nm, not shown) is superposed on that of the B_g. Figure 2 (b) shows the d_{001} peaks of clay mineral and pillared samples B_g (1.81nm) and A_{200} (1.45 and 1.80 nm).

Table 3.
Textural parameters of the pillared clays

		A		B			
		A_{200}	A_g	B_{200}	B_g	$B_{g\,(powder)}$	$B_{g\,(core)}$
BET s.s.a.	(m²/g)	199	137	431	423	430	436
Langmuir s.s.a.	(m²/g)	214	152	463	456	464	472
V_{total} p/po=0.95	(mL/g)	0.121	0.105	0.256	0.238	0.246	0.249
DR							
$V_{\mu pore\,DR}$	(mL/g)	0.083	0.060	0.179	0.176	0.179	0.182
t-plot							
$V_{\mu pore}$	(mL/g)	0.070	0.05	0.194	0.194	0.196	0.187
S_{ext}	(m²/g)	31	32	37	27	30	37
HK pore size	(nm)	0.68	0.62	0.72	0.72	0.72	0.72
CEC	(meq/g)	0.42	0.40	0.34	0.33	0.33	0.33
XRD	(nm)	1.45 - 1.80	-	1.85	1.81	1.81	1.81

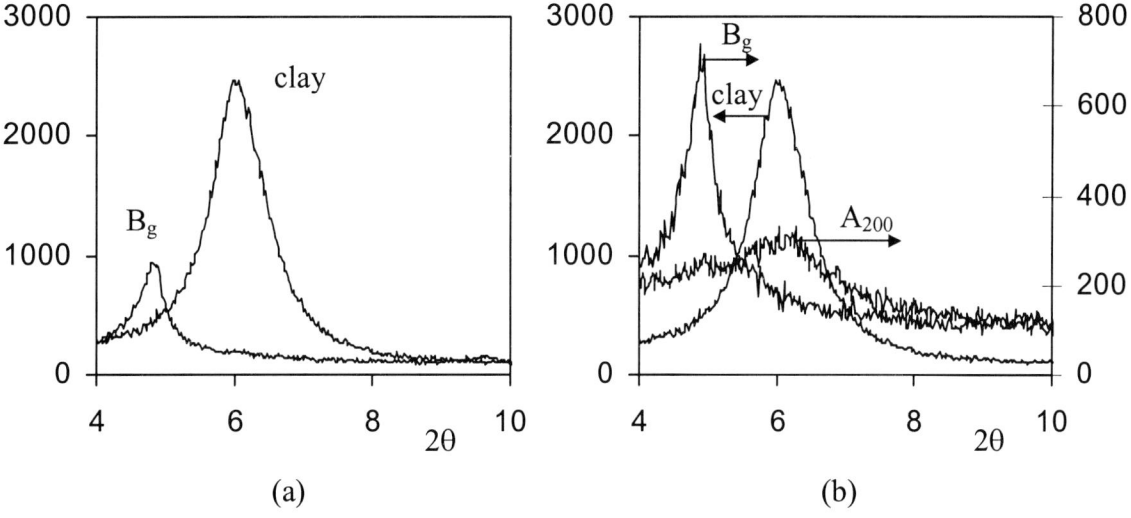

Figure 2. d_{001} of: (a) clay mineral and B_g sample; (b) clay mineral and samples A_{200} and B_g referred as shown to ordinate axis in cps.

4. DISCUSSION

The Cerro Largo montmorillonite from Uruguay has Ca^{+2} as its main exchange cation in accordance with the d001 XRD peak (1.47 nm) and good crystallinity.

The pillaring procedure presented here uses the clay mineral with any other previous pretreatment than milling. Two different mills were used and two different samples were selected from each of the milling processes. The pillaring agent was the Al_{13} oxocation and it was fixed to the clay layer at 550°C. Pillared samples were characterized by nitrogen adsorption isotherms and XRD.

For all samples, nitrogen adsorption isotherms were type I in the BDDT classification (17). A very small or quite non-existent hysteresis loop suggests the absence of a defined structure of mesopores for A and B samples; this observation is in accordance with the BJH pore size distribution diagrams that show no peaks. The increase in the adsorption with relative pressure can be assigned to pores in interparticle spaces.

The adsorption at low values of p/p_o, characteristic of microporous structures, was higher for B samples and does not depend on the particle size of the mineral selected for pillaring. For A samples instead this adsorption was higher when the small particle size fraction of the mineral was pillared. These observations are in accordance with the values of the different textural parameters analyzed here: specific micropore volume and specific surface area.

A good correlation was observed for specific micropore volumes determined after DR and t-plots when the appropriated p/p_o ranges were selected.

The d_{001} peaks from XRD data show two peaks for sample A_{200}, the higher of them corresponding to the ground mineral and a smaller one to the pillared clay, as if only a partial pillaring was attained. Instead, the B_{200} and the B_g samples show d_{001} peaks for pillared clay and none for the starting mineral as if most of the interlayer spaces were pillared. The thickness of the montmorillonite layer reported by van Olphen (18) was 0.96 nm; therefore the interlayer space derived from XRD data should be 0.85 nm.

The analysis of the nitrogen adsorption isotherms at low relative pressures by the Horvath-Kawazoe theory gives pore size in the range of 0.72 nm for B samples and 0.62-0.68 nm for the A samples. These values are in accordance with those derived from XRD data.

The comparison of the data obtained for the B samples in the three right hand columns in Table 3, shows that the products display a good homogeneity.

The surface area of clay layers that is introduced by the pillaring process can be calculated after Diano et al (19) as:

$$S_L = S_{N_2} + S_P - S_{ext}$$

where S_L is the surface area of the layers, S_{N2} is the total area computed, for example by BET, S_P is the area occupied by the bonded pillars to the clay sheets and S_{ext} the external surface area calculated from nitrogen adsorption isotherm data using the t-plot. The total amount of aluminium taken up during pillaring was determined for B samples to be 3 mmol/g. Considering that 13 aluminium atoms form a pillar and that each pillar covers an area of 1 nm^2 (20, 21) S_L was 539 m^2/g. This result compared with the maximum crystallographic surface area of 750 m^2/g (22) indicates that 72% of the theoretical s.s.a. was exposed to nitrogen adsorption by pillaring.

The cation exchange capacity of the starting montmorillonite (1.1 meq/g) was sharply reduced after pillaring to 0.3-0.4 meq/g. Grim (23) states that 20% of the CEC of a montmorillonite can be assigned to the broken bonds in the particle edges and the remaining to the substitutions within the lattice.

The higher specific micropore volume and the smaller hysteresis loop for B samples can be explained in relation to the higher shear stress of the mill used in the pretreatment of the mineral.

The BET specific surface area values for B samples were twice that of A_{200} sample. The s.s.a. values of 430 m^2/g obtained for B samples are higher than those reported by Sergio and Diano (24), i.e. 370 m^2/g when a clay fraction of a homoionic sodium montmorillonite (< 2μm) was pillared at 550°C.

5. CONCLUSIONS

It has been shown that the grinding process used to prepare the clay mineral for pillaring determines the textural properties of the resulting solid. When a high shear stress mill was used, textural properties of pillared montmorillonites were not related to the particle size of the parent clay.

Using the method described here, it is possible to prepare solids in a reproducible way with high s.s.a. and a porous structure in the range of micropores.

The present paper describes a preparation process that is more economic and less time-consuming than others but provides a high yield of the product.

6. ACKNOWLEDGEMENTS

The authors are indebted to the following Institutions of Uruguay for supporting the present work: CONICYT from the Ministry of Education, CSIC from the University of the Republic and the Program for Supporting Basic Science (PEDECIBA).

REFERENCES

1. K.Suzuki, M. Horio, and T. Mori, Mat. Res. Bull. 23 (1988) 1711.
2. G.M. Brindley and R.E. Sempels, Clay Miner. 12 (1970) 229.
3. N. Lahav, U. Shani and J. Shabtai, Clays and Clay Miner. 26 (1978) 107.
4. D.E.W. Vaughan, Catal. Today, 2, (1988) 187.
5. R. Rubino, Thesis, Universidad de la República, Montevideo, Uruguay, (1992).
6. R.A. Schoonheydt and H.Leeman, Clay Miner. 27 (1992) 249.
7. R. Molina, A. Vieira-Coelho and G. Poncelet, Clays Clay Miner. 40 (1992) 480.
8. W. Diano, R. Rubino and M. Sergio, Microporous Materials. 2 (1994) 179.
9. R.M. Taylor, in H. van Olphen (eds), Proc. International Clay Conference, Elsevier (1981) 749.
10. E. Quagliatta and R.Valentín, Personal Comunication (1989).
11. J.Y.Bottero, J.M.Cases, F.Fiessinger and J.E.Poirier, J.Phys.Chem. 84 (1980) 2933.

12. S.J. Gregg and K.S.W. Sing, Adsorption, Surfase Area and Porosity. Academic Press, New York, NY, 2nd Ed. (1982).
13. M.M. Dubinin and L.V. Radushkevich, Proc. Acad. Sci. USSR 55 (1947) 331.
14. F. Rouquerol, J. Rouquerol and K. Sing, Adsorption by Powders and Porous Solids. Principles, Methodology and Applications. Academic Press (1999) 345.
15. E.P. Barret, L.G. Joyner and P.P. Halenda, Amer.Chem.Soc. 73 (1951) 373.
16. G. Horvath and K. Kawazoe, J.Chem.Eng.Jpn. 16 (1983), 470.
17. K.S.W. Sing, D.H. Everett, R.A.W. Haul, L. Moscou, R.A. Pierotti, J. Rouquerol, T. Siemieniewska, Pure Applied Chem., 57, 603, 1985
18. H. Van Olphen, An introduction to Clay Colloid Chemistry, Willey, NY (1977).
19. W.Diano, R. Rubino and M. Sergio, Microporous Materials, 2 (1994) 179.
20. T.J.Pinnavaia, V. Rainey, Tson Ming-Sing and J.M. White, J.Mol.Catal. 27 (1984) 213.
21. J.Y. Bottero, M. Burant and J.M. Cases, Clay Miner., 23 (1988) 213.
22. H.van Olphen, in D.H. Everett and R.H. Ottewill (eds), Proceedings of the International Symposium on Surface Area Determination, Bristol 1969, Butterworth Science Publishers (1970) 255
23. R.E. Grim, Clay Mineralogy. Mc.GRAW-HILL (1953)
24. M.Sergio and W. Diano, International Workshop on Activated Clays, (1998) 77.

Synthesis and characterization of pillared clays from Argentinean bentonites

C. Solar[a], E. Fernández[b], E. Perino[b], E. Strasser[c] and K. Sapag[a],*

[a]Laboratorio de Ciencias de Superficies y Medios Porosos *LacSuMP*. Departamento de Física, Universidad Nacional de San Luis (UNSL). 5700 San Luis. Argentina
*email: sapag@unsl.edu.ar

[b]Área de Química Analítica, Facultad de Química Bqca y Farmacia. UNSL

[c]Departamento de Geología, Fac. de Cs. Fco Matem. y Naturales. UNSL

The synthesis and characterization of pillared interlayer clays (PILCs) made from a natural clay, is studied. The source material was an Argentinean bentonite with a high percentage of montmorillonite, called *SG*, obtained from Minas Santa Gema, San Juan. The intercalated clays were prepared with oligocations of Al, Zr, and Fe. The samples were characterized by means of AA, ICP, DTA, TGA, XRD and N_2-Sorption techniques. These materials present important variations in their texture and structure, depending on the oligocations incorporated and the preparation methods. These characteristics give to them a potential application in different processes, like adsorption and catalysis, because of their surface and porosity.

1. INTRODUCTION

In adsorption and catalysis, great efforts are made with a view to obtaining new materials that could improve the processes [1]. In most of these processes, the appropriate materials are those that present their porosity in the range of micro and mesoporous, which have the adequate size for the formation of many molecules of interest [2].

Laminar clays, as montmorillonite, show good physical-chemistry properties that are useful for a great number of applications [3,4] mainly because of their capacity of sorption and their acidic properties [5]. As the weakness of these materials is the low stability of their porous structure and their low specific area, a proper variation in these properties would lead them to a better application [6].

The montmorillonite has interlayered cations with weak bonds. These cations are easily interchangeable and hydrolyzed producing a swelling in the structure and, as a result, the opening of the layer, which permits the interchange with bigger cations [7], like poliyhidroximetallic cations [8, 9, 10, 11]. Having these cations introduced between the layers, after a calcination process, they are transformed into oxides that act as molecular props (pillar) avoiding the collapse of the structure. These materials, denominated Pillared Interlayered Clays -PILCs- are structures that form galleries with dimensions given by the size of the incorporated oligocation and by the distance between them. This distance is determined by their Cationic Exchange Capacity (CEC) and their interlayer surface. The importance of the PILCs is given by the possibility of controlling their structure and texture by means of the nature and density of the pillaring cations.

Since the preliminary work of Vaughan et al [12] many articles have been published [13,14] concerning the synthesis of this kind of materials with different oligocations, specially metallic ions from water soluble salts. Those materials were synthesized using either natural or synthetic clays. Variations in the method of synthesis provide different characteristics to the resulting materials, being the most important variation parameters those corresponding to the formation, intercalation and later fixing between the layers of polinuclear cations [15,16]. In spite of that, there are not general rules about the best conditions for the synthesis.

In this work, it is introduced the synthesis and characterization of aluminum, zircon and iron pillared clays, using a natural clay from San Juan, Argentina. The effect of the different oligocations on the texture and structure of the clay is analyzed by different techniques.

2. EXPERIMENTAL

2.1 Methods and Techniques

The analysis of the previously disintegrated sample was made by Atomic Absorption Spectrophotometry (AAS), with an equipment Metrolab 315 aa/ae, and by Induced Coupled Plasma (ICP) with a Baird ICP 2070 equipment.

The study of the properties for different temperatures were made by thermal gravimetric analysis (TGA) and by differential thermal analysis (DTA). The TGA were realised with a TGA 51 Shimadzu equipment, and DTA with DTA 50 Shimadzu equipment.

The study of the changes produced in the structural and mineral composition was made by the X-Ray Diffraction technique (XRD) called "dust method" with a diffractometer RIGAKU GEIGERFLEX with Cu anode and Ni filter. The diffractograms were obtained from $\theta:2\theta$ with steps of $0.02°$ and a retention time of 1 second in each angle.

The BET-method was used to analyse the texture of the samples. Nitrogen isotherms were obtained with a nitrogen sorptometer Micromeritics® ASAP-2000. The data corresponding to the microporous region were calculated with the equation of Harkins and Jura, and the total pore volume (V_{Tp}) according to the Gursvitch rule (applied at $p/p°=0.98$). The samples (0.3 to 0.8 g) were previously degassed at 300°C during 18 hours approximately and then studied at 77K (-196°C).

2.2 Preparation of the Samples

2.2.1 Former Material

The approximated mineralogist composition of the former material obtained by XRD was montmorillonite (85%), quartz (10%) and feldspar (5%). After a purification process, by Stokes method, the clay denominated *SG* remains free of quartz, and shows the following characteristics: 1.60 meq. g^{-1} of cationic exchange capacity (CEC); 12.4 Å of basal distance calculated by XRD; 40 m^2/g of BET specific area, micropore volume ($V\mu_p$)=0.006137cc/g; total volume (V_{Tp})= 0.091cc/g and an approximate structural formulae determined by chemical analysis:

$$(Al_{2,760} + Fe_{0,161} + Mg_{1,078} + Mn_{0,017})_{VI} + (Al_{0,323} + Si_{7,677})_{IV}\ O_{20}\ (OH)_4\ M_{1,359}$$

where VI and IV are referred to the coordination number in the octahedral and tetrahedral position respectively, and M corresponds to the interchangeable cations, Ca, Na, K, and Mg.

2.2.2. The Oligocations Synthesis:

a) Aluminum oligocations:

The aluminum oligocations were prepared with solution 0.2M of $AlCl_3 6H_2O$ and solution 0.5M of NaOH, with different basicity relationships OH/Al = 1; 1.5; 2 and 2.5. For all the samples, the sodium hydroxide solution was incorporated drop by drop into the aluminium chloride solution, which was maintained under stirring at 50°C. The resultant solution was aged during 16 h. The polymeric solution was incorporated drop by drop into a suspension made with 5% of natural clay into deionized water. The amount of the incorporated oligocation had a ratio of 20 meq of Al per gram of clay.

The samples were washed in a dialysis membrane, dried at 60°C and calcinated at 500°C for 1 hour. The resulting materials were denominated Al20-1; Al20-1,5; Al20-2 and Al20-2,5.

b) *Zirconium oligocations*

The Zirconium oligocations were prepared from $ZrClO_2.8H_2O$, 0.1M. This solution was incorporated drop by drop into a suspension made with 5% of natural clay into deionized water, which was maintained under stirring at 25 °C. Different samples were prepared with: 5; 10 and 15 meq. of Zr per gram of clay. The samples were washed in a dialysis membrane, dried at 60°C and calcinated at 500°C for 1 hour. The resulting materials were denominated Zr5; Zr10 and Zr15.

c) Iron Oligocations

The iron oligocations were prepared from $[Fe_3(OCOCH_3)_7OH.2H_2O]NO_3$, 0.04M. The trinuclear acetato hydroxy-iron (III) nitrate (TNAFe) was prepared following the procedure described By Yamanaka et al [17]. The solution of TNAFe was incorporated drop by drop into a suspension of clay, then filtered, and the residue dried at 60°C and calcinated at 500°C for 1 hour. Different samples were prepared with:5; 10 and 15meq of Fe per gram of clay. The resulting materials were denominated Fe5; Fe10 and Fe15.

3. RESULTS AND DISCUSSION

Thermogravimetric and differential thermal analysis applied to clays [18,19], were used to study the initial samples (before the calcination process).

In Fig. 1, are shown the TGA and DTA studies corresponding to: the natural clay SG and two samples interchanged with Al oligocations before calcinating. For SG, the TGA shows in the curve (up to 120°C approximately) a first loss of mass, that corresponds to the loss of adsorbed water and interlayered water coordinated to the exchange cations. The second loss of mass (at 600°C approximately) is related to the loss of water that proceeds from the superficial OH^- (dehydroxilation) and also determined by the exchange cations. Finally, the third loss of mass at temperatures from 750°C is related to the loss of structural HO^-. By DTA, the SG sample shows a first wide peak, which is associated to the water corresponding to mono and divalent cations. The endothermic peak corresponding to dehydroxilation appears at 600°C and the peak corresponding to destruction of crystalline structure appears at 850°C approximately. In TGA in the aluminium samples, the temperature of the first loss of mass is different comparing with SGA, which is attributed to the different hydration capacity

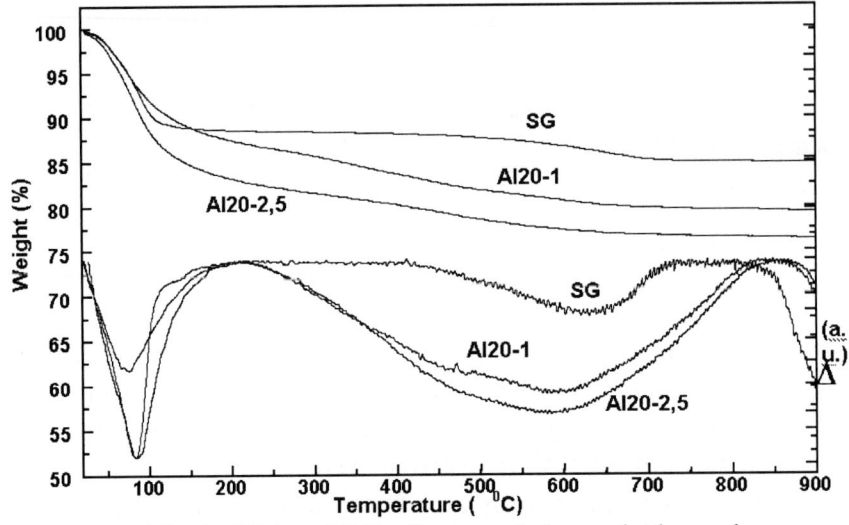

Fig. 1. TGA and DTA for natural clay and Al samples

of the Al cations comparing with the exchange cations. Although the amount of Al incorporated is the same, the difference in the basicity relationship of the solutions have lead to the formation of different aluminum cations.

The sample Al20.2,5 shows higher hydration and weaker bonds in the hydrated water, resulting in higher losses at lower temperatures (in DTA). By comparing with the thermogram corresponding to SGA, it is notorious in both Al samples a smooth loss of composition water

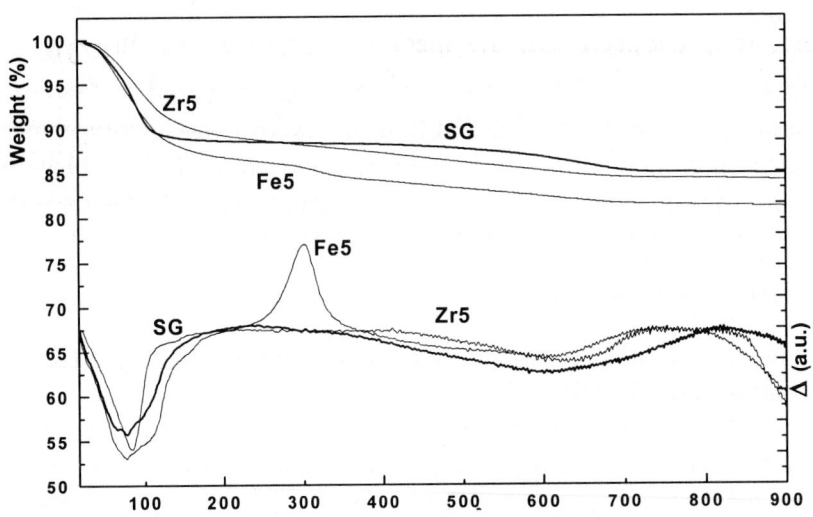

Fig.2 TGA and DTA for Fe and Zr samples comparing to SG

due to the process of dehydration of Al cations. For Al samples, in DTA, the endothermic peak corresponding to the loss of physisorbed water is wider than in SG, indicating an important solvation of the pillaring cations and the presence of polyvalent cations.

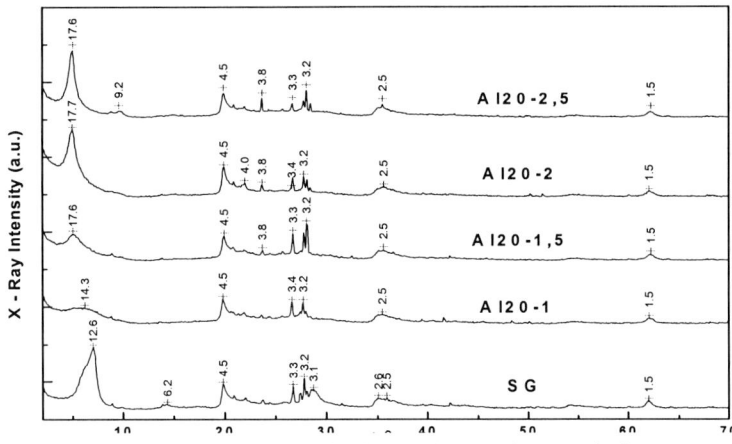

Fig 3. DRX of Al samples and natural clay

variation that appears in SG at 645°C corresponds to the loss of hydroxyls of composition. In Al samples this variation appears at temperatures about 500-600 °C, in which the pillars are "anchored".

In Fig. 2 are shown the studies for Zr and Fe samples, compare to SG. The sample with Zr has a similar behaviour than Al with low basicity according to TGA. In DTA, the sample with

Fe shows an exothermic peak, corresponding to the combustion of ion acetate to CO_2, obtaining Fe_2O_3, which is corroborated by TGA. The endothermic variations are not very important, giving and idea of the low hydration of the formed ions. As a consequence of the thermal analysis, we found that to anchor the pillars, the samples can be calcinated at 500°C. The XRD technique was used to study the presence of pillared materials in the samples.

In Fig. 3 are shown the diffractograms obtained for the Al-PILCs and the natural clay. From this figure, it is deduced that the incorporation of Al produces an increase in the distance between the planes 001 (d_{001}) whereas the rest of the structure was not appreciatively affected. This is a characteristic of the formation of the pillared clays. The increase in the basal space is related to the increase in the basicity relationship until OH/Al=2. For Al20-2 and Al20-2,5 the intensities are similar in the first peak, being a bit smaller in the second sample. As a consequence, we can assure the obtaining of good Al pillared clays from SG, if the basicity relationship is in the range of 1,5 and 2,5.

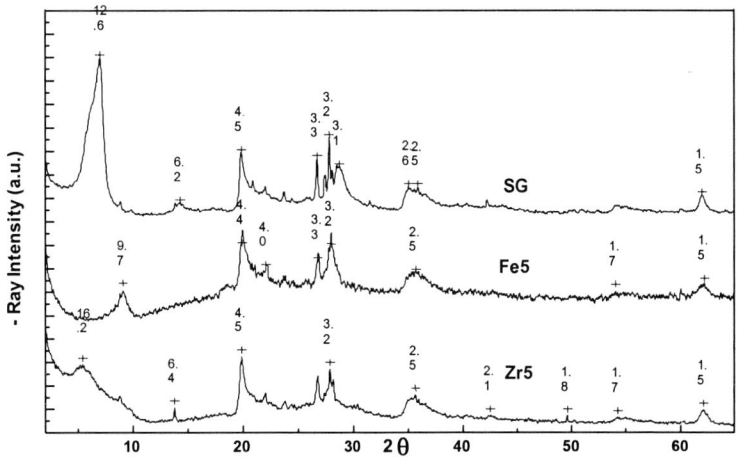

Fig. 4 DRX of Fe and Zr samples comparing to SG

To complete the study of these materials, their texture was analysed by N_2 adsorption-desorption at 77K. Since Zr and Fe samples have a low porosity and specific surface, a special attention were paid to the Al samples.

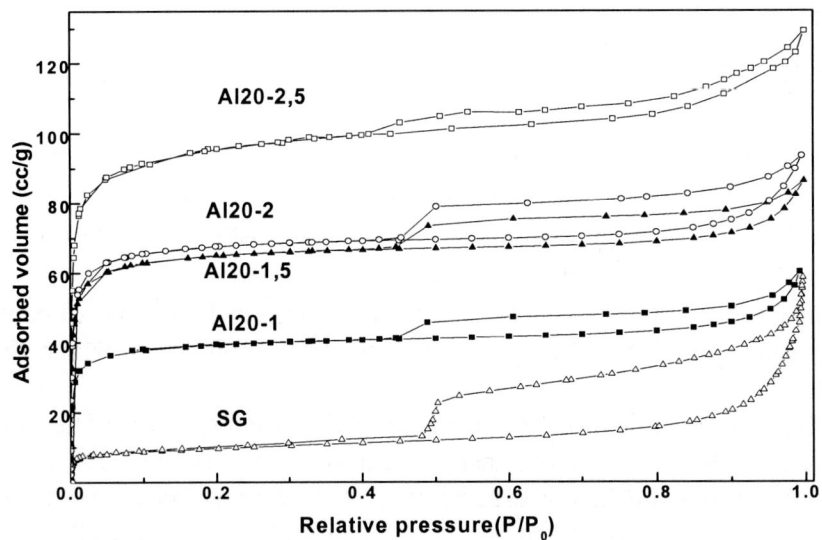

Fig. 5 Adsoprtion-desorption isotherms from Al samples and SG

In Fig.5 are represented the N_2 adsorption-desorption isotherms of the Al samples, which resemble a IV type isotherm in the BDDT classification, with hysteresis H4, from IUPAC, typical of porous between parallel plates. In this figure, it is detectable that due to the pillaring, an increase in the surface area was obtained, emphasizing that the pillared material kept the same layered structure as the source clay. The Al-polimeric species that can be formed from the oligocations under study differ principally in their relationship metal/charge and consequently influences the incorporation between the layers and the formation of different textures, despite of the similar basal spaces. The increase in the surface area corresponds to the increase in the microporosity and is according to the increase in the basicity relationship. This fact is directly related to the pillaring process. The Al20-1, Al20-1,5 and Al20-2 samples show a similar mesoporous texture, given by their hysteresis cycle, with a small variation for Al20-2,5.

In table 1, it is shown the data obtained by XRD(d_{001}) and N_2 adsorption-desoption (S_{BET}, $V_{\mu p}$, V_{Tp}) for the clays intercalated with aluminum, compared to the natural clay.

Table 1: Characterization data of Al samples and the natural clay

Sample	D_{001} (Å)	S_{BET} (m^2/g)	$V_{\mu p}$(m^2/g)	V_{Tp}(cc/g)
SG	12,6	34	0,0061	0,0908
Al20-1	14,3	144	0,0483	0,0933
Al20-1,5	17,6	239	0,0836	0,1336
Al20-2	17,7	250	0,0885	0,1556
Al20-2,5	17,6	345	0,0938	0,1955

It is noticeable that in relation to the former SG, the pillaring process has increased almost ten times the surface area (S_{BET}) and up to fifteen times the microporous volume ($V_{\mu p}$) when the basicity is increased. The basal space (d_{001}) increases from 12 to 18 Å approximately, for all basicity relationships except for OH/Al=1 in which the increase is up to 14,3 Å. The characteristics obtained demonstrate a pillarization process in Al samples for each relationship of basicity.

4. CONCLUSIONS

Different intercalated materials were synthesised and characterised from a natural clay from San Juan, Argentina. Al, Zr and Fe from hydrolysing metallic salts were used as intercalation oligocations. The samples were analysed by thermal techniques, obtaining the adequate temperature for the possible formation of the oxide form of the metals. The influence in the basal space of the oligocations was studied by DRX, from where we conclude that the Al oligocation produces PILCs more easily, under optimum preparation conditions. The changes in the texture of the samples intercalated with Al, were studied by N_2-sorption, obtaining an increase in the microporosity and specific area, typical of pillared clays.

To conclude, starting from a material with low porosity -the SG clay- we obtained a microporous material with an adequate specific surface to be used in adsorption and catalysis.

REFERENCES

1. K. Sapag, PhD Thesis. Universidad Autónoma de Madrid, Madrid, España 1997.
2. K. Sapag, S. Rojas, M. López Granados, J.L.G. Fierro and S. Mendioroz. J. of Mol. Catalysis A: Chemical, Vol. 167, No. 1-2, 2001, p. 81-89.
3. N. Lahav, D.White and S.Chang. Science 201 (1978) 67.
4. L.B.Ryland, M.W.Tanale and J.N.Wilson. "Catalysis", P.H.Emmett Ed., N.Y. 7:1 (1960).
5. J.Shabtai. Chem. Ind. 61 (1979) 734.
6. K. Sapag, R. L. Girón and S. Mendioroz, Granular Matter 3 (2001) 1/2, p.131-135.
7. T.J.Pinnavia. Science, 220 N4595 (1983).
8. G.W.Brindley and R.E.Sempels. Clay Minerals 12 (1977) 229.
9. N.Lahav, V.Shani and J.Shabtai. Clays and Clay Minerals 26 (1978) 107.
10. S.Yamanaka and G.W. Brindley. Clays and Clay Minerals 26 (1978) 21.
11. D.E.W.Vaughan, R.J.Lussier and J.S.Magee. U.S.Patent 4,176,090 (1979).
12. D.E.W. Vaughan, P.K. Maher, E.W. Alberx. Zl.S.Patent 3.775.345 (1973).
13. R. Burch. Catalysis Today 3 (1988) 2-3.
14. F.Figueras. Cat. Rev. Sci. Eng. 30 (1988) 3.
15. S.L.Jones. Catalysis Today 2 (1988) 209.
16. G.J.J.Bartley. Catalysis Today 2 (1988) 233.
17. S. Yamanaka, T. Doi, S. Sako and M. Hattori. Mat. Res. Bull., 19 (1984)161-168.
18. H. Van Olphen and J.J.Fripiat. Data Handbook for Clay Materials and other Non-Metallic Minerals, Pergamon Press, Oxford, U.K. (1979).
19. S.B. Hendricks, R.A. Nelson and L.T. Alexander. J. Amer. Chem. Soc. 62:1 (1940) 457.

Influence of the smectite type on the basal spacing after polyhydroxy-aluminum pillaring in some Brazilian and Argentinian clays

A.C. Vieira Coelho[a]; P. Souza Santos[a]; C. Volzone [b]; L.D.V. Abreu[a]

[a] Laboratório de Matérias-Primas Particuladas e Sólidos Não-Metálicos - Depto de Engenharia Metalúrgica e de Materiais - Escola Politécnica da USP - Av. Prof. Luciano Gualberto, Travessa 3, 380 CEP 05508-970 - São Paulo, SP, Brasil – Email : acvcoelh@usp.br

[b] CETMIC - Centro de Tecnología de Recursos Minerales y Cerámica – CONICET. CC.49, Cno. Centenario y 506, (1897) M.B.Gonnet. Prov. Buenos Aires. Argentina.

This paper presents the results of a study, using six Brazilian and five Argentinean smectitic clays, on the effects of their beidellite contents and the proportions of the montmorillonite types (Wyoming or Cheto) intercalation and pillaring with polyhydroxy-aluminum cations. The desirable qualities of intercalation and subsequent pillaring were evaluated by the values of d(001) and the width at half-height of the *(001)* reflection. It was observed that the higher the beidellite content of the clay, the more difficult it was to obtain good quality intercalated and pillared smectites. No direct relationship was found between the montmorillonite type, Wyoming or Cheto, and the quality of Al intercalation and pillaring in the group of clays studied.

1. INTRODUCTION

The synthesis and characterization processes of pillared clays have been the aim of much research since 1980 due to the interest shown in these materials for heterogeneous catalysis. The most important characteristics of these materials, when prepared from smectitic clays, is their specific surface area and the microporosity developed between the 2:1 layers of the crystalline structure of the smectitic clay mineral. The simplest way to characterize these materials is by measurement of the basal interlayer spacing (d_{001}), after firing at 400°C or higher temperatures, using X-ray diffraction (XRD). The increase in the original specific surface area and the development of microporosity in the fired clay is directly correlated to the increase of the value of d_{001} of the smectitic clay mineral.

The basal spacing in pillared clays may be influenced by the crystalline structure of the smectite species in the smectitic clays used in their preparation [1,2].

The aim of this paper is to present the results of an evaluation of the beidellitic content of source Brazilian and Argentine smectitic clays and to compare the effects of Wyoming or Cheto types of montmorillonites on the quality of the intercalation and later pillaring of polyhydroxy-aluminum, as evaluated by XRD.

2. MATERIALS AND METHODS

Fourteen clays were used: commercial samples from Wyoming (USA) i.e. Sigma, KWK and Volclay; from Brazil, Ca-smectitic clay from Campina Grande (PB); commercial smectites from Campina Grande (State of Paraíba): natural calcium clays that had been exchanged with sodium, using sodium carbonate, i.e. Brasgel Aço, FF and PA; Ca/Mg-smectitic clay from Campina Grande, locally named "Róseo Bofe"; gray colored smectitic clay from Franca (State of São Paulo) [3]; and commercial smectitic clays from Patagonia, Argentina, i.e. SIB, G2A, MCM, JB and SIEC. The samples were air dried at 110°C and sieved through ABNT sieve #200 (74 μm aperture); the fraction smaller than 74 μm was used in the experiments. The clays were dispersed in water (2%, w/w), and the suspensions were aged for seven days, prior to intercalation; they were stirred once a day.

The clays were intercalated with aqueous solutions of aluminum polyhydroxycations from the partial hydrolysis of aluminum nitrate solution with sodium hydroxide solution, following the procedure described by Vieira Coelho [1], with OH/Al ratio equal to 1.6. The aluminum solution was added dropwise to the clay suspensions in the proportion 20 miliequivalents of aluminum per gram of clay. After time for intercalation, washing and drying at 110°C, the Al-intercalated clays were heated in a programmed electric furnace for 3 hours at 400°C, for the dehydroxylation of the Al-olygomer and pillared clay formation.

The (*001*) basal spacing of the clays were measured in oriented specimens in an X-ray diffraction Philips equipment model X'PERT MPD (at LMPSol-USP) between 3°(2θ) – 15°(2θ). Cu-Kα radiation (λ = 0.15416 nm) was used. Step scanning was used; each step was 0.02°(2θ). The accumulation time for each step was 10 seconds.

The presence of beidellitic clay mineral was characterized in the original 74 μm samples by a variation of Greene-Kelly's procedure [4], using lithium bromide instead of lithium chloride, heating at 250°-300°C, followed by ethylene glycol treatment [5]. Measuring the basal spacing of the ethylene glycol treated specimen, the value of 0.95 nm (9.5Å) characterizes the montmorillonite structure while the beidellitic structure is characterized by values near 1.7-1.8 nm, indicating the non-contraction of the original basal space. The percentage of beidellite is evaluated by the quotient between the areas of the two peaks [4,5].

The characterization of the Wyoming or Cheto montmorillonite types was carried out with the K^+ saturation test [6]: the montmorillonite clay was treated with an 3M aqueous solution of potassium chloride; followed by centrifugation, washing and air drying, when the K-clay was examined by XRD; a 1.7 nm peak characterizes an Wyoming smectite while a 1.5 nm peak is indicative of a Cheto smectite [6]. These tests were made with a Philips PW1140 equipment operating between 3° and 15°(2θ); continuous scanning at a 0,02°(2θ) / s rate; 40kV and 20mA; Cu-Kα radiation; with a Ni filter to eliminate Kβ radiation.

The "quality" of the intercalated and pillared clays can be evaluated by the width at half-height of the (*001*) reflexion and by the value of d(001). For an intercalated smectite, high values of d(001) and small width values are desirable. Also the stability of the values of these parameters is desirable for pillared smectites; the good performance of a pillared smectite, in a technological use, such as in adsorption and catalysis, is dependent of the accessibility of the 2:1 interlayer spacing in temperature conditions in which a contraction of the interlayer space would normally happen, such as above 300°C [1,2].

3. EXPERIMENTAL RESULTS

The values of the d(001) spacing and widths at half-heights of the *(001)* reflexions in the intercalated and pillared clays (average of three measurements), as well as the basal spacing of the K-clays and their evaluated beidellite contents are shown in Table 1 and in Figures 1 to 4.

Table 1
Values of beidellite content; d(001) for K-clay; and d(001) and width at half-height (FWHM expressed in °2θ) for intercalated and pillared clays.

Clay	% beidellite	d_{001} K$^+$ (nm)	Intercalated Clay		Pillared Clay	
			d_{001} (nm)	FWHM(°2θ)	d_{001} (nm)	FWHM(°2θ)
Sigma	0	1.700	1.910	0.98	1.786	1.08
KWK	0	1.700	1.902	0.70	1.788	1.00
Volclay	0	1.700	1.945	1.05	1.788	1.02
Brasgel PA	19	1.635	1.879	2.09	1.750	1.87
Brasgel FF	18	1.591	1.900	1.54	1.789	1.68
Brasgel Aço	13	1.685	1.917	1.51	1.804	1.73
Bras. NT-25	23	1.590	1.885	1.76	1.722	1.95
Róseo Bofe	20	1.605	1.815	1.30	1.724	1.70
Franca	7	1.520	1.898	1.56	1.731	1.33
SIB	0	1.641	1.910	1.15	1.789	1.93
G2A	3.2	1.655	1.932	0.70	1.827	1.14
MCM	0	1.541	1.921	0.70	1.847	1.10
SIEC	10.2	1.638	1.910	1.19	1.786	1.94
JB	57	1.633	1.899	1.24	1.789	1.77

4. DISCUSSION

Figure 1 shows a tentative plot to establish a linear correlation between one of the mineralogical characteristics of the studied smectitic clays (their beidellite content) and the d(001) values for the intercalated and pillared clays. It can be observed in Figure 1 that an increase of the beidellite content would correspond to smaller d(001) values either for the intercalated or the corresponding pillared clays. Straight lines adjusted by the minimum squares method are shown in the Figure 1, representing that tendency (in the adjustment, the sample with the highest beidellite content was not considered). The use of "tendency" instead of "linear correlation" is due to the fact that the linear coefficient (r) is low; it is 0.44 for intercalated and 0.40 for pillared clays.

Figure 2 shows the tentative plot to establish a linear correlation between the beidellite content and the width at half-height of the *(001)* peak. Again, higher beidellite contents tend to give higher FWHM values (in 2θ) and, therefore, lower "quality" of intercalation and pillaring. The values of the linear correlation coefficient were slightly better: r = 0.65 for intercalated and 0.52 for pillared clays, using the same assumptions as from Figure 1.

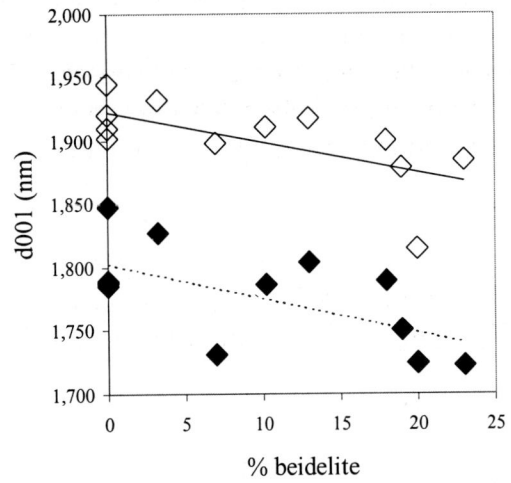

Figure 1. Plot of d(001) spacing vs % beidellite; Al-intercalated (◊) and pillared (♦) clays.

Figure 2. Plot of FWHM values vs % beidellite; Al-intercalated (◊) and pillared (♦) clays.

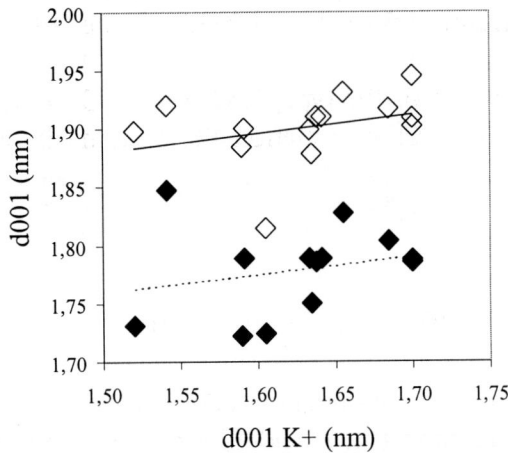
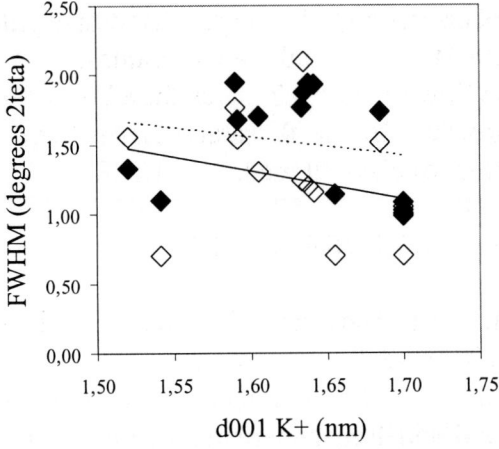

Figure 3. Plot of d(001) values vs d(001) values of K-saturated clays; Al-intercalated (◊) and pillared (♦) clays.

Figure 4. Plot of FWHM values vs d(001) values of K-saturated clays; Al-intercalated (◊) and pillared (♦) clays.

The type of tendency shown by these experimental data could be explained by the fact that the beidellite structure has a greater charge density in the tetrahedral sheet than the montmorillonite structure; such high spatial charge density could hinder the reversible exchange between the small cations originally present in the clay and the larger aluminum polyhydroxycations, such as Al_{13}^{7+}; the proximity between the opposite charges would lead to stronger electrical interactions, thus hindering the exchange for intercalation, and, as a consequence, for the pillaring of the beidellite richer clays.

Figures 3 and 4 show plots of the d(001) spacing values of the K-clays, indicating the montmorillonite types, either Wyoming or Cheto, against the d(001) spacings and FWHM values respectively for the same intercalated and respectively pillared clays. Examining both figures, no apparent linear relationship exists between the montmorillonite type and the d(001) spacing and FWHM values for either intercalated or pillared clays. That observation is in agreement with the very low values for the linear correlation coefficient (r) : in Figure 3, r = 0.09 for intercalated and r = 0.06 for pillared clays; in Figure 4, r = 0.08 for intercalated and r = 0.04 for pillared clays.

These observations indicated that, in the studied group of clays, the increasing presence of beidellite structure hinders the preparation of good quality pillared smectites; on the other hand, the presence of the montmorillonite type, Wyoming or Cheto, does not affect that quality.

5. CONCLUSIONS

a) Studying a group of six Brazilian an five Argentinian smectitic clays, it was possible to show that the beidellite content (as measured by the Li test) is a good index for an evaluation of the quality of the intercalation and pillaring these clays by the use of polyhydroxy-aluminium cations.
b) It was observed that the higher the beidellite content of the clay, the more difficult it was to obtain good quality intercalated and pillared smectites, as evaluated by the values of the d(001) spacing and FWHM values.
c) No direct relationship was found between the montmorillonite type, Wyoming or Cheto (as measured by the K^+ saturation test) and the quality of Al intercalation and pillaring in the group of clays studied.

ACKNOWLEDGEMENTS

This paper is part of Projeto Temático FAPESP 1995/00544-0 and Projeto FAPESP 2001/04681-5.

REFERENCES

1. A.C. Vieira Coelho, Caractérisation d'Argiles Nickelifères du Brésil et Étude du Pontage de ces Argiles et d'une Montmorillonite par des Solutions Hydroxy-Galliques et/ou Aluminiques. Université Catholique de Louvain, Louvain-la-Neuve. Belgium. 1991.
2. L.D.V. de Abreu, Estudo da Pilarização, com Alumina, de Argilas Esmectíticas Brasileiras. Dissertação de Mestrado. Departamento de Engenharia Química. Escola Politécnica da Universidade de São Paulo. São Paulo. 1997.
3. P. de Souza Santos, Ciência e Tecnologia de Argilas. Edgard Blücher. São Paulo. 1992.
4. R. Greene-Kelly, J. Soil Sci. **4** (1953), 233.
5. C. Volzone, J. Mater. Letters **10** (1991), 957.
6. R. E. Grim and G. Kulbicki, Amer. Mineral. **46** (1961), 1329.